List of the Elements

Name	Symbol	Atomic Number	Atomic Weight	Name	Symbol	Atomic Number	Atomic Weight
Actinium	Ac	89	227.028	Mercury	Hg	80	200.59
Aluminum	Al	13	26.982	Molybdenum	Mo	42	95.94
Americium	Am	95	243	Neodymium	Nd	60	144.24
Antimony	Sb	51	121.76	Neon	Ne	10	20.180
Argon	Ar	18	39.948	Neptunium	Np	93	237.048
Arsenic	As	33	74.922	Nickel	Ni	28	58.69
Astatine	At	85	210	Niobium	Nb	41	92.906
Barium	Ba	56	137.327	Nitrogen	N	7	14.007
Berkelium	Bk	97	247	Nobelium	No	102	259
Beryllium	Be	4	9.012	Osmium	Os	76	190.23
Bismuth	Bi	83	208.980	Oxygen	O	8	15.999
Bohrium	Bh	107	262	Palladium	Pd	46	106.42
Boron	B	5	10.811	Phosphorus	P	15	30.974
Bromine	Br	35	79.904	Platinum	Pt	78	195.08
Cadmium	Cd	48	112.411	Plutonium	Pu	94	244
Calcium	Ca	20	40.078	Polonium	Po	84	209
Californium	Cf	98	251	Potassium	K	19	39.098
Carbon	C	6	12.011	Praseodymium	Pr	59	140.908
Cerium	Ce	58	140.115	Promethium	Pm	61	145
Cesium	Cs	55	132.905	Protactinium	Pa	91	231.036
Chlorine	Cl	17	35.453	Radium	Ra	88	226.025
Chromium	Cr	24	51.996	Radon	Rn	86	222
Cobalt	Co	27	58.933	Rhenium	Re	75	186.207
Copper	Cu	29	63.546	Rhodium	Rh	45	102.906
Curium	Cm	96	247	Rubidium	Rb	37	85.468
Dubnium	Db	105	262	Ruthenium	Ru	44	101.07
Dysprosium	Dy	66	162.5	Rutherfordium	Rf	104	261
Einsteinium	Es	99	252	Samarium	Sm	62	150.36
Erbium	Er	68	167.26	Scandium	Sc	21	44.956
Europium	Eu	63	151.964	Seaborgium	Sg	106	263
Fermium	Fm	100	257	Selenium	Se	34	78.96
Fluorine	F	9	18.998	Silicon	Si	14	28.086
Francium	Fr	87	223	Silver	Ag	47	107.868
Gadolinium	Gd	64	157.25	Sodium	Na	11	22.990
Gallium	Ga	31	69.723	Strontium	Sr	38	87.62
Germanium	Ge	32	72.61	Sulfur	S	16	32.066
Gold	Au	79	196.967	Tantalum	Ta	73	180.948
Hafnium	Hf	72	178.49	Technetium	Tc	43	98
Hassium	Hs	108	265	Tellurium	Te	52	127.60
Helium	He	2	4.003	Terbium	Tb	65	158.925
Holmium	Ho	67	164.93	Thallium	Tl	81	204.383
Hydrogen	H	1	1.0079	Thorium	Th	90	232.038
Indium	In	49	114.82	Thulium	Tm	69	168.934
Iodine	I	53	126.905	Tin	Sn	50	118.71
Iridium	Ir	77	192.22	Titanium	Ti	22	47.88
Iron	Fe	26	55.845	Tungsten	W	74	183.84
Krypton	Kr	36	83.8	Uranium	U	92	238.029
Lanthanum	La	57	138.906	Vanadium	V	23	50.942
Lawrencium	Lr	103	262	Xenon	Xe	54	131.29
Lead	Pb	82	207.2	Ytterbium	Yb	70	173.04
Lithium	Li	3	6.941	Yttrium	Y	39	88.906
Lutetium	Lu	71	174.967	Zinc	Zn	30	65.39
Magnesium	Mg	12	24.305	Zirconium	Zr	40	91.224
Manganese	Mn	25	54.938	—	Uun	110	269
Meitnerium	Mt	109	266	—	Uuu	111	272
Mendelevium	Md	101	258	—	Uub	112	277

Free Student Aid.

Log on.

Tune in.

Succeed.

To help you succeed in your Conceptual Chemistry course, your professor has arranged for you to enjoy access to a great media resource, The Chemistry Place. You'll find that The Chemistry Place website that accompanies your textbook will enhance your course materials.

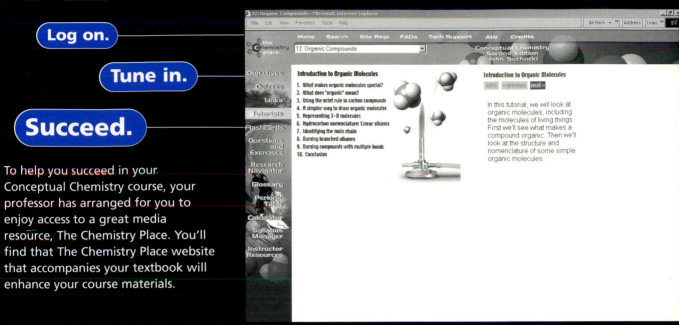

What your system needs to use these media resources:

WINDOWS
266 MHz Minimum CPU
Windows 98 or Higher
64 MB Minimum RAM
800 x 600 screen resolution
Thousands of colors
Internet Explorer 5.0 or Higher
Netscape 4.7 or Higher
Plug-Ins: Flash Player
NOTE: Use of Netscape 6.0 is not recommended due to a known compatibility issue between Netscape 6.0 and the Flash and Shockwave plug-ins.

MACINTOSH
266 MHz Minimum CPU
OS 9.2 or Higher
64 MB Minimum RAM
800 x 600 screen resolution
Thousands of colors
Internet Explorer 5.0 or Higher
Netscape 4.7 or Higher
Plug-Ins: Flash Player
NOTE: Use of Netscape 6.0 is not recommended due to a known compatibility issue between Netscape 6.0 and the Flash and Shockwave plug-ins.

Got technical questions?

For technical support, please visit www.aw.com/techsupport.

Here's your personal ticket to success:

How to log on to www.aw.com/chemplace

1. Go to www.aw.com/chemplace.
2. Click **Conceptual Chemistry, 2nd Edition**.
3. Click "Register"
4. Enter your pre-assigned access code exactly as it appears below.
5. Complete the online registration form to create your own personal login name and password.
6. Once your personal login name and password are confirmed by email, go back to www.aw.com/chemplace, click **Conceptual Chemistry, 2nd Edition**, type in your new login name and password, and click "Log In."

Your Access Code is:

Record your new login name and password on the back of this card.

Cut out this card and keep it handy. It's your ticket to valuable information.

Important: Please read the License Agreement located on the log in screen before using **The Chemistry Place**. By using the website, you indicate that you have read, understood, and accepted the terms of this agreement.

0-8053-3228-6

Conceptual Chemistry

Understanding Our World of Atoms and Molecules

Conceptual Chemistry
Understanding Our World of Atoms and Molecules

Second Edition

John Suchocki

Saint Michael's College

Benjamin Cummings

San Francisco Boston New York
Capetown Hong Kong London Madrid Mexico City
Montreal Munich Paris Singapore Sydney Tokyo Toronto

Publisher: Jim Smith
Associate Editor: Lisa Leung
Develomental Editor: Hilair Chism
Senior Marketing Manager: Christy Lawrence
Market Development Manager: Susan Winslow
Media Managing Editor: Claire Masson
Production Coordination: Joan Marsh
Production Management and Composition: Joan Keyes, Dovetail Publishing Services
Cover Design/Illustration: Blakeley Kim
Text Design: Joan Keyes and Emiko-Rose Koike
Artists: Emiko-Rose Koike, J. B. Woolsey and Associates
Photo Research: Stuart Kenter, Yvos Riezebos
Manufacturing Supervisor: Vivian McDougal
Prepress House: H&S Graphics
Printer and Binder: Von Hoffmann Press

Library of Congress Cataloging-in-Publication Data

Suchocki, John.
    Conceptual chemistry : understanding our world of atoms and molecules /
John A. Suchocki. — 2nd ed.
        p. cm.
    Includes index.
    ISBN 0-8053-3228-6
    1. Chemistry. I. Title.

QD33.2.S83 2003
540—dc21   2002041603

1 2 3 4 5 6 7 8 9 10—VH—03 02 01 00
www.aw.com

To
Tracy Suchocki

for more than I had ever dreamed,
my first thought is with you

Brief Contents

Detailed Contents

Conceptual Chemistry Photo Album

Conceptual Chemistry is personalized with photographs of my family and friends. A photo of my uncle and mentor Paul Hewitt, appears on page xxx. On Uncle Paul's lap is my son Evan Suchocki (pronounced Su-hock'-ee, with a silent c) who, as a toddler, sums up the book with his optimistic message.

Taking advantage of water's high heat of vaporization is my wife, Tracy, who is seen fearlessly walking over hot coals on page 264. You will also find her on page 294 taking a close look at a set of marbles as their kinetic energy disperses into heat. Demonstrating the potential energy of a drawn bow and arrow on page 17 is our precious oldest son, Ian, who is also seen as a baby with his mom on page 88 letting us know that the closeness between us is in the heart. Our third child, Maitreya Rose, is proudly showcased both as a fetus and as a baby on page 445, as one of the models of Figure 13.18, on page 416, highlighting the value of proteins, and as a two-year-old holding the cellulose and color-rich Vermont autumn leaves on page 408. About to enjoy his favorite beverage—by the liter—is son Evan on page 11. The inverted image of Evan and his mom enjoying the balmy beaches of Hawaii can be seen on page 339 in the discussion on the chemistry of photography. Those are Ian's hands holding the mineral fluorite on page 181 and my fingers on page 151 lightly touching the strings of Betsy, my guitar since childhood. (FYI, there's an "Easter egg" of Betsy in action hidden on the *Conceptual Chemistry Alive!* CD-ROM.) Also of our immediate family is Rusty Cat, whom you will find on page 627 helping provide perspective for the propane tank to the side of our home. Our dog Sam demonstrates his panting skills on page 252.

A few members of our extended family have also made their way into *Conceptual Chemistry*. My nephew Graham Orr is seen on page 51 enjoying one of the most valuable resources this planet has to offer—fresh water. Exploring the microscopic realm with the uncanny resolution of electron waves is my cousin George Webster, who is seen on page 146 alongside his own scanning electron microscope. George's son, Christian, is the cute kid in the Chapter 3 opening photo. Christian is holding a model of the amino acid glycine in front of the multitude of stars from which most all atoms, other than hydrogen, arise. Friend and former housemate Rinchen Trashi is seen looking through the spectroscope on page 139. Tracy's brother, Peter Elias, is found on page 593 smelling the camphorous odor of a freshly cut Ping-Pong ball. Look carefully on page 612 and you will see Peter again along with his mom (my mother-in-law), Sharon Hopwood, as they perch on the branch of a tree made strong by its composite nature. Both Peter and Sharon were key players in the development of the *Conceptual Chemistry Alive!* CD-ROM tutorial.

In addition to family photographs, the photographs of many of our friends' children grace this book. Ayano Jeffers-Fabro is the adorable girl hugging the tree on page 10. Jill Rabinov and her daughter Michaela appear on page 45 demonstrating the chemical nature of biological growth. Cole Stevens, who is seen on page 238, helps us to be amazed by what happens to the volume of water as it freezes. Cole's sister, Maya Stevens, is seen pondering the organic chemicals within vanilla and chocolate ice cream on page 361. In the Chapter 13 opener on page 399 are Daniel Glassman-Vinci and his twin brother Jacob. I'll leave it to you to decide whether they are two people at the same time or two people at different times—read the opener carefully. Last, but certainly not least, is Makani Nelson, who on page 400 provides us with a fine example of a human body full of cells and biomolecules. Look also for Makani's cameo appearance in the opening montage video of *Conceptual Chemistry Alive!* We are born with the desire to learn about our environment and our place in it. Let the sparkle of curiosity in the eyes of our sons and daughters portrayed in this textbook serve as a reminder of this important fact.

To the Student

Welcome to the world of chemistry—a world where everything around you can be traced to those incredibly tiny particles called atoms. Chemistry is the study of how atoms combine to form materials. By learning chemistry, you gain a unique perspective on what things are made of and why they behave as they do.

Chemistry is a science with a very practical outlook. By understanding and controlling the behavior of atoms, chemists have been able to produce a broad range of new and useful materials—alloys, fertilizers, pharmaceuticals, polymers, computer chips, recombinant DNA, and more. These materials have raised our standards of living to unprecedented levels. Learning chemistry, therefore, is worthwhile simply because of the impact this field has on society. More important, with a background in chemistry you can judge for yourself whether or not available technologies are in harmony with the environment and with what you believe to be right.

This book presents chemistry *conceptually*, focusing on the concepts of chemistry with little emphasis on calculations. Though sometimes wildly bizarre, the concepts of chemistry are straightforward and accessible—all it takes is the desire to learn. What you will gain from your efforts, however, may

be more than new knowledge about your environment and your personal relation to it—you may improve your learning skills and become a better thinker! But remember, just as with any other form of training, you'll get out of your study of chemistry only as much as you put in.

I enjoy chemistry, and I know you can, too. So put on your boots and let's go explore this world from the perspective of its fundamental building blocks.

Good chemistry to you!

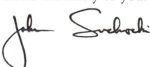

To the Instructor

As instructors, we share a common desire for our teaching efforts to have a long-lasting positive impact on our students. We focus, therefore, on what we think is most important for the student to learn. For students taking liberal arts chemistry courses, certain learning goals are clear. They should become familiar with and, perhaps, even interested in the basic concepts of chemistry, especially the ones that apply to their daily lives. They should understand, for example, how soap works and why ice floats on water. They should be able to distinguish between stratospheric ozone depletion and global warming, and also know what it takes to ensure a safe drinking water supply. Along the way, they should learn how to think about matter from the perspective of atoms and molecules. Furthermore, by studying chemistry, students should come to understand the methods of scientific inquiry and become better equipped to pass this knowledge along to future generations. In short, these students should become citizens of above-average scientific literacy.

These are noble goals, and it is crucial that we do our best to achieve them. Judging from my encounters with former liberal arts students in the midst of their daily lives, however, I have come to conclude that this is not what they usually cherish most from having taken a course in chemistry. Rather, it is the personal development they experienced through the process.

As all science educators know, chemistry—with its many abstract concepts—is fertile ground for the development of higher-thinking skills. Thus, it seems reasonable for us to share this valuable scientific offering—tempered to an appropriate level—with all students. Liberal arts students, like all other students, come to college not just to learn about specific subjects but for personal growth as well. This growth should include improvements in their analytical and verbal-reasoning skills along with a boost in self-confidence from having successfully met well-placed challenges. The value of our teaching, therefore, rests not only on our ability to help students learn chemistry but also on our ability to help them learn about themselves.

These are the premises upon which *Conceptual Chemistry* was written. You will find the standard discussions of the applications of chemistry, as shown in the table of contents. True to its title, this textbook also builds a conceptual base from which nonscience students may view nature more perceptively by helping them visualize the behavior of atoms and molecules and showing them how this behavior gives rise to our macroscopic environment.

Numerical problem-solving skills and memorization are not stressed. Instead, chemistry concepts are developed in a story-telling fashion with the frequent use of analogies and tightly integrated illustrations and photographs. Follow-up exercises are designed to challenge the students' understanding of concepts and their ability to synthesize and articulate conclusions. Concurrent with helping students learn chemistry, *Conceptual Chemistry* aims to be a tool by which students can learn how to become better thinkers and reach their personal goals of self-discovery.

Organization

The basic concepts of chemistry are developed in the first 12 chapters of *Conceptual Chemistry*. Threaded into the development, real-life applications facilitate the understanding and appreciation of chemistry concepts. In the remaining seven chapters, students have the opportunity to exercise their understanding of earlier material as they explore numerous chemistry-related topics.

Features

Key features of *Conceptual Chemistry* include the following:

- A conversational and clear writing style aimed at engaging student interest.

- In-text **Concept Checks** that pose a question and provide an answer immediately. These questions primarily reinforce ideas just presented before the student moves on to new concepts.

- **Hands-On Chemistry** activities that allow students to witness chemistry outside a formal laboratory setting. These activities can be performed using common household ingredients and equipment. Most chapters have two or three Hands-On features, which lend themselves well to distance learning or to in-class activities.

- **Calculation Corners** appear in selected chapters. They are included so that students can practice the quantitative-reasoning skills needed to perform chemical calculations. In each Calculation Corner, an example problem and answer show students how to perform a specific calculation; then their understanding is tested in a Your Turn section. None of the calculations involve skills beyond fractions, percentages, or basic algebra.

Extensive end-of-chapter material includes:

- **Key Terms and Matching Definitions** providing a short summary of important terms that appear boldfaced in the text.

- **Review Questions** designed to guide the student through the essentials of the chapter. They are grouped by chapter section to help the student stay focused while reviewing the material.

- **Hands-On Chemistry Insights** that follow up on the Hands-On Chemistry activities. These Insights are designed to ensure that the student is getting the most out of performing the Hands-On activities and also to clear up any misconceptions that may have developed.

- **Exercises** are designed to challenge student understanding of the chapter material and to emphasize critical thinking rather than mere recall. In many cases, the Exercises link chemistry concepts to familiar situations. The solutions to all odd-numbered exercises and problems appear in Appendix C. Thus, you can consider assigning even-numbered exercises for group studies, in-class discussions, or exams.

- **Problems** featuring concepts that are more clearly understood with numerical values and straightforward calculations. They are based on information presented in the Calculation Corners and therefore appear only in chapters containing this feature.

- **Discussion Topics** in the topical chapters (13–19) prompt students to express their opinions on issues that have no definitive answers. These topics may promote student debate about controversial ideas.

- **Exploring Further** references appear in every chapter, but are particularly important for the topical chapters for which you may be more inclined to assign research papers or poster presentations.

New to the Second Edition

The many positive comments on *Conceptual Chemistry*'s readability have been inspiring. Building on this strength, a major focus of this second edition was placed on streamlining the narrative and altering sentence structures to take *Conceptual Chemistry*'s user-friendliness to an even higher level. Another major focus was on correcting a number of content inaccuracies discovered by users of the first edition.

This second edition maintains the organizational structure of the first edition, with but one exception. Each chapter now ends with a section entitled "In Perspective." This feature serves both as a chapter summary and as food for thought. Many examples of this feature were present in the first edition, but for this second edition more were developed, and all are now highlighted.

Content changes include the introduction of breeder reactors in Chapter 4, The Atomic Nucleus. Also, the first section of Chapter 6, Chemical Bonding, was substantially reworked to support a course syllabus in which the atomic models of Chapter 5 are not emphasized. Perhaps the most significant change is the addition of a section on the second law of thermodynamics, which you will find as the last section of Chapter 9, An Overview of Chemical Reactions. This section was developed with the great assistance of Frank L. Lambert, Professor Emeritus of Occidental College. Frank aptly points out that entropy is merely a gauge of the natural tendency of energy to disperse. For more on this fresh and simple approach, be sure to explore www.secondlaw.com.

Other modifications to this edition include an updating of the Web references found at the end of each chapter. The topical chapters, such as Chapter 15, Optimizing Food Production, have also been updated based on current events. Also, you will find that each chapter opens on an odd-numbered page. This feature facilitates the production of custom-published versions of *Conceptual Chemistry*. Let your sales representative know which chapters you would like your students to have, and four-color custom copies will quickly be available.

The ancillaries that accompany the second edition have also been reworked. Most notably, the test bank has been expanded to include a vast number of new questions as well as multiple-choice versions of all the Exercises and Problems that appear at the end of each chapter. These multiple-choice versions of the Exercises and Problems also appear on the textbook's Web site. The intent is to provide a mechanism for rewarding students who have taken the time to work through the short-answer Exercises and Problems in the textbook.

Last but not least, you will find that each copy of the second edition comes with the first CD-ROM of *Conceptual Chemistry Alive!* which is a self-guided tutorial designed to help your students come to class prepared for a student-centered learning experience.

Support Package

The *Conceptual Chemistry* instructional package provides complete support materials for both students and faculty.

For the Student

- *Conceptual Chemistry Alive!* is a semester-length student tutorial presented by the author through a series of 12 CD-ROMs—one for each of the first 12 chapters. This tutorial features mini-lectures, demonstrations, animations, home chemistry projects, and explorations of chemistry in the community. Students browse through Quicktime movies in an interactive environment that follows the *Conceptual Chemistry* table of contents. After viewing a segment, students answer Concept Checks that encourage them to test their understanding of key material before progressing further. A student's answers to these Concept Checks are recorded in an electronic notebook that can be submitted to an instructor for assessment. More than a study supplement, *Conceptual Chemistry Alive!* is a textbook companion suitable for distance-learning programs and for instructors seeking to free up class time for student-centered curricula.

  A complimentary demonstration copy of the first CD is included in the back of each textbook. The full set of 12 discs is available to the student at a price that is affordable when bundled with a new textbook. A full set may also be purchased separately from the textbook through www.ConceptChem.com, which is the technical support Web site for all-*Conceptual Chemistry Alive!* users.

  For further information, please visit www.ConceptChem.com or contact your local sales representative, whom you may find through www.awl.com/replocator.

- *The Chemistry Place* Web site (http://www.aw.com/chemplace) is a unique study tool that offers detailed learning objectives, practice quizzes, flash cards, and Web links for each chapter of the text. The Chemistry Place also includes 30 new interactive tutorials, for a total of 41, featuring simulations, animations, and 3-D visualization tools. Also new at The Chemistry Place is the Research Navigator, a special portal to topical chemistry news and Web links.

- *Student Laboratory Manual for Conceptual Chemistry* (ISBN 0-8053-3238-3) co-authored with Donna Gibson, Chabot College, features laboratory activities tightly correlated to the chapter content.

For the Instructor

- The *Instructor's Manual* for *Conceptual Chemistry* is very different from most others. More than 400 pages in length and written by the author, it includes a variety of sample syllabi, lecture ideas and topics not treated in the book, teaching tips, and suggested step-by-step lectures and demonstrations. Answers to the Matching Key Terms and Review Questions and complete solutions to the Exercises and Problems are available to instructors in a format suitable for photocopying and posting for students to review. (ISBN 0-8053-3232-4)

- *The Chemistry Place* Web site (http://www.aw.com/chemplace) contains areas accessible only by the course instructor. These areas provide an on-line syllabus manager and a link to the complete *Instructor's Manual* on-line.

- A set of 250 four-color acetates of figures and tables from the text is available (ISBN 0-8053-3233-2).

- A CD-ROM contains the book's art library in a high-resolution format for electronic presentation (ISBN 0-8053-3231-6).

- A test bank comes in both printed format (ISBN 0-8053-3234-0) and on a cross-platform CD-ROM (ISBN 0-8053-3230-8).

- Course management technologies:

  WebCT: cms.aw.com/webct

  Blackboard: http://cms.aw.com/blackboard

  CourseCompass™: www.coursecompass.com

  In addition to offering Blackboard, we also offer CourseCompass™— a nationally hosted on-line course management system. All CourseCompass™ and Blackboard courses offer preloaded content, including testing and assessment, interactive Web-based activities, animations, Web links, illustrations, and photos. To view a demonstration of any course, go to http://cms.awlonline.com.

Acknowledgments

As every author knows, no words can express the debt one owes to the members of one's immediate family. My family's tireless support and belief in me have been an invaluable gift. To my wife, Tracy, to whom this book is dedicated, I am grateful for her endless patience and for the love and time she gives to me daily. Our late-night conversations and her countless paragraph edits have helped shape the scope and focus of this book. To Ian, Evan, and Maitreya, who have grown knowing only a dad who pours hours over his computer, thank you for reminding me of the important things in life.

There are numerous other individuals I am grateful and indebted to for their assistance in the development of *Conceptual Chemistry*. Standing at the head of this crowd is my uncle, mentor, and friend, Paul G. Hewitt. He planted the seed for this book in the early 1980s and has lovingly nurtured its growth ever since. Forever encouraging, patient, and inspirational, I thank you, Uncle Paul, for your guidance and more.

I am also blessed with a very capable extended family. Thanks to Uncle Paul and cousin Leslie Hewitt Abrams for allowing me to use their material from *Conceptual Physical Science*. Thanks to my pharmacist sister, Joan Lucas, for helping with the early drafts of Chapter 14 and to my chemical engineer brother-in-law, Rick Lucas, for consultations regarding the petroleum industry. To Bill Candler, rock collector and husband to my dear sister Cathy, thank you for supplying various minerals for photographing. I would like to thank my molecular geneticist step-brother, Nicholas Kellar, for assistance in the Chapter 13 laboratory for isolating DNA from plant material. Thanks also to my electron microscopist cousin, George Webster, and his wife, Lolita, for supplying photos of their SEM.

Special thanks are extended to my mother, Marjorie Hewitt Suchocki, and my father, John M. Suchocki, for their love and for instilling in me a positive attitude about life's work. Personal thanks are also extended to all my friends and other relatives for their support throughout the years, most notably to my mother-in-law, Sharon Hopwood, for photo research and for being a wonderful grandmother. Personal thanks also go to my past mentors, Professors Everette May and Albert Sneden of Virginia Commonwealth University.

I am particularly grateful for the past support of the faculty and staff of Leeward Community College, especially during my frequent book-writing sabbatical leaves. I am notably indebted to Michael Reese, Bob Asato,

George Shiroma, Patricia Domingo, Manny Cabral, Mike Lee, Kakkala Mohanan, Irwin Yamamoto, Stacy Thomas, Sharon Narimatsu, and Mark Silliman. To Alayne Schroll and the other faculty of the chemistry department at St. Michael's College, and to John Kenney, Dean of the College, thank you for welcoming me to your campus. Special thanks are extended to Frank L. Lambert, Professor Emeritus Occidental College, for his much-appreciated assistance in the development of *Conceptual Chemistry*'s presentation of the second law of thermodynamics.

For developing the *Conceptual Chemistry Laboratory Manual,* I am forever grateful to Donna Gibson of Chabot College. It has been a privilege to work with Donna—a remarkable woman and highly skilled instructor. For developing the *Conceptual Chemistry* test bank, I am indebted to Pam Marks of Arizona State University. To Pam's graduate students, Debbie Leedy and Rachel Morgan, I give thanks for creating answers to the matching terms and review questions in the *Instructor's Manual.*

The first chapter of this textbook is graced by the research efforts of Professors Jim McClintock of the University of Alabama, Birmingham, and Bill Baker of the University of South Florida, who were quick to provide not only their permissions but some beautiful photographs of the Antarctic. A big *mahalo* to them both.

There were many individuals in the publishing world who played vital and often pivotal roles in the early development of *Conceptual Chemistry.* These people include Doug Humphrey, Shelly McCarthy, Cathleen Petree, Mary Castellion, Richard Stratton, Paul Corey, and John Challice, and the late John Vondeling. I am grateful for their generous support and sound advice.

For making the publication of the first edition of *Conceptual Chemistry* possible, I owe special thanks to Ben Roberts, who has been all things to this textbook from acquisitions editor to senior editor. Ben's commitment to providing students with the best possible learning tools is unwavering and runs bone deep. We've been through a lot together, and I regard him as a close friend. Thanks, Ben, for your vision and for these opportunities.

Working with me on the nitty-gritty of each paragraph over the course of many years was my developmental editor, Hilair Chism, whose brillliant mind was able to keep track of the many details that my own mind was quick to forget. To Hilair I send my deepest appreciation, especially for her focus on integrating the text with the art program. Working with Hilair over the years has been a tremendous learning experience as well as a joy. I will always treasure our kinship.

Conceptual Chemistry has benefitted from the efforts of not just one, but two great developmental editors. In addition to Hilair's input, I had the input of the very talented Irene Nunes, my developmental editor for *Conceptual Physical Science.*

For this second edition, I am grateful for the support of my chemistry editor, Jim Smith, and senior editor, Frank Ruggirello. Thanks to Jim and Frank, as well as to Linda Baron Davis, for their trust and confidence in me as an author. I am also indebted to Lisa Leung and Sharon Hopwood, who were of great assistance in the preparation of the manuscript.

For overseeing the production of *Conceptual Chemistry,* I send a heartfelt thanks to Joan Marsh. For rendering the art, thanks to Emi Koike,

Blakeley Kim, and the artists at J. B. Woolsey and Associates. Special thanks to Jean Lake and Tony Asaro for managing the ancillary materials and to Margot Otway for providing valuable input throughout the project. Joan Keyes and Jonathan Peck of Dovetail Publishing Services did a superior job of composing pages and keeping track of the work flow. My deep gratitude goes to Stuart Kenter and Rachel Epstein of Stuart Kenter Associates for their remarkable photo research. As for marketing, I am excited to know that *Conceptual Chemistry* is in the very capable hands of Stacy Treco, Christy Lawrence, and Chalon Bridges. Thanks to you all. An author couldn't possibly ask for more support than this.

The development of *Conceptual Chemistry* relied heavily on the comments and criticisms of numerous reviewers. These people should know that their input was carefully considered and most often incorporated. A tremendous thanks goes to the reviewers listed here, who contributed immeasurably to the development of the first and second editions of the book:

Pamela M. Aker, University of Pittsburgh
Edward Alexander, San Diego Mesa College
Sandra Allen, Indiana State University
Susan Bangasser, San Bernardino Valley College
Ronald Baumgarten, University of Illinois, Chicago
Stacey Bent, New York University
Richard Bretz, University of Toledo
Benjamin Bruckner, University of Maryland, Baltimore County
Kerry Bruns, Southwestern University
Patrick E. Buick, Florida Atlantic University
John Bullock, Central Washington University
Barbara Burke, California State Polytechnical University, Pomona
Robert Byrne, Illinois Valley Community College
Richard Cahill, De Anza College
David Camp, Eastern Oregon University
Jefferson Cavalieri, Dutchess Community College
William J. Centobene, Cypress College
Ana Ciereszko, Miami Dade Community College
Jerzy Croslowski, Florida State University
Richard Clarke, Boston University
Cynthia Coleman, SUNY Potsdam
Virgil Cope, University of Michigan-Flint
Kathryn Craighead, University of Wisconsin/River Falls
Jack Cummini, Metropolitan State College of Denver
William Deese, Louisiana Tech University
Rodney A. Dixon, Towson University
Jerry A. Driscoll, University of Utah
Melvyn Dutton, California State University, Bakersfield
J. D. Edwards, University of Southwestern Louisiana
Karen Eichstadt, Ohio University
David Farrelly, Utah State University
Ana Gaillat, Greenfield Community College
Patrick Garvey, Des Moines Area Community College
Shelley Gaudia, Lane Community College

Donna Gibson, Chabot College
Palmer Graves, Florida International University
Jan Gryko, Jacksonville State University
William Halpern, University of West Florida
Marie Hankins, University of Southern Indiana
Alton Hassell, Baylor University
Barbara Hillery, SUNY Old Westbury
Angela Hoffman, University of Portland
John Hutchinson, Rice University
Mark Jackson, Florida Atlantic University
Kevin Johnson, Pacific University
Stanley Johnson, Orange Coast College
Joe Kirsch, Butler University
Louis Kuo, Lewis and Clark College
Frank Lambert, Occidental College
Carol Lasko, Humboldt State University
Joseph Lechner, Mount Vernon Nazarene College
Robley Light, Florida State University
Maria Longas, Purdue University
David Lygre, Central Washington University
Art Maret, University of Central Florida
Vahe Marganian, Bridgewater State College
Irene Matusz, Community College of Baltimore County—Essex
Robert Metzger, San Diego State University
Luis Muga, University of Florida
B. I. Naddy, Columbia State Community College
Donald R. Neu, St. Cloud State University
Larry Neubauer, University of Nevada, Las Vegas
Frazier Nyasulu, University of Washington
Frank Palocsay, James Madison University
Robert Pool, Spokane Community College
Brian Ramsey, Rollins College
Kathleen Richardson, University of Central Florida
Ronald Roth, George Mason University
Elizabeth Runquist, San Francisco State University
Maureen Scharberg, San Jose State University
Francis Sheehan, John Jay College of Criminal Justice
Mee Shelley, Miami University
Vincent Sollimo, Burlington County College
Ralph Steinhaus, Western Michigan University
Mike Stekoll, University of Alaska
Dennis Stevens, University of Nevada, Las Vegas
Anthony Tanner, Austin College
Joseph C. Tausta, State University College at Oneonta
Bill Timberlake, Los Angeles Harbor College
Margaret A. Tolbert, University of Colorado
Anthony Toste, Southwest Missouri State University
Carl Trindle, University of Virginia
Everett Turner, University of Massachusetts Amherst
George Wahl, North Carolina State University

M. Rachel Wang, Spokane Community College
Karen Weichelman, University of Southwestern Louisiana
Bob Widing, University of Illinois at Chicago
Ted Wood, Pierce University
Sheldon York, University of Denver

To the struggling student, thank you for your learning efforts—you are on the road to making this world a better place.

Much effort has gone into keeping this textbook error-free and accurate. It is possible, however, that some errors or inaccuracies may have escaped our notice. Your forwarding such errors or inaccuracies to me would be greatly appreciated. Your questions, general comments, and criticisms are also welcome. I look forward to hearing from you.

John Suchocki
ConceptChem@aol.com

Wow, Great Uncle Paul! Before this chickie exhausted its inner space resources and poked out of its shell, it must have thought it was at its last moments. But what seemed like its end was a new beginning. Are we like chickies, ready to poke through to a new environment and new understanding of our place in the universe?

1

Chemistry Is a Science

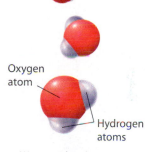

Oxygen atom

Hydrogen atoms

Water molecule, H_2O

Looking at the World of Atoms and Molecules

From afar, a sand dune looks to be made of a smooth, continuous material. Up close, however, the dune reveals itself to be made of tiny particles of sand. In a similar fashion, everything around us—no matter how smooth it may appear—is made of basic units called atoms. Atoms are so small, however, that a single grain of sand contains on the order of 125 million trillion of them. There are roughly 250,000 times more atoms in a single grain of sand than there are grains of sand in the dunes of this chapter's opening photograph.

As small as atoms are, there is much we have learned about them. We know, for example, that there are more than 100 different types of atoms, and they are listed in a widely recognized chart known as the *periodic table*. Some atoms link together to form larger but still incredibly small basic units of matter called *molecules*. As shown to the right, for example, two hydrogen atoms and one oxygen atom link together to form a single molecule of water, which you may know as H_2O. Water molecules are so small that an 8-oz glass of water contains about a trillion trillion of them.

Our world can be studied at different levels of magnification. At the *macroscopic* level, matter is large enough to be seen, measured, and handled. A handful of sand and a glass of water are macroscopic samples of matter. At the *microscopic* level, physical structure is so fine that it can be seen only with a microscope. A biological cell is microscopic, as is the detail on a dragonfly's wing. Beyond the microscopic level is the **submicroscopic**—the realm of atoms and molecules and an important focus of chemistry.

1

1.1   Chemistry Is a Central Science Useful to Our Lives

When you wonder what the Earth, sky, or ocean is made of, you are thinking about chemistry. When you wonder how a rain puddle dries up, how a car gets energy from gasoline, or how your body gets energy from the food you eat, you are again thinking about chemistry. By definition, **chemistry** is the study of matter and the transformations it can undergo. **Matter** is anything that occupies space. It is the stuff that makes up all material things—anything you can touch, taste, smell, see, or hear is matter. The scope of chemistry, therefore, is very broad.

Chemistry is often described as a central science because it touches all the other sciences. It springs from the principles of physics, and it serves as the foundation for the most complex science of all—biology. Indeed, many of the great advances in the life sciences today, such as genetic engineering, are applications of some very exotic chemistry. Chemistry sets the foundation for the Earth sciences—geology, volcanology, oceanography, meteorology—as well as for such related branches as archeology. It is also an important component of space science. Just as we learned about the origin of the moon from the chemical analysis of moon rocks in the early 1970s, we are now learning about the history of Mars and other planets from the chemical information gathered by space probes.

(a)   (b)   (c)   (d)   (e)

Figure 1.1
Chemistry is a foundation for many other disciplines. (a) Biochemists analyzing DNA profiles. (b) Meteorologist relasing weather balloon to study the chemistry of the upper atmosphere. (c) Technicians conducting DNA research. (d) Paleontologists preparing fossilized dinosaur bones for transport to laboratory for chemical analysis. (e) Astronomer studying the composition of asteroids.

Transparent matrix of processed silicon dioxide (Chapter 18)

Chemically disinfected drinking water (Chapter 16)

Caffeine solution (Chapter 14)

Thermoset polymer (Chapter 12)

Prescription medicines stored in refrigerator (Chapter 14)

Chlorofluorocarbon-free refrigerating fluids (Chapter 17)

Electrical energy from a fossil fuel or nuclear power plant (Chapter 19)

Metal alloy (Chapter 18)

Roasting carbohydrates, fats, proteins, and vitamins (Chapter 13)

Natural gas laced with odoriferous sulfur compounds (Chapter 12)

Fertilizer grown vegetables (Chapter 15)

Figure 1.2
Most of the material items in any modern house are shaped by some human-devised chemical process.

Progress in science, including chemistry, is made by scientists as they conduct research, which is any activity aimed at the systematic discovery and interpretation of new knowledge. **Basic research** leads us to a greater understanding of how the natural world operates. Many scientists focus on basic research. The foundation of knowledge laid down by basic research frequently leads to useful applications. Research that focuses on developing these applications is known as **applied research**. The majority of chemists have applied research as their major focus. Applied research in chemistry has provided us with medicine, food, water, shelter, and so many of the material goods that characterize modern life. Just a few of a myriad of examples are shown in Figure 1.2.

Over the course of the 20th century, we excelled at manipulating atoms and molecules to create materials to suit our needs. At the same time, however, mistakes were made when it came to caring for the environment. Waste products were dumped into rivers, buried in the ground, or vented into the air without regard for possible long-term consequences. Many people believed that the Earth was so large that its resources were virtually unlimited and that it could absorb wastes without being significantly harmed.

Most nations now recognize this as a dangerous attitude. As a result, government agencies, industries, and concerned citizens are involved in extensive efforts to clean up toxic-waste sites. Such regulations as the international ban on ozone-destroying chlorofluorocarbons have been enacted to protect the environment. Members of the American Chemistry Council, who as a group produce 90 percent of the chemicals manufactured in the United States, have adopted a program called Responsible Care, in which they have pledged to manufacture without causing environmental damage. The Responsible Care program—its emblem is shown in Figure 1.3—is based on the understanding that just as modern technology can be

Figure 1.3
The Responsible Care symbol of the American Chemistry Council.

Figure 1.4
More than 70 percent of all legislation placed before the Congress of the United States addresses science-related questions and issues, and many of these issues pertain to chemistry. Learning about science is an important endeavor for all citizens, particularly those destined to become leaders.

used to harm the environment, it can also be used to protect the environment. For example, by using chemistry wisely, most waste products can be minimized, recycled, engineered into sellable commodities, or rendered environmentally benign.

Chemistry has influenced our lives in profound ways and will continue to do so in the future. For this reason, it is in everyone's interest to become acquainted with the concepts of chemistry. A knowledge of chemistry gives us a handle on many of the questions and issues we face as a society. Are generic medicines really just as effective as brand-name ones (Chapter 14)? Should food supplements be federally regulated (Chapter 15)? Is genetically modified food safe (Chapter 15)? Should fluoride be added to local water supplies (Chapter 16)? What is happening to stratospheric ozone, and how does this problem differ from the problem of global warming (Chapter 17)? Why is it important to recycle (Chapter 18)? What should be our primary energy resources in the future (Chapter 19)? At some point, either we or the people we elect will be considering questions such as these, as the scene in Figure 1.4 illustrates. The more informed we are, the greater the likelihood that the decisions we make will be good ones.

Concept Check ✔

Chemists have learned how to produce aspirin using petroleum as a starting material. Is this an example of basic or applied research?

Was this your answer? This is an example of applied research because the primary goal was to develop a useful commodity. However, the ability to produce aspirin from petroleum depended on an understanding of atoms and molecules, an understanding that came from many years of basic research.

1.2   Science Is a Way of Understanding the Universe

Science has given us much. Our modern world is built on it. Nearly all forms of technology—from medicine to space travel—are applications of science. But what exactly is this amazing thing called *science*? How should science be used? Where did science come from? And what would the world be like without it?

Science is an organized body of knowledge about nature. It is the product of observations, common sense, rational thinking, and (sometimes) brilliant insights. Science has grown out of group efforts as well as individuals' discoveries. It has been built up over thousands of years and gathered from places all around the Earth. It is a huge gift to us today from the thinkers and experimenters of the past.

Yet science is not just a body of knowledge. It is also a method, a way of exploring nature and discovering the order within it. Importantly, science is also a tool for solving problems.

Science began back before recorded history, when people first discovered repeating patterns in nature, such as star patterns in the night sky, weather patterns, and patterns in animal migration. From these patterns, people learned to make predictions that gave them some control over their surroundings.

Although there are many paths scientists can follow in doing science, regardless of which path is taken, a number of key elements traditionally arise: observations, questions, scientific hypotheses, predictions, and tests. A **scientific hypothesis** is a testable assumption, or guess, often used to explain an observed phenomenon. As Figure 1.5 suggests, the results of tests invariably lead to further observations, questions, and scientific hypotheses, meaning that the scientific process can never have any end.

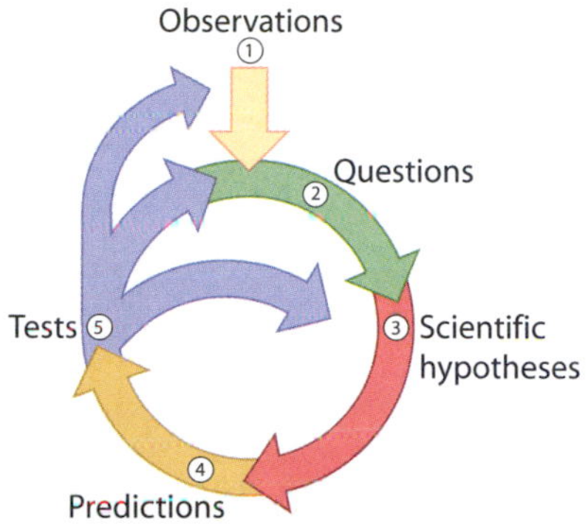

Observations
①
Questions
②
Tests ⑤
③ Scientific hypotheses
④
Predictions

Figure 1.5
The scientific process often—but not always—proceeds in the following order: observations, questions, scientific hypotheses, predictions, tests. Tests lead to more observations, more questions, more scientific hypotheses, and so on. From this cyclic process comes a greater understanding of the universe.

A Study of Sea Butterflies Illustrates the Process of Science

The scientific process is aptly illustrated by the efforts of an Antarctic research team headed by James McClintock, professor of biology at the University of Alabama at Birmingham, and Bill Baker, professor of chemistry at the University of South Florida, both shown in Figure 1.6. One

Figure 1.6
The Chemical Ecology of Antarctic Marine Organisms Research Project was initiated in 1988 by James McClintock, shown here (fifth from left) with his team of colleagues and research assistants. In 1992, he was joined by Bill Baker (second from right). Baker is shown in the inset dressing for a dive into the icy Antarctic water. Like many other science projects, this one was interdisciplinary, involving the efforts of scientists from a variety of backgrounds.

aspect of their research involved studying the toxic chemicals Antarctic marine organisms secrete to defend themselves against predators. McClintock and Baker observed an unusual relationship between two animal species, a sea butterfly and an amphipod—a relationship that led to a question, a scientific hypothesis, a prediction, and tests about the chemistry involved in the relationship.

1. **Observation**. The sea butterfly *Clione antarctica* is a brightly colored, shell-less snail with winglike extensions used in swimming (Figure 1.7a), and the amphipod *Hyperiella dilatata* resembles a small shrimp. McClintock and Baker observed a large percentage of amphipods carrying sea butterflies on their backs, with the sea butterflies held tightly by the hind legs of the amphipods (Figure 1.7b). Any amphipod that lost its sea butterfly would quickly seek another—the amphipods were actively abducting the sea butterflies!

(a)

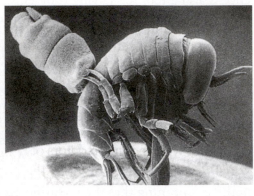
(b)

Figure 1.7
(a) The graceful Antarctic sea butterfly is a species of snail that does not have a shell. (b) The shrimp-like amphipod attaches a sea butterfly to its back even though doing so limits the amphipod's mobility.

(a) Sea butterfly

(b) Amphipod

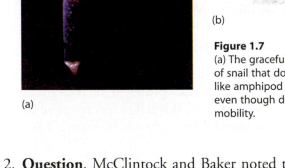
(c) Sea butterfly and amphipod

Figure 1.8
In McClintock and Baker's initial experiment, a predatory fish (a) rejected the sea butterfly, (b) ate the free-swimming amphipod, and (c) rejected the amphipod coupled with a sea butterfly.

2. **Question**. McClintock and Baker noted that amphipods carrying sea butterflies were slowed considerably, making the amphipods more vulnerable to predators and less adept at capturing prey. Why then did the amphipods abduct the sea butterflies?

3. **Scientific Hypothesis**. Given their experience with the chemical defense systems of various sea organisms, the research team hypothesized that amphipods carry sea butterflies because the sea butterflies produce a chemical that deters a predator of the amphipod.

4. **Prediction**. Based on their hypothesis, they predicted (a) that they would be able to isolate this chemical and (b) that an amphipod predator would be deterred by it.

5. **Tests**. To test their hypothesis and prediction, the researchers captured several predator fish species and conducted the test shown in Figure 1.8. The fish were presented with solitary sea butterflies, which they took

into their mouths but promptly spit back out. The fish readily ate uncoupled amphipods but spit out any amphipod coupled with a sea butterfly. These are the results expected if the sea butterfly was secreting some sort of chemical deterrent. The same results would be obtained, however, if a predator fish simply didn't like the feel of the sea butterfly in its mouth. The results of this simple test were therefore ambiguous.

All scientific tests need to minimize the number of possible conclusions. Often this is done by running an experimental test alongside a **control test**. Ideally, the two tests should differ by only one variable. Any differences in results can then be attributed to how the experimental test differed from the control test.

To confirm that the deterrent was chemical and not physical, the researchers made one set of food pellets containing both fish meal and sea-butterfly extract (the experimental pellets). For their control test, they made a physically identical set containing only fish meal (the control pellets). As shown in Figure 1.9, the predator fish readily ate the control pellets but not the experimental ones. These results strongly supported the chemical-deterrent hypothesis.

Further processing of the sea-butterfly extract yielded five major chemical compounds, only one of which deterred predator fish from eating the pellets. Chemical analysis of this compound revealed it to be the previously unknown molecule shown in Figure 1.10, which they named pteroenone.

As frequently happens in science, McClintock and Baker's results led to new questions. What are the properties of pteroenone? Does this substance have potential for treating human disease? In fact, a majority of the drugs prescribed in the United States were developed by chemists working with naturally occurring materials. As we explore further in Chapter 14, this is an important reason to preserve marine habitats and tropical rainforests, which house countless yet-to-be-discovered substances.

Reproducibility and an Attitude of Inquiry Are Essential Components of Science

In addition to running control tests, scientists confirm experimental results by repeated testing. The Antarctic researchers, for example, made many food pellets, both experimental and control, so that each test could be repeated many times. Only after obtaining consistent results can a scientist begin to decide whether the hypothesis in question is supported or not.

If there is an undetected variable or flaw in an experiment, it doesn't matter how many times the tests are repeated. Measuring your weight on a broken scale is a good example of a flawed procedure—no matter how many times you step on the scale, the weight you measure will be wrong every time. Similarly, had the Antarctic researchers not been careful to make sure the fish in the experimental and control tests were equally hungry and of the same species, their results and conclusions would have been unreliable.

Because of the great potential for unseen error in any procedure, the results of a scientific experiment are considered valid only if they can be reproduced by other scientists working in similarly equipped laboratories. This restriction helps to confirm the experimental results and lends more

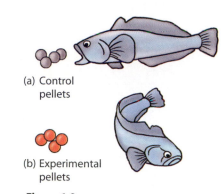

(a) Control pellets

(b) Experimental pellets

Figure 1.9
The predator fish (a) ate the control pellets but (b) rejected the experimental pellets, which contained sea-butterfly extract.

Pteroenone

Figure 1.10
Pteroenone is a molecule produced by sea butterflies as a chemical deterrent against predators. Its name is derived from *ptero-*, which means "winged" (for the sea butterfly), and *-enone*, which describes information about the chemical structure. The black spheres represent carbon atoms, the white hydrogen atoms, and the red oxygen atoms.

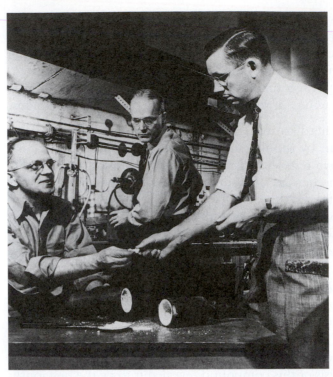

Figure 1.11
Roy Plunkett (right) and his colleagues pose for this reenactment photograph of their discovery of Teflon. Their success was due in great part to their curiosity.

credence to an interpretation. Reproducibility is an essential component of science. Without it, the understandings we gain through science become questionable.

Although the traditional methods of science are powerful, the success of science has more to do with an *attitude* common to scientists rather than with a particular method. This attitude is one of inquiry, experimentation, honesty, and a faith that all natural phenomena can be explained. Accordingly, many scientific discoveries have involved trial and error, experimentation without guessing, or just plain accidental discovery. In the late 1930s, for example, the DuPont researcher Roy Plunkett and his colleagues, shown in Figure 1.11, filled a pressurized cylinder with a gas called tetrafluoroethylene. The next morning they were surprised to discover that the cylinder appeared empty. Not believing that the contents had simply vanished, they hacksawed the cylinder apart and discovered a white solid coating the inner surface. Driven by curiosity, they continued to investigate the material, which eventually came to be known and marketed as the highly useful polymer Teflon.

Concept Check ✔

Why is it important for a scientist to be honest?

Was this your answer? Any discovery made by a scientist is subject to the scrutiny and testing of other scientists. Sooner or later, mistakes or wishful thinking or even outright deceptions are found out. Honesty, so important to the progress of science, thus becomes a matter of self-interest.

A Theory Is a Single Idea That Has Great Explanatory Power

Periodically, science moves to a point where a wide range of observations can be explained by a single comprehensive idea that has stood up to repeated scrutiny. Such an idea is what scientists call a **theory**. Biologists, for instance, speak of the *theory of natural selection* and use it to explain both the unity and the diversity of life. Physicists speak of the *theory of relativity* and use it to explain how we are held to the Earth by gravity. Chemists speak of the *theory of the atom* and use it to explain how one material can transform into another.

Theories are a foundation of science, but they are not fixed. Rather, they evolve as they go through stages of redefinition and refinement. Since it was first proposed 200 years ago, for example, the theory of the atom has been repeatedly refined as new evidence about atomic behavior has been gathered. Those who know little about science may argue that scientific theories have little value because they are always being modified. Those who understand science, however, see it differently—theories grow stronger as they are modified.

Science Has Limitations

Science is a powerful means of gaining knowledge about the natural world, but it is not without limitations. No experiment can ever prove definitively that a scientific hypothesis is correct. What happens instead is that we gain more and more confidence in a hypothesis as it continues to be supported by the results of many different experiments conducted by many different investigators. If any experimental result contradicts the hypothesis, and if this result is reproducible, the hypothesis must be either discarded or revised. Even the most firmly planted theories are subject to this same scrutiny.

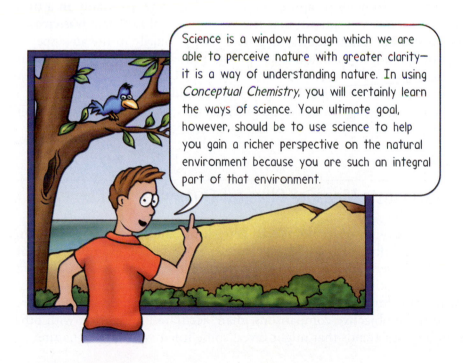

Science deals only with hypotheses that are testable. As such, its domain is restricted to the observable natural world. While scientific methods can be used to debunk various claims, science has no way of verifying testimonies involving the supernatural. The term *supernatural* literally means "above nature." Science works within nature, not above it. Likewise, science is unable to answer such philosophical questions as "What is the purpose of life?" or such religious questions as "What is the nature of the human spirit?" Though these questions are valid and have great importance to us, they rely on subjective personal experience and do not lead to testable hypotheses.

Concept Check ✓

Which statement is a *scientific* hypothesis?
   a. The moon is made of Swiss cheese.
   b. Human consciousness arises from an essence that is undetectable.

Was this your answer? Both statements attempt to explain observed phenomena, and so both are hypotheses. Only statement a is testable, however, and therefore only statement a is a *scientific* hypothesis.

While we shall never be able to detect the undetectable, we have traveled to the moon and found that it is not made of Swiss cheese (the last manned flight to the moon was in 1974). Analyses of moon samples have shown that the moon has a chemical composition very similar to that of the Earth, a finding that led to more questions and scientific hypotheses. For example, why are the moon and the Earth of similar chemical composition? Were they once bound together as a single body that split apart billions of years ago? Continued experiments suggest that the answer to this question is yes!

Science Helps Us Learn the Rules of Nature

Just as you can't enjoy a ball game, computer game, or party game until you know its rules, so it is with nature. Because science helps us learn the rules of nature, it also helps us appreciate nature. You may see beauty in a tree, but you'll see more beauty in that tree when you realize that it was created from substances found not in the ground but primarily in the air—specifically, the carbon dioxide and water put into the air by respiring organisms such as yourself (Figure 1.12).

Learning science builds new perspectives and is not unlike climbing a mountain. Each step builds on the previous step while the view grows ever more astounding.

Figure. 1.12
The tree Ayano hugs is made primarily from carbon dioxide and water, the very same chemicals Ayano releases through her breath. In return, the tree releases oxygen, which Ayano uses to sustain her life. We are one with our environment down to the level of atoms and molecules.

1.3 Scientists Measure Physical Quantities

Science starts with observations. To get a firm handle on an observation, however, the observer should take measurements. For example, it's not enough to observe that a material has mass. A more complete observation will include both a measurement of how much mass and a measurement of how much volume this mass occupies. By quantifying observations, we are able to make objective comparisons, share accurate information with others, or look for trends that might reveal some inner workings of nature.

Scientists measure *physical quantities*. Some examples of physical quantities you will be learning about and using in this book are length, time, mass, weight, volume, energy, temperature, heat, and density. Any measurement of a physical quantity must always include a number followed by a *unit* that tells us what was measured. It would be meaningless, for instance, to say that an animal is 3 because the number by itself does not give us enough information. The animal could be 3 meters tall, 3 kilograms in mass, or even 3 seconds old. Meters, kilograms, and seconds are all units that tell us the significance of the physical quantity, and they must be included to complete the description.

There are two major unit systems used in the world today. One is the United States Customary System (USCS, formerly called the British System of Units), used in the United States, primarily for nonscientific purposes.* The other is the Système International (SI), which is used in most other nations. This system is also known as the International System of Units or as the metric system. The orderliness of this system makes it useful for scientific work, and it is used by scientists all over the world, including those in the United States. (And the International System is beginning to be used for nonscientific work in the United States, as Figure 1.13 shows.)

This book uses the SI units given in Table 1.1. On occasion, USCS units are also used to help you make comparisons.

Figure 1.13
The metric system is finally making some headway in the United States, where various commercial goods, such as Evan's favorite soda, are now sold in metric quantities.

Table 1.1

Metric Units for Physical Quantities and Their USCS Equivalents

Physical Quantity	Metric Unit	Abbreviation	USCS Equivalent
length	kilometer	km	1 km = 0.621 miles (mi)
	meter	m	1 m = 1.094 yards (yd)
	centimeter	cm	1 cm = 0.3937 inches (in.) 1 in. = 2.54 cm
	millimeter	mm	none commonly used
time	second	s	second also used in USCS
mass	kilogram	kg	1 kg = 2.205 pounds (lb)
	gram	g	1 g = 0.03528 ounces (oz) 1 oz = 28.345 g
	milligram	mg	none commonly used
volume	liter	L	1 L = 1.057 quarts (qt)
	milliliter	mL	1 mL = 0.0339 fl oz
	cubic centimeter	cm^3	1 cm^3 = 0.0339 fl oz
energy	kilojoule	kJ	1 kJ = 0.239 kilocalories (kcal)
	joule	J	1 J = 0.239 calories (cal) 1 cal = 4.184 J
temperature	degree Celsius	°C	(°C × 1.8) + 32 = degrees Fahrenheit, °F
	kelvin	K	°C + 273 = K

*Two other countries that continue to use the USCS are Liberia and Myanmar.

Calculation Corner: Unit Conversion

Welcome to Calculation Corner! *Conceptual Chemistry* focuses on visual models and qualitative understandings. As with any other science, however, chemistry has its quantitative aspects. In fact, it is only by the interpretation of quantitative data obtained through laboratory experiments that chemical concepts can be reliably deduced. Thus, it is only natural that there are times when your conceptual understanding of chemistry can be nicely reinforced by some simple, straightforward calculations.

Often in chemistry, and especially in a laboratory setting, it is necessary to convert from one unit to another. To do so, you need only multiply the given quantity by the appropriate *conversion factor*. All conversion factors can be written as ratios in which the numerator and denominator represent the equivalent quantity expressed in different units. Because any quantity divided by itself is equal to 1, all conversion factors are equal to 1. For example, the following two conversion factors are both derived from the relationship 100 centimeters = 1 meter:

$$\frac{100 \text{ centimeters}}{1 \text{ meter}} = 1 \qquad \frac{1 \text{ meter}}{100 \text{ centimeters}} = 1$$

Because all conversion factors are equal to 1, multiplying a quantity by a conversion factor does not change the value of the quantity. What does change are the units. Suppose you measured an item to be 60 centimeters in length. You can convert this measurement to meters by multiplying it by the conversion factor that allows you to cancel centimeters.

Example
Convert 60 centimeters to meters.

Answer

$$(60 \text{ centimeters})\frac{(1 \text{ meter})}{(100 \text{ centimeters})} = 0.6 \text{ meter}$$

$$\underset{\substack{\text{quantity} \\ \text{in centimeters}}}{\uparrow} \qquad \underset{\substack{\text{conversion} \\ \text{factor}}}{\uparrow} \qquad \underset{\substack{\text{quantity} \\ \text{in meters}}}{\uparrow}$$

To derive a conversion factor, consult a table that presents unit equalities, such as Table 1.1. Then multiply the given quantity by the conversion factor, and voilà, the units are converted. Always be careful to write down your units. They are your ultimate guide, telling you what numbers go where and whether you are setting up the equation properly.

Your Turn
Multiply each physical quantity by the appropriate conversion factor to find its numerical value in the new unit indicated. You will need paper, pencil, a calculator, and Tables 1.1 and 1.2.

  a. 7320 grams to kilograms.
  b. 235 kilograms to pounds.
  c. 4500 milliliters to liters.
  d. 2.0 liters to quarts.
  e. 100 calories to kilocalories.
  f. 100 calories to joules.

The answers for Calculation Corners appear at the end of each chapter.

One major advantage of the metric system is that it uses a decimal system, which means all units are related to smaller or larger units by a factor of 10. Some of the more commonly used prefixes along with their decimal equivalents are shown in Table 1.2. From this table, you can see that 1 kilometer is equal to 1000 meters, where the prefix *kilo-* indicates

Table 1.2

Metric Prefixes

Prefix	Symbol	Decimal Equivalent	Exponential Form	Example
tera-	T	1,000,000,000,000.	10^{12}	1 terameter (Tm) = 1 trillion meters
giga-	G	1,000,000,000.	10^{9}	1 gigameter (Gm) = 1 billion meters
mega-	M	1,000,000.	10^{6}	1 megameter (Mm) = 1 million meters
kilo-	k	1,000.	10^{3}	1 kilometer (km) = 1 thousand meters
hecto-	h	100.	10^{2}	1 hectometer (hm) = 1 hundred meters
deka-	da	10.	10^{1}	1 dekameter (dam) = ten meters
no prefix	—	1.	10^{0}	1 meter (m) = 1 meter
deci-	d	0.1	10^{-1}	1 decimeter (dm) = 1 tenth of a meter
centi-	c	0.01	10^{-2}	1 centimeter (cm) = 1 hundredth of a meter
milli-	m	0.001	10^{-3}	1 millimeter (mm) = 1 thousandth of a meter
micro-	μ	0.000 001	10^{-6}	1 micrometer (μm) = 1 millionth of a meter
nano-	n	0.000 000 001	10^{-9}	1 nanometer (nm) = 1 billionth of a meter
pico-	p	0.000 000 000 001	10^{-12}	1 picometer (pm) = 1 trillionth of a meter

1000. Likewise, 1 millimeter is equal to 0.001 meter, where the prefix *milli-* indicates ¹⁄₁₀₀₀. You need not memorize this table, but you will find it a useful reference when you come across these prefixes in your course of study.

The remaining sections of this chapter introduce some physical quantities important to the study of chemistry. In the spirit of the atomic and molecular theme of *Conceptual Chemistry*, these physical quantities are described from the point of view of atoms and molecules. In addition to physical quantities, the various phases of matter—solid, liquid, gas—are also described from the point of view of atoms and molecules.

1.4 Mass Is How Much and Volume Is How Spacious

To describe a material object, we can quantify any number of properties, but perhaps the most fundamental property is mass. **Mass** is the quantitative measure of how much matter a material object contains. The greater the mass of an object, the greater amount of matter in it. A gold bar that is twice as massive as another gold bar, for example, contains twice as many gold atoms.

Mass is also a measure of an object's *inertia*, which is the resistance the object has to any change in its motion. A cement truck, for example, has a lot of mass (inertia), which is why it requires a powerful engine to get moving and powerful brakes to come to a stop.

The standard unit of mass is the *kilogram*, and a replica of the primary cylinder used to determine exactly what mass "1 kilogram" describes is shown in Figure 1.14. An average-sized human male has a mass of about 70 kilograms (154 pounds). For smaller quantities, we use the *gram*. Table 1.2 tells us that the prefix *kilo-* means "1000," and so we see that 1000 grams are needed to make a single kilogram (1000 grams = 1 kilogram). For even smaller quantities, the *milligram* is used (1000 milligrams = 1 gram).

Figure 1.14
The standard kilogram is defined as the mass of a platinum–iridium cylinder kept at the International Bureau of Weights and Measures in Sevres, France. The cylinder is removed from its very safe location only once a year for comparison with duplicates, such as the one shown here, which is housed at the National Institute of Standards and Technology in Washington, D.C.

Welcome to "Hands-On Chemistry," the interactive corner of your *Conceptual Chemistry* textbook. This feature provides you the opportunity to apply chemistry concepts outside a formal laboratory setting. Each activity is guaranteed to be meaningful, interesting, and, at times, surprising.

At the end of each chapter, you will find follow-up discussions, called Insights, for each Hands-On Chemistry activity. Ideally, you should look over the Insights only after you have attempted the activity. The job of the Insights is to correct any misconceptions you may have about the results of the activity and also to provide food for further thought—sort of a "Minds-On Chemistry." So let's begin!

Hands-On Chemistry: Penny Fingers

Conceptual Chemistry Alive!

Pennies dated 1982 or earlier are nearly pure copper, each having a mass of about 3.5 grams. Pennies dated after 1982 are made of copper-coated zinc, each having a mass of about 2.9 grams. Hold a pre-1982 penny on the tip of your right index finger and a post-1982 penny on the tip of your left index finger. Move your forearms up and down to feel the difference in inertia—the difference of 0.6 grams (600 milligrams) is subtle but not beyond a set of well-tuned senses. If one penny on each finger is below your threshold, try two pre-1982s stacked on one finger and two post-1982s stacked on the other. Share this activity with a friend.

Mass is easy to understand. It is simply a measure of the amount of matter in a sample, which is a function of how many atoms the sample contains. Accordingly, the mass of an object remains the same no matter where it is located. A one-kilogram gold bar, for example, has the same mass whether it is on the Earth, on the moon, or floating "weightless" in space. This is because it contains the same number of atoms in each location.

Weight is more complicated. By definition, **weight** is the gravitational force exerted on an object by the nearest most massive body, such as the Earth. The weight of an object, therefore, depends entirely upon its location, as is shown in Figure 1.15. On the moon, a gold bar weighs less than it does on the Earth. This is because the moon is much less massive than the Earth; hence, the gravitational force exerted by the moon on the bar is much less. On Jupiter, the gold bar would weigh more than it does on the Earth because of the greater gravitational force exerted on the bar by this very massive planet.

Because mass is independent of location, it is customary in science to measure matter by its mass rather than its weight. *Conceptual Chemistry* adheres to this convention by presenting matter in units of mass, such as kilograms, grams, and milligrams. Such weight units as pounds and tons (1 ton = 2000 pounds) are occasionally provided as a reference because of their familiarity.

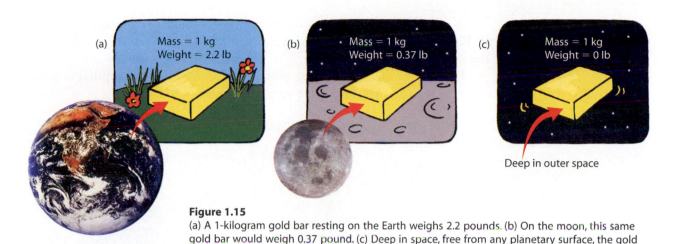

Figure 1.15
(a) A 1-kilogram gold bar resting on the Earth weighs 2.2 pounds. (b) On the moon, this same gold bar would weigh 0.37 pound. (c) Deep in space, free from any planetary surface, the gold bar would weigh 0 pounds but still have a mass of 1 kilogram.

Concept Check ✓

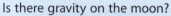

Is there gravity on the moon?

Was this your answer? Yes, absolutely! The moon exerts a downward gravitational pull on any body near its surface, as evidenced by the fact that astronauts were able to land and walk on the moon. This NASA photograph shows an astronaut jumping. Without gravity, this jump would have been his last.

So, since there is gravity, why doesn't the flag droop downward? Look carefully and you'll see that it is held up by a stick across its top edge. There is no wind on the moon (because there is no atmosphere), and so the crew used the support stick to make the wrinkled flag display nicely in photographs.

The amount of space a material object occupies is its **volume**. The SI unit of volume is the liter, which is only slightly larger than the USCS unit of volume, the quart. A liter is the volume of space marked off by a cube measuring 10 centimeters by 10 centimeters by 10 centimeters, which is 1000 cubic centimeters. A smaller unit of volume is the milliliter, which is ¹⁄₁₀₀₀ of a liter, or 1 cubic centimeter.

A convenient way to measure the volume of an irregular object is shown in Figure 1.16. The volume of water displaced is equal to the volume of the object.

Figure 1.17 gives you a sense of the relative sizes of some familiar objects, some very large ones, and some very small ones.

Figure 1.16
The volume of an object, no matter what its shape, can be measured by the displacement of water. When this rock is immersed in the water, the rise in the water level equals the volume of the rock, which in this example measures about 90 mL.

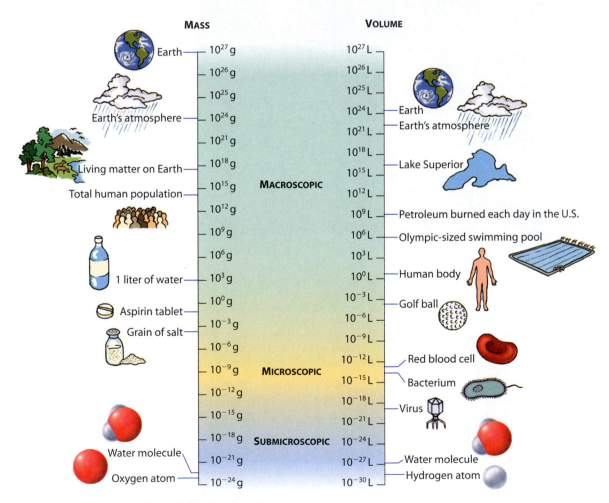

Figure 1.17
Range of masses and volumes in the universe.

Hands-On Chemistry: Decisive Dimensions

This activity challenges any misconceptions you might have about what determines volume as it answers the question "Can one object have a greater mass than another and yet have the same volume?" Try this activity and find out.

What You Need

Tall, narrow glass; masking tape; empty film canister with lid; bunch of pennies

Procedure

① Fill the glass two-thirds full with water and mark the water level with masking tape.

② Fill the canister with pennies. Cap the canister and place it in the water. Note the new water level with a second strip of tape.

③ Remove the canister, being careful not to splash water out of the glass. If the water level after you remove the canister is below the original level, add water until the volume is again at that level.

④ Remove half the pennies from the canister so as to decrease its mass. Cap the canister, predict how much the water level will rise when you submerge the canister, and then submerge it.

Which statement do your results support:

    a. The volume of water an object displaces depends only on the dimensions of the object and not on its mass.

    b. The volume of water an object displaces depends on both the dimensions and the mass of the object.

1.5   Energy Is the Mover of Matter

Matter is substance, and energy is that which can move substance. The concept of energy is abstract and therefore not as easy to define as the concepts of mass and volume. One definition of **energy** is the capacity to do work. If something has energy, it can do work on something else—it can exert a force and move that something else. Accordingly, energy is not something we observe directly. Rather, we only witness its effects.

There are two principal forms of energy: potential and kinetic. **Potential energy** is stored energy. A boulder perched on the edge of a cliff has potential energy due to the force of gravity, just as the poised arrow in Figure 1.18 has potential energy due to the tension force of the bow. The potential energy of an object increases as the distance over which the force is able to act increases. The higher a boulder is positioned above level ground, the more potential energy it has to do work as it falls downward under the pull of gravity. Similarly, an arrow in a fully drawn bow, for example, has more potential energy than does one in a half-drawn bow.

Figure 1.18
Much of the potential energy in Ian's drawn bow will be converted to the kinetic energy of the arrow upon its release.

Kinetic energy is the energy of motion. Both a falling boulder and a flying arrow have kinetic energy. The faster a body moves, the more kinetic energy it has and therefore the more work it can do. For example, the faster an arrow flies, the more work it can do to a target, as evidenced by its deeper penetration.

Concept Check ✓

How does a flying arrow have potential energy as well as kinetic energy?

Was this your answer? It has potential energy as long as it remains above the ground. Once it reaches the ground, its potential energy is zero because it no longer has any potential to do work.

Chemical substances possess what is known as *chemical potential energy*, which is the energy that is stored within atoms and molecules. Any material that can burn has chemical potential energy. The firecracker in Figure 1.19, for example, has chemical potential energy. This energy gets released when the firecracker is ignited. During the explosion, some of the chemical potential energy is transformed to the kinetic energy of flying particles. Much of the chemical potential energy is also transformed to light and heat. We explore the relationship between energy and chemical reactions in Chapter 9.

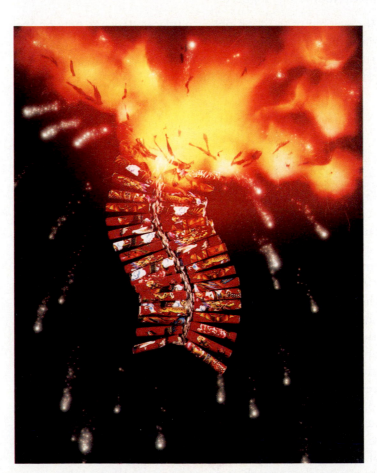

Figure 1.19
A firecracker is a mixture of solids that possess chemical potential energy. When a firecracker explodes, the solids react to form gases that fly outward and so possess a great deal of kinetic energy. Light and heat (both of which are forms of energy) are also formed.

Concept Check ✓

How does a wooden arrow lying on the ground have potential energy?

Was this your answer? The arrow contains chemical potential energy because it can burn.

The SI unit of energy is the *joule*, which is about the amount of energy released from a candle burning for only a moment. In the United States, a common unit of energy is the *calorie*. One calorie is by definition the amount of energy required to raise the temperature of 1 gram of water by 1 degree Celsius. One calorie is 4.184 times larger than 1 joule. Put differently, 4.184 joules of energy is equivalent to 1 calorie (4.184 joules = 1 calorie). So, a joule is about one-fourth of a calorie.

In the United States, the energy content of food is measured by the *Calorie* (note the uppercase C). One Calorie equals 1 kilocalorie, which is 1000 calories (note the lowercase c). The candy bar in Figure 1.20 offers 230 Calories (230 kilocalories), bestowing a total of 230,000 calories to the consumer.

Figure 1.20
The energy content of this candy bar (230 Calories = 230,000 calories), when released through burning, is enough to heat up 230,000 grams (about 507 pounds) of water by 1 degree Celsius.

1.6 Temperature Is a Measure of How Hot—Heat It Is Not

The atoms and molecules that form matter are in constant motion, jiggling to and fro or bouncing from one position to another. By virtue of their motion, these particles possess kinetic energy. Their average kinetic energy is directly related to a property you can sense: how hot something is. Whenever something becomes warmer, the kinetic energy of its submicroscopic particles increases. For example, strike a penny with a hammer and the penny becomes warm because the hammer's blow causes its atoms to jostle faster, increasing their kinetic energy (the hammer becomes warm for the same reason). Put a flame to a liquid and the liquid becomes warmer because the energy of the flame causes the particles of the liquid to move faster, increasing their kinetic energy. For example, the molecules in the hot coffee in Figure 1.21 are moving faster, on average, than those in the cold coffee.

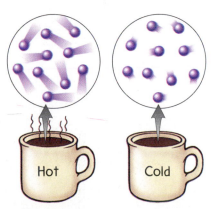

Figure 1.21
The difference between hot coffee and cold coffee is the average speed of the molecules. In the hot coffee, the molecules are moving faster, on average, than they are in the cold coffee. (The longer "motion trails" on the molecules of hot coffee indicate their higher speed.)

Figure 1.22
Can we trust our sense of hot and cold? Will both fingers feel the same temperature when they are put in the warm water? Try this yourself, and you will see why we use a thermometer for an objective measurement.

The quantity that tells us how warm or cold an object is relative to some standard is called **temperature**. We express temperature by a number that corresponds to the degree of hotness on some chosen scale. Just touching an object certainly isn't a good way of measuring its temperature, as Figure 1.22 illustrates. To measure temperature, therefore, we take advantage of the fact that nearly all materials expand when their temperature is raised and contract when it is lowered. With increasing temperature, the particles move faster and are on average farther apart—the material expands. With decreasing temperature, the particles move more slowly and are on average closer together—the material contracts. A **thermometer** exploits this characteristic of matter, measuring temperature by means of the expansion and contraction of a liquid, usually mercury or colored alcohol.

Concept Check ✓

You may have noticed telephone wires sagging on a hot day. This happens because the wires are longer in hot weather than in cold. What is happening on the atomic level to cause such changes in wire length?

Was this your answer? On a hot day, the atoms in the wire are moving faster, and as a result the wire expands. On a cold day, those same atoms are moving more slowly, which causes the wire to contract.

The most common thermometer in the world is the Celsius thermometer, named in honor of the Swedish astronomer Anders Celsius (1701–1744), who first suggested the scale of 100 degrees between the freezing point and boiling point of fresh water. In a Celsius thermometer, the number 0 is assigned to the temperature at which pure water freezes and the number 100 is assigned to the temperature at which it boils (at standard atmospheric pressure), with 100 equal parts called *degrees* between these two points.

In the United States, we use a Fahrenheit thermometer, named after its originator, the German scientist G. D. Fahrenheit (1686–1736), who chose to assign 0 to the temperature of a mixture containing equal weights of snow and common salt and 100 to the body temperature of a human. Because these reference points are not dependable, the Fahrenheit scale has since been modified such that the freezing point of pure water is designated 32°F and the boiling point of pure water is designated 212°F. On this recalibrated scale, normal human body temperature is around 98.6°F.

A temperature scale favored by scientists is the Kelvin scale, named after the British physicist Lord Kelvin (1824–1907). This scale is calibrated not in terms of the freezing and boiling points of water but rather in terms of the motion of atoms and molecules. On the Kelvin scale, zero is the temperature at which there is no atomic or molecular motion. This is a theoretical limit called **absolute zero**, which is the temperature at which the particles of a substance have absolutely no kinetic energy to give up. Absolute zero corresponds to −459.7°F on the Fahrenheit scale

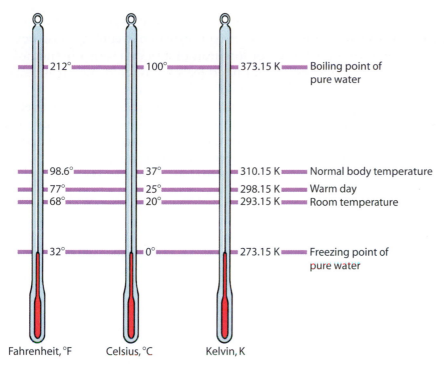

Figure 1.23
Some familiar temperatures measured on the Fahrenheit, Celsius, and Kelvin scales.

and −273.15°C on the Celsius scale. On the Kelvin scale, this temperature is simply 0 K, which is read "zero kelvin" or "zero K." Marks on the Kelvin scale are the same distance apart as those on the Celsius scale, and so the temperature of freezing water is +273 kelvin. (Note that the word *degree* is not used with the Kelvin scale. To say "273 degrees kelvin" is incorrect. To say "273 kelvin" is correct.) The three scales are compared in Figure 1.23.

It is important to understand that temperature is a measure of the *average* amount of energy in a substance, not the *total* amount of energy, as Figure 1.24 shows. The total energy in a swimming pool full of boiling water is much more than the total energy in a cupful of boiling water even though both are at the same temperature. Your utility bill after heating the swimming pool water to 100°C would show this. Whereas the total amount of energy in the pool is much more than in the cup, the *average* molecular motion is the same in both water samples. The water molecules in the swimming pool are moving on average just as fast as the water molecules in the cup. The only difference is that the swimming pool contains more water molecules and hence a greater total amount of energy.

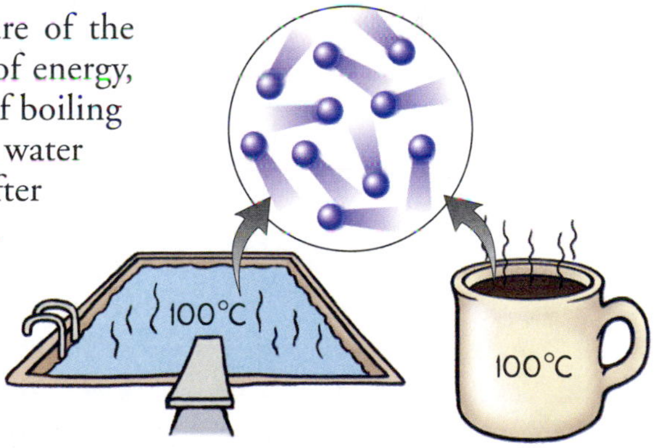

Heat is energy that flows from a higher-temperature object to a lower-temperature object. If you touch a hot stove, heat enters your hand because the stove is at a higher temperature than your hand. When you touch a piece of ice, energy passes out of your hand and into the ice because the ice is at a lower temperature than your hand. From a human

Figure 1.24
Bodies of water at the same temperature have the same average molecular kinetic energies. The *volume* of the water has nothing to do with its temperature.

perspective, if you are receiving heat you experience warmth; if you are giving away heat, you experience cooling. The next time you touch the hot forehead of a sick, feverish friend, ask her or him whether your hand feels hot or cold. Whereas temperature is absolute, hot and cold are relative.

In general, the greater the temperature difference between two bodies in contact with each other, the greater the rate of heat flow. This is why a hot clothes iron can cause much more damage to your skin than a warm clothes iron.

Because heat is a form of energy, its unit is the joule.

Concept Check ✔

When you enter a swimming pool, the water may feel quite cold. After a while, though, your body "gets used to it," and the water no longer feels so cold. Use the concept of heat to explain what is going on.

Was this your answer? Heat flows because of a temperature difference. When you enter the water, your skin temperature is much higher than the water temperature. The result is a significant flow of heat from your body to the water, which you experience as cold. Once you have been in the water awhile, your skin temperature is much closer to the water temperature (due to the cooling effects of the water and your body's ability to conserve heat), and so the flow of heat from your body is less. With less heat flowing from your body, the water no longer feels so cold.

1.7 The Phase of a Material Depends on the Motion of Its Particles

One of the most apparent ways we can describe matter is by its physical form, which may be in one of three phases (also sometimes described as physical states): *solid*, *liquid*, or *gas*. A **solid** material, such as a rock, occupies a constant amount of space and does not readily deform upon the application of pressure. In other words, a solid has both definite volume and definite shape. A **liquid** also occupies a constant amount of space (it has a definite volume), but its form changes readily (it has an indefinite shape). A liter of milk, for example, may take the shape of its carton or the shape of a puddle, but its volume is the same in both cases. A **gas** is diffuse, having neither definite volume nor definite shape. Any sample of gas assumes both the shape and the volume of the container it occupies. A given amount of air, for example, may assume the volume and shape of a toy balloon or the volume and shape of a bicycle tire. Released from its container, a gas diffuses into the atmosphere, which is a collection of various gases held to our planet only by the force of gravity.

On the submicroscopic level, the solid, liquid, and gaseous phases are distinguished by how well the submicroscopic particles hold together. This is illustrated in Figure 1.25. In solid matter, the attractions between particles are strong enough to hold all the particles together in some fixed three-dimensional arrangement. The particles are able to vibrate about fixed positions, but they cannot move past one another. Adding heat causes these vibrations to increase until, at a certain temperature, the vibrations are rapid enough to disrupt the fixed arrangement. The particles can then slip past one another and tumble around much like a bunch of marbles in a bag. This is the liquid phase of matter, and it is the mobility of the submicroscopic particles that gives rise to the liquid's fluid character—its ability to flow and take on the shape of its container.

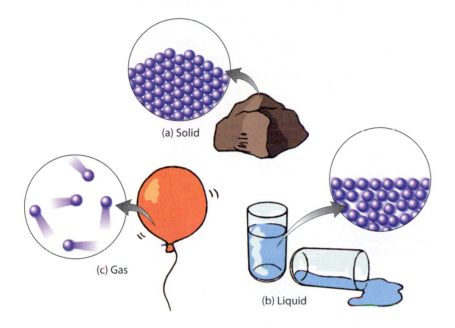

(a) Solid

(c) Gas

(b) Liquid

Figure 1.25
The familiar bulk properties of a solid, liquid, and gas. (a) The submicroscopic particles of the solid phase vibrate about fixed positions. (b) The submicroscopic particles of the liquid phase slip past one another. (c) The fast-moving submicroscopic particles of the gaseous phase are separated by large average distances.

Further heating causes the submicroscopic particles in the liquid to move so fast that the attractions they have for one another are unable to hold them together. They then separate from one another, forming a gas. Moving at an average speed of 500 meters per second (1100 miles per hour), the particles of a gas are widely separated from one another. Matter in the gaseous phase therefore occupies much more volume than it does in the solid or liquid phase, as Figure 1.26 shows. Applying pressure to a gas squeezes the gas particles closer together, which makes for a smaller volume.

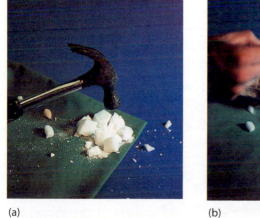

(a)

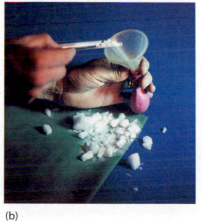

(b)

(c)

Figure 1.26
The gaseous phase of any material occupies significantly more volume than either its solid or liquid phase. (a) Solid carbon dioxide (dry ice) is broken up into powder form. (b) The powder is funneled into a balloon. (c) The balloon expands as the contained carbon dioxide becomes a gas as the powder warms up.

Figure 1.27
In traveling from point A to point B, the typical gas particle travels a circuitous path because of numerous collisions with other gas particles—about eight billion collisions every second! The changes in direction shown here represent only a few of these collisions. Although the particle travels at very high speeds, it takes a relatively long time to cross between two distant points because of these numerous collisions.

Enough air for an underwater diver to breathe for many minutes, for example, can be squeezed (compressed) into a tank small enough to be carried on the diver's back.

Although gas particles move at high speeds, the speed at which they can travel from one side of a room to the other is relatively slow. This is because the gas particles are continually hitting one another, and the path they end up taking is circuitous. At home, you get a sense of how long it takes for gas particles to migrate each time someone opens the oven door after baking, as Figure 1.27 shows. A shot of aromatic gas particles escapes the oven, but there is a notable delay before the aroma reaches the nose of someone sitting in the next room.

Concept Check

Why are gases so much easier to compress into smaller volumes than are solids and liquids?

Was this your answer? Because there is a lot of space between gas particles. The particles of a solid or liquid, on the other hand, are already close to one another, meaning there is little room left for a further decrease in volume.

Familiar Terms Are Used to Describe Changing Phases

Figure 1.28 illustrates that you must either add heat to a substance or remove heat from it if you want to change its phase. The process of a solid transforming to a liquid is called **melting**. To visualize what happens when heat begins to melt a solid, imagine you are holding hands with a group of people and each of you start jumping around randomly. The more violently you jump, the more difficult it is to hold onto one another. If everyone jumps violently enough, keeping hold is impossible. Something like this happens to the submicroscopic particles of a solid when it is heated. As heat is added to the solid, the particles vibrate more and more violently. If enough heat is added, the attractive forces between the particles are no longer able to hold them together. The solid melts.

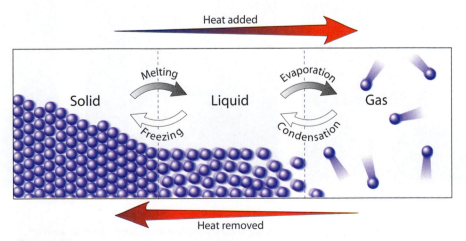

Figure 1.28
Melting and evaporation involve the addition of heat; condensation and freezing involve the removal of heat.

A liquid can be changed to a solid by the removal of heat. This process is called **freezing**, and it is the reverse of melting. As heat is withdrawn from the liquid, particle motion diminishes until the particles, on average, are moving slowly enough for attractive forces between them to take permanent hold. The only motion the particles are capable of then is vibration about fixed positions, which means the liquid has solidified, or frozen.

A liquid can be heated so that it becomes a gas—a process called **evaporation**. As heat is added, the particles of the liquid acquire more kinetic energy and move faster. Particles at the liquid surface eventually gain enough energy to jump out of the liquid and enter the air. In other words, they enter the gaseous phase. As more and more particles absorb the heat being added, they too acquire enough energy to escape from the liquid surface and become gas particles. Because a gas results from evaporation, this phase is also sometimes referred to as *vapor*. Water in the gaseous phase, for example, may be referred to as water vapor.

The rate at which a liquid evaporates increases with temperature. A puddle of water, for example, evaporates from a hot pavement more quickly than it does from your cool kitchen floor. When the temperature is hot enough, evaporation occurs beneath the surface of the liquid. As a result, bubbles form and are buoyed up to the surface. We say that the liquid is **boiling**. A substance is often characterized by its *boiling point*, which is the temperature at which it boils. At sea level, the boiling point of fresh water is 100°C.

The transformation from gas to liquid—the reverse of evaporation—is called **condensation**. This process can occur when the temperature of a gas decreases. The water vapor held in the warm daylight air, for example, may condense to form a wet dew in the cool of the night.

Hands-On Chemistry: Hot-Water Balloon

See for yourself that a material in its gaseous phase occupies much more space than it does in its liquid phase.

What You Need

2 teaspoons of water, 9-inch rubber balloon, microwave oven, oven mitt, safety glasses

Procedure

① Pour the water into the balloon. Squeeze out as much air as you can and knot the balloon.

② Put the balloon in the microwave oven and cook at full power for however many seconds it takes for boiling to begin, which is indicated by a rapid growth in the size of the balloon. It may take only about 10 seconds for the balloon to reach full size once it starts expanding. (The balloon will pop if you add too much water or if you cook it for too long.)

③ Remove the heated balloon with the oven mitt, shake the balloon around, and listen for the return of the liquid phase. You should be able to hear it raining inside the balloon.

What happens if you submerge the inflated balloon in a pot of ice-cold water?

1.8 Density Is the Ratio of Mass to Volume

The relationship between an object's mass and the amount of space it occupies is the object's density. Density is a measure of compactness, of how tightly mass is squeezed into a given volume. A block of lead has much more mass squeezed into its volume than does a same-sized block of aluminum. The lead is therefore more dense. We think of density as the "lightness" or "heaviness" of objects of the same size, as Figure 1.29 shows.

Same-sized block of aluminum

Block of lead

Figure 1.29
The amount of mass in a block of lead far exceeds the amount of mass in a block of aluminum of the same size. Hence, the lead weighs much more and is more difficult to lift.

Density is the amount of mass contained in a sample divided by the volume of the sample:

$$\text{density} = \frac{\text{mass}}{\text{volume}}$$

An object having a mass of 1 gram and a volume of 1 milliliter, for example, has a density of

$$\text{density} = \frac{1 \text{ g}}{1 \text{ mL}} = 1\frac{\text{g}}{\text{mL}}, \text{ which reads "one gram per milliliter"}$$

An object having a mass of 2 grams and a volume of 1 milliliter is denser; its density is

$$\text{density} = \frac{2 \text{ g}}{1 \text{ mL}} = 2\frac{\text{g}}{\text{mL}}, \text{ which reads "two grams per milliliter"}$$

Other units of mass and volume besides gram and milliliters may be used in calculating density. The densities of gases, for example, because they are so low, are often given in grams per liter. In all cases, however, the units are a unit of mass divided by a unit of volume.

Concept Check ✓

Which occupies a greater volume: 1 kilogram of lead or 1 kilogram of aluminum?

Was this your answer? The aluminum. Think of it this way. Because lead is so dense, you need only a little bit in order to have 1 kilogram. Aluminum, by contrast, is far less dense, and so 1 kilogram of aluminum occupies much more volume than the same mass of lead.

Calculation Corner: Manipulating an Algebraic Equation

With a little algebraic manipulation, it is easy to change the equation for density around so that it solves either for mass or for volume. The first step is to multiply both sides of the density equation by volume. Then canceling the volumes that appear in the numerator and denominator results in the equation for the mass of an object:

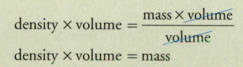

$$\text{density} \times \text{volume} = \frac{\text{mass} \times \cancel{\text{volume}}}{\cancel{\text{volume}}}$$

$$\text{density} \times \text{volume} = \text{mass}$$

Example

A pre-1982 penny has a density of 8.92 grams per milliliter and a volume of 0.392 milliliters. What is its mass?

Answer

$$D \times V = M = 8.92 \frac{g}{\cancel{mL}} \times 0.392 \; \cancel{mL} = 3.50 \; g$$

To solve for the volume of an object, divide both sides of the equation for the mass by the density. Canceling the densities results in the equation for volume:

$$\frac{\cancel{\text{density}} \times \text{volume}}{\cancel{\text{density}}} = \frac{\text{mass}}{\text{density}}$$

$$\text{volume} = \frac{\text{mass}}{\text{density}}$$

Example

A post-1982 penny has a density of 7.40 grams per milliliter and a mass of 2.90 grams. What is its volume?

Answer

$$V = \frac{M}{D} = \frac{2.90 \; g}{7.40 \frac{g}{mL}} = 0.392 \; mL$$

In summary, the three equations expressing the relationship among density, mass, and volume are

Density	**Mass**	**Volume**
$D = \dfrac{M}{V}$	$M = D \times V$	$V = \dfrac{M}{D}$

A good way to remember these relationships is to use the digram shown below. Use your finger to cover the quantity you want to know, and that quantity's relationship to the other quantities is revealed. For example, covering the M shows that mass is equal to density times volume, $D \times V$.

Your Turn

1. What is the average density of a loaf of bread that has a mass of 500 grams and a volume of 1000 milliliters?

2. The loaf of bread in the previous problem loses all its moisture after being left out for several days. Its volume remains at 1000 milliliters, but its density has been reduced to 0.4 grams per milliliter. What is its new mass?

3. A sack of groceries accidently set on a 500-gram loaf of bread increases the average density of the loaf to 5 grams per milliliter. What is its new volume?

Table 1.3

Densities of Some Solids, Liquids, and Gases

Substance	Density (g/mL)	Density (g/L)
Solids		
osmium	22.5	22,500
gold	19.3	19,300
lead	11.3	11,300
copper	8.92	8,920
iron	7.86	7,860
zinc	7.14	7,140
aluminum	2.70	2,700
ice	0.92	920
Liquids		
mercury	13.6	13,600
sea water	1.03	1,030
water at 4°C	1.00	1,000
ethyl alcohol	0.81	810
*Gases**		
dry air		
0°C	0.00129	1.29
20°C	0.00121	1.21
helium at 0°C	0.000178	0.178
oxygen at 0°C	0.00143	1.43

*All values at sea-level atmospheric pressure.

The densities of some substances are given in Table 1.3. Which would be more difficult to pick up: a liter of water or a liter of mercury?

Gas densities are much more affected by pressure and temperature than are the densities of solids and liquids. With an increase in pressure, gas molecules are squeezed closer together. This makes for less volume and therefore greater density. The density of the air inside a diver's breathing tank, for example, is much greater than the density of air at normal atmospheric pressure. With an increase in temperature, gas molecules are moving faster and thus have a tendency to push outward, thereby occupying a greater volume. Thus, hot air is less dense than cold air, which is why hot air rises and the balloon in Figure 1.30 can take its passengers for a breathtaking ride.

Concept Check ✔

1. Which has a greater density: 1 gram of water or 10 grams of water?
2. Which has a greater density: 1 gram of lead or 10 grams of aluminum?

Was this your answer?

1. The density is the same for any amount of water. Whereas 1 gram of water occupies a volume of 1 milliliter, 10 grams occupies a volume of 10 milliliters. The ratio 1 gram/1 milliliter is the same as the ratio 10 grams/10 milliliters.
2. The lead. Density is mass per volume, and this ratio is greater for any amount of lead than for any amount of aluminum.

Figure 1.30
The hot air inside this hot-air balloon is less dense than the surrounding colder air, which is why the balloon rises.

In Perspective

In this chapter you were introduced to how chemistry is the study of matter and how the scope of chemistry is very broad—including anything you can touch, taste, smell, see, or hear. Chemistry is a major branch of science, which this chapter defined as an organized body of knowledge resulting from our observations, common sense, rational thinking, and insights into nature. After centuries of development, science has become a powerful tool for helping us to perceive our environment with greater clarity.

To help you in your study of chemistry, it's important that you be familiar with some basic physical quantities. These include mass, volume, energy, temperature, and density. *Mass* is a measure of how much, whereas *volume* is a measure of how spacious. *Energy* is an abstract concept but best understood as that which is required to move matter. The higher the *temperature* of a material, the greater the average kinetic energy of its submicroscopic particles.

The phase of a material is a function of its temperature. In the solid phase, the atoms or molecules only vibrate about fixed positions. In the liquid phase, these submicroscopic particles are able to tumble over one another. In the gaseous phase, the atoms or molecules have enough kinetic energy to be separated from one another by relatively large distances. Finally, this chapter introduced *density*, which is the ratio of a material's mass to its volume. Materials are often characterized by their densities.

A Word About Chapter Endmatter from the Author

Each chapter in this book concludes with a list of Key Terms and Matching Definitions, Review Questions, Hands-On Chemistry Insights, Exercises, and in some chapters, a few mathematical Problems.

Key Terms are listed alphabetically and followed by **Matching Definitions** listed in order of appearance in the chapter. For a guided review of the Key Terms, find the term that matches each definition. To get the most out of this format, match as many definitions as you can before checking back in the chapter. The more familiar you are with these terms, the easier it will be for you to apply the concepts.

The **Review Questions** help you fix ideas more firmly in your mind and catch the essentials of the chapter material. Like the Key Terms and Matching Definitions, they are not meant to be difficult. You can find the answers to the Review Questions in the chapter. If you study only the Key Terms and Review Questions and nothing else, you are minimizing your chance for success in this class.

The **Hands-On Chemistry Insights** are follow-up discussions designed to make sure you are getting the most out of performing these activities and also to clear up any misconceptions that may develop. Ideally, you should read these follow-ups after performing the activities.

In contrast to the Review Questions, the **Exercises** are designed to challenge your understanding of the chapter material. They emphasize thinking rather than mere recall and should be attempted only after you are well acquainted with the chapter through the Matching Definitions and Review Questions. In many cases, the intention of a particular Exercise is to help

you to apply the ideas of chemistry to familiar situations. Your answers should be in complete sentences, with an explanation or sketch when applicable. Exercises are a favorite among instructors for exam material. If you do well with the Exercises, you can expect to do well on your exams. To help keep you on track, the answers to all odd-numbered exercises and problems are provided in Appendix C.

Problems feature concepts that are more clearly understood with numerical values and straightforward calculations. Most are based on information given in the chapter's *Calculation Corners*. Be sure to include units in your answers. The Problems are relatively few in number to avoid an emphasis on problem solving that could obscure the primary goal of *Conceptual Chemistry*—a focus on the concepts of chemistry and how they relate to everyday living.

Want to succed in your chemistry course? For every hour you spend in class, you should spend about three hours outside of class studying. This is an investment in yourself that will pay off in more ways than just a good grade, so don't delay. Find a comfortable place to study and go to it! Better yet, study with a partner and practice articulating what you've learned. You know you haven't really learned something until you can express it using your own voice.

Key Terms and Matching Definitions

_____ absolute zero
_____ applied research
_____ basic research
_____ boiling
_____ chemistry
_____ condensation
_____ control test
_____ density
_____ energy
_____ evaporation
_____ freezing
_____ gas
_____ heat
_____ kinetic energy
_____ liquid
_____ mass
_____ matter
_____ melting
_____ potential energy
_____ science
_____ scientific hypothesis
_____ solid
_____ submicroscopic
_____ temperature
_____ theory
_____ thermometer
_____ volume
_____ weight

1. An organized body of knowledge resulting from our observations, common sense, rational thinking, and insights into nature.
2. The realm of atoms and molecules, where objects are smaller than can be detected by optical microscopes.
3. The study of matter and the transformations it can undergo.
4. Anything that occupies space.
5. Research dedicated to the discovery of the fundamental workings of nature.
6. Research dedicated to the development of useful products and processes.
7. A testable assumption often used to explain an observed phenomenon.
8. A test performed by scientists to increase the conclusiveness of an experimental test.
9. A comprehensive idea that can be used to explain a broad range of phenomena.

10. The quantitative measure of how much matter an object contains.
11. The gravitational force of attraction between two bodies (where one body is usually the Earth).
12. The quantity of space an object occupies.
13. The capacity to do work.
14. Stored energy.
15. Energy due to motion.
16. How warm or cold an object is relative to some standard. Also, a measure of the average kinetic energy per molecule of a substance, measured in degrees Celsius, degrees Fahrenheit, or kelvins.
17. An instrument used to measure temperature.
18. The lowest possible temperature any substance can have; the temperature at which the atoms of a substance have no kinetic energy: 0 K = −273.15°C = −459.7°F.
19. The energy that flows from one object to another because of a temperature difference between the two.
20. Matter that has a definite volume and a definite shape.
21. Matter that has a definite volume but no definite shape, assuming the shape of its container.
22. Matter that has neither a definite volume nor a definite shape, always filling any space available to it.
23. A transformation from a solid to a liquid.
24. A transformation from a liquid to a solid.
25. A transformation from a liquid to a gas.
26. Evaporation in which bubbles form beneath the liquid surface.
27. A transformation from a gas to a liquid.
28. The ratio of an object's mass to its volume.

Review Questions

Chemistry Is a Central Science Useful to Our Lives

1. Are atoms made of molecules, or are molecules made of atoms?
2. What is the difference between basic research and applied research?
3. Why is chemistry often called the central science?
4. What do members of the Chemical Manufacturers Association pledge in the Responsible Care program?

Science Is a Way of Understanding the Universe

5. Why were McClintock and Baker exploring the oceans off Antarctica?

6. What evidence supported McClintock and Baker's hypothesis that amphipods abducted sea butterflies for chemical defense against predators?

7. What is the purpose of a control test?

8. Why is reproducibility such a vital component of science?

9. Does a theory become stronger or weaker the more it is modified to account for experimental evidence?

10. What kinds of questions is science unable to answer?

Scientists Measure Physical Quantities

11. Why are the units of a measurement just as important as the number?

12. What are the two major systems of measurement used in the world today?

13. Why are prefixes used in the metric system?

14. Which is greater: a micrometer or a decimeter?

15. A kilogram is equal to how many grams?

16. A milligram is equal to how many grams?

Mass Is How Much and Volume Is How Spacious

17. What is inertia, and how is it related to mass?

18. Which is the more complicated concept: mass or weight?

19. Which can change from one location to another: mass or weight?

20. What is volume?

21. What is the difference between an object's mass and its volume?

Energy Is the Mover of Matter

22. Why is energy hard to define?

23. What do we call the energy an object has because of its position?

24. What do we call the energy an object has because of its motion?

25. Which represents more energy: a joule or a calorie?

26. Which represents more energy: a calorie or a Calorie?

Temperature Is a Measure of How Hot—Heat It Is Not

27. In which is the average speed of the molecules less: in cold coffee or in hot coffee?

28. What happens to the volume of most materials when they are heated?

29. Which temperature scale has its zero point as the point of zero atomic and molecular motion?

30. Which has more total energy: a cup of boiling water at 100°C or a swimming pool of boiling water at 100°C?

31. Is it natural for heat to travel from a cold object to a hotter object?

32. What determines the direction of heat flow?

The Phase of a Material Depends on the Motion of Its Particles

33. How are the particles in a solid arranged differently from those in a liquid?

34. How does the arrangement of particles in a gas differ from the arrangements in liquids and solids?

35. Which occupies the greatest volume: 1 gram of ice, 1 gram of liquid water, or 1 gram of water vapor?

36. Gas particles travel at speeds of up to 500 meters per second. Why, then, does it take so long for gas molecules to travel the length of a room?

37. As liquid water evaporates, what is happening to the molecules?

38. Which requires the extraction of energy: melting or freezing?

39. Which requires the input of energy: evaporation or condensation?

40. What is it called when evaporation takes place beneath the surface of a liquid?

Density Is the Ratio of Mass to Volume

41. Is a more massive object necessarily more dense than a less massive object?

42. Is a denser object necessarily more massive than a less dense object?

43. The units of density are a ratio of what two quantities?

44. What happens to the density of a gas as it is compressed into a smaller volume?

Hands-On Chemistry Insights

Penny Fingers

You have to be moving the pennies up and down in order to optimize your threshold of detection. What you are sensing here is the *difference* in inertia. Recall that inertia is an object's resistance to any change in its motion. If you minimize the motion, you minimize your ability to detect any difference in inertia the two coins may have. Switch pennies between your left and right index fingers (with your eyes closed) to confirm your ability to detect a difference.

With or without the motion, the pennies exert a downward pressure that your nerve endings sense. To feel this pressure, repeat the experiment with the pre-1982s on one index finger and the post-1982s on the other index finger, but this time keep your hands still. How many do you need to stack before you can sense a difference in pressure? If you did this pressure experiment on the moon, would you need to stack more or fewer? Why?

Decisive Dimensions

Don't feel bad if your prediction was wrong—you are in good company. But now you understand that the volume of water displaced by an object is equal to the object's volume, not its mass or its weight.

Hot-Water Balloon

A marble hitting your hand pushes against your hand. In a similar fashion, a gaseous water molecule hitting the inside of the balloon pushes against the balloon. The force of a single water molecule is not that great, but the combined forces of the billions and billions of them in this activity is sufficient to inflate the balloon

as the liquid water evaporates to the gaseous phase. Thus, you saw how the gaseous phase occupies much more volume than the liquid phase. If you observed the balloon carefully, you noticed it continues to inflate (although not so rapidly) after all the water has been converted to water vapor. This occurs because the microwaves continue to heat the gaseous water molecules, making them move faster and faster, pushing harder and harder against the balloon's inner surface.

After you take the balloon out of the microwave, the balloon is in contact with air molecules, which, being cooler, move more slowly than the water molecules. Gaseous water molecules colliding with the inner surface of the balloon pass their kinetic energy to the slower air molecules, and the air molecules get warmer because their kinetic energy increases. (This is similar to how the kinetic energy of a hammer pounding a nail into a flimsy wall can be transferred to a picture frame hanging on the opposite side of the wall.) You can feel this warming by holding your hand close to the balloon.

As the gaseous water molecules lose kinetic energy, they begin to condense into the liquid phase, a noisy process amplified by the balloon (listen carefully).

From a molecular point of view, why does the balloon shrink more quickly in ice water? How is this activity similar to the demonstration depicted in Figure 1.26?

Exercises

1. Why is it important to work through the Review Questions before attempting the Exercises?

2. In what sense is a color computer monitor or television screen similar to our view of matter? Place a drop (and only a drop) of water on your computer monitor or television screen for a closer look.

3. Of the three sciences physics, chemistry, and biology, which is the most complex?

4. Is chemistry the study of the submicroscopic, the microscopic, the macroscopic, or all three? Defend your answer.

5. Some politicians take pride in maintaining a particular point of view. They think a change of mind would be seen as a sign of weakness. How is a change of mind viewed differently in science?

6. Why is the process of science not restricted to any one particular method?

7. Distinguish between a scientific hypothesis and a theory.

8. How might the demand for reproducibility in science have the long-run effect of compelling honesty?

9. McClintock and Baker worked together on scientific research projects involving marine organisms of the Antarctic seas, yet they have different scientific backgrounds—McClintock in biology and Baker in chemistry. Is this unusual? Explain.

10. Which of the following are scientific hypotheses?
    a. Stars are made of the lost teeth of children.
    b. Albert Einstein was the greatest scientist ever to have lived.
    c. The planet Mars is reddish because it is coated with cotton candy.
    d. Aliens from outer space have transplanted themselves into the minds of government workers.
    e. Tides are caused by the moon.
    f. You were Abraham Lincoln in a past life.
    g. A human remains self-aware while sleeping.
    h. A human remains self-aware after death.

11. In answer to the question "When a plant grows, where does the material come from?" the ancient Greek philosopher Aristotle (384–322 B.C.) hypothesized that all material came from the soil. Do you consider his hypothesis to be correct, incorrect, or partially correct? What experimental tests do you propose to support your choice?

12. The great philosopher and mathematician Bertrand Russell (1872–1970) wrote, "I think we must retain the belief that scientific knowledge is one of the glories of man. I will not maintain that knowledge can never do harm. I think such general propositions can almost always be refuted by well-chosen examples. What I will maintain—and maintain vigorously—is that knowledge is very much more often useful than harmful and that fear of knowledge is very much more often harmful than useful." Think of examples to support this statement.

13. Name two physical quantities discussed in this chapter that change when a junked car is neatly crushed into a compact cube.

14. Which would you rather have: a decigram of gold or a kilogram of gold?

15. Can an object have mass without having weight? Can it have weight without having mass?

16. Why do we use different units for mass and weight?

17. Gravity on the moon is only one-sixth as strong as gravity on the Earth. What is the mass of a 6-kilogram object on the moon, and what is its mass on the Earth?

18. Does a 2-kilogram solid iron brick have twice as much mass as a 1-kilogram solid iron brick? Twice as much weight? Twice as much volume?

19. Does a 2-kilogram solid iron brick have twice as much mass as a 1-kilogram solid block of wood? Twice as much volume? Explain.

20. Which is more evident: potential or kinetic energy? Explain.

21. Will your body possess energy after you die? If so, what kind?

22. What is temperature a measure of?

23. An old remedy for separating two nested drinking glasses stuck together is to run water at one temperature into the inner glass and then run water at a different temperature over the surface of the outer glass. Which water should be hot and which cold?

24. A Concorde supersonic airplane heats up considerably when traveling through the air at speeds greater than the speed of sound. As a result, the Concorde in flight is about 20 centimeters longer than when it is on the ground. Offer an explanation for this length change from a submicroscopic perspective.

25. Which has more total energy: a cup of boiling water at 100°C or a swimming pool of slightly cooler water at 90°C?

26. If you drop a hot rock into a pail of water, the temperature of the rock and that of the water both change until the two are equal. The rock cools, and the water warms. Does this hold true if the hot rock is dropped into the Atlantic Ocean?

27. Which has stronger attractions among its sub-microscopic particles: a solid at 25°C or a gas at 25°C? Explain.

28. The leftmost diagram below shows the moving particles of a gaseous material within a rigid container. Which of the three boxes on the right (a, b, or c) best represents this material upon the addition of heat.

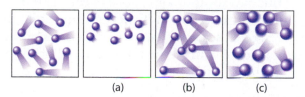

(a)   (b)   (c)

29. The leftmost diagram below shows two phases of a single substance. In the middle box, draw what these particles would look like if heat were taken away. In the box on the right, show what they would look like if heat were added. If each particle represents a water molecule, what is the temperature in the box on the left?

30. Humidity is a measure of the amount of water vapor in the atmosphere. Why is humidity always very low inside your kitchen freezer?

31. What happens to the density of a gas as the gas is compressed into a smaller volume?

32. A post-1982 penny is made with zinc, but its density is greater than that of zinc. Why?

33. The following three boxes represent the number of submicroscopic particles in a given volume of a particular substance at different temperatures. Which box represents the highest density? Which box represents the highest temperature? Why would this be a most unusual substance if box a represented the liquid phase and box b represented the solid phase?

  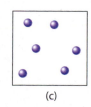

(a)   (b)   (c)

Problems

1. What is the mass in kilograms of a 130-pound human standing on the Earth?

2. What is the mass in kilograms of a 130-pound human standing on the moon?

3. How many joules are there in a candy bar containing 230,000 calories?

4. How many milliliters of dirt are there in a hole that has a volume of 5 liters? How many milliliters of air?

5. Someone wants to sell you a piece of gold and says it is nearly pure. Before buying the piece, you measure its mass to be 52.3 grams and find that it displaces 4.16 milliliters of water. Calculate its density and consult Table 1.3 to assess its purity.

6. What volume of water will a 52.3-gram sample of pure gold displace?

Answers to Calculation Corners

Unit Conversion

a. 7.32 kg	b. 518 lb	c. 4.5 L
d. 2.1 qt	e. 0.1 kcal	f. 400 J

Perhaps you are wondering about how many digits to include in your answers. Were you perplexed, for example, that the answer to f is 400 J and not 418.4 J? There are specific procedures to follow in figuring which digits from your calculator to write down. The digits you are supposed to write down are called *significant figures*. Because there are few calculations in the chapter portions of *Conceptual Chemistry*, however, a full discussion of significant figures is left to Appendix B. It is there for those of you looking for a little more quantitative depth, which is certainly often needed when performing experiments in the laboratory.

Manipulating an Algebraic Equation

1. 0.5 grams per milliliter

2. 400 grams

3. 100 milliliters

Exploring Further

Ronald Breslow, *Chemistry Today and Tomorrow: The Central, Useful, and Creative Science.* Sudbury, MA: Jones and Bartlett, 1997.

Written by a past president of the American Chemical Society, this paperback analyzes the role chemistry has played in the development of our modern society and how chemistry is sure to play an increasing role in our future.

Roald Hoffmann, *The Same and Not the Same.* New York: Columbia University Press, 1995.

Written by a noted Nobel Prize winner, researcher, and teacher, this book seeks to explain the workings of chemistry to the general public, with an emphasis on the beauty of molecular design.

John Allen Paulos, *Innumeracy: Mathematical Illiteracy and Its Consequences.* New York: Vintage Books, 1990.

A delightful account of the statistical misunderstandings of the common citizen. What *illiteracy* is to reading, *innumeracy* is to mathematics.

Royston Roberts, *Serendipity: Accidental Discoveries in Science.* New York: John Wiley & Sons, 1989.

Fascinating stories illustrating how many scientific advances are made by chance discovery when an investigator is keen enough to recognize the significance of the discovery.

Carl Sagan, *The Demon-Haunted World: Science as a Candle in the Dark.* New York: Random House, 1995.

An eloquent account of the present status of society's attitudes toward science and the dangers we face when reason gives way to the myths of pseudoscience, New Age thinking, and fundamentalism.

http://www.awis.org

Home page for the Association for Women in Science, an organization dedicated to achieving equity and full participation for women in science, mathematics, engineering, and technology.

http://www.chemcenter.org

Maintained by the American Chemical Society, this site is an excellent starting point for searching out such chemistry-related information as current events or the status of a particular avenue of research.

http://www.csicop.org

Home page for the Committee for the Scientific Investigation of Claims of the Paranormal. This organization of Nobel laureates and other respected scientists takes on the claims of pseudoscience with all the rigor required of any scientific claim.

http://www.newscientist.com

Web site for the British weekly science and technology newsmagazine. Current events and many "hot issues" in science are presented.

http://www.rsc.org/lap/rsccom/wcc/wccindex.htm

Home page for the Women Chemists Committee of The Royal Society of Chemistry (UK). This organization promotes the entry and re-entry of women into the profession of chemistry and collects and disseminates information about women in chemistry.

http://www.sciencenews.org/sn_arch

Archives of *ScienceNews*, a widely read weekly magazine covering current developments in science.

Chemistry Is a Science
Visit The Chemistry Place at:
www.aw.com/chemplace

2

Elements of Chemistry

Understanding Chemistry Through Its Language

As you progress through this chemistry course, you will note an accumulating list of terms. Chapter 1 introduced 28 key terms, and in this chapter you'll find another 32! Why all these new terms? In the laboratory, chemists perform experiments, make many observations, and then draw conclusions. Over time, the result is a growing body of new knowledge that inevitably exceeds the capacity of everyday language. For example, in the language of chemistry we say that there are more than 100 kinds of *atoms* and that any material consisting of a single kind of atom is an *element*. (Some examples of elements are shown in this chapter's opening photographs.) Atoms can link together to form a *molecule*, and a molecule consisting of atoms from different elements is a *compound*. And on and on, one term building on another as we attempt to describe the nature of matter beyond its casual appearance.

Rather than memorizing all the chemistry-related terms in this text, however, you will serve yourself far better by focusing on the underlying concept each term represents. Based on what you learned in Chapter 1, for example, why is it that hot coffee can burn your tongue? One answer might be that hot coffee contains a large amount of molecular kinetic energy. While true, this answer assumes an understanding of the term *kinetic energy* (Section 1.5). A term is only a label, however. It is possible to know the term without understanding the chemistry—just as it is possible to understand the chemistry without knowing the term. If you truly understand the chemistry but forget the term, you may find yourself comparing the energy of molecular motion to that of a speeding bullet. The faster the bullet—in other words, the greater its kinetic energy—the greater its capacity to cause harm. Similarly, the faster the speed of the molecules in the coffee, the greater their capacity to harm your tongue. So, while *kinetic energy* and other chemistry-related terms are useful for communication, they do not guarantee conceptual understanding. If you focus first on the concepts, the language used to describe them will come to you much more naturally.

Because this chapter focuses on how chemists describe and classify matter, it lays the foundation for all future chapters. Take special note of the **boldfaced** terms, and be sure to practice articulating and paraphrasing the concepts they represent. Do this by describing these concepts aloud to yourself (or to a friend) without looking at the book. When you are able to express these concepts in your own words, you will have the insight to do well in this course and beyond.

We begin by looking at how chemists describe matter by its physical and chemical properties.

2.1 Matter Has Physical and Chemical Properties

Properties that describe the look or feel of a substance, such as color, hardness, density, texture, and phase, are called **physical properties**. Every substance has its own set of characteristic physical properties that we can use to identify that substance (Figure 2.1).

The physical properties of a substance can change when conditions change, but that does not mean a different substance is created. Cooling liquid water to below 0°C causes the water to transform to solid ice,

Gold
Opacity: opaque
Color: yellowish
Phase at 25˚C: solid
Density: 19.3 g/mL

Diamond
Opacity: transparent
Color: colorless
Phase at 25˚C: solid
Density: 3.5 g/mL

Water
Opacity: transparent
Color: colorless
Phase at 25˚C: liquid
Density: 1.0 g/mL

Figure 2.1
Gold, diamond, and water can be identified by their physical properties. If a substance has all the physical properties listed under gold, for example, it must be gold.

but the substance is still water because it is still H_2O no matter which phase it is in. The only difference is how the H_2O molecules are oriented relative to one another. In the liquid, the water molecules tumble around one another, whereas in the ice they vibrate about fixed positions. The freezing of water is an example of what chemists call a **physical change**. During a physical change, a substance changes its phase or some other physical property but *not* its chemical composition, as Figure 2.2 shows.

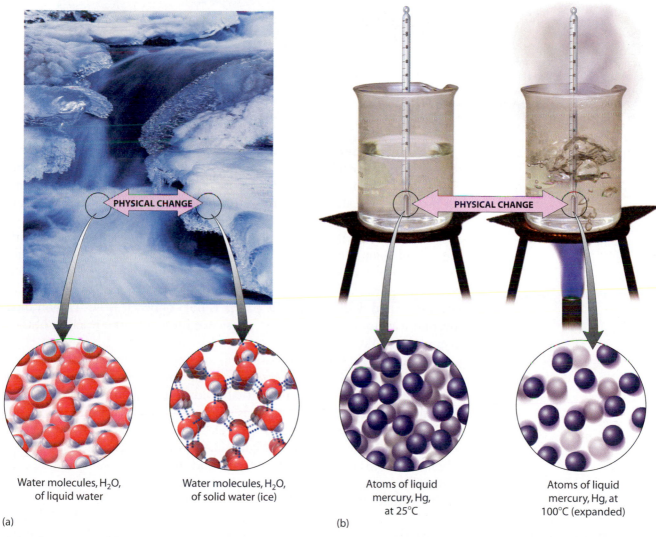

PHYSICAL CHANGE

PHYSICAL CHANGE

Water molecules, H_2O, of liquid water

Water molecules, H_2O, of solid water (ice)

Atoms of liquid mercury, Hg, at 25°C

Atoms of liquid mercury, Hg, at 100°C (expanded)

(a)

(b)

Figure 2.2
Two physical changes. (a) Liquid water and ice might look like different substances, but at the submicroscopic level, it is evident that both consist of water molecules. (b) At 25°C, the atoms in a sample of mercury are a certain distance apart, yielding a density of 13.53 grams per milliliter. At 100°C, the atoms are farther apart, meaning that each milliliter now contains fewer atoms than at 25°C, and the density is now 13.35 grams per milliliter. The physical property we call density has changed with the temperature, but the identity of the substance remains unchanged: mercury is mercury.

Concept Check ✔

The melting of gold is a physical change. Why?

Was this your answer? During a physical change, a substance changes only one or more of its *physical* properties; its *identity* does not change. Because melted gold is still gold but in a different form, this change is a physical change.

Figure 2.3
The chemical properties of substances allow them to transform to new substances. Natural gas and baking soda transform to carbon dioxide, water, and heat. Copper transforms to patina.

Methane
Reacts with oxygen to form carbon dioxide and water, giving off lots of heat during the reaction.

Baking soda
Reacts with vinegar to form carbon dioxide and water, absorbing heat during the reaction.

Copper
Reacts with carbon dioxide and water to form the greenish-blue substance called patina.

Chemical properties are those that characterize the ability of a substance to react with other substances or to transform from one substance to another. Figure 2.3 shows three examples. The methane of natural gas has the chemical property of reacting with oxygen to produce carbon dioxide and water, along with lots of heat energy. Similarly, it is a chemical property of baking soda to react with vinegar to produce carbon dioxide and water while absorbing a small amount of heat energy. Copper has the chemical property of reacting with carbon dioxide and water to form a greenish-blue solid known as patina. Copper statues exposed to the carbon dioxide and water in the air become coated with patina. The patina is not copper, it is not carbon dioxide, and it is not water. It is a new substance formed by the reaction of these chemicals with one another.

All three of these transformations involve a change in the way the atoms in the molecules are *chemically bonded* to one another. (A *chemical bond* is the attraction between two atoms that holds them together in a molecule.) A methane molecule, for example, is made of a single carbon atom bonded to four hydrogen atoms, and an oxygen molecule is made of two oxygen atoms bonded to each other. Figure 2.4 shows the chemical change in which the atoms in a methane molecule and those in two oxygen molecules first pull apart and then form new bonds with different partners, resulting in the formation of molecules of carbon dioxide and water.

Figure 2.4
The chemical change in which molecules of methane and oxygen transform to molecules of carbon dioxide and water as atoms break old bonds and form new ones. The actual mechanism of this transformation is more complicated than depicted here; however, the idea that new materials are formed by the rearrangement of atoms is accurate.

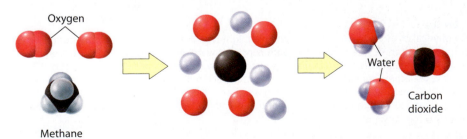

Any change in a substance that involves a rearrangement of the way atoms are bonded is called a **chemical change**. Thus the transformation of methane to carbon dioxide and water is a chemical change, as are the other two transformations shown in Figure 2.3.

The chemical change shown in Figure 2.5 occurs when an electric current is passed through water. The energy of the current causes the water molecules to split into atoms that then form new chemical bonds. Thus, water molecules are changed into molecules of hydrogen and oxygen, two substances that are very different from water. The hydrogen and oxygen are both gases at room temperature, and they can be seen as bubbles rising to the surface.

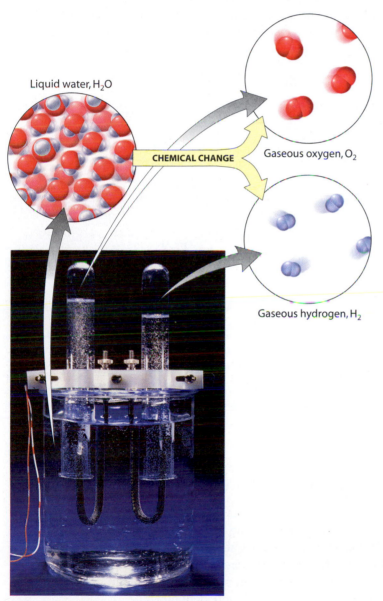

Liquid water, H_2O

CHEMICAL CHANGE

Gaseous oxygen, O_2

Gaseous hydrogen, H_2

Figure 2.5
Water can be transformed to hydrogen gas and oxygen gas by the energy of an electric current. This is a chemical change because new materials (the two gases) are formed as the atoms originally in the water molecules are rearranged.

In the language of chemistry, materials undergoing a chemical change are said to be *reacting*. Methane *reacts* with oxygen to form carbon dioxide and water. Water *reacts* when exposed to electricity to form hydrogen gas and oxygen gas. Thus the term *chemical change* means the same thing as *chemical reaction*. During a **chemical reaction**, new materials are formed by a change in the way atoms are bonded together. We shall explore chemical bonds and the reactions in which they are formed and broken in later chapters.

Concept Check ✔

Each sphere in the following diagrams represents an atom. Joined spheres represent molecules. One set of diagrams shows a physical change, and the other shows a chemical change. Which is which?

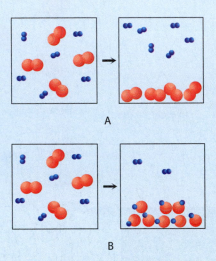

Was this your answer? Remember that a chemical change (also known as a chemical reaction) involves molecules breaking apart so that the atoms are free to form new bonds with new partners. You must be careful to distinguish this breaking apart from a mere change in the relative positions of a group of molecules. In set A, the molecules before and after the change are the same. They differ only in their positions relative to one another. Set A therefore represents a physical change. In set B, new molecules consisting of bonded red and blue spheres appear after the change. These molecules represent a new material, and so B is a chemical change.

Hands-On Chemistry: Fire Water

This activity is for those of you with access to a gas stove. Place a large pot of cool water on top of the stove, and set the burner on high. What product from the combustion of the natural gas do you see condensing on the outside of the pot? Where did it come from? Would more or less of this product form if the pot contained ice water? Where does this product go as the pot gets warmer? What physical and chemical changes can you identify?

Conceptual **Chemistry** *Alive!*

Determining Whether a Change Is Physical or Chemical Can Be Difficult

How can you tell whether a change you observe is physical or chemical? It can be tricky because in both cases there are changes in physical appearance. Water, for example, looks quite different after it freezes, just as a car looks quite different after it rusts (Figure 2.6). The freezing of water results from a change in how water molecules are oriented relative to one another. This is a physical change because liquid water and frozen water are both forms of water. The rusting of a car, by contrast, is the result of the transformation of iron to rust. This is a chemical change because iron and rust are two different materials, each consisting of a different arrangement of atoms. As we shall see in the next two sections, iron is an *element* and rust is a *compound* consisting of iron and oxygen atoms.

Figure 2.6
The transformation of water to ice and the transformation of iron to rust both involve a change in physical appearance. The formation of ice is a physical change, and the formation of rust is a chemical change.

By studying this chapter, you can expect to learn the difference between a physical change and a chemical change. However, you cannot expect to have a firm handle on how to categorize an observed change as physical or chemical because doing so requires a knowledge of the chemical identity of the materials involved as well as an understanding of how their atoms and molecules behave. This sort of insight builds over many years of study and laboratory experience.

There are, however, two powerful guidelines that can assist you in assessing physical and chemical changes. First, in a physical change, a change in appearance is the result of a new set of conditions imposed on the *same* material. Restoring the original conditions restores the original appearance: frozen water melts upon warming. Second, in a chemical change, a change in appearance is the result of the formation of a *new* material that has its own unique set of physical properties. The more evidence you have suggesting that a different material has been formed, the greater the likelihood that the change is a chemical change. Iron is a material that can be used to build cars. Rust is not. This suggests that the rusting of iron is a chemical change.

Figure 2.7 shows potassium chromate, a material whose color depends on temperature. At room temperature, potassium chromate is a bright canary yellow. At higher temperatures, it is a deep reddish orange. Upon cooling, the canary color returns, suggesting that the change is physical. With a chemical change, reverting to original conditions does not restore the original appearance. Ammonium dichromate, shown in Figure 2.8, is an orange material that when heated explodes into ammonia, water vapor, and green chromium(III) oxide. When the test tube is returned to the original temperature, there is no trace of orange ammonium dichromate. In its place are new substances having completely different physical properties.

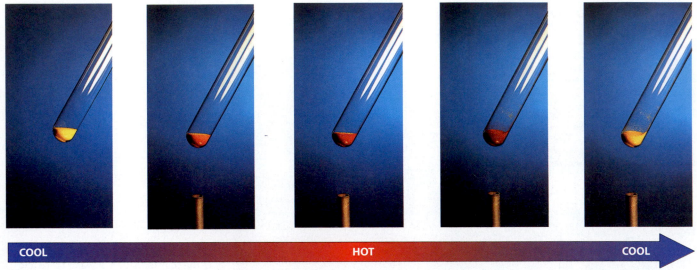

COOL — HOT — COOL

Figure 2.7
Potassium chromate changes color as its temperature changes. This change in color is a physical change. A return to the original temperature restores the original bright yellow color.

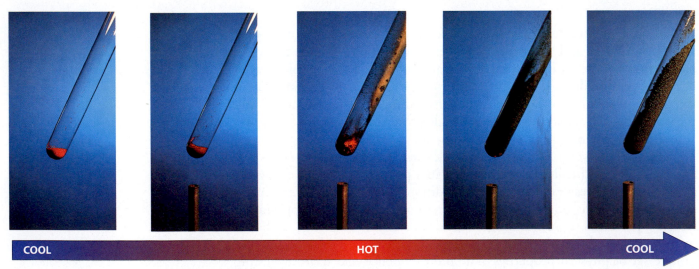

COOL — HOT — COOL

Figure 2.8
When heated, orange ammonium dichromate undergoes a chemical change to ammonia, water vapor, and chromium(III) oxide. A return to the original temperature does not restore the orange color because the ammonium dichromate is no longer there.

Concept Check ✓

Michaela has grown an inch in height over the past year. Is this best described as a physical or a chemical change?

Was this your answer? Are new materials being formed as Michaela grows? Absolutely—created out of the food she eats. Her body is very different from, say, the peanut butter sandwich she ate yesterday. Yet through some very advanced chemistry, her body is able to take the atoms of that peanut butter sandwich and rearrange them into new materials. Biological growth, therefore, is best described as a chemical change.

2.2 Atoms Are the Fundamental Components of Elements

You know that atoms make up the matter around you, from stars to steel to chocolate ice cream. You might think that there must be many different kinds of atoms to account for the huge diversity of matter, but the number of different kinds of atoms is surprisingly small. The great variety of substances results from the many ways a few kinds of atoms can be combined. Just as the three colors red, green, and blue can be combined to form any color on a television screen or the 26 letters of the alphabet make up all the words in a dictionary, only a few kinds of atoms combine in different ways to produce all substances. To date, we know of slightly more than 100 distinct atoms. Of these, about 90 are found in nature. The remaining atoms have been created in the laboratory.

Any material that is made up of only one type of atom is classified as an **element**. A few examples are shown in Figure 2.9. Pure gold, for example, is an element—it contains only gold atoms. Similarly, one of the gases in air is nitrogen, an element. Nitrogen gas is an element because it contains only nitrogen atoms. Likewise, the graphite in your pencil is an element—carbon. Graphite is made up solely of carbon atoms. All of the elements are listed in a chart called the **periodic table**, shown in Figure 2.10.

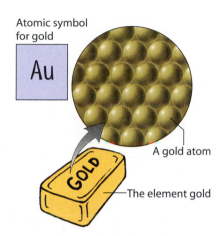

Atomic symbol for gold

Au

A gold atom

The element gold

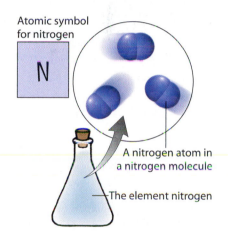

Atomic symbol for nitrogen

N

A nitrogen atom in a nitrogen molecule

The element nitrogen

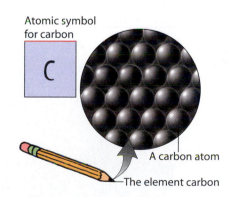

Atomic symbol for carbon

C

A carbon atom

The element carbon

Figure 2.9
Any element consists of only one kind of atom. Gold consists of only gold atoms, a flask of gaseous nitrogen consists of only nitrogen atoms, and the carbon of a graphite pencil consists of only carbon atoms.

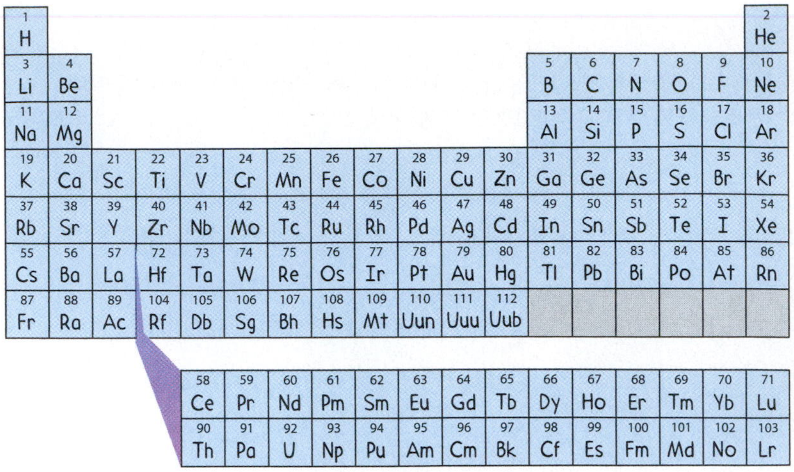

Figure 2.10
The periodic table lists all the known elements.

1 H																	2 He
3 Li	4 Be											5 B	6 C	7 N	8 O	9 F	10 Ne
11 Na	12 Mg											13 Al	14 Si	15 P	16 S	17 Cl	18 Ar
19 K	20 Ca	21 Sc	22 Ti	23 V	24 Cr	25 Mn	26 Fe	27 Co	28 Ni	29 Cu	30 Zn	31 Ga	32 Ge	33 As	34 Se	35 Br	36 Kr
37 Rb	38 Sr	39 Y	40 Zr	41 Nb	42 Mo	43 Tc	44 Ru	45 Rh	46 Pd	47 Ag	48 Cd	49 In	50 Sn	51 Sb	52 Te	53 I	54 Xe
55 Cs	56 Ba	57 La	72 Hf	73 Ta	74 W	75 Re	76 Os	77 Ir	78 Pt	79 Au	80 Hg	81 Tl	82 Pb	83 Bi	84 Po	85 At	86 Rn
87 Fr	88 Ra	89 Ac	104 Rf	105 Db	106 Sg	107 Bh	108 Hs	109 Mt	110 Uun	111 Uuu	112 Uub						

58 Ce	59 Pr	60 Nd	61 Pm	62 Sm	63 Eu	64 Gd	65 Tb	66 Dy	67 Ho	68 Er	69 Tm	70 Yb	71 Lu
90 Th	91 Pa	92 U	93 Np	94 Pu	95 Am	96 Cm	97 Bk	98 Cf	99 Es	100 Fm	101 Md	102 No	103 Lr

As you can see from the periodic table, each element is designated by its **atomic symbol**, which comes from the letters of the element's name. For example, the atomic symbol for carbon is C, and that for chlorine is Cl. In many cases, the atomic symbol is derived from the element's Latin name. Gold has the atomic symbol Au after its Latin name, *aurum*. Lead has the atomic symbol Pb after its Latin name, *plumbum* (Figure 2.11). Elements having symbols derived from Latin names are usually those discovered earliest.

Note that only the first letter of an atomic symbol is capitalized. The symbol for the element cobalt, for instance, is Co, while CO is a combination of two elements: carbon, C, and oxygen, O.

Figure. 2.11
A plumb bob, a heavy weight attached to a string and used by carpenters and surveyors to establish a straight vertical line, gets it name from the lead (plumbum, Pb) that is still sometimes used as the weight. Plumbers got their name because they once worked with lead pipes.

The terms *element* and *atom* are often used in a similar context. You might hear, for example, that gold is an element made of gold atoms. Generally, *element* is used in reference to an entire macroscopic or microscopic sample, and *atom* is used when speaking of the submicroscopic particles in the sample. The important distinction is that elements are made of atoms and not the other way around.

How many atoms are bound together in an element is shown by an **elemental formula**. For elements in which the basic units are individual atoms, the elemental formula is simply the chemical symbol: Au is the elemental formula for gold, and Li is the elemental formula for lithium, to name just two examples. For elements in which the basic units are two or more atoms bonded into molecules, the elemental formula is the chemical symbol followed by a subscript indicating the number of atoms in each molecule. For example, elemental nitrogen, as was shown in Figure 2.9, commonly consists of molecules containing two nitrogen atoms per molecule. Thus N_2 is the usual elemental formula given for nitrogen. Similarly, O_2 is the elemental formula for oxygen, and S_8 is the elemental formula for sulfur.

Concept Check ✓

> The oxygen we breathe, O_2, is converted to ozone, O_3, in the presence of an electric spark. Is this a physical or chemical change?

Was this your answer? When atoms regroup, the result is an entirely new substance, and that is what happens here. The oxygen we breathe, O_2, is odorless and life-giving. Ozone, O_3, can be toxic and has a pungent smell commonly associated with electric motors. The conversion of O_2 to O_3 is therefore a chemical change. However, both O_2 and O_3 are elemental forms of oxygen.

2.3 Elements Can Combine to Form Compounds

When atoms of *different* elements bond to one another, they make a **compound**. Sodium atoms and chlorine atoms, for example, bond to make the compound sodium chloride, commonly known as table salt. Nitrogen atoms and hydrogen atoms join to make the compound ammonia, a common household cleaner.

A compound is represented by its **chemical formula**, in which the symbols for the elements are written together. The chemical formula for sodium chloride is NaCl, and that for ammonia is NH_3. Numerical subscripts indicate the ratio in which the atoms combine. By convention, the subscript 1 is understood and omitted. So the chemical formula NaCl tells us that in the compound sodium chloride there is one sodium for every one chlorine, and the chemical formula NH_3 tells us that in the compound ammonia there is one nitrogen atom for every three hydrogen atoms, as Figure 2.12 shows.

Compounds have physical and chemical properties that are different from the properties of their elemental components. The sodium chloride,

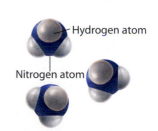

Sodium atom

Chlorine atom

Sodium chloride, NaCl

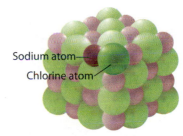

Hydrogen atom

Nitrogen atom

Ammonia, NH_3

Figure 2.12
The compounds sodium chloride and ammonia are represented by their chemical formulas, NaCl and NH_3. A chemical formula shows the ratio of atoms used to make the compound.

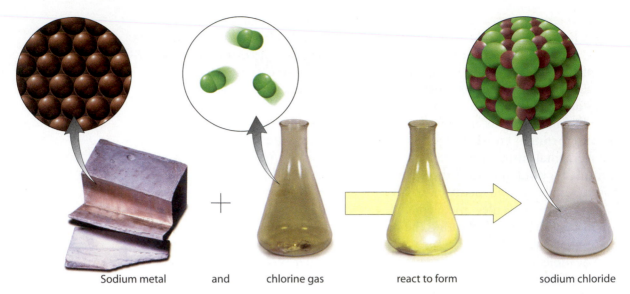

Sodium metal   and   chlorine gas   react to form   sodium chloride

Figure 2.13
Sodium metal and chlorine gas react together to form sodium chloride. Although the compound sodium chloride is composed of sodium and chlorine, the physical and chemical properties of sodium chloride are very different from the physical and chemical properties of either sodium metal or chlorine gas.

NaCl, shown in Figure 2.13 is very different from the elemental and elemental chlorine used to form it. Elemental sodium, Na, c nothing but sodium atoms, which form a soft, silvery metal that can be cut easily with a knife. Its melting point is 97.5°C, and it reacts violently with water. Elemental chlorine, Cl_2, consists of chlorine molecules. This material, a yellow-green gas at room temperature, is very toxic and was used as a chemical warfare agent during World War I. Its boiling point is −34°C. The compound sodium chloride, NaCl, is a translucent, brittle, colorless crystal having a melting point of 800°C. Sodium chloride does not chemically react with water the way sodium does, and not only is it not toxic to humans the way chlorine is, but the very opposite is true: it is an essential component of all living organisms. Sodium chloride is not sodium, nor is it chlorine; it is uniquely sodium chloride, a tasty chemical when sprinkled lightly over popcorn.

Concept Check ✔

Hydrogen sulfide, H_2S, is one of the smelliest compounds. Rotten eggs get their characteristic bad smell from the hydrogen sulfide they release. Can you infer from this information that elemental sulfur, S_8, is just as smelly?

Was this your answer?   No, you cannot. In fact, the odor of elemental sulfur is negligible compared with that of hydrogen sulfide. Compounds are truly different from the elements from which they are formed. Hydrogen sulfide, H_2S, is as different from elemental sulfur, S_8, as water, H_2O, is from elemental oxygen, O_2.

Hands-On Chemistry: Oxygen Bubble Bursts

Conceptual
Chemistry
Alive!

Compounds can be broken down to their component elements. For example, when you pour a solution of the compound hydrogen peroxide, H_2O_2, over a cut, an enzyme in your blood decomposes it to produce oxygen gas, O_2, as evidenced by the bubbling that takes place. It is this oxygen at high concentrations at the site of injury that kills off microorganisms. A similar enzyme is found in baker's yeast.

What You Need

Packet of baker's yeast; 3% hydrogen peroxide solution; short, wide drinking glass; tweezers; matches

Safety Note

Wear safety glasses, and remove all combustibles, such as paper towels, from the area. Keep your fingers well away from the flame because it will glow more brightly as it is exposed to the oxygen.

Procedure

① Pour the yeast into the glass. Add a couple of capfuls of the hydrogen peroxide and watch the oxygen bubbles form.

② Test for the presence of oxygen by holding a lighted match with the tweezers and putting the flame near the bubbles. Look for the flame to glow more brightly as the escaping oxygen passes over it.

Describe oxygen's physical and chemical properties.

Compounds Are Named According to the Elements They Contain

A system for naming the countless number of possible compounds has been developed by the International Union for Pure and Applied Chemistry (IUPAC). This system is designed so that a compound's name reflects the elements it contains and how those elements are put together. Anyone familiar with the system, therefore, can deduce the chemical identity of a compound from its *systematic name*.

As you might imagine, this system is very intricate. However, you need not learn all its rules. At this point, learning some guidelines will prove most helpful. These guidelines alone will not enable you to name every compound; however, they will acquaint you with how the system works for many simple compounds consisting of only two elements.

Guideline 1   The name of the element farther to the left in the periodic table is followed by the name of the element farther to the right, with the suffix *-ide* added to the name of the latter:

NaCl	Sodium chloride	HCl	Hydrogen chloride
Li_2O	Lithium oxide	MgO	Magnesium oxide
CaF_2	Calcium fluoride	Sr_3P_2	Strontium phosphide

Guideline 2 When two or more compounds have different numbers of the same elements, prefixes are added to remove the ambiguity. The first four prefixes are *mono-* ("one"), *di-* ("two"), *tri-* ("three"), and *tetra-* ("four"). The prefix *mono-*, however, is commonly omitted from the beginning of the first word of the name:

Carbon and oxygen

CO Carbon monoxide
CO_2 Carbon dioxide

Nitrogen and oxygen

NO_2 Nitrogen dioxide
N_2O_4 Dinitrogen tetroxide

Sulfur and oxygen

SO_2 Sulfur dioxide
SO_3 Sulfur trioxide

Guideline 3 Many compounds are not usually referred to by their systematic names. Instead, they are assigned *common names* that are more convenient or have been used traditionally for many years. Some common names we use in *Conceptual Chemistry* are *water* for H_2O, *ammonia* for NH_3, and *methane* for CH_4. Pteroenone, the name of the compound extracted from the sea butterfly referred to in Chapter 1, is a common name. The systematic name for this compound, though more descriptive of the elements it contains, is much longer: 5(S)-methyl-6(R)-hydroxy-7,9-dimethyl-7,9-diene-4-undecanone.

Concept Check ✔

What is the systematic name for NaF?

Was this your answer? This compound is a cavity-fighting substance added to some toothpastes—sodium fluoride.

2.4   Most Materials Are Mixtures

A **mixture** is a combination of two or more substances in which each substance retains its properties. Most materials we encounter are mixtures: mixtures of elements, mixtures of compounds, or mixtures of elements and compounds. Stainless steel, for example, is a mixture of the elements iron, chromium, nickel, and carbon. Seltzer water is a mixture of the liquid compound water and the gaseous compound carbon dioxide. Our atmosphere, as Figure 2.14 illustrates, is a mixture of the elements nitrogen, oxygen, and argon plus small amounts of such compounds as carbon dioxide and water vapor.

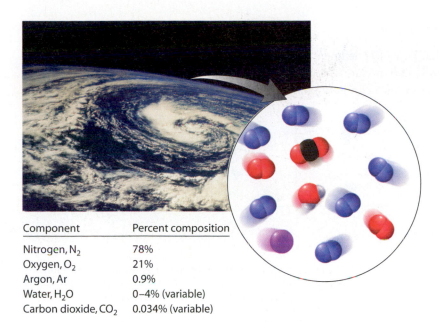

Component	Percent composition
Nitrogen, N_2	78%
Oxygen, O_2	21%
Argon, Ar	0.9%
Water, H_2O	0–4% (variable)
Carbon dioxide, CO_2	0.034% (variable)

Figure 2.14
The Earth's atmosphere is a mixture of gaseous elements and compounds. Some of them are shown here.

Tap water is a mixture containing mostly water but also many other compounds. Depending on your location, your water may contain compounds of calcium, magnesium, fluorine, iron, and potassium; chlorine disinfectants; trace amounts of compounds of lead, mercury, and cadmium; organic compounds; and dissolved oxygen, nitrogen, and carbon dioxide. While it is surely important to minimize any toxic components in your drinking water, it is unnecessary, undesirable, and impossible to remove all other substances from it. Some of the dissolved solids and gases give water its characteristic taste, and many of them promote human health: fluoride compounds protect teeth, chlorine destroys harmful bacteria, and as much as 10 percent of our daily requirement for iron, potassium, calcium, and magnesium is obtained from drinking water (Figures 2.15 and 2.16).

Figure 2.15
Tap water provides us with water as well as a large number of other compounds, many of which are flavorful and healthful. Bottoms up!

Figure 2.16
Most of the oxygen in the air bubbles produced by an aquarium aerator escapes into the atmosphere. Some of the oxygen, however, mixes with the water. It is this oxygen the fish depend on to survive. Without this dissolved oxygen, which they extract with their gills, the fish would promptly drown. So fish don't "breathe" water. They breathe the oxygen, O_2, dissolved in the water.

Concept Check ✔

So far, you have learned about three kinds of matter: elements, compounds, and mixtures. Which box below contains only an element? Which contains only a compound? Which contains a mixture?

A	B	C

Was this your answer? The molecules in box A each contain two different types of atoms and so are representative of a compound. The molecules in box B each consist of the same atoms and so are representative of an element. Box C is a mixture of the compound and the element.

Note how the molecules of the compound and those of the element remain intact in the mixture. That is, upon the formation of the mixture, there is no exchange of atoms between the components.

There is a difference between the way substances—either elements or compounds—combine to form mixtures and the way elements combine to form compounds. Each substance in a mixture retains its chemical identity. The sugar molecules in the teaspoon of sugar in Figure 2.17, for example, are identical to the sugar molecules already in the tea. The only difference is that the sugar molecules in the tea are mixed with other substances, mostly water. The formation of a mixture, therefore, is a physical change. As was discussed in Section 2.3, when elements join to form compounds, there is a change in chemical identity. Sodium chloride is not a mixture of sodium and chlorine atoms. Instead, sodium chloride is a compound, which means it is entirely different from the elements used to make it. The formation of a compound is therefore a chemical change.

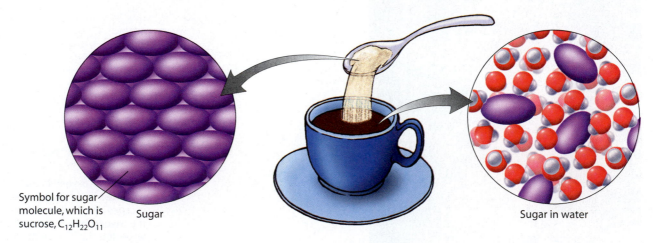

Symbol for sugar molecule, which is sucrose, $C_{12}H_{22}O_{11}$

Sugar

Sugar in water

Figure 2.17
Table sugar is a compound consisting of only sucrose molecules. Once these molecules are mixed into hot tea, they become interspersed among the water and tea molecules and form a sugar–tea–water mixture. No new compounds are formed, and so this is an example of a physical change.

Mixtures Can Be Separated by Physical Means

The components of mixtures can be separated from one another by taking advantage of differences in the components' physical properties. A mixture of solids and liquids, for example, can be separated using filter paper through which the liquids pass but the solids do not. This is how coffee is often made: the caffeine and flavor molecules in the hot water pass through the filter and into the coffee pot while the solid coffee grounds remain behind. This method of separating a solid–liquid mixture is called *filtration* and is a common technique used by chemists.

Mixtures can also be separated by taking advantage of a difference in boiling or melting points. Seawater is a mixture of water and a variety of compounds, mostly sodium chloride. Whereas pure water boils at 100°C, sodium chloride doesn't even *melt* until 800°C. One way to separate pure water from the mixture we call seawater, therefore, is to heat the seawater to about 100°C. At this temperature, the liquid water readily transforms to water vapor but the sodium chloride stays behind dissolved in the remaining water. As the water vapor rises, it can be channeled into a cooler container, where it condenses to a liquid without the dissolved solids. This process of collecting a vaporized substance, called *distillation*, is illustrated in Figure 2.18. After all the water has been distilled from seawater, what remain are dry solids. These solids, also a mixture of compounds, contain a variety of valuable materials, including sodium chloride, potassium bromide, and a small amount of gold! (For details on why this gold is not recoverable, see Section 19.3.) Further separation of the components of this mixture is of significant commercial interest (Figure 2.19).

(a)

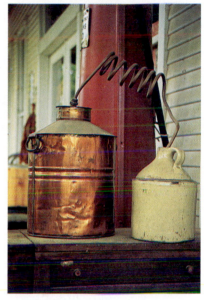

(b)

Figure 2.18
(a) A simple distillation setup used to separate one component—water—from the mixture we call seawater. The seawater is boiled in the flask on the left. The rising water vapor is channeled into a downward-slanting tube kept cool by cold water flowing across its outer surface. The water vapor inside the cool tube condenses and collects in the flask on the right. (b) A whiskey still works on the same principle. A mixture containing alcohol is heated to the point where the alcohol, some flavoring molecules, and some water are vaporized. These vapors travel through the copper coils, where they condense to a liquid ready for collection.

Figure 2.19
At the southern end of San Francisco Bay are areas where the seawater has been partitioned off. These are evaporation ponds where the water is allowed to evaporate, leaving behind the solids that were dissolved in the seawater. These solids are further refined for commercial sale. The remarkable color of the ponds results from suspended particles of iron oxide and other minerals, which are easily removed during refining.

Hands-On Chemistry: Bottoms Up and Bubbles Out

What's in a glass of water? Separate the components of your tap water to find out.

What You Need

Tap water, sparkling clean cooking pot, stove, knife

Safety Note

Wear safety glasses for step 1 because some splattering may occur.

Procedure

① Put on your safety glasses and add the tap water to the cooking pot. Boil the water to dryness. (Turn off the burner before the water is all gone. The heat from the pot will finish the evaporation.)

② Examine the resulting residue by scraping it with the knife. These are the solids you ingest with every glass of water you drink.

③ To see the gases dissolved in your water, fill a clean cooking pot with water and let it stand at room temperature for several hours. Note the bubbles that adhere to the inner sides of the pot.

Where did the bubbles of step 3 come from? What do you suppose they contain?

2.5   Chemists Classify Matter as Pure or Impure

If a material is **pure**, it consists of only a single element or a single compound. In pure gold, for example, there is nothing but the element gold. In pure table salt, there is nothing but the compound sodium chloride. If a material is **impure**, it is a mixture and contains two or more elements or compounds. This classification scheme is shown in Figure 2.20.

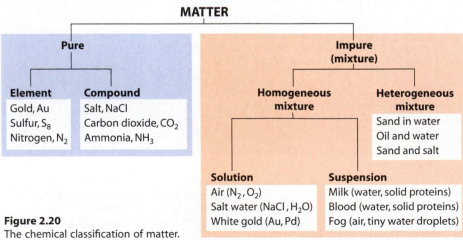

Figure 2.20
The chemical classification of matter.

Because atoms and molecules are so small, it is impractical to prepare a sample that is truly pure—that is, truly 100 percent of a single material. For example, if just one atom or molecule out of a trillion trillion were different, then the 100 percent pure status would be lost. Samples can be "purified" by various methods, however, such as distillation. When we say *pure*, it is understood to be a relative term. Comparing the purity of two samples, the purer one contains fewer impurities. A sample of water that is 99.9 percent pure has a greater proportion of impurities than does a sample of water that is 99.9999 percent pure.

Sometimes naturally occurring mixtures are labeled as being pure, as in "pure orange juice." Such a statement merely means that nothing artificial has been added. According to a chemist's definition, however, orange juice is anything but pure, as it contains a wide variety of materials, including water, pulp, flavorings, vitamins, and sugars.

Mixtures may be heterogeneous or homogeneous. In a **heterogeneous mixture**, the different components can be seen as individual substances, such as pulp in orange juice, sand in water, or oil globules dispersed in vinegar. The different components are visible. **Homogeneous mixtures** have the same composition throughout. Any one region of the mixture has the same ratio of substances as does any other region, and the components cannot be seen as individual identifiable entities. The distinction is shown in Figure 2.21.

| Granite | "Snow" in snow globe | Pizza |

(a) Heterogeneous mixtures

| Air | Clear seawater | White gold |

(b) Homogeneous mixtures

Figure 2.21
(a) In heterogeneous mixtures, the different components can be seen with the naked eye.
(b) In homogeneous mixtures, the different components are mixed at a much finer level and so are not readily distinguished.

Figure 2.22
The path of light becomes visible when the light passes through a suspension.

A homogeneous mixture may be either a solution or a suspension. In a **solution**, all components are in the same phase. The atmosphere we breathe is a gaseous solution consisting of the gaseous elements nitrogen and oxygen as well as minor amounts of other gaseous materials. Salt water is a liquid solution because both the water and the dissolved sodium chloride are found in a single liquid phase. An example of a solid solution is white gold, which is a homogeneous mixture of the elements gold and palladium. We shall be discussing solutions in more detail in Chapter 7.

A **suspension** is a homogeneous mixture in which the different components are in different phases, such as solids in liquids or liquids in gases. In a suspension, the mixing is so thorough that the different phases cannot be readily distinguished. Milk is a suspension because it is a homogeneous mixture of proteins and fats finely dispersed in water. Blood is a suspension composed of finely dispersed blood cells in water. Another example of a suspension is clouds, which are homogeneous mixtures of tiny water droplets suspended in air. Shining a light through a suspension, as is done in Figure 2.22, results in a visible cone as the light is reflected by the suspended components.

The easiest way to distinguish a suspension from a solution in the laboratory is to spin a sample in a centrifuge. This device, spinning at thousands of revolutions per minute, separates the components of suspensions but not those of solutions, as Figure 2.23 shows.

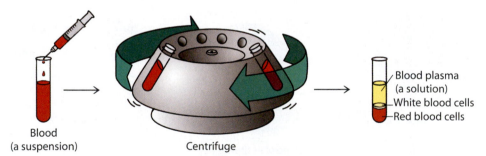

Blood
(a suspension)

Centrifuge

Blood plasma
(a solution)
White blood cells
Red blood cells

Figure 2.23
Blood, because it is a suspension, can be centrifuged into its components, which include the blood plasma (a yellowish solution) and white and red blood cells. The components of the plasma cannot be separated from one another here because a centrifuge has no effect on solutions.

Concept Check ✔

Impure water can be purified by
a. removing the impure water molecules.
b. removing everything that is not water.
c. breaking down the water to its simplest components.
d. adding some disinfectant such as chlorine.

Was this your answer? Water, H_2O, is a compound made of the elements hydrogen and oxygen in a 2-to-1 ratio. Every H_2O molecule is exactly the same as every other, and there's no such thing as an impure H_2O molecule. Just about anything, including you, beach balls, rubber ducks, dust particles, and bacteria, can be found in water. When something other than water is found in water, we say that the water is impure. It is important to see that the impurities are in the water and not part of the water, which means that it is possible to remove them by a variety of physical means, such as filtration or distillation. The answer to this Concept Check is b.

2.6 Elements Are Organized in the Periodic Table by Their Properties

As was mentioned in Section 2.2, the periodic table is a listing of all the known elements. There is so much more to this table, however. Most notably, the elements are organized in the table based on their physical and chemical properties. One of the most apparent examples is how the elements are grouped as metals, nonmetals, and metalloids.

As shown in Figure 2.24, most of the known elements are **metals**, which are defined as those elements that are shiny, opaque, and good conductors of electricity and heat. Metals are *malleable*, which means they can be hammered into different shapes or bent without breaking. They are also *ductile*, which means they can be drawn into wires. All but a few metals are solid at room temperature. The exceptions include mercury, Hg; gallium, Ga; cesium, Cs; and francium, Fr; which are all liquids at a warm room temperature of 30°C (86°F). Another interesting exception is hydrogen, H,

Please put to rest any fear you may have about needing to memorize the periodic table, or even parts of it—better to focus on the many great concepts behind its organization.

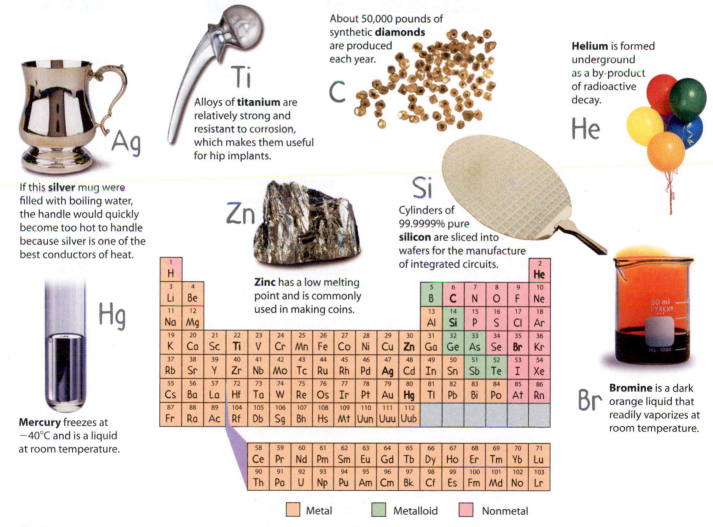

Alloys of **titanium** are relatively strong and resistant to corrosion, which makes them useful for hip implants.

About 50,000 pounds of synthetic **diamonds** are produced each year.

Helium is formed underground as a by-product of radioactive decay.

If this **silver** mug were filled with boiling water, the handle would quickly become too hot to handle because silver is one of the best conductors of heat.

Zinc has a low melting point and is commonly used in making coins.

Cylinders of 99.9999% pure **silicon** are sliced into wafers for the manufacture of integrated circuits.

Mercury freezes at −40°C and is a liquid at room temperature.

Bromine is a dark orange liquid that readily vaporizes at room temperature.

Metal   Metalloid   Nonmetal

Figure 2.24
The periodic table color-coded to show metals, nonmetals, and metalloids.

Figure 2.25
Geoplanetary models suggest that hydrogen exists as a liquid metal deep beneath the surfaces of Jupiter (shown here) and Saturn. These planets are composed mostly of hydrogen. Inside them, the pressure exceeds 3 million times the Earth's atmospheric pressure. At this tremendously high pressure, hydrogen is pressed to a liquid-metal phase. Back here on the Earth at our relatively low atmospheric pressure, hydrogen exists as a nonmetallic gas of hydrogen molecules, H_2.

which takes on the properties of a liquid metal only at very high pressures (Figure 2.25). Under normal conditions, hydrogen atoms combine to form hydrogen molecules, H_2, which behave as a nonmetallic gas.

The nonmetallic elements, with the exception of hydrogen, are on the right of the periodic table. **Nonmetals** are very poor conductors of electricity and heat, and may also be transparent. Solid nonmetals are neither malleable nor ductile. Rather, they are brittle and shatter when hammered. At 30°C (86°F), some nonmetals are solid (carbon, C), others are liquid (bromine, Br), and still others are gaseous (helium, He).

Six elements are classified as **metalloids**: boron, B; silicon, Si; germanium, Ge; arsenic, As; antimony, Sb; and tellurium, Te. Situated between the metals and the nonmetals in the periodic table, the metalloids have both metallic and nonmetallic characteristics. For example, these elements are weak conductors of electricity, which makes them useful as semiconductors in the integrated circuits of computers. Note from the periodic table how germanium, Ge (number 32), is closer to the metals than to the nonmetals. Because of this positioning, we can deduce that germanium has more metallic properties than silicon, Si (number 14), and is a slightly better conductor of electricity. So we find that integrated circuits fabricated with germanium operate faster than those fabricated with silicon. Because silicon is much more abundant and less expensive to obtain, however, silicon computer chips remain the industry standard.

A Period Is a Horizontal Row, a Group a Vertical Column

Two other important ways in which the elements are organized in the periodic table are by horizontal rows and vertical columns. Each horizontal row is called a **period**, and each vertical column is called a **group**

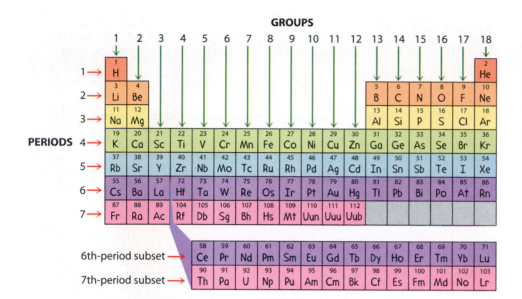

GROUPS

Figure 2.26
The 7 periods (horizontal rows) and 18 groups (vertical columns) of the periodic table. Note that not all periods contain the same number of elements. Also note that, for reasons explained later, the sixth and seventh periods each include a subset of elements, which are listed apart from the main body.

(or sometimes a *family*). As shown in Figure 2.26, there are 7 periods and 18 groups.

Across any period, the properties of elements gradually change. This gradual change is called a **periodic trend**. As is shown in Figure 2.27, one periodic trend is that atomic size tends to decrease as you move from left to right across any period. Note that the trend repeats from one horizontal row to the next. This phenomenon of repeating trends is called *periodicity*, a term used to indicate that the trends recur in cycles. Each horizontal row is called a *period* because it corresponds to one full cycle of a trend. As we explore in further detail in Section 5.8, there are many other properties of elements that change gradually in moving from left to right across the periodic table.

Figure 2.27
The size of atoms gradually decreases in moving from left to right across any period. Atomic size is a periodic (repeating) property.

GROUPS

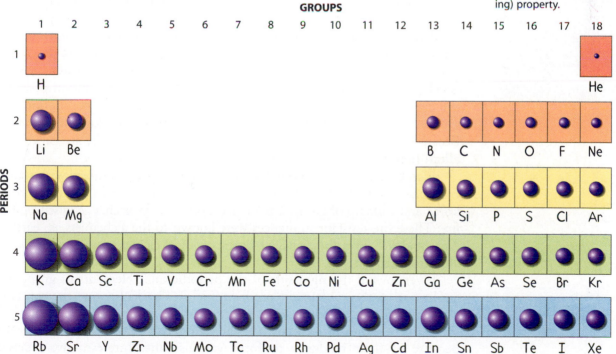

Concept Check ✔

Which are larger: atoms of cesium, Cs (number 55), or atoms of radon, Rn (number 86)?

Was this your answer? Perhaps you tried looking to Figure 2.27 to answer this question and quickly became frustrated because the sixth-period elements are not shown. Well, relax. Look at the trends and you'll see that, in any one period, all atoms to the left are larger than those to the right. Accordingly, cesium is positioned at the far left of period 6, and so you can reasonably predict that its atoms are larger than those of radon, which is positioned at the far right of period 6. The periodic table is a road map to understanding the elements.

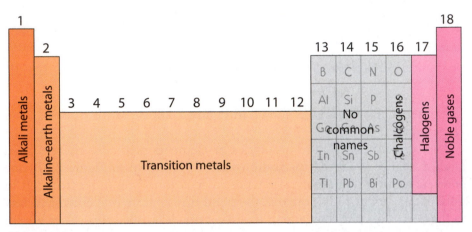

Figure 2.28
The common names for various groups of elements.

Figure 2.29
Ashes and water make a slippery alkaline solution once used to clean hands.

Down any group (vertical column), the properties of elements tend to be remarkably similar, which is why these elements are said to be "grouped" or "in a family." As Figure 2.28 shows, several groups have traditional names that describe the properties of their elements. Early in human history, people discovered that ashes mixed with water produce a slippery solution useful for removing grease. By the Middle Ages, such mixtures were described as being *alkaline*, a term derived from the Arabic word for ashes, *al-qali*. Alkaline mixtures found many uses, particularly in the preparation of soaps (Figure 2.29). We now know that alkaline ashes contain compounds of group 1 elements, most notably potassium carbonate, also known as *potash*. Because of this history, group 1 elements, which are metals, are called the **alkali metals**.

Elements of group 2 also form alkaline solutions when mixed with water. Furthermore, medieval alchemists noted that certain minerals (which we now know are made up of group 2 elements) do not melt or change when put in fire. These fire-resistant substances were known to the alchemists as "earth." As a holdover from these ancient times, group 2 elements are known as the **alkaline-earth metals**.

Over toward the right side of the periodic table elements of group 16 are known as the *chalcogens* ("ore-forming" in Greek) because the top two elements of this group, oxygen and sulfur, are so commonly found in ores. Elements of group 17 are known as the **halogens** ("salt-forming" in Greek) because of their tendency to form various salts. Interestingly, a small amount of the halogen iodine or the halogen bromine inside a lamp allows the lamp's tungsten filament to glow more brightly without burning out so quickly. Such lamps are commonly referred to as halogen lamps. Group 18 elements are all unreactive gases that tend not to combine with other elements. For this reason, they are called the **noble gases**, presumably because the nobility of earlier times were above interacting with common folk.

The elements of groups 3 through 12 are all metals that do not form alkaline solutions with water. These metals tend to be harder than the alkali metals and less reactive with water; hence they are used for structural purposes. Collectively they are known as the **transition metals**, a name that denotes their central position in the periodic table. The transition metals include some of the most familiar and important elements—iron, Fe; copper, Cu; nickel, Ni; chromium, Cr; silver, Ag; and gold, Au. They also include many lesser-known elements that are nonetheless important in modern technology. Persons with hip implants appreciate the transition metals titanium, Ti; molybdenum, Mo; and manganese, Mn, because these noncorrosive metals are used in implant devices.

Concept Check ✓

> The elements copper, Cu; silver, Ag; and gold, Au, are three of the few metals that can be found naturally in their elemental state. These three metals have found great use as currency and jewelry for a number of reasons, including their resistance to corrosion and their remarkable colors. How is the fact that these metals have similar properties reflected in the periodic table?

Was this your answer? Copper (number 29), silver (number 47), and gold (number 79) are all in the same group in the periodic table (group 11), which suggests they should have similar—though not identical—physical and chemical properties.

In the sixth period is a subset of 14 metallic elements (numbers 58 to 71) that are quite unlike any of the other transition metals. A similar subset (numbers 90 to 103) is found in the seventh period. These two subsets are the **inner transition metals**. Inserting the inner transition metals into the main body of the periodic table as in Figure 2.30 results in a long and cumbersome table. So that the table can fit nicely on a standard paper size, these elements are commonly placed below the main body of the table, as shown in Figure 2.31.

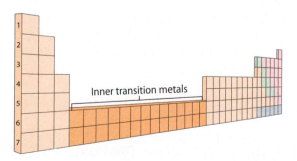

Inner transition metals

Figure 2.30
Inserting the inner transition metals between atomic groups 3 and 4 results in a periodic table that is not easy to fit on a standard sheet of paper.

Figure 2.31
The typical display of the inner transition metals. The count of elements in the sixth period goes from lanthanum (La, 57) to cerium (Ce, 58) on through to lutetium (Lu, 71) and then back to hafnium (Hf, 72). A similar jump is made in the seventh period.

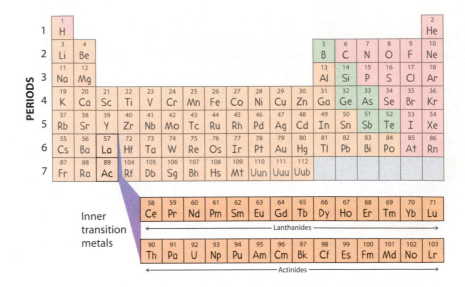

The sixth-period inner transition metals are called the **lanthanides** because they fall after lanthanum, La. Because of their similar physical and chemical properties, they tend to occur mixed together in the same locations in the Earth. Also because of their similarities, lanthanides are unusually difficult to purify. Recently, the commercial use of lanthanides has increased. Several lanthanide elements, for example, are used in the fabrication of the light-emitting diodes (LEDs) of laptop computer monitors.

The seventh-period inner transition metals are called the **actinides** because they fall after actinium, Ac. They, too, all have similar properties and hence are not easily purified. The nuclear power industry faces this obstacle because it requires purified samples of two of the most publicized actinides: uranium, U, and plutonium, Pu. Actinides heavier than uranium are not found in nature but are synthesized in the laboratory.

In Perspective

In this chapter we explored many of the rudiments of chemistry, including how matter is described by its physical and chemical properties and denoted by elemental and chemical formulas. We saw how compounds are different from the elements from which they are formed and how mixtures can be separated by taking advantage of differences in the physical properties of the components. Also addressed was what a chemist means by *pure* and how matter can be classified as element, compound, or mixture. Lastly, we saw how elements are organized in the periodic table by their physical and chemical properties. Along the way, you were introduced to some of the most important key terms of chemistry. With an understanding of these fundamental concepts and of the language used to describe them, you are well equipped to continue your study of nature's submicroscopic realm.

Key Terms and Matching Definitions

_____ actinide
_____ alkali metal
_____ alkaline-earth metal
_____ atomic symbol
_____ chemical change
_____ chemical formula
_____ chemical property
_____ chemical reaction
_____ compound
_____ element
_____ elemental formula
_____ group
_____ halogen
_____ heterogeneous mixture
_____ homogeneous mixture
_____ impure
_____ inner transition metal
_____ lanthanide
_____ metal
_____ metalloid
_____ mixture
_____ noble gas
_____ nonmetal
_____ period
_____ periodic table
_____ periodic trend
_____ physical change
_____ physical property
_____ pure
_____ solution
_____ suspension
_____ transition metal

1. Any physical attribute of a substance, such as color, density, or hardness.
2. A change in which a substance changes its physical properties without changing its chemical identity.
3. A type of property that characterizes the ability of a substance to change its chemical identity.
4. During this kind of change, atoms in a substance are rearranged to give a new substance having a new chemical identity.
5. Synonymous with chemical change.
6. A fundamental material consisting of only one type of atom.

7. A chart in which all known elements are organized by physical and chemical properties.
8. An abbreviation for an element or atom.
9. A notation that uses the atomic symbol and (sometimes) a numerical subscript to denote how atoms are bonded in an element.
10. A material in which atoms of different elements are bonded to one another.
11. A notation used to indicate the composition of a compound, consisting of the atomic symbols for the different elements of the compound and numerical subscripts indicating the ratio in which the atoms combine.
12. A combination of two or more substances in which each substance retains its properties.
13. The state of a material that consists of a single element or compound.
14. The state of a material that is a mixture of more than one element or compound.
15. A mixture in which the various components can be seen as individual substances.
16. A mixture in which the components are so finely mixed that the composition is the same throughout.
17. A homogeneous mixture in which all components are in the same phase.
18. A homogeneous mixture in which the various components are in different phases.
19. An element that is shiny, opaque, and able to conduct electricity and heat.
20. An element located toward the upper right of the periodic table that is neither a metal nor a metalloid.
21. An element that exhibits some properties of metals and some properties of nonmetals.
22. A horizontal row in the periodic table.
23. A vertical column in the periodic table, also known as a family of elements.
24. The gradual change of any property in the elements across a period.
25. Any group 1 element.
26. Any group 2 element.
27. Any "salt-forming" element.
28. Any unreactive element.
29. Any element of groups 3 through 12.
30. Any element in the two subgroups of the transition metals.
31. Any sixth-period inner transition metal.
32. Any seventh-period inner transition metal.

Review Questions

Matter Has Physical and Chemical Properties

1. What is a physical property?

2. What is a chemical property?

3. What doesn't change during a physical change?

4. Why is it sometimes difficult to decide whether an observed change is physical or chemical?

5. What are some of the clues that help us determine whether an observed change is physical or chemical?

Atoms Are the Fundamental Components of Elements

6. How many types of atoms can you expect to find in a pure sample of any element?

7. Distinguish between an atom and an element.

8. How many atoms are in a sulfur molecule that has the elemental formula S_8?

Elements Can Combine to Form Compounds

9. What is the difference between an element and a compound?

10. How many atoms are there in one molecule of H_3PO_4? How many atoms of each element are there in one molecule of H_3PO_4?

11. Are the physical and chemical properties of a compound necessarily similar to those of the elements from which it is composed?

12. What is the IUPAC systematic name for the compound KF?

13. What is the chemical formula for the compound titanium dioxide?

14. Why are common names often used for chemical compounds instead of systematic names?

Most Materials Are Mixtures

15. What defines a material as being a mixture?

16. How can the components of a mixture be separated from one another?

17. How does distillation separate the components of a mixture?

18. Oxygen, O_2, has a boiling point of 90 K ($-183°C$), and nitrogen, N_2, has a boiling point of 77 K ($-196°C$). Which is a liquid and which is a gas at 80 K ($-193°C$)?

Chemists Classify Matter as Pure or Impure

19. Why is it not practical to have a macroscopic sample that is 100 percent pure?

20. Classify the following as (a) homogeneous mixture, (b) heterogeneous mixture, (c) element, or (d) compound:

   milk _____   steel _____

   ocean water _____   blood _____

   sodium _____   planet Earth _____

21. How is a solution different from a suspension?

22. How can a solution be distinguished from a suspension?

Elements Are Organized in the Periodic Table by Their Properties

23. How is the periodic table more than just a listing of the known elements?

24. Are most elements metallic or nonmetallic?

25. Why is hydrogen, H, most often considered a nonmetallic element?

26. How do the physical properties of nonmetals differ from the physical properties of metals?

27. Where are metalloids located in the periodic table?

28. How many periods are there in the periodic table? How many groups?

29. What happens to the properties of elements across any period of the periodic table?

30. Why are group 1 elements called alkali metals?

31. Why are group 17 elements called halogens?

32. Which group of elements are all gases at room temperature?

33. Why are the inner transition metals not listed in the main body of the periodic table?

34. Why is it difficult to purify an inner transition metal?

Hands-On Chemistry Insights

Fire Water

As you can see in Figure 2.4, the two primary products when natural gas burns are carbon dioxide and water. Because of the heat generated by the burning, the water is released as water vapor. When it comes into contact with the relatively cool sides of the pot, this water vapor condenses to the liquid phase and is seen as "sweat." If the pot contained ice water, more vapor would condense, enough to form drops that roll off the bottom edge. As the pot gets warmer, this liquid water is heated and returns to the gaseous phase.

The only chemical change is the conversion of natural gas to carbon dioxide and water vapor. There are two physical changes—condensation of the water vapor created in the methane combustion and evaporation of this water once the pot gets sufficiently hot. (Of course, the evaporation of the water in the pot is another physical change.)

Oxygen Bubble Bursts

Hydrogen peroxide, H_2O_2, is a relatively unstable compound. In solution with water, it slowly decomposes, producing oxygen gas. In describing oxygen's physical properties, you should have noted that it is an invisible gas having no odor detectable over that of the yeast. Oxygen is light enough to rise out of the glass once it is released from the bubbles. What is your evidence of this? A chemical property of oxygen is that it intensifies burning.

Bottoms Up and Bubbles Out

It would be humorous to scrape the residue from your boiled-down drinking water into sealable containers labeled as drinking water from your particular region, such as "Rocky Mountain Drinking Water." Think of the potential market. You could ship these containers to customers around the world, and because the containers are not weighted down with water, shipping costs would be very low. Of course, each bottle would have to come with the instruction "Just add distilled water." Would you or would you

not want to push it by adding the word *Pure* to your label? With your classmates, discuss the science and ethics of such a venture.

As we explore in Chapter 7, gases do not dissolve well in hot liquids. Air that is dissolved in room-temperature water, for example, will bubble out when the water is heated. Thus you can speed up step 3 by using warm water.

For further experimentation, perform step 3 in two pots side by side. In one pot, use warm water from the kitchen faucet. In the second pot, use boiled water that has cooled down to the same temperature. You'll find that boiling *deaerates* the water, that is, removes the atmospheric gases. Chemists sometimes need to use deaerated water, which is made by allowing boiled water to cool in a sealed container. Why don't fish live very long in deaerated water?

Exercises

1. Each night you measure your height just before going to bed. When you arise each morning, you measure your height again and consistently find that you are 1 inch taller than you were the night before but only as tall as you were 24 hours ago! Is what happens to your body in this instance best described as a physical change or a chemical change? Be sure to try this activity if you haven't already.

2. Classify the following changes as physical or chemical. Even if you are incorrect in your assessment, you should be able to defend why you chose as you did.
   a. grape juice turns to wine   _____
   b. wood burns to ashes   _____
   c. water begins to boil   _____
   d. a broken leg mends itself   _____
   e. grass grows   _____
   f. an infant gains 10 pounds   _____
   g. a rock is crushed to powder   _____

3. Is the following transformation representative of a physical change or a chemical change?

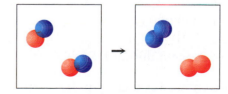

4. Each sphere in the diagrams below represents an atom. Joined spheres represent molecules. Which box contains a liquid phase? Why can you not assume that box B represents a lower temperature?

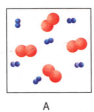
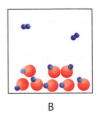

A   B

5. Based on the information given in the following diagrams, which substance has the lower boiling point, 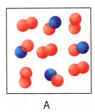 or 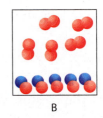 ?

A   B

6. What physical and chemical changes occur when a wax candle burns?

7. Which elements are some of the oldest known? What is your evidence?

8. Oxygen atoms are used to make water molecules. Does this mean that oxygen, O_2, and water, H_2O, have similar properties? Why do we drown when we breathe in water despite all the oxygen atoms present in this material?

9. A sample of water that is 99.9999 percent pure contains 0.0001 percent impurities. Consider from Chapter 1 that a glass of water contains on the order of a trillion trillion (1×10^{24}) molecules. If 0.0001 percent of these molecules were the molecules of some impurity, about how many impurity molecules would this be?
   a. 1000 (one thousand: 1×10^3)
   b. 1,000,000 (one million: 1×10^6)
   c. 1,000,000,000 (one billion: 1×10^9)
   d. 1,000,000,000,000,000,000 (one million trillion: 1×10^{18}) (One million trillion is the same as one quintillion.)

   How does your answer make you feel about drinking water that is 99.9999 percent free of

some poison, such as a pesticide? (See Appendix A for a discussion of scientific notation.)

10. Read carefully. Twice as much as one million trillion is two million trillion. One thousand times as much is 1000 million trillion. One million times as much is 1,000,000 million trillion, which is the same as one trillion trillion. Thus, one trillion trillion is one million times greater than a million trillion. Got that? So how many more water molecules than impurity molecules are there in a glass of water that is 99.9999 percent pure?

11. Someone argues that he or she doesn't drink tap water because it contains thousands of molecules of some impurity in each glass. How would you respond in defense of the water's purity, if it indeed does contain thousands of molecules of some impurity per glass?

12. Explain what chicken noodle soup and garden soil have in common without using the phrase *heterogeneous mixture*.

13. Classify the following as element, compound, or mixture, and justify your classifications: table salt, stainless steel, tap water, table sugar, vanilla extract, butter, maple syrup, aluminum, ice, milk, cherry-flavored cough drops.

14. If you eat metallic sodium or inhale chlorine gas, you stand a strong chance of dying. Let these two elements react with each other, however, and you can safely sprinkle the compound on your popcorn for better taste. What is going on?

15. Which of the following boxes contains an element? A compound? A mixture? How many different types of molecules are shown altogether in all three boxes?

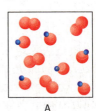
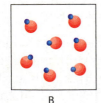
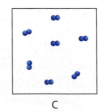

A   B   C

16. Common names of chemical compounds are generally much shorter than the corresponding systematic names. The systematic names for water, ammonia, and methane, for example, are dihydrogen monoxide, H_2O; trihydrogen nitride, NH_3; and tetrahydrogen carbide, CH_4.

For these compounds, which would you rather use: common names or systematic names? Which do you find more descriptive?

17. What is the difference between a compound and a mixture?

18. How might you separate a mixture of sand and salt? How about a mixture of iron and sand?

19. Mixtures can be separated into their components by taking advantage of differences in the chemical properties of the components. Why might this separation method be less convenient than taking advantage of differences in the physical properties of the components?

20. Why can't the elements of a compound be separated from one another by physical means?

21. Germanium, Ge (number 32), computer chips operate faster than silicon, Si (number 14), computer chips. So how might a gallium, Ga (number 31), chip compare with a germanium chip?

22. Is the air in your house a homogeneous or heterogeneous mixture? What evidence have you seen?

23. Helium, He, is a nonmetallic gas and the second element in the periodic table. Rather than being placed adjacent to hydrogen, H, however, helium is placed on the far right of the table. Why?

24. Name ten elements you have access to macroscopic samples of as a consumer here on the Earth.

25. Strontium, Sr (number 38), is especially dangerous to humans because it tends to accumulate in calcium-dependent bone marrow tissues (calcium, Ca, number 20). How does this fact relate to what you know about the organization of the periodic table?

26. With the periodic table as your guide, describe the element selenium, Se (number 34), using as many of this chapter's key terms as you can.

27. Many dry cereals are fortified with iron, which is added to the cereal in the form of small iron particles. How might these particles be separated from the cereal?

28. Why is half-frozen fruit punch always sweeter than the same fruit punch completely melted?

Exploring Further

The CRC Handbook of Chemistry and Physics. Boca Raton, FL: CRC Press, 1996.
  Toward the front of this classic reference book, you'll find a section on the history and general properties of each element.

http://www.chemsoc.org
  Chemistry news updates, an on-line chemistry magazine, and much more at this site run by the Royal Society of Chemistry.

http://www.chemsoc.org/viselements/pages/periodic_table.html
  The Visual Elements project of the Royal Society of Chemistry, providing animations of almost all the elements. A high-speed Internet connection is required.

http://www.csc.fi/lul/chem/graphics.html
  A virtual art gallery of molecular animations and other cool things, maintained by Finland's Center for Scientific Computing.

http://www.gsi.de
  Web site for the heavy-ion research facility in Darmstadt, Germany, where many of the heaviest but shortest-lived elements are being created.

http://newton.dep.anl.gov
  The Division of Educational Programs of the Argonne National Laboratory presents the Newton Bulletin Board Service, which features "Ask a Scientist." Explore the chemistry archives for the answers to more than 1500 student questions compiled since 1991.

http://www.shef.ac.uk/~chem/web-elements
  There are a large number of periodic tables posted on the Web, and this is one of the most popular ones.

the Chemistry place

Elements of Chemistry
Visit The Chemistry Place at:
www.aw.com/chemplace

Discovering the Atom and Subatomic Particles

Where We've Been and What We Know Now

The origin of most atoms goes back to the birth of the universe. Hydrogen, H, the lightest atom, was probably the original atom, and hydrogen atoms make up more than 90 percent of the atoms in the known universe. Heavier atoms are produced in stars, which are massive collections of hydrogen atoms pulled together by gravitational forces. The great pressures deep in a star's interior cause hydrogen atoms to fuse to heavier atoms. With the exception of hydrogen, therefore, all the atoms that occur naturally on the Earth—including those in your body—are the products of stars. A tiny fraction of these atoms came from our own star, the sun, but most are from stars that ran their course long before our solar system came into being. You are made of stardust, as is everything that surrounds you.

So most atoms are ancient. They have existed through imponderable ages, recycling through the universe in innumerable forms, both nonliving and living. In this sense, you don't "own" the atoms that make up your body—you are simply their present caretaker. There will be many caretakers to follow.

Atoms are so small that there are more than 10 billion trillion of them in each breath you exhale. This is more than the number of breaths in the Earth's atmosphere. Within a few years, the atoms of your breath are uniformly mixed throughout the atmosphere. What this means is that anyone anywhere on the Earth inhaling a breath of air takes in numerous atoms that were once part of you. And, of course, the reverse is true: you inhale atoms that were once part of everyone who has ever lived. We are literally breathing one another.

In this chapter, we trace the history of the discovery of the atom, which is perhaps the most important discovery humans have ever made. We also look at how researchers discovered that atoms are made of even smaller units of matter known as subatomic particles. Along the way, you will see how progress in science depends not only on keen observations and interpretations but also on an open-mindedness so frequently found in each new generation of investigators.

3.1 Chemistry Developed Out of Our Interest in Materials

We humans have long tinkered with the materials around us and used them to our advantage. Once we learned how to control fire, we were able to create many new substances. Moldable wet clay, for example, was found to harden to ceramic when heated by fire. By 5000 B.C., pottery fire pits gave way to furnaces hot enough to convert copper ores to metallic copper. By 1200 B.C., even hotter furnaces were converting iron ores to iron. This technology allowed for the mass production of metal tools and weapons and made possible the many achievements of ancient Chinese, Egyptian, and Greek civilizations.

In the fourth century B.C., the influential Greek philosopher Aristotle (384–322 B.C.) described the composition and behavior of matter in terms of the four qualities shown in Figure 3.1: hot, cold, moist, and dry. Although we know today that Aristotle's model is wrong, it was nonetheless a remarkable achievement for its day, and people using it in Aristotle's time found it made sense. When pottery was made, for example, wet clay was converted to ceramic because the heat of the fire drove out the moist quality of the wet clay and replaced it with the dry quality of the ceramic. Likewise, warm air caused ice to melt by replacing the dry quality characteristic of ice with the moist quality characteristic of water.

Aristotle's views on the nature of matter made so much sense to people that less obvious views were difficult to accept. One alternative view was the forerunner of our present-day model: matter is composed of a finite number of incredibly small but discrete units we call atoms. This model was advanced by several Greek philosophers, including Democritus (460–370 B.C.), who coined the term *atom* from the Greek phrase *a tomos*, which means "not cut" or "that which is indivisible." According to the atomic model of Democritus, the texture, mass, and color of a material were a function of the texture, mass, and color of its atoms, as illustrated in Figure 3.2. So compelling was Aristotle's reputation, however, that the atomic model would not reappear for 2000 years.

According to Aristotle, it was theoretically possible to transform any substance to another substance simply by altering the relative proportions of the four basic qualities. This meant that, under the proper conditions, a metal like lead could be transformed to gold. This concept laid the foundation of **alchemy**, a field of study concerned primarily with finding potions that would produce gold or confer immortality. Alchemists from the time of Aristotle to as late as the 1600s tried in vain to convert various metals to gold.

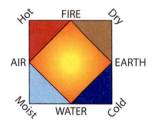

Figure 3.1
Aristotle thought that all materials were made of various proportions of four fundamental qualities: hot, dry, cold, and moist. Various combinations of these qualities gave rise to the four basic *elements*: hot and dry gave fire, moist and cold gave water, hot and moist gave air, and dry and cold gave earth. A hard substance like rock contained mostly the dry quality, for example, and a soft substance like clay contained mostly the moist quality.

Figure 3.2
In his atomic model, Democritus imagined that atoms of iron were shaped like coils, making iron rigid, strong, and malleable, and that atoms of fire were sharp, lightweight, and yellow.

Despite the futility of their efforts, the alchemists learned much about the behavior of many chemicals, and many useful laboratory techniques were developed.

3.2   Lavoisier Laid the Foundation of Modern Chemistry

In the 1400s, the printing press was invented in Europe, and an explosion of information, including scientific information, followed. Evidence against Aristotle's model of matter began to accumulate. In 1661, a well-known English experimentalist, Robert Boyle (1627–1691), departed from Aristotelian thought by proposing that a substance was not an element if it was made of two or more components. He published these and related thoughts in a text entitled *The Skeptical Chymist*, which had a significant impact on future generations of chemical thinkers.

About a century after Boyle, huge steps toward our present understanding of elements and compounds were taken by the French chemist Antoine Lavoisier (1743–1794). Embracing Boyle's views, Lavoisier accepted the idea of an *element* as any material made of only one component. Taking this model a step further, as shown in Figure 3.3, he identified a *compound* as any material composed of two or more elements. As you may recall from Chapter 2, these definitions are in line with our present understanding. Hydrogen, for example, is an element because it is made of only hydrogen atoms, and water is a compound because it is made of atoms of the elements hydrogen and oxygen.

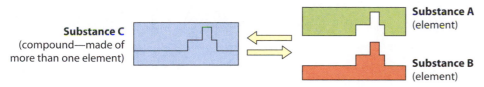

Figure 3.3
A representation of Lavoisier's notion of elements and compounds. Substances A and B cannot be broken down to smaller components and so are classified as elements. These two elements can react together to form the more complex substance C, which is classified as a compound because it is made of more than one element.

Antoine Lavoisier, shown here with his wife, Marie-Anne, who assisted him in many of his experiments, was a concerned citizen as well as a first-rate scientist. He established free schools, advocated the use of fire hydrants, and designed street lamps to make travel through urban neighborhoods safer at night. To help finance his scientific projects, Lavoisier took part-time employment as a tax collector. Because of this employment, he was beheaded in 1794 during the French Revolution. Soon after his execution, however, the French government was erecting statues in his honor.

The significance of Lavoisier's definitions is that they required experimentation. This was counter to the old ways of the Greek philosophers, who formulated their ideas based on logic and reason. Because he focused on the results of laboratory research, Lavoisier was a key player in the development of modern chemistry. To many, he is known as the "father of modern chemistry."

Mass Is Conserved in a Chemical Reaction

In the important experiment illustrated in Figure 3.4, Lavoisier carefully measured the mass of a sealed glass vessel that contained tin. When he heated the vessel, a chemical reaction occurred and the tin turned to a white powder. Lavoisier again measured the vessel's mass and found it had not changed. From the results of his experiments and the results of similar experiments performed by other investigators, Lavoisier hypothesized that mass is always conserved during a chemical reaction, where *conserved* in this context means that the amount of the mass does not change—the number of grams of mass present after the reaction is the same as the number of grams present before the reaction. This hypothesis is now considered to be a **scientific law**, which is any scientific hypothesis that has been tested over and over again and has not been contradicted. (A scientific law is sometimes referred to as a *scientific principle*.) Formally, the **law of mass conservation** states the following:

There is no detectable change in the total mass of materials when they react chemically to form new materials.

The law of mass conservation remains one of the most important laws in chemistry today. It is easy to see, however, why it took earlier investigators so long to formulate this law. After all, when wood burns, the mass of the ashes is always less than the mass of the original wood. Also, it was known that some substances, such as hardening cement, tend to gain mass as they undergo chemical change. What early investigators failed to recognize, however, is the role gases play in many chemical reactions. When

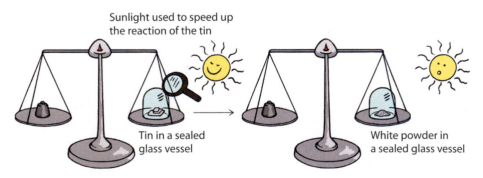

Figure 3.4
Lavoisier measured the mass of a sealed glass vessel containing tin and the mass of the same vessel containing a white powder left after the tin underwent a chemical reaction. He found the mass to be the same before and after the reaction.

wood burns, gaseous carbon dioxide and water vapor are released. The ashes have less mass because they are only one of the products of the reaction. Cement gains mass as it absorbs atmospheric carbon dioxide.

Lavoisier was exceedingly careful in attending to details. In recognizing the role that gases might play, he knew the importance of sealing his apparatus before performing a chemical reaction.

Concept Check

Had Lavoisier been a follower of ancient Greek philosophy, what might have prevented him from discovering the law of mass conservation?

Was this your answer? Using only logic and reason, it is difficult to conclude that mass is conserved in a chemical reaction. In most reactions, the total amount of mass appears to change because some of the products of the reaction are invisible atmospheric gases. Thus the law of mass conservation would have likely escaped Lavoisier's notice had he relied on the "common sense" logic and reason used by Greek philosophers rather than on precise measurements and experimentation.

When Lavoisier opened the sealed vessel in which the white powder had formed, he observed that air rushed in. He hypothesized that, as it formed the white powder, the tin absorbed either the air inside the vessel or perhaps only some component of the air. To find out what percentage of the air had reacted with the tin, he performed the same experiment using the arrangement shown in Figure 3.5. When the tin was done reacting, the water level in the jar had risen to about one-fifth of the total volume of the jar. Lavoisier realized that the only possible explanation was that air originally in the jar (which kept water out before the reaction began) had somehow been consumed in the chemical reaction with the tin. Because water replaced about 20 percent of the original volume of air, he reasoned that air is a mixture of at least two gaseous materials. One gas, making up about 20 percent of the air, disappeared from its gaseous phase by combining with the tin. The second gas, making up the other 80 percent, must have remained in the gaseous phase because it did not combine with the metal.

Soon after Lavoisier completed these experiments, he learned that an English chemist, Joseph Priestley (1733–1804), had prepared and isolated a

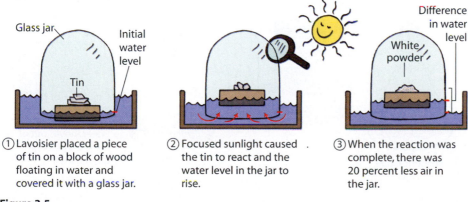

① Lavoisier placed a piece of tin on a block of wood floating in water and covered it with a glass jar.

② Focused sunlight caused the tin to react and the water level in the jar to rise.

③ When the reaction was complete, there was 20 percent less air in the jar.

Figure 3.5
Lavoisier directed sunlight at tin floating on a block of wood in a glass jar inverted over water. As the tin reacted to form a powder, the water level in the jar rose, indicating that some of the air originally in the jar had taken part in the reaction.

Hands-On Chemistry: Air Out

You can witness the involvement of a gas in a reaction by performing an experiment similar to the ones Lavoisier performed with tin.

What You Need

Nonsoapy steel-wool pad; narrow, straight-sided jar, such as an olive jar; wide-mouth jar (or shallow cooking pot); water

Procedure

① Follow the setup shown in the photograph. Stuff the steel-wool pad into the bottom of the narrow jar. A pad that doesn't fit through the jar opening can be cut into strips with scissors. Pour water into the wide-mouth jar to a depth of about 2 inches. Then invert the narrow jar into the water. Place the lip of the inverted jar on a coin to prevent sealing with the bottom of the wide-mouth jar.

② Note the water level inside the inverted jar.

③ Leave the setup alone until the steel wool, which is primarily iron, begins to look rusty (a couple of hours should do it).

What has happened to the water level in the inverted jar? Why?

Joseph Priestley was a self-trained scientist. He was the first to recognize the nature of carbonated beverages and began the study of photosynthesis with his discovery that plants absorb carbon dioxide when exposed to sunlight. A radical in many of his political views, Priestley was regarded with much suspicion, especially after he publicly sympathized with the French Revolution. After he was harassed and a mob had burned his home and library, he took the advice of his good friend Benjamin Franklin and moved to America, where he spent the last few years of his life in self-imposed exile.

gas that had remarkable properties. This gas caused candles to burn more brightly and charcoal to burn hotter. Lavoisier found that this gas couldn't be broken down to simpler substances and so recognized it to be an element. Because the gas produced acidic solutions when bubbled through water, Lavoisier gave it the name *oxygen*, which means "acid former." He found that oxygen was the reacting component of the air in his tin experiments.

See Figure 3.6 for a description of how Priestley first prepared and isolated oxygen.

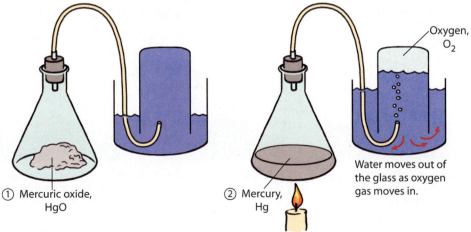

Figure 3.6
Priestley collected oxygen gas by heating the highly toxic metallic compound known today as mercuric oxide, HgO. When heated, mercuric oxide decomposes to liquid mercury and oxygen gas. Priestley collected the oxygen gas in an apparatus similar to the one illustrated here. As the gas is formed, it displaces water in the submerged inverted glass.

Hands-On Chemistry: Collecting Bubbles

A common method of collecting a gas produced in a chemical reaction is by the displacement of water. In this activity, you will collect the carbon dioxide produced from the reaction between baking soda and vinegar.

What You Need

Large pot (or sink), water, small plastic soda bottle (the 20-ounce size works nicely), film canister with lid, sharp knife, baking soda, vinegar, long wooden match, assistant

Safety Note

Wear safety glasses while using the knife and while the chemical reaction is taking place.

Procedure

① Cut a hole no wider than a pencil in the lid of the film canister. Add a teaspoon of baking soda to the canister and leave it uncapped.

② Fill the large pot (or the sink) three-fourths full with water. Fill the soda bottle with water, and with your hand over the opening, invert it into the water in the pot. Have your assistant hold the soda bottle upright in the pot with the mouth of the bottle not touching the pot bottom.

③ Pour a capful of vinegar into the film canister. As the bubbles begin to form, quickly cap the canister, placing your thumb over the hole in the lid to hold in as much gas as possible. Immediately submerge the canister right side up directly below the mouth of the soda bottle. Release your thumb, and bubbles of carbon dioxide will rise and be captured in the bottle.

④ To collect more carbon dioxide, drain the water from the film canister and repeat the procedure while your assistant continues to hold the inverted bottle.

⑤ To examine the properties of the gas generated in the baking soda–vinegar reaction, seal your hand over the mouth of the inverted soda bottle, carefully place the bottle upright on a table, and then uncover the mouth. There may still be some water in the bottle, but the gas above the water is carbon dioxide, which stays in the bottle because it is heavier than air. Light the match and place the flame into the carbon dioxide. Because there is no oxygen (there is only carbon dioxide), the flame is quickly extinguished because carbon dioxide does not support burning the way oxygen does.

Proust Proposed the Law of Definite Proportions

In 1766, the English chemist Henry Cavendish (1731–1810) isolated a gas that could be ignited in air to produce water and heat. Lavoisier was the first to recognize this gas as an element, which he named hydrogen—a Greek word that means "water former." Lavoisier was also the first to recognize that in forming water, the hydrogen was reacting with atmospheric oxygen. Thus, water must be a compound (not an element as Aristotle professed) made of the two elements hydrogen and oxygen.

By the 1790s, the French chemist Joseph Proust (1754–1826) had noted that, in forming water, hydrogen and oxygen always react in a particular mass ratio. He found, for example, that 8 grams of oxygen reacts with 1 gram of hydrogen (no more and no less) to produce 9 grams of water. Equivalently, 32 grams of oxygen reacts with 4 grams of hydrogen (no more and no less) to produce 36 grams of water. In all cases, the ratio of oxygen mass to hydrogen mass is 8:1. Even if oxygen and hydrogen are not mixed in an 8:1 ratio, they react as though they were. If 10 grams of oxygen, for example, is mixed with 1 gram of hydrogen, only 8 grams of oxygen reacts, leaving 2 grams of oxygen untouched. The result is still 9 grams of water, as Figure 3.7 shows.

				Leftover unreacted chemicals	
	Oxygen +	Hydrogen ⟶	Water +	Oxygen +	Hydrogen
(a)	8 g	1 g	9 g		
(b)	10 g	1 g	9 g	2 g	
(c)	8 g	2 g	9 g		1 g

Figure 3.7
(a) In forming water, oxygen and hydrogen always react in an 8:1 mass ratio. (b) When an excess of oxygen is present, the reaction still occurs in an 8:1 ratio, with the excess oxygen—2 grams in this case—remaining unreacted. (c) When an excess of hydrogen is present, the reaction still occurs in an 8:1 ratio, with the excess hydrogen remaining unreacted.

These and similar results with other chemical reactions, especially those involving metallic compounds, led Proust to propose the **law of definite proportions**:

Elements combine in definite mass ratios to form compounds.

Another example of the law of definite proportions involves nitrogen and hydrogen, which react in a 14:3 mass ratio to form ammonia. This means that 14 grams of nitrogen reacts with 3 grams of hydrogen to produce 17 grams of ammonia. Accordingly, mixing 14 grams of nitrogen with 14 grams of hydrogen also produces 17 grams of ammonia, not 28 grams. Somehow the 14 grams of ammonia "knows" to react with only 3 grams of hydrogen, despite the presence of 14 grams of hydrogen. How elements "knew" to react in particular mass ratios was a great mystery. The law of definite proportions, however, turned out to be one of the greatest clues to the discovery of the atom.

Concept Check ✓

How many grams of water can be produced when 16 grams of oxygen is mixed with 2 grams of hydrogen?

Was this your answer? Oxygen and hydrogen react in an 8:1 mass ratio to form water, but that doesn't mean you have to have exactly 8 grams of oxygen and exactly 1 gram of hydrogen in the reaction vessel. What this ratio means is that the mass of oxygen taking part in the reaction is always exactly eight times the mass of hydrogen taking part. Because here the oxygen mass—16 grams—is eight times the hydrogen mass—2 grams—the 16 grams of oxygen reacts fully with the 2 grams of hydrogen to produce 18 grams of water. Note that a 16:2 ratio is mathematically the same as an 8:1 ratio.

Calculation Corner: Finding Out How Much of a Chemical Reacts

When the definite proportions for a given reaction are known, unit conversion (see the Calculation Corner on page 12) can help you determine the amount of a chemical that takes part in the reaction. For instance, we can express the law of definite proportions for the formation of water as:

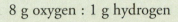

8 g oxygen : 1 g hydrogen

From this relationship, we can derive two conversion factors:

$$\frac{8 \text{ g oxygen}}{1 \text{ g hydrogen}} \quad \text{and} \quad \frac{1 \text{ g hydrogen}}{8 \text{ g oxygen}}$$

If you are given a certain amount of one element and want to know how much of the other element is needed for a complete reaction, you need only multiply by the appropriate conversion factor to find the answer.

Example

How much hydrogen is needed in order for 64 grams of oxygen to react completely in the formation of water?

Answer

$$64 \text{ g oxygen} \times \frac{1 \text{ g hydrogen}}{8 \text{ g oxygen}} = 8 \text{ g hydrogen}$$

↑ Quantity of oxygen in grams   ↑ Conversion factor   ↑ Quantity of hydrogen in grams

Knowing how much of each element must be present for complete reaction then allows you to determine how much product forms. If 64 grams of oxygen reacts with 8 grams of hydrogen, then a total of 64 grams + 8 grams = 72 grams of water is formed.

Your Turn

1. Nitrogen and hydrogen react in a 14:3 mass ratio to form ammonia. How much hydrogen is needed in order for 7.0 grams of nitrogen to react completely?

2. How much ammonia forms in the reaction between 7.0 grams of nitrogen and 6.0 grams of hydrogen?

3.3   Dalton Deduced That Matter Is Made of Atoms

The observations of Lavoisier, Proust, and others led John Dalton (1766–1844), a self-educated English schoolteacher, to reintroduce the atomic ideas of Democritus. In 1803 Dalton wrote a series of postulates—claims he assumed to be true based on experimental evidence—that can be summarized as follows:

1. Each element consists of indivisible, minute particles called atoms.

2. Atoms can be neither created nor destroyed in chemical reactions.

3. All atoms of a given element are identical.

4. Atoms chemically combine in definite whole-number ratios to form compounds.

5. Atoms of different elements have different masses.

John Dalton was born into a very poor family. Although his formal schooling ended at age 11, he continued to learn on his own and even began teaching others when he was only 12. His primary research interest was weather, which led him to conduct many experiments with gases. Soon after publishing his conclusions on the atomic nature of matter, his reputation as a first-rate scientist increased rapidly. In 1810, he was elected into Britain's premiere scientific organization, the Royal Society.

Even though Dalton was not correct about atoms being indivisible or about all atoms of a given element being identical (we'll see why later in the chapter), his postulates answered many questions about the nature of elements and compounds. Postulate 2, for example, which described the *indestructible* nature of atoms, accounted for Lavoisier's mass-conservation principle. During a chemical reaction, atoms may be exchanged, but never are they created out of nothing nor do they simply vanish. Postulate 4 explained compounds as the combination of the atoms of different elements. Oxygen and hydrogen atoms, for example, combine to form water.

Dalton concluded that because 8 grams of oxygen always combines with 1 gram of hydrogen, the oxygen atom must be eight times more massive than the hydrogen atom. In drawing this conclusion, he assumed that a single oxygen atom joins with a single hydrogen atom. According to Dalton, therefore, the most fundamental unit of water was HO rather than the familiar H_2O we know today.

Dalton Defended His Atomic Hypothesis Against Experimental Evidence

In 1808, the French chemist Joseph Gay-Lussac (1778–1850) reported that when gases react, the volumes that react—in a manner similar to what Proust described for elements in his law of definite proportions—are in the ratio of small whole numbers. Gay-Lussac's experiments showed, for example, that 2 liters of hydrogen completely reacts with 1 liter of oxygen (no more and no less) to form 2 liters of water vapor:

2 liters hydrogen gas + **1** liter oxygen gas ⟶ **2** liters water vapor

Dalton, however, was highly critical of Gay-Lussac's experiments. Dalton had already firmly established in his own mind that the formula for water was HO. If water contained twice as much hydrogen as oxygen, the formula would have to be H_2O. In addition, Dalton could not understand how 2 liters of water formed and not just 1 liter. Assuming that hydrogen gas and oxygen gas consisted of individual atoms, and assuming that equal volumes of two gases contain equal numbers of particles, as shown in Figure 3.8, then 2 liters of hydrogen should react with 1 liter of oxygen to produce 1 liter of water vapor, as shown in Figure 3.8a.

Gay-Lussac's results, illustrated in Figure 3.8b, showed that *2* liters of water vapor formed. So where did the atoms needed to create the additional liter of water vapor come from? Did each hydrogen atom and each oxygen atom split in half? This would effectively double the number of atoms, allowing for a second volume of water. The notion of an atom splitting in half, however, was counter to Dalton's well-received atomic hypothesis.

In 1811, the Italian physicist and lawyer Amadeo Avogadro (1776–1856) gave an accurate explanation for Gay-Lussac's experimental results. Avogadro hypothesized that the fundamental particles of hydrogen and oxygen were not atoms but rather diatomic molecules, where the term *diatomic* indicates two atoms per molecule. Thus, the formula for hydrogen

In addition to exploring the chemical identity of gases and the nature of chemical reactions, Joseph Gay-Lussac was one of the first balloonists. In one of his balloon flights to test hypotheses on the composition of air and the extent of the Earth's magnetic field, Gay-Lussac reached an altitude of 7000 meters (23,000 feet). This record remained unbroken for the next 50 years.

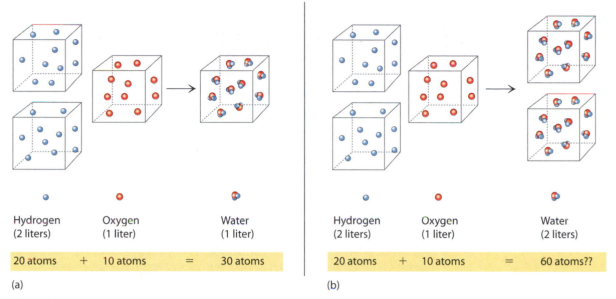

| Hydrogen (2 liters) | | Oxygen (1 liter) | | Water (1 liter) | | | Hydrogen (2 liters) | | Oxygen (1 liter) | | Water (2 liters) |

| 20 atoms | + | 10 atoms | = | 30 atoms | | 20 atoms | + | 10 atoms | = | 60 atoms?? |

(a)   (b)

Figure 3.8
(a) Dalton pointed out that if water had the formula H_2O, then 2 liters of hydrogen (shown here as 20 atoms) and 1 liter of oxygen (shown here as 10 atoms) should yield 1 liter of water vapor (shown here as 10 molecules containing a total of 30 atoms). (b) Gay-Lussac's experiments showed that 2 liters of water vapor formed. Where did the atoms for this second liter of water vapor come from? Questions such as this led Dalton to distrust Gay-Lussac's experimental results.

is H_2, and the formula for oxygen is O_2. Diatomic particles of hydrogen and oxygen would have double the number of atoms in a given volume, thus allowing for the formation of a second volume of water, as Figure 3.9 shows.

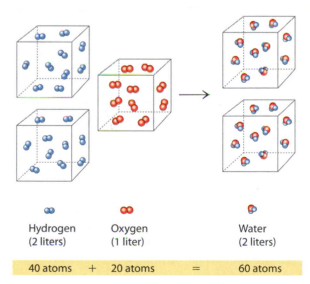

| Hydrogen (2 liters) | Oxygen (1 liter) | Water (2 liters) |

| 40 atoms | + | 20 atoms | = | 60 atoms |

Figure 3.9
Because each particle of gaseous hydrogen and gaseous oxygen is diatomic, 2 liters of hydrogen and 1 liter of oxygen form 2 liters of water vapor. In this way, molecules of hydrogen, H_2, and oxygen, O_2, are split rather than *atoms* of hydrogen, H, and oxygen, O, and Dalton's atomic theory is not violated.

Amadeo Avogadro received a doctor of law degree when he was 20 years old. He enjoyed practicing law but was more interested in science, which eventually became his life's occupation. As a professor of physics and mathematics in Italy, he was geographically and intellectually isolated from the chemical community developing on the other side of the Alps in northern France and in England. This isolation made it difficult for him to defend his views on the nature of atoms. In addition, he valued his privacy and preferred focusing his energies on family matters.

Concept Check ✔

How many molecules of hydrogen chloride, HCl, form when ten molecules of diatomic hydrogen, H_2, react with ten molecules of diatomic chlorine, Cl_2? If ten molecules represent one volume, how many volumes of hydrogen chloride form?

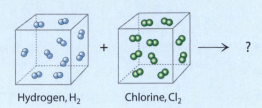

Hydrogen, H_2 Chlorine, Cl_2 ?

Were these your answers? In this reaction, 20 hydrogen chloride molecules are formed. One volume of hydrogen plus one volume of chlorine react to form two volumes of hydrogen chloride.

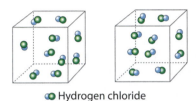

Hydrogen chloride

Stanislao Cannizzaro's main research interests were in the chemistry of carbon compounds found in living organisms. Cannizzaro did much to dispel the then widely held belief that the laws governing those chemicals were different from the laws governing chemicals not found in living organisms.

Dalton understood Avogadro's creative argument but found it unacceptable because it failed to explain how two atoms of the same element could bond to each other. Based on his own research, Dalton had come to the erroneous conclusion that atoms of the same kind always have a natural repulsion for one another. Because of Dalton's authority in the scientific community, Avogadro's hypothesis was discarded and did not reappear for another half-century.

In 1860, an international conference of chemists convened to discuss how the masses of atoms of different elements could be measured and compared with one another. (As we explore in Section 9.2, knowing the relative masses of atoms helps chemists understand and control chemical reactions.) At the time, there was little agreement because different chemists using different experimental procedures and assuming different theories came up with different results. Progress in chemistry was stymied by this problem.

At this conference, a pamphlet written by the Italian chemist Stanislao Cannizzaro (1826–1910) was presented. In this pamphlet, which he had used with his students for several years, Cannizzaro explained and justified Avogadro's hypothesis and showed how correct atomic masses and formulas could be obtained through easy calculations. The concept was simple: provided equal volumes of gases contain equal numbers of atoms or molecules, the relative masses of these particles can be obtained by weighing equal volumes of gases that are at the same temperature and pressure. As shown in Figure 3.10, for example, 1 liter of oxygen is 16 times heavier (more massive) than 1 liter of hydrogen, suggesting that, assuming the same number of atoms per molecule, an oxygen molecule is 16 times heavier

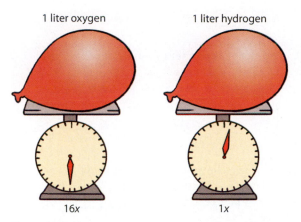

1 liter oxygen   1 liter hydrogen

16*x*   1*x*

Figure 3.10
One liter of oxygen is 16 times more massive than the same volume of hydrogen. Assuming 1 liter of oxygen and 1 liter of hydrogen contain the same number of particles, the mass of an oxygen particle must be 16 times greater than the mass of a hydrogen particle.

(more massive) than a hydrogen molecule. Analysis of equal volumes of gases for many other elements resulted in the first tables of accurate relative atomic masses, and from these tables grew the all-important periodic table.

Despite the critical attacks by Dalton, the contributions of Gay-Lussac and Avogadro turned out to be most important to the development of atomic theory. Their efforts, however, were not widely recognized until after their deaths. Gay-Lussac died 10 years prior to the 1860 conference, and Avogadro died 4 years prior.

Mendeleev Used Known Relative Atomic Masses to Create the Periodic Table

By the 1860s, many scientists working independently had noted that when they listed elements in order of relative masses, a number of interesting patterns arose. Many physical and chemical properties, for example, tended to change gradually in moving from one element to the next. At regular intervals, however, an element had properties that were vastly different from those of the preceding element and more like those of a much lighter element. In other words, the properties of elements tended to recur in *cycles*, exhibiting the *periodicity* we looked at in Section 2.6.

In 1869, a Russian chemistry professor, Dmitri Mendeleev (1834–1907), produced a chart summarizing the properties of known elements for his students. Mendeleev's chart was unique in that it resembled a calendar. Across each horizontal row, he placed all the elements that appeared in one interval of repeating properties. Down each vertical column he placed elements of similar properties. He found, however, that in order to align elements properly in a column, he had to shift elements left or right occasionally. This left gaps—blank spaces that could not be filled by any known element (Figure 3.11). Instead of looking on these gaps as defects, Mendeleev boldly predicted the existence of elements that had not yet been discovered. Furthermore, his predictions about the properties of some of those missing elements led to their discovery.

Dmitri Mendeleev was a devoted and highly effective teacher. Students adored him and would fill lecture halls to hear him speak about chemistry. Much of his work on the periodic table occurred in his spare time following his lectures. Mendeleev taught not only in the university classrooms but anywhere he traveled. During his journeys by train, he would travel third class with peasants to share his findings about agriculture.

— 70 —

но въ ней, мнѣ кажется, уже ясно выражается примѣнимость вы ставляемаго мною начала ко всей совокупности элементовъ, пай которыхъ извѣстенъ съ достовѣрностію. На этотъ разъ я и желалъ преимущественно найдти общую систему элементовъ. Вотъ этотъ опытъ:

```
                                    Ti=50    Zr=90    ?=180.
                                    V=51     Nb=94    Ta=182.
                                    Cr=52    Mo=96    W=186.
                                    Mn=55    Rh=104,4 Pt=197,4
                                    Fe=56    Ru=104,4 Ir=198.
                             Ni=Co=59        Pl=106,6, Os=199.
H=1                                 Cu=63,4  Ag=108   Hg=200.
         Be=9,4    Mg=24    Zn=65,2  Cd=112
         B=11      Al=27,4  ?=68     Ur=116   Au=197?
         C=12      Si=28    ?=70     Sn=118
         N=14      P=31     As=75    Sb=122   Bi=210
         O=16      S=32     Se=79,4  Te=128?
         F=19      Cl=35,5  Br=80    I=127
Li=7 Na=23        K=39      Rb=85,4  Cs=133   Tl=204
                  Ca=40     Sr=57,6  Ba=137   Pb=207.
                  ?=45      Ce=92
                  ?Er=56    La=94
                  ?Yt=60    Di=95
                  ?In=75,6  Th=118?
```

а потому приходится въ разныхъ рядахъ имѣть различное измѣненіе разностей, чего нѣтъ въ главныхъ числахъ предлагаемой таблицы. Или же придется предпо- лагать при составленіи системы очень много недостающихъ членовъ. То и другое мало выгодно. Мнѣ кажется притомъ, наиболѣе естественнымъ составить кубическую систему (предлагаемая есть плоскостная), но и попытки для ея образо- ванія не повели къ надлежащимъ результатамъ. Слѣдующія двѣ попытки могутъ по- казать то разнообразіе сопоставленій, какое возможно при допущеніи основнаго начала, высказаннаго въ этой статьѣ.

Li	Na	K	Cu	Rb	Ag	Cs	—	Tl
7	23	39	63,4	85,4	108	133		204
Be	Mg	Ca	Zn	Sr	Cd	Ba	—	Pb
B	Al	—	—	—	Ur	—	—	Bi?
C	Si	Ti	—	Zr	Sn	—	—	—
N	P	V	As	Nb	Sb	—	Ta	—
O	S	—	Se	—	Te	—	W	—
F	Cl	—	Br	—	J	—	—	—
19	35,5	58	80	190	127	160	190	220.

Figure 3.11
An early draft of Mendeleev's periodic table.

That Mendeleev was able to predict the properties of new elements helped convince many scientists of the accuracy of Dalton's atomic hypothesis. This in turn helped promote Dalton's proposed atomic nature of matter from a hypothesis to a more widely accepted theory. Mendeleev's chart, which ultimately led to our modern periodic table with its horizontal periods and vertical groups, also helped lay the groundwork for our understanding of atomic behavior and is recognized as one of the most important achievements of modern science.

Concept Check ✔

The following statements summarize the scientific discoveries presented in Sections 3.2 and 3.3. Place them in chronological order.
a. Elements are made of atoms.
b. Chemicals react in definite whole-number ratios.
c. Relative masses of atoms can be measured.
d. Hydrogen gas and oxygen gas consist of diatomic molecules.
e. The periodic table can be used to predict the properties of elements.
f. Mass is conserved during a chemical reaction.

Was this your answer? Lavoisier discovered that *mass is conserved during a chemical reaction*, which led Proust to discover that *chemicals react in definite whole-number ratios*. Based on this information, Dalton proposed that *elements are made of atoms*. This was followed by Gay-Lussac's experiments, which suggested to Avogadro that *hydrogen gas and oxygen gas consist of diatomic molecules*. Using this as a premise, Cannizzaro was able to show how Avogadro's hypothesis could be used to *measure the relative masses of atoms*. Knowing relative masses allowed Mendeleev to *devise the periodic table and use it to predict the properties of elements* yet to be discovered. The correct sequence is therefore f, b, a, d, c, e.

Today, the results of many scientific experiments confirm the atomic nature of matter. Contrary to Dalton's notion of the indivisible atom, however, an accumulation of evidence tells us that atoms are in fact divisible and that they are made of smaller particles called *electrons*, *protons*, and *neutrons*. For the remainder of this chapter, we explore these *subatomic particles* in detail, continuing with our historical perspective.

3.4 The Electron Was the First Subatomic Particle Discovered

In 1752, Benjamin Franklin (1706–1790) learned from experiments with thunderstorms that lightning is a flow of electrical energy through the atmosphere. This discovery prompted other scientists to explore whether or not electrical energy could travel through gases other than the atmosphere. To find out, they applied a voltage across glass tubes in which they had sealed various gases. (To apply a voltage means to connect each end of a tube to a wire and then connect the free ends of the two wires to a battery.)

In every case, the result was a brightly glowing ray (Figure 3.12a). This meant that electrical energy was able to travel through different types of gases. To the surprise of these early investigators, a ray was also produced when the voltage was applied across a glass tube that had been evacuated and was thus empty of any gas (Figure 3.12b). This implied that the ray was not a consequence of the gas but rather an entity in and of itself.

Experiments showed that the ray emerged from the end of the tube that was negatively charged. Because this negatively charged end was called the *cathode*, the apparatus, shown in Figure 3.13a, was named a **cathode ray tube**. Magnetic fields caused the ray to deflect, as did small electrically charged metal plates. When such plates were used, the ray was always

Benjamin Franklin invented the lightning rod, which is a sharp point of metal placed on a rooftop and connected to the ground by a long wire. Houses equipped with such rods are protected from the danger of lightning bolts. A popular myth holds that Franklin discovered the electrical nature of lightning by flying a kite during a lightning storm. Franklin, however, was smart enough to know the extreme danger posed by such a foolish act.

(a)

Figure 3.12
(a) Electrical energy passing through a glass tube filled with neon gas generates a bright red glowing ray. (b) The ray passing through an evacuated glass tube is not usually visible. In the tube shown here, however, the ray is highlighted by a fluorescent backing that glows green as the ray passes over it.

(b)

deflected toward the positively charged plate and away from the negatively charged plates. Because like signs of electric charge repel each other, this meant the cathode ray was negatively charged. The speed of the ray was found to be considerably less than the speed of light. Because of these characteristics, it appeared that the ray behaved more like a beam of particles than a beam of light.

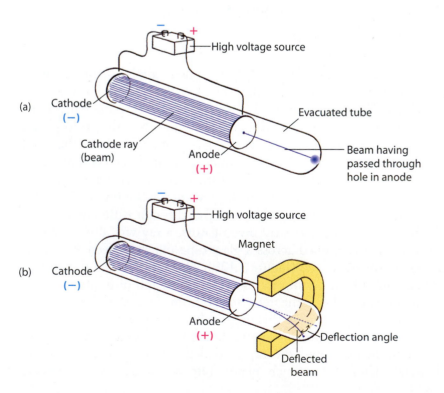

Figure 3.13
(a) A simple cathode ray tube. The small hole in the positvely charged end of the tube, the *anode*, permits the passage of a narrow beam that strikes the end of the evacuated tube, producing a glowing dot as the beam interacts with the glass. (b) The cathode ray is deflected by a magnetic field.

In 1897, J. J. Thomson (1856–1940) measured the deflection angles of cathode ray particles in a magnetic field, using a magnet positioned as shown in Figure 3.13b. He reasoned that the deflection of the particles depended on their mass and electric charge. The greater a particle's mass, the greater its resistance to a change in motion and therefore the *smaller* the deflection. The greater a particle's charge, the stronger the magnetic interactions and therefore the *larger* the deflection. The angle of deflection, he concluded, was equal to the ratio of the particle's charge to its mass:

$$\text{angle of deflection} = \frac{\text{charge}}{\text{mass}}$$

Knowing only the angle of deflection, however, Thomson was unable to calculate either the charge or the mass of each particle. In order to calculate the mass, he needed to know the charge, but in order to calculate the charge, he needed to know the mass.

Concept Check ✔

For which equation is it *not* possible to calculate one specific value for *x*:

$$4 = \frac{x}{2} \qquad 3 = \frac{x}{y}$$

Was this your answer? In the first equation, it's possible to figure that $x = 8$ (because $8 \div 2 = 4$). In the second equation, x would equal 15 if y equaled 5 (becasue $15 \div 5 = 3$) but would equal 9 if y equaled 3 (because $9 \div 3 = 3$). In other words, one specific value for x cannot be determined unless the value of y is known. Likewise, the value of y cannot be determined unless the value of x is known. Similarly, Thomson could not calculate the electron's mass without knowing its charge.

Joseph John Thomson, known to his colleagues as J. J., was one of the first directors of the famous Cavendish Laboratory of Cambridge University in England, where almost all the discoveries concerning subatomic particles and their behavior were made. Seven of Thomson's students went on to receive Nobel prizes for their scientific work. Thomson himself won a Nobel prize in 1906 for his work with the cathode ray tube.

In 1909, the American physicist Robert Millikan (1868–1953) calculated the numerical value of a single increment of electric charge on the basis of the innovative experiment shown in Figure 3.14. Millikan sprayed tiny oil

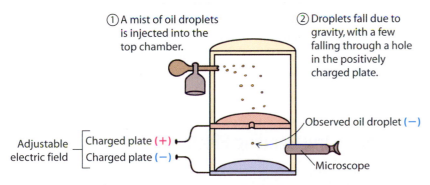

① A mist of oil droplets is injected into the top chamber.

② Droplets fall due to gravity, with a few falling through a hole in the positively charged plate.

Observed oil droplet (−)

Adjustable electric field — Charged plate (+) Charged plate (−)

Microscope

③ The electric field is adjusted until a droplet hovers. The upward electric force exerted on the droplet by the positively charged plate is exactly balanced by the downward force of gravity exerted on the droplet.

Figure 3.14
Millikan determined the charge of an electron with this oil-drop experiment.

The European science community of the 1800s viewed most American scientists as inventors—clever, but not profound in their thinking or discoveries. This attitude began to change at the turn of the 20th century, principally because of the work of the American scientist Robert Millikan, who excelled in his experimental designs and conclusions. In addition to research, he also spent much time preparing textbooks so that his students did not have to rely so much on lectures. He won a Nobel prize in 1923 and served as the president of Caltech from 1921 to 1945.

droplets into a specially designed chamber in which droplets could be suspended in air by the application of an electric field. (This is similar to the way a person's hair can be made to stand straight up by placing a statically charged balloon near the hair.) When the field was strong on Millikan's apparatus, some of the droplets moved upward, indicating they had a very slight negative charge and so were attracted to the upper, positively charged plate. Millikan adjusted the field so that some of the droplets would hover motionless. He knew that the downward force of gravity on these motionless droplets was exactly balanced by the upward electric force. By altering the field strength, he could make other droplets, of different masses, hover motionless. Repeated measurements showed that the electric charge on any droplet was always some multiple of a single very small value, 1.60×10^{-19} coulomb, which Millikan proposed to be the fundamental increment of all electric charge. (The *coulomb* is a unit of electric charge.) Using this value and the charge-to-mass ratio discovered by Thomson, Millikan calculated the mass of a cathode ray particle to be considerably less than that of the smallest known atom, hydrogen. This was remarkable because it provided strong evidence that the atom was not the smallest particle of matter.

Concept Check ✔

What do the numbers 45, 30, 60, 75, 105, 35, 80, 55, 90, 20, 65 have in common?

Was this your answer? They are all multiples of 5. In a similar fashion, Millikan noted that all the readings from his electronic equipment were multiples of a very small number, which he calculated to be 1.60×10^{-19} coulomb.

The cathode ray particle is known today as the **electron**, a name that comes from the Greek word for amber (*electrik*), which is a material the early Greeks used to study the effects of static electricity. The electron is a fundamental component of all atoms. All electrons are identical, each having a negative electric charge and an incredibly small mass of 9.1×10^{-31} kilograms. Electrons determine many of a material's properties, including chemical reactivity and such physical attributes as taste, texture, appearance, and color.

The cathode ray—a stream of electrons—has found a great number of applications. Most notably, a traditional television set (not the modern thin LCD screens) is a cathode ray tube with one end widened out into a phosphor-coated screen. Signals from the television station cause electrically

Hands-On Chemistry: Bending Electrons

Stare at a non-LCD television set or computer monitor, and you stare down the barrel of a cathode ray tube. You can find evidence for this by holding a magnet up to the screen. Note the distortion. *Important: use only a small magnet and hold it up to the screen only briefly; otherwise the distortion may become permanent.*

charged plates in the tube to control the direction of the ray such that images are traced onto the screen.

3.5   The Mass of an Atom Is Concentrated in Its Nucleus

It was reasoned that if atoms contained negatively charged particles, some balancing positively charged matter must also exist. From this, Thomson put forth what he called a *plum-pudding model* of the atom, shown in Figure 3.15. Further experimentation, however, soon proved this model to be wrong.

Around 1910, a more accurate picture of the atom came to one of Thomson's former students, the New Zealand physicist Ernest Rutherford (1871–1937). Rutherford oversaw the now-famous gold-foil experiment, which was the first experiment to show that the atom is mostly empty space and that most of its mass is concentrated in a tiny central core called the **atomic nucleus**.

In Rutherford's experiment, shown in Figure 3.16, a beam of positively charged particles, called alpha particles, was directed through an ultrathin sheet of gold foil. Since alpha particles were known to be thousands of times more massive than electrons, it was expected that the alpha-particle stream would not be impeded as it passed through the "atomic pudding" of gold foil. This was indeed observed to be the case—for the most part. Nearly all alpha particles passed through the gold foil undeflected and produced spots of light when they hit a fluorescent screen positioned around the gold foil. However, some particles were deflected from their straight-line path as they passed through the foil. A few of them were widely deflected, and a very few were even deflected straight back toward the source! These alpha particles must have hit something relatively massive, but what? Rutherford reasoned that undeflected particles traveled through regions of the gold foil that were empty space, as Figure 3.17 shows, and the deflected ones were repelled from extremely dense positively charged centers. Each atom, he concluded, must contain one of these centers, which he named the *atomic nucleus*.

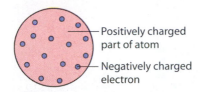

Figure 3.15
Thomson's plum-pudding model of the atom. Thomson proposed that the atom might be made of thousands of tiny, negatively charged particles swarming within a cloud of positive charge, much like plums and raisins in an old-fashioned Christmas plum pudding.

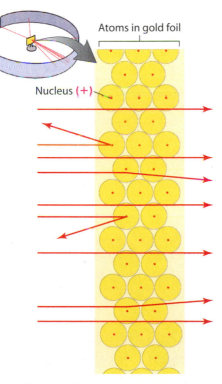

Figure 3.17
Rutherford's interpretation of the results from his gold-foil experiment. Most alpha particles passed through the empty space of the gold atoms undeflected, but a few were deflected by an atomic nucleus.

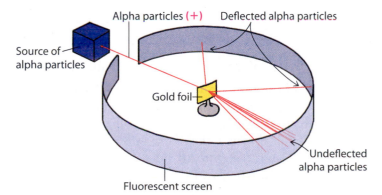

Figure 3.16
Rutherford's gold-foil experiment. A beam of positively charged alpha particles was directed at a piece of gold foil. Most of the particles passed through the foil undeflected, but some were deflected. This result implied that each gold atom was mostly empty space with a concentration of mass at its center—the atomic nucleus.

When Ernest Rutherford was 24, he placed second in a New Zealand scholarship competition to attend Cambridge University in England, but the scholarship was awarded to Rutherford after the winner decided to stay home and get married. In addition to discovering the atomic nucleus, Rutherford was also first to characterize and name many of the nuclear phenomena discussed in the following chapter. He won a Nobel prize in 1908 for showing how elements such as uranium can become different elements through the process of radioactive decay. At the time, the idea of one element transforming to another was shocking and met with great skepticism because it seemed reminiscent of alchemy.

Rutherford guessed that the atomic nucleus must have a positive electric charge to balance the negative charge of the electrons in the atom. He also guessed that the electrons were not part of this nucleus and so must be outside it but still somewhere in the atom. Today we know that, as Figure 3.18 illustrates, the electrons do indeed exist outside the nucleus, swirling around it at ultrahigh speeds. Figure 3.18 also shows that an atom is mostly empty space, with the diameter of the whole atom being about 10,000 times greater than the diameter of its nucleus. If a nucleus were the size of the period at the end of this sentence, the outer edges of the atom would be located 3.3 meters (11 feet) away. Also, because the nucleus is so dense, the mass of such a period-sized nucleus would be on the order of 2500 kilograms—the mass of a large pick-up truck.

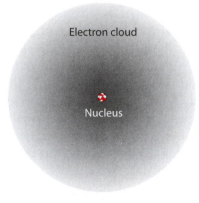

Electron cloud

Nucleus

Figure 3.18
Electrons whiz around the atomic nucleus, forming what can be best described as a cloud. If this illustration were drawn to scale, the atomic nucleus would be too small to be seen. An atom is mostly empty space.

Figure 3.19
As close as Tracy and Ian are in this photograph, none of their atoms meet. The closeness between us is in our hearts.

We and all materials around us are mostly empty space because the atoms we are made of are mostly empty space. So why don't atoms simply pass through one another? How is it that we are supported by the floor despite the empty nature of its atoms? Although subatomic particles are much smaller than the volume of the atom, the range of their electric field is several times larger than that volume. In the outer regions of any atom are electrons, which repel the electrons of neighboring atoms. Two atoms therefore can get only so close to each other before they start repelling (provided they don't join in a chemical bond, as is discussed in Chapter 6).

When the atoms of your hand push against the atoms of a wall, electrical repulsions between electrons in your hand and electrons in the wall prevent your hand from passing through the wall. These same electrical repulsions prevent us from falling through the solid floor. They also allow us the sense of touch. Interestingly, when you touch someone, your atoms and those of the other person do not meet. Instead, atoms from the two of you get close enough so that you sense an electrical repulsion. There is still a tiny, though imperceptible, gap between the two of you (Figure 3.19).

3.6 The Atomic Nucleus Is Made of Protons and Neutrons

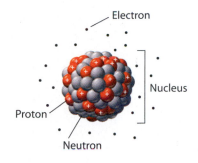

The positive charge of any atomic nucleus was found to be equal in magnitude to the combined negative charge of all the electrons in the atom. It was thus reasoned, and then experimentally confirmed, that positively charged subatomic particles make up the nucleus. Today we call these positively charged particles **protons**. The proton is nearly 2000 times more massive than the electron. The electric charge on the proton is numerically equal to the electric charge on the electron, but, as just mentioned, the charge on the proton is positive. Thus each electron has an electric charge of -1.60×10^{-19} coulomb, and each proton has an electric charge of $+1.60 \times 10^{-19}$ coulomb. The number of protons in the nucleus of any atom is equal to the number of electrons whirling about the nucleus, and so the positive charge and negative charge cancel each other, which means the atom is electrically balanced. For example, an electrically balanced oxygen atom has eight electrons and eight protons.

Scientists have agreed to identify elements by **atomic number**, which is the number of protons each atom of a given element contains. The modern periodic table lists the elements in order of increasing atomic number. Hydrogen, with one proton per atom, has atomic number 1; helium, with two protons per atom, has atomic number 2; and so on.

Concept Check ✔

How many protons are there in an iron atom, Fe (atomic number 26)?

Was this your answer? The atomic number of an atom and its number of protons are the same. Thus, there are 26 protons in an iron atom. Another way to put this is that all atoms that contain 26 protons are, by definition, iron atoms.

If we compare the electric charges and masses of different atoms, we see that the atomic nucleus must be made up of more than just protons. Helium, for example, has twice the electric charge of hydrogen but *four* times the mass. The added mass is due to another subatomic particle found in the nucleus, the **neutron**, which was first detected in 1932 by the British physicist James Chadwick (1891–1974). The neutron has about the same mass as the proton, but it has no electric charge. Any object that has no net electric charge is said to be *electrically neutral*, and that is where the neutron got its name. We discuss the important role that neutrons play in holding the atomic nucleus together in the following chapter.

Both protons and neutrons are called **nucleons**, a term that denotes their location in the atomic nucleus.

A neutron goes into a restaurant and asks the waiter, "How much for a drink?" The waiter replies, "For you, no charge."

Table 3.1 summarizes the basic facts about our three subatomic particles.

Table 3.1

Subatomic Particles

	Particle	Charge	Relative Mass	Actual Mass* (kg)
	Electron	−1	1	9.11×10^{-31} **
Nucleons	Proton	+1	1836	1.673×10^{-27}
	Neutron	0	1841	1.675×10^{-27}

* Not measured directly but calculated from experimental data.

** 9.11×10^{-31} kg = 0.000000000000000000000000000000911 kg (see Appendix A).

For any element, there is no set number of neutrons in the nucleus. For example, most hydrogen atoms (atomic number 1) have no neutrons. A small percentage, however, have one neutron, and a smaller percentage have two neutrons. Similarly, most iron atoms (atomic number 26) have 30 neutrons, but a small percentage have 29 neutrons. Atoms of the same element that contain different numbers of neutrons are **isotopes** of one another.

We identify isotopes by their **mass number**, which is the total number of protons and neutrons (in other words, the number of nucleons) in the nucleus. As Figure 3.20 shows, a hydrogen isotope with only one proton is called hydrogen-1, where 1 is the mass number. A hydrogen isotope with one proton and one neutron is therefore hydrogen-2, and a hydrogen isotope with one proton and two neutrons is hydrogen-3. Similarly, an iron isotope with 26 protons and 30 neutrons is called iron-56, and one with only 29 neutrons is iron-55.

Figure 3.20
Isotopes of an element have the same number of protons but different numbers of neutrons and hence different mass numbers. The three hydrogen isotopes have special names: protium for hydrogen-1, deuterium for hydrogen-2, and tritium for hydrogen-3. Of these three isotopes, hydrogen-1 is most common. For most elements, such as iron, the isotopes have no special names and are indicated merely by mass number.

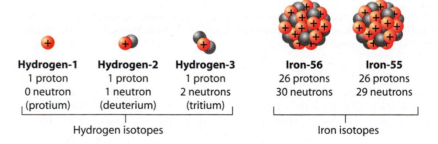

Hydrogen-1	**Hydrogen-2**	**Hydrogen-3**	**Iron-56**	**Iron-55**
1 proton	1 proton	1 proton	26 protons	26 protons
0 neutron	1 neutron	2 neutrons	30 neutrons	29 neutrons
(protium)	(deuterium)	(tritium)		

Hydrogen isotopes   Iron isotopes

An alternative method of indicating isotopes is to write the mass number as a superscript and the atomic number as a subscript to the left of the atomic symbol. For example, an iron isotope with a mass number of 56 and atomic number of 26 is written

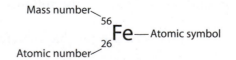

Mass number $\diagdown$ $^{56}_{26}$Fe — Atomic symbol

Atomic number $\diagup$

The total number of neutrons in an isotope can be calculated by subtracting its atomic number from its mass number:

$$
\begin{array}{r}
\text{mass number} \\
-\ \underline{\text{atomic number}} \\
\text{number of neutrons}
\end{array}
$$

For example, uranium-238 has 238 nucleons. The atomic number of uranium is 92, which tells us that 92 of these 238 nucleons are protons. The remaining 146 nucleons must be neutrons:

Nucleons — $^{238}_{92}$U — Protons

$$
\begin{array}{r}
238 \text{ protons and neutrons} \\
-\ \underline{\ 92 \text{ protons}} \\
146 \text{ neutrons}
\end{array}
$$

Atoms interact with one another electrically. Therefore the way any atom behaves in the presence of other atoms is determined largely by the charged particles it contains, especially its electrons. Isotopes of an element differ only by mass, not by electric charge. For this reason, isotopes of an element share many characteristics—in fact, as chemicals they cannot be distinguished from one another. For example, a sugar molecule containing seven neutrons per carbon nucleus is digested no differently from a sugar molecule containing six neutrons per carbon nucleus. In fact, about 1 percent of the carbon we eat is the carbon-13 isotope containing seven neutrons per nucleus. The remaining 99 percent of the carbon in our diet is the more common carbon-12 isotope containing six neutrons per nucleus.

The total mass of an atom is called its **atomic mass**. This is the sum of the masses of all the atom's components (electrons, protons, and neutrons). Because electrons are so much less massive than protons and neutrons, their contribution to atomic mass is negligible. As we explore further in Section 9.2, a special unit has been developed for atomic masses. This is the *atomic mass unit*, amu, where 1 atomic mass unit is equal to 1.661×10^{-24} gram, which is slightly less than the mass of a single proton. As shown in Figure 3.21, the atomic masses listed in the periodic table are in atomic mass units. As is explored in the Calculation Corner on page 92, the atomic mass of an element as presented in the periodic table is actually the average atomic mass of its various isotopes.

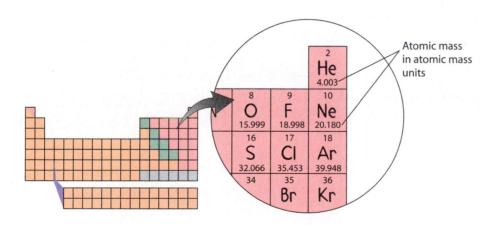

Figure 3.21
Helium, He, has an atomic mass of 4.003 atomic mass units, and neon, Ne, has an atomic mass of 20.180 atomic mass units.

Concept Check ✔

Distinguish between mass number and atomic mass.

Was this your answer? Both terms include the word *mass* and so are easily confused. Focus your attention on the second word of each term, however, and you'll get it right every time. Mass *number* is a count of the *number* of nucleons in an isotope. An atom's mass number requires no units because it is simply a count. Atomic *mass* is a measure of the total *mass* of an atom, which is given in atomic mass units. If necessary, atomic mass units can be converted to grams using the relationship 1 atomic mass unit $= 1.661 \times 10^{-24}$ gram.

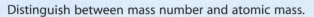

Calculation Corner: Calculating Atomic Mass

Most elements have a variety of isotopes, each with its own atomic mass. For this reason, the atomic mass listed in the periodic table for any given element is the average of the masses of all the element's isotopes based on their relative abundance.

About 99 percent of all carbon atoms, for example, are the isotope carbon-12, and most of the remaining 1 percent are the heavier isotope carbon-13. This small amount of carbon-13 raises the *average* mass of carbon from 12.000 atomic mass units to the slightly greater value 12.011 atomic mass units.

To arrive at the atomic mass presented in the periodic table, you first multiply the mass of each naturally occurring isotope of an element by the fraction of its abundance and then add up all the fractions.

Example

Carbon-12 has a mass of 12.0000 atomic mass units and makes up 98.89 percent of naturally occurring carbon. Carbon-13 has a mass of 13.0034 atomic mass units and makes up 1.11 percent of naturally occurring carbon. Use this information to show that the atomic mass of carbon shown in the periodic table, 12.011 atomic mass units, is correct.

Answer

Recognize that 98.89 percent and 1.11 percent expressed as fractions are 0.9889 and 0.0111, respectively.

	Contributing Mass of ^{12}C	Contributing Mass of ^{13}C
Fraction of Abundance	0.9889	0.0111
Mass (amu)	× 12.0000	× 13.0034
	11.867	0.144

⎫ step 1

atomic mass $= 11.867 + 0.144 = 12.011$ step 2

Your Turn

Chlorine-35 has a mass of 34.97 atomic mass units, and chlorine-37 has a mass of 36.95 atomic mass units. Determine the atomic mass of chlorine, Cl (atomic number 17), if 75.53 percent of all chlorine atoms are the chlorine-35 isotope and 24.47 percent are the chlorine-37 isotope.

In Perspective

You may recognize that the atomic masses given in the periodic table are the relative masses attendees at the 1860 chemistry conference were so avidly working toward. From the atomic masses, for example, we can easily calculate that neon atoms are 20.18/4.003 = 5.041 times more massive than helium atoms. In this and many other regards, the periodic table is the culmination of the efforts of many talented individuals, only some of whom were discussed in this chapter. Table 3.2 summarizes all the atomic history we have covered.

When the initial discoveries about atoms were being made, scientists based their conclusions on experimental evidence, and such evidence is always open to critical review. Investigators were thus able to look beyond the biases of the past to conceive new and more accurate models of nature.

Table 3.2

Democritus (460–370 B.C.)		Proposed an atomic model for matter
Aristotle (384–322 B.C.)		Proposed that matter is continuous
Boyle (1627–1691)	1661	Identified an element as that which cannot be broken down to simpler parts
Franklin (1706–1790)	1752	Investigated the nature of electricity
Cavendish (1731–1810)	1766	Discovered hydrogen
Lavoisier (1743–1794)	1774	Developed the law of mass conservation
Priestley (1733–1804)	1774	Discovered oxygen but did not identify it
Proust (1754–1826)	1797	Proposed the law of definite proportions
Dalton (1766–1844)	1803	Developed five postulates describing the atomic model of matter
Gay-Lussac (1778–1850)	1808	Showed that gases react in definite volume ratios
Avogadro (1776–1856)	1811	Explained Gay-Lussac's observations by proposing that the particles of a gas exist as diatomic molecules
Cannizzaro (1826–1910)	1860	Reintroduced the work of Avogadro as a reliable means of measuring relative atomic masses
Mendeleev (1834–1907)	1869	Developed a chart—forerunner to our modern periodic table—that organized the elements by properties
Thomson (1856–1940)	1897	Measured the charge-to-mass ratio for a beam of electrons
Millikan (1868–1953)	1909	Calculated the mass of an electron and found it to be smaller than the mass of the smallest known atom
Rutherford (1871–1937)	1910	Discovered the atomic nucleus
Chadwick (1891–1974)	1932	Discovered the neutron

Key Terms and Matching Definitions

_____ alchemy
_____ atomic mass
_____ atomic nucleus
_____ atomic number
_____ cathode ray tube
_____ electron
_____ isotope
_____ law of definite proportions
_____ law of mass conservation
_____ mass number
_____ neutron
_____ nucleon
_____ proton
_____ scientific law

1. A medieval endeavor concerned with turning other metals to gold.
2. Any scientific hypothesis that has been tested over and over again and has not been contradicted. Also known as a scientific principle.
3. A law stating that there is no detectable change in the amount of mass present before and after a chemical reaction.
4. A law stating that elements combine in definite mass ratios to form compounds.
5. A device that emits a beam of electrons.
6. An extremely small, negatively charged subatomic particle found outside the atomic nucleus.
7. The dense, positively charged center of every atom.
8. A positively charged subatomic particle of the atomic nucleus.
9. A count of the number of protons in the atomic nucleus.
10. An electrically neutral subatomic particle of the atomic nucleus.
11. Any subatomic particle found in the atomic nucleus. Another name for either proton or neutron.
12. Any member of a set of atoms of the same element whose nuclei contain the same number of protons but different numbers of neutrons.
13. The number of nucleons (protons and neutrons) in the atomic nucleus. Used primarily to identify isotopes.
14. The mass of an element's atoms listed in the periodic table as an *average* value based on the relative abundance of the element's isotopes.

Review Questions

Chemistry Developed Out of Our Interest in Materials

1. In what ways was Aristotle's erroneous model of matter a remarkable achievement?

2. According to Aristotle's model, how is clay converted to ceramic?

3. How did chemistry benefit from alchemy?

Lavoisier Laid the Foundation of Modern Chemistry

4. How did Lavoisier define an element and a chemical compound?

5. Why did Lavoisier's mass-conservation law escape earlier investigators?

6. Who named the element oxygen?

7. What is the meaning of the word *hydrogen*?

8. How many grams of water can be formed from the reaction between 10 grams of oxygen and 1 gram of hydrogen?

Dalton Deduced That Matter Is Made of Atoms

9. How did Dalton define an element?

10. How did Dalton explain the fact that elements combine in whole-number ratios to form chemical compounds?

11. Which of Dalton's five postulates accounts for Lavoisier's mass-conservation principle?

12. According to Dalton, how do the atoms of different elements differ from one another?

13. What was Dalton's proposed formula for water?

14. In what volume ratio do hydrogen gas and oxygen gas react to form water?

15. How did Avogadro account for the formation of two volumes of water from two volumes of hydrogen and one volume of oxygen?

16. When was Avogadro's hypothesis finally accepted by the scientific community?

17. What observation led Mendeleev to develop his early version of the periodic table?

The Electron Was the First Subatomic Particle Discovered

18. What is a cathode ray?

19. Why is a cathode ray deflected by a nearby electric charge or magnet?

20. What did Thomson discover about the electron?

21. Why couldn't Thomson calculate the mass of the electron?

22. What did Millikan discover about the electron?

The Mass of an Atom Is Concentrated in Its Nucleus

23. What did Rutherford discover about the atom?

24. What was the fate of the vast majority of alpha particles in Rutherford's gold-foil experiment?

25. To Rutherford's surprise, what was the fate of a tiny fraction of alpha particles in the gold-foil experiment?

26. What kind of force prevents atoms from squishing into one another?

The Atomic Nucleus Is Made of Protons and Neutrons

27. A proton is how much more massive than an electron?

28. Compare the electric charge on the proton with the electric charge on the electron.

29. What is the definition of atomic number?

30. What role does atomic number play in the periodic table?

31. What effect do isotopes of a given element have on the atomic mass calculated for that element?

32. Name two nucleons.

33. Distinguish between atomic number and mass number.

34. Distinguish between mass number and atomic mass.

Hands-On Chemistry Insights

Air Out

As the iron rusts, it absorbs oxygen molecules from the air in the inverted jar. This allows the water to rise into the jar. How far the water rises is a function of the amount of oxygen removed. You can find out how much oxygen was removed by rubber-banding a ruler to the jar such that the zero mark on the ruler is at the initial water level inside the inverted jar. When the water stops rising, divide the water height inside the inverted jar by the height of the air that was initially inside the jar. The fraction you obtain gives a rough estimate of the percentage of air removed from the jar, which corresponds to the percentage of oxygen in the atmosphere. How close do you come to the accepted value of 21 percent?

You might also make a graph plotting the water height in the inverted jar at successive 10-minute intervals. Why does the graph gradually level off? What effect does the volume of the steel wool have on your data?

Collecting Bubbles

Don't restrict yourself to the setup given in this hands-on activity. Improvise with available household items, and you may well devise a more successful way of collecting the carbon dioxide. Consider, for example, using either rubber tubing or a drinking straw to connect the CO_2 source to the inverted bottle. You can shape one end of the tubing into a J shape by inserting a straightened paper clip into the tubing and then bending the clip until the tubing end remains curved. Then slip the curved end into the inverted bottle.

In this activity, always be wary of the pressure that builds up when baking soda and vinegar are mixed in a closed container.

You can pour the carbon dioxide over the flame of a birthday candle. As the carbon dioxide flows out of the bottle, it falls onto the candle and extinguishes the flame. (Don't tilt the bottle too far or some water will pour out.) At times, you may see the motion of the carbon dioxide by the streaming of the smoke from the extinguished candle. This is all evidence that carbon dioxide is heavier than air.

Bending Electrons

If you can find one, a black-and-white television or computer monitor shows this effect most vividly. On a color screen, you'll see color changes in addition to the distortions. Most televisions and monitors today are equipped with automatic "degaussers" that alleviate distortion from the magnets in a nearby audio speaker or even from the Earth's magnetic field.

Are the distortions you see the same for both ends of the magnet?

Exercises

1. A cat strolls across your backyard. An hour later, a dog with its nose to the ground follows the trail of the cat. Explain what is going on from a molecular point of view.

2. If all the molecules of a body remained part of that body, would the body have any odor?

3. Which are older, the atoms in the body of an elderly person or those in the body of a baby?

4. Where did the atoms that make up a newborn baby originate?

5. In what sense can you truthfully say that you are a part of every person around you?

6. Considering how small atoms are, what are the chances that at least one of the atoms exhaled in your first breath will be in your last breath?

7. Describe how Lavoisier used the scientific approach (observation, questions, hypothesis, predictions, tests, theory) in his development of the principle of mass conservation.

8. Lavoisier heated a piece of tin on a floating block of wood covered by a glass jar. As the tin decomposed, the water level inside the jar rose. How did Lavoisier explain this result?

9. According to Proust, why is only 9 grams of water formed when 10 grams of oxygen reacts with 1 gram of hydrogen?

10. Substances A and B combine to make substance C. Substances C and B combine to make substances A and D. Place the letter of each substance next to the symbol that best describes its atomic or molecular structure:

11. Proust noted that oxygen and hydrogen react in an 8:1 ratio, whereas Gay-Lussac noted that they react in a 1:2 ratio. Who was right? Defend your answer.

12. Two of the substances in Exercise 10 are elements and two are compounds. Which are which?

13. A sample of iron weighs more after it rusts. Why?

14. The following diagram depicts the reaction between gaseous oxygen, O_2, and gaseous hydrogen, H_2, to form water vapor, H_2O. What symbols and how many of them should be drawn in the empty box? How many grams of water are formed under these conditions? How many grams of which chemical remain unreacted?

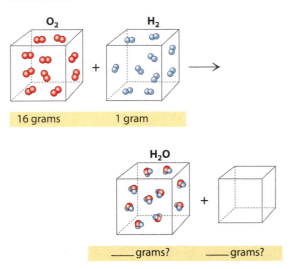

15. How did Avogadro account for Dalton's observation that a given volume of oxygen gas has more mass than an equal volume of water vapor?

16. If all atoms had the same mass, and 8 grams of oxygen still reacted with 1 gram of hydrogen, what would be the formula for water?

17. The following diagrams depict the reaction between gaseous chlorine, Cl_2, and gaseous hydrogen, H_2, to form gaseous hydrogen chloride, HCl. What should be drawn in the empty boxes? Specify the quantities beneath each box.

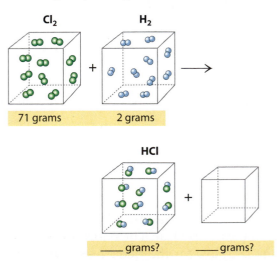

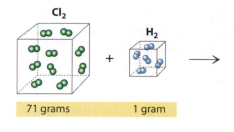

71 grams   1 gram

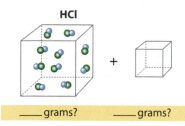

_____ grams?   _____ grams?

18. Gas A is composed of diatomic molecules (two atoms per molecule) of a pure element. Gas B is composed of triatomic molecules (three atoms per molecule) of another pure element. A volume of gas B is found to be three times more massive than an equal volume of gas A. How does the mass of an atom of gas B compare with the mass of an atom of gas A?

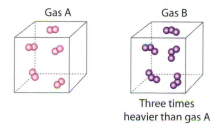

Three times heavier than gas A

19. Max Planck, a famous physicist of the early 20th century, is quoted as saying, "A new scientific truth does not triumph by convincing its opponents and making them see the light, but rather because its opponents eventually die and a new generation grows up." Cite a case where this statement applies to the development of modern chemistry.

20. How might Planck's statement apply to politics or religion?

21. Of all the investigators presented in this chapter, who was the youngest at the time of his discovery besides Democritus and Aristotle? (See Table 3.2.)

22. Why is it important for a chemist to know the relative masses of atoms? Why do we refer to relative masses rather than absolute masses?

23. If the particles of a cathode ray had a greater electric charge, would the ray be bent more or less in a magnetic field?

24. Why did Rutherford assume that the atomic nucleus was positively charged?

25. Why does the ray of light in a neon sign bend when a magnet is held up to it?

26. How does Rutherford's model of the atom explain why some of the alpha particles directed at the gold foil were deflected straight back toward the source?

27. Which of the following diagrams best represents the size of the atomic nucleus relative to the size of the atom:

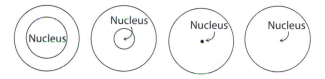

28. If two protons and two neutrons are removed from the nucleus of an oxygen atom, a nucleus of which element remains?

29. You could swallow a capsule of germanium, Ge (atomic number 32), without ill effects. If a proton were added to each germanium nucleus, however, you would not want to swallow the capsule. Why? (Consult a periodic table of the elements.)

30. If an atom has 43 electrons, 56 neutrons, and 43 protons, what is its approximate atomic mass? What is the name of this element?

31. The nucleus of an electrically neutral iron atom contains 26 protons. How many electrons does this iron atom have?

32. Evidence for the existence of neutrons did not come until many years after the discoveries of the electron and the proton. Give a possible explanation.

33. Which has more atoms: a 1-gram sample of carbon-12 or a 1-gram sample of carbon-13? Explain.

34. Why are the atomic masses listed in the periodic table not whole numbers?

Problems

1. How many grams of water can be produced by the combination of 8 grams of oxygen and 8 grams of hydrogen?

2. How many grams of water can be produced from the combination of 25 grams of hydrogen and 225 grams of oxygen? How much of which element will be left over?

3. The isotope lithium-7 has a mass of 7.0160 atomic mass units, and the isotope lithium-6 has a mass of 6.0151 atomic mass units. Given the information that 92.58 percent of all lithium atoms found in nature are lithium-7 and 7.42 percent are lithium-6, calculate the atomic mass of lithium, Li (atomic number 3).

4. The element bromine, Br (atomic number 35), has two major isotopes of similar abundance, both around 50 percent. The atomic mass of bromine is reported in the periodic table as 79.904 atomic mass units. Choose the most likely set of mass numbers for these two bromine isotopes: (a) ^{80}Br, ^{81}Br; (b) ^{79}Br, ^{80}Br; (c) ^{79}Br, ^{81}Br.

Answers to Calculation Corner

Finding Out How Much of a Chemical Reacts

1. The fact that 14 grams nitrogen reacts fully with 3 grams of hydrogen gives you two conversion factors:

$$\frac{14 \text{ g nitrogen}}{3 \text{ g hydrogen}} \quad \text{and} \quad \frac{3 \text{ g hydrogen}}{14 \text{ g nitrogen}}$$

Use the second conversion factor to convert the 7.0 grams of nitrogen to grams of hydrogen:

$$7.0 \text{ g nitrogen} \times \frac{3.0 \text{ g hydrogen}}{14 \text{ g nitrogen}} = 1.5 \text{ g hydrogen}$$

(See Appendix B for why 3.0 g is used rather than 3 g.)

2. From the preceding answer, you know that 7.0 grams of nitrogen reacts with 1.5 grams of hydrogen to form 7.0 grams + 1.5 grams = 8.5 grams of ammonia. If 6.0 grams of hydrogen is mixed with 7.0 grams of nitrogen, only 1.5 grams of that 6.0 grams reacts, so still only 8.5 grams of ammonia is formed. A total of 6.0 grams − 1.5 grams = 4.5 grams of hydrogen remains unreacted.

Calculating Atomic Mass

	Contributing Mass of ^{35}C	Contributing Mass of ^{37}C
Fraction of Abundance	0.7553	0.2447
Mass (amu)	× 34.97	× 36.95
	26.41	9.04

atomic mass = 26.41 + 9.04 = 35.45

Exploring Further

Jean-Pierre Poirer (translated by Rebecca Balinski), *Lavoisier: Chemist, Biologist, Economist*. Philadelphia: University of Pennsylvania Press, 1997.

An authoritative and detailed look at the life and times of the father of modern chemistry. With an intriguing account of the dynamics leading to his execution, this book explores Lavoisier's life not only as a chemist but as an accountant, administrator, educator, and tax collector.

Hugh Salzburg, *From Caveman to Chemist: Circumstances and Achievements*. Washington, DC: American Chemical Society, 1991.

An easy-to-read, informative, absorbing account of the history of chemistry.

http://www.aip.org/history/electron/jjthomson.htm

An in-depth presentation of the discovery of the electron by J. J. Thomson.

http://www.woodrow.org/teachers/ci/1992/

A series of insightful biographies of historical figures in chemistry, written by participants at the 1992 Institute on the History of Chemistry and sponsored by the Woodrow Wilson National Fellowship Foundation.

Discovering the Atom and Subatomic Particles

Visit The Chemistry Place at:
www.aw.com/chemplace

Uranium-235
nucleus

Neutron

The Atomic Nucleus

Know Nukes

Nuclear power plants generate electricity in much the same way fossil-fuel power plants do. Water is heated to create steam that can be used to turn electricity-generating turbines. The fundamental difference between these two types of power plants is the fuel used to heat the water. A fossil-fuel plant burns fossil fuel, such as coal or petroleum, but a nuclear plant, such as the one shown in this chapter's opening photograph, uses the heat created by nuclear fission to heat the water.

The burning of fossil fuel is a chemical reaction, which, as you recall from Section 2.1, is a reaction that involves changes in the way atoms are bonded and results in the formation of new materials. For fossil fuels, these new materials are mostly carbon dioxide and water vapor. As we explore in future chapters, the only thing that determines the ability of atoms to form new materials in a chemical reaction is the atoms' ability to share or exchange *electrons*—the atomic nuclei are not directly involved. The chemistry of an atom is therefore more a function of its electrons than of its nucleus. Nuclear fission, by contrast, involves *nuclear reactions*, which, as shown in the chapter opening photograph, involve the atomic nucleus. In this sense, the study of the atomic nucleus is not a primary focus of chemistry.

Nuclear processes, however, have certainly impacted society and have raised many issues regarding our health, energy sources, and national security. At the same time, the atomic nucleus is one of the most misunderstood areas of science. Public fears about anything *nuclear* are much like the fears people had about electricity a century ago. Since that time, however, society has determined that the benefits of electricity outweigh the risks. Today, we are making similar decisions about the risks and benefits of nuclear technology. In order that we make the best possible decisions, everyone should have an adequate understanding of the atomic nucleus and its processes. So, as part of our study of the atom, we briefly turn to the atomic nucleus and the related concept of radioactivity. We shall then be set to revisit the nucleus in Chapter 19 when we study energy sources.

4.1 The Cathode Ray Led to the Discovery of Radioactivity

In 1896, the German physicist Wilhelm Roentgen (1845–1923) discovered a "new kind of ray" emanating from a point where cathode rays hit the glass surface of a high-voltage cathode ray tube. (Recall from Chapter 3 that a cathode ray is a beam of electrons.) Unlike cathode rays, these new rays were not deflected by either an electric field or a magnetic field. Furthermore, they could pass through opaque materials.

Roentgen discovered this latter property when he let the rays fall on a photographic plate wrapped in black paper thick enough to keep all visible light from falling on the plate. A photographic plate is coated with light-sensitive chemicals, and when light falls on the chemicals, the plate is said to have been *exposed* to the light. Light is one form of radiation, as we shall learn in Chapter 5, and Roentgen's rays are another form of radiation. Because the rays were able to pass through the lightproof paper in which Roentgen's plate was wrapped, the rays exposed the plate, as Figure 4.1 shows. Not able to deduce the nature of these rays, Roentgen called them X rays.

Figure 4.1
X rays can pass through solid materials. The denser a material, however, the greater its ability to block X rays. Bones, for example, are more effective at blocking X rays than is soft tissue. For this reason, the region of the plate lying below the bone parts of the hand are less exposed than are the regions lying below the tissue parts. As a result, the shadow of bones shows up clearly on the plate.

X rays

Photographic film enclosed in lightproof holder

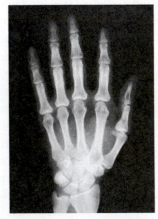

Exposed and developed photographic film

A few months after Roentgen announced his discovery of X rays, the French physicist Antoine Henri Becquerel (1852–1908) experimented to see if they were emitted by phosphorescent substances—those that glow in the dark after being exposed to bright light. One substance that appeared to confirm the idea that phosphorescence resulted in X rays was uranium. When placed in the sunlight and on top of a photographic plate wrapped in dark paper, uranium exposed the photographic plate much the way X rays from the cathode ray tube did. When cloudy weather forced Becquerel to suspend his research, he stored the uranium and a photographic plate together in a closed drawer. Several days later on a whim, he thought to develop the plate, and to his amazement, he saw something like what is shown in Figure 4.2—the plate had been exposed to some sort of rays without sunlight or any other source of energy. The rays must have originated from the uranium! Subsequent experiments revealed that these rays emanating from the uranium had nothing to do with X rays or phosphorescence.

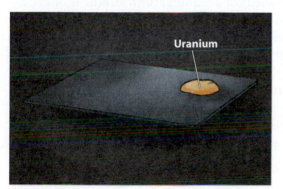

Uranium

Photographic film enclosed in lightproof paper and stored in total darkness

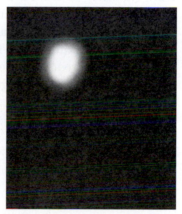

Developed photographic film

Figure 4.2
Becquerel noted that a piece of uranium left on a photographic plate wrapped in opaque black paper exposed the plate even in the absence of light. From this he deduced that the uranium was giving off some sort of radiation.

A couple of years later, one of Becquerel's students, Marie Sklodowska Curie (1867–1934), shown in Figure 4.3, became keenly interested in this strange form of radiation. She showed that the radiation was also emitted by several other elements known at the time and suggested that it should be possible to isolate yet undiscovered elements by studying any radiation they might be emitting. Using chemical techniques, she and her husband, Pierre Curie (1859–1906), laboriously divided an 8-ton pile of uranium ore

(a)   (b)

Figure 4.3
For their work on radioactivity, (a) Becquerel and (b) the Curies shared the 1903 Nobel Prize in Physics.

into fractions, keeping those fractions giving off high levels of radiation and discarding the rest. The Curies used the term **radioactivity** to describe the tendency of these elements to emit radiation. Ultimately, they succeeded in isolating purified samples of two new radioactive elements. Marie named the first element *polonium* after her native Poland and the second *radium* because of its intense radioactivity.

The Three Major Products of Radioactivity Are Alpha, Beta, and Gamma Rays

At about the time the Curies were isolating new radioactive elements, Ernest Rutherford discovered that there are at least two major forms of radioactivity, which he identified as alpha rays and beta rays. Alpha rays, he found, consist of positively charged particles he called alpha particles. As discussed in Section 3.5, these are the particles he used in his discovery of the atomic nucleus. An **alpha particle** is a combination of two protons and two neutrons (in other words, it is the nucleus of a helium atom, atomic number 2). Beta rays he found to be identical to cathode rays. A **beta particle** therefore is simply another name for an electron ejected from a nucleus.

Shortly after Rutherford had identified alpha and beta rays, a third major form of radioactivity, gamma rays, was discovered by other investigators. Unlike alpha and beta rays, **gamma rays** carry no electric charge and have no mass. Instead, they are an extremely energetic form of nonvisible light.

As is shown in Figure 4.4, the three major types of radiation given off by radioactive materials can be separated by putting a magnetic field across their paths.

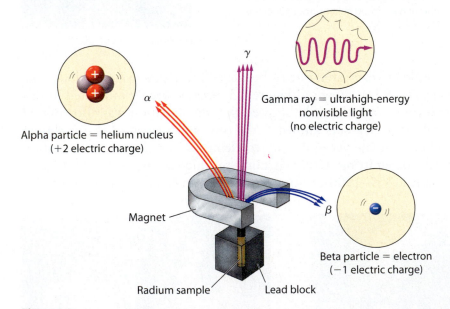

Alpha particle = helium nucleus (+2 electric charge)

Gamma ray = ultrahigh-energy nonvisible light (no electric charge)

Beta particle = electron (−1 electric charge)

Magnet

Radium sample   Lead block

Figure 4.4
The three most common forms of radiation coming from a radioactive substance are called by the first three letters of the Greek alphabet, α, β, γ—*alpha, beta,* and *gamma*. In a magnetic field, alpha rays bend one way, beta rays bend the other way, and gamma rays do not bend at all. Note that the alpha rays bend less than do the beta rays. This happens because the alpha particles have more inertia (because they have more mass) than the beta particles. The source of all three radiations is a radioactive material placed at the bottom of a hole drilled in a lead block.

Alpha particles do not easily penetrate solid material because of their relatively large size and their double positive charge (+2). Because of their great kinetic energies, however, alpha particles can cause significant damage to the surface of a material, especially living tissue. As they travel through air, even through distances as short as a few centimeters, alpha particles pick up electrons, slow down, and become harmless helium. Almost all the Earth's helium atoms, including those in a helium balloon, were once energetic alpha particles ejected from radioactive elements.

Beta particles are normally faster than alpha particles and not as easy to stop. For this reason, they are able to penetrate light materials such as paper and clothing. They can penetrate fairly deeply into skin, where they have the potential for harming or killing cells. They are not able to penetrate deeply into denser materials, however, such as aluminum. Beta particles, once stopped, become part of the material they are in, like any other electron.

Like visible light, a gamma ray is pure energy. The amount of energy in a gamma ray is much greater than the amount of energy in visible light. Because they have no mass or electric charge and because of their high energies, gamma rays are able to penetrate through most materials. However, they cannot penetrate unusually dense materials such as lead, which absorbs them. The delicate molecules in body cells exposed to gamma rays suffer structural damage. Hence, gamma rays are generally more harmful to us than alpha or beta rays.

Figure 4.5 shows the relative penetrating power of the three types of radiation, and Figure 4.6 shows an interesting practical use for gamma radiation.

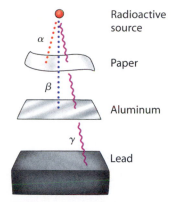

Figure 4.5
Alpha particles are the least penetrating form of radiation and can be stopped by a sheet of paper. Beta particles readily pass through paper but not through a sheet of aluminum. Gamma rays penetrate several centimeters into solid lead.

Figure 4.6
The shelf life of fresh strawberries and other perishables is markedly increased when the food is subjected to gamma rays from a radioactive source. The strawberries on the right were treated with gamma radiation, which kills the microorganisms that normally lead to spoilage. The food is only a receiver of radiation and is in no way transformed to an emitter of radiation, as can be confirmed with a radiation detector.

Concept Check ✓

> Pretend you are given three radioactive rocks—one an alpha emitter, one a beta emitter, and one a gamma emitter. You can throw one away, but of the remaining two, you must hold one in your hand and place the other in your pocket. What can you do to minimize your exposure to radiation?

Was this your answer? Ideally you should get as far from all the rocks as possible. If you must hold one and put one in your pocket, however, hold the alpha emitter because the skin on your hand will shield you. Put the beta emitter in your pocket because its rays might be stopped by the combined thickness of clothing and skin. Throw away the gamma emitter because its rays would penetrate deep into your body from either of these places.

4.2   Radioactivity Is a Natural Phenomenon

A common misconception is that radioactivity is new in the environment, but it has been around far longer than the human race. It is as much a part of our environment as the sun and the rain. It has always occurred in the soil we walk on and in the air we breathe, and it warms the interior of the Earth and makes it molten. The energy released by radioactive substances in the Earth's interior heats the water that spurts from a geyser and the water that wells up from a natural hot spring.

As Figure 4.7 shows, most of the radiation we encounter is natural background radiation that originates in the Earth and in space and was present long before we humans got here. Even the cleanest air we breathe is somewhat radioactive as a result of bombardment by cosmic rays. At sea level, the protective blanket of the atmosphere reduces background radiation, but at higher altitudes radiation is more intense. In Denver, the "Mile-High City," a person receives more than twice as much radiation from cosmic rays as at sea level. A couple of round-trip flights between New York and San Francisco exposes us to as much radiation as we receive in a chest X ray at the doctor's office. The air time of airline personnel is limited because of this extra radiation.

Cells are able to repair most kinds of molecular damage caused by radiation if the damage is not too severe. A cell can survive an otherwise lethal dose of radiation if the dose is spread over a long period of time to allow intervals for healing. When radiation is sufficient to kill cells, the dead cells can be replaced by new ones. Sometimes radiation alters the genetic information of a cell by damaging its DNA molecules (see Section 13.5). New cells arising from the damaged cell retain the altered genetic information, which is called a *mutation*. Usually the effects of a mutation are insignificant, but occasionally the mutation results in cells that do not function as well as unaffected ones, sometimes leading to a cancer. If the damaged DNA is in an individual's reproductive cells, the genetic code of the individual's offspring may retain the mutation.

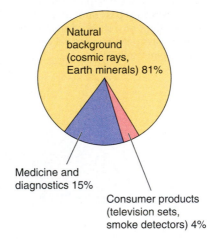

Figure 4.7
Origins of radiation exposure for an average individual in the United States.

Natural background (cosmic rays, Earth minerals) 81%

Medicine and diagnostics 15%

Consumer products (television sets, smoke detectors) 4%

Rems Are Units of Radiation

We measure the ability of radiation to cause harm in **rems**. Lethal doses of radiation begin at 500 rems. A person has about a 50 percent chance of surviving a dose of this magnitude received over a short period of time.

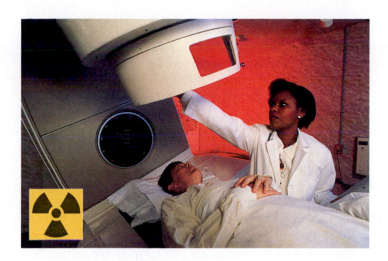

Figure 4.8
Nuclear radiation is focused on harmful tissue, such as a cancerous tumor, to selectively kill or shrink the tissue in a technique known as *radiation therapy*. This application of nuclear radiation has saved millions of lives—a clear-cut example of the benefits of nuclear technology. The inset shows the international symbol indicating an area where radioactive material is being handled or produced.

During radiation therapy, a patient may receive localized doses in excess of 200 rems each day for a period of weeks (Figure 4.8).

All the radiation we receive from natural sources and medical procedures is only a fraction of 1 rem. For convenience, the smaller unit *millirem* is used, where 1 millirem (mrem) is 1/1000 of a rem.

The average person in the United States is exposed to about 360 millirems a year, as Table 4.1 indicates. About 80 percent of this radiation comes from natural sources, such as cosmic rays (radiation from our sun as well as other stars) and the Earth. A typical diagnostic X ray exposes a person to between 5 and 30 millirems (0.005 and 0.030 rem), less than 1/10,000 of the lethal dose. Interestingly, the human body is a significant source of natural radiation, primarily from the potassium we ingest. Our bodies contain about 2 kilograms of potassium. Of this quantity, about 20 milligrams is the radioactive isotope potassium-40, a beta-ray emitter. In a human body, about 60,000 potassium-40 atoms emit pulses of radioactivity in the time it takes the heart to beat once.

Table 4.1

Annual Radiation Exposure

Source	Typical Amount Received in One Year (millirems)
Natural Origin	
Cosmic radiation	26
Ground	33
Air (radon-222)	198
Human tissues (potassium-40; radium-226)	35
Human Origin	
Medical procedures	
Diagnostic X rays	40
Nuclear medicine	15
Television tubes, other consumer products	11
Weapons-test fallout	1

Figure 4.9
A commercially available radon test kit for the home.

The leading source of naturally occurring radiation, however, is radon-222, an inert gas arising from uranium deposits. Radon is heavier than air and therefore tends to accumulate in basements after it seeps up through cracks in the floor. Levels of radon vary from region to region, depending on local geology. You can check the radon level in your home with a radon detection kit like the one shown in Figure 4.9. If levels are abnormally high, corrective measures, such as sealing the basement floor and walls and maintaining adequate ventilation, are recommended. The U.S. Environmental Protection Agency projects that anywhere from 7000 to 30,000 cases of lung cancer each year are attributed to radon exposure. Smokers who inhale the radon that occurs naturally in tobacco smoke are at particularly high risk.

About one-fifth of our annual exposure to radiation comes from non-natural sources, primarily medical procedures. Television sets, fallout from nuclear testing, and the coal and nuclear power industries are minor but significant non-natural sources. Interestingly, the coal industry far outranks the nuclear power industry as a source of radiation. The global combustion of coal annually releases into the atmosphere about 13,000 tons of radioactive thorium and uranium. Worldwide, the nuclear power industries generate about 10,000 tons of radioactive waste each year. Most of this waste is contained, however, and is *not* released into the environment. As we explore in Chapter 19, where to bury this contained radioactive waste is a heated issue yet to be resolved.

4.3 Radioactive Isotopes Are Useful as Tracers and for Medical Imaging

Radioactive isotopes can be incorporated into molecules whose location can then be traced by the radiation they emit. When used in this way, radioactive isotopes are called *tracers*, and Figure 4.10 shows one use. To check the action of a fertilizer, researchers incorporate radioactive isotopes into the molecules of the fertilizer and then apply the fertilizer to plants. The amount taken up by the plants can be measured with radiation detectors. From such measurements, scientists can tell farmers how much fertilizer to use because fertilizer uptake is a physical and chemical process that is not affected by the radioactivity of the materials involved.

Fertilizer with radioactive isotope applied to crop

Radioactivity detected in plant

Figure 4.10
Tracking fertilizer uptake with a radioactive isotope.

Tracers are also used in industry. Motor oil manufacturers can quantify the lubricating qualities of their products by running oil in engines containing small but measurable amounts of radioactive isotopes. As the engine runs and the pistons rub against the inner chambers, some of the metal from the engine invariably makes its way into the motor oil, and this metal carries with it the embedded radioactive isotopes. The better the lubricating qualities of a motor oil, the fewer radioactive isotopes it will contain after running in the engine for a given length of time.

In a technique known as *medical imaging*, tracers are used in medicine for the diagnosis of internal disorders. Small amounts of a radioactive material,

Hands-On Chemistry: Personal Radiation

We all live with radiation. It was in our environment well before the discovery of the atom and even before the first human civilization. Recent technologies, however, have increased our exposure. Use the following worksheet to estimate your annual exposure.

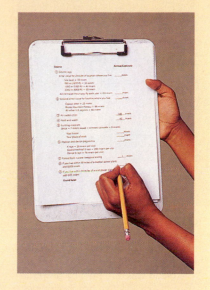

Source		Annual Exposure
1. Cosmic rays		
Enter value for altitude of location where you live:	30	mrem
Sea level = 30 mrem		
500 m (1650 ft) = 35 mrem		
1000 m (3300 ft) = 40 mrem		
2000 m (6600 ft) = 60 mrem		
Airline travel: Hours you fly each year × 0.6 mrem	3	mrem
2. Ground: enter value for location where you live:	23	mrem
Coastal state = 23 mrem		
Rocky Mountain Plateau = 90 mrem		
All other U.S. regions = 46 mrem		
3. Air (radon-222):	198	mrem
4. Food and water:	40	mrem
5. Building materials (brick = 7 mrem; wood = 4 mrem; concrete = 8 mrem)		
Your house	7	mrem
Your place of work	7	mrem
6. Medical and dental diagnostics		mrem
X rays = 20 mrem per visit		
Gastrointestinal X rays = 200 mrem per visit		
Dental X rays = 10 mrem per visit		
Radiation therapy (ask your radiologist)		
7. Fallout from nuclear weapons testing	1	mrem
8. If you live within 50 miles of a nuclear power plant, add 0.009 mrem	0.009	mrem
9. If you live within 50 miles of a coal power plant, add 0.03 mrem	0.03	mrem
Grand total	309	**mrem**

Data from the U.S. Environmental Protection Agency and the National Council for Radiation Protection. See the Web page http://www.epa.gov/rpdweb00/students/calculate.html for a more detailed calculation.

such as sodium iodide, NaI, which contains the radioactive isotope iodine-131, are administered to a patient and traced through the body with a radiation detector. The result, shown in Figure 4.11, is an image that shows how

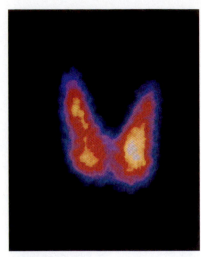

Figure 4.11
The thyroid gland, located in the neck, absorbs much of the iodine that enters the body in food and drink. This image of the gland was obtained by giving a patient the radioactive isotope iodine-131. Such images are useful in diagnosing metabolic disorders.

the material is distributed in the body. This technique works because the path the tracer material takes is influenced only by its physical and chemical properties, not by its radioactivity. The tracer may be introduced alone or along with some other chemical, known as a *carrier compound*, that helps target the isotope to a particular type of tissue in the body.

Table 4.2 lists the uses of a number of radioactive isotopes.

Table 4.2

Uses for various radioactive isotopes

Isotope	Usage
Calcium-47	Used in the study of bone formation in mammals
Californium-252	Used to inspect airline luggage for explosives
Hydrogen-3 (tritium)	Used for life-science and drug-metabolism studies to ensure safety of potential new drugs
Iodine-131	Used to diagnose and treat thyroid disorders
Iridium-192	Used to test integrity of pipeline welds, boilers, and aircraft parts
Thallium-201	Used in cardiology and for tumor detection
Xenon-133	Used in lung-ventilation and blood-flow studies

Source: Nuclear Regulatory Council

4.4 Radioactivity Results from an Imbalance of Forces in the Nucleus

We know that electric charges of like sign repel one another. So how is it possible that all the positively charged protons of the nucleus can stay clumped together? This question led to the discovery of an attractive force called the **strong nuclear force**, which acts between all nucleons. This force is very strong but only over extremely short distances (about 10^{-15} meter, the diameter of a typical atomic nucleus). Repulsive electrical interactions, on the other hand, are relatively long-ranged. Figure 4.12 compares the strength of these two forces over distance. For protons that are close together, as in a

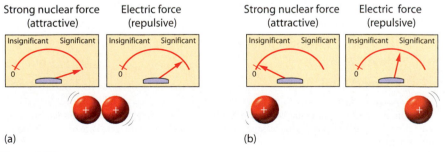

Figure 4.12
(a) Two protons near each other experience both an attractive strong nuclear force and a repulsive electric force. At this tiny separation distance, the strong nuclear force overcomes the electric force, and as a result the protons stay close together. (b) When the two protons are relatively far from each other, the electric force is more significant than the strong nuclear force, and as a result the protons repulse each other. It is this proton–proton repulsion in large atomic nuclei that causes radioactivity.

small atomic nucleus, the attractive strong nuclear force easily overcomes the repulsive electric force. For protons that are far apart, like those on opposite edges of a large nucleus, the attractive strong nuclear force may be smaller than the repulsive electric force.

Because the strong nuclear force decreases over distance, a large nucleus is not as stable as a small one, as shown in Figure 14.13. In other words, a large atomic nucleus is more susceptible to falling apart and emitting either high-energy particles or gamma rays. This process is radioactivity, which, because it involves the decay of the atomic nucleus, is sometimes also called *radioactive decay*.

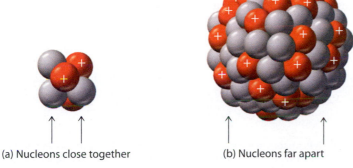

(a) Nucleons close together   (b) Nucleons far apart

Figure 4.13
(a) All nucleons in a small atomic nucleus are close to one another; hence, they experience an attractive strong nuclear force. (b) Nucleons on opposite sides of a large nucleus are not as close to one another, and so the attractive strong nuclear forces holding them together are much weaker. The result is that the large nucleus is less stable.

Neutrons serve as "nuclear cement" holding the atomic nucleus together. Protons attract both other protons and neutrons by the strong nuclear force, but they also repel other protons by the electric force. Neutrons, on the other hand, have no electric charge and so only attract protons and other neutrons by the strong nuclear force. The presence of neutrons therefore adds to the attraction among nucleons and helps hold the nucleus together, as illustrated in Figure 4.14.

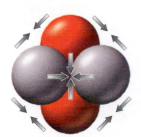 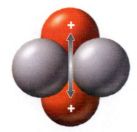

All nucleons, both protons and neutrons, attract one another by the strong nuclear force.   Only protons repel one another by the electric force.

Figure 4.14
The presence of neutrons helps hold the atomic nucleus together by increasing the effect of the attractive strong nuclear force, represented by the single-headed arrows.

The more protons there are in a nucleus, the more neutrons are needed to help balance the repulsive electric forces. For light elements, it is sufficient to have about as many neutrons as protons. The most common isotope of carbon, carbon-12, for instance, has six protons and six neutrons. For large nuclei, more neutrons than protons are required. Remember that the strong nuclear force diminishes rapidly with increasing distance between nucleons. Nucleons must be practically touching in order for the strong nuclear force to be effective. Nucleons on opposite sides of a large atomic nucleus are not as attracted to one another. The electric force, however, does not diminish by much across the diameter of a large nucleus and so begins to win out over the strong nuclear force. To compensate for the weakening of the strong nuclear force across the diameter of the nucleus, large nuclei have more neutrons than protons. Lead, for example, has about one and a half times as many neutrons as protons.

Concept Check ✔

Two protons in an atomic nucleus repel each other, but they are also attracted to each other. Explain.

Was this your answer? Two protons in a nucleus repel each other by the electric force, true, but they also attract each other by the strong nuclear force. Both forces act simultaneously. So long as the attractive strong nuclear force is more influential than the repulsive electric force, the protons remain together. Under conditions where the electric force overcomes the strong nuclear force, the protons fly apart.

Neutrons are stabilizing, and large nuclei require an abundance of them. Neutrons, however, are not always successful in keeping a nucleus intact, for two reasons. First, neutrons are not stable when they are by themselves. A lone neutron will spontaneously transform to a proton and an electron, as shown in Figure 4.15a. A neutron seems to need protons around to keep this from happening. After the size of a nucleus reaches a

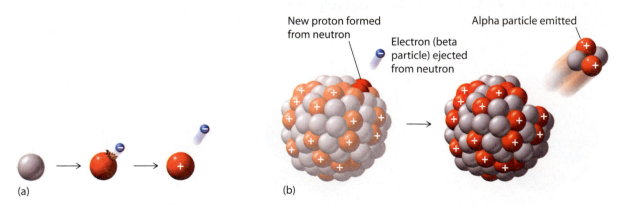

Figure 4.15
(a) A neutron near a proton is stable, but a neutron by itself is unstable and decays to a proton by emitting an electron. (b) Destabilized by an increase in the number of protons, the nucleus begins to shed fragments, such as alpha particles.

certain point, there are so many more neutrons than protons that there are not enough protons in the mix to prevent the neutrons from turning into protons. As neutrons in a nucleus change to protons, the stability of the nucleus decreases because the repulsive electric force becomes more and more significant. The result is that pieces of the nucleus fragment away in the form of radiation, as Figure 4.15b shows.

The second reason the stabilizing effect of neutrons is limited is that any proton in the nucleus is attracted by the strong nuclear force only to adjacent protons but is electrically repelled by all other protons in the nucleus. As more and more protons are squeezed into the nucleus, the repulsive electric forces increase substantially. For example, each of the two protons in a helium nucleus feels the repulsive effect of the other. Each proton in a nucleus containing 84 protons, however, feels the repulsive effects of 83 protons! The attractive nuclear force exerted by each neutron, however, extends only to its immediate neighbors. The size of the atomic nucleus is therefore limited. This in turn limits the number of possible elements in the periodic table. It is for this reason that all nuclei having more than 83 protons are radioactive. Also, the nuclei of the heaviest elements produced in the laboratory are so unstable (radioactive) that they exist for only fractions of a second.

Concept Check ✓

Which is more sensitive to distance: the strong nuclear force or the electric force?

Was this your answer? The strong nuclear force weakens rapidly over relatively short distances, but the electric force remains powerful over such distances.

Small nuclei also have the potential for being radioactive. This generally occurs when a nucleus contains more neutrons than protons. The nucleus of carbon-14, for example, contains eight neutrons but only six protons. With not enough protons to go around, one of the neutrons inevitably transforms to a proton, releasing an electron (beta radiation) in the process.

4.5   A Radioactive Element Can Transmute to a Different Element

When a radioactive nucleus emits an alpha or beta particle, the identity of the nucleus is changed because there is a change in atomic number. The changing of one element to another is called **transmutation**. Consider a uranium-238 nucleus, which contains 92 protons and 146 neutrons. When an alpha particle is ejected, the nucleus loses two protons and two neutrons. Because an element is defined by the number of protons in its nucleus, the

90 protons and 144 neutrons left behind are no longer identified as being uranium. What we have now is a nucleus of a different element—thorium. This transmutation can be written as a nuclear equation:

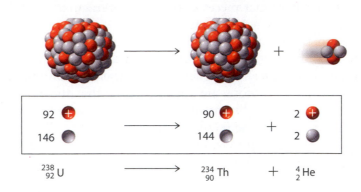

$$^{238}_{92}\text{U} \longrightarrow \, ^{234}_{90}\text{Th} + \, ^{4}_{2}\text{He}$$

This equation shows that $^{238}_{92}\text{U}$ transmutes to the two elements written to the right of the arrow. When this transmutation happens, energy is released, partly in the form of gamma radiation and partly in the form of kinetic energy in the alpha particle ($^{4}_{2}\text{He}$) and the thorium atom. In this and all other nuclear equations, the mass numbers balance ($238 = 234 + 4$) and the atomic numbers also balance ($92 = 90 + 2$).

Thorium-234 is also radioactive. When it decays, it emits a beta particle. Recall that a beta particle is an electron emitted by a neutron as the neutron transforms to a proton. So with thorium, which has 90 protons, beta emission leaves the nucleus with one fewer neutron and one more proton. The new nucleus has 91 protons and is no longer thorium; now it is the element protactinium. Although the atomic number has increased by 1 in this process, the mass number (protons + neutrons) remains the same. The nuclear equation is

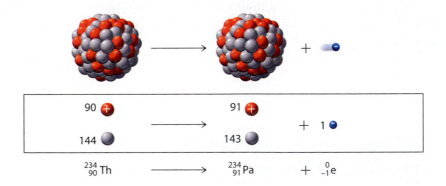

$$^{234}_{90}\text{Th} \longrightarrow \, ^{234}_{91}\text{Pa} + \, ^{0}_{-1}\text{e}$$

We write an electron as $^{0}_{-1}\text{e}$. The superscript 0 indicates that the electron's mass is insignificant relative to that of protons and neutrons. The subscript −1 is the electric charge of the electron.

So we see that when an element ejects an alpha particle from its nucleus, the mass number of the remaining atom is decreased by 4 and its atomic number is decreased by 2. The resulting atom is an atom of the element two spaces back in the periodic table because this atom has two fewer protons. When an element ejects a beta particle from its nucleus, the mass of the atom is practically unaffected, meaning there is no change in mass number, but its atomic number increases by 1. The resulting atom is an atom of the element one place forward in the periodic table because it has one more proton.

The decay of $^{238}_{92}$U to $^{206}_{82}$Pb, an isotope of lead, is shown in Figure 4.16. Each blue arrow shows an alpha decay, and each red arrow shows a beta decay.

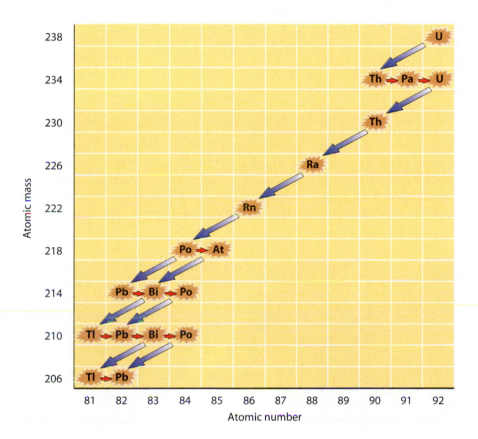

Figure 4.16
Uranium-238 decays to lead-206 through a series of alpha (blue) and beta (red) decays.

Concept Check ✓

1. Complete the nuclear reactions (a) $^{218}_{84}$Ra $\longrightarrow$ $^{?}_{?}$? + $_{-1}^{0}$e and (b) $^{210}_{84}$Po $\longrightarrow$ $^{206}_{82}$Pb + $^{?}_{?}$?.
2. What finally becomes of all the uranium that undergoes radioactive decay?

Were these your answers?

1. (a) $^{218}_{84}$Ra $\longrightarrow$ $^{218}_{85}$At + $_{-1}^{0}$e;   and   (b) $^{210}_{84}$Po $\longrightarrow$ $^{206}_{82}$Pb + $^{4}_{2}$He.
2. All uranium ultimately becomes lead. On the way, it exists as the elements shown in Figure 4.16.

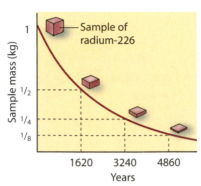

Figure 4.17
Radium-226 has a half-life of 1620 years, meaning that every 1620 years the amount of radium decreases by half as the radium transmutes to other elements.

4.6   The Shorter the Half-Life, the Greater the Radioactivity

The rate of decay of a radioactive isotope is measured in terms of a characteristic time called the **half-life**. This is the time it takes for half of the material in a radioactive sample to decay. For example, Figure 4.17 shows that radium-226 has a half-life of 1620 years. This means that half of a sample of radium has decayed to other elements by the end of 1620 years. In the next 1620 years, half of the remaining radium decays, leaving only one-fourth the original amount.

Half-lives are remarkably constant and not affected by external conditions. Some radioactive isotopes have half-lives that are less than a millionth of a second, while others have half-lives of more than a billion years. For example, uranium-238 has a half-life of 4.5 billion years, which means that in 4.5 billion years, half the uranium in the Earth today will be lead.

It is not necessary to wait through the duration of a half-life in order to measure it. The half-life of an element can be accurately estimated by measuring the rate of decay of a known quantity of the element. This is easily done using a radiation detector. In general, the shorter the half-life of a substance, the faster it disintegrates and the more radioactivity per minute is detected. Figure 4.18 shows a Geiger counter being used by environmental workers.

Figure 4.18
A Geiger counter detects incoming radiation by the way the radiation affects a gas enclosed in the tube that the technician is holding in his right hand.

Hands-On Chemistry: Radioactive Paper Clips

You can simulate radioactive decay with a bunch of paper clips representing atoms of a radioactive element. The "atoms" are thrown onto a flat surface. The ones that land in a certain orientation are imagined to have decayed and so are now atoms of a different element. Decayed atoms are removed from the pile and not used for successive throws. This process is continued until all the atoms have decayed.

What You Need

At least 20 metal paper clips, paper, pencil

Procedure

① Unfold the two loops of each clip by 90 degrees so that the loops are at right angles to each other. Pull the end of the larger loop outward by 90 degrees to form a structure that looks like the drawing at right.

  The orientation drawn on the top we'll call the leg-up orientation. Rotate the upward-pointing "leg" 90 degrees, and you'll have the orientation drawn on the bottom, which we'll call the head-up orientation. You will be investigating the half-life of two pretend elements. The first, "legonium," decays when it lands in the leg-up orientation. The second, "headonium," decays when it lands in the head-up orientation.

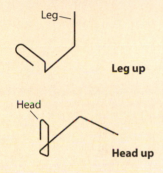

② You'll do two simulations similar to the one at right, one for legonium and one for headonium, and each simulation will involve five trials. For each trial, create a data table consisting of two columns, the first labeled "Throw" and the second labeled "Number of atoms remaining." In the row numbered zero, write the number of paper clips you start with (at least 20) in the column labeled "Number of atoms remaining."

LEGONIUM TRIAL 1	
Throw	Number of Atoms Remaining
0	20
1	___
2	___
.	.
.	.
.	.

③ Pretending each paper clip is a legonium atom, toss all of them onto a flat surface. Remove all the atoms in the leg-up orientation and write down the number remaining in the row numbered 1 of the data table, for trial 1. Continue until all the legonium atoms have been removed. This completes one trial, and so you need to run four more because there will be a lot of statistical variation. Count the number of throws required to remove all atoms in each trial and calculate an average.

④ Repeat the procedure, now pretending the paper clips are headonium atoms. The paper clips removed will be the ones that fall in the head-up orientation.

  Compare the legonium trials with the headonium trials. Half-life is given in units of time, but for this simulation the units are different. What are they? Estimate the half-life of legonium and headonium. Which element is more radioactive?

  If you were an atom in a sample of a radioactive element that has a half-life of 5 minutes, would you necessarily be decayed to another type of atom after 5 minutes?

Concept Check ✓

1. If you have a sample of a radioactive isotope that has a half-life of one day, how much of the original sample is left at the end of the second day? The third day?
2. What becomes of the atoms of the sample that decay?
3. With equal quantities of material, which gives a higher counting rate on a radiation detector, radioactive material that has a short half-life or radioactive material that has a long half-life?

Were these your answers?

1. At the end of two days, one-fourth of the original sample is left—one-half disappears by the end of the first day, and one-half of that one-half ($1/2 \times 1/2 = 1/4$) disappears by the end of the second day. At the end of three days, one-eighth of the original sample is left.
2. The atoms that decay are now atoms of a different element.
3. The material with the shorter half-life is more active and so gives a higher counting rate.

4.7 Isotopic Dating Measures the Age of a Material

The Earth's atmosphere is continuously bombarded by cosmic rays, and this bombardment causes many atoms in the upper atmosphere to transmute. These transmutations result in many protons and neutrons being "sprayed out" into the environment. Most of the protons are stopped as they collide with the atoms of the upper atmosphere. By stripping electrons from these atoms, the colliding protons become hydrogen atoms. The neutrons, however, keep going for longer distances because they have no electric charge and therefore do not interact electrically with matter. Eventually, many of them collide with atomic nuclei in the lower atmosphere. A nitrogen that captures a neutron, for instance, becomes an isotope of carbon by emitting a proton:

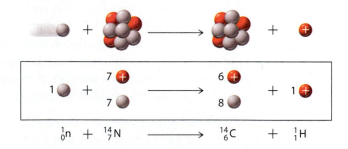

$$^{1}_{0}n + {}^{14}_{7}N \longrightarrow {}^{14}_{6}C + {}^{1}_{1}H$$

This carbon-14 isotope, which makes up less than one-millionth of 1 percent of the carbon in the atmosphere, is radioactive and has eight neutrons. (The most common isotope, carbon-12, has six neutrons and is not radioactive.). Because both carbon-12 and carbon-14 are forms of carbon, they have the same chemical properties. Both of these isotopes, for example, form carbon dioxide, which is taken in by plants. This means that all plants contain a tiny bit of radioactive carbon-14. All animals eat either plants or plant-eating animals, and therefore all animals have a little carbon-14 in them. In short, all living things on the Earth contain some carbon-14.

Carbon-14 is a beta emitter and decays back to nitrogen:

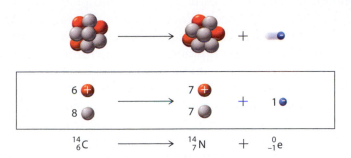

$$^{14}_{6}\text{C} \longrightarrow {}^{14}_{7}\text{N} + {}^{0}_{-1}\text{e}$$

Because plants take in carbon dioxide as long as they live, any carbon-14 lost to decay is immediately replenished with fresh carbon-14 from the atmosphere. In this way, a radioactive equilibrium is reached where there is a constant ratio of about one carbon-14 atom to every 100 billion carbon-12 atoms. When a plant dies, replenishment of carbon-14 stops. Then the percentage of carbon-14 decreases at a constant rate given by its half-life, but the amount of carbon-12 does not change because this isotope does not undergo radioactive decay. The longer a plant or other organism is dead, therefore, the less carbon-14 it contains relative to the constant amount of carbon-12.

The half-life of carbon-14 is about 5730 years. This means that half of the carbon-14 atoms now present in a plant or animal that dies today will decay in the next 5730 years. Half of the remaining carbon-14 atoms will then decay in the following 5730 years, and so on.

With this knowledge, scientists are able to calculate the age of carbon-containing artifacts, such as wooden tools or the skeleton shown in Figure 4.19, by measuring their current level of radioactivity. This process, known as **carbon-14 dating**, enables us to probe as much as 50,000 years into the past. Beyond this time span, there is too little carbon-14 remaining to permit an accurate analysis. (Understanding the local geology is another important tool used by archeologists in the dating of ancient relics.)

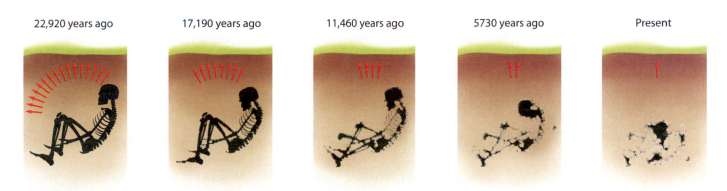

Figure 4.19
The amount of radioactive carbon-14 in the skeleton diminishes by one-half every 5730 years, with the result that today the skeleton contains only a fraction of the carbon-14 it originally had. The red arrows symbolize relative amounts of carbon-14.

Figure 4.20
Carbon-14 dating was developed by the American chemist William F. Libby (1908–1980) at the University of Chicago in the 1950s. For this work he received the Nobel Prize in Chemistry in 1960.

Carbon-14 dating would be an extremely simple and accurate dating method if the amount of radioactive carbon in the atmosphere had been constant over the ages, but it hasn't been. Fluctuations in the magnetic field of the sun and in that of the Earth cause fluctuations in cosmic-ray intensity in the Earth's atmosphere. These ups and downs in cosmic-ray intensity in turn produce fluctuations in the amount of carbon-14 in the atmosphere at any given time. In addition, changes in the Earth's climate affect the amount of carbon dioxide in the atmosphere. As we explore in Chapter 18, the oceans are great reservoirs of carbon dioxide. When the oceans are cold, they release less carbon dioxide into the atmosphere than when they are warm. Because of all these fluctuations in the carbon-14 production rate through the centuries, carbon-14 dating has an uncertainty of about 15 percent. This means, for example, that the straw of an old adobe brick dated to be 500 years old may really be only 425 years old on the low side or 575 years old on the high side. For many purposes, this is an acceptable level of uncertainty.

Concept Check ✔

Suppose an archeologist extracts 1.0 g of carbon from an ancient ax handle and finds that carbon to be one-fourth as radioactive as 1.0 g of carbon extracted from a freshly cut tree branch. About how old is the ax handle?

Was this your answer? The age of the ax handle is equal to two half-lives of ^{14}C; that's 2 × 5730 years ≈11,000 years old.

Scientists use radioactive minerals to date very old nonliving things. The naturally occurring mineral isotopes uranium-238 and uranium-235 decay very slowly and ultimately become lead—but not the common isotope lead-208. Instead, as was shown in Figure 14.16, uranium-238 decays to lead-206. Uranium-235, on the other hand, decays to lead-207. Thus the lead-206 and lead-207 that now exist in a uranium-bearing rock were at one time uranium. The older the rock, the higher the percentage of these remnant isotopes.

If you know the half-lives of uranium isotopes and the percentage of lead isotopes in some uranium-bearing rock, you can calculate the date the rock was formed. Rocks dated in this way have been found to be as much as 3.7 *billion* years old. Samples from the moon have been dated at 4.2 billion years, which is close to the estimated age of our solar system: 4.6 billion years.

4.8   Nuclear Fission Is the Splitting of the Atomic Nucleus

In 1938, two German scientists, Otto Hahn (1879–1968) and Fritz Strassmann (1902–1980), made a discovery that was to change the world. While bombarding a sample of uranium with neutrons in the hopes of creating heavier elements, they were astonished to find chemical evidence for the production of barium, an element having about half the mass of uranium. Hahn wrote of this news to his former colleague Lise Meitner (1878–1968), who had fled from Nazi Germany to Sweden because of her Jewish ancestry. From Hahn's evidence, Meitner concluded that the uranium nucleus, activated by neutron bombardment, had split in half. Soon thereafter, Meitner, working with her nephew Otto Frisch (1904–1979), published a paper in which the term *nuclear fission* was coined.

In the nucleus there exist both the attractive strong nuclear forces between nucleons and repulsive electric forces between protons. In all known nuclei, the strong nuclear forces dominate. As was discussed in Section 4.4, in many large nuclei this domination is easily lost and radioactive decay may occur. For a select number of large nuclei, however, another possibility exists. For example, a uranium-235 nucleus hit with a neutron elongates as shown in Figure 4.21. In a nucleus stretched into this elongated shape, the strong nuclear force weakens substantially because of the increased distance between opposite ends. The repulsive electric forces between protons remain powerful, however, and these forces may elongate the nucleus even more. If the elongation passes a certain point, the electric forces overwhelm the distance-sensitive strong nuclear forces and the nucleus splits into *fragments*. Typically, there are two large fragments accompanied by several smaller ones. This splitting of a nucleus into fragments is **nuclear fission**.

The energy released by the fission of one uranium-235 nucleus is enormous—about seven million times the energy released by the explosion of one TNT molecule. This energy is mainly in the form of kinetic energy of the fission fragments, which fly apart from one another. A much smaller amount of energy is released as gamma radiation.

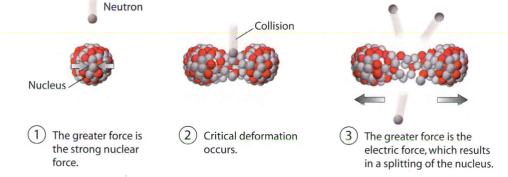

① The greater force is the strong nuclear force.

② Critical deformation occurs.

③ The greater force is the electric force, which results in a splitting of the nucleus.

Figure 4.21
Nuclear deformation may result in repulsive electric forces overcoming attractive strong nuclear forces, in which case fission occurs.

Here is the equation for a typical uranium fission reaction:

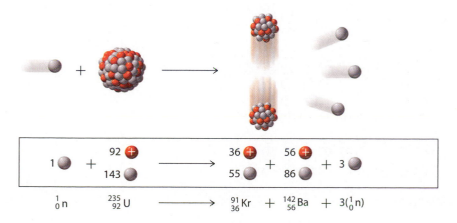

$$^{1}_{0}n + ^{235}_{92}U \longrightarrow ^{91}_{36}Kr + ^{142}_{56}Ba + 3(^{1}_{0}n)$$

Note in this reaction that one neutron starts the fission of the uranium nucleus and that the fission produces three neutrons. (It is also possible for a given fission event to produce either fewer than three neutrons or more than three.) These product neutrons can cause the fissioning of three other uranium atoms, releasing nine more neutrons. If each of these 9 neutrons succeeds in splitting a uranium atom, the next step in the reaction produces 27 neutrons, and so on. Such a sequence, illustrated in Figure 4.22, is called a **chain reaction**—a self-sustaining reaction in which the products of one reaction event stimulate further reaction events.

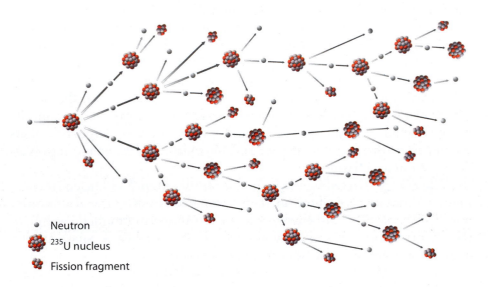

- • Neutron
- 235U nucleus
- • Fission fragment

Figure 4.22
A chain reaction.

Chain reactions do not occur to any great extent in naturally occurring uranium ore because not all uranium atoms fission so easily. Fission occurs mainly in the isotope uranium-235, which is rare and makes up only 0.7 percent of the uranium in pure uranium metal (Figure 4.23). When the more abundant isotope uranium-238 absorbs neutrons created by fission of

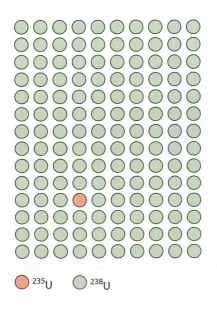

235U   238U

Figure 4.23
Only 1 part in 140 of naturally occurring uranium is uranium-235.

a uranium-235 atom, the uranium-238 typically does not undergo fission. So any chain reaction getting set up in the uranium-235 atoms in an ore sample is snuffed out by the neutron-absorbing uranium-238, as well as by other neutron-absorbing elements in the rock in which the ore is imbedded.

If a chain reaction occurred in a baseball-sized chunk of pure uranium-235, an enormous explosion would result. If the chain reaction were started in a smaller chunk of pure uranium-235, however, no explosion would occur. This is because of geometry: the ratio of surface area to mass is larger in a small piece than in a large piece. Just as there is more skin on six small potatoes having a combined mass of 1 kilogram than there is on a single 1-kilogram potato, there is more surface area on a bunch of smaller pieces of uranium-235 than on a large piece. In a small piece of uranium-235, therefore, neutrons have a greater chance of reaching the surface and escaping before they cause additional fission events, as Figure 4.24 illustrates. In a bigger piece, the chain reaction builds up to enormous energies before the neutrons get to the surface and escape. For masses greater than a certain amount, called the **critical mass**, an explosion of enormous magnitude may take place.

Consider a large quantity of uranium-235 divided into two pieces, each having a mass smaller than critical. The units are *subcritical.* Neutrons in either piece readily reach the surface and escape before a sizable chain reaction builds up. If the pieces are suddenly pushed together, however, the total surface area decreases. If the timing is right and the combined mass is greater than critical, a violent explosion takes place. This is what happens in a nuclear fission bomb, as Figure 4.25 shows.

Constructing a fission bomb is a formidable task. The difficulty is in separating enough uranium-235 from the more abundant uranium-238. Scientists took more than two years to extract enough of the 235 isotope from uranium ore to make the bomb detonated at Hiroshima, Japan, in 1945. To this day, uranium isotope separation remains a difficult process.

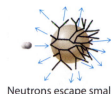

Neutrons escape small lump of uranium-235

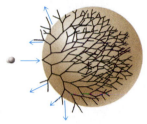

Neutrons trigger more reactions within large lump of uranium-235

Figure 4.24
This exaggerated view shows that a chain reaction in a small piece of pure uranium-235 runs its course before it can cause a large explosion because neutrons leak from the surface too soon. The surface area of the small piece is large relative to the mass. In a larger piece, more uranium and less surface are presented to the neutrons.

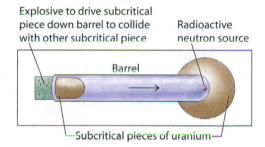

Explosive to drive subcritical piece down barrel to collide with other subcritical piece

Radioactive neutron source

Barrel

Subcritical pieces of uranium

Figure 4.25
Simplified diagram of a uranium fission bomb.

Concept Check

A 1-kilogram ball of uranium-235 has critical mass, but the same ball broken up into small chunks does not. Explain.

Was this your answer? The small chunks have more combined surface area than the ball from which they came. Neutrons escape via the surface of each small chunk before a sustained chain reaction can build up.

Nuclear Fission Reactors Convert Nuclear Energy to Electrical Energy

The awesome energy of nuclear fission was introduced to the world in the form of nuclear bombs, and this violent image still colors our thinking about nuclear power, making it difficult for many people to recognize its potential usefulness. Currently, about 20 percent of electrical energy in the United States is generated by *nuclear fission reactors*, which are simply nuclear boilers, as Figure 4.26 shows. Like fossil-fuel furnaces, reactors do nothing more elegant than boil water to produce steam for a turbine. The

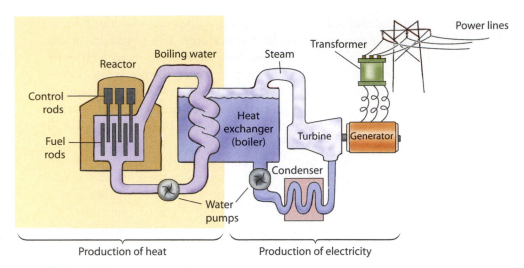

Figure 4.26
Diagram of a nuclear fission power plant. Note that the water in contact with the fuel rods is completely contained, and radioactive materials are not involved directly in the generation of electricity. The details of the production of electricity are covered in Chapter 19.

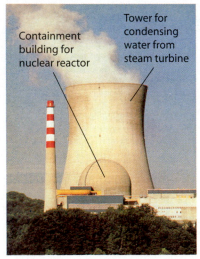

Figure 4.27
A nuclear reactor is housed within a dome-shaped containment building designed to prevent the release of radioactive isotopes in the event of an accident.

greatest practical difference is the amount of fuel involved: a mere 1 kilogram of uranium fuel yields more energy than 30 freightcar loads of coal.

A fission reactor contains three components: nuclear fuel rods, control rods, and a liquid (usually water) to transfer the heat created by fission from the reactor to the turbine. The nuclear fuel is primarily uranium-238 plus about 3 percent uranium-235. Because the uranium-235 atoms are so highly diluted with uranium-238 atoms, an explosion like that of a nuclear bomb is not possible. The reaction rate, which depends on the number of neutrons available to initiate fission of uranium-235 nuclei, is controlled by rods inserted into the reactor. The control rods are made of a neutron-absorbing material, such as cadmium or boron.

Water surrounding the nuclear fuel is kept under high pressure to keep it at a high temperature without boiling. Heated by fission, this water transfers heat to a second, lower-pressure water system, which operates a turbine and an electric generator. Two separate water systems are used so that no radioactivity reaches the turbine, and the entire setup resides inside a building like the one shown in Figure 4.27, designed to keep any radioactive material from ever being released into the environment.

One disadvantage of fission power is the generation of waste products that are radioactive. Smaller atomic nuclei are most stable when composed of equal numbers of protons and neutrons, as we learned earlier, and it is mainly heavy nuclei that need more neutrons than protons for stability. For example, there are 143 neutrons but only 92 protons in uranium-235. When this uranium fissions into two medium-sized elements, the extra neutrons in their nuclei make them unstable. These fragments are therefore radioactive. Most of them have very short half-lives, but some of them have half-lives of thousands of years. Safely disposing of these waste products as well as materials made radioactive in the production of nuclear fuels requires special storage casks and procedures. Although fission power goes back nearly a half-century, the technology of radioactive waste disposal is still in the developmental stage. You can read further details on the subject in Section 19.3.

The Breeder Reactor Breeds Its Own Fuel

One of the fascinating features of fission power is the *breeding* of fission fuel from nonfissionable uranium-238. Breeding occurs when small amounts of fissionable isotopes are mixed with uranium-238 in a reactor. Fission liberates neutrons that convert the relatively abundant nonfissionable uranium-238 to uranium-239, which beta-decays to neptunium-239, which in turn beta-decays to fissionable plutonium-239. So in addition to the abundant energy produced, fission fuel is bred from relatively abundant uranium-238 in the process.

Breeding occurs to some extent in all fission reactors, but a reactor specifically designed to breed more fissionable fuel than is put into it is called a *breeder reactor*. Using a breeder reactor is like filling your car's gas tank with water, adding some gasoline, then driving the car and having more gasoline after the trip than at the beginning! The basic principle of the breeder reactor is very attractive, for after a few years of operation a breeder-reactor power plant can produce vast amounts of power while at the same time breeding twice the amount of fuel it started with.

The downside of breeder reactors is their enormous complexity. The United States gave up on breeders more than a decade ago, and only France and Germany are still investing in them. Officials in these countries point out that supplies of naturally occurring uranium-235 are limited. At present rates of consumption, all natural sources of uranium-235 may be depleted within a century. If countries then decide to turn to breeder reactors, they may well find themselves digging up the radioactive wastes they once buried.

The benefits of fission power are plentiful electricity, conservation of many billions of tons of fossil fuels annually, and the elimination of the megatons of sulfur oxides and other poisons put into the air each year by the burning of fossil fuels.

4.9   Nuclear Energy Comes from Nuclear Mass and Vice Versa

In the early 1900s, Albert Einstein (1879–1955) discovered that mass is actually "congealed" energy. He realized that mass and energy are two sides of the same coin, as stated in his celebrated equation $E = mc^2$. In this equation, E stands for the energy that any mass at rest has, m stands for mass, and c is the speed of light. This relationship between energy and mass is the key to understanding why and how energy is released in nuclear reactions. Any time a nucleus fissions to two smaller nuclei, the combined mass of all nucleons in the smaller nuclei is less than the combined mass of all nucleons in the original nucleus. The mass "missing" after the fission event has been converted to energy and given off to the surroundings. Let's see how.

Figure 4.28
Much work is required to pull a nucleon from an atomic nucleus.

From physics we know that energy is the capacity to do work (Section 1.5) and that work is equal to the product of force times distance:

work = force × distance

Think of the enormous external force required to pull a nucleon out of the nucleus through a distance sufficient to overcome the attractive strong nuclear force, comically represented in Figure 4.28. As per the word equation for work just given, enormous *force* exerted through a *distance* means that enormous work is required. This work is energy that has been added to the nucleon.

According to Einstein's equation, this newly acquired energy reveals itself as an increase in the nucleon's mass—the mass of a nucleon outside a nucleus is greater than the mass of the same nucleon locked inside a nucleus. For example, a carbon-12 atom—the nucleus of which is made up of six protons and six neutrons—has a mass of exactly 12.00000 atomic mass units. Therefore, each proton and each neutron contributes a mass of 1 atomic mass unit. However, outside the nucleus, a proton has a mass of 1.00728 atomic mass units and a neutron has a mass of 1.00867 atomic mass units. Thus we see that the combined mass of six free protons and six free neutrons—(6 × 1.00728) + (6 × 1.00867) = 12.09570—is greater than the mass of one carbon-12 nucleus. The greater mass reflects the energy that was required to pull the nucleons apart from one another. Thus, what mass a nucleon has depends on where the nucleon is.

The graph shown in Figure 4.29 results when we plot average mass *per nucleon* for the elements hydrogen through uranium. This graph is the key to understanding the energy released in nuclear processes. To obtain the average mass per nucleon, you divide the total mass of a nucleus by the number of nucleons in the nucleus. (Similarly, if you divide the total mass of a roomful of people by the number of people in the room, you get the average mass per person.)

Figure 4.29
This graph shows that the average mass of a nucleon depends on which nucleus it is in. Individual nucleons have the most mass in the lightest nuclei, the least mass in iron, and intermediate mass in the heaviest nuclei.

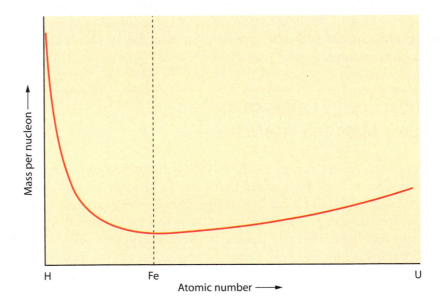

From Figure 4.29 you can see how energy is released when a uranium nucleus splits into two nuclei of lower atomic number. Uranium, being at the right of the graph, has a relatively large amount of mass per nucleon.

When a uranium nucleus splits, however, smaller nuclei of lower atomic numbers are formed. As shown in Figure 4.30, these nuclei are lower on the graph than uranium, which means they have a smaller amount of mass per nucleon. Thus, nucleons lose mass as they go from being in a uranium nucleus to being in the nucleus of one of its fragments. All the mass lost by the nucleons as they change from being nucleons of uranium to being nucleons of atoms such as barium and krypton is converted to energy, and this energy is what we harness and use as "nuclear power." If you wanted to calculate exactly how much energy is released in each fission event, you'd use Einstein's equation: multiply the decrease in mass by the speed of light squared (c^2 in the equation), and the product is the amount of energy yielded by each uranium nucleus as it undergoes fission.

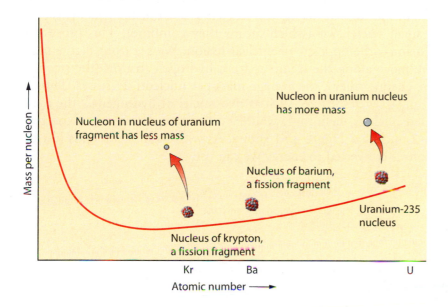

Figure 4.30
The mass of each nucleon in a uranium nucleus is greater than the mass of each nucleon in any one of its fission fragments. This lost mass has been converted to energy, which is why nuclear fission is an energy-releasing process.

Interestingly, Einstein's mass/energy relationship applies to chemical reactions as well as to nuclear reactions. For nuclear reactions, the energies involved are so great that the change in mass is measureable, corresponding to about 1 part in 1000. In chemical reactions, the energy involved is so small that the change in mass, about 1 part in 1,000,000,000, is not detectable. This is why the mass conservation law (Section 3.2) states that there is no *detectable* change in the total mass of materials as they chemically react to form new materials. In truth, there are changes in the mass of atoms during a chemical reaction. These changes are too small to be of any concern to the working chemist, however.

Concept Check ✓

Correct this statement: When a heavy element undergoes fission, there are fewer nucleons after the reaction than before.

Was this your answer? When a heavy element undergoes fission, there aren't fewer nucleons after the reaction. Instead, there's *less mass* in the same number of nucleons.

We can think of the mass-per-nucleon graph shown in Figure 4.29 as an energy valley that starts at hydrogen (the highest point) and slopes steeply to the lowest point (iron), then slopes gradually up to uranium. Iron is at the bottom of the energy valley and is therefore the most stable nucleus. It is also the most tightly bound nucleus; more energy per nucleon is required to separate nucleons from an iron nucleus than from any other nucleus.

4.10   Nuclear Fusion Is the Combining of Atomic Nuclei

As mentioned earlier, a drawback to nuclear fission is the production of radioactive waste products. A more promising long-range source of nuclear energy is to be found with the lightest elements. In a nutshell, energy is produced as small nuclei *fuse* (which means they combine). This process is **nuclear fusion**—the opposite of nuclear fission. We see from Figure 4.29 that, as we move along the elements from hydrogen to iron (the steepest part of the energy valley), the average mass per nucleon decreases. Thus if two small nuclei were to fuse, such as two nuclei of hydrogen-2, the mass of the fused nucleus, helium-4, would be less than the mass of the two hydrogen-2 nuclei, as Figure 4.31 shows. As with fission, the mass lost by the nucleons is converted to energy we can use.

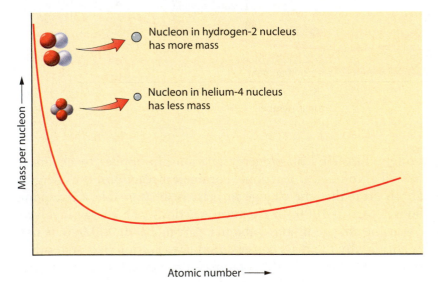

Figure 4.31
The mass of each nucleon in a hydrogen-2 nucleus is greater than the mass of each nucleon in a helium-4 nucleus, which results from the fusion of two hydrogen-2 nuclei. This lost mass has been converted to energy, which is why nuclear fusion is an energy-releasing process.

If a fusion reaction is to occur, the nuclei must be traveling at extremely high speeds when they collide in order to overcome their mutual electrical repulsion. The required speeds correspond to the extremely high temperatures found deep in the sun and in other stars. Fusion brought about by high temperatures is called **thermonuclear fusion**. In the high temperatures of the sun, approximately 657 million tons of hydrogen is fused to 653 million tons of helium *each second.* The 4 million tons of nucleon mass lost is discharged as radiant energy.

Concept Check ✓

To get energy from the element iron, should iron be fissioned or fused?

Was this your answer? Neither, because iron is at the very bottom of the energy-valley curve of Figure 4.29. If you fuse two iron nuclei, the product lies somewhere to the right of iron on the curve, which means the product has a higher mass per nucleon. If you split an iron nucleus, the products lie to the left of iron on the curve, which again means a higher mass per nucleon. Because no mass decrease occurs in either reaction, no mass is available to be converted to energy, and as a result no energy is released.

Prior to the development of the atomic bomb, the temperatures required to initiate nuclear fusion on the Earth were unattainable. When researchers found that the temperature inside an exploding atomic bomb is four to five times the temperature at the center of the sun, the thermonuclear bomb was but a step away. This first thermonuclear bomb, a hydrogen bomb, was detonated in 1952. Whereas the critical mass of fissionable material limits the size of a fission bomb (atomic bomb), no such limit is imposed on a fusion bomb (thermonuclear or hydrogen bomb). A typical thermonuclear bomb stockpiled by the United States today, for example, is about 1000 times more destructive than the atomic bomb detonated over Hiroshima at the end of World War II.

The hydrogen bomb is another example of a discovery used for destructive rather than constructive purposes. The potential constructive possibility is the controlled release of vast amounts of clean energy.

The Holy Grail of Nuclear Research Today Is Controlled Fusion

Carrying out fusion reactions under controlled conditions requires temperatures of millions of degrees. As you can imagine, this poses many technical difficulties, especially when it comes to the large-scale production of energy. For example, one major problem is that any reaction vessel being used would melt and vaporize long before these temperatures were reached.

One proposed technique is to aim an array of laser beams at a common point and drop solid pellets of hydrogen isotopes through the synchronous crossfire, as Figure 4.32 on page 128 shows. The energy of the multiple beams should crush the pellets to densities 20 times that of lead. Such a fusion could produce several hundred times more energy than the amount delivered by the laser beams. Like the succession of fuel/air explosions in an automobile engine's cylinders that convert to a smooth flow of mechanical power, the successive ignition of pellets in a laser fusion device may similarly produce a steady stream of electric power. A plant equipped with the device could produce 1000 million watts of electric power, enough to supply a city of 600,000 people. High-power lasers that work reliably, however, have yet to be developed.

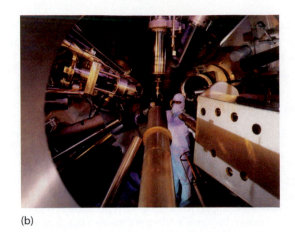

(a)   (b)

Figure 4.32
(a) Fusion with multiple laser beams. Pellets of hydrogen isotopes are rhythmically dropped into synchronized laser crossfire in this planned device. The resulting heat is carried off by molten lithium to produce steam. (b) The pellet chamber at Lawrence Livermore Laboratory. The laser source is Nova, the most powerful laser in the world, which directs ten beams into the target region.

Also under development are techniques that use magnetic fields to confine fusing materials. Examples are presented in Section 19.3.

Concept Check ✔

Fission and fusion are opposite processes, yet each releases energy. Isn't this contradictory?

Was this your answer? No, no, no! As Figure 4.29 shows, only the fusion of light elements and the fission of heavy elements result in a decrease in nucleon mass and therefore a release of energy.

In Perspective

If people are one day to dart about the universe the way we jet about the Earth today, their supply of fuel is assured. The fuel for fusion—hydrogen—is found in every part of the universe, not only in the stars but also in the space between them. About 91 percent of the atoms in the universe are estimated to be hydrogen. For people of the future, the supply of raw materials is also assured because all the elements known to exist result from the fusing of more and more hydrogen nuclei. Simply put, if you fuse 8 hydrogen-2 nuclei, you have oxygen; 26, you have iron; and so forth. Future humans might synthesize their own elements and produce energy in the process, just as the stars have always done.

Key Terms and Matching Definitions

_____ alpha particle
_____ beta particle
_____ carbon-14 dating
_____ chain reaction
_____ critical mass
_____ gamma ray
_____ half-life
_____ nuclear fission
_____ nuclear fusion
_____ radioactivity
_____ rem
_____ strong nuclear force
_____ thermonuclear fusion
_____ transmutation

1. The tendency of some elements, such as uranium, to emit radiation as a result of changes in the atomic nucleus.
2. A helium atom nucleus, which consists of two neutrons and two protons and is ejected by certain radioactive elements.
3. An electron ejected from an atomic nucleus during the radioactive decay of certain nuclei.
4. High-energy radiation emitted by the nuclei of radioactive atoms.
5. A unit for measuring the ability of radiation to harm living tissue.
6. The force of interaction between all nucleons, effective only at very, very, very close distances.
7. The conversion of an atomic nucleus of one element to an atomic nucleus of another element through a loss or gain of protons.
8. The time required for half the atoms in a sample of a radioactive isotope to decay.
9. The process of estimating the age of once-living material by measuring the amount of a radioactive isotope of carbon present in the material.
10. The splitting of a heavy nucleus into two lighter nuclei, accompanied by the release of much energy.
11. A self-sustaining reaction in which the products of one fission event stimulate further events.
12. The minimum mass of fissionable material needed to sustain a chain reaction.
13. The joining together of light nuclei to form a heavier nucleus, accompanied by the release of much energy.
14. Nuclear fusion produced by high temperature.

Review Questions

The Cathode Ray Led to the Discovery of Radioactivity

1. What did Wilhelm Roentgen discover in 1896?
2. How did Henri Becquerel determine that phosphorescence was not responsible for the emission of radiation by uranium?
3. Who coined the term *radioactivity*?
4. How do the electric charges of alpha particles, beta particles, and gamma rays differ from one another?
5. Which has the greatest penetrating power—alpha partciles, beta particles, or gamma rays?

Radioactivity Is a Natural Phenomenon

6. What is the origin of most of the radiation you encounter?
7. Which is worse: having cells in your body damaged by radiation or killed by radiation?
8. Is radioactivity on the Earth something relatively new? Defend your answer.
9. What is a rem?

Radioactive Isotopes Are Useful as Tracers and for Medical Imaging

10. What is a radioactive tracer?
11. How are radioactive isotopes used in medical imaging?

Radioactivity Results from an Imbalance of Forces in the Nucleus

12. How are the strong nuclear force and the electric force different from each other?
13. What role do neutrons play in the atomic nucleus?
14. Why is there a limit to the number of neutrons a nucleus can contain?

A Radioactive Element Can Transmute to a Different Element

15. When thorium, atomic number 90, decays by emitting an alpha particle, what is the atomic number of the resulting nucleus?

16. When thorium-90 decays by emitting a beta particle, what is the atomic number of the resulting nucleus?

17. What change in atomic number occurs when a nucleus emits an alpha particle? A beta particle?

18. What is the long-range fate of all the uranium that exists in the world today?

The Shorter the Half-Life, the Greater the Radioactivity

19. What is meant by the half-life of a radioactive sample?

20. What is the half-life of radium-226?

21. How does the decay rate of an isotope relate to its half-life?

Isotopic Dating Measures the Age of a Material

22. What do cosmic rays have to do with transmutation?

23. How is carbon-14 produced in the atmosphere?

24. Which is radioactive, carbon-12 or carbon-14?

25. Why is there more carbon-14 in living bones than in once-living ancient bones of the same mass?

26. Why is carbon-14 dating useless for dating old coins but not old pieces of cloth?

27. Why is lead found in all deposits of uranium ores?

28. What does the proportion of lead and uranium in rock tell us about the age of the rock?

Nuclear Fission Is the Splitting of the Atomic Nucleus

29. Why does a chain reaction not occur in uranium mines?

30. Is a chain reaction more likely to occur in two separate pieces of uranium-235 or in the same pieces stuck together?

31. How is a nuclear reactor similar to the furnace in a fossil-fuel power plant? How is it different?

32. What is the function of control rods in a nuclear reactor?

Nuclear Energy Comes from Nuclear Mass and Vice Versa

33. Is work required to pull a nucleon out of an atomic nucleus? Does the nucleon, once outside the nucleus, have more mass than it had inside the nucleus?

34. Does the mass of a nucleon after it has been pulled from an atomic nucleus depend on which nucleus it was extracted from?

35. How does the mass per nucleon in uranium compare with the mass per nucleon in the fission fragments of uranium?

36. If an iron nucleus split in two, would its fission fragments have more mass per nucleon or less mass per nucleon?

37. If a pair of iron nuclei were fused, would the product nucleus have more mass per nucleon or less mass per nucleon?

Nuclear Fusion Is the Combining of Atomic Nuclei

38. When two hydrogen isotopes are fused, is the mass of the product nucleus more or less than the total mass of the hydrogen nuclei?

39. From where does the sun gets its energy?

40. How do the products of fusion reactions differ from the products of fission reactions?

Hands-On Chemistry Insights

Personal Radiation

Will the effects of the radiation you receive be passed on to your children? Only radiation received by your reproductive organs (testes in males and ovaries in females) has the potential of causing effects that might be passed on to future generations. All other radiation you receive is for your body only.

What percentage of your estimated annual radiation comes from natural sources? What adjustments might you be willing to make in order to decrease your annual exposure?

Radioactive Paper Clips

The unit of half-life in this simulation is *number of throws*. The element with the shorter half-life is considered to be the more radioactive one because it decays faster and in the process emits more radiation per unit time. For most students, this turns out to be legonium.

Uranium-238 has a half-life of 4.5 billion years, while polonium-214 has a half-life of 0.00016 second. So, which would you rather hold in your hands: 1 gram of uranium-238 or 1 gram of polonium-214?

Perhaps the most important point of this activity is that radioactive decay is a statistical phenomenon. Follow any one of the paper clips, and you'll find that it may decay well before or after the half-life. Half-life is a predictable quantity only when a large number of particles are involved.

Exercises

1. Why is a sample of radium always a little warmer than its surroundings?

2. Is it possible for a hydrogen nucleus to emit an alpha particle? Defend your answer.

3. Why are alpha particles and beta particles deflected in opposite directions in a magnetic field? Why are gamma rays undeflected?

4. The alpha particle has twice the electric charge of the beta particle but deflects less in a magnetic field. Why?

5. Which type of radiation—alpha, beta, or gamma—results in the greatest change in mass number? The greatest change in atomic number?

6. Which type of radiation—alpha, beta, or gamma—results in the least change in mass number? The least change in atomic number?

7. Which type of radiation—alpha, beta, or gamma—predominates on the inside of a high-flying commercial airplane? Why?

8. In bombarding atomic nuclei with proton "bullets," why must the protons be given large amounts of kinetic energy in order to make contact with the target nuclei?

9. Why would you expect alpha particles to be less able to penetrate materials than beta particles?

10. What evidence supports the hypothesis that, at short intranuclear distances, the strong nuclear force is stronger than the electric force?

11. The isotope cesium-137, which has a half-life of 30 years, is a product of nuclear power plants. How long will it take this isotope to decay to one-sixteenth its original amount?

12. When the isotope bismuth-213 emits an alpha particle, what new element results? What new element results if it instead emits a beta particle?

13. When $^{226}_{88}$Ra decays by emitting an alpha particle, what is the atomic number of the resulting nucleus? What is the resulting atomic mass?

14. What are the atomic number and atomic mass of the element formed when $^{218}_{84}$Po emits a beta particle? What are they if the polonium emits an alpha particle?

15. How is it possible for an element to decay "forward in the periodic table"—that is, decay to an element of higher atomic number?

16. Elements above uranium in the periodic table do not exist in any appreciable amounts in nature because they have short half-lives. Yet there are several elements below uranium in the table that have equally short half-lives but do exist in appreciable amounts in nature. How can you account for this?

17. You and a friend journey to the mountain foothills to get closer to nature and escape such things as radioactivity. While bathing in the warmth of a natural hot spring, she wonders aloud how the spring gets its heat. What do you tell her?

18. People who work around radioactivity wear film badges to monitor the amount of radiation that reaches their bodies. Each badge consists of a small piece of photographic film enclosed in a lightproof wrapper. What kind of radiation do these devices monitor, and how can they determine the amount of radiation the people receive?

19. Coal contains only minute quantities of radioactive materials, and yet there is more environmental radiation surrounding a coal-fired power plant than a fission power plant. What does this indicate about the shielding that typically surrounds these two types of plants?

20. A friend checks the local background radiation with a Geiger counter, which ticks audibly. Another friend, who normally fears most that which is understood least, makes an effort to keep away from the region of the Geiger counter and looks to you for advice. What do you say?

21. Why is carbon-14 dating not accurate for estimating the age of materials more than 50,000 years old?

22. The age of the Dead Sea Scrolls was determined by carbon-14 dating. Could this technique have worked if they had been carved on stone tablets? Explain.

23. A certain radioactive element has a half-life of 1 hour. If you start with a 1-gram sample of the element at noon, how much is left at 3:00 P.M.? At 6:00 P.M.? At 10:00 P.M.?

24. Why will nuclear fission probably never be used directly for powering automobiles? How could it be used indirectly?

25. Why does a neutron make a better nuclear bullet than a proton or an electron?

26. Does the average distance a neutron travels through fissionable material before escaping increase or decrease when two pieces of fissionable material are assembled into one piece? Does this assembly increase or decrease the probability of an explosion?

27. Why does plutonium not occur in appreciable amounts in natural ore deposits?

28. Uranium-235 releases an average of 2.5 neutrons per fission, while plutonium-239 releases an average of 2.7 neutrons per fission. Which of these elements might you therefore expect to have the smaller critical mass?

29. Why, after a uranium fuel rod reaches the end of its fuel cycle (typically three years), does most of its energy come from plutonium fission?

30. If a nucleus of $^{232}_{90}$Th absorbs a neutron and the resulting nucleus undergoes two successive beta decays, which nucleus results?

31. To predict the approximate energy release of either a fission or a fusion reaction, explain how a physicist uses a table of nuclear masses and the equation $E = mc^2$.

32. Which process would release energy from gold, fission or fusion? From carbon? From iron?

33. If a uranium nucleus were to fission into three fragments of approximately equal size instead of two, would more energy or less energy be released? Defend your answer using Figures 4.29 and 4.30.

34. Explain how radioactive decay has always warmed the Earth from the inside and how nuclear fusion has always warmed the Earth from the outside.

35. Speculate about some worldwide changes likely to follow the advent of successful fusion reactors.

Exploring Further

M. Mitchell Waldrop, "The Shroud of Turin: An Answer Is at Hand." *Science*, September 30, 1988. Describes the events leading up to the dating of the controversial Shroud of Turin.

http://www.friendsofpast.org
A nonprofit organization dedicated to promoting and advancing the rights of scientists and the public to learn about America's past through archeology. Emphasis is given to the Kennewick Man debate, in which Native Americans claim kinship to 9000-year-old skeletal remains.

http://www.iaea.or.at/worldatom
The Web site for the International Atomic Energy Agency, which monitors almost all issues related to nuclear technology. A good starting point for exploring applications of many of the concepts discussed in this chapter.

http://www.iter.org
The Web site for the International Thermonuclear Experimental Reactor project. Explore this site for the latest on the science and politics of this important project.

http://www.rw.doe.gov/homejava/homejava.htm
Home page for the Office of Civilian Radioactive Waste Management, established in 1982 to develop and manage a federal system for disposing of spent nuclear fuel resulting from atomic energy defense activities. It's here that you'll find the official position of the U.S. government regarding Yucca Mountain, Nevada, as a potential nuclear waste repository.

the **Chemistry** place

The Atomic Nucleus
Visit The Chemistry Place at:
www.aw.com/chemplace

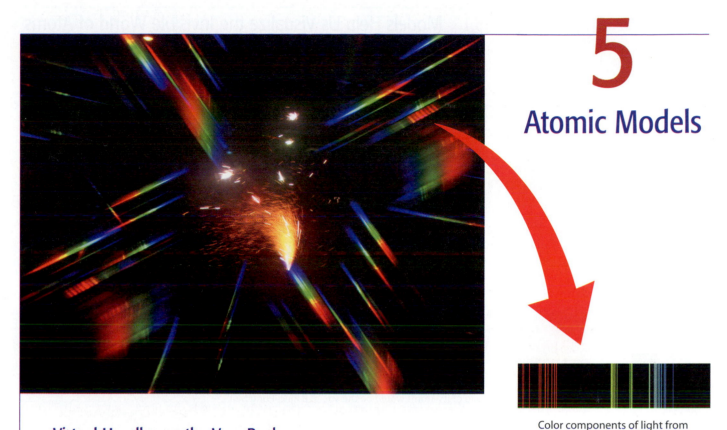

5

Atomic Models

Color components of light from glowing strontium (high resolution)

Virtual Handles on the Very Real

The elements of many chemical compounds glow with color when heated. Strontium, for example, glows red, sodium glows yellow, and barium glows yellow-green. Package such elements with burning gun powder, and the result is a brilliant fireworks display. Interestingly, the glow of a single element consists of a number of overlapping colors, which can be separated from one another with a prism. Such a separation is shown in the opening photograph of this chapter. In the center of the photograph are glowing sparks of strontium. The adjacent diagonal stripes, created by a prismlike filter on the camera, reveal this element's many hues, called a *spectral pattern*.

Each element emits its own characteristic spectral pattern, which can be used to identify the element just as a fingerprint can be used to identify a person. As we discuss in this chapter, scientists of the early 1900s saw these spectral patterns as clues to the internal structure and dynamics of atoms. By studying spectral patterns and by conducting experiments, these scientists were able to develop models of the atom. Through these models, which continue to be refined even today, chemists gain a powerful understanding of how atoms behave.

This chapter explores the development of atomic models. It is one of the more challenging chapters of this tetbook. The background it provides, however, will give you both a deeper understanding of the periodic table and a foundation for understanding how atoms react with one another to form new materials—an important topic of subsequent chapters.

5.1   Models Help Us Visualize the Invisible World of Atoms

Atoms are so small that the number of them in a baseball is roughly equal to the number of Ping-Pong balls that could fit inside a hollow sphere as big as the Earth, as Figure 5.1 illustrates. This number is incredibly large—

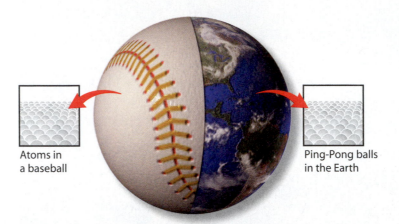

Figure 5.1
If the Earth were filled with nothing but Ping-Pong balls, the number of balls would be roughly equal to the number of atoms in a baseball. Put differently, if a baseball were the size of the Earth, one of its atoms would be the size of a Ping-Pong ball.

beyond our intuitive grasp. Atoms are so incredibly small that we can never *see* them in the usual sense. This is because light travels in waves, and atoms are smaller than the wavelengths of visible light, which is the light that allows the human eye to see things. We could stack microscope on top of microscope and still not see an individual atom. As illustrated in Figure 5.2, the diameter of an object visible under the highest magnification must be larger than the wavelengths of visible light.

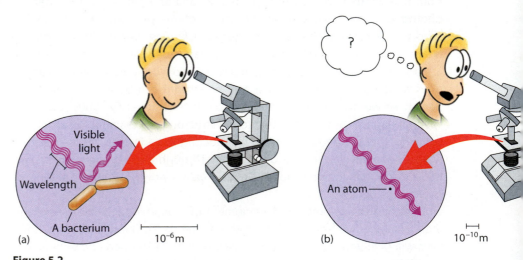

Figure 5.2
Microscopic objects can be seen through a microscope that works with visible light, but submicroscopic particles cannot. (a) A bacterium is visible because it is larger than the wavelengths of visible light. We can see the bacterium through the microscope because the bacterium reflects visible light. (b) An atom is invisible because it is smaller than the wavelengths of visible light and so does not reflect the light toward our eyes.

Although we cannot see atoms *directly*, we can generate images of them *indirectly*. In the mid 1980s, researchers developed the *scanning tunneling microscope* (STM), which produces images by dragging an ultrathin needle back and forth over the surface of a sample. Bumps the size of atoms on the surface cause the needle to move up and down. This vertical motion is detected and translated by a computer into a topographical image that corresponds to the positions of atoms on the surface (Figure 5.3). An STM can also be used to push individual atoms into desired positions. This ability opened up the field of nanotechnology, in which incredibly small electronic circuits and motors are built atom by atom.

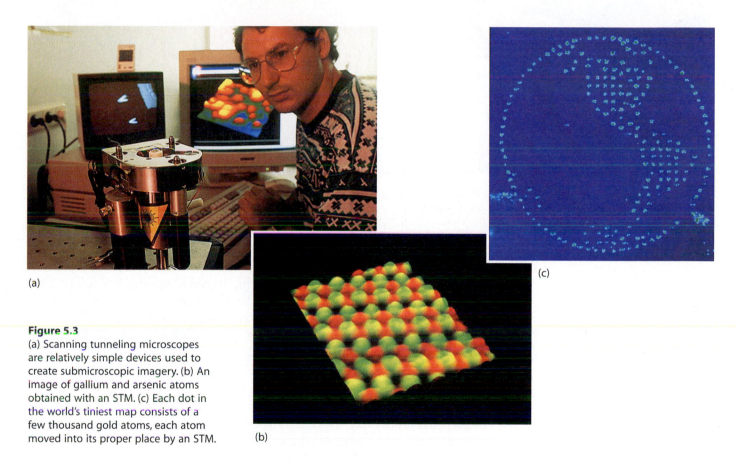

(a)

(c)

(b)

Figure 5.3
(a) Scanning tunneling microscopes are relatively simple devices used to create submicroscopic imagery. (b) An image of gallium and arsenic atoms obtained with an STM. (c) Each dot in the world's tiniest map consists of a few thousand gold atoms, each atom moved into its proper place by an STM.

Concept Check ✓

Why are atoms invisible?

Was this your answer? An individual atom is smaller than the wavelengths of visible light and so is unable to reflect that light. Atoms are invisible, therefore, because visible light passes right by them. The atomic images generated by STMs are not photographs taken by a camera. Rather, they are computer renditions generated from the movements of an ultrathin needle.

A very small or very large visible object can be represented with a **physical model**, which is a model that replicates the object at a more convenient scale. Figure 5.4a, for instance, shows a large-scale physical model

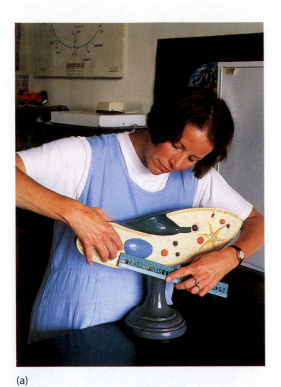

(a)

(b)

Figure 5.4
(a) This large-scale model of a microorganism is a physical model. (b) Weather forecasters rely on conceptual models such as this one to predict the behavior of weather systems.

of a microorganism that a biology student uses to study the microorganism's internal structure. Because atoms are invisible, however, we cannot use a physical model to represent them. In other words, we cannot simply scale up the atom to a larger size, as we might with a microorganism. (An STM merely shows the *positions* of atoms and not actual images of atoms, which do not have the solid surfaces implied in the STM images of Figure 5.3.) So, rather than describing the atom with a physical model, chemists use what is known as a **conceptual model,** which describes a *system.* The more accurate a conceptual model, the more accurately it predicts the behavior of the system. The weather is best described using a conceptual model like the one shown in Figure 5.4b. Such a model shows how the various components of the system—humidity, atmospheric pressure, temperature, electric charge, the motion of large masses of air—interact with one another.

Other systems that can be described by conceptual models are the economy, population growth, the spread of diseases, and team sports.

Concept Check ✓

A basketball coach describes a playing strategy to her team by way of sketches on a game card. Do the illustrations represent a physical model or a conceptual model?

Was this your answer? The sketches are a conceptual model the coach uses to describe a system (the players on the court), with the hope of predicting an outcome (winning the game).

Like the weather, the atom is a complex system of interacting components, and it is best described with a conceptual model. You should therefore be careful not to interpret any visual representation of an atomic conceptual model as a re-creation of an actual atom. In Section 5.4, for example, you will be introduced to the planetary model of the atom, wherein electrons are shown orbiting the atomic nucleus much as planets orbit the sun. This planetary model is limited, however, in that it fails to explain many properties of atoms. Thus newer and more accurate (and more complicated) conceptual models of the atom have since been introduced. In these models, electrons appear as a cloud hovering around the atomic nucleus, but even these models have their limitations. Ultimately, the best models of the atom are ones that are purely mathematical.

In this textbook, our focus is on conceptual atomic models that are easily represented by visual images, including the planetary model, the electron-cloud model, and a model in which electrons are grouped in units called *shells*. Despite their limitations, such images are excellent guides to learning chemistry, especially for the beginning student. These models were developed by scientists to help explain how atoms emit light. We begin our study of atomic models, therefore, by reviewing the fundamental nature of light.

5.2 Light Is a Form of Energy

Light is a form of energy known as *electromagnetic radiation*. It travels in waves that are analogous to the waves produced by a pebble dropped into a pond. Electromagnetic waves, however, are oscillations (vibrations) of electric and magnetic fields, not oscillations of a material medium such as water. Most of the electromagnetic radiation we encounter is generated by electrons, which can oscillate at exceedingly high rates because of their small size.

The distance between two crests of an electromagnetic wave is called the **wavelength** of the wave. Electromagnetic wavelengths range from less than 1 nanometer (10^{-9} meter) for high-energy gamma rays to more than 1 kilometer (10^3 meters) for radio waves, which are low-energy electromagnetic radiation. Figure 5.5 labels two wavelengths—one very long, the other very short—on a fictitious wave drawn for illustration only.

Electromagnetic waves can also be characterized by their **wave frequency**, a measure of how rapidly they oscillate. The shorter the wavelength of an electromagnetic wave, the greater its wave frequency. Gamma rays, for example, have very short wavelengths, which means their wave frequencies are very high. Radio waves have very long wavelengths, which means their wave frequencies are very low.

The basic unit of wave frequency is the *hertz* (abbreviated Hz), where 1 hertz equals 1 cycle per second and *cycle* refers to one complete oscillation. Wave frequencies for electromagnetic radiation range from 10^{24} hertz for gamma rays to less than 10^3 hertz for radio waves. The higher the frequency of a wave, the greater its energy, which means that gamma rays are far more energetic than radio waves.

As I idly tap this stick on the water surface, I generate waves that emanate outward from the point of contact. Similarly, as electrons oscillate back and forth in an atom, they generate electromagnetic waves that emanate from the atom. Interestingly, the faster I tap, the closer together the waves are to one another.

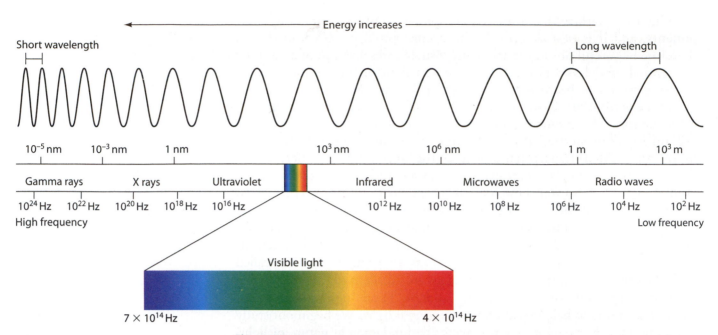

Figure 5.5
The electromagnetic spectrum is a continuous band of wave frequencies extending from high-energy gamma rays, which have short wavelengths and high frequencies, to low-energy radio waves, which have long wavelengths and low frequencies. The descriptive names of these regions are merely a historical classification, for all waves are the same in nature, differing only in wavelength and frequency.

Figure 5.5 shows a full range of frequencies and wavelengths of electromagnetic radiation in a display known as the **electromagnetic spectrum**. The most energetic region of the electromagnetic spectrum consists of gamma rays. Next is the region of slightly lower energy where we find X rays, and next is the electromagnetic radiation we call ultraviolet light. Within a narrow region from about 7×10^{14} (700 trillion) hertz to about 4×10^{14} (400 trillion) hertz are the frequencies of electromagnetic radiation known as visible light. This region includes the rainbow of colors our eyes are able to detect, from violet at 700 trillion hertz to red at 400 trillion hertz. Lower in energy than visible light are infrared waves (detected by our skin as "heat waves"), then microwaves (used to cook foods), and finally radio waves (through which radio and television signals are sent), the waves of lowest energy.

Concept Check ✓

Can you see radio waves? Can you hear them?

Was this your answer? Your eyes are equipped to see only the narrow range of frequencies of electromagnetic radiation from about 700 trillion to 400 trillion hertz—the range of visible light. Radio waves are one type of electromagnetic radiation, but their frequency is much lower than what your eyes can detect. Thus, you can't see radio waves. Neither can you hear them. You can, however, turn on an electronic gizmo called a radio, which translates radio waves into signals that drive a speaker to produce sound waves your ears can hear.

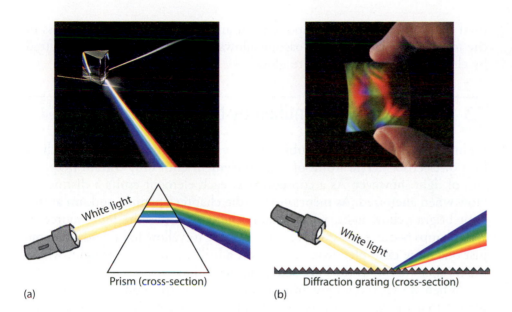

Figure 5.6
White light is separated into its color components by (a) a prism and (b) a diffraction grating.

White light

Prism (cross-section)

(a)

White light

Diffraction grating (cross-section)

(b)

We see white light when all frequencies of visible light reach our eye at the same time. By passing white light through a prism or through a diffraction grating, which is a glass plate or plastic sheet with microscopic lines etched into it, the color components of the light can be separated, as shown in Figure 5.6. (Remember—each color of visible light corresponds to a different frequency.) A **spectroscope**, shown in Figure 5.7, is an instrument

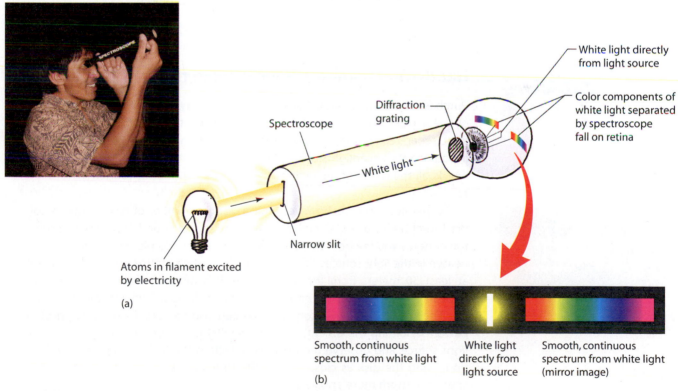

White light directly from light source

Color components of white light separated by spectroscope fall on retina

Diffraction grating

Spectroscope

White light

Narrow slit

Atoms in filament excited by electricity

(a)

Smooth, continuous spectrum from white light

White light directly from light source

Smooth, continuous spectrum from white light (mirror image)

(b)

Figure 5.7
(a) In a spectroscope, light emitted by atoms passes through a narrow slit before being separated into particular frequencies by a prism or (as shown here) a diffraction grating. (b) This is what the eye sees when the slit of a diffraction-grating spectroscope is pointed toward a white-light source. Spectra of colors appear to the left and right of the slit.

used to observe the color components of any light source. As we discuss in the following section, a spectroscope allows us to analyze the light emitted by elements as they are made to glow.

5.3   Atoms Can Be Identified by the Light They Emit

Light is given off by atoms subjected to various forms of energy, such as heat or electricity. The atoms of a given element emit only certain frequencies of light, however. As a consequence, each element emits a distinctive glow when energized. As mentioned in the chapter opening, sodium atoms emit bright yellow light, which makes them useful as the light source in street lamps because our eyes are very sensitive to yellow light, and, to name just one more example, neon atoms emit a brilliant red-orange light, which makes them useful as the light source in neon signs.

When we view the light from glowing atoms through a spectroscope, we see that the light consists of a number of discrete (separate from one another) frequencies rather than a continuous spectrum like the one shown in Figures 5.5, 5.6, and 5.7. The pattern of frequencies formed by a given element—some of which are shown in Figure 5.8—is referred to as that element's **atomic spectrum**. The atomic spectrum is an element's fingerprint. You can identify the elements in a light source by analyzing the light through a spectroscope and looking for characteristic patterns. If you don't have the opportunity to work with a spectroscope in your laboratory, check out the Hands-On Chemistry *Spectral Patterns* below.

Hands-On Chemistry: Spectral Patterns

Purchase some "rainbow" glasses from a nature, toy, or hobby store. The lenses of these glasses are diffraction gratings. Looking through them, you will see light separated into its color components. Certain light sources, such as the moon or a car's headlights, are separated into a continuous spectrum—in other words, all the colors of the rainbow appear in a continuous sequence from red to violet.

Other light sources, however, emit a distinct number of discontinuous colors. Examples include streetlights, neon signs, sparklers, and fireworks. The spectral patterns you see from these light sources are the atomic spectra of elements heated in the light sources. You'll be able to see the patterns best when you are at least 50 meters from the light source. This distance makes the spectrum appear as a series of dots similar to the series of lines shown in Figure 5.8.

The rainbow side of a compact disk can also be used for viewing spectral patterns. Holding the disk at eye level parallel to the ground, look over it at a light source, and observe the rainbow reflection. While focusing on the reflection, bring the disk as close as possible to your eye. Doing so will make the spectral pattern more apparent.

Share your rainbow glasses and disk with a friend on your next "night on the town." You'll find each type of light has its own signature pattern. How many different patterns are you able to observe?

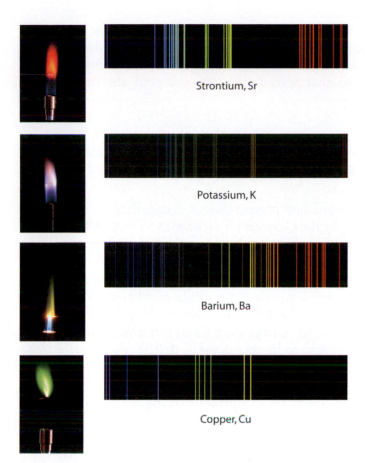

Figure 5.8
Elements heated by a flame glow their characteristic color. This is commonly called a *flame test* and is used to test for the presence of an element in a sample. When viewed through a spectroscope, the color of each element is revealed to consist of a pattern of distinct frequencies known as an atomic spectrum.

Strontium, Sr

Potassium, K

Barium, Ba

Copper, Cu

Concept Check ✔

How might you deduce the elemental composition of a star?

Was this your answer?   Aim a well-built spectroscope at the star, and study its spectral patterns. In the late 1800s, this was done with our own star, the sun. Spectral patterns of hydrogen and some other known elements were observed, in addition to one pattern that could not be identified. Scientists concluded that this unidentified pattern belonged to an element not yet discovered on the Earth. They named this element *helium* after the Greek word for "sun," *helios*.

Researchers in the 1800s noted that the lightest element, hydrogen, has a far more orderly atomic spectrum than the other elements. Figure 5.9 shows a portion of the hydrogen spectrum. Note that the spacing between successive lines decreases in a regular way. A Swiss schoolteacher, Johann Balmer (1825–1898), expressed these line positions by a mathematical formula. Another regularity in hydrogen's atomic spectrum was noticed by Johannes Rydberg (1854–1919)—the sum of the frequencies of two lines sometimes equals the frequency of a third line. For example,

Figure 5.9
A portion of the atomic spectrum for hydrogen. These frequencies are higher than those of visible light, which is why they are not shown in color.

First spectral line	1.6×10^{14} Hz
Second spectral line	$+ \underline{4.6 \times 10^{14} \text{ Hz}}$
Third spectral line	6.2×10^{14} Hz

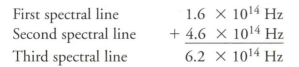

The orderliness of hydrogen's atomic spectrum was most intriguing to Balmer, Rydberg, and other investigators of the time. However, as to why such orderliness should exist, these early workers were unable to formulate any hypothesis that agreed with any accepted atomic model of the day.

5.4 Niels Bohr Used the Quantum Hypothesis to Explain Atomic Spectra

An important step toward our present-day understanding of atoms and their spectra was taken by the German physicist Max Planck (1858–1947). In 1900, Planck hypothesized that light energy is *quantized* in much the same way matter is.

To understand what this interesting new term means, consider a gold brick. The mass of the brick equals some whole-number multiple of the mass of a single gold atom. Similarly, an electric charge is always some whole-number multiple of the charge on a single electron. Mass and electric charge are therefore said to be *quantized* in that they consist of some number of fundamental units.

What Planck did with his **quantum hypothesis** was to recognize that a beam of light energy is not the continuous (nonquantized) stream of energy we think it is. Instead the beam consists of zillions of small, discrete *packets* of energy, each packet called a **quantum**, as represented in Figure 5.10. A few years later, in 1905, Einstein recognized that these quanta of light behave much like tiny particles of matter. To emphasize their *particulate* nature, each quantum of light was called a **photon**, a name coined because of its similarity to the words *electron*, *proton*, and *neutron*.

Take a moment to let this amazing fact sink in: light as a stream of tiny little bullets! And if that is so, why does Section 5.2 say light is an electromagnetic *wave*? Was that an error? A lie? Neither, for evidence tells us that light behaves as *both a wave and a particle*, and this is where the idea of conceptual models comes into play. When scientists study visible light (or any other electromagnetic radiation), they are free to choose the model that best fits their needs—light as wave or light as stream of particles. Depending on

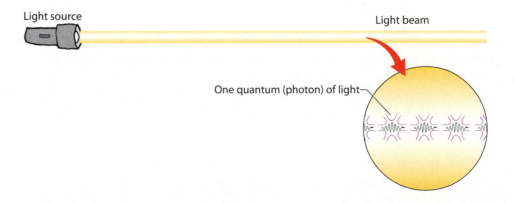

Figure 5.10
Light is quantized, which means it consists of a stream of energy packets. Each packet is called a quantum, also known as a photon.

the model chosen, light has the properties of a wave or the properties of a particle. To represent this duality, photons are illustrated in this text as a burst of light with a wave drawn inside the burst.

As shown in Figure 5.11, the amount of energy in a photon increases with the frequency of the light. One photon of ultraviolet light, for example, possesses more energy than one photon of infrared light because ultraviolet light has higher frequency than infrared light (as Figure 5.5 shows).

Using Planck's quantum hypothesis, the Danish scientist Niels Bohr (1885–1962) explained the formation of atomic spectra as follows. First, Bohr recognized that the potential energy of an electron in an atom depends on the electron's distance from the nucleus. This is analogous to the potential energy of an object held some distance above the Earth's surface. The object has more potential energy when it is held high above the ground than when it is held close to the ground. Likewise, an electron has more potential energy when it is far from the nucleus than when it is close to the nucleus. Second, Bohr recognized that when an atom absorbs a photon of light, it is absorbing *energy*. This energy is acquired by one of the electrons surrounding the atom's nucleus. Because this electron has gained energy, it must move away from the nucleus. In other words, absorption of a photon causes a low-potential-energy electron in an atom to become a high-potential-energy electron.

Bohr also realized that the opposite is true: when a high-potential-energy electron in an atom loses some of its energy, the electron moves closer to the nucleus and the energy lost from the electron is emitted from the atom as a photon of light. Both absorption and emission are illustrated in Figure 5.12.

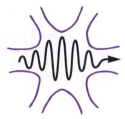

High-frequency, high-energy photon

Low-frequency, low-energy photon

Figure 5.11
The greater the frequency of a photon of light, the greater the energy packed into that photon.

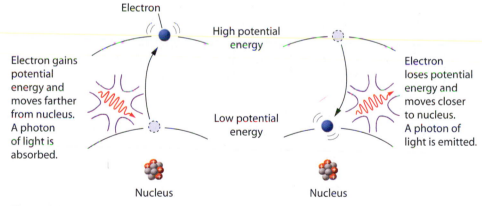

Electron

High potential energy

Low potential energy

Electron gains potential energy and moves farther from nucleus. A photon of light is absorbed.

Electron loses potential energy and moves closer to nucleus. A photon of light is emitted.

Nucleus

Nucleus

Figure 5.12
An electron is lifted away from the nucleus as the atom it is in absorbs a photon of light and drops closer to the nucleus as the atom releases a photon of light.

Concept Check ✔

Which has more energy: a photon of red light or a photon of infrared light?

Was this your answer? As shown in Figure 5.5, red light has a higher frequency than infrared light, which means a photon of red light has more energy than a photon of infrared light. Recall that a photon is a single discrete packet (a *quantum*) of radiant energy.

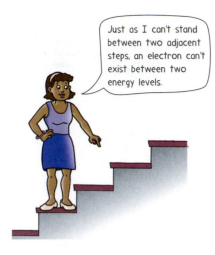

Just as I can't stand between two adjacent steps, an electron can't exist between two energy levels.

Bohr reasoned that because light energy is quantized, the energy of an electron in an atom must also be quantized. In other words, an electron cannot have just any amount of potential energy. Rather, within the atom there must be a number of distinct *energy levels*, analogous to steps on a staircase. Where you are on a staircase is restricted to where the steps are—you cannot stand at a height that is, say, halfway between any two adjacent steps. Similarly, there are only a limited number of permitted energy levels in an atom, and an electron can never have an amount of energy between these permitted energy levels. Bohr gave each energy level a **principal quantum number n**, where n is always some integer. The lowest energy level has a principal quantum number $n = 1$. An electron for which $n = 1$ is as close to the nucleus as possible, and an electron for which $n = 2$, $n = 3$, and so forth is farther away from the nucleus.

Using these ideas, Bohr developed a conceptual model in which an electron moving around the nucleus is restricted to certain distances from the nucleus, with these distances determined by the amount of energy the electron has. Bohr saw this as similar to how the planets are held in orbit around the sun at given distances from the sun. The allowed energy levels for any atom, therefore, could be graphically represented as orbits around the nucleus, as shown in Figure 5.13. Bohr's quantized model of the atom thus became known as the *planetary model*.

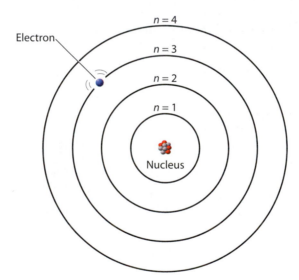

Figure 5.13
Bohr's planetary model of the atom, in which electrons orbit the nucleus much like planets orbit the sun, is a graphical representation that helps us understand how electrons can possess only certain quantities of energy.

Bohr used his planetary model to explain why atomic spectra contain only a limited number of light frequencies. According to the model, photons are emitted by atoms as electrons move from higher-energy outer orbits to lower-energy inner orbits. The energy of an emitted photon is equal to the *difference* in energy between the two orbits. Because an electron is restricted to discrete orbits, only particular light frequencies are emitted, as atomic spectra show.

Interestingly, any transition between two orbits is always instantaneous. In other words, the electron doesn't "jump" from a higher to lower orbit the way a squirrel jumps from a higher branch in a tree to a lower one. Rather, it takes no time for an electron to move between two orbits.

Bohr was serious when he stated that electrons could *never* exist *between* permitted energy levels!

Bohr was also able to explain why the sum of two frequencies of light emitted by an atom often equals a third emitted frequency. If an electron is raised to the third energy level—that is, the third highest orbit, the one for which $n = 3$—it can return to the first orbit by two routes. As shown in Figure 5.14, it can return by a single transition from the third to the first orbit, or it can return by a double transition from the third orbit to the second and then to the first. The single transition emits a photon of frequency C, and the double transition emits two photons, one of frequency A and one of frequency B. These three photons of frequencies A, B, and C are responsible for three spectral lines. Note that the energy transition for A plus B is equal to the energy transition for C. Because frequency is proportional to energy, frequency A plus frequency B equals frequency C.

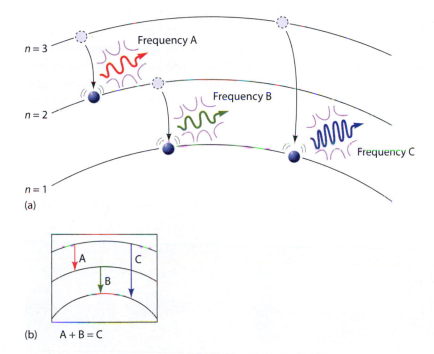

Figure 5.14
(a) The frequency of light emitted (or absorbed) by an atom is proportional to the energy difference between electron orbits. Because the energy differences between orbits are discrete, the frequencies of light emitted (or absorbed) are also discrete. The electron here can emit only three discrete frequencies of light—A, B, and C. The greater the transition, the higher the frequency of the photon emitted. (b) The sum of the energies (and frequencies) for transitions A and B equals the energy (and frequency) of transition C.

Concept Check ✔

Suppose the frequency of light emitted in Figure 5.14 is 5 billion hertz along path A and 7 billion hertz along path B. What frequency of light is emitted when an electron makes a transition along path C?

Was this your answer? Add the two known frequencies to get the frequency of path C: 5 billion hertz + 7 billion hertz = 12 billion hertz.

Bohr's planetary atomic model proved to be a tremendous success. By utilizing Planck's quantum hypothesis, Bohr's model solved the mystery of atomic spectra. Despite its successes, though, Bohr's model was limited because it did not explain why energy levels in an atom are quantized. Bohr himself was quick to point out that his model was to be interpreted only as a crude beginning, and the picture of electrons whirling about the nucleus like planets about the sun was not to be taken literally (a warning to which popularizers of science paid no heed).

5.5   Electrons Exhibit Wave Properties

If light has both wave properties and particle properties, why can't a material particle, such as an electron, also have both? This question was posed by the French physicist Louis de Broglie (1892–1987) while he was still a graduate student in 1924. His revolutionary answer was that every particle of matter is somehow endowed with a wave to guide it as it travels. The more slowly an electron moves, the more its behavior is that of a particle with mass. The more quickly it moves, however, the more its behavior is that of a wave of energy. This duality is an extension of Einstein's famous equation $E = mc^2$, which tells us that matter and energy are interconvertible (Section 4.9).

A practical application of the wave properties of fast-moving electrons is the electron microscope, which focuses not visible-light waves but rather electron waves. Because electron waves are much shorter than visible-light waves, electron microscopes are able to show far greater detail than optical microscopes, as Figure 5.15 shows.

(a)

(b)

Figure 5.15
(a) An electron microscope makes practical use of the wave nature of electrons. The wavelengths of electron beams are typically thousands of times shorter than the wavelengths of visible light, and so the electron microscope is able to distinguish detail not visible with optical microscopes. (b) Detail of a female mosquito head as seen with an electron microscope at a "low" magnification of 200 times. Note the remarkable resolution.

In an atom, an electron moves at very high speeds—on the order of 2 million meters per second—and therefore exhibits many of the properties of a wave. An electron's wave nature can be used to explain why electrons in an atom are restricted to particular energy levels. Permitted energy levels are a natural consequence of electron waves closing in on themselves in a synchronized manner.

As an analogy, consider the wire loop shown in Figure 5.16. This loop is affixed to a mechanical vibrator that can be adjusted to create waves of different wavelengths in the wire. Waves passing through the wire that meet up with themselves, as shown in Figure 5.16b, form a stationary wave pattern called a *standing wave*. This pattern results because the peaks and valleys of successive waves are perfectly matched, which makes the waves reinforce one another. With other wavelengths, as shown in Figure 5.16c, successive waves are not synchronized. As a result, the waves do not build to great amplitude.

The only waves that an electron exhibits while confined to an atom are those that are self-reinforcing. These are the ones that resemble a standing wave centered on the atomic nucleus. Each standing wave corresponds to one of the permitted energy levels. Only the frequencies of light that match

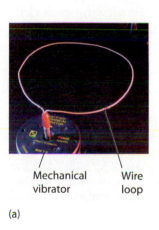

Mechanical vibrator   Wire loop

(a)

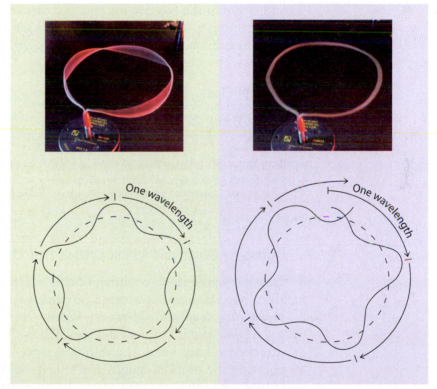

(b) Wavelength is self-reinforcing.

(c) Wavelength produces chaotic motion.

Figure 5.16
For the fixed circumference of a wire loop, only some wavelengths are self-reinforcing. (a) The loop affixed to the post of a mechanical vibrator at rest. Waves are sent through the wire when the post vibrates. (b) Waves created by vibration at particular rates are self-reinforcing. (c) Waves created by vibration at other rates are not self-reinforcing.

Hands-On Chemistry: Rubber Waves

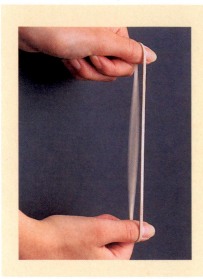

Stretch a rubber band between your two thumbs and pluck one length of it. Note that no matter where along the length you pluck, the area of greatest oscillation is always at the midpoint. This is a self-reinforcing wave that occurs as overlapping waves bounce back and forth from thumb to thumb.

Under regular light, it is difficult to see the waves traveling back and forth. For a better view, pluck the rubber band in front of a computer monitor or a television screen that uses a cathode ray tube. The light from these devices, which acts like a strobe light, makes the waves appear to slow down.

Vary the tension in the rubber band to see different effects.

the difference between any two of these permitted energy levels can be absorbed or emitted by an atom.

The wave nature of electrons also explains why they do not spiral closer and closer to the positive nucleus that attracts them. By viewing each electron orbit as a self-reinforcing wave, we see that the circumference of the smallest orbit can be no smaller than a single wavelength.

Concept Check ✔

What must an electron be doing in order to have wave properties?

Was this your answer? According to de Broglie, particles of matter behave like waves by virtue of their motion. An electron must therefore be *moving* in order to have wave properties. In atoms, electrons move at speeds of about 2 million meters per second, and so their wave nature is most pronounced.

Probability Clouds and Atomic Orbitals Help Us Visualize Electron Waves

Electron waves are three-dimensional, which makes them difficult to visualize, but scientists have come up with two ways of visualizing them: as *probability clouds* and as *atomic orbitals*.

As you saw if you just did the Hands-On exercise, when you pluck a stretched rubber band, the resulting waves are most intense at the midpoint of the plucked length and much weaker at the ends. In a similar fashion, standing electron waves in an atom are more intense in some regions than in others. In 1926, the Austrian–German scientist Erwin Schrodinger (1887–1961) formulated a remarkable equation from which the intensities of electron waves in an atom could be calculated. It was soon recognized that the intensity at any given location determined the probability of finding the electron at that location. In other words, the electron was most likely to be found where its wave intensity was greatest and least likely to be found where its wave intensity was smallest.

If we could plot the positions of an electron of a given energy over time as a series of tiny dots, the resulting pattern would resemble what is called a **probability cloud**. Figure 5.17a shows a probability cloud for hydrogen's electron. The denser a region of the cloud, the greater the probability of finding the electron in that region. The densest regions correspond to where the electron's wave intensity is greatest. A probability cloud is therefore a close approximation of the actual shape of an electron's three-dimensional wave.

An **atomic orbital**, like a probability cloud, specifies a volume of space where the electron is most likely to be found. By convention, atomic orbitals are drawn to delineate the volume inside which the electron is located 90 percent of the time. This gives the atomic orbital an apparent border, as shown in Figure 5.17b. This border is arbitrary, however, because the electron may exist on either side of it. Most of the time, though, the electron remains within the border.

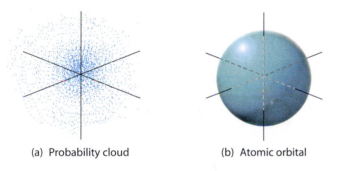

(a) Probability cloud   (b) Atomic orbital

Figure 5.17
(a) The probability cloud for hydrogen's electron. The more concentrated the dots, the greater the chance of finding the electron at that location. (b) The atomic orbital for hydrogen's electron. The electron is somewhere inside this spherical volume 90 percent of the time.

Probability clouds and atomic orbitals are essentially the same thing. They differ only in that atomic orbitals specify an outer limit, which makes them easier to depict graphically.

As Table 5.1 shows, the first four atomic orbitals are classified by the letters *s*, *p*, *d*, and *f* and they come in a variety of shapes, some quite exquisite. The simplest is the spherical *s* orbital. The *p* orbital consists of two lobes and resembles an hourglass. There are three kinds of *p* orbitals, and they differ from one another only by their orientation in three-dimensional space. The more complex *d* orbitals have five possible shapes, and the *f* orbitals have seven. Please do not feel compelled to memorize all the orbital shapes, especially the *d* and *f* ones. However, you should understand that each orbital represents a different region in which an electron of a given energy is most likely to be found.

Concept Check ✔

What is the relationship between an electron wave and an atomic orbital?

Was this your answer? The atomic orbital is an approximation of the shape of the standing electron wave surrounding the atomic nucleus.

In addition to a variety of shapes, atomic orbitals also come in a variety of sizes that correspond to different energy levels. In general, highly

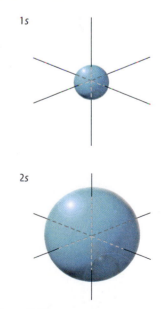

1s

2s

Figure 5.18
The 2s orbital is larger than the 1s orbital because the 2s accommodates electrons of greater energy.

Table 5.1

The First Four Atomic Orbitals: *s*, *p*, *d*, *f*

Orbital Type	Spatial Orientations

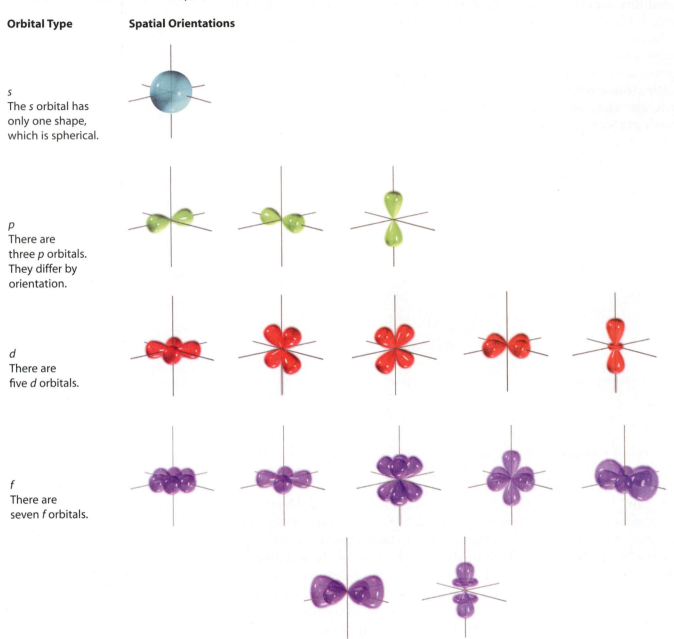

s
The *s* orbital has only one shape, which is spherical.

p
There are three *p* orbitals. They differ by orientation.

d
There are five *d* orbitals.

f
There are seven *f* orbitals.

energized electrons are able to extend farther away from the attracting nucleus, which means they are distributed over a greater volume of space. The higher the energy of the electron, therefore, the larger its atomic orbital. Because electron energies are quantized, however, the possible sizes of the atomic orbitals are quantized. The size of an orbital is thus indicated by Bohr's principal quantum number $n = 1, 2, 3, 4, 5, 6, 7$, or greater.

The first two *s* orbitals are shown in Figure 5.18. The smallest *s* orbital is the 1*s* (pronounced one-ess), where 1 is the principal quantum number. The next largest *s* orbital is the 2*s*, and so forth.

An atomic orbital is simply a volume of space within which an electron may reside. Orbitals may therefore overlap one another in an atom. As shown in Figure 5.19, the electrons of a fluorine atom are distributed among its 1*s*, 2*s*, and three 2*p* orbitals.*

The hourglass-shaped *p* orbital illustrates the significance of the wave nature of the electron. Unlike the case with a real hourglass, the two lobes of this orbital are not open to each other, and yet an electron freely moves from one lobe to the other. To understand how this can happen, consider an analogy from the macroscopic world. A guitar player can gently tap a guitar string at its midpoint (the 12th fret) and pluck it elsewhere at the same time to produce a high-pitched tone called a harmonic. Close inspection of this string, shown in Figure 5.20, reveals that it oscillates everywhere along the string except at the point directly above the 12th fret. This point of zero oscillation is called a *node*. Although there is no motion at the node, waves nonetheless travel through it. Thus, the guitar string oscillates on both sides of the node when only one side is plucked. Similarly, the point between the two lobes of a *p* orbital is a node through which the electron may pass—but only by virtue of its ability to take on the form of a wave.

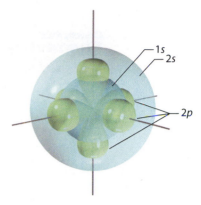

Figure 5.19
The fluorine atom has five overlapping atomic orbitals that contain its nine electrons, which are not shown.

Node

Figure 5.20
The guitar string can oscillate on both sides of the 12th-fret node even when the string is plucked on only one side of the node. This occurs because waves can pass through the node.

Concept Check ✔

Distinguish between an *orbital* and one of Bohr's *orbits*.

Was this your answer? An orbit is a *distinct path* followed by an object in its revolution around another object. In Bohr's planetary model of the atom, he proposed an analogy between electrons orbiting the atomic nucleus and planets orbiting the sun.

An atomic orbital is a *volume of space* around an atomic nucleus where an electron of a given energy will most likely be found. What orbits and orbitals have in common is that they both use Bohr's principal quantum number to indicate energy levels in an atom.

A disadvantage of Bohr's planetary atomic model is that the restriction that electrons in an atom can have only discrete energy values is introduced arbitrarily in order to account for spectral data. Atomic models based on the electron's wave behavior, on the other hand, show that discrete electron

---

* For reasons that are beyond the scope of this text, the 1*p* orbital does not exist. The smallest *p* orbital is therefore the 2*p*. Other nonexistent orbitals are the 1*d*, 2*d*, 1*f*, 2*f*, and 3*f*.

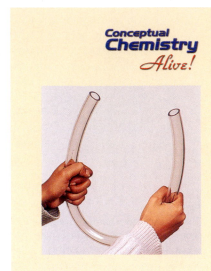

Hands-On Chemistry: Quantized Whistle

You can "quantize" your whistle by whistling down a long tube, such as the tube from a roll of wrapping paper. First, without the tube, whistle from a high pitch to a low pitch. Do it in a single breath and as loud as you can. Next, try the same thing while holding the tube to your lips. Ah, ha! Note that some frequencies simply cannot be whistled, no matter how hard you try. These frequencies are forbidden because their wavelengths are not a multiple of the length of the tube.

Try experimenting with tubes of different lengths. To hear yourself more clearly, use a flexible plastic tube and twist the outer end toward your ear.

When your whistle is confined to the tube, the consequence is a quantization of its frequencies. When an electron wave is confined to an atom, the consequence is a quantization of the electron's energy.

energy values are a natural consequence of the electron's confinement to the atom. While Bohr's planetary model accounts for the generation of light quanta, the wave model takes things a step further by treating light and matter in the same way—both behaving sometimes like a wave and sometimes like a particle. As abstract as the wave model may be, these successes indicate that it presents a more fundamental description of the atom than does Bohr's planetary model.

5.6   Energy-Level Diagrams Describe How Orbitals Are Occupied

Each orbital has a capacity for two, but no more than two, electrons. To understand how two electrically repelling electrons may coexist in the same region of space, we turn to physics. From physics we learn that electrons have a property called *spin*. Spin can have two states, and these states are analogous to a ball spinning either clockwise or counterclockwise, as shown in Figure 5.21. Two electrons that have opposite spins also have oppositely

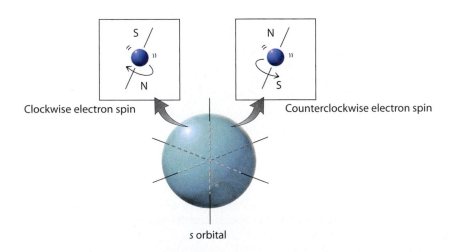

Clockwise electron spin   Counterclockwise electron spin

s orbital

Figure 5.21
Two electrons having opposite spins may pair together in an atomic orbital.

aligned magnetic fields, which are mutually attractive. This compensates for the electrical repulsion between the electrons.*

We can combine this spin concept and our orbital model to "build" atoms electron by electron. This is done by using what is called an **energy-level diagram**, shown in Figure 5.22. Each box represents an orbital, each electron is represented by an arrow, and two electrons spinning in opposite directions in the same orbital are shown as two arrows pointing in opposite directions.

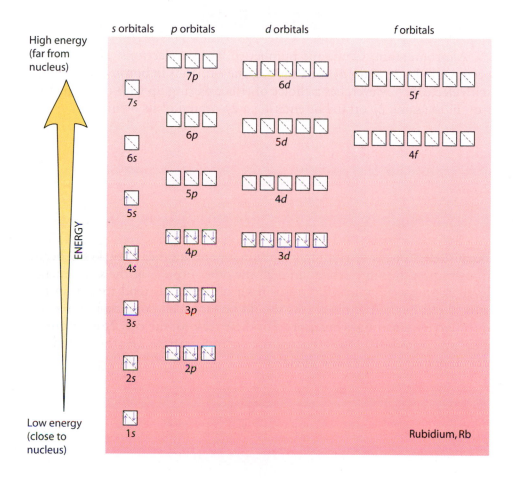

Figure 5.22
This energy-level diagram shows the relative energy levels of atomic orbitals in a multielectron atom (in this case rubidium, Rb, atomic number 37).

Consider the lithium atom, Li (atomic number 3), which has three electrons. In which orbitals do you suppose lithium's electrons are most likely to be found? As indicated in Figure 5.22, the lower a box is located in the diagram, the lower the energy of the orbital it represents. Low-energy orbitals are the ones that allow the electrons to get closest to the nucleus. Thus, these low-energy orbitals are the ones that tend to get filled first. Accordingly, a lithium atom in its lowest energy state, as depicted at left, has two electrons filling the 1s orbital and the third electron in the 2s orbital.

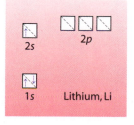

---

* This explanation is not fully accurate. A more accurate explanation, however, is beyond the scope of this text.

A boron atom, B (atomic number 5), in its lowest energy state has four of its five electrons filling the 1*s* and 2*s* orbitals. Its fifth electron may reside in any one of the 2*p* orbitals, all of which are at same energy level:

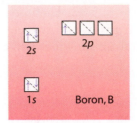

A carbon atom, C (atomic number 6), has six electrons. Five of them occupy the 1*s*, 2*s*, and 2*p* orbitals just as the electrons in boron do. Carbon's sixth electron, however, has a choice of either pairing up with the fifth electron in the same 2*p* orbital or entering a 2*p* orbital of its own:

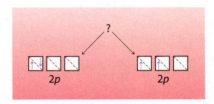

Because electrons have a natural repulsion for one another, they do not begin to pair up in the same orbital until all the other orbitals at the same energy level are singly occupied. Electrons in separate orbitals tend to spin in the same direction, and so the arrows should be shown all pointing in the same direction until pairing is necessary. For these reasons, the two 2*p* electrons of a carbon atom in its lowest energy state are in separate 2*p* orbitals and are drawn pointing in the same direction:

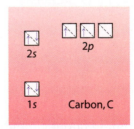

There is no pairing of 2*p* electrons in the nitrogen atom, N (atomic number 7), which has seven electrons. The oxygen atom, O (atomic number 8), however, has eight electrons, two of which are forced to pair up in one 2*p* orbital (it doesn't matter which one).

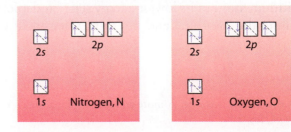

How electrons enter orbitals of the same energy level, such as the three 2*p* orbitals, is not unlike a bunch of strangers boarding a bus with double seats. Pretend these strangers prefer to occupy double seats alone. Only after all the seats are singly occupied do the strangers begin to pair up.

The arrangement of electrons in the orbitals of an atom is called the atom's **electron configuration**. The electron configuration of any atom can be shown in an energy-level diagram like Figure 5.22 by placing that atom's electrons in the orbitals in order of increasing energy level. Also, remember to pair electrons of opposite spin in an orbital only when necessary.

Concept Check ✓

1. How many 3*d* orbitals are there?
2. Fill in this energy-level diagram for sodium, Na (atomic number 11):

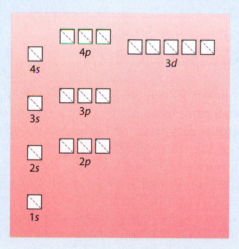

Were these your answers?

1. There are five 3*d* orbitals, each represented by one box in an energy-level diagram. These 3*d* orbitals differ from one another by their spatial orientations and shapes, as shown in Table 5.1.
2. Begin with the lowest-energy orbital, which is the 1*s*. Place two electrons in each orbital, using oppositely oriented arrows to represent the opposite spins of the electrons. Sodium's 11 electrons fill the 1*s*, 2*s*, and all three 2*p* orbitals. The 11th electron resides alone in the 3*s* orbital.

An abbreviated way of presenting electron configuration is to write the principal quantum number and letter of each occupied orbital and then use a superscript to indicate the number of electrons in each orbital. The orbitals of each atom are then written in order of increasing energy levels. For the group 1 elements, this notation is

Hydrogen, H	$1s^1$
Lithium, Li	$1s^2 2s^1$
Sodium, Na	$1s^2 2s^2 2p^6 3s^1$
Potassium, K	$1s^2 2s^2 2p^6 3s^2 3p^6 4s^1$
Rubidium, Rb	$1s^2 2s^2 2p^6 3s^2 3p^6 4s^2 3d^{10} 4p^6 5s^1$
Cesium, Cs	$1s^2 2s^2 2p^6 3s^2 3p^6 4s^2 3d^{10} 4p^6 5s^2 4d^{10} 5p^6 6s^1$
Francium, Fr	$1s^2 2s^2 2p^6 3s^2 3p^6 4s^2 3d^{10} 4p^6 5s^2 4d^{10} 5p^6 6s^2 4f^{14} 5d^{10} 6p^6 7s^1$

Note that all the superscripts for an atom must add up to the total number of electrons in the atom—1 for hydrogen, 3 for lithium, 11 for sodium, and so forth. Also note that the orbitals are not always listed in order of principal quantum number. The $4s$ orbital, for example, is lower in energy than the $3d$ orbitals, as is indicated on the energy-level diagram of Figure 5.22. The $4s$ orbital, therefore, appears *before* the $3d$ orbital.

The properties of an atom are determined mostly by its outermost electrons, the ones farthest from the nucleus. These are the electrons toward the "outer surface" of the atom and hence the ones in direct contact with the external environment. Elements that have similar electron configurations in their outermost orbitals, therefore, have similar properties. For example, in the alkaline metals of group 1, shown above, the outermost occupied orbital (shown in blue) is an *s* orbital containing a single electron. In general, elements in the same group of the periodic table have similar electron configurations in the outermost orbitals, which explains why elements in the same group have similar properties—a concept first presented in Section 2.6.

5.7   Orbitals of Similar Energies Can Be Grouped into Shells

Orbitals having comparable energies can be grouped together. As shown in Figure 5.23, no other orbital has energy similar to that of the $1s$ orbital, and so this orbital is grouped by itself. The energy level of the $2s$ orbital, however, is very close to the energy level of the three $2p$ orbitals, and so these four orbitals are grouped together. Likewise, the $3s$ and three $3p$ orbitals, the $4s$, five $3d$, and three $4p$ orbitals, and so on. The result is a set of seven distinct horizontal rows of orbitals.

The seven rows in Figure 5.23 correspond to the seven periods in the periodic table, with the bottom row corresponding to the first period, the next row up from the bottom corresponding to the second period, and so on. Furthermore, the maximum number of electrons each row can hold is equal to the number of elements in the corresponding period. The bottom row in Figure 5.23 can hold a maximum of two electrons, and so there are only two elements, hydrogen and helium, in the first period of the periodic table. The second and third rows up from the bottom each have a capacity for eight electrons, and so eight elements are found in both the second and

High energy

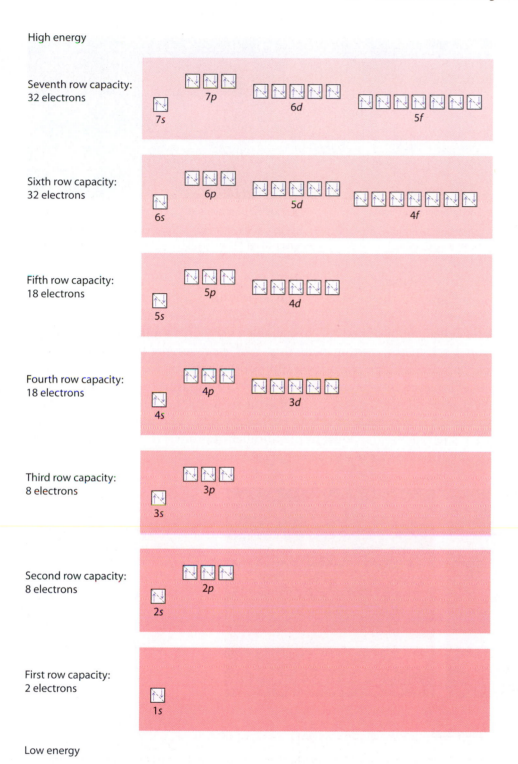

Seventh row capacity:
32 electrons

Sixth row capacity:
32 electrons

Fifth row capacity:
18 electrons

Fourth row capacity:
18 electrons

Third row capacity:
8 electrons

Second row capacity:
8 electrons

First row capacity:
2 electrons

Low energy

Figure 5.23
Orbitals of comparable energy levels can be grouped together to give rise to a set of seven rows of orbitals.

third periods. Continue analyzing Figure 5.23 in this way, and you will find 18 elements in the fourth and fifth periods, and 32 elements in the sixth and seventh periods. (As of this writing, the discovery of only 26 seventh-period elements has been confirmed.)

Recall from Section 5.5 that the higher the energy level of an orbital, the farther away an electron in that orbital is located from the nucleus. Electrons in the same row of orbitals in Figure 5.23, therefore, are roughly

Figure 5.24
The second row of orbitals, which consists of the 2s and three 2p orbitals, can be represented either as a single smooth spherical shell or as a cross-section of such a shell.

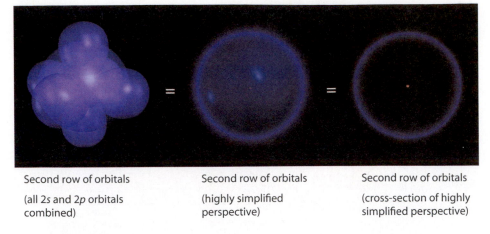

Second row of orbitals

(all 2s and 2p orbitals combined)

Second row of orbitals

(highly simplified perspective)

Second row of orbitals

(cross-section of highly simplified perspective)

the same distance from the nucleus. Graphically, this can be represented by converging all the orbitals in a given row into a single three-dimensional hollow shell, as shown in Figure 5.24. Each shell is a graphic representation of a collection of orbitals of comparable energy in a multielectron atom. As you'll see in the next section, this *shell model* of the atom allows us to explain much about the organization of the periodic table.

The seven rows of orbitals in Figure 5.23 can thus be represented either by a series of seven concentric shells or by a series of seven cross-sectional circles of these shells, as shown in Figure 5.25. The number of electrons each shell can hold is equal to the number of orbitals it contains multiplied by two (because there can be two electrons per orbital).

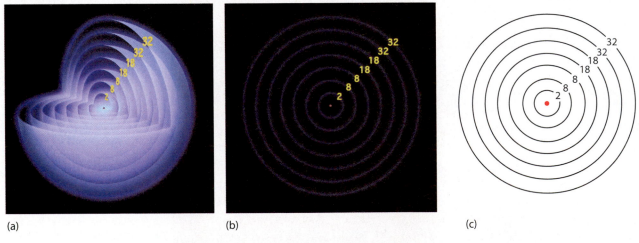

(a) (b) (c)

Figure 5.25
(a) A cutaway view of the seven shells, with the number of electrons each shell can hold indicated. (b) A two-dimensional, cross-sectional view of the shells. (c) An easy-to-draw cross-sectional view that resembles Bohr's planetary model.

You fill in electrons in a shell diagram just as in an energy-level diagram— electrons first fill the shells closest to the nucleus. Also, in accordance with the strangers-on-a-bus analogy, electrons do not begin to pair in a shell until the shell is half filled. Figure 5.26 shows how this works for the first three periods. As with energy-level diagrams, there is one shell for each period, and the number of elements in a period is equal to the maximum number of electrons the shells representing that period can hold.

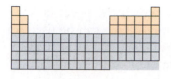

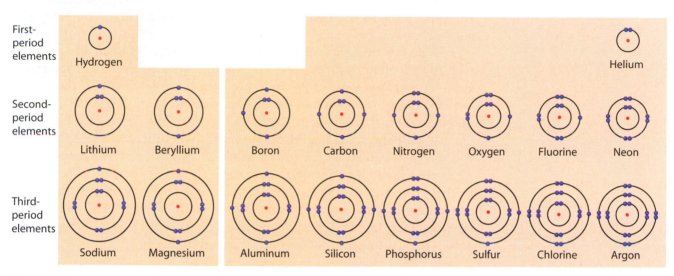

Figure 5.26
The first three periods of the periodic table according to the shell model. Elements in the same period have electrons in the same shells. Elements in the same period differ from one another by the number of electrons in the outermost shell.

Concept Check ✓

How many orbitals make up the fourth shell? What is the electron capacity of this shell?

Were these your answers? There are nine orbitals in the fourth shell. In order of increasing energy level, they are the one $4s$ orbital, the five $3d$ orbitals, and the three $4p$ orbitals. Because each orbital can hold two electrons, the total electron capacity of the fourth shell is $2 \times 9 = 18$ electrons, which is the same number of elements found in the fourth period of the periodic table.

In the next section, we explore how the shell model can be used to explain periodic trends. An even further simplified shell model, known as *electron-dot structure*, is then developed in Chapter 6 to assist you in understanding chemical bonding. As you use these models, please keep in mind that electrons are not really confined to the "surface" of one shell or another. Instead, any thorough description of electrons in an atom must involve the orbitals these shells represent. For the purpose of an elementary understanding of chemistry, however, the simplified shell model is very useful.

Two-time Nobel laureate Linus Pauling (1901–1994) was an early proponent of teaching beginning chemistry students a shell model from which the organization of the periodic table could be described. In this model, as is described here, orbitals are grouped according to energy level. However, this shell model differs from that found in advanced physics and chemistry textboks, which identify a shell as a group of orbitals that all have the same principal quantum number.

5.8   The Periodic Table Helps Us Predict Properties of Elements

Larger particles of sand can be separated from smaller particles by tossing the sand up and down over a wire-mesh screen. All particles larger than the holes in the screen stay above the screen, and all particles smaller than the holes pass through. Now imagine a membrane in which the pores (analogous to the holes in a screen) are so tiny that the membrane can be used to separate two different-sized molecules, say nitrogen, N_2, and oxygen, O_2. This would be an incredible feat because the diameters of these molecules differ by no more than 0.02 nanometer (2×10^{-11} meter). Which would you expect to pass through such a membrane more readily: nitrogen molecules or oxygen molecules?

To answer this question, you need look no further than the periodic table, which allows you to make fairly accurate predictions about the properties both of atoms and of the molecules they form. For instance, the farther to the left and lower down in the table an element is, the larger the atoms of that element are. Conversely, the farther to the right and higher up in the table an element is, the smaller the atoms of that element are. Knowing this, you can predict that oxygen atoms, being in the same row *but farther to the right*, are smaller than nitrogen atoms, which means oxygen molecules are smaller than nitrogen molecules. The smaller oxygen molecules, therefore, would pass through the membrane more readily. In fact, such membranes exist, and they are being developed as a cost-effective means of separating atmospheric nitrogen from oxygen, as shown in Figure 5.27.

Figure 5.27
The composition of air passing through this membrane changes from 21 percent oxygen to up to 44 percent oxygen because the larger nitrogen molecules get left behind.

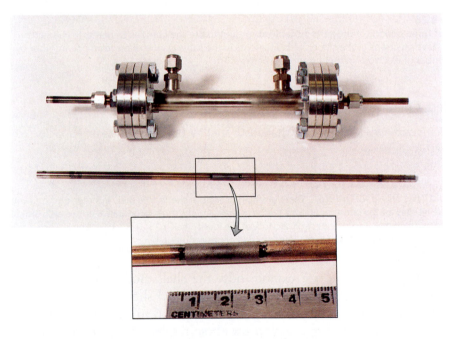

Recall from Chapter 2 that a gradual change in properties as we move in any direction in the periodic table is called a *periodic trend*. Most periodic trends can be understood from the perspective of the simplified shell model, and underlying most trends are two important concepts: *inner-shell shielding* and *effective nuclear charge*.

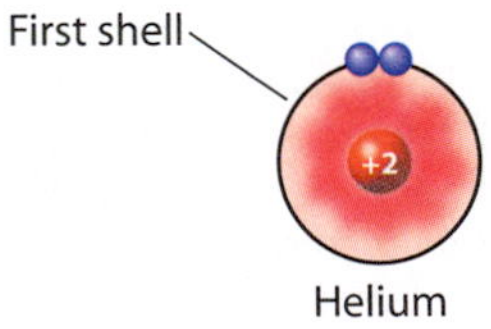

Imagine you are one of the two electrons in the shell of a helium atom. You share this shell with one other electron, but that electron doesn't affect your attraction to the nucleus because you both have the same "line of sight" to the nucleus. As shown in Figure 5.28, you and your neighbor electron sense a nucleus of two protons, and the two of you are equally attracted to it.

The situation is different for atoms beyond helium, which have more than one shell occupied by electrons. In these cases, inner-shell electrons

First shell

Helium

Figure 5.28
The two electrons in a helium atom have equal exposure to the nucleus; hence, they experience the same degree of attraction, represented by the pink shading in the space between the nucleus and the shell boundary.

weaken the attraction between outer-shell electrons and the nucleus. Imagine, for example, you are that second-shell electron in the lithium atom shown in Figure 5.29. Looking toward the nucleus, what do you sense? Not just the nucleus but also the two electrons in the first shell. These two inner electrons, with their negative charge repelling your negative charge, have the effect of weakening your electrical attraction to the nucleus. This is **inner-shell shielding**—inner-shell electrons shield electrons farther out from some of the attractive pull exerted by the positively charged nucleus.

Because inner-shell electrons diminish the attraction outer-shell electrons have for the nucleus, the nuclear charge sensed by outer-shell electrons is always less than the actual charge of the nucleus. This diminished nuclear charge experienced by outer-shell electrons is called the **effective nuclear charge** and is abbreviated Z^* (pronounced zee-star), where Z stands for the nuclear charge and the asterisk indicates this charge appears to be less than it actually is. The second-shell electron in lithium, for example, does not sense the full effect of lithium's $+3$ nuclear charge (there are three protons in the nucleus of lithium). Instead, the total charge on the first-shell electrons, -2, subtracts from the charge of the nucleus, $+3$, to give an effective nuclear charge of $+1$ sensed by the second-shell electron.

For most elements, subtracting the total number of inner-shell electrons from the nuclear charge provides a convenient estimate of the effective nuclear charge, as Figure 5.30 illustrates.

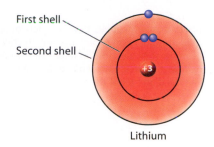

First shell
Second shell

Lithium

Figure 5.29
Lithium's two first-shell electrons shield the second-shell electron from the nucleus. The nuclear attraction, again represented by pink shading, is less intense in the second shell.

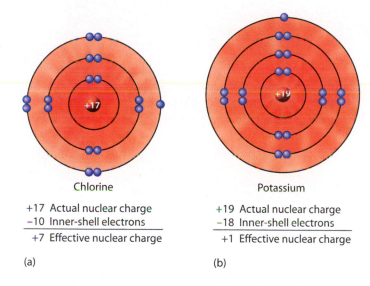

Chlorine

+17 Actual nuclear charge
−10 Inner-shell electrons
―――――――――――
+7 Effective nuclear charge

(a)

Potassium

+19 Actual nuclear charge
−18 Inner-shell electrons
―――――――――――
+1 Effective nuclear charge

(b)

Figure 5.30
(a) A chlorine atom has three occupied shells. The $2 + 8 = 10$ electrons of the inner two shells shield the 7 electrons of the third shell from the $+17$ nucleus. The third-shell electrons therefore experience an effective nuclear charge of $17 - 10 = +7$. (b) In a potassium atom, the fourth-shell electron experiences an effective nuclear charge of $19 - 18 = +1$.

The Smallest Atoms Are at the Upper Right of the Periodic Table

From left to right across any row of the periodic table, the atomic diameters get *smaller*. Let's look at this trend from the point of view of effective nuclear charge. Consider lithium's outermost electron, which experiences an effective nuclear charge of $+1$. Then look across period 2 to neon, where each outermost electron experiences an effective nuclear charge of $+8$, as Figure 5.31 shows. Because the outer-shell neon electrons experience a greater attraction to the nucleus, they are pulled in closer to it than is the outer-shell electron in lithium. So neon, although nearly three times as massive as

Figure 5.31
Lithium's outermost electron experiences an effective nuclear charge of +1, while those of neon experience an effective nuclear charge of +8. As a result, the outer-shell electrons in neon are closer to the nucleus than is the outer-shell electron in lithium, and so the diameter of the neon atom is smaller than the diameter of the lithium atom.

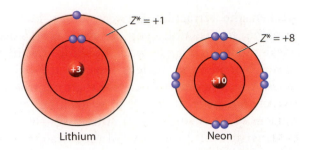

Lithium   Neon

lithium, has a considerably smaller diameter. In general, across any period from left to right, atomic diameters become smaller because of an increase in effective nuclear charge. Look back to Figure 5.26 and you will see this trend illustrated for the first three periods. In addition, Figure 5.32 shows relative atomic diameters as estimated from experimental data. Note there are some exceptions to this trend, especially between groups 12 and 13.

Moving down a group, atomic diameters get larger because of an increasing number of occupied shells. Whereas lithium has a small diameter because it has only two occupied shells, francium has a much larger diameter because it has seven occupied shells.

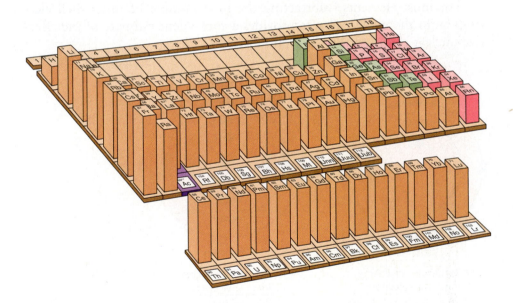

Figure 5.32
Relative atomic diameters indicated by height. Note that atomic size generally decreases in moving to the upper right of the periodic table.

Concept Check ✔

Which is larger, a sulfur atom, S (atomic number 16), or an arsenic atom, As (atomic number 33)? Consult the periodic table that appears on the inside front cover of this textbook.

Was this your answer? The arsenic atom is larger because it is positioned closer to the lower left corner of the periodic table. Note that you didn't need to memorize some long list of atomic sizes nor look to Figure 5.32 in order to answer this question. Instead, you were able to use a common periodic table as a tool to help you find the answer.

The Smallest Atoms Have the Most Strongly Held Electrons

How strongly electrons are bound to an atom is another property that changes gradually across the periodic table. In general, the trend is that the smaller the atom, the more tightly bound its electrons.

As discussed earlier, effective nuclear charge increases in moving from left to right across any period. Thus not only are atoms toward the right in any period smaller, but their electrons are held more strongly. It takes about four times as much energy to remove an outer electron from a neon atom, for example, than to remove the outer electron from a lithium atom.

Moving down any group, the effective nuclear charge generally stays the same. The effective nuclear charge for all group 1 elements, for example, is about +1. Because of their greater number of shells, however, elements toward the bottom of a group are larger than elements toward the top of the group. The electrons in the outermost shell are therefore farther from the nucleus by an appreciable distance. From physics we learn that the electric force weakens rapidly with increasing distance. As Figure 5.33 illustrates, an outer-shell electron in a larger atom, such as cesium, is not held as tightly as an outer-shell electron in a smaller atom, such as lithium. As a consequence, the energy needed to remove the outer electron from a cesium atom is about half the energy needed to remove the outer electron from a lithium atom.

The combination of increasing effective nuclear charge from left to right and increasing number of shells from top to bottom creates a periodic

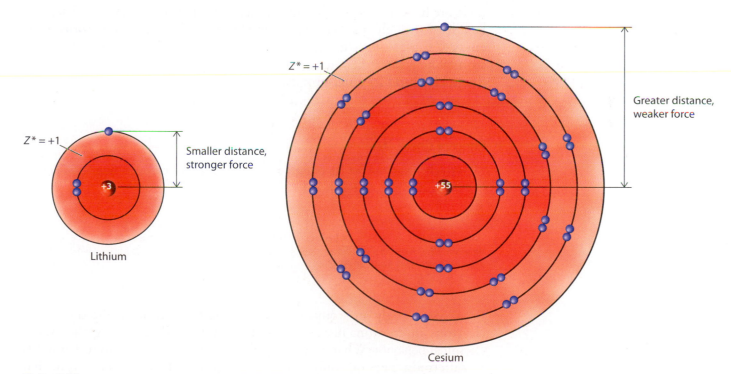

Figure 5.33
In both lithium and cesium, the outermost electron experiences an effective nuclear charge of +1. The outermost electron in a cesium atom, however, is not held as strongly to the nucleus because of its greater distance from the nucleus.

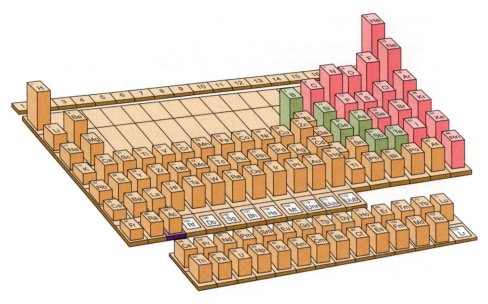

Figure 5.34
Trends in ionization energy. The attraction an atomic nucleus has for the outermost electrons in an atom indicated by height. Note that atoms at the upper right tend to have the greatest ionization energy and those at the lower left the least.

trend in which the electrons in atoms at the upper right of the periodic table are held most strongly and the electrons in atoms at the lower left are held least strongly. This is reflected in Figure 5.34, which shows **ionization energy**, the amount of energy needed to pull an electron away from an atom. The greater the ionization energy, the greater the attraction between the nucleus and its outermost electrons.

Concept Check ✔

Which loses one of its outermost electrons more easily: a francium, Fr, atom, or a helium, He, atom?

Was this your answer? A francium, Fr, atom loses electrons much more easily than does a helium, He, atom. Why? Because a francium atom's electrons are not held so tightly by its nucleus, which is buried deep beneath many layers of shielding electrons.

How strongly an atomic nucleus is able to hold on to the outermost electrons in an atom plays an important role in determining the atom's chemical behavior. What do you suppose happens when an atom that holds its outermost electrons only weakly comes into contact with an atom that has a very strong pull on its outermost electrons? As we explore in Chapter 6, either the atom that pulls strongly may swipe one or more electrons from the other atom or the two atoms may share electrons.

In Perspective

In this chapter we have gone into a fair amount of detail regarding atomic models. We discussed how electrons are arranged around an atomic nucleus. Rather than moving in neat orbits like planets around the sun, electrons are wavelike entities that swarm in various volumes of space called *atomic orbitals*. Furthermore, atomic orbitals of comparable energy can be grouped together and represented by a single *shell*. Such a shell should not be taken literally. Rather, the shell represents a region of space within which electrons of similar energy are most likely to be found.

Remember that these models are not to be interpreted as actual representations of the atom's physical structure. Rather, they serve as tools to help us understand and predict how atoms behave in various circumstances. These models, therefore, are the foundation of chemistry and the key to a richer understanding of the atomic and molecular environment that surrounds us.

Key Terms and Matching Definitions

_____ atomic orbital
_____ atomic spectrum
_____ conceptual model
_____ effective nuclear charge
_____ electromagnetic spectrum
_____ electron configuration
_____ energy-level diagram
_____ inner-shell shielding
_____ ionization energy
_____ photon
_____ physical model
_____ principal quantum number *n*
_____ probability cloud
_____ quantum
_____ quantum hypothesis
_____ spectroscope
_____ wave frequency
_____ wavelength

1. A representation of an object on some convenient scale.
2. A representation of a system that helps us predict how the system behaves.
3. The distance between two crests of a wave.
4. A measure of how rapidly a wave oscillates. The higher this value, the greater the amount of energy in the wave.
5. The complete range of waves, from radio waves to gamma rays.
6. A device that uses a prism or diffraction grating to separate light into its color components.
7. The pattern of frequencies of electromagnetic radiation emitted by the atoms of an element, considered to be an element's "fingerprint."
8. The idea that light energy is contained in discrete packets called quanta.
9. A small, discrete packet of light energy.
10 Another term for a single quantum of light, a name chosen to emphasize the particulate nature of light.
11. An integer that specifies the quantized energy level of an atomic orbital.
12. The pattern of electron positions plotted over time to show the likelihood of an electron's being at a given position at a given time.
13. A region of space in which an electron in an atom has a 90 percent chance of being located.

14. Drawing used to arrange atomic orbitals in order of energy levels.
15. The arrangement of electrons in the orbitals of an atom.
16. The tendency of inner-shell electrons to partially shield outer-shell electrons from the nuclear charge.
17. The nuclear charge experienced by outer-shell electrons, diminished by the shielding effect of inner-shell electrons.
18. The amount of energy required to remove an electron from an atom.

Review Questions

Models Help Us Visualize the Invisible World of Atoms

1. If a baseball were the size of the Earth, about how large would its atoms be?
2. When we use a scanning tunneling microscope, do we see atoms directly or do we see them only indirectly?
3. Why are atoms invisible to visible light?
4. What is the difference between a physical model and a conceptual model?
5. What is the function of an atomic model?

Light Is a Form of Energy

6. Does visible light constitute a large or small portion of the electromagnetic spectrum?
7. Why does ultraviolet light cause more damage to our skin than visible light?
8. As the frequency of light increases, what happens to its energy?
9. What does a spectroscope do to the light coming from an atom?

Atoms Can Be Identified by the Light They Emit

10. What causes an atom to emit light?
11. Why do we say atomic spectra are like fingerprints of the elements?
12. What did Rydberg note about the atomic spectrum of hydrogen?

Niels Bohr Used the Quantum Hypothesis to Explain Atomic Spectra

13. What was Planck's quantum hypothesis?

14. Which has more potential energy: an electron close to an atomic nucleus or one far from an atomic nucleus?

15. What happens to an electron as it absorbs a photon of light?

16. What is the relationship between the light emitted by an atom and the energies of the electrons in the atom?

17. Did Bohr think of his planetary model as an accurate representation of what an atom looks like?

Electrons Exhibit Wave Properties

18. About how fast does an electron travel around the atomic nucleus?

19. How does the speed of an electron change its fundamental nature?

20. Who developed the equation that relates the intensity of an electron's wave to the electron's most probable location?

21. How is an atomic orbital similar to a probability cloud?

Energy-Level Diagrams Describe How Orbitals Are Occupied

22. How many electrons may reside in a single orbital?

23. How many $2p$ orbitals are there, and what is the total number of electrons they can hold?

24. What atom has the electron configuration $1s^2\ 2s^2\ 2p^6$?

25. Which electrons are most responsible for the physical and chemical properties of an atom?

26. Using the abbreviated notation, give the electron configuration for strontium, Sr (atomic number 38).

Orbitals of Similar Energies Can Be Grouped into Shells

27. What do the orbitals in a shell have in common?

28. The shell model presented in this book is not very accurate. Why then is it presented?

29. How many orbitals are there in the third shell?

30. How is the number of shells an atom of a given element contains related to the row of the periodic table in which that element is found?

31. What is the relationship between the maximum number of electrons each shell can hold and the number of elements in each period of the periodic table?

The Periodic Table Helps Us Predict Properties of Elements

32. How would you know from looking at the periodic table that oxygen, O (atomic number 8), molecules are smaller than nitrogen, N (atomic number 7), molecules?

33. The nucleus of a carbon atom, C (atomic number 6), has a charge of +6, but this is not the charge sensed by electrons in carbon's outer shell. Why?

34. How many shells are occupied by electrons in a gold atom, Au (atomic number 79)?

35. Based on the periodic trend of atomic diameter, which should be larger, an atom of technetium, Tc (atomic number 43), or an atom of tantalum, Ta (atomic number 73)?

36. What is the effective nuclear charge for an electron in the outermost shell of a fluorine atom, F (atomic number 9)? How about one in the outermost shell of a sulfur atom, S (atomic number 16)?

37. Why is it more difficult for fluorine to lose an electron than for sulfur to do so?

Hands-On Chemistry Insights

Spectral Patterns

The diffraction gratings used in rainbow glasses have lines etched vertically and horizontally, which makes the colors appear to the left and right, above and below, and in all corners as well. A compact disk behaves as a diffraction grating because its surface contains many rows of microscopic pits.

To the naked eye, a glowing element appears as only a single color. However, this color is an average of the many different visible frequencies the element is emitting. Only with a device such as a spectroscope are you able to discern the different frequencies. So when you look at an atomic spectrum, don't get confused and think that each frequency of light (color) corresponds to a different element. Instead, remember that what you are looking at is all the frequencies of light emitted by a single element as its electrons make transitions back and forth between energy levels.

Not all elements produce discrete-line patterns in the visible spectrum. Tungsten, for example, produces the full spectrum of colors (white light), which makes it useful as the glowing component of a car's headlights, as shown in the photograph below. Also, the sunlight reflecting off the moon, also shown below, is so bright and contains the glow of so many different elements that it, too, appears as a broad spectrum.

Rubber Waves

A self-reinforcing wave may sound beautiful on a guitar, but it can spell disaster for a bridge. In 1940, light winds across the Tacoma Narrows in the state of Washington caused the newly constructed Tacoma Narrows Bridge to start oscillating at a frequency that allowed the waves to be self-reinforcing. As the energy of the wind was absorbed by the bridge, the waves grew stronger (over the course of several days), to the point where the bridge collapsed. One of the tasks of building a durable structure, therefore, is to design it such that self-reinforcing waves are not likely to form.

Quantized Whistle

People watching you perform this activity may not believe that the audible "steps" of your whistling down the tube are not intentional. Explain quantization to them before allowing them to attempt this activity for themselves. Try to count the number of steps in your tubular whistle, understanding that each step is analogous to an energy level in an atom. Does a longer tube create fewer or more steps than a shorter tube? Why is it so difficult to whistle down a garden hose?

If you punch a few holes along the tube, you alter the frequencies of the standing waves that can form in the tube, with the result that different pitches are produced. This is the underlying principle in such musical instruments as flutes and saxophones.

Exercises

1. With scanning tunneling microscopy (STM) technology, we see not actual atoms, but rather images of them. Explain.

2. Why is it not possible for an STM to make images of the inside of an atom?

3. Would you use a physical model or a conceptual model to describe the following: brain, mind, solar system, birth of universe, stranger, best friend, gold coin, dollar bill, car engine, virus, spread of sexually transmitted disease?

4. How might you distinguish a sodium-vapor lamp from a mercury-vapor lamp?

5. How can a hydrogen atom, which has only one electron, create so many spectral lines?

6. Suppose a certain atom has four energy levels. Assuming that all transitions between levels are possible, how many spectral lines will this atom exhibit? Which transition corresponds to the highest-energy light emitted? Which corresponds to the lowest-energy light emitted?

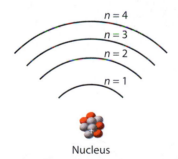

Nucleus

7. An electron drops from the fourth energy level in an atom to the third level and then to the first level. Two frequencies of light are emitted. How does their combined energy compare with the energy of the single frequency that would be emitted if the electron dropped from the fourth level directly to the first level?

8. Figure 5.14 shows three energy-level transitions that produce three spectral lines in a spectroscope. Note that the distance between the $n = 1$ and $n = 2$ levels is greater than the distance between the $n = 2$ and $n = 3$ levels. Would the number of spectral lines produced change if the distance between the $n = 1$ and $n = 2$ levels were exactly the same as the distance between the $n = 2$ and $n = 3$ levels?

9. Which color of light comes from a greater energy transition, red or blue?

10. How does the wave model of electrons orbiting the nucleus account for the fact that the electrons can have only discrete energy values?

11. What might the spectrum of an atom look like if the atom's electrons were not restricted to particular energy levels?

12. How does an electron get from one lobe of a p orbital to the other?

13. Light is emitted as an electron transitions from a higher-energy orbital to a lower-energy orbital. How long does it take for the transition to take place? At what point in time is the electron found between the two orbitals?

14. Why is there only one spatial orientation for the s orbital?

15. In which lobe of a p orbital is an electron more likely found?

16. Fill in these three energy-level diagrams. Why do these three elements have such similar chemical properties?

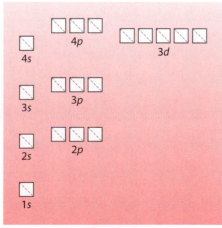

Oxygen, O

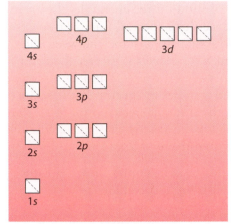

Sulfur, S

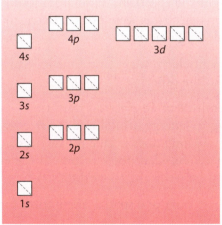

Selenium, Se

17. When does a carbon atom contain more energy—when its electrons are in the configuration on the left or when they are in the configuration on the right:

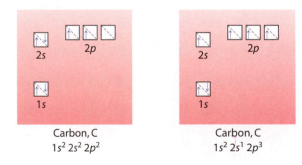

Carbon, C
$1s^2\,2s^2\,2p^2$

Carbon, C
$1s^2\,2s^1\,2p^3$

18. Write the electron configuration for uranium, U (atomic number 92), in abbreviated notation.

19. List these electron configurations for fluorine in order of increasing energy for the atom:

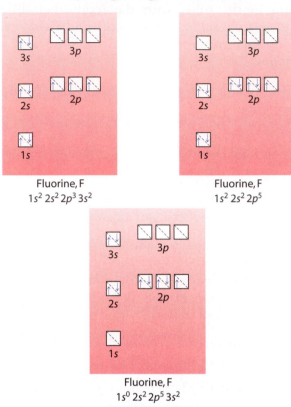

Fluorine, F
$1s^2\,2s^2\,2p^3\,3s^2$

Fluorine, F
$1s^2\,2s^2\,2p^5$

Fluorine, F
$1s^0\,2s^2\,2p^5\,3s^2$

20. What do the electron configurations for the group 18 noble gases have in common?

21. Place the proper number of electrons in each shell:

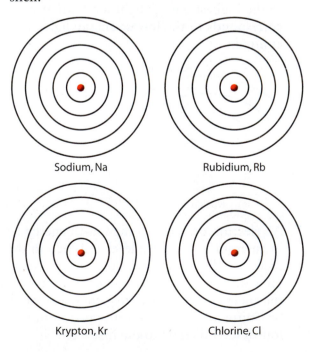

Sodium, Na

Rubidium, Rb

Krypton, Kr

Chlorine, Cl

22. Which element is represented in Figure 5.25 if all seven shells are filled to capacity?

23. Does an orbital or shell have to contain electrons in order to exist?

24. Why does an electron in a 7s orbital have more energy than one in a 1s orbital?

25. Neon, Ne (atomic number 10), has a relatively large effective nuclear charge, and yet it cannot attract any additional electrons. Why?

26. Which experiences a greater effective nuclear charge, an electron in the outermost occupied shell of neon or one in the outermost occupied shell of sodium? Why?

27. An electron in the outermost occupied shell of which element experiences the greatest effective nuclear charge?
    a. sodium, Na
    b. potassium, K
    c. rubidium, Rb
    d. cesium, Cs
    e. all experience the same effective nuclear charge

28. List the following atoms in order of increasing atomic size: thallium, Tl; germanium, Ge; tin, Sn; phosphorus, P:

    _____ < _____ < _____ < _____
    (smallest)   (largest)

29. Arrange the following atoms in order of increasing ionization energy: tin, Sn; lead, Pb; phosphorus, P; arsenic, As:

    _____ < _____ < _____ < _____
    (least)   (most)

30. Which of the following concepts underlies all the others: ionization energy, effective nuclear charge, atomic size?

31. It is relatively easy to pull one electron away from a potassium atom but very difficult to remove a second one. Use the shell model and the idea of effective nuclear charge to explain why.

32. Another interesting periodic trend is density. Osmium, Os (atomic number 76), has the greatest density of all elements, and, with some exceptions, the closer an element is to osmium in the periodic table, the greater its density. Use this trend to list the following elements in order of increasing density: copper, Cu; gold, Au; platinum, Pt; and silver, Ag:

    _____ < _____ < _____ < _____
    (least dense)   (most dense)

33. How is the following graphic similar to the energy-level diagram of Figure 5.23? Use it to explain why a gallium atom, Ga (atomic number 31), is larger than a zinc atom, Zn (atomic number 30):

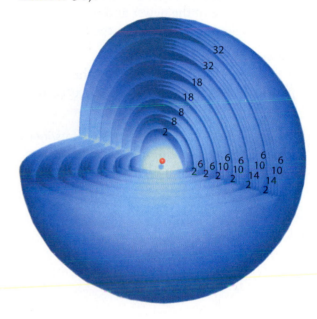

Exploring Further

George Gamow, *Thirty Years That Shook Physics.*
New York: Dover, 1985.

> A historical tracing of quantum theory by
> someone who was part of its development.

G. J. Milburn, *Schrodinger's Machines.* New York:
W. H. Freeman, 1997.

> Our understanding of quantum theory has
> already led to a number of society-shaping inven-
> tions, such as the transistor, a basic component
> of all computers, and the laser, which scans every-
> thing from our groceries to our music. This book
> presents some of the newer and fantastic quantum
> technologies we can expect in the next 50 years.

http://www.achilles.net/~jtalbot

> A superb site for learning about the spectral
> patterns of stars and how they are used to study
> the universe.

http://www.achilles.net/~jtalbot/data/elements/index.html

> Here is where you will find high-resolution
> spectral patterns of a variety of elements.

http://www.physics.purdue.edu/nanophys

> Site of the nanoscale physics laboratory of
> Purdue University, where they look at things
> that are really, really, really, really, really, really,
> really small. Lots of pretty pictures.

http://www.superstringtheory.com

> If you think the wave nature of the electron
> is bizarre, explore this site for information on
> and references to the potentially revolutionary
> theory that particles, forces, space, and time are
> merely manifestations of incredibly tiny strings
> that exist in 11 dimensions.

Atomic Models

Visit The Chemistry Place at:
www.aw.com/chemplace

6

Chemical Bonding and Molecular Shapes

How Atoms Connect to One Another

Millions of years ago, the Great Plains of the United States were ocean. As sea levels fell and at the same time the North American continent rose, many isolated pockets of seawater, called saline lakes, formed. Over time these lakes evaporated, leaving behind the solids that had been dissolved in the seawater. Most abundant was sodium chloride, which collected in cubic crystals referred to by mineralogists as the mineral *halite*. When conditions were right, halite crystals like the ones in this chapter's opening photograph would grow to be several centimeters across.

Why do halite crystals have such a distinct shape? As we explore in this chapter, the macroscopic properties of any substance can be traced to how its submicroscopic parts are held together. The sodium and chloride ions in a halite crystal, for example, are held together in a cubic orientation, and as a result the macroscopic object we know as a halite crystal is also cubic.

Similarly, the macroscopic properties of substances made of molecules are a result of how the atoms in the molecules are held together. For example, many of the properties of water are the result of how the hydrogen and oxygen atoms of each water molecule are held together at an angle. Because of this angled orientation, one side of the molecule has a slight negative charge and the opposite side has a slight positive charge. This charge separation in water molecules gives rise to such phenomena as the inability of water and oil to mix and water's high boiling temperature.

173

The force of attraction that holds ions or atoms together is the electric force, which is the force that occurs between oppositely charged particles. Chemists refer to this ion-binding or atom-binding force as a chemical bond. In this chapter we explore two types of chemical bonds: the *ionic bond,* which holds ions together in a crystal, and the *covalent bond,* which holds atoms together in a molecule.

6.1 An Atomic Model Is Needed to Understand How Atoms Bond

In Chapter 5, we discussed how electrons are arranged around an atomic nucleus. Rather than moving in neat orbits like planets around the sun, electrons are wavelike entities that swarm in various volumes of space called *shells.*

As was shown in Figure 5.25, there are seven shells available to the electrons in any atom, and the electrons fill these shells in order, from innermost to outermost. Furthermore, the maximum number of electrons allowed in the first shell is 2, and for the second and third shells it is 8. The fourth and fifth shells can each hold 18 electrons, and the sixth and seventh shells can each hold 32 elecrons.* These numbers match the number of elements in each period (horizontal row) of the periodic table. Figure 6.1 shows how this model applies to the first four elements of group 18.

Electrons in the outermost occupied shell of any atom may play a significant role in that atom's chemical properties, including its ability to form chemical bonds. To indicate their importance, these electrons are called **valence electrons** (from the Latin *valentia,* "strength"), and the shell they occupy is called the **valence shell**. Valence electrons can be conveniently represented as a series of dots surrounding an atomic symbol. This notation is called an **electron-dot structure** or, sometimes, a *Lewis dot symbol* in honor of the American chemist G. N. Lewis, who first proposed the concepts of shells and valence electrons.

Figure 6.2 shows the electron-dot structures for the atoms important in our discussions of ionic and covalent bonds. (Atoms of elements in groups 3 through 12 form *metallic bonds,* which we'll study in Chapter 18.)

When you look at the electron-dot structure of an atom, you immediately know two important things about that element. You know how many valence electrons it has and how many of these are paired. Chlorine, for example, has three sets of paired electrons and one unpaired electron, and carbon has four unpaired electrons:

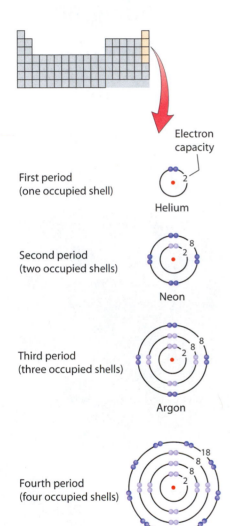

First period
(one occupied shell)

Electron capacity

Helium

Second period
(two occupied shells)

Neon

Third period
(three occupied shells)

Argon

Fourth period
(four occupied shells)

Krypton

Figure 6.1
Occupied shells in the group 18 elements helium through krypton. Each of these elements has a filled outermost occupied shell, and the number of electrons in each outermost occupied shell corresponds to the number of elements in the period to which a particular group 18 element belongs.

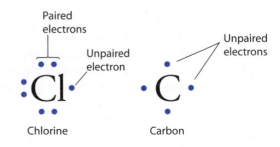

Paired electrons

Unpaired electron

Unpaired electrons

Chlorine   Carbon

---

* These are shells of orbitals grouped by similar energy levels rather than by principal quantum number. See page 159.

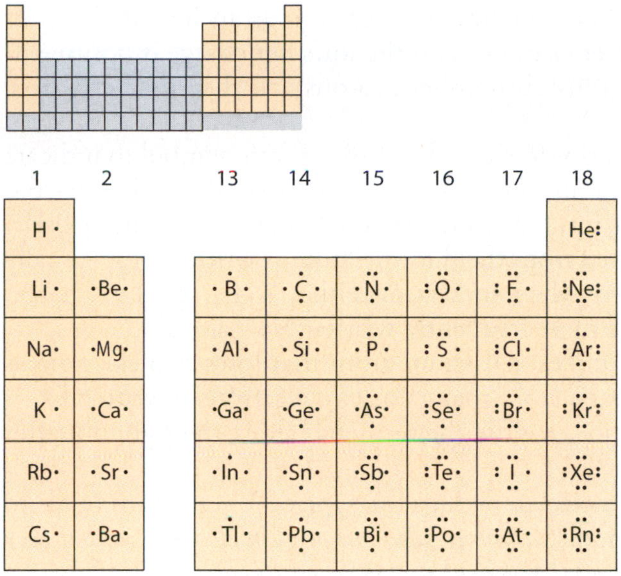

Figure 6.2
The valence electrons of an atom are shown in its electron-dot structure. Note that the first three periods here parallel Figure 5.26. Also note that for larger atoms, not all the electrons in the valence shell are valence electrons. Krypton, Kr, for example, has 18 electrons in its valence shell, as shown in Figure 6.1, but only 8 of these are classified as valence electrons.

Paired valence electrons are relatively stable. In other words, they usually do not form chemical bonds with other atoms. For this reason, electron pairs in an electron-dot structure are called **nonbonding pairs**. (Do not take this name literally, however, for in Chapter 10 you'll see that, under the right conditions, even "nonbonding" pairs can form a chemical bond.) As we discuss in Section 6.5, nonbonding pairs can have a significant influence on the shape of any molecule containing them.

Valence electrons that are unpaired, by contrast, have a strong tendency to participate in chemical bonding. By doing so, they become paired with an electron from another atom. The chemical bonds discussed in this chapter all result from either a transfer or a sharing of unpaired valence electrons.

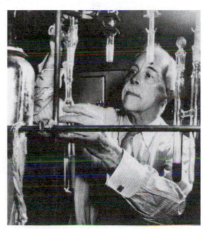

Figure 6.3
Gilbert Newton Lewis (1875–1946) revolutionized chemistry with his theory of chemical bonding, which he published in 1916. He worked most of his life in the chemistry department of the University of California at Berkeley, where he was not only a productive researcher but also an exceptional teacher. Among his teaching innovations was the idea of providing students with problem sets as a follow-up to lectures and readings.

Concept Check ✓

Where are valence electrons located, and why are they important?

Was this your answer? Valence electrons are located in the outermost occupied shell of an atom. They are important because they play a leading role in determining the chemical properties of the atom.

6.2 Atoms Can Lose or Gain Electrons to Become Ions

When the number of protons in the nucleus of an atom equals the number of electrons in the atom, the charges balance and the atom is electrically neutral. If one or more electrons are lost or gained, as illustrated in Figures 6.4 and 6.5, the balance is upset and the atom takes on a net electric charge.

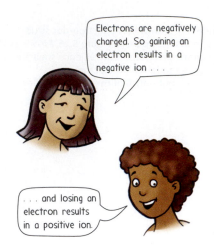

Electrons are negatively charged. So gaining an electron results in a negative ion . . .

. . . and losing an electron results in a positive ion.

Any atom having a net electric charge is referred to as an **ion**. If electrons are lost, protons outnumber electrons and the ion's net charge is positive. If electrons are gained, electrons outnumber protons and the ion's net charge is negative.

Chemists use a superscript to the right of the atomic symbol to indicate the magnitude and sign of an ion's charge. Thus, as shown in Figures 6.4 and 6.5, the positive ion formed from the sodium atom is written Na^{1+} and the negative ion formed from the fluorine atom is written F^{1-}. Usually the numeral *1* is omitted when indicating either a $1+$ or $1-$ charge. Hence, these two ions are most frequently written Na^+ and F^-.

To give two more examples, a calcium atom that loses two electrons is written Ca^{2+}, and an oxygen atom that gains two electrons is written O^{2-}. (Note that the convention is to write the numeral before the sign, not after it: $2+$, not $+2$.)

We can use the shell model to deduce the type of ion an atom tends to form. According to this model, *atoms tend to lose or gain electrons so that they end up with an outermost occupied shell that is filled to capacity.* Let's take a moment to consider this point, looking to Figures 6.4 and 6.5 as visual guides.

If an atom has only one or only a few electrons in its valence shell, it tends to lose these electrons so that the next shell inward, which is already filled, becomes the outermost occupied shell. The sodium atom of Figure 6.4, for example, has one electron in its valence shell, which is the third shell. In forming an ion, the sodium atom loses this electron, thereby making the second shell, which is already filled to capacity, the outermost occupied shell. Because the sodium atom has only one valence electron to lose, it tends to form the $1+$ ion.

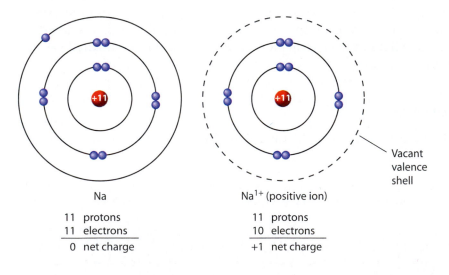

Figure 6.4
An electrically neutral sodium atom contains 11 negatively charged electrons surrounding the 11 positively charged protons of the nucleus. When this atom loses an electron, the result is a positive ion.

Na

11	protons
11	electrons
0	net charge

Na^{1+} (positive ion)

11	protons
10	electrons
+1	net charge

Vacant valence shell

If the valence shell of an atom is almost filled, that atom attracts electrons from another atom and so forms a negative ion. The fluorine atom of Figure 6.5, for example, has one space available in its valence shell for an additional electron. After this additional electron is gained, the fluorine achieves a filled valence shell. Fluorine therefore tends to form the $1-$ ion.

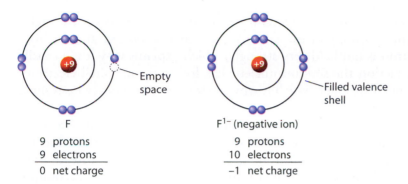

Figure 6.5
An electrically neutral fluorine atom contains nine protons and nine electrons. When this atom gains an electron, the result is a negative ion.

You can use the periodic table as a quick reference when determining the type of ion an atom tends to form. As Figure 6.6 shows, each atom of any group 1 element, for example, has only one valence electron and so tends to form the 1+ ion. Each atom of any group 17 element has room for one additional electron in its valence shell and therefore tends to form the 1– ion. Atoms of the noble-gas elements tend not to form any type of ion because their valence shells are already filled to capacity.

Concept Check ✓

What type of ion does the magnesium atom, Mg, tend to form?

Was this your answer? The magnesium atom (atomic number 12) is found in group 2 and has two valence electrons to lose (see Figure 6.2). It therefore tends to form the 2+ ion.

As was discussed in Chapter 5 and is indicated in Figure 6.6, the attraction an atom's nucleus has for its valence electrons is weakest for elements

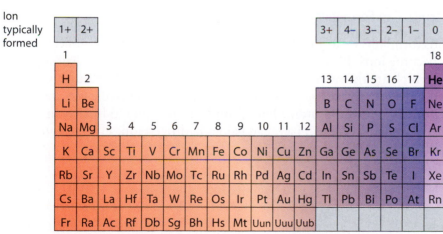

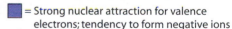
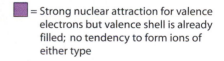

Figure 6.6
The periodic table is your guide to the types of ions atoms tend to form.

on the left in the periodic table and strongest for elements on the right. From sodium's position in the table, we see that a sodium atom's single valence electron is not held very strongly, which explains why it is so easily lost. The attraction the sodium nucleus has for its second-shell electrons, however, is much stronger, which is why the sodium atom rarely loses more than one electron.

At the other side of the periodic table, the nucleus of a fluorine atom holds on strongly to its valence electrons, which explains why the fluorine atom tends *not* to lose any electrons to form a positive ion. Instead, fluorine's nuclear pull on the valence electrons is strong enough to accommodate even an additional electron "imported" from some other atom.

The nucleus of a noble-gas atom pulls so strongly on its valence electrons that they are very difficult to lose. Because there is no room left in the valence shell of a noble-gas atom, no additional electrons are gained. Thus, a noble-gas atom tends not to form an ion of any sort.

Concept Check ✓

Why does the magnesium atom tend to form the 2+ ion?

Was this your answer? Magnesium is on the left in the periodic table, and so atoms of this element do not hold on to the two valence electrons very strongly. The details of why this is so were explained in Section 5.8 using the concept of *inner-shell shielding*. For now, you need recognize only that, because these electrons are not held very tightly, they are easily lost, which is why the magnesium atom tends to form the 2+ ion.

Using our shell model to explain how ions form works well for groups 1 and 2 and 13 through 18. This model is too simplified to work well for the transition metals of groups 3 through 12, however, or for the inner transition metals. In general, these metal atoms tend to form positive ions, but the number of electrons lost varies. Depending on conditions, for example, an iron atom may lose two electrons to form the Fe^{2+} ion, or it may lose three electrons to form the Fe^{3+} ion.

Molecules Can Form Ions

Atoms form ions by losing or gaining electrons. Interestingly, molecules can also become ions. In most cases they do so by either losing or gaining protons, which are the same thing as hydrogen ions, H^+. (Recall that a hydrogen atom is a proton together with an electron. The hydrogen ion, H^+, therefore, is simply a proton.) For example, as is explored further in Chapter 10, a water molecule, H_2O, can gain a hydrogen ion, H^+ (a proton), to form the hydronium ion, H_3O^+:

Water   Hydrogen ion   Hydronium ion
                            (proton)

Similarly, the carbonic acid molecule H_2CO_3 can lose two protons to form the carbonate ion CO_3^{2-}:

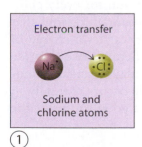

Carbonic acid → Carbonate ion + 2 H⁺ Hydrogen ions (protons)

How these reactions occur will be explored in later chapters. For now, you should understand that the hydronium and carbonate ions are examples of **polyatomic ions**, which are molecules that carry a net electric charge. Table 6.1 lists some commonly encountered polyatomic ions.

Table 6.1

Common Polyatomic Ions

Name	Formula
Hydronium ion	H_3O^+
Ammonium ion	NH_4^+
Bicarbonate ion	HCO_3^-
Acetate ion	$CH_3CO_2^-$
Nitrate ion	NO_3^-
Cyanide ion	CN^-
Hydroxide ion	OH^-
Carbonate ion	CO_3^{2-}
Sulfate ion	SO_4^{2-}
Phosphate ion	PO_4^{3-}

6.3 Ionic Bonds Result from a Transfer of Electrons

When an atom that tends to lose electrons is placed in contact with an atom that tends to gain them, the result is an electron transfer and the formation of two oppositely charged ions. This is what happens when sodium and chlorine are combined. As shown in Figure 6.7, the sodium atom loses one of its electrons to the chlorine atom, resulting in the formation of a positive sodium ion and a negative chloride ion. The two oppositely charged ions are attracted to each other by the electric force, which holds them close together. This electric force of attraction between two oppositely charged ions is called an **ionic bond**.

Figure 6.7
① An electrically neutral sodium atom loses its valence electron to an electrically neutral chlorine atom. ② This electron transfer results in two oppositely charged ions. ③ The ions are then held together by an ionic bond. The spheres drawn around the electron-dot structures here and in subsequent illustrations indicate the relative sizes of the atoms and ions. Note that the sodium ion is smaller than the sodium atom because the lone electron in the third shell is gone once the ion forms, leaving the ion with only two occupied shells. The chloride ion is larger than the chlorine atom because adding that one electron to the third shell makes the shell expand as a result of the repulsions among the electrons.

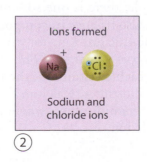

Electron transfer

Sodium and chlorine atoms

①

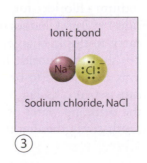

Ions formed

Sodium and chloride ions

②

Ionic bond

Sodium chloride, NaCl

③

A sodium ion and a chloride ion together make the chemical compound sodium chloride, commonly known as table salt. This and all other chemical compounds containing ions are referred to as **ionic compounds**. All ionic compounds are completely different from the elements from which they are made. As discussed in Section 2.3, sodium chloride is not sodium, nor is it chlorine. Rather, it is a collection of sodium and chloride ions that form a unique material having its own physical and chemical properties.

Concept Check ✔

Is the transfer of an electron from a sodium atom to a chlorine atom a physical change or a chemical change?

Was this your answer? Recall from Chapter 2 that only a chemical change involves the formation of new material. Thus this or any other electron transfer, because it results in the formation of a new substance, is a chemical change.

As Figure 6.8 shows, ionic compounds typically consist of elements found on opposite sides of the periodic table. Also, because of how the metals and nonmetals are organized in the periodic table, positive ions are generally derived from metallic elements and negative ions are generally derived from nonmetallic elements.

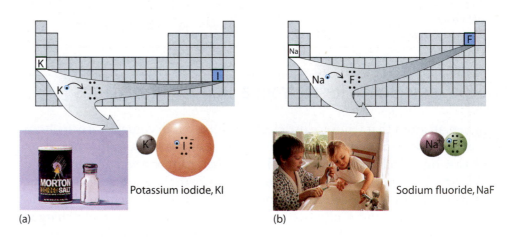

Figure 6.8
(a) The ionic compound potassium iodide, KI, is added in minute quantities to commercial salt because the iodide ion, I⁻, it contains is an essential dietary mineral. (b) The ionic compound sodium fluoride, NaF, is often added to municipal water supplies and toothpastes because it is a good source of the tooth-strengthening fluoride ion, F⁻.

Potassium iodide, KI

Sodium fluoride, NaF

(a)

(b)

For all ionic compounds, positive and negative charges must balance. In sodium chloride, for example, there is one sodium 1+ ion for every chloride 1− ion. Charges must also balance in compounds containing ions that carry multiple charges. The calcium ion, for example, carries a charge of 2+, but the fluoride ion carries a charge of only 1−. Because two fluoride ions are needed to balance each calcium ion, the formula for calcium fluoride is CaF_2, as Figure 6.9 illustrates. Calcium fluoride occurs naturally in the drinking water of some communities, where it is a good source of the tooth-strengthening fluoride ion, F⁻.

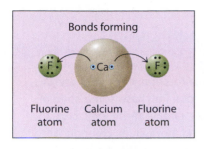

Ionic bonds formed

Calcium fluoride, CaF$_2$

Fluorite

Figure 6.9
A calcium atom loses two electrons to form a calcium ion, Ca^{2+}. These two electrons may be picked up by two fluorine atoms, transforming the atoms to two fluoride ions. Calcium ions and fluoride ions then join to form the ionic compound calcium fluoride, CaF$_2$, which occurs naturally as the mineral fluorite.

An aluminum ion carries a 3+ charge, and an oxide ion carries a 2− charge. Together, these ions make the ionic compound aluminum oxide, Al$_2$O$_3$, the main component of such gemstones as rubies and sapphires. Figure 6.10 illustrates the formation of aluminum oxide. The three oxide ions in Al$_2$O$_3$ carry a total charge of 6−, which balances the total 6+ charge of the two aluminum ions. Interestingly, rubies and sapphires differ in color because of the impurities they contain. Rubies are red because of minor amounts of chromium ions, and sapphires are blue because of minor amounts of iron and titanium ions.

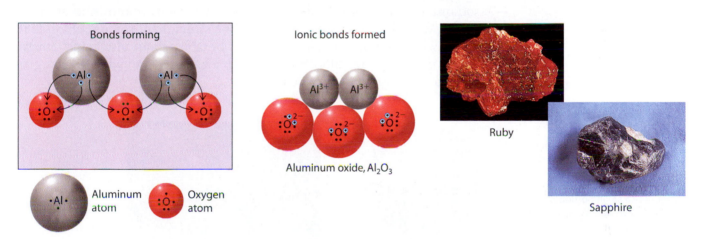

Ruby

Sapphire

Figure 6.10
Two aluminum atoms lose a total of six electrons to form two aluminum ions, Al^{3+}. These six electrons may be picked up by three oxygen atoms, transforming the atoms to three oxide ions, O^{2-}. The aluminum and oxide ions then join to form the ionic compound aluminum oxide, Al$_2$O$_3$.

Concept Check ✔

What is the chemical formula for the ionic compound magnesium oxide?

Was this your answer? Because magnesium is a group 2 element, you know a magnesium atom must lose two electrons to form an Mg^{2+} ion. Because oxygen is a group 16 element, an oxygen atom gains two electrons to form an O^{2-} ion. These charges balance in a one-to-one ratio, and so the formula for magnesium oxide is MgO.

An ionic compound typically contains a multitude of ions grouped together in a highly ordered three-dimensional array. In sodium chloride, for example, each sodium ion is surrounded by six chloride ions and each

Hands-On Chemistry: Up Close with Crystals

View crystals of table salt with a magnifying glass or, better yet, a microscope if one is available. If you do have a microscope, crush the crystals with a spoon and examine the resulting powder. Purchase some sodium-free salt, which is potassium chloride, KCl, and examine these ionic crystals, both intact and crushed. Sodium chloride and potassium chloride both form cubic crystals, but there are significant differences. What are they?

chloride ion is surrounded by six sodium ions (Figure 6.11). Overall there is one sodium ion for each chloride ion, but there are no identifiable sodium–chloride pairs. Such an orderly array of ions is known as an *ionic crystal*. On the atomic level, the crystalline structure of sodium chloride is cubic, which is why macroscopic crystals of table salt are also cubic. Smash a large cubic sodium chloride crystal with a hammer, and what do you get? Smaller cubic sodium chloride crystals!

Figure 6.11
(a) Sodium chloride, as well as other ionic compounds, forms ionic crystals in which every internal ion is surrounded by ions of the opposite charge. (For simplicity, only a small portion of the ion array is shown here. A typical NaCl crystal involves millions and millions of ions.) (b) A view of crystals of table salt through a microscope shows their cubic structure. The cubic shape is a consequence of the cubic arrangement of sodium and chloride ions.

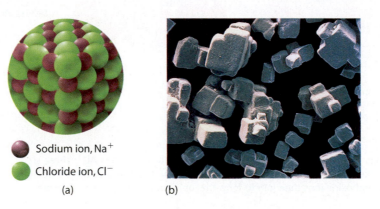

● Sodium ion, Na$^+$

● Chloride ion, Cl$^-$

(a)　　　　　　　　(b)

Similarly, the crystalline structures of other ionic compounds, such as calcium fluoride and aluminum oxide, are a consequence of how the ions pack together.

6.4 Covalent Bonds Result from a Sharing of Electrons

Imagine two children playing together and sharing their toys. A force that keeps the children together is their mutual attraction to the toys they share. In a similar fashion, two atoms can be held together by their mutual attraction for electrons they share. A fluorine atom, for example, has a strong attraction for one additional electron to fill its outermost occupied shell. As shown in Figure 6.12, a fluorine atom can obtain an additional electron by holding on to the unpaired valence electron of another fluorine atom. This

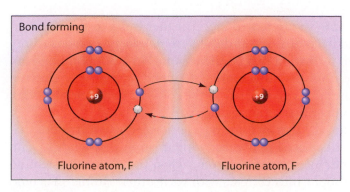

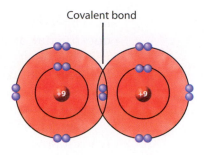

Figure 6.12
The effect of the positive nuclear charge (represented by red shading) of a fluorine atom extends beyond the atom's outermost occupied shell. This positive charge can cause the fluorine atom to become attracted to the unpaired valence electron of a neighboring fluorine atom. Then the two atoms are held together in a fluorine molecule by the attraction they both have for the two shared electrons. Each fluorine atom achieves a filled valence shell.

results in a situation in which the two fluorine atoms are mutually attracted to the same two electrons. This type of electrical attraction in which atoms are held together by their mutual attraction for shared electrons is called a **covalent bond**, where *co-* signifies sharing and *-valent* refers to the fact that it is valence electrons that are being shared.

A substance composed of atoms held together by covalent bonds is a **covalent compound**. The fundamental unit of most covalent compounds is a **molecule**, which we can now formally define as any group of atoms held together by covalent bonds. Figure 6.13 uses the element fluorine to illustrate this principle.

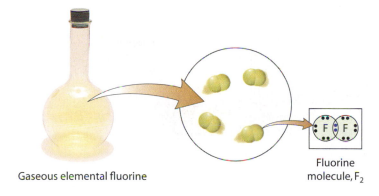

Gaseous elemental fluorine

Fluorine
molecule, F_2

Figure 6.13
Molecules are the fundamental units of the gaseous covalent compound fluorine, F_2. Notice that in this model of a fluorine molecule, the spheres overlap, whereas the spheres shown earlier for ionic compounds do not. Now you know that this difference in representation is because of the difference in bond types.

When writing electron-dot structures for covalent compounds, chemists often use a straight line to represent the two electrons involved in a covalent bond. In some representations, the nonbonding electron pairs are left out. This is done in instances where these electrons play no significant role in the process being illustrated. Here are two frequently used ways of showing the electron-dot structure for a fluorine molecule without using spheres to represent the atoms:

$$:\ddot{F} - \ddot{F}: \qquad F - F$$

Remember—the straight line in both versions represents two electrons, *one from each atom.* Thus we now have two types of electron pairs to keep track

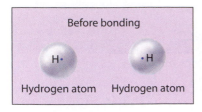

Covalent bond formed

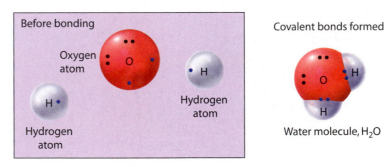

Hydrogen molecule, H₂

Figure 6.14
Two hydrogen atoms form a covalent bond as they share their unpaired electrons.

of. The term *nonbonding pair* refers to any pair that exists in the electron-dot structure of an individual atom, and the term *bonding pair* refers to any pair that results from formation of a covalent bond. In a nonbonding pair, both electrons come from the same atom; in a bonding pair, one electron comes from one of the atoms taking part in the covalent bond and the other electron comes from the other atom taking part in the bond.

Recall from Section 6.3 that an ionic bond is formed when an atom that tends to lose electrons is placed in contact with an atom that tends to gain them. A covalent bond, by contrast, is formed when two atoms that tend to gain electrons are brought into contact with each other. Atoms that tend to form covalent bonds are therefore primarily atoms of the nonmetallic elements in the upper right corner of the periodic table (with the exception of the noble-gas elements, which are very stable and tend not to form bonds).

Hydrogen tends to form covalent bonds because, unlike the other group 1 elements, it has a fairly strong attraction for an additional electron. Two hydrogen atoms, for example, covalently bond to form a hydrogen molecule, H_2, as shown in Figure 6.14.

The number of covalent bonds an atom can form is equal to the number of additional electrons it can attract, which is the number it needs to fill its valence shell. Hydrogen attracts only one additional electron, and so it forms only one covalent bond. Oxygen, which attracts two additional electrons, finds them when it encounters two hydrogen atoms and reacts with them to form water, H_2O, as Figure 6.15 shows. In water, not only does the oxygen atom have access to two additional electrons by covalently bonding to two hydrogen atoms, but each hydrogen atom has access to an additional electron by bonding to the oxygen atom. Each atom thus achieves a filled valence shell.

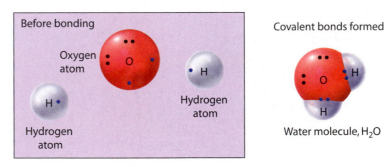

Figure 6.15
The two unpaired valence electrons of oxygen pair with the unpaired valence electrons of two hydrogen atoms to form the covalent compound water.

Nitrogen attracts three additional electrons and is thus able to form three covalent bonds, as occurs in ammonia, NH_3, shown in Figure 6.16. Likewise, a carbon atom can attract four additional electrons and is thus able to form four covalent bonds, as occurs in methane, CH_4. Note that the number of covalent bonds formed by these and other nonmetallic elements parallels the type of negative ions they tend to form (see Figure 6.6). This makes sense because covalent bond formation and negative ion formation are both applications of the same concept: nonmetallic atoms tend to gain electrons until their valence shells are filled.

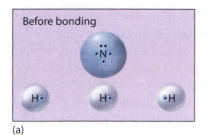

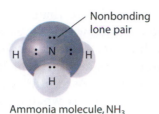

(a)

Nonbonding lone pair

Ammonia molecule, NH₃

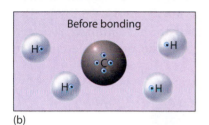

(b)

Methane molecule, CH₄

Figure 6.16
(a) A nitrogen atom attracts the three electrons in three hydrogen atoms to form ammonia, NH₃, a gas that can dissolve in water to make an effective cleanser. (b) A carbon atom attracts the four electrons in four hydrogen atoms to form methane, CH₄, the primary component of natural gas. In these and most other cases of covalent bond formation, the result is a filled valence shell for all the atoms involved.

Diamond is a most unusual covalent compound consisting of carbon atoms covalently bonded to one another in four directions. The result is a *covalent crystal,* which, as shown in Figure 6.17, is a highly ordered three-dimensional network of covalently bonded atoms. This network of carbon atoms forms a very strong and rigid structure, which is why diamonds are so hard. Also, because a diamond is a group of atoms held together only by covalent bonds, it can be characterized as a single molecule! Unlike most other molecules, a diamond molecule is large enough to be visible to the naked eye, and so it is more appropriately referred to as a *macromolecule.*

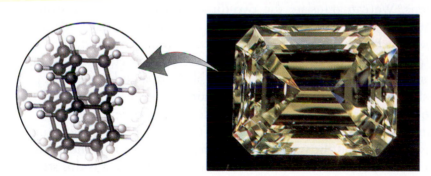

Figure 6.17
The crystalline structure of diamond is best illustrated by using sticks to represent the covalent bonds. It is the molecular nature of diamond that is responsible for this material's unusual properties, such as its extreme hardness.

Concept Check

How many electrons make up a covalent bond?

Was this your answer? Two—one from each participating atom.

It is possible to have more than two electrons shared between two atoms, and Figure 6.18 shows a few examples. Molecular oxygen, O_2, which is what we breathe, consists of two oxygen atoms connected by four shared electrons. This arrangement is called a *double covalent bond* or, for short, a *double bond.* As another example, the covalent compound carbon dioxide, CO_2, which is what we exhale, consists of two double bonds connecting two oxygen atoms to a central carbon atom.

Some atoms can form *triple covalent bonds*, in which six electrons—three from each atom—are shared. One example is molecular nitrogen, N_2. Most of the air surrounding you right now (about 78%) is gaseous molecular nitrogen, N_2.

Any double or triple bond is often referred to as a *multiple covalent bond.* Multiple bonds higher than these, such as the quadruple covalent bond, are not commonly observed.

Oxygen, O_2　　　　Carbon dioxide, CO_2　　　　Nitrogen, N_2

Figure 6.18
Double covalent bonds in molecules of oxygen, O_2, and carbon dioxide, CO_2, and a triple covalent bond in a molecule of nitrogen, N_2.

6.5 Valence Electrons Determine Molecular Shape

Molecules are three-dimensional entities and therefore best depicted in three dimensions. We can translate the two-dimensional electron-dot structure representing a molecule into a more accurate three-dimensional rendering by using the model known as **valence-shell electron-pair repulsion**, also called VSEPR (pronounced ves-per). According to this model, electron pairs in a valence shell strive to get as far away as possible from all other electron pairs in the shell. This includes nonbonding pairs and any bonding pairs or groups of bonding pairs held together in a double or triple bond.

Note that the VSEPR model talks about the repulsions between pairs of electrons, not between the two electrons in a pair. (Recall that the electrons in a pair can stay together because of their opposite spins.) It is this striving for maximum separation distance between electron pairs that determines the geometry of any molecule.

The two-dimensional electron-dot structure for methane, CH_4, is

$$
\begin{array}{c}
\text{H} \\
| \\
\text{H}-\text{C}-\text{H} \\
| \\
\text{H}
\end{array}
$$

90°

In this structure, the bonding electron pairs (shown as straight lines representing one electron from each atom) are set 90 degrees apart because that is the farthest apart they can be shown in two dimensions. When we extend to three dimensions, however, we can create a more accurate rendering in which the four bonding pairs are 109.5 degrees apart:

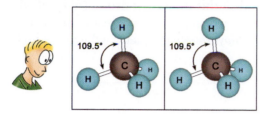

Stereo image

These two renderings of methane are *stereo images*—you can see them in three dimensions by looking at them cross-eyed so that they appear to overlap. You can think of this three-dimensional structure as follows: the central carbon atom has one hydrogen atom sticking out of its top and is supported on a tripod who legs are formed by the three lower C–H bonds.

Draw the four triangles defined by the hydrogen atoms in the above stereo images of CH_4 (one triangle being the base, the other three being the three vertical faces) and you'll see that the shape of the methane molecule is a pyramid that has a triangular base supporting three other triangles that meet at the pyramid apex. In geometry, a pyramid that has a triangular base is given the special name *tetrahedron*, and so chemists say that the methane molecule is *tetrahedral*:

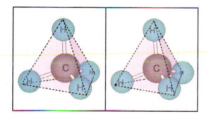

Stereo image of tetrahedral methane molecule

The VSEPR model allows us to use electron-dot structures to predict the three-dimensional geometry of simple molecules. This geometry is determined by considering the number of *substituents* surrounding the central atom. A **substituent** is any atom or nonbonding pair of electrons surrounding some central atom. The carbon of the methane molecule, for example, has four substituents—the four hydrogen atoms. The oxygen atom of a water molecule also has four substituents—two hydrogen atoms and two nonbonding pairs of electrons:

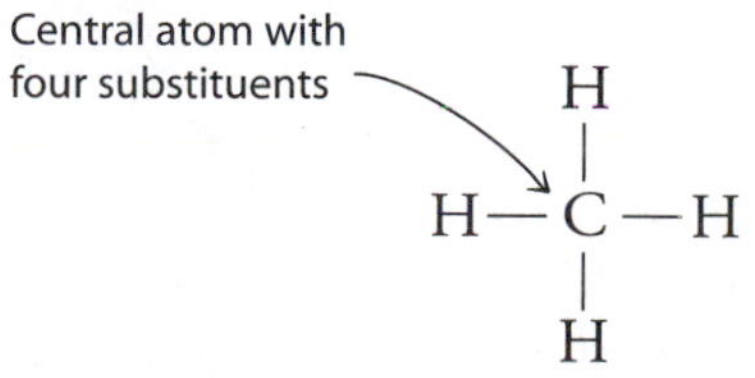

Central atom with four substituents

$$H-\overset{\displaystyle H}{\underset{\displaystyle H}{C}}-H$$

Methane, CH_4

Central atom with four substituents

Lobes used to indicate space occupied by nonbonding pair

$$\cdot\cdot O-H$$
$$|$$
$$H$$

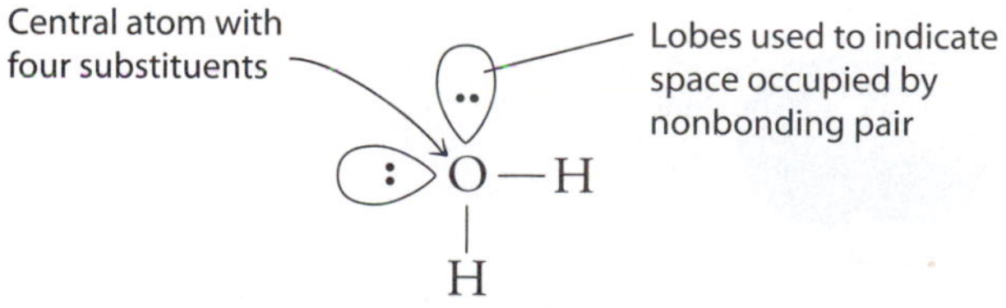

Water, H_2O

As shown in Table 6.2, when a central atom has only two substituents, the geometry of the molecule is *linear*, meaning a single straight line may be drawn passing through both substituents and the central atom. Three substituents arrange themselves in a triangle the plane of which passes through the central atom, and so this molecular geometry is called *triangular planar*. Four substituents form a tetrahedron, as already discussed. Five substituents result in a *triangular bipyramidal* geometry, which, as you'll see when you do

Table 6.2

Molecular Geometries

Number of Substituents	Three-Dimensional Geometry	Examples		
2	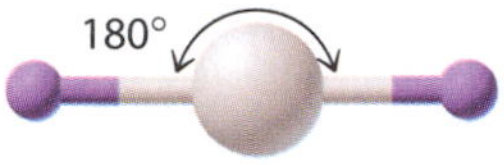 Linear — 180°	H—Be—H BeH$_2$	O=C=O CO$_2$	H—C≡N HCN
3	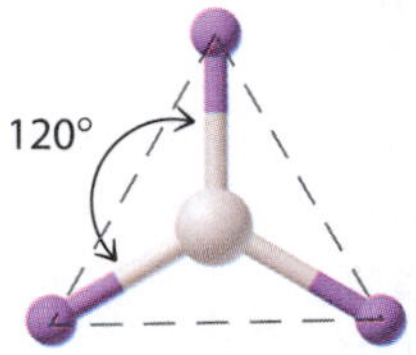 Triangular planar — 120°	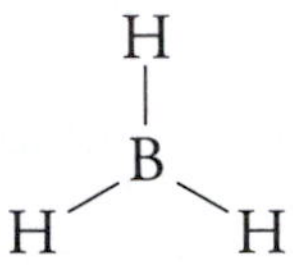 H$_2$, B(H)(H) BH$_3$	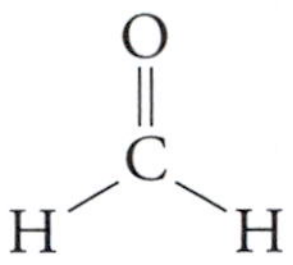 O=C(H)(H) H$_2$CO	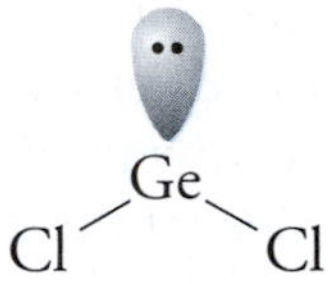 Ge(Cl)(Cl) GeCl$_2$
4	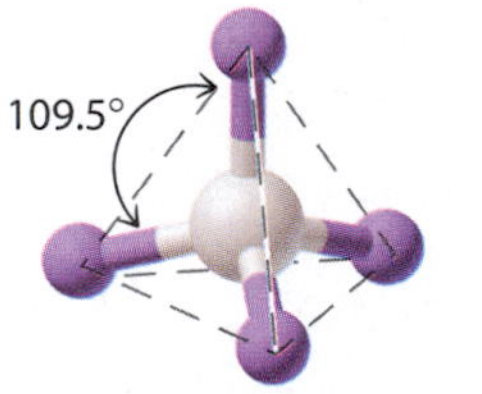 Tetrahedral — 109.5°	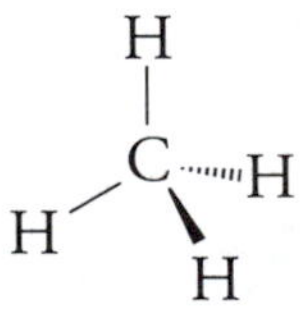 CH$_4$	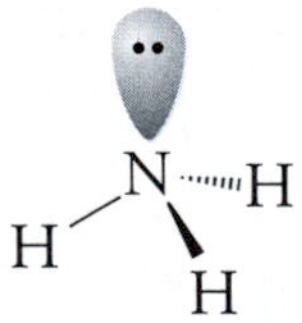 NH$_3$	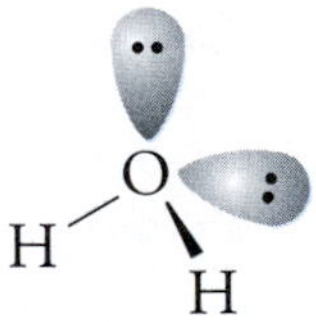 H$_2$O
5	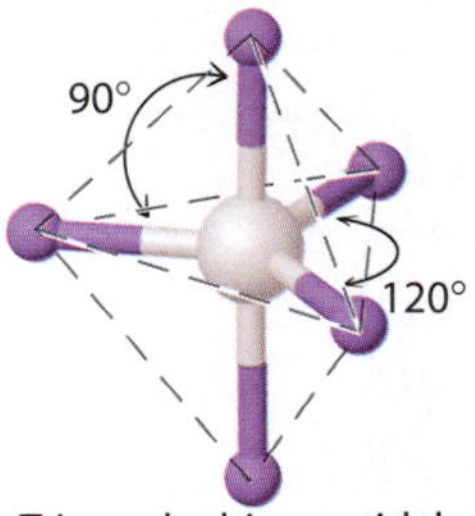 Triangular bipyramidal — 90°, 120°	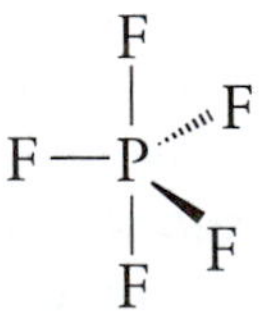 PF$_5$	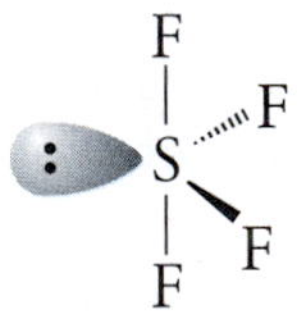 SF$_4$	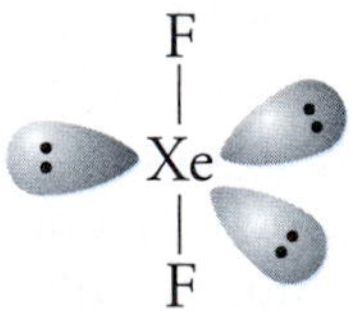 XeF$_2$
6	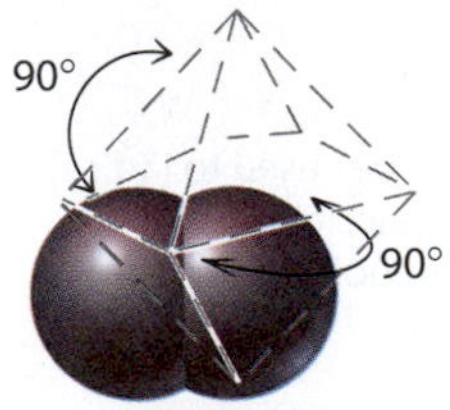 Octahedral — 90°, 90°	 SF$_6$	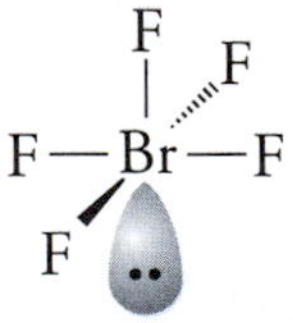 BrF$_5$	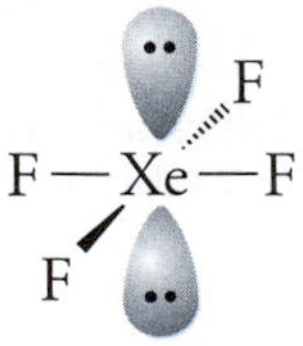 XeF$_4$

the Hands-On Chemistry activity on page 190, is two triangle-based pyramids sharing a base and having the two apexes pointing in opposite directions. Six substituents arrange themselves around the central atom in a geometry that, if it had a surface, would show eight sides. To indicate this eight-sided geometry, this structure is called *octahedral*.

Why these geometries? Simply put, these are the geometries that allow for maximum distance between substituents.

Concept Check ✔

Why are the two oxygen atoms in carbon dioxide, CO_2, spaced 180 degrees apart?

Was this your answer? If the two oxygen atoms were on the same side of the carbon atom, the bonding electrons would be relatively close to each other:

Incorrect geometry for
carbon dioxide, CO_2

Because electron pairs repel one another, this is not a stable situation. Instead, the oxygen atoms position themselves so that the bonding pairs of the two double bonds are as far from each other as possible, which is on opposite sides of the carbon atom, 180 degrees apart, as shown in Table 6.2.

Molecular Shape Is Defined by Where the Substituent Atoms Are

Now that you have learned how to use VSEPR to determine molecular *geometry*, you are ready to see how chemists figure out molecular *shape*. What's the difference, you ask? Just this: when chemists talk about molecular geometry, they are talking about the relative positions of *everything* surrounding a central atom in the molecule, both atoms and nonbonding pairs of electrons. When they talk about molecular shape, they are talking about the relative positions of *only the atoms surrounding a central atom.*

Figuring out a molecular shape is a two-step process. The first step is to use VSEPR to position all substituents, both atoms and nonbonding pairs, around a central atom. The second step is to "freeze" the orientations you've come up with so that no atom can change its position, remove all nonbonding pairs, and then decide what three-dimensional shape the atoms form. Let's work through a few examples from Table 6.2 to see what all this means.

In any molecule in which there are no nonbonding pairs around the central atom, the molecular shape is the same as the molecular geometry. Thus, to use the examples from Table 6.2, all three two-substituent molecules have both a linear geometry and a linear shape. Both BH_3 and H_2CO have a triangular planar shape, CH_4 has a tetrahedral shape, PF_5 a triangular bipyramidal shape, and SF_6 a square bipyramidal shape.

Now let's look at molecules that have nonbonding pairs, beginning with germanium chloride, $GeCl_2$, and its one nonbonding pair. The geometry is triangular planar, and to get the shape we ignore the nonbonding pair. This reveals the germanium and two chlorine atoms held together at an angle—a shape known as *bent*. Similarly, ignoring the two nonbonding pairs in a water molecule also reveals three atoms forming a bent shape. Now you know why water molecules are always depicted with the two hydrogen atoms close to each other like a set of mouse ears rather than as far apart as possible on opposite sides of the oxygen atom—there are two nonbonding pairs pushing them into this orientation.

Ignoring the nonbonding pair of the ammonia molecule, NH_3, in Table 6.2 means the shape is not tetrahedral because in a tetrahedron all four corners must be equally distant from the central atom. Ammonia's shape is thus more accurately defined as triangular pyramidal.

This same process of ignoring the nonbonding electron pairs reveals the shapes of the remaining molecules of Table 6.2, which are shown in Figure 6.19.

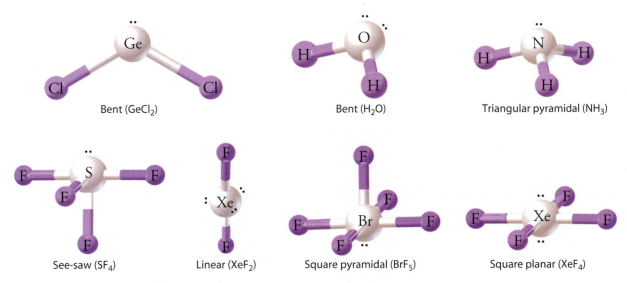

Bent ($GeCl_2$)   Bent (H_2O)   Triangular pyramidal (NH_3)

See-saw (SF_4)   Linear (XeF_2)   Square pyramidal (BrF_5)   Square planar (XeF_4)

Figure 6.19
The shapes of molecules from Table 6.2.

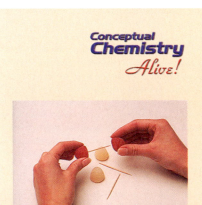

Conceptual **Chemistry** *Alive!*

Hands-On Chemistry: Gumdrop Molecules

Use toothpicks and gumdrops or jelly beans of different colors to build models of the molecules shown in Figure 6.19, letting the different colors represent different elements.

Once you have become proficient at building these models, test your expertise by building models for difluoromethane, CH_2F_2; ethane, C_2H_6; hydrogen peroxide, H_2O_2; and acetylene, C_2H_2. Keep in mind that each carbon atom must have four covalent bonds, each oxygen must have two, and each fluorine and hydrogen must have only one.

No fair peeking at the Hands-On Chemistry Insights at the end of this chapter until you have made an honest attempt to build these molecules.

Concept Check ✓

What is the shape of a chlorine trifluoride molecule, ClF₃, which has a triangular bipyramidal geometry:

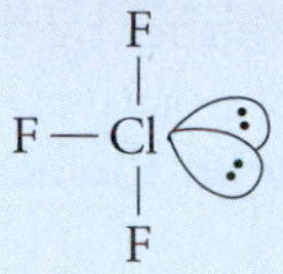

Was this your answer? Ignore the two nonbonding pairs, and the shape of the molecule is all four atoms in the same plane. They form a triangle having a fluorine atom at each corner and the chlorine atom sitting at the midpoint of one side:

Call it what you like—most chemists call it T-shaped. There are even more molecular shapes that can be derived from the geometries in Table 6.2. How many can you find? How might you name them? Curious? Talk with your instructor.

6.6 Polar Covalent Bonds Result from an Uneven Sharing of Electrons

If the two atoms in a covalent bond are identical, their nuclei have the same positive charge, and therefore the electrons are shared *evenly*. We can represent these electrons as being centrally located by using an electron-dot structure in which the electrons are situated exactly halfway between the two atomic symbols. Alternatively, we can draw a probability cloud (Section 5.5) in which the positions of the two bonding electrons over time are shown as a series of dots. Where the dots are most concentrated is where the electrons have the greatest probability of being located:

H : H   H H

In a covalent bond between nonidentical atoms, the nuclear charges are different, and consequently the bonding electrons may be shared *unevenly*. This occurs in a hydrogen–fluorine bond, where electrons are more attracted to fluorine's greater nuclear charge:

H : F   H F

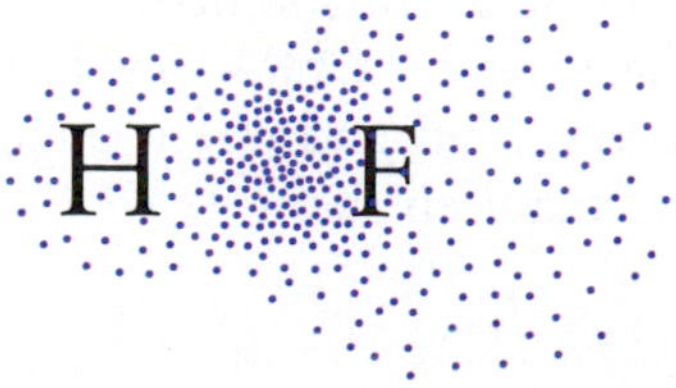

The bonding electrons spend more time around the fluorine atom. For this reason, the fluorine side of the bond is slightly negative and, because the bonding electrons have been drawn away from the hydrogen atom, the hydrogen side of the bond is slightly positive. This separation of charge is called a **dipole** (pronounced die-pole) and is represented either by the characters $\delta-$ and $\delta+$, read "slightly negative" and "slightly positive," respectively, or by a crossed arrow pointing to the negative side of the bond:

$$\overset{\delta+}{\mathrm{H}}-\overset{\delta-}{\mathrm{F}} \qquad \overset{+\!\!\longrightarrow}{\mathrm{H}-\mathrm{F}}$$

So, atoms forming a chemical bond engage in a tug-of-war for electrons. How strongly an atom is able to tug on bonding electrons has been measured experimentally and quantified as the atom's **electronegativity**. The range of electronegativities runs from 0.7 to 3.98, as Figure 6.20 shows. The greater an atom's electronegativity, the greater its ability to pull electrons toward itself when bonded. Thus in hydrogen fluoride, fluorine has a greater electronegativity, or pulling power, than does hydrogen.

H 2.2																	He —
Li 0.98	Be 1.57											B 2.04	C 2.55	N 3.04	O 3.44	F 3.98	Ne —
Na 0.93	Mg 1.31											Al 1.61	Si 1.9	P 2.19	S 2.58	Cl 3.16	Ar —
K 0.82	Ca 1.0	Sc 1.36	Ti 1.54	V 1.63	Cr 1.66	Mn 1.55	Fe 1.83	Co 1.88	Ni 1.91	Cu 1.90	Zn 1.65	Ga 1.81	Ge 2.01	As 2.18	Se 2.55	Br 2.96	Kr —
Rb 0.82	Sr 0.95	Y 1.22	Zr 1.33	Nb 1.6	Mo 2.16	Tc 1.9	Ru 2.2	Rh 2.28	Pd 2.20	Ag 1.93	Cd 1.69	In 1.78	Sn 1.96	Sb 2.05	Te 2.1	I 2.66	Xe —
Cs 0.79	Ba 0.89	La 1.10	Hf 1.3	Ta 1.5	W 2.36	Re 1.9	Os 2.2	Ir 2.20	Pt 2.8	Au 2.54	Hg 2.00	Tl 2.04	Pb 2.33	Bi 2.02	Po 2.0	At 2.2	Rn —
Fr 0.7	Ra 0.9	Ac 1.1	Rf —	Db —	Sg —	Bh —	Hs —	Mt —	Uun —	Uuu —	Uub —						

Figure 6.20
The experimentally measured electronegativities of elements.

Electronegativity is greatest for elements at the upper right of the periodic table and lowest for elements at the lower left. Noble gases are not considered in electronegativity discussions because, with only a few exceptions, they do not participate in chemical bonding.

When the two atoms in a covalent bond have the same electronegativity, no dipole is formed (as is the case with H_2) and the bond is classified as a **nonpolar bond**. When the electronegativities of the atoms differ, a dipole may form (as with HF) and the bond is classified as a **polar bond**. Just how polar a bond is depends on the difference between the electronegativity values of the two atoms—the greater the difference, the more polar the bond.

As can be seen in Figure 6.20, the farther apart two atoms are in the periodic table, the greater the difference in their electronegativities, and hence the greater the polarity of the bond between them. So a chemist need not even read the electronegativities to predict which bonds are more polar than others. To find out, he or she need only look at the relative positions of the atoms in the periodic table—the farther apart they are, especially when one is at the lower left and one is at the upper right, the greater the polarity of the bond between them.

Concept Check ✓

List these bonds in order of increasing polarity: P–F, S–F, Ga–F, Ge–F (F, fluorine, atomic number 9; P, phosphorus, atomic number 15; S, sulfur, atomic number 16; Ga, gallium, atomic number 31; Ge, germanium, atomic number 32):

(least polar) _____ < _____ < _____ < _____ < (most polar)

Was this your answer? *If you answered the question, or attempted to, before reading this answer, hooray for you! You're doing more than reading the text—you're learning chemistry.* The greater the *difference* in electronegativities between two bonded atoms, the greater the polarity of the bond, and so the order of increasing polarity is S–F < P–F < Ge–F < Ga–F.

Note that this answer can be obtained by looking only at the relative positions of these elements in the periodic table rather than by calculating the differences in their electronegativities.

The magnitude of bond polarity is sometimes indicated by the size of the crossed arrow or $\delta+/\delta-$ symbol used to depict a dipole, as shown in Figure 6.21.

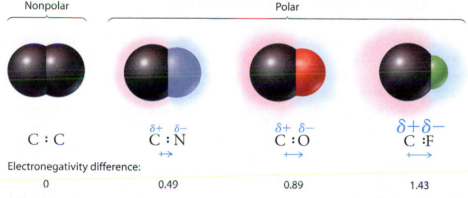

Figure 6.21
These bonds are in order of increasing polarity from left to right, a trend indicated by the larger and larger crossed arrows and $\delta+/\delta-$ symbols. Which of these pairs of elements are farthest apart in the periodic table?

Note that the electronegativity difference between atoms in an ionic bond can also be calculated. For example, the bond in NaCl has an electronegativity difference of 2.23, far greater than the difference of 1.43 shown for the C−F bond in Figure 6.21.

What is important to understand here is that there is no black-and-white distinction between ionic and covalent bonds. Rather, there is a gradual change from one to the other as the atoms that bond are located farther and farther apart in the periodic table. This continuum is illustrated in Figure 6.22. Atoms on opposite sides of the periodic table have great differences in electronegativity, and hence the bonds between them are highly polar—in other words, ionic. Nonmetallic atoms of the same type have the same electronegativities, and so their bonds are nonpolar covalent. The polar covalent bond with its uneven *sharing* of electrons and slightly *charged* atoms is between these two extremes.

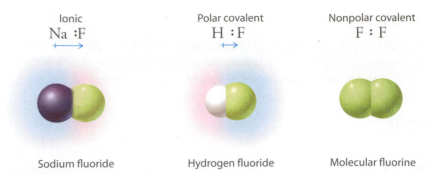

Ionic	Polar covalent	Nonpolar covalent
Na :F	H :F	F : F
Sodium fluoride	Hydrogen fluoride	Molecular fluorine

Figure 6.22
The ionic bond and the nonpolar covalent bond represent the two extremes of chemical bonding. The ionic bond involves a *transfer* of one or more electrons, and the nonpolar covalent bond involves the equitable *sharing* of electrons. The character of a polar covalent bond falls between these two extremes.

6.7 Molecular Polarity Results from an Uneven Distribution of Electrons

If all the bonds in a molecule are nonpolar, the molecule as a whole is also nonpolar—as is the case with H_2, O_2, and N_2. If a molecule consists of only two atoms and the bond between them is polar, the polarity of the molecule is the same as the polarity of the bond—as with HF, HCl, and ClF.

Complexities arise when assessing the polarity of a molecule containing more than two atoms. Consider carbon dioxide, CO_2, shown in Figure 6.23. The cause of the dipole in either one of the carbon–oxygen bonds is oxygen's greater pull (because oxygen is more electronegative than carbon) on the bonding electrons. At the same time, however, the oxygen atom on the opposite side of the carbon pulls those electrons back to the carbon. The net result is an even distribution of bonding electrons around the whole molecule. So, dipoles that are of equal strength but pull in opposite directions in a molecule effectively cancel each other, with the result that the molecule as a whole is nonpolar.

Figure 6.23
There is no net dipole in a carbon dioxide molecule, and so the molecule is nonpolar. This is analogous to two people in a tug-of-war. As long as they pull with equal forces but in opposite directions, the rope remains stationary.

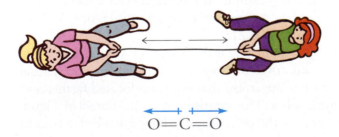

$$O = C = O$$

Figure 6.24 illustrates a similar situation in boron trifluoride, BF_3, where three fluorine atoms are oriented 120 degrees from one another around a central boron atom. Because the angles are all the same, and because each fluorine atom pulls on the electrons of its boron–fluorine bond with the same force, the resulting polarity of this molecule is zero.

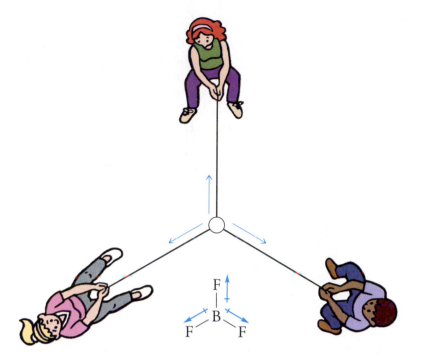

Figure 6.24
The three dipoles of a boron trifluoride molecule oppose one another at 120-degree angles, which makes the overall molecule nonpolar. This is analogous to three people pulling with equal force on ropes attached to a central ring. As long as they all pull with equal force and all maintain the 120-degree angles, the ring remains stationary.

Nonpolar molecules have only relatively weak attractions to other nonpolar molecules. The covalent bonds in a carbon dioxide molecule, for example, are many times stronger than any forces of attraction that might occur between two adjacent carbon dioxide molecules. This lack of attraction between nonpolar molecules explains the low boiling points of many nonpolar substances. Recall from Section 1.7 that boiling is a process wherein the molecules of a liquid separate from one another as they go into the gaseous phase. When there are only weak attractions between the molecules of a liquid, less heat energy is required to liberate the molecules from one another and allow them to enter the gaseous phase. This translates into a relatively low boiling point for the liquid, as, for instance, in the nitrogen, N_2, shown in Figure 6.25. The boiling points of hydrogen, H_2; oxygen, O_2; carbon dioxide, CO_2; and boron trifluoride, BF_3, are also quite low for the same reason.

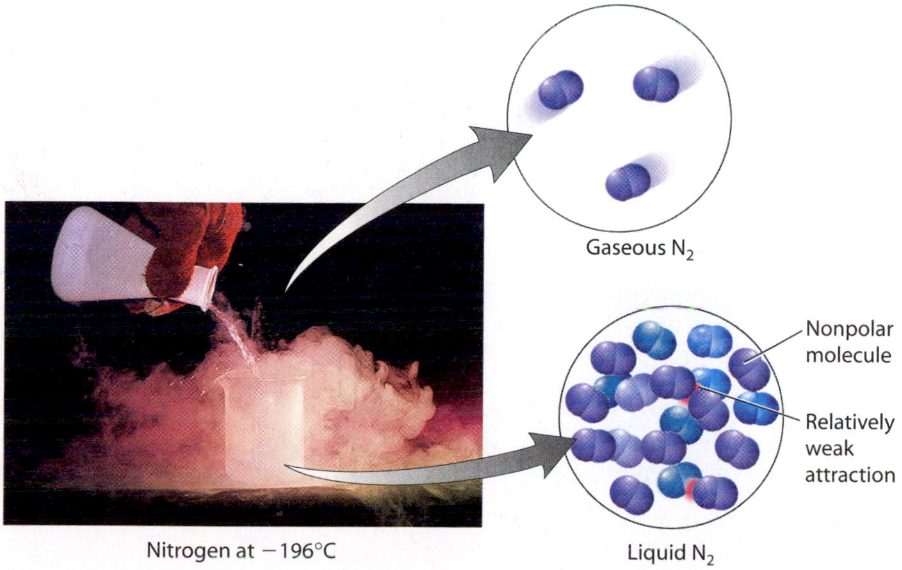

Gaseous N_2

Nonpolar molecule

Relatively weak attraction

Nitrogen at $-196°C$

Liquid N_2

Figure 6.25
Nitrogen is a liquid at temperatures below its chilly boiling point of $-196°C$. Nitrogen molecules are not very attracted to one another because they are nonpolar. As a result, the small amount of heat energy available at $-196°C$ is enough to separate them and allow them to enter the gaseous phase.

There are many instances in which the dipoles of different bonds in a molecule do not cancel each other. Reconsider the rope analogy of Figure 6.24. As long as everyone pulls equally hard, the ring stays put. Imagine, however, that one person begins to ease off on the rope. Now the pulls are no longer balanced, and the ring begins to move away from the person who is slacking off, as Figure 6.26 shows. Likewise, if one person began to pull harder, the ring would move away from the other two people.

Figure 6.26
If one person eases off in a three-way tug-of-war but the other two continue to pull, the ring moves in the direction of the purple arrow.

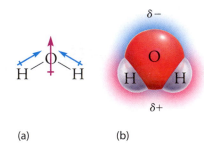

(a)   (b)

Figure 6.27
(a) The individual dipoles in a water molecule add together to give a large overall dipole for the whole molecule, shown in purple. (b) The region around the oxygen atom is therefore slightly negative, and the region around the two hydrogens is slightly positive.

A similar situation occurs in molecules where polar covalent bonds are not equal and opposite. Perhaps the most relevant example is water, H_2O. Each hydrogen–oxygen covalent bond has a relatively large dipole because of the great electronegativity difference. Because of the bent shape of the molecule, however, the two dipoles, shown in blue in Figure 6.27, do not cancel each other the way the $C{=}O$ dipoles in Figure 6.23 do. Instead, the dipoles in the water molecule work together to give an overall dipole, shown in purple, for the molecule.

Concept Check ✓

Which of these molecules is polar and which is nonpolar:

Was this your answer? Symmetry is often the greatest clue for determining polarity. Because the molecule on the left is symmetrical, the dipoles on the two sides cancel each other. This molecule is therefore nonpolar:

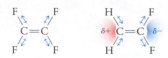

The molecule on the right is less symmetrical (more "lopsided") and so is the polar molecule. Because carbon is more electronegative than hydrogen, the dipoles of the two hydrogen–carbon bonds point toward the carbon. Because fluorine is more electronegative than carbon, the dipoles of the carbon–fluorine bonds point toward the fluorines. Because the general direction of all dipole arrows is toward the fluorines, so is the average distribution of the bonding electrons. The fluorine side of the molecule is therefore slightly negative, and the hydrogen side is slightly positive.

Figure 6.28 illustrates how polar molecules electrically attract one another and as a result are relatively difficult to separate. In other words, polar molecules can be thought of as being "sticky," which is why it takes more energy to separate them and let them enter the gaseous phase. For this reason, substances composed of polar molecules typically have higher boiling points than substances composed of nonpolar molecules, as Table 6.3 shows.

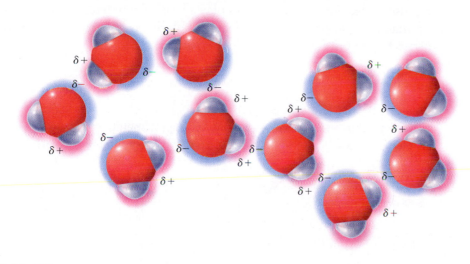

Figure 6.28
Water molecules attract one another because each contains a slightly positive side and a slightly negative side. The molecules position themselves such that the positive side of one faces the negative side of a neighbor.

Table 6.3

Boiling Points of Some Polar and Nonpolar Substances

Substance	Boiling Point (°C)
Polar	
Hydrogen fluoride, HF	20
Water, H_2O	100
Ammonia, NH_3	−33
Nonpolar	
Hydrogen, H_2	−253
Oxygen, O_2	−183
Nitrogen, N_2	−196
Boron trifluoride, BF_3	−100
Carbon dioxide, CO_2	−79

Figure 6.29
Oil and water are difficult to mix, as is evident from this 1989 oil spill of the Exxon Valdez oil tanker in Alaska's Prince William Sound. It's not, however, that oil and water repel each other. Rather, water molecules are so attracted to themselves because of their polarity that they pull themselves together. The nonpolar oil molecules are thus excluded and left to themselves. Being less dense than water, oil floats on the surface, where it poses great danger to wildlife.

Water, for example, boils at 100°C, whereas carbon dioxide boils at −79°C. This 179 C° difference is quite dramatic when you consider that a carbon dioxide molecule is more than twice as massive as a water molecule.

Because molecular "stickiness" can play a lead role in determining a substance's macroscopic properties, molecular polarity is a central concept of chemistry. Figure 6.29 describes an interesting example.

Concept Check ✓

Substance A boils at 150°C, and substance B boils at 30°C. The molecules of these substances, represented below, are approximately the same size, but their shapes are different. Which substance is likely the more polar?

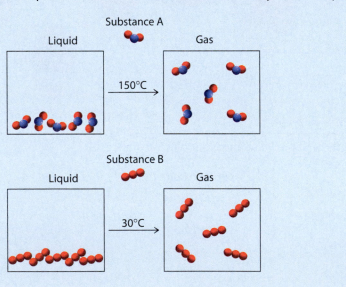

Was this your answer? There are two reasons to believe substance A is more polar. First, the A molecules are bent, which suggests they might have a dipole, much like water has a dipole. Second, assuming similar sizes, polar molecules tend to stick to one another more than nonpolar molecules do, which means the boiling points of polar substances tend to be higher. In other words, more heat energy is required for polar molecules to separate from one another. (Note that boiling is a physical change because the molecules remain intact.

In Perspective

In this chapter, we explored two types of chemical bonds: ionic and covalent. Ionic bonds are formed when one or more electrons move from one atom to another. In this way, the atoms become ions—one positive, the other negative—and are held together by the resulting electrical attraction. Covalent bonds form when atoms share electrons. When the sharing is completely equitable, the bond is nonpolar covalent. When one atom pulls more strongly on the electrons because of its greater electronegativity, the bond is polar covalent and a dipole may be formed.

We also looked at how the shape of a molecule can play a role in determining its polarity and how molecular polarity has a great influence on macroscopic behavior. Consider what the world would be like if the oxygen atom in a water molecule did not have its two nonbonding pairs of electrons. Instead of being bent, each water molecule would be linear, much like carbon dioxide. The dipoles of the two hydrogen–oxygen bonds would cancel each other, which would make water a nonpolar substance and give it a relatively low boiling point. Water would not be a liquid at the ambient temperatures of our planet, and we in turn would not be here discussing these concepts. Hooray for the two nonbonding pairs on the oxygen atom! Hooray for the insights we gain by thinking about the molecular realm!

Key Terms and Matching Definitions

_____ covalent bond
_____ covalent compound
_____ dipole
_____ electron-dot structure
_____ electronegativity
_____ ion
_____ ionic bond
_____ ionic compound
_____ molecule
_____ nonbonding pair
_____ nonpolar bond
_____ polar bond
_____ polyatomic ion
_____ substituent
_____ valence electron
_____ valence shell
_____ valence-shell electron-pair repulsion

1. An electron that is located in the outermost occupied shell in an atom and can participate in chemical bonding.
2. The outermost occupied shell of an atom.
3. A shorthand notation of the shell model of the atom in which valence electrons are shown around an atomic symbol.
4. Two paired valence electrons that don't participate in a chemical bond and yet influence the shape of the molecule.
5. An electrically charged particle created when an atom either loses or gains one or more electrons.
6. An ionically charged molecule.
7. A chemical bond in which an attractive electric force holds ions of opposite charge together.
8. Any chemical compound containing ions.
9. A chemical bond in which atoms are held together by their mutual attraction for two or more electrons they share.
10. An element or chemical compound in which atoms are held together by covalent bonds.
11. A group of atoms held tightly together by covalent bonds.
12. A model that explains molecular geometries in terms of electron pairs striving to be as far apart from one another as possible.
13. An atom or nonbonding pair of electrons surrounding a central atom.

14. A separation of charge that occurs in a chemical bond because of differences in the electronegativities of the bonded atoms.
15. The ability of an atom to attract a bonding pair of electrons to itself when bonded to another atom.
16. A chemical bond that has no dipole.
17. A chemical bond that has a dipole.

Review Questions

An Atomic Model Is Needed to Understand How Atoms Bond

1. How many shells are needed to account for the seven periods of the periodic table?

2. How many electrons can fit in the first shell? How many in the second shell?

3. How many shells are completely filled in an argon atom, Ar (atomic number 18)?

4. Which electrons are represented by an electron-dot structure?

5. How do the electron-dot structures of elements in the same group in the periodic table compare with one another?

6. How many nonbonding pairs are there in the valence shell of an oxygen atom? How many unpaired valence electrons?

Atoms Can Lose or Gain Electrons to Become Ions

7. How does an ion differ from an atom?

8. To become a negative ion, does an atom lose or gain electrons?

9. Do metals more readily gain or lose electrons?

10. How many electrons does the calcium atom tend to lose?

11. Why does the fluorine atom tend to gain only one electron?

12. What do molecules lose or gain to become polyatomic ions?

Ionic Bonds Result from a Transfer of Electrons

13. Which elements tend to form ionic bonds?

14. Is an ionic compound an example of a chemical compound, or is a chemical compound an example of an ionic compound?

15. What is the electric charge on the calcium ion in the compound calcium chloride, $CaCl_2$?

16. What is the electric charge on the calcium ion in the compound calcium oxide, CaO?

17. Suppose an oxygen atom gains two electrons to become an oxygen ion. What is its electric charge?

18. What is an ionic crystal?

Covalent Bonds Result from a Sharing of Electrons

10. Which elements tend to form covalent bonds?

20. What force holds two atoms together in a covalent bond?

21. How many electrons are shared in a double covalent bond?

22. How many electrons are shared in a triple covalent bond?

23. How many valence electrons is an oxygen atom able to attract from other atoms?

24. How many covalent bonds is an oxygen atom able to form?

Valence Electrons Determine Molecular Shape

25. What does *VSEPR* stand for?

26. How many faces are there on a tetrahedron?

27. What is meant by the term *substituent*?

28. When is the geometry of a molecule not the same as its shape?

29. How many substituents does the oxygen atom in a water molecule have?

Polar Covalent Bonds Result from an Uneven Sharing of Electrons

30. What is a dipole?

31. Which element of the periodic table has the greatest electronegativity? Which has the smallest?

32. Which is more polar: a carbon–oxygen bond or a carbon–nitrogen bond?

33. How is a polar covalent bond similar to an ionic bond?

Molecular Polarity Results from an Uneven Distribution of Electrons

34. How can a molecule be nonpolar when it consists of atoms that have different electronegativities?

35. Why do nonpolar substances tend to boil at relatively low temperatures?

36. Which tends to have a greater degree of symmetry: a polar molecule or a nonpolar molecule?

37. Why don't oil and water mix?

38. Which would you describe as "stickier": a polar molecule or a nonpolar one?

Hands-On Chemistry Insights

Up Close with Crystals

One thing you probably noticed under the magnifying glass when you compared uncrushed crystals was sharp, angular edges in NaCl and rounded edges in KCl. Then you probably found it easier to grind the KCl crystals to powder. These differences have the same origin: a potassium ion, K^+, is larger than a sodium ion, Na^+.

The positive and negative ions in a crystal are attracted to one another by the attractive electric force between oppositely charged particles. The negative charge on a negative ion is, as we saw in Chapter 5, outside the nucleus, distributed among all the electrons. The positive charge on a positive ion, however, is all in the nucleus. This means the positive and negative charges of the ionic bond are farther apart in the compound containing the larger positive ion:

Shorter distance between positive and negative charges

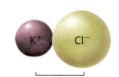

Longer distance between positive and negative charges

Because the electric force weakens with increasing distance between the opposite charges, the KCl ionic bond is weaker than the NaCl ionic bond. Weaker ionic bonds mean that KCl crystals are less resilient to stress and impact than are NaCl crystals, accounting for the rounder edges you observed in the KCl crystals and for the fact that it was easier to grind the KCl to a powder.

This difference in bond strength is also responsible for many other differences in the physical properties of these two substances. For instance, whereas the melting point of NaCl is 801°C, that of KCl is "only" 770°C. The 31 C° difference is easy to explain in terms of what happens when a solid-to-liquid phase change occurs: the particles of the solid have to be pried apart from one another. The weaker ionic bonds in KCl mean the ions separate more easily, and the macroscopic evidence of this is the lower melting point of KCl.

Gumdrop Molecules

Your molecular models should look like this:

Difluoromethane,
CH_2F_2,
tetrahedron

Ethane,
C_2H_6,
two tetrahedrons

Hydrogen peroxide,
H_2O_2,
two bent shapes
stuck together

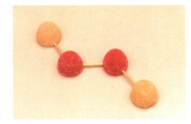

$$H - C \equiv C - H$$

Acetylene,
C_2H_2,
linear

Exercises

1. An atom loses an electron to another atom. Is this an example of a physical change or a chemical change?

2. Why is it so easy for a magnesium atom to lose two electrons?

3. Why doesn't the sodium atom gain seven electrons so that its third shell becomes the filled outermost occupied shell?

4. Magnesium ions carry a 2+ charge, and chloride ions carry a 1− charge. What is the chemical formula for the ionic compound magnesium chloride?

5. Barium ions carry a 2+ charge, and nitrogen ions carry a 3− charge. What is the chemical formula for the ionic compound barium nitride?

6. Does an ionic bond have a dipole?

7. Why doesn't a neon atom tend to gain electrons?

8. Why doesn't a neon atom tend to lose electrons?

9. Why doesn't a hydrogen atom form more than one covalent bond?

10. What drives an atom to form a covalent bond: its nuclear charge or the need to have a filled valence shell? Explain.

11. Is there an abrupt change or a gradual change between ionic and covalent bonds? Explain.

12. Classify the following bonds as ionic, polar covalent, or nonpolar covalent (O, atomic number 8; F, atomic number 9; Na, atomic number 11; Cl, atomic number 17; Ca, atomic number 20; U, atomic number 92):

    O with F _____

    Ca with Cl _____

    Na with Na _____

    U with Cl _____

13. Atoms of nonmetallic elements form covalent bonds, but they can also form ionic bonds. How is this possible?

14. Atoms of metallic elements can form ionic bonds, but they are not very good at forming covalent bonds. Why?

15. Phosphine is a covalent compound of phosphorus, P, and hydrogen, H. What is its chemical formula?

16. Why is a germanium chloride molecule, $GeCl_2$, bent even though there are only two atoms surrounding the central germanium atom?

17. Write the electron-dot structure for the ionic compound calcium chloride, $CaCl_2$.

18. Write the electron-dot structure for the covalent compound ethane, C_2H_6.

19. Write the electron-dot structure for the covalent compound hydrogen peroxide, H_2O_2.

20. Write the electron-dot structure for the covalent compound acetylene, C_2H_2.

21. In two dimensions, sulfuric acid, H_2SO_4, is often written

$$HO-\overset{\overset{\displaystyle O}{\|}}{\underset{\underset{\displaystyle O}{\|}}{S}}-OH$$

What three-dimensional shape does this molecule most likely have?

22. Examine the three-dimensional geometries of PF_5 and SF_4 in Table 6.2. Which do you think is the more polar compound?

23. What is the source of an atom's electronegativity?

24. Which bond is most polar: H–N, N–C, C–O, C–C, O–H, C–H?

25. Which molecule is most polar: S=C=S, O=C=O, O=C=S?

26. In each molecule, which atom carries the greater positive charge: H–Cl, Br–F, C≡O, Br–Br?

27. List these bonds in order of increasing polarity: N–N, N–F, N–O, H–F

_____ < _____ < _____ < _____
(least polar)   (most polar)

28. Which is more polar: a sulfur–bromine bond, S–Br, or a selenium–chlorine bond, Se–Cl?

29. Water, H_2O, and methane, CH_4, have about the same mass and differ by only one type of atom. Why is the boiling point of water so much higher than that of methane?

30. An individual carbon–oxygen bond is polar. Yet carbon dioxide, CO_2, which has two carbon–oxygen bonds, is nonpolar. Explain.

31. In each pair, which compound probably has the higher boiling point (atomic numbers: Cl 17, S 16, O 8, C 6, H 1):

(a) $\underset{\displaystyle H}{\overset{\displaystyle Cl}{\diagdown}}C=C\underset{\displaystyle H}{\overset{\displaystyle Cl}{\diagup}}$ $\underset{\displaystyle Cl}{\overset{\displaystyle H}{\diagdown}}C=C\underset{\displaystyle H}{\overset{\displaystyle Cl}{\diagup}}$

(b) S=C=O   O=C=O

(c) $\underset{\displaystyle Cl}{\overset{\displaystyle Cl}{\diagdown}}C=O$ $\underset{\displaystyle Cl}{\overset{\displaystyle Cl}{\diagdown}}C=C\underset{\displaystyle H}{\overset{\displaystyle H}{\diagup}}$

32. Why is ammonia, NH_3, more polar than borane, BH_3?

Exploring Further

http://www.ada.org/public/topics/fluoride/fluoride.html
Fluoride page of the American Dental Association, with many links to information regarding fluorides and fluoridation of drinking water and toothpastes.

http://www.google.com
Fluoride ions at concentrations of about 1 milligram per liter have proved most effective at preventing tooth decay. At greater concentrations, fluoride ions are toxic. A 10-gram dose of sodium fluoride, for example, is enough to kill an adult. Use the Google search engine to

explore the controversies of fluoride ions in our environment, using such phrases as *fluoride ion and tooth decay* and *fluoride ion toxicity* to do your search. Remember during your Internet searches that the loudest voices you encounter are not necessarily the most accurate.

http://www.saltinstitute.org/idd.htm

Numerous reports in the literature demonstrate the effectiveness of iodized salt in controlling the medical condition called goiter. Check out this site for historical case studies that first pointed to this conclusion.

http://www.soils.wisc.edu/virtual_museum

Home page of the Virtual Museum of Minerals and Molecules, curated by Phillip Barak of the University of Minnesota and Ed Nater of the University of Wisconsin. Through this site, you will find molecular models that you can manipulate in three dimensions. To do so, your browser will need to be equipped with the Chime plug-in, which you may download by following the hyperlinks to http://www.mdli.com/download/chimedown.html.

Chemical Bonding and Molecular Shapes

Visit The Chemistry Place at:
www.aw.com/chemplace

7

Molecular Mixing

How Molecules Attract One Another

Can fish drown? To many people, this question may sound silly. Recall from Section 2.4, however, that fish do not "breathe" water. Rather, their gills are equipped to extract oxygen molecules mixed in with water. Fish therefore can drown if they are in water that contains an insufficient number of oxygen molecules. This is what happens when excessive amounts of organic wastes are discharged into a lake or river. As we explore in Chapter 16, the organic wastes are consumed by microorganisms that also use molecular oxygen. As these microorganisms thrive, the amount of molecular oxygen in the water drops to the point where fish and many other aquatic organisms drown.

The number of oxygen molecules that can mix with a given volume of water is amazingly low. Water that has been fully aerated at room temperature, for example, contains only about 1 oxygen molecule for every 200,000 water molecules, a ratio represented pictorially in the illustration to the right. The gills of a fish, therefore, must be highly efficient at extracting molecular oxygen from water.

This chapter explains how many of the physical properties of materials are a consequence of attractions among the submicroscopic particles making up the materials. Why only small amounts of oxygen can mix with water, for example, can be explained by the fact that the attractive forces between water molecules and oxygen molecules are very weak. We begin by looking at four types of electrical attractions that occur between submicroscopic particles.

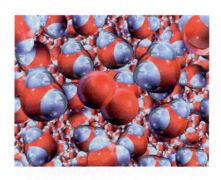

7.1 Submicroscopic Particles Electrically Attract One Another

We can think of any pure substance as being made up of one type of sub-microscopic particle. For an ionic compound, that particle is an ion; for a covalent compound, it is a molecule; and for an element, it is an atom.

Table 7.1 lists four types of electrical attractions that can occur between these particles. The strength of even the strongest of these attractions is many times weaker than any chemical bond, however. The attraction between two adjacent water molecules, for example, is about 20 times weaker than the chemical bonds holding the hydrogen and oxygen atoms together in the water molecules. Although particle-to-particle attractions are relatively weak, you can see their profound effect on the substances around you.

We now explore these interparticle attractions in order of relative strength, beginning with the strongest.

Table 7.1

Electrical Attractions Between Submicroscopic Particles

Attraction	Relative Strength
Ion–dipole	Strongest
Dipole–dipole	
Dipole–induced dipole	
Induced dipole–induced dipole	Weakest

Ions and Polar Molecules Attract One Another

You probably remember from Chapter 6 that a polar molecule is one in which the bonding electrons are unevenly distributed. One side of the molecule carries a slight negative charge, and the opposite side carries a slight positive charge. This separation of charge is a dipole.

So what happens to polar molecules, such as water molecules, when they are near an ionic compound, such as sodium chloride? The opposite charges electrically attract one another. The positive sodium ions attract the negative side of the water molecules, and the negative chloride ions attract the positive side of the water molecules. This is illustrated in Figure 7.1. Such an attraction between an ion and the dipole of a polar molecule is called an *ion–dipole attraction*.

Ion–dipole attractions are much weaker than ionic bonds. However, a large number of ion–dipole attractions can act collectively to disrupt an ionic bond. This is what happens to sodium chloride in water. Attractions

Figure 7.1
Electrical attractions are shown as a series of overlapping arcs. The blue arcs indicate negative charge, and the red arcs indicate positive charge.

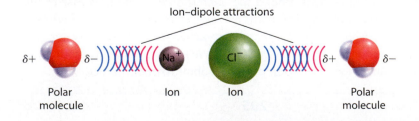

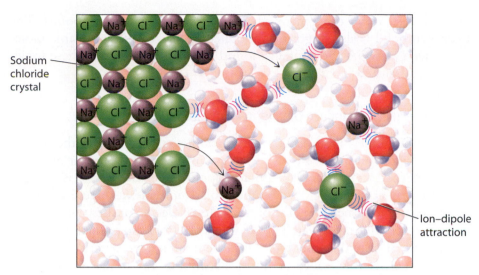

Aqueous solution of sodium chloride

Figure 7.2
Sodium and chloride ions tightly bound in a crystal lattice are separated from one another by the collective attraction exerted by many water molecules to form an aqueous solution of sodium chloride.

exerted by the water molecules break the ionic bonds and pull the ions away from one another. The result, represented in Figure 7.2, is a solution of sodium chloride in water. (A solution in water is called an *aqueous solution.*)

Polar Molecules Attract Other Polar Molecules

An attraction between two polar molecules is called a *dipole–dipole attraction.* An unusually strong dipole–dipole attraction is the **hydrogen bond**. This attraction occurs between molecules that have a hydrogen atom covalently bonded to a highly electronegative atom, usually nitrogen, oxygen, or fluorine. Recall from Chapter 6 that the electronegativity of an atom describes how well that atom is able to pull bonding electrons toward itself. The greater the atom's electronegativity, the better it is able to gain electrons and thus the more negative is its charge.

Look at Figure 7.3 to see how hydrogen bonding works. The hydrogen side of a polar molecule (water in this example) has a positive charge because the more electronegative oxygen the hydrogen is bonded to tugs on the hydrogen's electron. This hydrogen is therefore electrically attracted to a pair of nonbonding electrons on the negatively charged atom of another molecule (in this case, another water molecule). This mutual attraction between hydrogen and the negatively charged atom of another molecule is a hydrogen bond.

The strength of a hydrogen bond depends on two things: (1) the strength of the dipoles involved (this in turn depends on the difference in electronegativity of the atoms in the polar molecules) and (2) how strongly nonbonding electrons on one molecule can attract a hydrogen atom on a nearby molecule.

Even though the hydrogen bond is much weaker than any covalent or ionic bond, the effects of hydrogen bonding can be very pronounced.

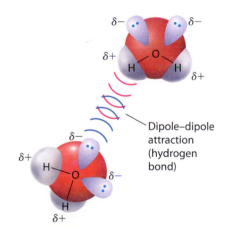

Figure 7.3
The dipole–dipole attraction between two water molecules is a hydrogen bond because it involves hydrogen atoms bonded to highly electronegative oxygen atoms.

For example, water owes many of its properties to hydrogen bonds. The hydrogen bond is also of great importance in the chemistry of large molecules, such as DNA and proteins, found in living organisms. These molecules are discussed in Chapter 13.

Polar Molecules Can Induce Dipoles in Nonpolar Molecules

In many molecules, the electrons are distributed evenly, and so there is no dipole. The oxygen molecule, O_2, is an example. Such a nonpolar molecule can be induced to become a temporary dipole, however, when it is brought close to a water molecule or any other polar molecule, as Figure 7.4 illustrates. The slightly negative side of the water molecule pushes the electrons in the oxygen molecule away. Thus oxygen's electrons are pushed to the side that is farthest from the water molecule. The result is a temporary uneven distribution of electrons called an **induced dipole**. The resulting attraction between the permanent dipole (water) and the induced dipole (oxygen) is a *dipole–induced dipole attraction.*

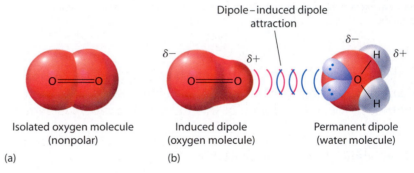

Figure 7.4
(a) An isolated oxygen molecule has no dipole; its electrons are distributed evenly. (b) An adjacent water molecule induces a redistribution of electrons in the oxygen molecule. (The slightly negative side of the oxygen molecule is shown larger than the slightly positive side because the slightly negative side contains more electrons.)

Concept Check ✔

How does the electron distribution in an oxygen molecule change when the hydrogen side of a water molecule is nearby?

Was this your answer? Because the hydrogen side of the water molecule is slightly positive, the electrons in the oxygen molecule are pulled *toward* the water molecule, inducing in the oxygen molecule a temporary dipole in which the larger side is nearest the water molecule (rather than as far away as possible as it was in Figure 7.4).

Remember, induced dipoles are only temporary. If the water molecule in Figure 7.4b were removed, the oxygen molecule would return to its normal, nonpolar state. As a consequence, dipole–induced dipole attractions are weaker than dipole–dipole attractions. Dipole–induce dipole attractions are strong enough to hold relatively small quantities of oxygen dissolved in water, however. As this chapter's introduction discusses, this attraction between water and molecular oxygen is vital for fish and other forms of aquatic life that rely on molecular oxygen mixed in water.

Dipole–induced dipole attractions also occur between molecules of carbon dioxide, which are nonpolar, and water. It is these attractions that help keep carbonated beverages (which are mixtures of carbon dioxide in water) from losing their fizz too quickly after they've been opened. Dipole–induced dipole attractions are also responsible for holding plastic wrap to glass, as shown in Figure 7.5. These wraps are made of very long nonpolar molecules that are induced to have dipoles when placed in contact with glass, which is highly polar. As is discussed next, the molecules of a nonpolar material, such as plastic wrap, can also induce dipoles among themselves. This explains how plastic wrap sticks not only to polar materials such as glass but also to itself.

Figure 7.5
Temporary dipoles induced in the normally nonpolar molecules in plastic wrap makes it stick to glass.

Concept Check ✔

Distinguish between a dipole–dipole attraction and a dipole–induced dipole attraction.

Was this your answer? The dipole–dipole attraction is stronger and involves two permanent dipoles. The dipole–induced dipole attraction is weaker and involves a permanent dipole and a temporary one.

Atoms and Nonpolar Molecules Can Form Temporary Dipoles on Their Own

Nonpolar argon   Temporary dipole in argon

Individual atoms and nonpolar molecules, on average, have a fairly even distribution of electrons. Because of the randomness of electron motion, however, at any given moment the electrons in an atom or a nonpolar molecule may be bunched to one side. The result is a temporary dipole, as shown in Figure 7.6.

Just as the permanent dipole of a polar molecule can induce a dipole in a nonpolar molecule, a temporary dipole can do the same thing. This gives rise to the weakest of the particle-to-particle attractions: the *induced dipole–induced dipole attraction*, illustrated in Figure 7.7.

Figure 7.6
The electron distribution in an atom is normally even. At any given moment, however, the electron distribution may be somewhat uneven, resulting in a temporary dipole.

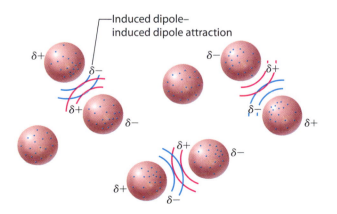

Figure 7.8
Because the normally even distribution of electrons in atoms can momentarily become uneven, atoms can be attracted to one another by induced dipole–induced dipole attractions.

Temporary dipoles are more significant for larger atoms. This is because the electrons in larger atoms have more space available for random motion and a greater likelihood of bunching together on one side. The electrons in smaller atoms are less able to bunch to one side because they are confined to a smaller space. The resulting greater electrical repulsion tends to keep these electrons evenly spread out. So it is larger atoms—and molecules made of larger atoms—that have the strongest induced dipole–induced dipole attractions.

As shown in Figure 7.8, nonpolar iodine molecules, I_2, are relatively large. Because of this, they have a greater attraction for one another than do relatively small nonpolar fluorine molecules, F_2. This explains why iodine molecules stick together as a solid at room temperature but at the same temperature fluorine molecules drift apart into the gaseous phase.

Fluorine is one of the smallest atoms, and nonpolar molecules made with fluorine atoms exhibit only very weak induced dipole–induced dipole attractions. This is the principle behind the Teflon nonstick surface. The Teflon molecule, part of which is shown in Figure 7.9, is a long chain of carbon atoms chemically bonded to fluorine atoms, and the fluorine atoms exert essentially no attractions on any material in contact with the Teflon surface—scrambled eggs in a frying pan, for instance.

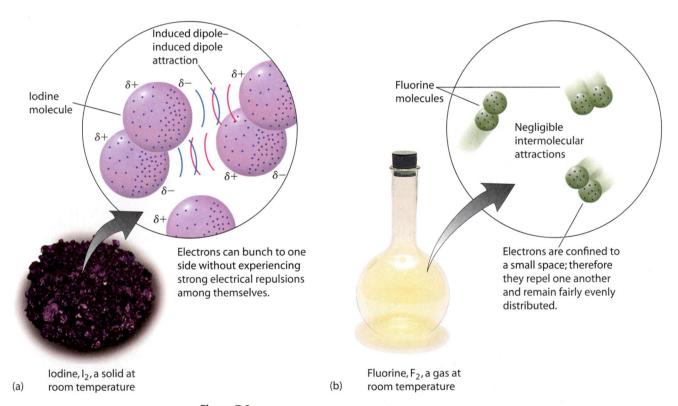

Figure 7.8
(a) Temporary dipoles more readily form in larger atoms, such as those in an iodine molecule, because in larger atoms electrons bunched to one side are still relatively far apart from one another and not so repelled by the electric force. (b) In smaller atoms, such as those in a fluorine molecule, electrons cannot bunch to one side as well because the repulsive electric force gets greater as the electrons get closer together.

Figure 7.9
Few things stick to Teflon because of the high proportion of fluorine atoms it contains. The structure depicted here is only a portion of the full length of the molecule.

Concept Check ✔

What is the distinction between a dipole–induced dipole attraction and an induced dipole–induced dipole attraction?

Was this your answer? The dipole–induced dipole attraction is stronger and involves a permanent dipole and a temporary one. The induced dipole–induced dipole attraction is weaker and involves two temporary dipoles.

Induced dipole–induced dipole attractions help explain why natural gas is a gas at room temperature but gasoline is a liquid. The major component of natural gas is methane, CH_4, and one of the major components of gasoline is octane, C_8H_{18}. We see in Figure 7.10 that the number of induced dipole–induced dipole attractions between two methane molecules is appreciably less than the number between two octane molecules. You know that two small pieces of Velcro are easier to pull apart than two long pieces. Like short pieces of Velcro, methane molecules can be pulled apart with little effort. That's why methane has a low boiling point, $-161°C$, and is a gas at room temperature. Octane molecules, like long strips of Velcro, are relatively hard to pull apart because of the larger number of induced dipole–induced dipole attractions. The boiling point of octane, 125°C, is therefore much higher than that of methane, and octane is a liquid at room temperature. (The greater mass of octane also plays a role in making its boiling point higher.)

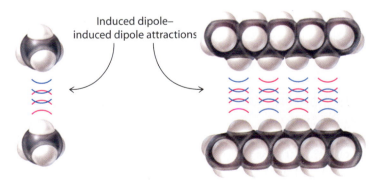

(a) Methane molecules   (b) Octane molecules

Figure 7.10
(a) Two nonpolar methane molecules are attracted to each other by induced dipole–induced dipole attractions, but there is only one attraction per molecule. (b) Two nonpolar octane molecules are similar to methane but longer. The number of induced dipole–induced dipole attractions between these two molecules is therefore greater.

Hands-On Chemistry: Circular Rainbows

Black ink contains pigments of many different colors. Acting together, these pigments absorb all the frequencies of visible light. Because no light is reflected, the ink appears black. We can use electrical attractions to separate the components of black ink with a technique called paper chromatography.

What You Need

Black felt-tip pen or black water-soluble marker; piece of porous paper, such as paper towel, table napkin, or coffee filter; solvent, such as water, acetone (fingernail polish remover), rubbing alcohol, or white vinegar.

Procedure

① Place a concentrated dot of ink at the center of the piece of porous paper.

② Carefully place one drop of solvent on top of the dot, and watch the ink spread radially with the solvent. Because the different components of the ink have different affinities for the solvent (based on the electrical attractions between component molecules and solvent molecules), they travel with the solvent at different rates.

③ Just after the drop of solvent is completely absorbed, add a second drop at the same place you put the first one, then a third, and so on until the ink components have separated to your satisfaction.

How the components separate depends on several factors, including your choice of solvent and your technique. It's also interesting to watch the leading edge of the moving ink under a strong magnifying glass or microscope.

Concept Check ✔

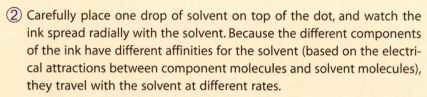

Methanol, CH_3OH, which can be used as a fuel, is not much larger than methane, CH_4, but is a liquid at room temperature. Suggest why.

Was this your answer? The polar oxygen–hydrogen covalent bond in each methanol molecule leads to hydrogen bonding between molecules. These relatively strong interparticle attractions hold methanol molecules together as a liquid at room temperature.

7.2   A Solution Is a Single-Phase Homogeneous Mixture

What happens when table sugar, known chemically as sucrose, is stirred into water? Is the sucrose destroyed? We know it isn't because it sweetens the water. Does the sucrose disappear because it somehow ceases to occupy space or because it fits within the nooks and crannies of the water? Not so, for the addition of sucrose changes the volume. This may not be noticeable at first, but continue to add sucrose to a glass of water and you'll see that the water level rises just as it would if you were adding sand.

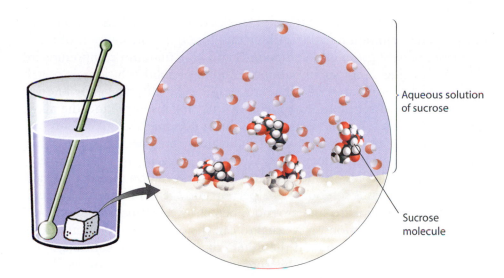

Aqueous solution of sucrose

Sucrose molecule

Figure 7.11
Water molecules pull the sucrose molecules in a sucrose crystal away from one another. This pulling away from the crystal does not, however, affect the covalent bonds within each sucrose molecule, which is why each dissolved sucrose molecule remains intact as a single molecule.

Sucrose stirred into water loses its crystalline form. Each sucrose crystal consists of billions upon billions of sucrose molecules packed neatly together. When the crystal is exposed to water, as was first shown in Figure 2.17 and is shown again here in Figure 7.11, an even greater number of water molecules pull on the sucrose molecules via hydrogen bonds formed between sucrose molecules and water molecules. With a little stirring, the sucrose molecules soon mix throughout the water. In place of sucrose crystals and water, we have a homogeneous mixture of sucrose molecules in water. As discussed in Section 2.5, homogeneous means that a sample taken from one part of a mixture is the same as a sample taken from any other part of the mixture. In our sucrose example, this means that the sweetness of the first sip of the solution is the same as the sweetness of the last sip.

Recall from Section 2.5 that a homogeneous mixture consisting of a single phase is called a *solution*. Sugar in water is a solution in the liquid phase. Solutions aren't always liquids, however. They can also be solid or gaseous, as Figure 7.12 shows. Gem stones are solid solutions. A ruby, for

(a)

(b)

(c)

Figure 7.12
Solutions may occur in (a) the solid phase, (b) the liquid phase, or (c) the gaseous phase.

example, is a solid solution of trace quantities of red chromium compounds in transparent aluminum oxide. A blue sapphire is a solid solution of trace quantities of light green iron compounds and blue titanium compounds in aluminum oxide. Another important example of solid solutions is metal alloys, which are mixtures of different metallic elements. The alloy known as brass is a solid solution of copper and zinc, for instance, and the alloy stainless steel is a solid solution of iron, chromium, nickel, and carbon.

An example of a gaseous solution is the air we breathe. By volume, this solution is 78 percent nitrogen gas, 21 percent oxygen gas, and 1 percent other gaseous materials, including water vapor and carbon dioxide. The air we *exhale* is a gaseous solution of 75 percent nitrogen, 14 percent oxygen, 5 percent carbon dioxide, and around 6 percent water vapor.

In describing solutions, it is usual to call the component present in the largest amount the **solvent** and the other component(s) the **solute(s)**. For example, when a teaspoon of table sugar is mixed with 1 liter of water, we identify the sugar as the solute and the water as the solvent.

The process of a solute mixing in a solvent is called **dissolving**. To make a solution, a solute must *dissolve* in a solvent; that is, the solute and solvent must form a homogeneous mixture. Whether or not one material dissolves in another is a function of electrical attractions.

Concept Check ✓

What is the solvent in the gaseous solution we call air?

Was this your answer? Nitrogen is the solvent because it is the component present in the greatest quantity.

There is a limit to how much of a given solute can dissolve in a given solvent, as Figure 7.13 illustrates. When you add table sugar to a glass of water, for example, the sugar rapidly dissolves. As you continue to add sugar, however, there comes a point when it no longer dissolves. Instead, it collects at the bottom of the glass, even after stirring. At this point, the water is *saturated* with sugar, meaning the water cannot accept any more sugar. When this happens, we have what is called a **saturated solution**, defined as one in which no more solute can dissolve. A solution that has not reached the limit of solute that will dissolve is called an **unsaturated solution**.

Figure 7.13
A maximum of 200 grams of sucrose dissolves in 100 milliliters of water at 20°C. (a) Mixing 150 grams of sucrose in 100 milliliters of water at 20°C produces an unsaturated solution. (b) Mixing 200 grams of sucrose in 100 milliliters of water at 20°C produces a saturated solution. (c) If 250 grams of sucrose is mixed with 100 milliliters of water at 20°C, 50 grams of sucrose remains undissolved. (As we discuss later, the concentration of a saturated solution varies with temperature.)

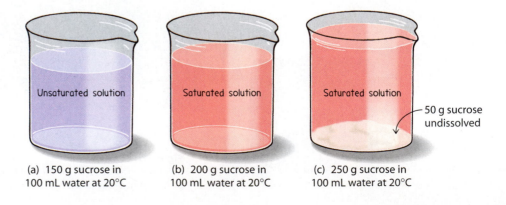

(a) 150 g sucrose in 100 mL water at 20°C — Unsaturated solution

(b) 200 g sucrose in 100 mL water at 20°C — Saturated solution

(c) 250 g sucrose in 100 mL water at 20°C — Saturated solution — 50 g sucrose undissolved

The quantity of solute dissolved in a solution is described in mathematical terms by the solution's **concentration**, which is the amount of solute dissolved per amount of solution:

$$\text{concentration of solution} = \frac{\text{amount of solute}}{\text{amount of solution}}$$

For example, a sucrose–water solution may have a concentration of 1 gram of sucrose for every liter of solution. This can be compared with concentrations of other solutions. A sucrose–water solution containing 2 grams of sucrose per liter of solution, for example, is more *concentrated*, and one containing only 0.5 gram of sucrose per liter of solution is less concentrated, or more *dilute*.

Chemists are often more interested in the number of solute particles in a solution rather than the number of grams of solute. Submicroscopic particles, however, are so very small that the number of them in any observable sample is incredibly large. To get around having to use awkwardly large numbers, scientists use a unit called the mole. One **mole** of any type of particle is, by definition, 6.02×10^{23} particles (this superlarge number is about 602 billion trillion):

$$
\begin{aligned}
1 \text{ mole} &= 6.02 \times 10^{23} \text{ particles} \\
&= 602{,}000{,}000{,}000{,}000{,}000{,}000{,}000{,}000 \text{ particles}
\end{aligned}
$$

One mole of pennies, for example, is 6.02×10^{23} pennies, 1 mole of marbles is 6.02×10^{23} marbles, and 1 mole of sucrose molecules is 6.02×10^{23} sucrose molecules.

Even if you've never heard the term *mole* in your life before now, you are already familiar with the basic idea. Saying "one mole" is just a shorthand way of saying "six point oh two times ten to the twenty-third." Just as *a couple of* means 2 of something and *a dozen of* means 12 of something, *a mole of* means 6.02×10^{23} of something. It's as simple as that:

- a couple of coconuts = 2 coconuts
- a dozen of donuts = 12 donuts
- a mole of mints = 6.02×10^{23} mints
- a mole of molecules = 6.02×10^{23} molecules

A stack containing 1 mole of pennies would reach a height of 860 quadrillion kilometers, which is roughly equal to the diameter of our Milky Way Galaxy. A mole of marbles would be enough to cover the entire land area of the 50 United States to a depth greater than 1.1 kilometers. Sucrose molecules are so small, however, that there are 6.02×10^{23} of them in only 342 grams of sucrose, which is about a cupful. Thus because 342 grams of sucrose contains 6.02×10^{23} molecules of sucrose, we can use our shorthand wording and say that 342 grams of sucrose contains 1 mole of sucrose. As Figure 7.14 shows, therefore, an aqueous solution that has a concentration of 342 grams of sucrose per liter of solution also has a concentration of 6.02×10^{23} sucrose molecules per liter of solution or, by definition, a concentration of 1 mole of sucrose per liter of solution.

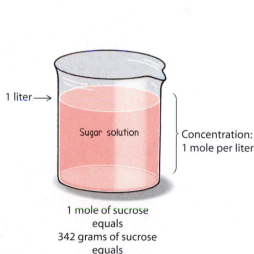

1 mole of sucrose
equals
342 grams of sucrose
equals
6.02×10^{23} molecules of sucrose

Figure 7.14
An aqueous solution of sucrose that has a concentration of 1 mole of sucrose per liter of solution contains 6.02×10^{23} sucrose molecules (342 grams) in every liter of solution.

Calculation Corner: Calculating for Solutions

From the formula for the concentration of a solution, we can derive equations for amount of solute and amount of solution:

$$\text{concentration of solution} = \frac{\text{amount of solute}}{\text{amount of solution}}$$

$$\text{amount of solute} =$$
$$\text{concentration of solution} \times \text{amount of solution}$$

$$\text{amount of solution} = \frac{\text{amount of solute}}{\text{concentration of solution}}$$

In solving for any of these values, the units must always match. If concentration is given in grams per liter, for example, the amount of solute must be in grams and the amount of solution must be in liters.

Note that these equations are set up for calculating amount of *solution* rather than amount of *solvent*. The amount of solution is greater than the amount of solvent because in addition to containing the solvent, the solution also contains the solute. As discussed at the beginning of this section, for example, the volume of an aqueous solution of sucrose depends not only on the volume of water but also on the volume of dissolved sucrose.

Example 1

How much sucrose, in grams, is there in 3 liters of an aqueous solution that has a concentration of 2 grams of sucrose per liter of solution?

Answer 1

This question asks for amount of solute, and so you should use the second of the three formulas given above:

$$\text{amount of solute} = \frac{2\text{ g}}{1\text{ L}} \times 3\text{ L} = 6\text{ g}$$

Example 2

A solution you are using in an experiment has a concentration of 10 grams of solute per liter of solution. If you pour enough of this solution into an empty laboratory flask to make the flask contain 5 grams of the solute, how many liters of the solution have you poured into the flask?

Answer 2

This question asks for amount of solution, and so what you want is the third formula:

$$\text{amount of solution} = \frac{5\text{ g}}{10\text{ g/L}} = 0.5\text{ L}$$

Your Turn

1. At 20°C, a saturated solution of sodium chloride in water has a concentration of about 380 grams of sodium chloride per liter of solution. How much sodium chloride, in grams, is required to make 3 liters of a saturated solution?

2. A student is told to use 20 grams of sodium chloride to make an aqueous solution that has a concentration of 10 grams of sodium chloride per liter of solution. How many liters of solution does she end up with?

The number of grams tells you the *mass* of solute in a given solution, and the number of moles tells you the actual *number* of molecules. Interestingly, the term *mole* is derived from the Latin word for "pile." A mole of marbles would be one amazingly large pile, wouldn't it! Finding the number of molecules in a given mass or the mass of a given number of molecules is something we explore in Chapter 9.

A common unit of concentration used by chemists is **molarity**, which is the solution's concentration expressed in moles of solute per liter of solution:

$$\text{molarity} = \frac{\text{number of moles of solute}}{\text{liters of solution}}$$

A solution that contains 1 mole of solute per liter of solution has a concentration of 1 *molar*, which is often abbreviated 1 *M*. A more concentrated, 2-molar (2 *M*) solution contains 2 moles of solute per liter of solution.

The difference between referring to the number of molecules of solute and referring to the number of grams of solute can be illustrated by the following question. A saturated aqueous solution of sucrose contains 200 grams of sucrose and 100 grams of water. Which is the solvent: sucrose or water?

As shown in Figure 7.15, there are 3.5×10^{23} molecules of sucrose in 200 grams of sucrose but almost 10 times as many molecules of water in 100 grams of water—3.3×10^{24} molecules. As defined earlier, the solvent is the component present in the largest amount, but what do we mean by *amount*? If amount means number of molecules, then water is the solvent. If amount means mass, then sucrose is the solvent. So, the answer depends on how you look at it. From a chemist's point of view, *amount* typically means the number of molecules, and so water is the solvent in this case.

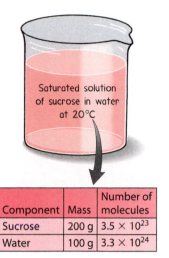

Figure 7.15
Although 200 grams of sucrose is twice as massive as 100 grams of water, there are about 10 times as many water molecules in 100 grams of water as there are sucrose molecules in 200 grams of sucrose. How can this be? Each water molecule is about 20 times less massive (and smaller) than each sucrose molecule, which means that about 10 times as many water molecules can fit within half the mass.

Component	Mass	Number of molecules
Sucrose	200 g	3.5×10^{23}
Water	100 g	3.3×10^{24}

Saturated solution of sucrose in water at 20°C

Concept Check ✔

1. How much sucrose, in moles, is there in 0.5 liter of a 2-molar solution? How many molecules of sucrose is this?
2. Does 1 liter of a 1-molar solution of sucrose in water contain 1 liter of water, less than 1 liter of water, or more than 1 liter of water?

Were these your answers?

1. First you need to understand that 2-molar means 2 moles of sucrose per liter of solution. Then you should multiply solution concentration by amount of solution to obtain amount of solute:

   (2 moles/L)(0.5 L) = 1 mole

   which is the same as 6.02×10^{23} molecules.
2. The definition of molarity refers to the number of liters of *solution*, not liters of solvent. When sucrose is added to a given volume of water, the volume of the solution increases. So, if 1 mole of sucrose is added to 1 liter of water, the result is more than 1 liter of solution. Therefore, 1 liter of a 1-molar solution requires less than 1 liter of water.

Hands-On Chemistry: Overflowing Sweetness

Just because a solid dissolves in a liquid doesn't mean the solid no longer occupies space.

What You Need

Tall glass, warm water, container larger than the tall glass, 4 tablespoons table sugar

Procedure

① Fill the glass to its brim with the warm water, and then carefully pour all the water into the larger container.

② Add the sugar to the empty glass.

③ Return half of the warm water to the glass and stir to dissolve all the sugar.

④ Return the remaining water, and as you get close to the top, ask a friend to predict whether the water level will be less than before, about the same as before, or more than before so that the water spills over the edge of the glass.

If your friend doesn't understand the result, ask him or her what would happen if you had added the sugar to the glass when the glass was full of water.

7.3 Solubility Is a Measure of How Well a Solute Dissolves

The **solubility** of a solute is its ability to dissolve in a solvent. As you might expect, this ability depends in great part on the submicroscopic attractions between solute particles and solvent particles. If a solute has any appreciable solubility in a solvent, then that solute is said to be **soluble** in that solvent.

Solubility also depends on attractions of solute particles for one another and attractions of solvent particles for one another. As shown in Figure 7.16,

Figure 7.16
A sucrose molecule contains many hydrogen–oxygen covalent bonds in which the hydrogen atoms are slightly positive and the oxygen atoms are slightly negative. These dipoles in any given sucrose molecule result in the formation of hydrogen bonds with neighboring sucrose molecules.

Sucrose

for example, there are many polar hydrogen–oxygen bonds in a sucrose molecule. Sucrose molecules, therefore, can form multiple hydrogen bonds with one another. These hydrogen bonds are strong enough to make sucrose a solid at room temperature and give it a relatively high melting point of 185°C. In order for sucrose to dissolve in water, the water molecules must first pull sucrose molecules away from one another. This puts a limit on the amount of sucrose that can dissolve in water—eventually a point is reached where there are not enough water molecules to separate the sucrose molecules from one another. As was discussed in Section 7.2, this is the point of saturation, and any additional sucrose added to the solution does not dissolve.

When the molecule-to-molecule attractions among solute molecules are comparable to the molecule-to-molecule attractions among solvent molecules, the result can be no practical point of saturation. As shown in Figure 7.17, for example, the hydrogen bonds among water molecules are about as strong as those between ethanol molecules. These two liquids therefore mix together quite well and in just about any proportion. We can even add ethanol to water until the ethanol rather than the water may be considered the solvent.

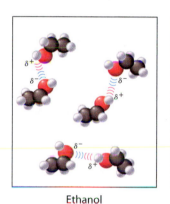
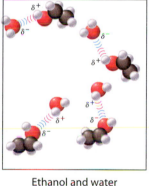
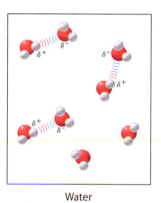

Ethanol   Ethanol and water   Water

Figure 7.17
Ethanol and water molecules are about the same size, and they both form hydrogen bonds. As a result, ethanol and water readily mix with each other.

A solute that has no practical point of saturation in a given solvent is said to be *infinitely soluble* in that solvent. Ethanol, for example, is infinitely soluble in water. Also, all gases are generally infinitely soluble in other gases because they can be mixed together in just about any proportion.

Let's now look at the other extreme of solubility, where a solute has very little solubility in a given solvent. An example is oxygen, O_2, in water. In contrast to sucrose, which has a solubility of 200 grams per 100 milliliters of water, only 0.004 gram of oxygen can dissolve in 100 milliliters of water. We can account for oxygen's low solubility in water by noting that the only electrical attractions that occur between oxygen molecules and water molecules are relatively weak dipole–induced dipole attractions. More important, however, is the fact that the stronger attraction of water molecules for one another—through the hydrogen bonds the water molecules form with one another—effectively excludes oxygen molecules from intermingling.

Figure 7.18
Glass is frosted by dissolving its
outer surface in hydrofluoric acid.

Figure 7.19
Is this cup melting or dissolving?

A material that does not dissolve in a solvent to any appreciable extent is said to be **insoluble** in that solvent. There are many substances we consider to be insoluble in water, including sand and glass. Just because a material is not soluble in one solvent, however, does not mean it won't dissolve in another. Sand and glass, for example, are soluble in hydrofluoric acid, HF, which is used to give glass the decorative frosted look shown in Figure 7.18. Also, although Styrofoam is insoluble in water, it is soluble in acetone, a solvent used in fingernail polish remover. Pour a little acetone into a Styrofoam cup, and the acetone soon dissolves the Styrofoam, as you can see in Figure 7.19.

Concept Check ✓

Why isn't sucrose infinitely soluble in water?

Was this your answer? The attraction between two sucrose molecules is much stronger than the attraction between a sucrose molecule and a water molecule. Because of this, sucrose dissolves in water only so long as the number of water molecules far exceeds the number of sucrose molecules. When there are too few water molecules to dissolve any additional sucrose, the solution is saturated.

Solubility Changes with Temperature

You probably know from experience that water-soluble solids usually dissolve better in hot water than in cold water. A highly concentrated solution of sucrose in water, for example, can be made by heating the solution almost to the boiling point. This is how syrups and hard candy are made.

Solubility increases with increasing temperature because hot water molecules have greater kinetic energy and therefore are able to collide with the solid solute more vigorously. The vigorous collisions facilitate the disruption of electrical particle-to-particle attractions in the solid.

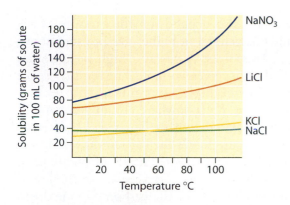

Figure 7.20
The solubility of many water-soluble solids increases with temperature, while the solubility of others is only very slightly affected by temperature.

Although the solubilities of some solid solutes—sucrose, to name just one example—are greatly affected by temperature changes, the solubilities of other solid solutes, such as sodium chloride, are only mildly affected, as Figure 7.20 shows. This difference has to do with a number of factors, including the strength of the chemical bonds in the solute molecules and the way those molecules are packed together.

When a solution saturated at a high temperature is allowed to cool, some of the solute usually comes out of solution and forms what is called a **precipitate**. When this happens, the solute is said to have *precipitated* from the solution. For example, at 165°C the solubility of sodium nitrate, $NaNO_3$, in water is 165 grams per 100 milliliters of water. As we cool this solution, the solubility of $NaNO_3$ decreases as shown in Figure 7.20, and this change in solubility causes some of the dissolved $NaNO_3$ to precipitate (come out of solution). At 20°C, the solubility of $NaNO_3$ is only 87 grams per 100 milliliters of water. So if we cool the 100°C solution to 20°C, 78 grams (165 grams − 87 grams) precipitates, as shown in Figure 7.21.

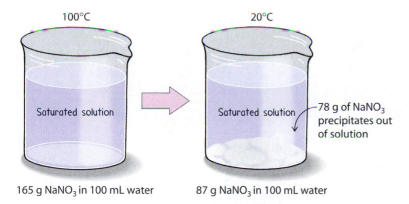

Figure 7.21
The solubility of sodium nitrate is 165 grams per 100 milliliters of water at 100°C but only 87 grams per 100 milliliters at 20°C. Cooling a 100°C saturated solution of $NaNO_3$ to 20°C causes 78 grams of the solute to precipitate.

Gases Are More Soluble at Low Temperatures and High Pressures

In contrast to the solubilities of most solids, the solubilities of gases in liquids *decrease* with increasing temperature, as Table 7.2 on page 222 shows. This is true because with an increase in temperature, the solvent molecules have more kinetic energy. This makes it more difficult for a gaseous solute to stay in solution because the solute molecules are literally being kicked out by the high-energy solvent molecules.

Perhaps you have noticed that warm carbonated beverages go flat faster than cold ones. The higher temperature causes the molecules of carbon dioxide gas to leave the liquid solvent at a higher rate.

Table 7.2

Temperature-Dependent Solubility of Oxygen Gas in Water at a Pressure of 1 Atmosphere

Temperature (°C)	O_2 Solubility (g O_2/L H_2O)
0	0.0141
10	0.0109
20	0.0092
25	0.0083
30	0.0077
35	0.0070
40	0.0065

The solubility of a gas in a liquid also depends on the pressure of the gas immediately above the liquid. In general, a higher gas pressure above the liquid means more of the gas dissolves. A gas at a high pressure has many, many gas particles crammed into a given volume. The "empty" space in an unopened soft drink bottle, for example, is crammed with carbon dioxide molecules in the gaseous phase. With nowhere else to go, many of these molecules dissolve in the liquid, as shown in Figure 7.22. Alternatively, we might say that the great pressure forces the carbon dioxide molecules into solution. When the bottle is opened, the "head" of highly pressurized carbon dioxide gas escapes. Now the gas pressure above the liquid is lower than it was. As a result, the solubility of the carbon dioxide drops and the carbon dioxide molecules once squeezed into the solution begin to escape into the air above the liquid.

The rate at which carbon dioxide molecules leave an opened soft drink is relatively slow. You can increase the rate by pouring in granulated sugar, salt, or sand. The microscopic nooks and crannies on the surface of the grains serve as *nucleation sites* where carbon dioxide bubbles are able to form rapidly and then escape by buoyant forces. Shaking the beverage also increases the surface area of the liquid-to-gas interface, making it easier for the carbon dioxide to escape from the solution. Once the solution is shaken, the rate at which carbon dioxide escapes becomes so great that the beverage froths over. You also increase the rate at which carbon dioxide escapes when you pour the beverage into your mouth, which abounds in nucleation sites. You can feel the resulting tingly sensation.

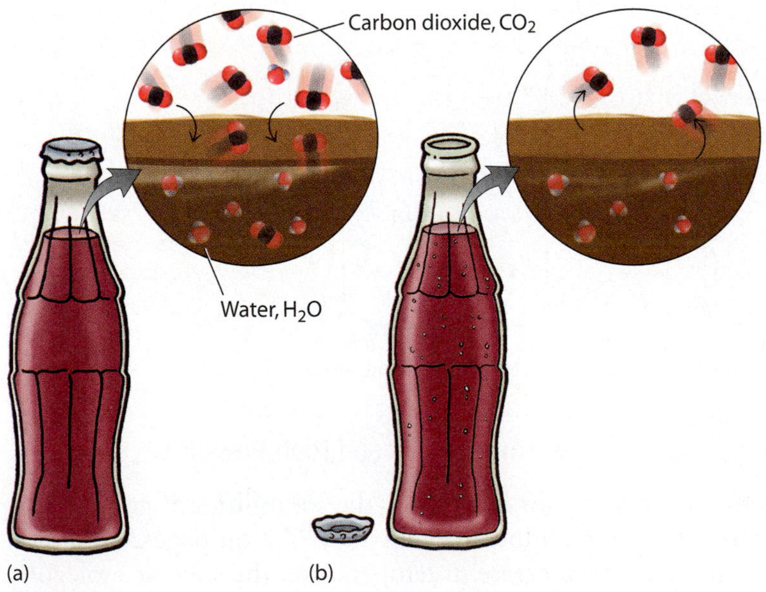

Figure 7.22
(a) The carbon dioxide gas above the liquid in an unopened soft drink bottle consists of many tightly packed carbon dioxide molecules that are forced by pressure into solution. (b) When the bottle is opened, the pressure is released and carbon dioxide molecules originally dissolved in the liquid can escape into the air.

Concept Check ✓

You open two cans of soft drink: one from a warm kitchen shelf, the other from the coldest depths of your refrigerator. Which provides more bubbles in the first gulp you take and why?

Was this your answer? The solubility of carbon dioxide in water decreases with increasing temperature. The warm drink will therefore fizz in your mouth more than does the cold one.

Hands-On Chemistry: Crystal Crazy

If a hot saturated solution is allowed to cool slowly and without disturbance, the solute may stay in solution. The result is a *supersaturated* solution. Supersaturated aqueous solutions of sucrose (table sugar) are fairly easy to make.

What You Need

Small cooking pot, water, table sugar, pencil that is longer than the diameter of the pot, string, weight (a nut or bolt works well), safety glasses to protect eyes from any hot liquid that may splatter

Procedure

① Fill the pot no more than 1 inch deep with water and heat the water to boiling.

② Lower the heat to medium–low. Slowly pour in sugar while carefully stirring to avoid splattering. Because sugar is very soluble in hot water, be prepared to add a volume of sugar equal to or greater than the volume of water you began with. Continue to add sugar until no more will dissolve even with persistent stirring.

③ Allow the solution to come back to a boil while stirring carefully. This should help dissolve any excess sugar added in step 2. Do not set the burner on high because doing so may make the sugar solution froth up and spill out of the pot. If the sugar still doesn't fully dissolve after the solution is brought to a slow boil, add more water 1 teaspoon at a time. If the sugar dissolves after being brought to a slow boil, add more sugar 1 tablespoon at a time. Ideally, you want a boiling-hot sugar solution that is just below saturation, which may be difficult to assess without prior experience.

④ Remove the clear (no undissolved sugar) boiling sugar solution from the heat. Tie some string to the weight and lower the weight into the hot solution. Support the string with the pencil set across the rim of the pot so that the weight does not touch the bottom.

⑤ Leave the mixture undisturbed for about a week, but check it periodically. You will see large sugar crystals, also known as rock candy, form on the string and also along the sides of the pot. The longer you wait, the larger the crystals.

Nonpolar Gases Readily Dissolve in Perfluorocarbons

As was discussed earlier, good solubility can result when the particle-to-particle attractions in a solute are comparable to those in a solvent. This was the case with ethanol and water, and it is also the case with oxygen and certain *perfluorcarbons*, such as perfluorodecalin, which are molecules consisting of only carbon and fluorine atoms. Oxygen and perfluorodecalin molecules are both nonpolar. Because of the large size of its molecules, perfluorodecalin is a liquid at room temperature. Because they are nonpolar, both perfluorodecalin molecules and oxygen molecules experience induced dipole–induced dipole attractions. At room temperature, consequently, a significant amount of oxygen gas is able to dissolve in liquid perfluorodecalin, as is demonstrated in Figure 7.23.

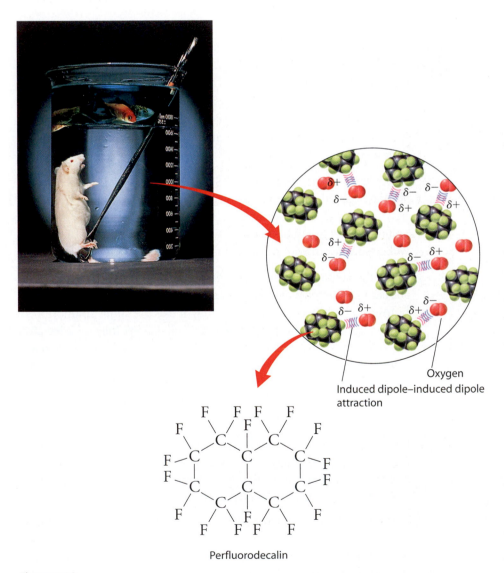

Oxygen
Induced dipole–induced dipole attraction

Perfluorodecalin

Figure 7.23
This mouse is alive and well, inhaling liquid perfluorodecalin saturated with oxygen gas.

Interestingly, a saturated solution of oxygen in a liquid perfluorocarbon contains about 20 percent more oxygen than does the atmosphere we breathe. When this perfluorocarbon solution is inhaled by a human or other animal, the lungs are able to absorb the oxygen in much the same way they absorb it from air. Because liquid perfluorocarbons are as inert as Teflon, which is a solid perfluorocarbon, negative side effects of having these liquids in the lungs are minimal.

Much research is currently being conducted on perfluorocarbons and their potential applications. For example, it is nearly impossible for babies born before seven months of gestation to breathe air. This is because their lungs have yet to develop an inner lining that prevents the moist walls from collapsing and sticking together like wet sheets of plastic food wrap. Researchers have found that premature infants can breathe oxygenated perfluorocarbons quite effectively. Adults may also benefit from inhaling perfluorocarbons because when the liquid is drained from the lungs, it carries with it foreign matter that has accumulated over time. Have you had your lungs cleaned lately?

Another exciting application of perfluorocarbons is their use as a blood substitute. Among the many advantages of *artificial blood* are that it can be stored for long periods of time without deteriorating and that it eliminates the transmission of such diseases as hepatitis and AIDS through blood transfusions. (Please note, however, that because of precautionary measures taken by blood banks, our current blood supply is safe from these diseases. For example, over the course of a year, the chance of dying from a blood transfusion is only about 1 in 100,000, whereas the chance of dying in a car accident is about 1 in 7000.)

The need for a reliable blood substitute arises from frequent blood bank shortages. Currently less than 5 percent of the population donates blood, and this percentage is dropping as demand increases worldwide by about 7.5 million liters each year. The shortfall could become critical sometime in the next 30 years. Because there is still much research needed on perfluorocarbons, donating blood is still a *very* worthwhile thing to do.

7.4 Soap Works by Being Both Polar and Nonpolar

Dirt and grease together make *grime*. Because grime contains many nonpolar components, it is difficult to remove from hands or clothing using just water. To remove most grime, we can use a nonpolar solvent such as turpentine or trichloroethane, which dissolves the grime because of strong induced dipole–induced dipole attractions. Turpentine, also known as a paint thinner, is good for removing the grime left on hands after such activities as changing a car's motor oil. Trichloroethane is the solvent used to "dry clean" clothes—a process whereby dirty clothes are churned in a container full of this nonpolar solvent, which removes the toughest of nonpolar stains without the use of water.

Rather than washing our dirty hands and clothes with nonpolar solvents, however, we have a more pleasant alternative—soap and water. Soap works because soap molecules have both nonpolar and polar properties.

A typical soap molecule has two parts: a long *nonpolar tail* of carbon and hydrogen atoms and a *polar head* containing at least one ionic bond:

$$H-\underset{\underset{H}{|}}{\overset{\overset{H}{|}}{C}}-\underset{\underset{H}{|}}{\overset{\overset{H}{|}}{C}}-\underset{\underset{H}{|}}{\overset{\overset{H}{|}}{C}}-\underset{\underset{H}{|}}{\overset{\overset{H}{|}}{C}}-\underset{\underset{H}{|}}{\overset{\overset{H}{|}}{C}}-\underset{\underset{H}{|}}{\overset{\overset{H}{|}}{C}}-\underset{\underset{H}{|}}{\overset{\overset{H}{|}}{C}}-\overset{\overset{O}{\|}}{C}-O^-\;Na^+$$

Nonpolar tail   Polar head

Because most of a soap molecule is nonpolar, it attracts nonpolar grime molecules via induced dipole–induced dipole attractions, as Figure 7.24 illustrates. In fact, grime quickly finds itself surrounded in three dimensions by the nonpolar tails of soap molecules. This attraction is usually enough to lift the grime away from the surface being cleaned. With the nonpolar tails facing inward toward the grime, the polar heads are all directed outward, where they are attracted to water molecules by relatively strong ion–dipole attractions. If the water is flowing, the whole conglomeration of grime and soap molecules flows with it, away from your hands or clothes and down the drain.

Figure 7.24
Nonpolar grime attracts and is surrounded by the nonpolar tails of soap molecules. The polar heads of the soap molecules are attracted by ion–dipole attractions to water molecules, which carry the soap–grime combination away.

For the past several centuries, soaps have been prepared by treating animal fats with sodium hydroxide, NaOH, also known as caustic lye. In this reaction, which is still used today, each fat molecule is broken down into three *fatty acid* soap molecules and one glycerol molecule:

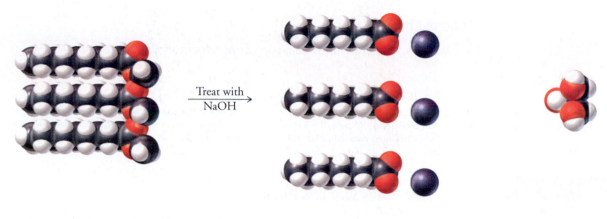

Fat molecule   Three fatty acid soap molecules   Glycerol molecule

Detergents Are Synthetic Soaps

In the 1940s, chemists began developing a class of synthetic soaps, known as *detergents*, that offer several advantages over soaps, such as stronger grease penetration and lower price.

The chemical structure of detergent molecules is similar to that of soap molecules in that both possess a polar head attached to a nonpolar tail. The polar head in a detergent molecule, however, typically consists of either a sulfate group, $-OSO_3^-$, or a sulfonate group, $-SO_3^-$, and the nonpolar tail can have an assortment of structures.

One of the most common sulfate detergents is sodium lauryl sulfate, a main ingredient of many toothpastes. A common sulfonate detergent is sodium dodecyl benzenesulfonate, also known as a linear alkylsulfonate, or LAS. You'll often find this compound in dishwashing liquids. Both these detergents are biodegradable, which means microorganisms can break down the molecules once they are released into the environment.

$$CH_3CH_2CH_2CH_2CH_2CH_2CH_2CH_2CH_2CH_2CH_2CH_2-O-\overset{\overset{\displaystyle O}{\|}}{\underset{\underset{\displaystyle O}{\|}}{S}}-O^-\ Na^+$$

Sodium lauryl sulfate

Sodium dodecyl benzenesulfonate

Concept Check ✓

What type of attractions hold soap or detergent molecules to grime?

Was this your answer? If you haven't yet formulated an answer, why not back up and re-read the question? You've got only four choices: ion–dipole, dipole–dipole, dipole–induced dipole, and induced dipole–induced dipole. The answer is induced dipole–induced dipole attractions, because the interaction is between two nonpolar entities—the grime and the nonpolar tail of a soap or detergent molecule.

Figure 7.25
Hard water causes calcium and magnesium compounds to build up on the inner surfaces of water pipes, especially those used to carry hot water.

Hard Water Makes Soap Less Effective

Water containing large amounts of calcium and magnesium ions is said to be *hard water*, and it has many undesirable qualities. For example, when hard water is heated, the calcium and magnesium ions tend to bind with negatively charged ions also found in the water to form solid compounds, like those shown in Figure 7.25, that can clog water heaters and boilers. You'll also find these calcium and magnesium compounds coated on the inside surface of a well-used tea kettle.

Hard water also inhibits the cleansing actions of soaps and, to a lesser extent, detergents. The sodium ions of soap and detergent molecules carry a 1+ charge, and calcium and magnesium ions carry a 2+ charge (note their positions in the periodic table). The negatively charged portion of the polar head of a soap or detergent molecule is more attracted to the double positive charge of calcium and magnesium ions than to the single positive charge of sodium ions. Soap or detergent molecules therefore give up their sodium ions to selectively bind with calcium or magnesium ions:

Soap or detergent molecules bound to calcium or magnesium ions tend to be insoluble in water. As they come out of solution, they form a scum that can appear as a ring around the bathtub. Because the soap or detergent molecules are tied up with calcium and magnesium ions, more of the cleanser must be added to maintain cleaning effectiveness.

Many detergents today contain sodium carbonate, Na_2CO_3, commonly known as washing soda. The calcium and magnesium ions in hard water are more attracted to the carbonate ion with its two negative charges than to a soap or a detergent molecule with its single negative charge. With the calcium and magnesium ions bound to the carbonate ion, as shown in Figure 7.26, the soap or detergent is free to do its job. Because it removes the ions that make water hard, sodium carbonate is known as a *water-softening agent*.

In some homes, the water is so hard that it must be passed through a water-softening unit. In a typical unit, illustrated in Figure 7.27, hard water is passed through a large tank filled with tiny beads of a water-insoluble resin known as an *ion-exchange resin*. The surface of the resin contains many

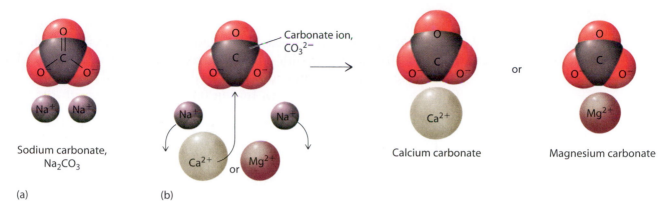

Figure 7.26
(a) Sodium carbonate is added to many detergents as a water-softening agent. (b) The doubly positive calcium and magnesium ions of hard water preferentially bind with the doubly negative carbonate ion, freeing the detergent molecules to do their job.

negatively charged ions bound to positively charged sodium ions. As the hard water with its calcium and magnesium ions passes over the resin, the ions displace the sodium ions and thereby become bound to the resin. The calcium and magnesium ions are able to do this because their positive charge (2+) is greater than that of the sodium ions (1+). The calcium and magnesium ions therefore have a greater attraction for the negative sites on the resin. The net result is that for every one calcium or magnesium ion that binds, two sodium ions are set loose. In this way, the resin *exchanges* ions. The water that exits the unit is free of calcium and magnesium ions but does contain sodium ions in their place.

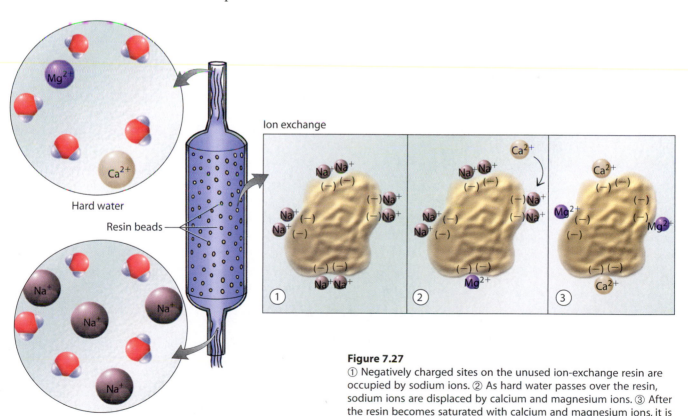

Figure 7.27
① Negatively charged sites on the unused ion-exchange resin are occupied by sodium ions. ② As hard water passes over the resin, sodium ions are displaced by calcium and magnesium ions. ③ After the resin becomes saturated with calcium and magnesium ions, it is no longer effective at softening water.

Eventually, all the sites for calcium and magnesium on the resin are filled, and then the resin needs to be either discarded or recharged. It is recharged by flushing it with a concentrated solution of sodium chloride, NaCl. The abundant sodium ions displace the calcium and magnesium ions (ions are *exchanged* once again), freeing up the binding sites on the resin.

In Perspective

We are now at a point in this textbook where you should have a firm understanding of how subatomic particles make atoms, how atoms make molecules, and how molecules interact with one another through relatively weak attractive electric forces. With this background, you are in a good position to understand and appreciate the real-world applications of chemistry, such as those that were discussed in the final section of this chapter. An important goal of *Conceptual Chemistry* is to give you an understanding of the submicroscopic basis of your macroscopic world. Toward this goal, the next chapter focuses on the macroscopic behavior of water and on how that behavior is determined by the properties of individual water molecules.

Key Terms and Matching Definitions

_____ concentration
_____ dissolving
_____ hydrogen bond
_____ induced dipole
_____ insoluble
_____ molarity
_____ mole
_____ precipitate
_____ saturated solution
_____ solubility
_____ soluble
_____ solute
_____ solvent
_____ unsaturated solution

1. A strong dipole–dipole attraction between a slightly positive hydrogen atom on one molecule and a pair of nonbonding electrons on another molecule.
2. A dipole temporarily created in an otherwise nonpolar molecule, induced by a neighboring charge.
3. The component in a solution present in the largest amount.
4. Any component in a solution that is not the solvent.
5. The process of mixing a solute in a solvent.
6. A solution containing the maximum amount of solute that will dissolve.
7. A solution that will dissolve additional solute if it is added.
8. A quantitative measure of the amount of solute in a solution.
9. 6.02×10^{23} of anything.
10. A unit of concentration equal to the number of moles of a solute per liter of solution.
11. The ability of a solute to dissolve in a given solvent.
12. Capable of dissolving to an appreciable extent in a given solvent.
13. Not capable of dissolving to any appreciable extent in a given solvent.
14. A solute that has come out of solution.

Review Questions

Submicroscopic Particles Electrically Attract One Another

1. What is the primary difference between a chemical bond and an attraction between two molecules?

2. Which is stronger, the ion–dipole attraction or the induced dipole–induced dipole attraction?
3. Why are water molecules attracted to sodium chloride?
4. How are ion–dipole attractions able to break apart ionic bonds, which are relatively strong?
5. Are electrons distributed evenly or unevenly in a polar molecule?
6. What is a hydrogen bond?
7. How are oxygen molecules attracted to water molecules?
8. Are induced dipoles permanent?
9. How can nonpolar atoms induce dipoles in other nonpolar atoms?
10. Why is it difficult to induce a dipole in a fluorine atom?
11. Why is the boiling point of octane, C_8H_{18}, so much higher than the boiling point of methane, CH_4?

A Solution Is a Single-Phase Homogeneous Mixture

12. What happens to the volume of a sugar solution as more sugar is dissolved in it?
13. Why is a ruby gemstone considered to be a solution?
14. Distinguish between a solute and a solvent.
15. What does it mean to say a solution is concentrated?
16. Distinguish between a saturated solution and an unsaturated solution.
17. How is the amount of solute in a solution calculated?
18. Is 1 mole of particles a very large number or a very small number of particles?

Solubility Is a Measure of How Well a Solute Dissolves

19. Why does oxygen have such a low solubility in water?
20. By what means are ethanol and water molecules attracted to each other?

21. What effect does temperature have on the solubility of a solid solute in a liquid solvent?

22. What effect does temperature have on the solubility of a gas solute in a liquid solvent?

23. How are supersaturated solutions made?

24. What does it mean to say that two materials are infinitely soluble in each other?

25. What kind of electrical attraction is responsible for oxygen's ability to dissolve in water?

26. What is the relationship between a precipitate and a solute?

27. Why does the solubility of a gas solute in a liquid solvent decrease with increasing temperature?

28. What do oxygen molecules and perfluorodecane molecules have in common?

Soap Works by Being Both Polar and Nonpolar

29. Which portion of a soap molecule is nonpolar?

30. Water and soap are attracted to each other by what type of electrical attraction?

31. Soap and grime are attracted to each other by what type of electrical attraction?

32. What is the difference between a soap and a detergent?

33. What component of hard water makes it hard?

34. Why are soap molecules so attracted to calcium and magnesium ions?

35. Calcium and magnesium ions are more attracted to sodium carbonate than to soap. Why?

Hands-On Chemistry Insights

Circular Rainbows

Paper chromatography was originally developed to separate plant pigments from one another. The separated pigments had different colors, which is how this technique got its name—*chroma* is Latin for "color." Mixtures need not be colored, however, to be separable by chromatography. All that's required is that the components have distinguishable affinities for the moving solvent and the stationary medium, such as paper, through which the solvent passes.

There are many other forms of chromatography besides paper chromatography. In *column chromatography*, the mixture to be separated is loaded at the top of a column of sandlike material. A solvent passing through the column pulls the components of the mixture through the material at different rates. As the purified components drip out the bottom of the column at different times, they can be collected in separate flasks.

In *gas chromatography*, a liquid mixture is injected into a long, narrow tube that has been heated to the point that the liquid mixture becomes a gaseous mixture. Each component of the gaseous mixture travels through the tube at its own rate, which is determined by the affinity that the component has for a stationary medium coating the inner surface of the tube. Gas chromatography can be used to isolate extremely small quantities, which makes it a valuable analytical tool for many purposes, such as drug testing.

Overflowing Sweetness

It can't be emphasized enough that a solute continues to occupy space whether or not it is dissolved in a liquid. The volume of water that spills over the edge of the glass in this activity is the volume of water displaced by the dissolved solid. As sugar dissolves in water, the sugar molecules are merely pulled out of the crystal lattice to become individual entities. Whether it is part of a lattice or free-floating, however, each sugar molecule occupies the same volume.

Crystal Crazy

Interesting crystals can also be made from supersaturated solutions of Epsom salts ($MgSO_4 \cdot 7 H_2O$) and alum ($KAl(SO_4)_2 \cdot 12 H_2O$), which is used for pickling and is available in the spice section of some grocery stores. Crystal shape directly relates to how the ions or molecules of a substance pack together. In fact, substances are often characterized by the shape of the crystals they form. *Crystallography* is the study of mineral crystals and their shapes and structure.

If you try any experiments with Epsom salts or alum, note how different solutes give rise to different crystal shapes.

Exercises

1. Why are ion–dipole attractions stronger than dipole–dipole attractions?

2. Chlorine, Cl_2, is a gas at room temperature, but bromine, Br_2, is a liquid. Why?

3. Plastic wrap is made of nonpolar molecules and is able to stick well to polar surfaces, such as glass, by way of dipole–induced dipole attractions. Why does plastic wrap also stick to itself so well?

4. Dipole–induced dipole attractions exist between molecules of water and molecules of gasoline, and yet these two substances do not mix because water has such a strong attraction for itself. Which compound might best help these two substances mix into a single liquid phase:

   (a)
   $$H-O-C-C-C-H$$
   (with three CH groups, each carbon bearing H above and below)

   (b) $Na^+ Cl^-$

   (c)
   $$H-C-H$$
   (carbon with H above and H below)

5. Explain why, for these three substances, the solubility in 20°C water goes down as the molecules get larger but the boiling point goes up:

Substance	Boiling point/ Solubility
CH_3-O with H	65°C infinite
$CH_3CH_2CH_2CH_2-O$ with H	117°C 8 g/100 mL
$CH_3CH_2CH_2CH_2CH_2-O$ with H	138°C 2.3 g/100 mL

6. The boiling point of 1,4-butanediol is 230°C. Would you expect this compound to be soluble or insoluble in room-temperature water? Explain.

$$H-O-CH_2CH_2CH_2CH_2-O-H$$

1,4-Butanediol

7. Based on atomic size, which would you expect to be more soluble in water: helium, He, or nitrogen, N_2?

8. If nitrogen, N_2, were pumped into your lungs at high pressure, what would happen to its solubility in your blood?

9. The air a scuba diver breathes is pressurized to counteract the pressure exerted by the water surrounding the diver's body. Breathing the high-pressure air causes excessive amounts of nitrogen to dissolve in body fluids, especially the blood. If a diver ascends to the surface too rapidly, the nitrogen bubbles out of the body fluids (much like the way carbon dioxide bubbles out of a soda immediately after the container is opened). This results in a painful and potentially lethal medical condition known as the *bends*. Why does breathing a mixture of helium and oxygen rather than air help divers avoid getting the bends?

10. Why are noble gases infinitely soluble in other noble gases?

11. Describe two ways to tell whether a sugar solution is saturated or not.

12. Which solute in Figure 7.20 has a solubility in water that changes the least with increasing temperature?

13. At 10°C, which is more concentrated: a saturated solution of sodium nitrate, $NaNO_3$, or a saturated solution of sodium chloride, NaCl? (See Figure 7.20.)

14. A saturated aqueous solution of compound X has a higher concentration than a saturated aqueous solution of compound Y at the same temperature. Does it follow that compound X is more soluble in water than compound Y is?

15. The volume of many liquid solvents expands with increasing temperature. What happens to the concentration of a solution made with such a solvent as the temperature of the solution is increased?

16. Suggest why sodium chloride, NaCl, is insoluble in gasoline. Consider the electrical attractions.

17. Recall from Chapter 3 that the isotopes of an atom differ only in the number of neutrons in the nucleus. Two isotopes of hydrogen are the more common *protium* isotope, which has no neutrons, and the less common *deuterium* isotope, which has one neutron. Either isotope can

be used to make water molecules. Water made with deuterium is known as *heavy water* because each molecule is about 11 percent more massive than water made with protium. Might you also expect the boiling point of heavy water to be about 11 percent greater than the boiling point of regular water? Draw a picture of these two molecules if you need help visualizing the difference between them.

18. Which would you expect to have a higher melting point: sodium chloride, NaCl, or aluminum oxide, Al_2O_3? Why?

19. Hydrogen chloride, HCl, is a gas at room temperature. Would you expect this material to be very soluble or not very soluble in water?

20. Would you expect to find more dissolved oxygen in ocean water around the North Pole or in ocean water close to the equator? Why?

21. Of the two structures shown below, one is a typical gasoline molecule and the other is a typical motor oil molecule. Which is which? Base your reasoning not on memorization but rather on what you know about electrical attractions between molecules and the various physical properties of gasoline and motor oil.

H H H H H H H H H H H H H H H H
H-C-C-C-C-C-C-C-C-C-C-C-C-C-C-C-C-H
H H H H H H H H H H H H H H H H

Structure A

H H H H H H H H
H-C-C-C-C-C-C-C-C-H
H H H H H H H H

Structure B

22. What is the boiling point of a single water molecule? Why does this question not make sense?

23. Account for the observation that ethanol, C_2H_5OH, dissolves readily in water but dimethyl ether, CH_3OCH_3, which has the same number and kinds of atoms, does not.

H   H
| |
H—C—C—O—H
| |
H   H

Ethanol

H   H H   H
\   / \   /
H—C   C—H
\   /
O
/   \
H   H

Dimethyl ether

24. Why are the melting points of most ionic compounds far higher than the melting points of most covalent compounds?

25. An inventor claims to have developed a perfume that lasts a long time because it doesn't evaporate. Comment on this claim.

26. How necessary is soap for removing salt from your hands? Why?

27. When you set a pot of tap water on the stove to boil, you'll often see bubbles start to form well before boiling temperature is reached. Explain this observation.

28. Fish don't live very long in water that has been boiled and brought back to room temperature. Why?

29. Why might softened water not be good for persons trying to reduce their dietary sodium-ion intake?

Problems

1. How much sucrose, in grams, is there in 5 liters of an aqueous solution of sucrose that has a concentration of 0.5 gram of sucrose per liter of solution?

2. How much sodium chloride, in grams, is needed to make 15 L of a solution that has a concentration of 3.0 grams of sodium chloride per liter of solution?

3. If water is added to 1 mole of sodium chloride in a flask until the volume of the solution is 1 liter, what is the molarity of the solution? What is the molarity when water is added to 2 moles of sodium chloride to make 0.5 liter of solution?

4. A student is told to use 20.0 grams of sodium chloride to make an aqueous solution that has a concentration of 10.0 grams of sodium chloride per liter of solution. Assuming that 20.0 grams of sodium chloride has a volume of 7.5 milliliters, about how much water will she use in making this solution?

Answers to Calculation Corner

Calculating for Solutions

1. Multiply the solution concentration by the amount of solution you must end up with to obtain the amount of solute required:

$$(380 \text{ g/L})(3 \text{ L}) = 1140 \text{ g}$$

2. Divide the amount of solute by the solution concentration to obtain the amount of solution she prepared:

$$\frac{20 \text{ g}}{10 \text{ g/L}} = 2 \text{ L}$$

Exploring Further

http://www.google.com

Use *hard water magnets* as a Web-search keyword to find a large number of Web sites that advertise the use of magnetic fields to prevent calcium buildup in plumbing. The proposed mechanism is that the magnetic field facilitates the formation of calcium carbonate crystals in the water rather than on the pipes. But wait! Calcium ions are not attracted to magnets the way metallic iron is. Does this method really work? Beware and be critical. The Web is chock full of misinformation.

http://www.med.umich.edu/liquid/Research.html

Use *perfluorocarbon* as a Web-search keyword and you will find references, most of them technical, to a variety of medical and other uses for liquid perfluorocarbons. The site listed here is that of the Liquid Ventilation Program at the University of Michigan. Scroll to the bottom of the home page for a list of useful links.

http://www.sugar.org

According to this site, the Sugar Association is proud to provide reliable, science-based information about pure, natural sugar. Reliance on sound science has positioned the Association as a leader in communicating accurate information about the nutritional and functional uses of sugar to consumers, professionals, and the media.

Molecular Mixing

Visit The Chemistry Place at:
www.aw.com/chemplace

Hydrogen bond

Macroscopic Consequences of Molecular Stickiness

You are made of water, born into a world of water, and then forevermore dependent on water. You can survive for more than a month without food, but without fresh water you would perish in a matter of days. Little wonder when you consider that water makes up about 60 percent of your body mass. It's the ideal solvent for transporting nutrients through your body and for supporting the countless biochemical reactions that keep you alive. All living organisms we know of depend on water. It is the medium of life on our planet and arguably our most vital natural resource.

Water is so common in our lives that its many unusual properties easily escape our notice. Consider, for example, that water is the only chemical substance on our planet's surface that can be found abundantly in all three phases—solid, liquid, and gas. Another unique property of water is its great resistance to any change in temperature. As a result, the water in you moderates your body temperature just as the oceans moderate global temperatures. Water's resistance to a change in temperature is also why it takes so long for a pot of water to boil and why firewalkers benefit by walking on wet grass before stepping on red-hot coals. Consider also that, unlike most other liquids, which freeze from the bottom up, liquid water freezes from the top down. To the trained eye of a chemist, water is far from a usual substance. Rather, it's downright bizarre and exotic.

Almost all of the amazing properties of water are a consequence of the ability of water molecules to cling tenaciously to one another by way of electrical attractions. Recall that each attraction, called a hydrogen bond, occurs between one of the positively charged hydrogen atoms of one water molecule and the negatively charged oxygen atom of another water molecule. In this chapter we explore the physical behavior of water while diving into the details and consequences of the "stickiness" of water molecules. We begin by exploring first the properties of solid water (ice), then those of liquid water, and finally those of gaseous water, also known as water vapor.

8.1 Water Molecules Form an Open Crystalline Structure in Ice

Experience tells us not to place a sealed glass jar of liquid water in the freezer, for we know that water expands as it freezes. Trapped in the jar, the freezing water expands outward with a force strong enough to shatter the glass into a hazardous mess, or to pop the lid from the jar, as shown in Figure 8.1. This expansion occurs because when the water freezes, the water molecules arrange themselves in a six-sided crystalline structure that contains many open spaces. As Figure 8.2 shows, a given number of water molecules in the liquid can get relatively close to one another and so occupy a certain volume. Water molecules in the crystalline structure of ice occupy a greater volume than in liquid water, however. Consequently, ice is less dense than liquid water, which is why ice floats in water. (Interestingly, the increase in volume that occurs when water freezes is equal to the volume of the ice floating above the liquid water's surface.)

Figure 8.1
The expansion of the freezing water inside the jar caused the lid to rise above the mouth of the jar.

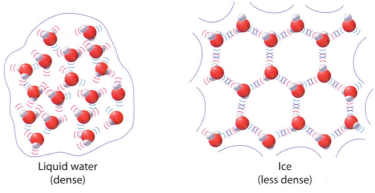

Liquid water
(dense)

Ice
(less dense)

Figure 8.2
Water molecules in the liquid phase are arranged more compactly than water molecules in the solid phase, where they form an open crystalline structure.

This property of expanding upon freezing is quite rare. The atoms or molecules of most frozen solids pack in such a way that the solid phase occupies a *smaller* volume than the liquid phase (Figure 8.3).

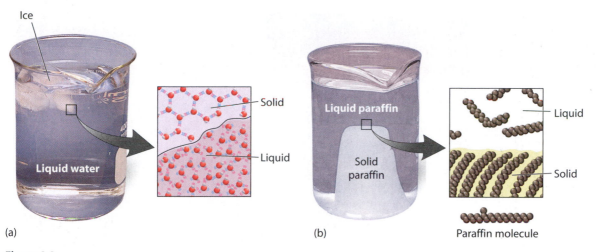

Ice

Solid

Liquid

Liquid water

(a)

Liquid paraffin

Solid paraffin

Liquid

Solid

Paraffin molecule

(b)

Figure 8.3
(a) Because water expands as it freezes, ice is less dense than liquid water and so floats in the water. (b) Like most other materials, paraffin is denser in its solid phase than in its liquid phase. Solid paraffin thus sinks in liquid paraffin.

Concept Check ✔

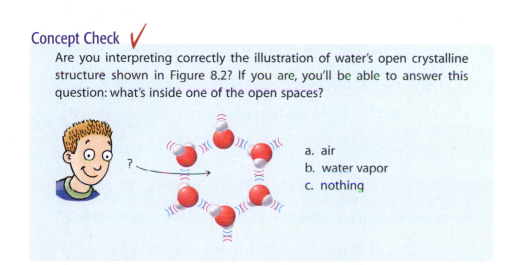

Are you interpreting correctly the illustration of water's open crystalline structure shown in Figure 8.2? If you are, you'll be able to answer this question: what's inside one of the open spaces?

?

a. air
b. water vapor
c. nothing

Was this your answer? If there were air in the spaces, the illustration would have to show the molecules that make up air, such as O_2 and N_2, which are comparable in size to water molecules. Any water vapor in the spaces would have to be shown as free-roaming water molecules spaced relatively far apart. The open spaces shown here represent nothing but empty space. The answer is c.

The hexagonal crystalline structure of H_2O molecules in ice has some interesting effects. Most snowflakes, like the one in Figure 8.4, share a similar hexagonal shape, which is the microscopic consequence of this molecular geometry.

Applying pressure to ice causes the open spaces to collapse, which transforms a small amount of ice to liquid water. The effect is only temporary because, as soon as the pressure is removed, the liquid water refreezes. The pressure applied by the blades of an ice skater or an ice-sailing craft is sufficient to generate a temporary thin film of liquid water over which the blades glide, as shown in Figure 8.5 on page 240. (The melting is also a

Figure 8.4
The six-sided geometry of ice crystals gives rise to the six-sided structure of snowflakes.

Figure 8.5
Skating blades cause a temporary melting of the ice, which generates a slippery thin film of liquid water.

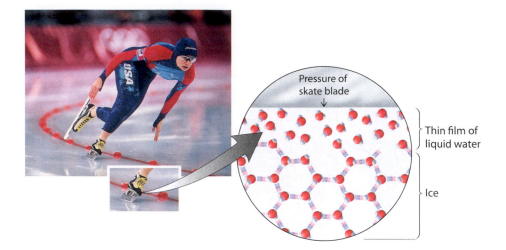

result of the heat generated from the friction of the blade.) At temperatures well below 0°C, the water molecules in ice are held so firmly in the hexagonal structure that there is no noticeable melting from an applied pressure. For this reason, skating blades do not glide well in extremely cold weather.

Hands-On Chemistry: A Slice of Ice

The principle behind ice skating can be used to pass a metal wire through a block of ice.

What You Need

Brick-sized block of ice; flat board, such as length of 2-by-4 or meterstick; two straight-back chairs; thin metal wire about 1 meter long (copper works best); two heavy weights, such as dumbbells or plastic milk jugs filled with water

Procedure

① Make the ice by freezing water in a plastic container of desired shape and size. To make a block that is fairly clear, use water that has just cooled down after having been boiled.

② Attach one weight to each end of the wire.

③ Support the board by laying it across the two chair backs, one chair back near one end of the board, the other near the other end.

④ Place the block of ice on the board and drape the wire across the top as shown in the drawing. The ice just beneath the wire will melt because of the pressure exerted by the wire. The melted ice above the wire then refreezes, trapping the downward-moving wire inside the ice.

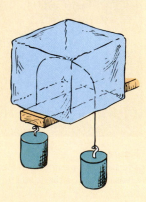

⑤ After a few minutes, the wire will have passed all the way through the ice block. Once that has happened, knock the ice with a hammer and see where it breaks. (In the days before refrigerators, this was the way large ice blocks were cut to size for the kitchen icebox.)

What do you suppose would happen if string were used instead of wire?

8.2   Freezing and Melting Go On at the Same Time

As Section 1.7 discussed, *melting* occurs when a substance changes from solid to liquid and *freezing* occurs when a substance changes from liquid to solid. When we view these processes from a molecular perspective, we see that melting and freezing occur simultaneously, as Figure 8.6 illustrates.

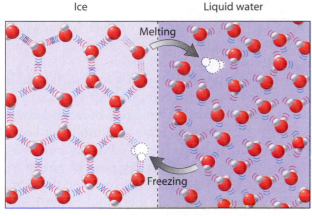

Ice   Liquid water

Melting

Freezing

Vibrating H$_2$O molecules   Slowly moving H$_2$O molecules
fixed in crystalline structure   near freezing temperature

Figure 8.6
At 0°C, ice crystals gain and lose water molecules simultaneously.

The temperature 0°C is both the melting temperature and the freezing temperature of water. At this temperature, water molecules in the liquid phase are moving slowly enough that they tend to clump together to form ice crystals—they freeze. At this same temperature, however, water molecules in ice are vibrating with great commotion, much more than they vibrate at colder temperatures. Many thus break loose from the crystalline structure to form liquid water—they melt. Thus melting and freezing occur simultaneously.

For water, 0°C is the special temperature at which the rate of ice formation equals the rate of liquid water formation. In other words, it is the temperature at which the opposite processes of melting and freezing counterbalance each other. This means that if a mixture of ice and liquid water is maintained at exactly 0°C, the two phases are able to coexist indefinitely.

Anytime we want a mixture of ice and liquid water at 0°C to freeze solid, we need to favor the rate of ice formation. This is accomplished by removing heat energy, a process that facilitates the formation of hydrogen bonds. As shown in Figure 8.7, when water molecules come together to form a hydrogen bond, heat energy is released. In order for the molecules to remain hydrogen-bonded, this heat energy must be removed—otherwise, the heat energy can be reabsorbed by the molecules, causing them to separate. The removal of heat energy therefore allows hydrogen bonds to remain intact after they have formed. As a result, there is the tendency for the ice crystals to grow.

Conversely, we can get a mixture of ice and liquid water at 0°C to melt completely by adding heat energy. This heat energy goes into breaking apart

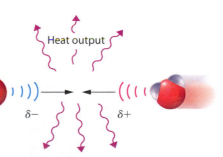

Heat output

$\delta-$ $\delta+$

Figure 8.7
As two water molecules come together to form a hydrogen bond, attractive electric forces cause them to accelerate toward each other. This results in an increase in their kinetic energies (the energy of motion), which is perceived on the macroscopic scale as heat energy.

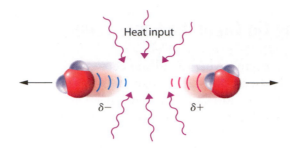

Figure 8.8
Heat energy must be added in order to separate two water molecules held together by a hydrogen bond. This heat energy causes the molecules to vibrate so rapidly that the hydrogen bond breaks.

the hydrogen bonds that hold the water molecules together, as shown in Figure 8.8. Because more hydrogen bonds between water molecules are breaking, there is the tendency for the ice crystals to melt.

Solutes tend to inhibit crystal formation. Anytime a solute, such as table salt or sugar, is added to water, the solute molecules take up space, as you learned in Section 7.2. When a solute is added to a mixture of ice and liquid water at 0°C, the solute molecules effectively decrease the number of liquid water molecules at the solid-liquid interface, as Figure 8.9 illustrates. With fewer liquid water molecules available to join the ice crystals, the rate of ice formation decreases. Because ice is a relatively pure form of water, the number of molecules moving from the solid phase to the liquid phase is not affected by the presence of solute. The net result is that the rate at which water molecules leave the solid phase is greater than the rate at which they enter the solid phase. This imbalance can be compensated for by decreasing the temperature to below 0°C. At lower temperatures, water molecules in the liquid phase move more slowly and have an easier time coalescing. Thus the rate of crystal formation is increased.

In general, adding anything to water lowers the freezing point. Antifreeze is a practical application of this process. The salting of icy roads is another.

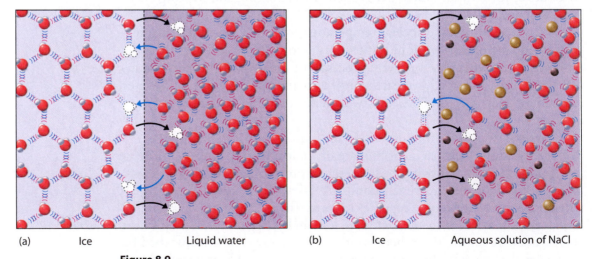

(a)   Ice   Liquid water   (b)   Ice   Aqueous solution of NaCl

Figure 8.9
(a) In a mixture of ice and liquid water at 0°C, the number of H_2O molecules entering the solid phase is equal to the number of H_2O molecules entering the liquid phase. (b) Adding a solute, such as sodium chloride, decreases the number of H_2O molecules entering the solid phase because now there are fewer liquid H_2O molecules at the interface.

Concept Check ✓

On a day with normal traffic, a certain parking lot has cars entering and leaving at the same rate. If the traffic in the streets around the lot were suddenly to change so that about half of it was trucks and limousines too large to park in the lot, what would happen to the rate at which cars enter the lot? What would happen to the number of cars parked in the lot if this situation persisted for a couple of hours? How is this scenario analogous to what happens when a solute is added to a mixture of ice and liquid water at 0°C?

Were these your answers? With fewer cars on the streets, the rate at which cars enter the lot decreases. The rate at which cars leave the lot, however, initially stays the same. Eventually, fewer emptied spaces get filled, and there is an overall decrease in the number of cars parked in the lot. This scenario is analogous to the ice–water case because the solute particles (trucks and limousines) lower the number of liquid water molecules (incoming cars) in contact with the ice (parking lot), thereby diminishing the rate at which water molecules (cars) enter the ice (parking lot). Freezing is thus deterred (fewer cars enter the lot), while melting continues unabated (cars leave the lot at the same rate).

Water Is Densest at 4°C

When the temperature of a substance is increased, its molecules vibrate faster and, on average, move farther apart. The result is that the substance expands. With few exceptions, all phases of matter—solids, liquids, and gases— expand when heated and contract when cooled. In many cases, these changes in volume are not very noticeable, but with careful observation you can usually detect them. Telephone wires, for instance, are longer and sag more on a hot summer day than on a cold winter day. Metal lids on glass jars can often be loosened by heating under hot water. If one part of a piece of glass is heated or cooled more rapidly than adjacent parts, the expansion or contraction that results may break the glass.

Within any given phase, water also expands with increasing temperature and contracts with decreasing temperature. This is true of all three phases—ice, liquid water, and water vapor. Liquid water at near-freezing temperatures, however, is an exception.

Liquid water at 0°C can flow just like any other liquid, but at 0°C the temperature is cold enough that microscopic crystals of ice are able to form. These crystals slightly "bloat" the liquid water's volume, as shown in Figure 8.10. As the temperature is increased to above 0°C, more and more of these microcrystals collapse, and as a result the volume of the liquid water *decreases*.

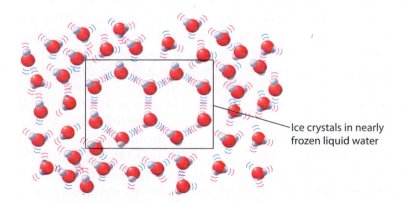

Ice crystals in nearly frozen liquid water

Figure 8.10
Within a few degrees of 0°C, liquid water contains crystals of ice. The open structure of these crystals makes the volume of the water slightly greater than it would be without the crystals.

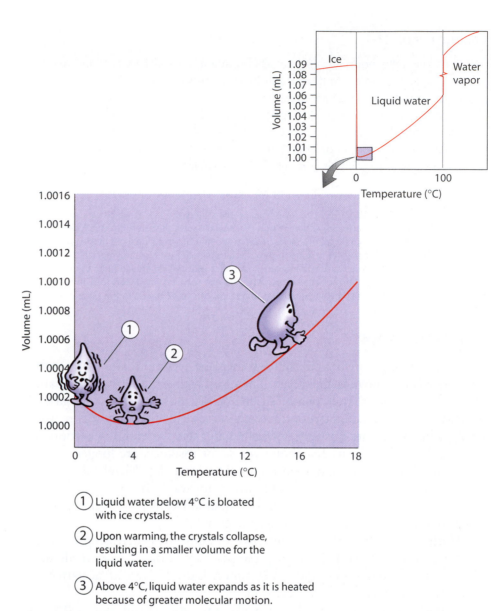

Figure 8.11
Between 0°C and 4°C, the volume of liquid water decreases as the temperature increases. Above 4°C, water behaves the way all other substances do: its volume increases as its temperature increases. The volumes shown here are for a 1-gram sample.

1. Liquid water below 4°C is bloated with ice crystals.

2. Upon warming, the crystals collapse, resulting in a smaller volume for the liquid water.

3. Above 4°C, liquid water expands as it is heated because of greater molecular motion.

Figure 8.11 shows that between 0°C and 4°C liquid water *contracts* as its temperature is raised. This contraction, however, continues only up to 4°C. As near-freezing water is heated, there is a simultaneous tendency for the water to expand due to greater molecular motion. Between 0°C and 4°C, the decrease in volume caused by collapsing ice crystals is greater than the increase in volume caused by the faster-moving molecules. As a result, the water volume continues to decrease. At temperatures just above 4°C, expansion overrides contraction because most of the ice crystals have collapsed.

So, because of the effect of collapsing microscopic ice crystals, liquid water has its smallest volume and thus its greatest density at 4°C. (Recall from Section 1.8 that *density* is the amount of mass contained in a sample of anything divided by the volume of the sample.) By definition, 1 gram of pure liquid water at this temperature has a volume of 1.0000 milliliter. As Figure 8.11 shows, 1 gram of liquid water at 0°C has an only slightly larger volume of 1.0002 milliliters. By comparison, 1 gram of ice at 0°C has a volume of 1.0870 milliliters. As can be seen in the small graph on the right in Figure 8.11, the volume of 1 gram of ice stays above 1.08 milliliters even below 0°C, meaning that even when ice is cooled to temperatures well below freezing, it is still always less dense than liquid water.

Although liquid water at 4°C is only slightly more dense than liquid water at 0°C, this small difference is of great importance in nature. Consider that if water were densest at its freezing point, as is true of most other liquids, the coldest water in a pond would settle to the bottom and the pond would freeze from the bottom up, destroying living organisms in winter months. Fortunately this does not happen. As winter comes on and the temperature of the water drops, its density also drops. The entire volume of water in the pond does not cool all at once, however. Surface water cools first because it is in direct contact with the cold air. Being cooler than the underlying water, this surface water is denser and so sinks, with warmer water rising to replace it. That new batch of surface water then cools to the air temperature, gets denser as it does so, and sinks, only to be replaced by warmer water that cools. This process continues on and on until the entire body of water has been cooled to 4°C. Then, if the air temperature remains below 4°C, the surface water also cools to below 4°C. This surface water does not sink, however, because at this colder temperature it is now *less* dense than the water below it. Thus, cooler less-dense water stays on the surface, where it can cool further, eventually reach 0°C, and turn to ice. While this ice forms at the surface, the organisms that require a liquid environment are happily swimming below the ice in liquid water at a "warm" 4°C, as Figure 8.12 illustrates.

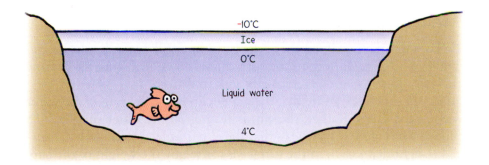

Figure. 8.12
As water cools to 4°C, it sinks. Then, as water at the surface is cooled to below 4°C, it floats on top and can freeze. Only after surface ice forms can temperatures lower than 4°C extend down into the pond. This does not happen very readily, however, because surface ice insulates the liquid water from the cold air.

An important effect of the vertical movement of water is the creation of vertical currents that, to the benefit of organisms living in the water, transport oxygen-rich surface water to the bottom and nutrient-rich bottom water to the surface. Marine biologists refer to this vertical cycling of water and nutrients as *upwelling*.

Very deep bodies of fresh water are not ice-covered even in the coldest of winters. This is because, as noted above, all the water must be cooled to 4°C before the temperature of the surface water can drop below 4°C. For deep water, the winter is not long enough for this to occur.

If only some of the water in a pond is 4°C, this water lies on the bottom. Because of water's ability to resist changes in temperature (Section 8.5) and its poor ability to conduct heat, the bottom of deep bodies of water in cold regions is a constant 4°C year round.

Concept Check ✔

What was the precise temperature at the bottom of Lake Michigan on New Year's Eve in 1901?

Was this your answer? If a body of water has 4°C water in it, then the temperature at the bottom of that body of water is 4°C, for the same reason that rocks are at the bottom. Both 4°C water and rocks are denser than water at any other temperature. If the body of water is deep and in a region of short summers, as is the case for Lake Michigan, the water at the bottom is 4°C year round.

8.3 The Behavior of Liquid Water Is the Result of the Stickiness of Water Molecules

In this section, we explore how water molecules in the liquid phase interact with one another via **cohesive forces**, which are forces of attraction between molecules of a single substance. For water, the cohesive forces are hydrogen bonds. We also explore how water molecules interact with other polar materials, such as glass, through **adhesive forces**, forces of attraction between molecules of two *different* substances.

Cohesive and adhesive forces involving water are dynamic. It is not one set of water molecules, for example, that holds a droplet of water to the side of a glass. Rather, the billions and billions of molecules in the droplet all take turns binding with the glass surface. Keep this in mind as you read this section and examine its illustrations, which, though informative, are merely freeze-frame depictions.

The Surface of Liquid Water Behaves like an Elastic Film

Gently lay a dry paper clip on the surface of some still water. If you're careful enough, the clip will rest on the surface, as shown in Figure 8.13. How can this be? Don't paper clips normally sink in water?

First, you should be aware that the paper clip is not floating *in* the water the way a boat floats. Rather, the clip is resting *on* the water surface. The closeup view in Figure 8.13 reveals that the clip is indeed resting on the surface. The slight depression in the surface is caused by the weight of the clip, which pushes down on the water much like the way the weight of a child

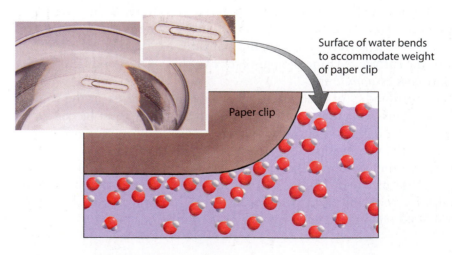

Surface of water bends to accommodate weight of paper clip

Paper clip

Figure 8.13
A paper clip rests on water, pushing the surface down slightly but not sinking.

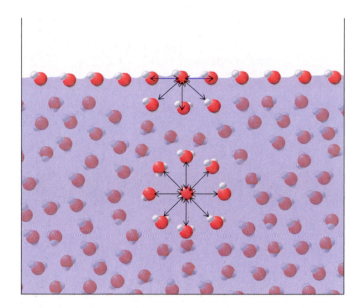

Figure 8.14
A molecule at the surface is pulled only sideways and downward by neighboring molecules. A molecule beneath the surface is pulled equally in all directions.

pushes down on a trampoline. This elastic tendency found at the surface of a liquid is known as **surface tension**.

Surface tension is caused by hydrogen bonds. As shown in Figure 8.14, beneath the surface, each water molecule is attracted in every direction by neighboring molecules, with the result that there is no tendency to be pulled in any preferred direction. A water molecule on the surface, however, is pulled only by neighbors to each side and those below; there is no pull upward. The combined effect of these molecular attractions is thus to pull the molecule from the surface into the liquid. This tendency to pull surface molecules into the liquid causes the surface to become as small as possible, and the surface behaves as if it were tightened into an elastic film. Lightweight objects that don't pierce the surface, such as a paper clip, are thus able to rest on the surface.

Surface tension accounts for the spherical shape of liquid drops. Raindrops, drops of oil, and falling drops of molten metal are all spherical because their surfaces tend to contract and this contraction forces each drop into the shape having the least surface area. This is a sphere, the geometric figure that has the least surface area for a given volume. In the weightless environment of an orbiting space shuttle, a blob of water takes on a spherical shape naturally, as is shown in Figure 8.15. Back on the Earth, the mist and dewdrops on spider webs or on the downy leaves of plants are also spherical, except for the distortions caused by the force of gravity.

(a)   (b)

Figure 8.15
(a) The surface tension in a freely floating blob of water causes the water to take on a spherical shape. (b) Small blobs of water resting on a surface would also be spheres if it weren't for the force of gravity, which squashes them into beads.

Figure 8.16
Soap or detergent molecules align themselves at the surface of liquid water so that their nonpolar tails can escape the polarity of the water. This arrangement disrupts the water's surface tension.

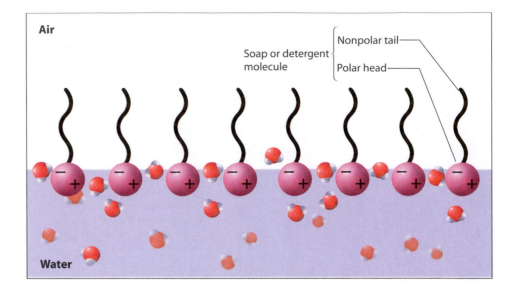

Surface tension is greater in water than in other common liquids because the hydrogen bonds in water are relatively strong. The surface tension in water is dramatically reduced, however, by the addition of soap or detergent. Figure 8.16 shows that soap or detergent molecules tend to aggregate at the surface of water, with their nonpolar tails sticking out away from the water. At the surface, these molecules interfere with the hydrogen bonds between neighboring water molecules, thereby reducing the surface tension. Get a metal paper clip floating on the surface of some water and then carefully touch the water a few centimeters away with the corner of a bar of wet soap or a dab of liquid detergent. You will be amazed at how quickly the surface tension is destroyed.

It is the strong surface tension of water that prevents it from wetting materials that have nonpolar surfaces, such as waxy leaves, umbrellas, and freshly polished automobiles. Rather than wetting (spreading out evenly), the water beads. This is good if the idea is to keep water away. If we want to clean, however, the idea is to get the object as wet as possible. This is another way in which soaps and detergents assist in cleaning, as Figure 8.17 illustrates. By destroying water's surface tension, they enhance its ability to wet. The nonpolar grime on dirty fabrics and dishes, for example, is penetrated by the water more rapidly, and cleaning is more efficient.

Figure 8.17
(a) Water beads on a surface that is clean and dry. (b) On a plate smeared with a thin film of detergent, water spreads evenly because the detergent has upset the water's surface tension.

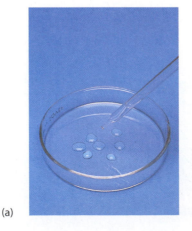

(a)

(b)

Capillary Action Results from the Interplay of Adhesive and Cohesive Forces

Because glass is a polar substance, there are adhesive forces between glass and water. These adhesive forces are relatively strong, and the many water molecules adjacent to the inner surface of a glass container compete to interact with the glass. They do so to the point of climbing up the inner surface of the glass above the water surface. Take a close look at the tube of colored water in Figure 8.18, and you'll see that the water is curved up the sides of the glass. We call the curving of the water surface (or the surface of any other liquid) at the interface between it and its container a **meniscus**.

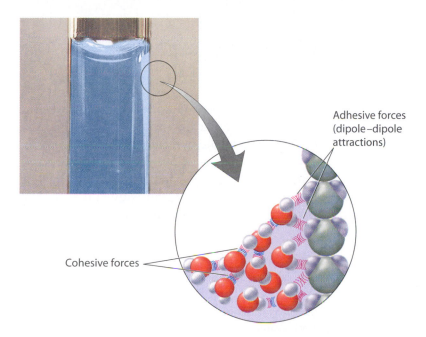

Adhesive forces (dipole–dipole attractions)

Cohesive forces

Figure 8.18
Adhesive forces between water and glass cause water molecules to creep up the sides of the glass, forming a meniscus.

Figure 8.19 illustrates what happens when a small-diameter glass tube is placed in water. ① Adhesive forces initially cause a relatively steep meniscus. ② As soon as the meniscus forms, the attractive cohesive forces among water molecules respond to the steepness by acting to minimize the surface area of the meniscus. The result is that the water level in the tube rises. ③ Adhesive forces will then cause the formation of another steep meniscus. ④ This is followed by the action of cohesive forces, which cause the steep meniscus to be "filled in." This cycle is repeated until the upward adhesive force equals the weight of the raised water in the tube. This rise of the liquid due to the interplay of adhesive and cohesive forces is called **capillary action**.

Figure 8.19
Water is drawn up a narrow glass tube by an interplay of adhesive and cohesive forces.

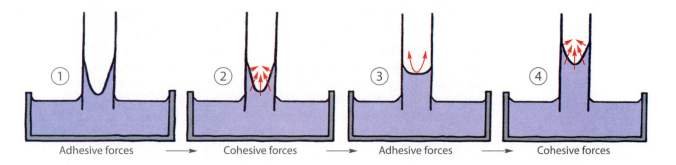

Adhesive forces ⟶ Cohesive forces ⟶ Adhesive forces ⟶ Cohesive forces

In a tube that has an internal diameter of about 0.5 millimeter, the water rises slightly higher than 5 centimeters. In a tube that has a smaller diameter, there is a smaller volume and less weight for a given height, and the water rises much higher, as Figure 8.20 illustrates.

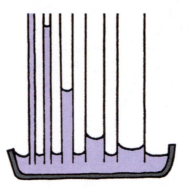

Figure 8.20
Capillary tubes. The smaller the diameter of the tube, the higher the liquid rises in it.

We see capillary action at work in many phenomena. If a paintbrush is dipped into water, the water rises into the narrow spaces between the bristles by capillary action. Hang your hair in the bathtub, and water seeps up to your scalp in the same way. This is how oil moves up a lamp wick and how water moves up a bath towel when one end hangs in water. Dip one end of a lump of sugar in coffee, and the entire lump is quickly wet. The capillary action occurring between soil particles is important in bringing water to the roots of plants.

Concept Check ✔

An astronaut sticks a narrow glass tube into a blob of floating water while in orbit, and the tube fills with water. Why?

Was this your answer? Capillary action causes the water to be drawn into the tube. In the free-fall environment of an orbiting spacecraft, however, there is no downward force to stop this capillary action. As a result, the water continues to creep along the inner surface of the tube until the tube is filled (and then starts spurting out the end).

8.4 Water Molecules Move Freely Between the Liquid and Gaseous Phases

Molecules of water in the liquid phase move about in all directions at different speeds. Some of these molecules may reach the liquid surface moving fast enough to overcome the hydrogen bonds and escape into the gaseous phase. As presented in Section 1.7, this process of molecules converting from the liquid phase to the gaseous phase is called *evaporation* (also sometimes called *vaporization*). The opposite of evaporation is *condensation*—the changing of a gas to a liquid. At the surface of any body of water, there is a constant exchange of molecules from one phase to the other, as illustrated in Figure 8.21.

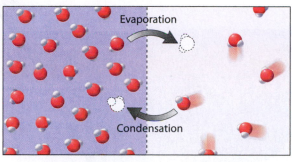
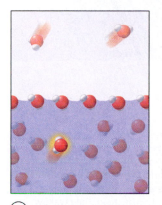
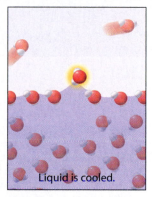
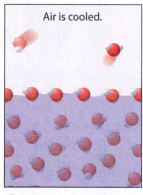

Liquid water   Water vapor

Figure 8.21
The exchange of molecules at the interface between liquid and gaseous water.

Air is cooled.

Liquid is cooled.

Figure 8.22
Evaporation is a cooling process.

① Liquid water molecule having **sufficient kinetic energy** to overcome surface hydrogen bonding approaches liquid surface.

② Liquid water cooled as it loses this high-speed water molecule.

③ Molecule enters gaseous phase, having lost kinetic energy in overcoming hydrogen bonding at the liquid surface. Air is cooled as it collects these slowly moving **gaseous particles.**

As evaporating molecules leave the liquid phase, they take their kinetic energy with them. This has the effect of lowering the average kinetic energy of all the molecules remaining in the liquid, and the liquid is cooled, as Figure 8.22 shows. Evaporation also has a cooling effect on the surrounding air because liquid molecules that escape into the gaseous phase are moving relatively slowly compared to other molecules in the air. This makes sense when you consider that these newly arrived molecules lost a fair amount of their kinetic energy in overcoming the hydrogen bonds of the liquid phase. Adding these slower molecules to the surrounding air effectively decreases the average kinetic energy of all the molecules making up the air, and the air is cooled. So no matter how you look at it, evaporation is a cooling process. Figure 8.23 shows a useful application of this cooling effect.

As water cools, the rate of evaporation slows down because fewer molecules have sufficient energy to escape the hydrogen bonds of the liquid phase. A higher rate of evaporation can be maintained if the water is in contact with a relatively warm surface, such as your skin. Body heat then flows from you into the water. In this way the water maintains a higher temperature and evaporation continues at a relatively high rate. This is why you feel cool as you dry off after getting wet—you are losing body heat to the energy-requiring process of evaporation.

Figure 8.23
When wet, the cloth covering on this canteen promotes cooling. As the faster-moving water molecules evaporate from the wet cloth, its temperature decreases and cools the metal, which in turn cools the water in the canteen.

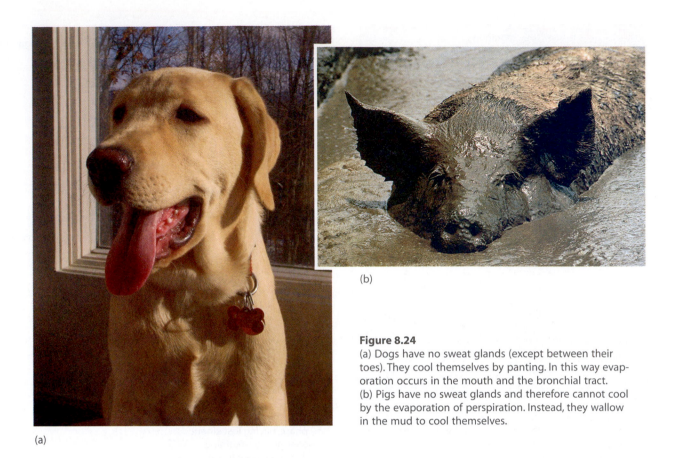

(a)

(b)

Figure 8.24
(a) Dogs have no sweat glands (except between their toes). They cool themselves by panting. In this way evaporation occurs in the mouth and the bronchial tract.
(b) Pigs have no sweat glands and therefore cannot cool by the evaporation of perspiration. Instead, they wallow in the mud to cool themselves.

When your body overheats, your sweat glands produce perspiration. The evaporation of perspiration cools you and helps maintain a stable body temperature. Many animals, such as the ones in Figure 8.24, do not have sweat glands and must cool themselves by other means.

Concept Check ✓

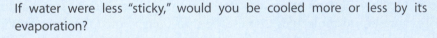

If water were less "sticky," would you be cooled more or less by its evaporation?

Was this your answer? Water molecules leave the liquid phase only when they have enough kinetic energy to overcome hydrogen bonding. It is hydrogen bonding that makes water sticky, and so to say water is less sticky is to say that hydrogen bonds are weaker than they actually are. Then at a given temperature, more molecules in the liquid phase would have sufficient kinetic energy to overcome the weaker hydrogen bonds and escape into the gaseous phase, carrying heat away from the liquid. The cooling power of evaporating water would therefore be greater. This is why less "sticky" substances, such as rubbing alcohol, have a noticeably greater cooling effect as they evaporate.

Warm water evaporates, but so does cool water. The only difference is that cool water evaporates at a slower rate. Even frozen water "evaporates." This form of evaporation, in which molecules jump directly from the solid phase to the gaseous phase, is called **sublimation**. Because water molecules are so firmly held in the solid phase, frozen water does not release molecules

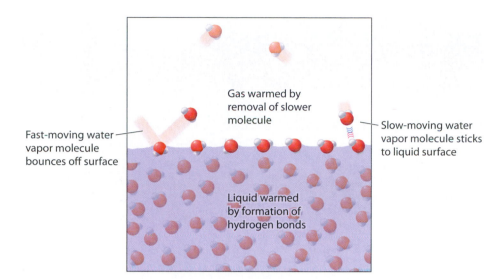

Figure 8.25
Condensation is a warming process.

into the gaseous phase as readily as liquid water does. Sublimation, however, does account for the loss of significant portions of snow and ice, especially on sunny, dry mountain tops. It's also why ice cubes left in the freezer for a long time tend to get smaller.

At the surface of any body of water, there is condensation as well as evaporation, as Figure 8.21 indicates. Condensation occurs as slow-moving water vapor molecules collide with and stick to the surface of a body of liquid water. Fast-moving water vapor molecules tend to bounce off each other or off the liquid surface, losing little of their kinetic energy. Only the slowest gas molecules condense into the liquid phase, as Figure 8.25 illustrates. As this happens, energy is released as hydrogen bonds are formed. This energy is absorbed by the liquid and increases its temperature. Condensation involves the removal of slower-moving water vapor molecules from the gaseous phase. The average kinetic energy of the remaining water vapor molecules is therefore increased, which means that the water vapor is warmer. So no matter how you look at it, condensation is a warming process.

A dramatic example of the warming that results from condensation is the energy given up by water vapor when it condenses—a painful experience if it condenses on you. That's why a burn from 100°C water vapor is much more damaging than a burn from 100°C liquid water; the water vapor gives up considerable energy when it condenses to a liquid and wets the skin. This energy released by condensation is utilized in heating systems, such as the household radiator shown in Figure 8.26.

The water vapor in our atmosphere also gives up energy as it condenses. This is the energy source for many weather systems, such as hurricanes, which derive much of their energy from the condensation of water vapor contained in humid tropical air, as Figure 8.27 on page 254 illustrates. The formation of 1 inch of rain over an area of 1 square mile yields the energy equivalent of about 32,000 tons of exploded dynamite.

After you take a shower, even a cold one, you are warmed by the heat energy released as the water vapor in the shower stall condenses. You quickly

Figure 8.26
Heat is given up by water vapor when the vapor condenses inside the radiator.

Figure 8.27
As it condenses, the water vapor in humid tropical air releases ample quantities of heat. Continued condensation can sometimes lead to powerful storm systems, such as hurricanes.

Figure 8.28
If you're chilly outside the shower stall, step back inside and be warmed by the condensation of the excess water vapor there.

sense the difference if you step out of the stall, as the chilly guy in Figure 8.28 is finding out. Away from the moisture, the rate of evaporation is much higher than the rate of condensation, and as a result you feel chilly. When you remain in the shower stall, where the humidity is higher, the rate of condensation is increased so that you feel that much warmer. So now you know why you can dry yourself with a towel much more comfortably if you remain in the shower stall. If you're in a hurry and don't mind the chill, dry yourself off in the hallway.

Spend a July afternoon in dry Tucson or Las Vegas, and you'll soon notice that the evaporation rate is appreciably greater than the condensation rate. The result of this pronounced evaporation is a much cooler feeling than you would experience on a same-temperature July afternoon in New York City or New Orleans. In these humid locations, condensation outpaces evaporation, and you feel the warming effect as water vapor in the air condenses on your skin.

Concept Check ✔

If the water level in a dish of water remains unchanged from one day to the next, can you conclude that no evaporation or condensation is taking place?

Was this your answer? Not at all, for there is much activity taking place at the molecular level. Both evaporation and condensation occur continuously and simultaneously. The fact that the water level remains constant indicates equal rates of evaporation and condensation—the number of H_2O molecules leaving the liquid surface by evaporation is equal to the number entering the liquid by condensation.

Boiling Is Evaporation Beneath a Liquid Surface

When liquid water is heated to a sufficiently high temperature, bubbles of water vapor form beneath the surface, as we saw in Section 1.7. These bubbles are buoyed to the surface, where they escape, and we say the liquid is *boiling*. As shown in Figure 8.29, bubbles can form only when the pressure of the vapor inside them is equal to or greater than the combined pressure

Figure 8.29
Boiling occurs when water molecules in the liquid are moving fast enough to generate bubbles of water vapor beneath the surface of the liquid.

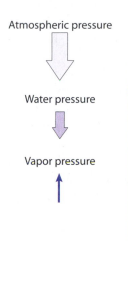

Atmospheric pressure

Water pressure

Vapor pressure

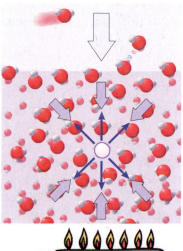

① As liquid water is heated, molecules gain enough energy to evaporate beneath the surface, forming bubbles of water vapor.

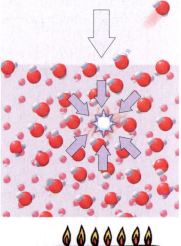

② Before the boiling point is reached, the pressure of the water vapor inside the bubbles is less than the sum of atmospheric pressure plus water pressure. As a result, the bubbles of water vapor collapse.

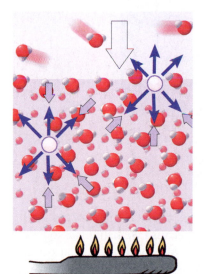

③ At the boiling point, the pressure of the water vapor inside the bubbles equals or exceeds the sum of atmospheric pressure plus water pressure. As a result, the bubbles of water vapor are buoyed to the surface and escape.

④ We see this evaporation as boiling.

Figure 8.30
The tight lid of a pressure cooker holds pressurized vapor above the water surface, and this inhibits boiling. In this way, the boiling temperature of the water is increased. Any food placed in this hotter water cooks more quickly than food placed in water boiling at 100°C.

exerted by the surrounding water and the atmosphere above. At the boiling point of the liquid, the pressure inside the bubbles equals or exceeds the combined pressure of the surrounding water and the atmosphere. At lower temperatures, the pressure inside the bubbles is not enough, and the surrounding pressure collapses any bubbles that form.

At what point boiling begins depends not only on temperature but also on pressure. As atmospheric pressure increases, the vapor molecules inside any bubbles that form must move faster in order to exert enough pressure from inside the bubble to counteract the additional atmospheric pressure. So increasing the pressure exerted on the surface of a liquid raises its boiling point. A cooking application of this effect of increased pressure is shown in Figure 8.30.

Conversely, lowered atmospheric pressure (as at high altitudes) decreases the boiling point of the liquid, as Figure 8.31 illustrates. In Denver, Colorado, the Mile-High City, for example, water boils at 95°C instead of the 100°C boiling temperature at sea level. If you try to cook food in boiling water that is cooler than 100°C, you must wait a longer time for proper cooking. A three-minute boiled egg in Denver is runny and undercooked. If the temperature of the boiling water were very low, food would not cook at all. As the German mountaineer Heinrich Harrer noted in his book *Seven Years in Tibet*, at an altitude of 4500 meters (15,000 feet) or higher, you can sip a cup of boiling tea without any danger of burning your mouth.

Figure 8.31
The boiling point of water (as well as other liquids) decreases with increasing altitude.

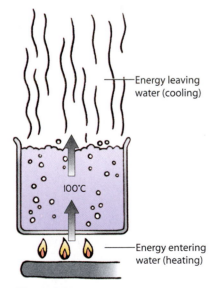

Figure 8.32
Heating warms the water from below, and boiling cools it from above. The net result is a constant temperature for the water.

Boiling, like evaporation, is a cooling process. At first thought, this may seem surprising—perhaps because we usually associate boiling with heating. But heating water is one thing; boiling it is another. As shown in Figure 8.32, boiling water is cooled by boiling as fast as it is heated by the energy from the heat source. So, boiling water remains at a constant temperature. If cooling did not take place, continued application of heat to a pot of boiling water would raise the temperature of the water. The reason the pressure cooker in Figure 8.30 reaches higher temperatures is because boiling is forestalled by increased pressure, which in effect prevents cooling.

Concept Check ✓

Is boiling a form of evaporation or is evaporation a form of boiling?

Was this your answer? Boiling is evaporation that takes place beneath the surface of a liquid.

A simple experiment that dramatically shows the cooling effect of evaporation and boiling consists of a shallow dish of room-temperature water in a vacuum jar. When the pressure in the jar is slowly reduced by a vacuum pump, the water starts to boil. The boiling process takes heat away from the water, which consequently cools. As the pressure is further reduced, more and more of the slower-moving liquid molecules boil away. Continued boiling results in a lowering of the temperature until the freezing point of approximately 0°C is reached. Continued cooling by boiling causes ice to form over the surface of the bubbling water. Boiling and freezing take place at the same time! The frozen bubbles of boiling water in Figure 8.33 are a remarkable sight.

Spray some drops of coffee into a vacuum chamber, and they, too, boil until they freeze. Even after they are frozen, the water molecules continue to evaporate into the vacuum until all that is left to be seen are little crystals of coffee solids. This is how freeze-dried coffee is made. The low temperature of this process tends to keep the chemical structure of the coffee solids from changing. When hot water is added, much of the original flavor of the coffee is retained.

The refrigerator also employs the cooling effect of boiling. A liquid coolant that has a low boiling point is pumped into the coils inside the refrigerator, where the liquid boils (evaporates) and draws heat from the food stored in the refrigerator. Then the coolant in its gas phase, along with its added energy, is directed outside the refrigerator to coils located in the back, appropriately called condensation coils, where heat is given off to the air as the coolant condenses back to a liquid. A motor pumps the coolant through the system as it undergoes the cyclic process of vaporization and condensation. The next time you're near a refrigerator, place your hand near the condensation coils in the back and you'll feel the heat that has been extracted from inside.

An air conditioner employs the same principle, pumping heat energy from inside a building to outside. Turn the air conditioner around so that cold air is pumped to the outside, and the air conditioner becomes a type of heater known as a heat pump.

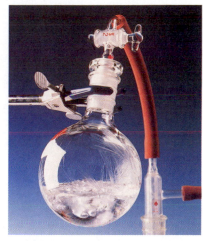

Figure 8.33
In a vacuum, water can freeze and boil at the same time.

8.5 It Takes a Lot of Energy to Change the Temperature of Liquid Water

Have you ever noticed that some foods stay hot much longer than others? The filling of a hot apple pie can burn your tongue while the crust will not, even when the pie has just been taken out of the oven. A piece of toast may be comfortably eaten a few seconds after coming from a hot toaster, whereas you must wait several minutes before eating hot soup.

Figure 8.34
It takes only 0.451 joule of heat to raise the temperature of 1 gram of iron by 1 C°. A 1-gram sample of water, by contrast, requires a whopping 4.184 joules for the same temperature change.

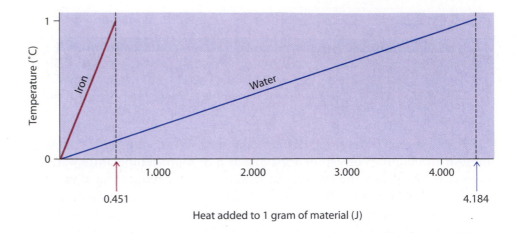

Different substances have different capacities for storing heat energy. This is because different materials absorb energy in different ways. The added energy may increase the jiggling motion of molecules, which raises the temperature, or it may pull apart the attractions among molecules and therefore go into potential energy, which does not raise the temperature. Generally there is a combination of the two ways.

It takes 4.184 joules of energy to raise the temperature of 1 gram of liquid water by 1 C°. As you can see in Figure 8.34, it takes only about one-ninth as much energy to raise the temperature of 1 gram of iron by the same amount. In other words, water absorbs more heat than iron for the same change in temperature. We say water has a higher **specific heat capacity** (sometimes shortened to *specific heat*), defined as the quantity of heat required to change the temperature of 1 gram of the substance by 1°C.

We can think of specific heat capacity as thermal inertia. As you learned in Section 1.4, *inertia* is a term used in physics to signify the resistance of an object to a change in its state of motion. Specific heat capacity is like a thermal inertia because it signifies the resistance of a substance to a change in temperature. Each substance has its own characteristic specific heat capacity, which may be used to assist in identification. Some typical values are given in Table 8.1.

Table 8.1

Specific Heat Capacities for Some Common Materials

Material	Specific Heat Capacity (J/g · °C)
Ammonia, NH_3	4.70
Liquid water, H_2O	4.184
Ethylene glycol, $C_2H_6O_2$ (antifreeze)	2.42
Ice, H_2O	2.01
Water vapor, H_2O	2.0
Aluminum, Al	0.90
Iron, Fe	0.451
Silver, Ag	0.24
Gold, Au	0.13

Guess why water has such a high specific heat capacity. Once again, the answer is hydrogen bonds. When heat is applied to water, much of the heat is consumed in breaking hydrogen bonds. Broken hydrogen bonds are a form of potential energy (just as two magnets pulled apart are a form of potential energy). Much of the heat added to water, therefore, is stored as this potential energy. Consequently, less heat is available to increase the kinetic energy of the water molecules. Since temperature is a measure of kinetic energy, we find that as water is heated, its temperature rises slowly. By the same token, when water is cooled, its temperature drops slowly—as the kinetic energy decreases, molecules slow down and more hydrogen bonds are able to re-form. This in turn releases heat that helps to maintain the temperature.

Concept Check ✔

Hydrogen bonds are not broken as heat is applied to ice (providing the ice doesn't melt) or water vapor. Would you therefore expect ice and water vapor to have specific heat capacities that are greater or less than that of liquid water?

Was this your answer? As Table 8.1 shows, the specific heat capacities of ice and water vapor are about half that of liquid water. Only liquid water has a remarkable specific heat capacity. This is because the liquid phase is the only phase in which hydrogen bonds are continually breaking and re-forming.

Global Climates Are Influenced by Water's High Specific Heat Capacity

The tendency of liquid water to resist changes in temperature improves the climate in many places. For example, notice the high latitude of Europe in Figure 8.35. If water did not have a high specific heat capacity, the countries of Europe would be as cold as the northeastern regions of Canada, for

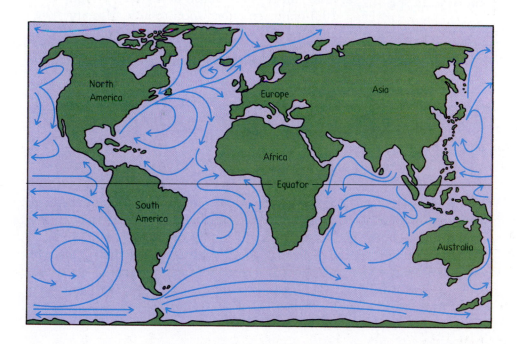

Figure 8.35
Many ocean currents, shown in blue, distribute heat from the warmer equatorial regions to the colder polar regions.

Calculation Corner: How Heat Changes Temperature

Heat must be applied to increase the temperature of a material. Conversely, heat must be withdrawn from a material in order to decrease its temperature. We can calculate the amount of heat required for a given temperature change from the equation

heat = specific heat capacity × mass
$$\times \text{ temperature change}$$

We can use this formula for any material provided there is no change of phase over the course of the temperature change. The value of the temperature change is obtained by subtracting the initial temperature T_i from the final temperature T_f:

$$\text{Temperature change} = T_f - T_i$$

Example 1
How much heat is required to increase the temperature of 1.00 gram of liquid water from an initial temperature of 30.0°C to a final temperature of 40.0°C?

Answer 1
The temperature change is $T_f - T_i = 40.0°C - 30.0°C = +10.0$ C°. To find the amount of heat needed for this temperature change, multiply this positive temperature change by the water's specific heat capacity and mass:

$$\text{heat} = (4.184 \text{ J/g} \cdot \text{C°})(1.00 \text{ g})(+10.0 \text{ C°})$$
$$= 41.8 \text{ J}$$

The temperature decrease that occurs when heat is removed from a material is indicated by a negative sign, as is shown in the next example.

Example 2
A glass containing 10.0 grams of water at an initial temperature of 25.0°C is placed in a refrigerator. How much heat does the refrigerator remove from the water as the water is brought to a final temperature of 10.0°C?

Answer 2
The temperature change is $T_f - T_i = 10.0°C - 25.0°C = -15.0$ C°. To find the heat removed, multiply this negative temperature change by the water's specific heat capacity and mass:

$$\text{heat} = (4.184 \text{ J/g} \cdot \text{C°})(10.0 \text{ g})(-15.0 \text{ C°})$$
$$= -628 \text{ J}$$

Your Turn
1. A residential water heater raises the temperature of 100,000 grams of liquid water (about 26 gallons) from 25.0°C to 55.0°C. How much heat was applied?

2. How much heat must be extracted from a 10.0-gram ice cube (specific heat capacity = 2.01 J/g · C°) in order to bring its temperature from a chilly −10.0°C to an even chillier −30.0°C?

both Europe and Canada get about the same amount of sunlight per square kilometer of surface area. An ocean current carries warm water northeast from the Caribbean. The water holds much of its heat long enough to reach the North Atlantic off the coast of Europe, where the water then cools. The energy released, 4.184 joules per Celsius degree for each gram of water that cools, is carried by the westerly winds (winds that blow west to east) over the European continent.

The winds in the latitudes of North America are westerly. On the western coast of the continent, therefore, air moves from the Pacific Ocean to the land. Because of water's high specific heat capacity, ocean temperatures do not vary much from summer to winter. In winter, the water warms the air, which is then blown eastward over the coastal regions. In summer, the water cools the air and the coastal regions are cooled. On the eastern coast of the continent, the temperature-moderating effects of the Atlantic Ocean are significant, but because the winds blow from the west—over land—temperature ranges in the east are much greater than in the west. San Francisco, for example, is warmer in winter and cooler in summer than Washington, D.C., which is at about the same latitude.

Islands and peninsulas, because they are more or less surrounded by water, do not have the extremes of temperatures observed in the interior of a continent. The high summer temperatures and low winter temperatures common in Manitoba and the Dakotas, for example, are largely due to the absence of large bodies of water. Europeans, islanders, and people living near ocean air currents should be glad that water has such a high specific heat capacity.

Hands-On Chemistry: Racing Temperatures

This activity is a qualitative measure of the specific heat capacity of two common kitchen ingredients: rice and salt.

What You Need

Uncooked rice, table salt, 1-cup measuring cup, aluminum foil, baking sheet, two identical ceramic coffee mugs, thermometer (optional)

Procedure

① Tear off two pieces of foil, each about half the size of the baking sheet. Place them side by side on the baking sheet.

② Measure out 1 cup of rice and pour onto one of the foil sheets. Measure out 1 cup of salt and pour onto the other foil sheet.

③ Heat the rice and salt for 10 minutes in an oven preheated to 250°C, then pour the rice into one of the mugs and the salt into the other.

④ Use a thermometer to note which comes out of the oven at the higher temperature and which cools down faster. If you don't have a thermometer, leave the heated rice and salt on the aluminum foil and judge their cooling rates by cautious touch.

Which has the higher specific heat capacity? Why does the heated rice adhere to the sides of the mug?

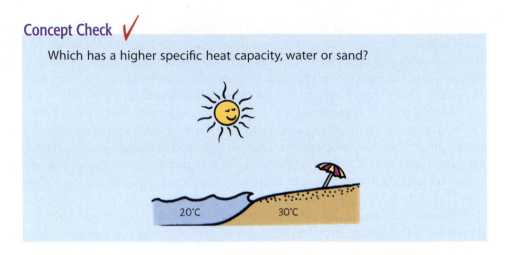

Which has a higher specific heat capacity, water or sand?

Was this your answer? As suggested by the illustration, the temperature of water increases less than the temperature of sand in the same sunlight. Water therefore has the higher specific heat capacity. Those who visit the beach frequently know that beach sand quickly turns hot on a sunny day while the water remains relatively cool. At night, however, the sand feels quite cool while the water's temperature feels about the same as it was during the day.

8.6   A Phase Change Requires the Input or Output of Energy

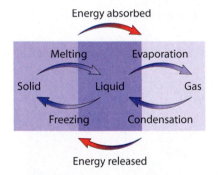

Figure 8.36
Energy changes with change of phase.

Any phase change involves the breaking or forming of molecular attractions. The changing of a substance from a solid to a liquid to a gas, for example, involves the breaking of molecular attractions. Phase changes in this direction therefore require the input of energy. Conversely, the changing of a substance from a gas to a liquid to a solid involves the forming of molecular attractions. Phase changes in this direction therefore result in the release of energy. Both these directions are summarized in Figure 8.36.

Consider a 1-gram piece of ice at −50°C put on a stove to heat. A thermometer in the container reveals a slow increase in temperature up to 0°C, as shown in Figure 8.37. At 0°C, the temperature stops rising, even though heat is still being added, for now all the added heat goes into melting the 0° ice, as indicated in Figure 8.38. This process of melting 1 gram of ice requires 335 joules. Only when all the ice has melted does the temperature begin to rise again. Then for every 4.184 joules absorbed, the water increases its temperature by 1 C° until the boiling temperature, 100°C, is reached. At

Figure 8.37
A graph showing the heat energy involved in converting 1 gram of ice initially at −50°C to water vapor. The horizontal portions of the graph represent regions of constant temperature.

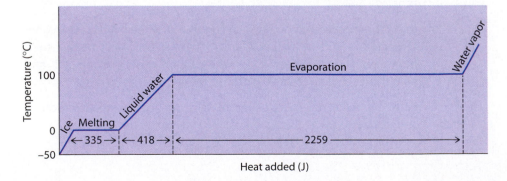

100°C, the temperature again stops rising even though heat is still being added, for now all the added heat is going into evaporating the liquid water to water vapor. The water must absorb a stunning 2259 joules of heat to evaporate all the liquid water. Finally, when all the liquid water has become vapor at 100°C, the temperature begins to rise once again and continues to rise as long as heat is added.

When the phase changes are in the opposite direction, the amounts of heat energy shown in Figure 8.37 are the amounts *released*—2259 joules per gram when water vapor condenses to liquid water and 335 joules per gram when liquid water turns to ice. The processes are reversible.

Figure 8.38
Add heat to melting ice and there is no change in temperature. The heat is consumed in breaking hydrogen bonds.

Concept Check ✔

From Figure 8.37, deduce how much energy (in joules) is transferred when 1 gram of
   a. water vapor at 100°C condenses to liquid water at 100°C.
   b. liquid water at 100°C cools to liquid water at 0°C.
   c. liquid water at 0°C freezes to ice at 0°C.
   d. water vapor at 100°C turns to ice at 0°C.

Were these your answers?
a.   2259 joules   b.   418 joules   c.   335 joules
d.   3012 joules (2259 joules + 418 joules + 335 joules)

The amount of heat energy required to change a solid to a liquid is called the **heat of melting**, and the amount of heat energy released when a liquid freezes is called the **heat of freezing**. Water has a heat of melting of +335 joules per gram. The positive sign indicates that this is the amount of heat energy that must be *added to* ice to melt it. Water's heat of freezing is −335 joules per gram. The negative sign indicates that this is the amount of heat energy that is *released* from liquid water as it freezes—the same amount that was required to melt it.

The amount of heat energy required to change a liquid to a gas is called the **heat of vaporization**. For water, this is +2259 joules per gram. The amount of heat energy released when a gas condenses is called the **heat of condensation**. For water, this is −2259 joules per gram. These values are high relative to the heats of vaporization and condensation for most other substances. This is due to the relatively strong hydrogen bonds between water molecules that must be broken or formed during these processes.

Concept Check ✔

Can you add heat to ice without melting it?

Was this your answer? A common misconception is that ice cannot have a temperature lower than 0°C. In fact, ice can have any temperature below 0°C, down to absolute zero, −273°C. Adding heat to ice below 0°C raises its temperature, say from −200°C to −100°C. As long as its temperature stays below 0°C, the ice does not melt.

Figure 8.39
Water extinguishes a flame by wetting but also by absorbing much of the heat that the fire needs to sustain itself.

Figure 8.40
Tracy Suchocki walks with wetted bare feet across red-hot wood coals without harm.

Although water vapor at 100°C and liquid water at 100°C have the same temperature, each gram of the vapor contains an additional 2259 joules of potential energy because the molecules are relatively far apart from one another. As the molecules bind together in the liquid phase, this 2259 joules per gram is released to the surroundings in the form of heat. In other words, the potential energy of the far-apart water vapor molecules transforms to heat as the molecules get closer together. This is like the potential energy of two attracting magnets separated from each other; when released, their potential energy is converted first to kinetic energy and then to heat as the magnets strike each other.

Water's high heat of vaporization allows you to *briefly* touch your *wetted* finger to a hot skillet or hot stove without harm. You can even touch it a few times in succession *as long as your finger remains wet*. This is because energy that ordinarily would go into burning your finger goes instead into changing the phase of the moisture on your finger from liquid to vapor. You can judge the hotness of a clothes iron in the same way—with a *wet* finger.

The firefighters in Figure 8.39 know that certain types of flames are best extinguished with a fine mist of water rather than a steady stream. The fine mist readily turns to water vapor and in doing so quickly absorbs heat energy and cools the burning material.

Water's high heat of vaporization makes walking barefooted on red-hot coals more comfortable, as shown in Figure 8.40. When your feet are wet, either from perspiration or because you are stepping off of wet grass, much

of the heat from the coals is absorbed by the water and not by your skin. (Firewalking also relies on the fact that wood is a poor conductor of heat, even when it is in the form of red-hot coals.)

In Perspective

Many of the properties of water we have explored in this chapter at the molecular level are nicely summarized in the scene shown in Figure 8.41. First, notice that the massive bear is supported by the floating ice. The ice floats because hydrogen bonds hold the H_2O molecules in the ice together in an open crystalline structure that makes the ice less dense than the liquid water. Because so much energy is required both to melt ice and to evaporate liquid water, most of the Arctic ice, which builds up primarily from snowfall, remains in the solid phase throughout the year. Where the seawater is in contact with the ice, the temperature is lower than 0°C because salts dissolved in the seawater inhibit the formation of ice crystals and thereby lower the freezing point of the seawater. The high specific heat capacity of the Arctic ocean beneath the bear moderates the Arctic climate. It gets cold in the Arctic in winter, yes, but not as cold as it gets in Antarctica, where the specific heat capacity of the mile-thick ice is only half as great as that of the liquid water that predominates in the Arctic. Covered by ice with its much lower specific heat capacity, Antarctica experiences much greater extremes in temperature than the Arctic.

The main focus of this chapter was the physical behavior of water molecules. In the previous chapter, the focus was the physical behavior of molecules and ions in general. For the next several chapters, we shift our attention to the chemical behavior of molecules and ions, which change their fundamental identities as they chemically react with one another.

Key Terms and Matching Definitions

_____ adhesive force
_____ capillary action
_____ cohesive force
_____ heat of condensation
_____ heat of freezing
_____ heat of melting
_____ heat of vaporization
_____ meniscus
_____ specific heat capacity
_____ sublimation
_____ surface tension

1. An attractive force between two identical molecules.
2. An attractive force between molecules of two different substances.
3. The elastic tendency found at the surface of a liquid.
4. The curving of the surface of a liquid at the interface between the liquid surface and its container.
5. The rising of liquid into a small vertical space due to the interplay of cohesive and adhesive forces.
6. The process of a material transforming from a solid directly to a gas, without passing through the liquid phase.
7. The quantity of heat required to change the temperature of 1 gram of a substance by 1 Celsius degree.
8. The heat energy absorbed by a substance as it transforms from solid to liquid.
9. The heat energy released by a substance as it transforms from liquid to solid.
10. The heat energy absorbed by a substance as it transforms from liquid to gas.
11. The energy released by a substance as it transforms from gas to liquid.

Review Questions

Water Molecules Form an Open Crystalline Structure in Ice

1. What accounts for the fact that ice is less dense than water?
2. What is inside one of the open spaces of an ice crystal?
3. What happens to ice when great pressure is applied to it?

Freezing and Melting Go On at the Same Time

4. How is it possible for a substance to melt and freeze at the same time?
5. What is released when a hydrogen bond forms between two water molecules?
6. Why does extracting heat from a mixture of ice and liquid water at 0°C increase the rate of ice formation?
7. Why does adding heat to a mixture of ice and liquid water at 0°C increase the rate of water formation?
8. Why does water not freeze at 0°C when either ions or molecules other than H_2O are present?
9. Is the density of near-freezing water, which contains ice crystals, greater or less than the density of liquid water containing no ice crystals?
10. When the temperature of 0°C liquid water is increased slightly, does the water undergo a net expansion or a net contraction?
11. What happens to the amount of molecular motion in water, no matter what its phase, when its temperature is increased?
12. At what temperature do the competing effects of contraction and expansion produce the smallest volume for liquid water?
13. Why does ice form at the surface of a body of water instead of at the bottom?

The Behavior of Liquid Water Is the Result of the Stickiness of Water Molecules

14. What is the difference between cohesive forces and adhesive forces?
15. In what direction is a water molecule on the surface not pulled?
16. Why do liquids in which the molecular interactions are strong have greater surface tension than those in which the molecular interactions are weak?
17. Does liquid water rise higher in a narrow tube or a wide tube?
18. What determines the height that liquid water rises by capillary action?

Water Molecules Move Freely Between the Liquid and Gaseous Phases

19. Do all the molecules in a liquid have about the same speed?

20. Why is evaporation a cooling process? What does evaporation cool?

21. What phases are involved in sublimation?

22. Why is condensation a warming process? What does condensation warm?

23. Why is a burn from water vapor at 100°C more damaging than a burn from liquid water at the same temperature?

24. Why do we feel uncomfortably warm on a hot, humid day?

25. Is it the pressure on the food or the higher temperature that cooks food faster in a pressure cooker?

26. What condition permits liquid water to boil at a temperature below 100°C?

It Takes a Lot of Energy to Change the Temperature of Liquid Water

27. Why does liquid water have such a high specific heat capacity?

28. Is it easy or difficult to change the temperature of a substance that has a low specific heat capacity?

29. Does a substance that heats up quickly have a high or a low specific heat capacity?

30. How does the specific heat capacity of liquid water compare with the specific heat capacities of other common materials?

31. Northeastern Canada and much of Europe receive about the same amount of sunlight per unit of surface area. Why then is Europe generally warmer in winter?

32. Why is the temperature fairly constant on islands and peninsulas?

A Phase Change Requires the Input or Output of Energy

33. When liquid water freezes, is heat released to the surroundings or absorbed from the surroundings?

34. Why doesn't the temperature of melting ice rise as the ice is heated?

35. How much heat is needed to melt 1 gram of ice? Give your answer in joules.

36. Is the food compartment in a refrigerator cooled by evaporation or condensation of the refrigerating coolant?

37. Why is it important that your finger be wet when you touch it briefly to a hot clothes iron?

38. Why does it take so much more energy to boil 10 grams of liquid water than to melt 10 grams of ice?

Hands-On Chemistry Insights

A Slice of Ice

Liquid water normally contains an appreciable amount of dissolved air. As the water freezes, this air comes out of solution and forms bubbles that can make the ice cloudy. Interestingly, liquid water in an ice tray begins to freeze along the perimeter of each cube. The dissolved gases are thus pushed inward toward the center of each cube, where freezing occurs last. This is why an ice cube is typically clear on its perimeter and cloudy in the middle. Water that has just been boiled contains only small amounts of dissolved air, which is why it can be used to create fairly clear ice cubes.

It's interesting to consider this activity in light of Section 8.6. Note that changes in phase are occurring as the ice melts below the wire and as the liquid water refreezes above. When the liquid water immediately above the wire refreezes, the water gives up energy. How much? Enough to melt an equal amount of ice immediately under the wire. This energy must be conducted through the wire. Hence this demonstration requires that the wire be an excellent conductor of heat. String is a poor conductor of heat, which is why it does not work as a substitute for metal wire.

Ice skaters know that the sharper their blades, the easier it is for them to glide. A sharper blade has a smaller surface area in contact with the ice and is thus able to apply a greater pressure. Similarly, a thin wire is able to slice through a block of ice more quickly than a thick wire. A thin wire, however, is also weaker and so might not be able to hold the anchoring weights without breaking.

Racing Temperatures

The first piece of evidence that the salt has a lower specific heat capacity is that it has a higher temperature when you take your samples out of the oven. The second piece of evidence is that, despite this initially higher temperature, the salt cools faster than the rice. One reason rice has the higher specific heat capacity is that each grain contains a fair amount of moisture. When you heat the rice, much of this moisture is released. Moisture continues to be released even after you take the rice out of the oven, which is why the grains adhere to the mug.

Some people exploit rice's ability to absorb moisture by placing grains of rice in their salt shakers. The rice absorbs any moisture that would otherwise cause the salt crystals to clump together. Most commercial salt contains water-absorbing silicates that achieve the same result. You can see these silicates as you try to dissolve commercial salt in water—the cloudiness you see is not from the salt but from the insoluble silicates.

You can put the high specific heat capacity of rice to practical use, either keeping warm on cold evenings or soothing painful cramps. Fill a clean sock three-quarters full with rice. Tie the open end closed with a string (don't use metal wire!) and cook in a microwave for a couple of minutes. (Don't use a conventional oven!!) The moisture in the grains becomes apparent when you take the sock out of the oven— the released moisture has made the sock slightly damp. Wrap the sock around your neck for instant gratification. Need a neck cooler? Store the rice-filled sock in the freezer. The moisture in the rice stays cold for a long time. These devices make great homemade gifts when a fancy fabric is used in place of the sock and a mild fragrance is added.

Exercises

1. Like water, hydrogen fluoride, HF, and ammonia, NH_3, have relatively high boiling points. Explain.

2. Ice floats in room-temperature water, but does it float in boiling water? Why or why not?

3. As an ice cube floating in a glass of water melts, what happens to the water level?

4. How does the combined volume of the billions and billions of hexagonal open spaces in the crystals in a piece of ice compare with the portion of the ice that floats above the water line?

5. Why is it important to protect water pipes in your home from freezing?

6. What happens to the freezing temperature of a solution of table salt in water as the solution becomes more concentrated?

7. Why does adding heat to an ice–water mixture decrease the rate of ice formation?

8. Suppose liquid water is used in a thermometer instead of mercury. If the temperature is initially 4°C and then changes, why can't the thermometer indicate whether the temperature is rising or falling?

9. Which graph most accurately represents the density of liquid water plotted against temperature:

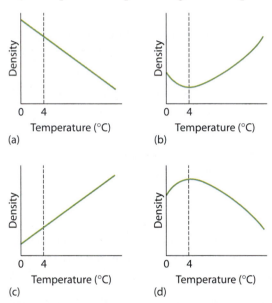

(a)   (b)   (c)   (d)

10. If cooling occurred at the bottom of a pond instead of at the surface, would a lake freeze from the bottom up? Explain.

11. Unlike fresh water, ocean water contracts as it cools to its freezing point, which is about −18°C. Why?

12. The polar ice cap that rests over the Arctic Ocean gets thicker during the winter. Does it grow from above or below? Explain.

13. Consider a lake in which the water is uniformly 10°C. What happens to the oxygen-rich surface water as it cools to 4°C? What concurrently happens to the nutrient-rich deeper water?

14. Why are polar oceans far more fertile in autumn?

15. Nutrient-rich water tends to be murky. Why does tropical water tend to be so clear?

16. Capillary action causes water to climb up the internal walls of narrow glass tubes. Why does the water not climb as high when the glass tube is wider?

17. Mercury forms a convex meniscus with glass rather than the concave meniscus shown in Figure 8.18. What does this tell you about the cohesive forces between mercury atoms versus the adhesive forces between mercury atoms and glass? Which forces are stronger?

 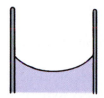

Convex meniscus   Concave meniscus

18. Can a glass be filled to above its brim with water without the water spilling over the edge? Try it and see. Explain your observations.

19. Would you expect the surface tension of water to increase or decrease with temperature? Defend your answer.

20. Dip a paper clip into water and then slowly pull it upward to the point where it is nearly free from the surface. You'll find that for a short distance, the water is brought up with the metal. Are these adhesive or cohesive forces at work?

21. Why does water bead on a freshly waxed surface?

22. Water coming out of volcanic sea vents at the bottom of the ocean can reach temperatures in excess of 300°C without boiling. Explain.

23. You can determine wind direction by wetting your finger and holding it up in the air. Explain.

24. Why does blowing over hot soup cool the soup?

25. Why does wetting a cloth in a bucket of cold water and then wrapping the wet cloth around a bottle produce a cooler bottle than placing the bottle directly in the bucket of cold water?

26. Could you cook an egg in water boiling in a vacuum? Explain.

27. An inventor friend proposes a design of cookware that will allow water to boil at a temperature lower than 100°C so that food can be cooked with less energy. Comment on this idea.

28. What is the gas inside a bubble of boiling water?

29. Your instructor hands you a closed flask partly filled with room-temperature water. When you hold it, the heat from your bare hands causes the water to boil. Quite impressive! How is this accomplished?

30. As an evaporating liquid cools, does something else get warm? If so, what?

31. A lid on a cooking pot filled with water shortens both the time it takes the water to come to a boil and the time the food takes to cook in the boiling water. Explain what is going on in each case.

32. If liquid water had a lower specific heat capacity, would ponds be more likely to freeze or less likely to freeze?

33. Should the specific heat capacity of your automobile's radiator liquid be as high as possible or as low as possible? Explain.

34. Would fevers run higher or lower if liquid water's specific heat capacity were not so high?

35. Bermuda is close to North Carolina, but unlike North Carolina it has a tropical climate year round. Why?

36. If the winds at the latitude of San Francisco and Washington, D.C., were from the east rather than from the west, why might San Francisco be able to grow cherry trees and Washington, D.C., palm trees?

37. To impart a hickory flavor to a roasted turkey, a cook places a pot of water containing hickory chips in the oven with the turkey. Why does the turkey take longer than expected to cook?

38. Why is it that in cold winters a large tub of liquid water placed in a farmer's canning cellar helps prevent canned food from freezing?

39. A great deal of heat is released when liquid water freezes. Why doesn't this heat simply remelt the ice?

40. Suppose 4 grams of liquid water at 100°C is spread over a large surface so that 1 gram evaporates rapidly. If evaporation of the 1 gram takes 2259 joules from the remaining 3 grams of water and no other heat transfer takes place, what are the temperature and phase of the remaining 3 grams once the 1 gram is evaporated?

Problems

1. A walnut stuck to a pin is burned beneath a can containing 100.0 grams of water at 21°C. After the walnut has completely burned, the water's final temperature is 28°C. How much heat energy came from the burning walnut?

2. How much heat is required to raise the temperature of 100,000 grams of iron by 30 C°?

3. By how much will the temperature of 5.0 grams of liquid water increase upon the addition of 230 joules of heat?

Answers to Calculation Corner

How Heat Changes Temperature

1. The temperature change is final temperature minus initial temperature: 55.0°C − 25.0°C = +30.0 C°. Multiply this positive temperature change by the water's specific heat capacity and mass:

   heat = (4.184 J/g · C°)(100,000 g)(+30.0 C°)
        = 12,552,000 J

   This large number helps to explain how an electric water heater consumes about 25 percent of all household electricity. With the proper number of significant figures (see Appendix B), this answer should be expressed as 10,000,000 joules.

2. The temperature change is −30.0°C − (−10.0°C) = −20.0 C°. Multiply this negative temperature change by the ice's specific heat capacity and mass:

   heat = (2.01 J/g · C°)(10.0 g)(−20.0 C°)
        = −402 J

   The next time you're near a refrigerator/freezer, place your hand near its back, and you'll feel the heat that has been extracted from the food inside.

Exploring Further

http://madsci.wustl.edu/posts/archives/mar97/852921998.As.r.html
   A discussion about the recent discovery of ice on the moon's south pole, with an explanation of how the ice can persist despite the absence of an atmosphere. See also http://www.nrl.navy.mil/clementine.

http://seawifs.gsfc.nasa.gov/OCEAN_PLANET/HTML/oceanography_recently_revealed1.html
   Volcanic vents at the bottom of the ocean spew out water at temperatures in excess of 300°C. In 1977, geologists exploring these vents discovered odd-looking animals surviving on the sunless sea floor.

the
Chemistry
place

Those Incredible Water Molecules
Visit The Chemistry Place at:
www.aw.com/chemplace

Nitrogen monoxide

Nitric acid

Nitrous acid

Nitrate ion

Nitrite ion

9

An Overview of Chemical Reactions

How Reactants React to Form Products

The heat of a lightning bolt causes many chemical reactions in the atmosphere, including one in which nitrogen and oxygen react to form nitrogen monoxide, NO. The nitrogen monoxide formed in this manner then reacts with atmospheric oxygen and water vapor to form nitric acid, HNO_3, and nitrous acid, HNO_2. These acids are carried by rain into the ground, where they form ions, which plants need for growth—a process that involves further chemical reactions.

Scientists have learned how to control chemical reactions to produce many useful materials—nitrates and other nitrogen-based fertilizers from atmospheric nitrogen, metals from rocks, plastics and pharmaceuticals from petroleum. These materials and the thousands of others produced by chemical reactions, as well as the abundant energy released when fossil fuels take part in the chemical reaction called combustion, have dramatically improved our living conditions.

The goal of this chapter is to give you a stronger handle on the basics of chemical reactions, which were introduced in Chapter 2. Then in the following chapters we'll look at specific classes of chemical reactions, such as acid-base reactions, oxidation-reduction reactions, and reactions involving organic chemicals.

9.1 Chemical Reactions Are Represented by Chemical Equations

During a chemical reaction, one or more new compounds are formed as a result of the rearrangement of atoms. To represent a chemical reaction, we can write a **chemical equation**, which shows the substances about to react, called **reactants**, to the left of an arrow that points to the newly formed substances, called **products**:

$$\text{reactants} \longrightarrow \text{products}$$

Typically, reactants and products are represented by their atomic or molecular formulas, but molecular structures or simple names may be used instead. Phases are also often shown: (s) for solid, (ℓ) for liquid, and (g) for gas. Compounds dissolved in water are designated (aq) for aqueous. Lastly, numbers are placed in front of the reactants or products to show the ratio in which they either combine or form. These numbers are called **coefficients**, and they represent numbers of individual atoms and molecules. For instance, to represent the chemical reaction in which coal (solid carbon) burns in the presence of oxygen to form gaseous carbon dioxide, we write the chemical equation

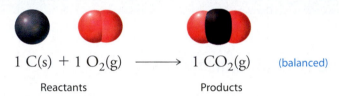

$$1 \text{ C(s)} + 1 \text{ O}_2\text{(g)} \longrightarrow 1 \text{ CO}_2\text{(g)} \qquad \text{(balanced)}$$

Reactants   Products

One of the most important principles of chemistry is the *law of mass conservation*, which states that matter is neither created nor destroyed during a chemical reaction (Section 3.2). The atoms present at the beginning of a reaction merely rearrange to form new molecules. This means that no atoms are lost or gained during any reaction. The chemical equation must therefore be *balanced*, which means each atom shown in the equation must appear on both sides of the arrow the same number of times. The preceding equation for the formation of carbon dioxide is balanced because each side shows one carbon atom and two oxygen atoms. You can count the number of atoms in the space-filling models to see this for yourself.

In another chemical reaction, two hydrogen gas molecules, H_2, react with one oxygen gas molecule, O_2, to produce two molecules of water, H_2O, in the gaseous phase:

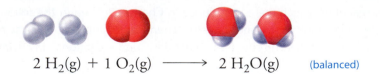

$$2 \text{ H}_2\text{(g)} + 1 \text{ O}_2\text{(g)} \longrightarrow 2 \text{ H}_2\text{O(g)} \qquad \text{(balanced)}$$

This equation for the formation of water is also balanced—there are four hydrogen and two oxygen atoms before and after the arrow.

A coefficient in front of a chemical formula tells us the number of times that element or compound must be counted. For example, $2 \text{ H}_2\text{O}$ indicates two water molecules, which contain a total of four hydrogen atoms and two oxygen atoms.

By convention, the coefficient 1 is omitted so that the above chemical equations are typically written

$C(s) + O_2(g) \longrightarrow CO_2(g)$ (balanced)

$2 H_2(g) + O_2(g) \longrightarrow 2 H_2O(g)$ (balanced)

Concept Check ✔

How many oxygen atoms are indicated by the balanced equation

$3 O_2(g) \longrightarrow 2 O_3(g)$

Was this your answer? Six. Before the reaction these six oxygen atoms are found in three O_2 molecules. After the reaction these same six atoms are found in two O_3 molecules.

An unbalanced chemical equation shows the reactants and products without the correct coefficients. For example, the equation

$NO(g) \longrightarrow N_2O(g) + NO_2(g)$ (not balanced)

is not balanced because there is one nitrogen atom and one oxygen atom before the arrow but three nitrogen atoms and three oxygen atoms after the arrow.

You can balance unbalanced equations by adding or changing coefficients to produce correct ratios. (It's important **not to change subscripts**, however, because to do so changes the compound's identity—H_2O is water, but H_2O_2 is hydrogen peroxide!) For example, to balance the above equation, add a 3 before the NO:

$3 NO(g) \longrightarrow N_2O(g) + NO_2(g)$ (balanced)

Now there are three nitrogen atoms and three oxygen atoms on each side of the arrow, and the law of mass conservation is not violated.

Concept Check ✔

Write a balanced equation for the reaction showing hydrogen gas and nitrogen gas forming ammonia gas:

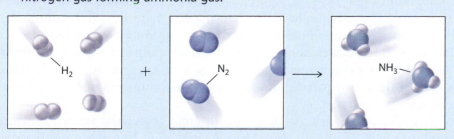

Was this your answer?

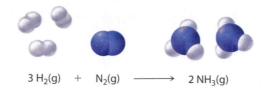

$$3\,H_2(g) \quad + \quad N_2(g) \quad \longrightarrow \quad 2\,NH_3(g)$$

You can see that there are equal numbers of each kind of atom before and after the arrow. For more practice balancing equations, see the Exercises at the end of this chapter.

Practicing chemists develop a skill for balancing equations. This skill involves creative energy and, like other skills, improves with experience. There are some useful tricks of the trade for balancing equations, and maybe your instructor will share some with you. More important than being an expert at balancing equations, however, is knowing why they need to be balanced. And the reason is the law of mass conservation, which tells us that atoms are neither created nor destroyed in a chemical reaction—they are simply rearranged. So every atom present before the reaction must be present after the reaction, even though the groupings of atoms are different.

9.2 Chemists Use Relative Masses to Count Atoms and Molecules

In any chemical reaction, a specific number of reactant atoms or molecules react to form a specific number of product atoms or molecules. For example, when carbon and oxygen combine to form carbon dioxide, they always combine in the ratio of one carbon atom to one oxygen molecule. A chemist who wants to carry out this reaction in the laboratory would be wasting chemicals and money if she were to combine, say, four carbon atoms for every one oxygen molecule. The excess carbon atoms would have no oxygen molecules to react with and would remain unchanged.

How is it possible to measure out a specific number of atoms or molecules? Rather than counting these particles individually, chemists can use a scale that measures the mass of bulk quantities. Because different atoms and molecules have different masses, however, a chemist can't simply measure out equal masses of each. Say, for example, he needs the same number of carbon atoms as oxygen molecules. Measuring equal masses of the two materials would not provide equal numbers.

You know that 1 kilogram of Ping-Pong balls contains more balls than 1 kilogram of golf balls, as Figure 9.1 illustrates. Likewise, because different atoms and molecules have different masses, there are different numbers of them in a 1-gram sample of each. Because carbon atoms are less massive than oxygen molecules, there are more carbon atoms in 1 gram of carbon than there are oxygen molecules in 1 gram of oxygen. So, clearly, equal masses of these two particles do not yield equal numbers of carbon atoms and oxygen molecules.

If we know the *relative masses* of different materials, we can measure equal numbers. Golf balls, for example, are about 20 times more massive

Equal masses

Figure 9.1
The number of balls in a given mass of Ping-Pong balls is very different from the number of balls in the same mass of golf balls.

than Ping-Pong balls, which is to say the relative mass of Ping-Pong balls to golf balls is 20 to 1. Measuring out 20 times as much mass of golf balls as Ping-Pong balls, therefore, gives equal numbers of each, as is shown in Figure 9.2.

The mass of one Ping-Pong ball is 2 grams.   The mass of one golf ball is 40 grams.

A Ping-Pong ball is 2/40, or 1/20, as massive as a golf ball.

Number of Ping-Pong balls   =   Number of golf balls

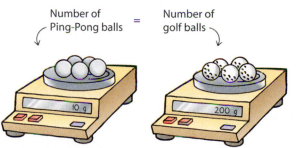

Figure 9.2
The number of golf balls in 200 grams of golf balls equals the number of Ping-Pong balls in 10 grams of Ping-Pong balls.

Concept Check ✔

A customer wants to buy a 1:1 mixture of blue and red jelly beans. Each blue bean is twice as massive as each red bean. If the clerk measures out 5 pounds of red beans, how many pounds of blue beans must she measure out?

Was this your answer? Because each blue jelly bean has twice the mass of each red one, the clerk needs to measure out twice as much mass of blues in order to have the same count, which means 10 pounds of blues. If the clerk did not know that the blue beans were twice as massive as the red ones, she would not know what mass of blues was needed for the 1:1 ratio. Likewise, a chemist would be at a loss in setting up a chemical reaction if she did not know the relative masses of the reactants.

The periodic table tells us the relative masses of carbon and molecular oxygen; therefore, we can measure out equal numbers of their fundamental particles—atoms for carbon and molecules for oxygen. Figure 9.3 illustrates this concept. The atomic mass of carbon is 12.011 atomic mass units. (As discussed in Section 3.6, 1 *atomic mass unit* (amu) $= 1.661 \times 10^{-24}$ gram.) The **formula mass** of a substance is the sum of the atomic masses of the elements in its chemical formula. Therefore, the formula mass of an oxygen molecule, O_2, is 15.999 atomic mass units $+$ 15.999 atomic mass units $\approx$ 32 atomic mass units. A carbon atom, therefore, is about $12/32 = 3/8$ as massive as an oxygen molecule. To measure out equal numbers of carbon atoms and oxygen molecules, we measure out only three-eighths as much carbon. If we started with 8 grams of oxygen, we need 3 grams of carbon to have the same number of particles (because 3 is three-eighths of 8). Alternatively, if we started with 32 grams of oxygen, we need 12 grams of carbon to have the same number of particles (because 12 is three-eighths of 32).

Atomic mass of O = 15.999 amu
+ Atomic mass of O = 15.999 amu
Formula mass of $O_2 \approx$ 32 amu

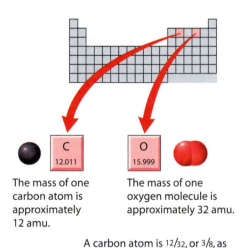

The mass of one carbon atom is approximately 12 amu.

The mass of one oxygen molecule is approximately 32 amu.

A carbon atom is 12/32, or 3/8, as massive as an oxygen molecule.

Number of   = Number of
carbon atoms   oxygen molecules

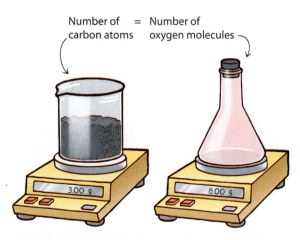

Figure 9.3
To have equal numbers of carbon atoms and oxygen molecules requires measuring out three-eighths as much carbon as oxygen.

Concept Check ✓

1. Reacting 3 grams of carbon, C, with 8 grams of molecular oxygen, O_2, results in 11 grams of carbon dioxide, CO_2. Does it follow that 1.5 grams of carbon will react with 4 grams of oxygen to form 5.5 grams of carbon dioxide?
2. Would reacting 5 grams of carbon with 8 grams of oxygen also result in 11 grams of carbon dioxide?

Were these your answers?

1. The quantities are only half as much, but their ratio is the same as when 11 grams of carbon dioxide is formed: 1.5:4:5.5 = 3:8:11.
2. It is a common error of many students to think that no reaction will occur if the proper ratios of reactants are not provided. You should understand, however, that in a 5-gram sample of carbon, 3 grams of carbon is available for reacting. This 3 grams will react with the 8 grams of oxygen to form 11 grams of carbon dioxide. There will be 2 grams of carbon unreacted after the reaction. Reacting this remaining 2 grams of carbon would require more oxygen.

The Periodic Table Helps Us Convert Between Grams and Moles

Atoms and molecules react in specific ratios. In the laboratory, however, chemists work with bulk quantities of materials, which are measured by mass. Chemists therefore need to know the relationship between the mass of a given sample and the number of atoms or molecules contained in that mass. The key to this relationship is the *mole*. Recall from Section 7.2 that the mole is a unit equal to 6.02×10^{23}. This number is known as **Avogadro's number**, in honor of Amadeo Avogadro (Section 3.3).

As Figure 9.4 illustrates, if you express the numeric value of the atomic mass of any element in grams, the number of atoms in a sample of the

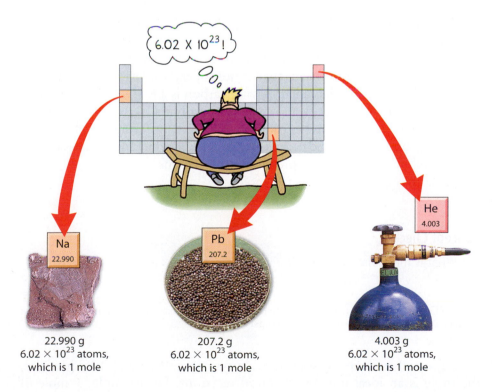

6.02 × 10²³!

Na
22.990

22.990 g
6.02×10^{23} atoms,
which is 1 mole

Pb
207.2

207.2 g
6.02×10^{23} atoms,
which is 1 mole

He
4.003

4.003 g
6.02×10^{23} atoms,
which is 1 mole

Figure 9.4
Express the numeric value of the atomic mass of any element in grams, and that many grams contains 6.02×10^{23} atoms.

element having this mass is always 6.02×10^{23}, which is 1 mole. For example, a 22.990-gram sample of sodium metal, Na (atomic mass 22.990 atomic mass units), contains 6.02×10^{23} sodium atoms, and a 207.2-gram sample of lead, Pb (atomic mass = 207.2 atomic mass units), contains 6.02×10^{23} lead atoms.

The same concept holds for compounds. Express the numeric value of the formula mass of any compound in grams, and a sample having that mass contains 6.02×10^{23} molecules of that compound. For example, there are 6.02×10^{23} O_2 molecules in 31.998 grams of molecular oxygen, O_2 (formula mass 31.998 atomic mass units), and 6.02×10^{23} CO_2 molecules in 44.009 grams of carbon dioxide, CO_2 (formula mass 44.009 atomic mass units).

An explanation of this amazing relationship is beyond the scope of this book but well within the range of questions you might ask your instructor.

Concept Check ✓

1. How many atoms are there in a 6.941-gram sample of lithium, Li (atomic mass 6.941 atomic mass units)?
2. How many molecules are there in an 18.015-gram sample of water, H_2O (formula mass 18.015 atomic mass units)?

Were these your answers?

1. Because this number of grams of lithium is numerically equal to the atomic mass, there are 6.02×10^{23} atoms in the sample, which is 1 mole of lithium atoms.
2. Because this number of grams of water is numerically equal to the formula mass, there are 6.02×10^{23} water molecules in the sample, which is 1 mole of water molecules.

The **molar mass** of any substance, be it element or compound, is defined as the mass of 1 mole of the substance. Thus the units of molar mass are grams per mole. For instance, the atomic mass of carbon is 12.011 atomic mass units, which means that 1 mole of carbon has a mass of 12.011 grams, and we say that the molar mass of carbon is 12.011 grams per mole. The molar mass of molecular oxygen (O_2, formula mass 31.998 atomic mass units) is 31.998 grams per mole. For convenience, values such as these are often rounded off to the nearest whole number. The molar mass of carbon, therefore, might also be presented as 12 grams per mole, and that of molecular oxygen as 32 grams per mole.

Concept Check ✓

What is the molar mass of water (formula mass = 18 atomic mass units)?

Was this your answer? From the formula mass, you know that 1 mole of water has a mass of 18 grams. Therefore the molar mass is 18 grams per mole.

Because 1 mole of any substance always contains 6.02×10^{23} particles, the mole is an ideal unit for chemical reactions. For example, 1 mole of carbon (12 grams) reacts with 1 mole of molecular oxygen (32 grams) to give 1 mole of carbon dioxide (44 grams).

In many instances, the ratio in which chemicals react is not 1:1. As shown in Figure 9.5, for example, 2 moles (4 grams) of molecular hydrogen react with 1 mole (32 grams) of molecular oxygen to give 2 moles (36 grams) of water. Note how the coefficients of the balanced chemical equation can be conveniently interpreted as the number of moles of reactants or products. A chemist therefore need only convert these numbers of moles to grams in order to know how much mass of each reactant he or she should measure out to have the proper proportions.

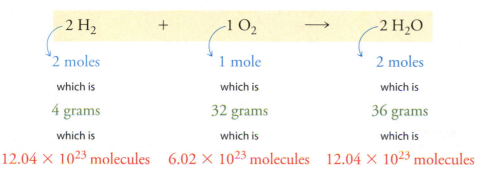

$$2\,H_2 \quad + \quad 1\,O_2 \quad \longrightarrow \quad 2\,H_2O$$

2 moles	1 mole	2 moles
which is	which is	which is
4 grams	32 grams	36 grams
which is	which is	which is
12.04×10^{23} molecules	6.02×10^{23} molecules	12.04×10^{23} molecules

Figure 9.5
Two moles of H_2 react with 1 mole of O_2 to give 2 moles of H_2O. This is the same as saying 4 grams of H_2 reacts with 32 grams of O_2 to give 36 grams of H_2O or, equivalently, that 12.04×10^{23} H_2 molecules react with 6.02×10^{23} O_2 molecules to give 12.04×10^{23} H_2O molecules.

Cooking and chemistry are similar in that both require measuring ingredients. Just as a cook looks to a recipe to find the necessary quantities measured by the cup or the tablespoon, a chemist looks to the periodic table to find the necessary quantities measured by the number of grams per mole for each element or compound.

9.3   Reaction Rate Is Influenced by Concentration and Temperature

A balanced chemical equation helps us determine the amount of products that might be formed from given amounts of reactants. The equation, however, tells us little about what is taking place on the submicroscopic level during the reaction. In this and the following section, we explore that level to show how the *rate* of a reaction can be changed either by changing the concentration or temperature of the reactants or by adding what is known as a *catalyst*.

Some chemical reactions, such as the rusting of iron, are slow, while others, such as the burning of gasoline, are fast. The speed of any reaction is indicated by its reaction rate, which is an indicator of how quickly the reactants transform to products. As shown in Figure 9.6 on page 280, initially a flask may contain only reactant molecules. Over time, these reactants form product molecules, and as a result, the concentration of product molecules increases. The **reaction rate**, therefore, can be defined either as how quickly the concentration of products increases or as how quickly the concentration of reactants decreases.

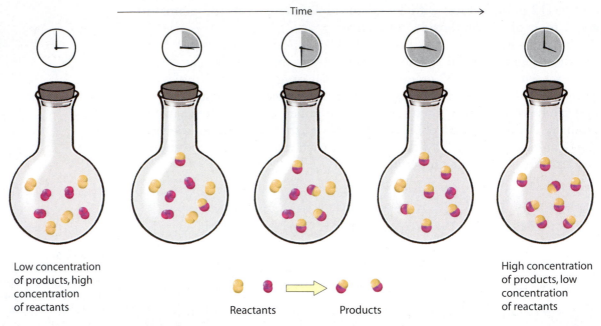

Time

Low concentration of products, high concentration of reactants

Reactants → Products

High concentration of products, low concentration of reactants

Figure 9.6
Over time, the reactants in this reaction flask may transform to products. If this happens quickly, the reaction rate is high. If this happens slowly, the reaction rate is low.

What determines the rate of a chemical reaction? The answer is complex, but one important factor is that reactant molecules must physically come together. Because molecules move rapidly, this physical contact is appropriately described as a collision. We can illustrate the relationship between molecular collisions and reaction rate by considering the reaction of gaseous nitrogen and gaseous oxygen to form gaseous nitrogen monoxide as shown in Figure 9.7.

Because reactant molecules must collide in order for a reaction to occur, the rate of a reaction can be increased by increasing the number of collisions. An effective way to increase the number of collisions is to increase the

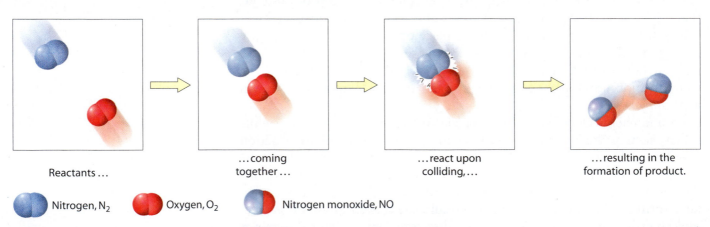

Reactants ...   ...coming together ...   ...react upon colliding, ...resulting in the formation of product.

Nitrogen, N_2 Oxygen, O_2 Nitrogen monoxide, NO

Figure 9.7
During a reaction, reactant molecules collide with each other.

Calculation Corner: Figuring Masses of Reactants and Products

Using conversion factors (Section 1.3) and the relationship between grams and moles, you can perform some very high-powered calculations.

Example

What mass of water is produced when 16 grams of methane, CH_4 (formula mass 16 atomic mass units), burns in the reaction

$$CH_4 + 2\,O_2 \longrightarrow CO_2 + 2\,H_2O$$

Step 1. Convert the given mass to moles:

Conversion factor

$$(16\ g\ CH_4)\left(\frac{1\ mole\ CH_4}{16\ g\ CH_4}\right) = 1\ mole\ CH_4$$

Step 2. Use the coefficients of the balanced equation to find out how many moles of H_2O are produced from this many moles of CH_4:

Conversion factor

$$(1\ mole\ CH_4)\left(\frac{2\ moles\ H_2O}{1\ mole\ CH_4}\right) = 2\ moles\ H_2O$$

Step 3. Now that you know how many moles of H_2O are produced, convert this value to grams of H_2O:

Conversion factor

$$(2\ moles\ H_2O)\left(\frac{18\ g\ H_2O}{1\ mole\ H_2O}\right) = 36\ g\ H_2O$$

This method of converting from grams to moles (step 1), then from moles to moles (step 2), and then from moles to grams (step 3) is an important aspect of what is called *stoichiometry*—the science of calculating the amount of reactants or products in any chemical reaction. It is a method that is developed much further in general chemistry courses. For this course, all you need to do is be familiar with what stoichiometry is all about, which is keeping tabs on atoms and molecules as they react to form products. Nonetheless, for a special assignment, you might try your analytical thinking skills on the following problems. First try to deduce the answer based on what you know about the law of mass conservation, and then follow the steps given here to check your answers.

Your Turn

1. How many grams of ozone (O_3, 48 amu) can be produced from 64 grams of oxygen (O_2, 32 amu) in the reaction

$$3\,O_2 \longrightarrow 2\,O_3$$

2. What mass of nitrogen monoxide (NO, 30 amu) is formed when 28 grams of nitrogen (N_2, 28 amu) reacts with 32 grams of oxygen (O_2, 32 amu) in the reaction

$$N_2 + O_2 \longrightarrow 2\,NO$$

concentration of the reactants. Figure 9.8 on page 282 shows that, with higher concentrations, there are more molecules in a given volume, which makes collisions between molecules more probable. As an analogy, consider a bunch of people on a dance floor—as the number of people increases, so does the rate at which they bump into one another. An increase in the concentration of nitrogen and oxygen molecules, therefore, leads to a greater number of collisions between these molecules and hence a greater number of nitrogen monoxide molecules formed in a given period of time.

Figure 9.8
The more concentrated a sample of nitrogen and oxygen, the greater the likelihood that N_2 and O_2 molecules will collide and form nitrogen monoxide.

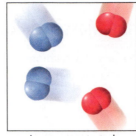
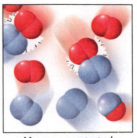

Less concentrated   More concentrated

Not all collisions between reactant molecules lead to products, however, because the molecules must collide in a certain orientation in order to react. Nitrogen and oxygen, for example, are much more likely to form nitrogen monoxide when the molecules collide in the parallel orientation shown in Figure 9.7. When they collide in the perpendicular orientation shown in Figure 9.9, nitrogen monoxide does not form. For larger molecules, which can have numerous orientations, this orientation requirement is even more restrictive.

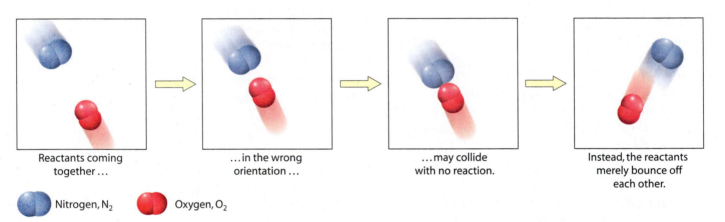

Reactants coming together …   …in the wrong orientation …   …may collide with no reaction.   Instead, the reactants merely bounce off each other.

Nitrogen, N_2 Oxygen, O_2

Figure 9.9
The orientation of reactant molecules in a collision can determine whether or not a reaction takes place. A perpendicular collision between N_2 and O_2 does not tend to result in formation of a product molecule.

A second reason not all collisions lead to product formation is that the reactant molecules must also collide with enough kinetic energy to break their bonds. Only then is it possible for the atoms in the reactant molecules to change bonding partners and form product molecules. The bonds in N_2 and O_2 molecules, for example, are quite strong. In order for these bonds to be broken, collisions between the molecules must contain enough energy to break the bonds. As a result, collisions between slow-moving N_2 and O_2 molecules, even those that collide in the proper orientation, may not form NO, as is shown in Figure 9.10.

The higher the temperature of a material, the faster its molecules are moving and the more forceful the collisions between them. Higher temperatures, therefore, tend to increase reaction rates. The nitrogen and oxygen molecules that make up our atmosphere, for example, are always colliding with one another. At the ambient temperatures of our atmosphere, however, these molecules do not generally have sufficient kinetic energy to allow for the formation of nitrogen monoxide. The heat of a lightning bolt, however,

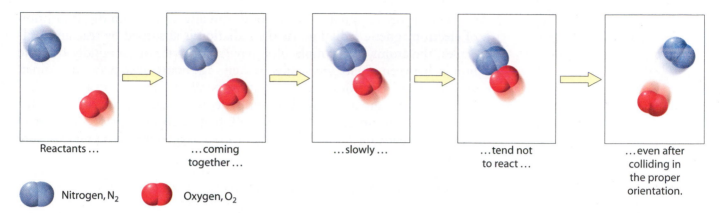

| Reactants ... | ...coming together ... | ...slowly ... | ...tend not to react ... | ...even after colliding in the proper orientation. |

🔵 Nitrogen, N$_2$ 🔴 Oxygen, O$_2$

Figure 9.10
Slow-moving molecules may collide without enough force to break bonds. In this case, they cannot react to form product molecules.

dramatically increases the kinetic energy of these molecules, to the point that a large portion of the collisions in the vicinity of the bolt result in the formation of nitrogen monoxide. As discussed in the opening of this chapter, the nitrogen monoxide formed in this manner undergoes further atmospheric reactions to form chemicals known as nitrates that plants depend on to survive. This is an example of *nitrogen fixation*, which we explore in Chapter 15.

Concept Check ✔

An internal-combustion engine works by drawing a mixture of air and gasoline vapors into a chamber. The action of a piston then compresses these gases into a smaller volume prior to ignition by the spark of a spark plug. What is the advantage of squeezing the vapors to a smaller volume?

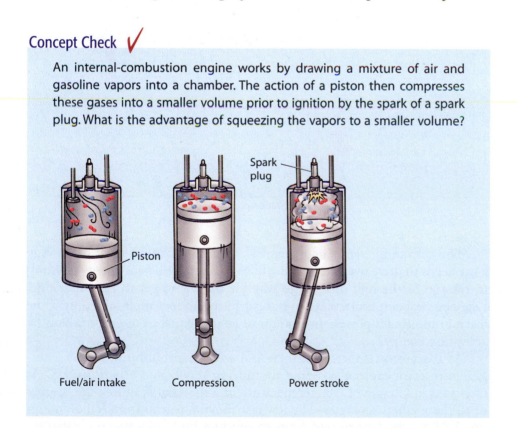

Spark plug

Piston

Fuel/air intake   Compression   Power stroke

Was this your answer? Squeezing the vapors to a smaller volume effectively increases their concentration and hence the number of collisions between molecules. This, in turn, promotes the chemical reaction.

The energy required to break bonds can also come from the absorption of electromagnetic radiation. As the radiation is absorbed by reactant molecules, the atoms in the molecules may start to vibrate so rapidly that the bonds between them are easily broken. In many instances, the direct absorption of electromagnetic radiation is all it takes to break chemical bonds and initiate a chemical reaction. As we discuss in Chapter 17, for example, the common atmospheric pollutant nitrogen dioxide, NO_2, may transform to nitrogen monoxide and atomic oxygen merely upon exposure to sunlight:

$$NO_2 + \text{sunlight} \longrightarrow NO + O$$

Whether the result of collisions, absorption of electromagnetic radiation, or both, broken bonds are a necessary first step in most chemical reactions. The energy required for this initial breaking of bonds can be viewed as an *energy barrier*. The minimum energy required to overcome this energy barrier is known as the **activation energy** E_a.

In the reaction between nitrogen and oxygen to form nitrogen monoxide, the energy barrier is so high (because the bonds in N_2 and O_2 are strong) that only the fastest-moving nitrogen and oxygen molecules possess sufficient energy to react. Figure 9.11 shows the energy barrier in this chemical reaction as a vertical hump.

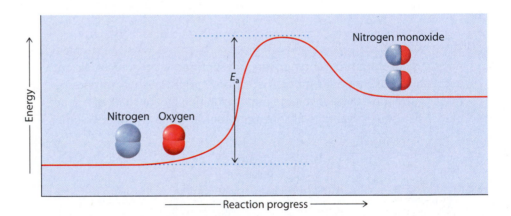

Figure 9.11
Reactant molecules must gain a minimum amount of energy, called the activation energy E_a, in order to transform to product molecules.

The activation energy of a chemical reaction is analogous to the energy a car needs to drive over the top of a hill. Without sufficient energy to climb to the top of the hill, there is no way for the car to get to the other side. Likewise, reactant molecules can transform to product molecules only if the reactant molecules possess an amount of energy equal to or greater than the activation energy.

At any given temperature, there is a wide distribution of kinetic energies in reactant molecules. Some are moving slowly and others quickly. As discussed in Chapter 1, the temperature of a material is simply the *average* of all these kinetic energies. The few fast-moving reactant molecules in Figure 9.12 are the first to transform to product molecules because these are the molecules that have enough energy to pass over the energy barrier.

When the temperature of reactants is increased, the number of reactant molecules having sufficient energy to pass over the barrier also increases,

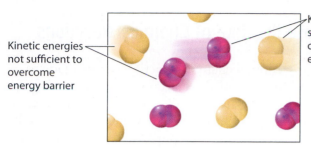

Kinetic energies not sufficient to overcome energy barrier

Kinetic energies sufficient to overcome energy barrier

Figure 9.12
Fast-moving reactant molecules possess sufficient energy to pass over the energy barrier and hence are the first ones to transform to product molecules.

which is why reactions are generally faster at higher temperatures. Conversely, at lower temperatures, there are fewer molecules having sufficient energy to pass over the barrier, which is why reactions are generally slower at lower temperatures.

Most chemical reactions are influenced by temperature in this manner, including those reactions occurring in living bodies. The body temperature of animals that regulate their internal temperature, such as humans, is fairly constant. However, the body temperature of some animals, such as the alligator shown in Figure 9.13, rises and falls with the temperature of the environment. On a warm day, the chemical reactions occurring in an alligator are "up to speed," and the animal can be most active. On a chilly day, however, the chemical reactions proceed at a lower rate, and as a consequence, the alligator's movements are unavoidably sluggish.

Figure 9.13
This alligator became immobilized on the pavement after being caught in the cold night air. By mid-morning, shown here, the temperature had warmed sufficiently to allow the alligator to get up and walk away.

Concept Check ✔

What kitchen device is used to lower the rate at which microorganisms grow on food?

Was this your answer? The refrigerator! Microorganisms, such as bread mold, are everywhere and difficult to avoid. By lowering the temperature of microorganism-contaminated food, the refrigerator decreases the rate of the chemical reactions that these microorganisms depend on for growth, thereby increasing the food's shelf life.

9.4 Catalysts Increase the Rate of Chemical Reactions

As discussed in the previous section, a chemical reaction can be made to go faster by increasing the concentration of the reactants or by increasing the temperature. A third way to increase the rate of a reaction is to add a **catalyst**, which is any substance that increases the rate of a chemical reaction by lowering its activation energy. The catalyst may participate as a reactant, but it is then regenerated as a product and is thus available to catalyze subsequent reactions.

The conversion of ozone, O_3, to oxygen, O_2, is normally sluggish because the reaction has a relatively high energy barrier, as shown in Figure 9.14a. However, when chlorine atoms act as a catalyst, the energy barrier is lowered, as shown in Figure 9.14b, and the reaction is able to proceed faster.

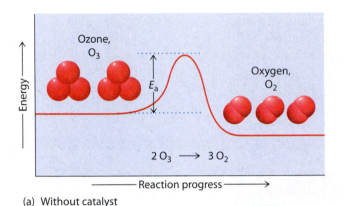

(a) Without catalyst

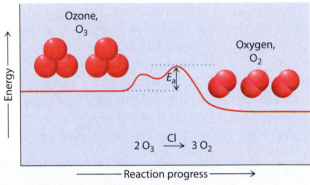

(b) With chlorine catalyst

Figure 9.14
(a) The relatively high energy barrier indicates that only the most energetic ozone molecules can react to form oxygen molecules. (b) Chlorine atoms lower the energy barrier, which means more reactant molecules have sufficient energy to form product. The chlorine allows the reaction to proceed in two steps, and the two smaller energy barriers correspond to these steps. (Note that the convention is to write the catalyst above the reaction arrow.)

Atomic chlorine lowers the energy barrier of this reaction by providing an alternate pathway involving intermediate reactions, each having a lower activation energy than the uncatalyzed reaction. This alternate pathway involves two steps. Initially, the chlorine reacts with the ozone to form chlorine monoxide and oxygen:

$$Cl + O_3 \longrightarrow ClO + O_2$$

Chlorine   Ozone   Chlorine   Oxygen
monoxide

The chlorine monoxide then reacts with another ozone molecule to re-form the chlorine atom as well as produce two additional oxygen molecules:

$$ClO + O_3 \longrightarrow Cl + 2 O_2$$

Chlorine   Ozone   Chlorine   Oxygen
monoxide

Although chlorine is used up in the first reaction, it is regenerated in the second reaction. As a result, there is no net consumption of chlorine. At the same time, however, a total of two ozone molecules are rapidly converted to three oxygen molecules. The chlorine is therefore a catalyst for the conversion of ozone to oxygen because the chlorine increases the speed of the reaction but is not consumed by the reaction.

Chlorine atoms in the stratosphere catalyze the destruction of the Earth's ozone layer. As we explore further in Chapter 17, evidence tells us that chlorine atoms are generated in the stratosphere as a by-product of human-made chlorofluorocarbons (CFCs), once widely produced as the cooling fluid of refrigerators and air-conditioners. Destruction of the ozone layer is a serious concern because of the role this layer plays in protecting us from the sun's harmful ultraviolet rays. One chlorine atom in the ozone layer, it is estimated, can catalyze the transformation of 100,000 ozone molecules to oxygen molecules in the one or two years before the chlorine atom is removed by natural processes.

Chemists have been able to harness the power of catalysts for numerous beneficial purposes. The exhaust that comes from an automobile engine, for example, contains a wide assortment of pollutants, such as nitrogen monoxide, carbon monoxide, and uncombusted fuel vapors (hydrocarbons). To reduce the amount of these pollutants entering the atmosphere, most automobiles are equipped with a *catalytic converter*, shown in Figure 9.15. Metal catalysts in a converter speed up reactions that convert exhaust pollutants to less toxic substances. Nitrogen monoxide is transformed to nitrogen and oxygen, carbon monoxide is transformed to carbon dioxide, and unburned fuel is converted to carbon dioxide and water vapor. Because catalysts are not consumed by the reactions they facilitate, a single catalytic converter may continue to operate effectively for the lifetime of the car.

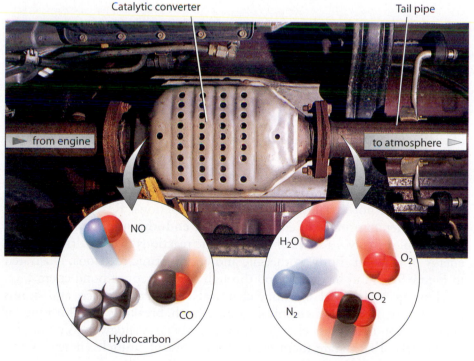

Catalytic converter

Tail pipe

from engine

to atmosphere

NO

Hydrocarbon

CO

H_2O

O_2

N_2

CO_2

Before it reaches the catalytic converter, the exhaust contains such pollutants as NO, CO, and hydrocarbons.

After it has passed through the catalytic converter, the exhaust contains water vapor, N_2, O_2, and CO_2.

Figure 9.15
A catalytic converter reduces the pollution caused by automobile exhaust by converting such harmful combustion products as NO, CO, and hydrocarbons to harmless N_2, O_2, and CO_2. The catalyst is typically platinum, Pt, palladium, Pd, or rhodium, Rd.

Figure 9.16
The exhaust from automobiles today is much cleaner than before the advent of the catalytic converter, but there are many more cars on the road. In 1960 there were 70 million registered motor vehicles in the United States. In 2000 there were more than 200 million.

Catalytic converters, along with microchip-controlled fuel–air ratios, have led to a significant drop in the per-vehicle emission of pollutants. A typical car in 1960 emitted about 11 grams of uncombusted fuel, 4 grams of nitrogen oxide, and 84 grams of carbon monoxide per mile traveled. An improved vehicle in 2000 emitted less than 0.5 gram of uncombusted fuel, less than 0.5 gram of nitrogen oxide, and only about 3 grams of carbon monoxide per mile traveled. This improvement, however, has been offset by an increase in the number of cars being driven, exemplified by the traffic jam shown in Figure 9.16.

The chemical industry depends on catalysts because they lower manufacturing costs by lowering required temperatures and providing greater product yields without being consumed. Indeed, more than 90 percent of all manufactured goods are produced with the assistance of catalysts. Without catalysts, the price of gasoline would be much higher, as would be the price of such consumer goods as rubber, plastics, pharmaceuticals, automobile parts, clothing, and food grown with chemical fertilizers.

Living organisms rely on special types of catalysts known as *enzymes*, which allow exceedingly complex biochemical reactions to occur with ease. The nature and behavior of enzymes are discussed in Chapter 13.

Concept Check ✔

How does a catalyst lower the energy barrier of a chemical reaction?

Was this your answer? The catalyst provides an alternate and easier-to-achieve pathway along which the chemical reaction can proceed.

9.5 Chemical Reactions Can Be Either Exothermic or Endothermic

As the preceding two sections discuss, reactants must have a certain amount of energy in order to overcome the energy barrier so that a chemical reaction can proceed. Once a reaction is complete, however, there may be either a net release or a net absorption of energy. Reactions in which there is a net release of energy are called **exothermic**. Rocketships lift off into space and campfires glow red hot as a result of exothermic reactions. Reactions in which there is a net absorption of energy are called **endothermic**. Photosynthesis, for example, involves a series of endothermic reactions that are driven by the energy of sunlight. Both exothermic and endothermic reactions, illustrated in Figure 9.17, can be understood through the concept of bond energy.

During a chemical reaction, chemical bonds are broken and atoms rearrange to form new chemical bonds. Such breaking and forming of chemical bonds involves changes in energy. As an analogy, consider a pair of magnets. To separate them requires an input of "muscle energy." Conversely, when the two separated magnets collide, they become slightly warmer than they were, and this warmth is evidence of energy released. Energy must be absorbed by the magnets if they are to break apart, and energy is released as they come together. The same principle applies to atoms. To pull bonded atoms apart requires an energy input. When atoms combine, there is an energy output, usually in the form of faster-moving atoms and molecules, electromagnetic radiation, or both.

The amount of energy required to pull two bonded atoms apart is the same as the amount released when they are brought together. This energy absorbed as a bond breaks or released as one forms is called **bond energy**. Each chemical bond has its own characteristic bond energy. The hydrogen–hydrogen bond energy, for example, is 436 kilojoules per mole. This means that 436 kilojoules of energy is absorbed as 1 mole of hydrogen–hydrogen bonds break apart, and 436 kilojoules of energy is released upon the formation of 1 mole of hydrogen–hydrogen bonds. Different bonds involving different elements have different bond energies, as Table 9.1 shows. You can refer to the table as you study this section, but please do not memorize these bond energies. Instead, focus on understanding what they mean.

Table 9.1

Selected Bond Energies

Bond	Bond Energy (kJ/mole)	Bond	Bond Energy (kJ/mole)
H–H	436	O–O	138
H–C	414	Cl–Cl	243
H–N	389	N–N	159
H–O	464	N=O	631
H–F	569	O=O	498
H–Cl	431	O=C	803
H–S	339	N≡N	946
C–C	347	C≡C	837

By convention, a positive bond energy represents the amount of energy absorbed as a bond breaks and a negative bond energy represents the amount of energy released as a bond forms. Thus when you are calculating the net energy released or absorbed during a reaction, you'll need to be careful about plus and minus signs. It is standard practice when doing such calculations to assign a plus sign to energy absorbed and a minus sign to energy released. For instance, when dealing with a reaction in which 1 mole of H–H bonds are broken, you'll write +436 kilojoules to indicate energy absorbed, and when dealing with the formation of 1 mole of H–H bonds, you'll write −436 kilojoules to indicate energy released. We'll do some sample calculations in a moment.

Figure 9.17
For the chemical reactions taking place in burning wood, there is a net release of energy. For those taking place in a photosynthetic plant, there is a net absorption of energy.

Concept Check ✔

Do all covalent single bonds have the same bond energy?

Was this your answer? Bond energy depends on the types of atoms bonding. The H–H single bond, for example, has a bond energy of 436 kilojoules per mole, but the H–O single bond has a bond energy of 464 kilojoules per mole. All covalent single bonds do not have the same bond energy.

An Exothermic Reaction Involves a Net Release of Energy

For any chemical reaction, the total amount of energy absorbed in breaking bonds in reactants is always different from the total amount of the energy released as bonds form in the products. Consider the reaction in which hydrogen and oxygen react to form water:

$$H\!-\!H + H\!-\!H + O\!=\!O \longrightarrow H\!-\!O\!\diagdown_H \;\; + \;\; \substack{H \diagdown\;\diagup H \\ O}$$

In the reactants, hydrogen atoms are bonded to hydrogen atoms and oxygen atoms are double-bonded to oxygen atoms. The total amount of energy absorbed as these bonds break is

Type of bond	Number of moles	Bond energy	Total energy
H–H	2	+436 kJ/mole	+872 kJ
O=O	1	+498 kJ/mole	+498 kJ
		Total energy absorbed	+1370 kJ

In the products there are four hydrogen–oxygen bonds. The total amount of energy released as these bonds form is

Type of bond	Number of moles	Bond energy	Total energy
H–O	4	−464 kJ/mole	−1856 kJ
		Total energy released	−1856 kJ

For this reaction the amount of energy released exceeds the amount of energy absorbed. The net energy of the reaction is found by adding the two quantities:

$$
\begin{aligned}
\text{net energy of reaction} &= \text{energy absorbed} + \text{energy released} \\
&= +1370\text{ kJ} + (-1856\text{ kJ}) \\
&= -486\text{ kJ}
\end{aligned}
$$

The negative sign on the net energy indicates that there is a net *release* of energy, and so the reaction is exothermic. For any exothermic reaction, energy can be considered a product and is thus sometimes included after the arrow of the chemical equation:

$$2\,H_2 + O_2 \longrightarrow 2\,H_2O + \text{energy}$$

In an exothermic reaction, the potential energy of atoms in the product molecules is lower than their potential energy in the reactant molecules. This is illustrated in the reaction profile shown in Figure 9.18. The potential energy of the atoms is lower in the product molecules because they are held more tightly together. This is analogous to two attracting magnets, whose potential energy decreases as they come closer together. The loss of potential energy is balanced by a gain in kinetic energy. As two free-floating

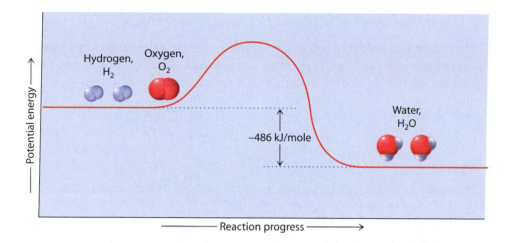

Figure 9.18
In an exothermic reaction, the product molecules are at a lower potential energy than the reactant molecules. The net amount of energy released by the reaction is equal to the difference in potential energies of the reactants and products.

magnets come together, they accelerate to higher speeds. Similarly, as reactants react to form products, the potential energy of the reactants is converted to kinetic energy, which can take the form of faster-moving atoms and molecules, electromagnetic radiation, or both. This kinetic energy is the energy released by the reaction, and it is equal to the difference between the potential energy of the reactants and the potential energy of the products, as is indicated in Figure 9.18.

It is important to understand that the energy released by an exothermic reaction is not created by the reaction. This is in accord with the *law of conservation of energy*, which states that energy is neither created nor destroyed in a chemical reaction. Instead, it is merely converted from one form to another. During an exothermic reaction, energy that was once in the form of the potential energy of chemical bonds is released as the kinetic energy of fast-moving molecules and/or as electromagnetic radiation.

The amount of energy released in an exothermic reaction depends on the amount of reactants. The reaction of large amounts of hydrogen and oxygen, for example, provides the energy to lift the space shuttle shown in Figure 9.19 into orbit. There are two compartments in the large central tank to which the orbiter is attached—one filled with liquid hydrogen and the other with liquid oxygen. Upon ignition, these two liquids mix and react chemically to form water vapor, which produces the needed thrust as it is expelled out the rocket cones. Additional thrust is obtained by a pair of solid-fuel rocket boosters containing a mixture of ammonium perchlorate, NH_4ClO_4, and powdered aluminum. Upon ignition, these chemicals react to form products that are expelled out the back of the rocket. The balanced equation representing this reaction is

Figure 9.19
A space shuttle uses exothermic chemical reactions to lift off from the Earth's surface.

$$3 \ NH_4ClO_4 + 3 \ Al \longrightarrow Al_2O_3 + AlCl_3 + 3 \ NO + 6 \ H_2O + energy$$

Concept Check

Where does the net energy released in an exothermic reaction go?

Was this your answer? This energy goes into making atoms and molecules move faster and/or into the formation of electromagnetic radiation.

An Endothermic Reaction Involves a Net Absorption of Energy

Many chemical reactions are endothermic, such that the amount of energy released as products form is *less* than the amount of energy absorbed in the breaking of bonds in the reactants. An example is the reaction of atmospheric nitrogen and oxygen to form nitrogen monoxide, which is the same reaction used for many of the discussions earlier in this chapter:

$$N{\equiv}N + O{=}O \longrightarrow N{=}O + N{=}O$$

The amount of energy absorbed as the chemical bonds in the reactants break is

Type of bond	Number of moles	Bond energy	Total energy
$N{\equiv}N$	1	$+946$ kJ/mole	$+946$ kJ
$O{=}O$	1	$+498$ kJ/mole	$+498$ kJ
		Total energy absorbed	$+1444$ kJ

The amount of energy released upon the formation of bonds in the products is

Type of bond	Number of moles	Bond energy	Total energy
$N{=}O$	2	-631 kJ/mole	-1262 kJ
		Total energy released	-1262 kJ

As before, the net energy of the reaction is found by adding the two quantities:

$$\begin{aligned} \text{net energy of reaction} &= \text{energy absorbed} + \text{energy released} \\ &= +1444\text{ kJ} + (-1262\text{ kJ}) \\ &= +182\text{ kJ} \end{aligned}$$

The positive sign indicates that there is a net *absorption* of energy, meaning the reaction is endothermic. For any endothermic reaction, energy can be considered a reactant and is thus sometimes included before the arrow of the chemical equation:

$$\text{energy} + N_2 + O_2 \longrightarrow 2\,NO$$

In an endothermic reaction, the potential energy of atoms in the product molecules is higher than their potential energy in the reactant molecules. This is illustrated in the reaction profile shown in Figure 9.20. Raising the potential energy of the atoms in the product molecules requires a net input of energy, which must come from some external source, such as electromagnetic radiation, electricity, or heat. Thus nitrogen and oxygen react to form nitrogen monoxide only with the application of much heat, as occurs adjacent to a lightning bolt or in an internal-combustion engine.

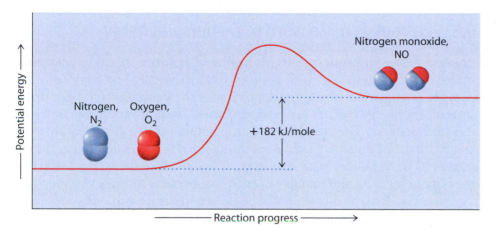

Figure 9.20
In an endothermic reaction, the product molecules are at a higher potential energy than the reactant molecules. The net amount of energy absorbed by the reaction is equal to the difference in potential energies of the reactants and products.

Hands-On Chemistry:
Warming and Cooling Water Mixtures

Recall from Section 7.1 that chemical bonds and intermolecular attractions are both consequences of the electric force, the difference being that chemical bonds are generally many times stronger than molecule-to-molecule attractions. So, just as the formation and breaking of chemical bonds involves energy, so does the formation and breaking of molecular attractions. For molecule-to-molecule attractions, the amount of energy absorbed or released per gram of material is relatively small. Physical changes involving the formation or breaking of molecule-to-molecule attractions, therefore, are much safer to perform, which makes them more suitable for a Hands-On Chemistry activity. Experience the exothermic and endothermic nature of physical changes for yourself by performing the following two activities.

① Hold some room-temperature water in the cupped palm of your hand over a sink. Pour an equal amount of room-temperature rubbing alcohol into the water. Is this mixing an exothermic or endothermic process? What's going on at the molecular level?

② Add lukewarm water to two plastic cups. (Do not use insulating Styrofoam cups.) Transfer the liquid back and forth between cups to ensure equal temperatures, ending up with the same amount of water in each cup. Add several tablespoons of table salt to one cup and stir. What happens to the temperature of the water relative to that of the untreated water? (Hold the cups up to your cheeks to tell.) Is this an exothermic or endothermic process? What's going on at the molecular level?

9.6 Entropy Is a Measure of Dispersed Energy

Energy tends to disperse. It flows from where it is concentrated to where it is spread out. This model fits with our everyday experience. The energy of a hot pan, for example, doesn't stay in the pan once the pan is taken off the stove. Instead, the energy radiates outward (disperses), away from the pan and into the surroundings. The energy found in gasoline burns explosively in a car engine. Some of this energy disperses through the transmission to get the car moving. The rest disperses as heat into the engine, its radiator fluid, and out the exhaust pipe. A third example of energy dispersal is shown in Figure 9.21.

Figure 9.21
Bouncing marbles don't bounce forever. Rather, they slow to a halt as frictional forces cause their kinetic energy to transform to heat energy, which disperses into the floor and throughout the room.

Scientists look to this tendency of energy to disperse as one of the central driving forces for physical and chemical processes. In other words, processes that result in the dispersion of energy tend to occur on their own—they are favored. This includes the cooling down of a hot pan and the burning of gasoline.

And the opposite holds true, too. Processes that result in the concentration of energy tend *not* to occur—they are not favored. Heat from the room, for example, will not spontaneously move back into the pan to heat it up. Likewise, the lower-energy molecules of the car's exhaust won't on their own come back together to re-form the higher-energy gasoline molecules. The natural flow of energy is always a one-way trip from where it is concentrated to where it is spread out.

Think about the *total* amount of energy held within a substance. The paper of this book, for example, is composed of atoms and molecules in constant motion, which is a form of kinetic energy. The paper can be burned easily, so we know it also possesses chemical potential energy. Energy within a substance is found in these and other forms. In general the total amount of energy within a substance depends on the substance's chemical identity, how much of it you have, and its temperature. Chemists measure this amount of energy as **entropy**, defined as the total amount of energy within a given amount of substance divided by the absolute temperature of the substance.

A way of interpreting this is to say that the entropy of a substance reflects the amount of energy that must go into that substance to raise its temperature from 0 kelvin up to some particular temperature, also measured in kelvins. The absolute temperature scale is used because, as discussed in Section 1.6, it is directly related to the motions of atoms and molecules.

The entropy of a substance can be measured by experiment, and some values are given in Table 9.2. Note that entropy is commonly represented by the letter S and is given in units of energy (joules) divided by absolute temperature (kelvins). In general, the higher the entropy value of a substance, the more energy the substance contains, as illustrated in Figure 9.22.

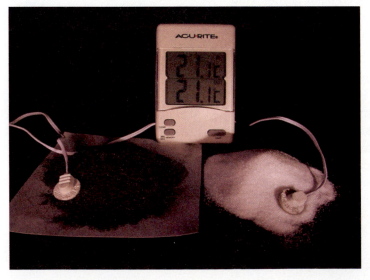

Figure 9.22
The 12.0 grams of graphite (1 mole) and 58.5 grams of sodium chloride (1 mole) shown here are at the same temperature, 294 K (21.1°C). The total amount of energy they have absorbed to get from 0 K to this temperature, however, is much greater for the sodium chloride ($S = 72.4$ J/K) than for the graphite ($S = 5.7$ J/K).

Table 9.2

Per Mole Entropies of Selected Substances at 298 K (25°C)

Substance	Entropy, S (J/K)
Carbon, C(s, diamond)	2.4
Carbon, C(s, graphite)	5.7
Sodium fluoride, NaF(s)	51.5
Water, $H_2O(\ell)$	69.9
Sodium chloride, NaCl(s)	72.4
Sodium bromide, NaBr(s)	86.8
Hydrogen, H(g)	114.7
Sodium chloride, NaCl(aq)	115.5
Methanol, $CH_3OH(\ell)$	126.8
Hydrogen, $H_2(g)$	130.7
Ammonium nitrate, $NH_4NO_3(s)$	151.1
Carbon, C(g)	158.1
Methane, $CH_4(g)$	186.3
Water, $H_2O(g)$	188.8
Nitrogen, $N_2(g)$	191.6
Ammonia, $NH_3(g)$	192.5
Oxygen, $O_2(g)$	205.1
Carbon dioxide, $CO_2(g)$	213.7
Nitrogen dioxide, $NO_2(g)$	240.1
Ammonium nitrate, $NH_4NO_3(aq)$	259.8

Entropy Changes as Chemicals React

During a chemical reaction there is a change in the identity of the substances. For example, 2 moles of atomic hydrogen, H, may react to form 1 mole of molecular hydrogen, H_2:

$$H(g) + H(g) \longrightarrow H_2(g)$$

Although H and H_2 are both made of hydrogen atoms, they are different substances and so have different properties, including different entropies, as Table 9.2 shows. Chemists calculate the difference in entropies for the chemicals in a reaction by adding up the entropies of all the products and subtracting the sum of the entropies of all the reactants:

entropy difference = entropies of products − entropies of reactants

For the formation of a molecular hydrogen, the entropy difference is

$$\begin{array}{ccccc} H & + & H & \longrightarrow & H_2 \\ \text{114.7 J/K} & & \text{114.7 J/K} & & \text{130.7 J/K} \end{array}$$

entropy difference = (130.7 J/K) − (114.7 J/K + 114.7 J/K)

= −98.7 J/K

Note that this is a negative entropy difference, which means the entropy *decreases* as the reaction proceeds. We can envision how entropy decreases by noting the number of particles. In general, more particles means energy has more opportunities to disperse (high entropy). Conversely, when there are fewer particles, there are fewer opportunities for energy to disperse (low entropy). For this hydrogen reaction, 2 moles of hydrogen atoms are transformed to 1 mole of hydrogen molecules, which means that after the reaction is over, there are fewer particles. Hence, the energy is less dispersed in the product than in the reactants.

As an analogy, consider two marbles dropped to the floor. You know that energy is dispersed as these marbles bounce off in different directions. What if these marbles were somehow combined into a larger single marble? When dropped, there may be the same release of energy, but because the marbles are combined, this energy is released over a smaller region. Two bouncing marbles, therefore, provide more entropy (energy dispersion) than a single bouncing marble of the same mass. Likewise, for a given temperature, two hydrogen atoms provide more entropy than a single hydrogen molecule of the same mass. Transforming two hydrogen atoms into a single hydrogen molecule, therefore, results in a *decrease* in entropy, which is seen above as a negative entropy difference.

Concept Check ✓

Two moles of hydrogen gas, H_2, react with 1 mole of oxygen gas, O_2, to form 2 moles of liquid water, H_2O. What is the difference in entropies between reactants and products?

$$H–H(g) + H–H(g) + O=O(g) \longrightarrow H–O–H(\ell) + H–O–H(\ell)$$

130.7 J/K   130.7 J/K   205.1 J/K   69.9 J/K   69.9 J/K

Was this your answer?

entropy difference = entropies of products − entropies of reactants
= (69.9 J/K + 69.9 J/K) − (130.7 J/K + 130.7 J/K + 205.1 J/K)
= −326.7 J/K

This is a negative entropy difference (entropy decreases), which makes sense because there are fewer molecules after the reaction has occurred. Furthermore, the reactants are gases and the products are liquids. Gases have higher entropy values than liquids because gas molecules are more dispersed.

According to the above analysis, hydrogen atoms should *not* transform to hydrogen molecules on their own because this would go against the natural tendency of energy to disperse. We do not yet have the whole story, however, and so let's look further. Although we have considered the energies products and reactants have within them, we have yet to consider the energy released or absorbed during a reaction. This is the energy of reaction that in Section 9.5 you learned to calculate by comparing the bond energies of the reactants and products.

The formation of molecular hydrogen involves only the formation of chemical bonds; no reactant bonds need to be broken. It is exothermic, releasing 436,000 J per mole of H_2 formed.* In being exothermic, the reac-

---

* When considered from the point of view of the chemical reaction, this is −436,000 J, where the minus sign indicates energy is *lost by the reaction*. Here we are looking at the energy *gained by the surroundings*, which we specify as +436,000 J.

tion spreads out energy into the environment. This fact indicates to us that hydrogen atoms *should* react to form hydrogen molecules. So we have two competing entropies: one in which less energy is dispersed to the hydrogen molecules and one involving much energy dispersed to the surroundings. Which dominates? We need only add these two competing entropies together to find out.

The entropy change resulting from the release or absorption of energy is calculated by dividing the energy of the reaction by its absolute temperature. Thus the 436,000 J of energy released in the formation of 1 mole of H_2 is 436,000 J/298 K, which equals +1463 J/K. The amount of entropy that goes into the universe from this reaction, therefore, is

$$\begin{array}{l} \text{change in} \\ S \text{ of universe} \end{array} = \begin{array}{l} \text{change in } S \text{ from} \\ \text{products} - \text{reactants} \end{array} + \begin{array}{l} \text{change in } S \text{ from} \\ \text{energy released/absorbed} \\ \text{during reaction} \end{array}$$

$$= -98.7 \text{ J/K} + 1463 \text{ J/K}$$

$$= +1364 \text{ J/K}$$

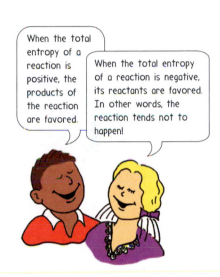

When the total entropy of a reaction is positive, the products of the reaction are favored.

When the total entropy of a reaction is negative, its reactants are favored. In other words, the reaction tends not to happen!

Thus, for the formation of molecular hydrogen, H_2, from atomic hydrogen, H, there is a net positive entropy change in the universe, which tells us that this reaction will proceed on its own. In other words, the smaller negative entropy change in the system of hydrogen molecules is compensated for by a larger positive entropy change in the surroundings.

Like the cooling of a hot pan, a chemical reaction proceeds on its own only if there is a net dispersal of energy (a positive net entropy change). If the opposite holds true (a negative net entropy change), the reaction won't proceed unless it is somehow driven by a hotter environment, which we discuss next.

Concept Check ✓

A total of 486,000 joules of energy is released as 2 moles of hydrogen, H_2, react with 1 mole of oxygen, O_2, to produce 2 moles of water, H_2O. Assume the reaction is run at a constant temperature of 298 K. Does the reaction involve a net increase or a net decrease in the entropy in the universe?

$$\text{H–H(g)} + \text{H–H(g)} + \text{O=O(g)} \longrightarrow \text{H–O–H}(\ell) + \text{H–O–H}(\ell)$$

130.7 J/K   130.7 J/K   205.1 J/K   69.9 J/K   69.9 J/K

Was this your answer? The entropy change associated with the energy released during this reaction is 486,000 J/298 K = 1631 J/K. Add this entropy change to the entropy difference between products and reactants, which was calculated in the previous concept check. The result is the amount of entropy that goes into the universe during this reaction:

$$\begin{array}{l} \text{change in} \\ S \text{ of universe} \end{array} = \begin{array}{l} \text{change in } S \text{ from} \\ \text{products} - \text{reactants} \end{array} + \begin{array}{l} \text{change in } S \text{ from} \\ \text{energy released/absorbed during reaction} \end{array}$$

$$= -326.7 \text{ J/K} + 1631 \text{ J/K} = +1304 \text{ J/K}$$

This net positive entropy change in the universe tells us this reaction proceeds on its own. Of course, as was discussed in Section 9.3, a tiny spark might be needed at first to overcome the activation energy barrier.

Endothermic Reactions Require the Absorption of Energy from the Surroundings

Exothermic reactions tend to proceed in the direction of reactants to products because these reactions release energy to the surroundings. What about endothermic reactions? Such reactions require that energy be absorbed from the surroundings. In other words, they cannot occur unless the surroundings have energy to give to the reaction.

Consider the decomposition of water, H_2O, to form hydrogen, H_2, and oxygen, O_2. At room temperature, 298 K, the total change in entropy for this reaction is the opposite of the entropy change for the formation of water, which means the decomposition reaction won't happen on its own. Note that here in the decomposition equation the entropy-change signs are all reversed from what they were in the previous two concept checks:

$$H_2O + H_2O \longrightarrow H_2 + H_2 + O_2$$

| change in S of universe | = | change in S from products − reactants | + | change in S from energy released/absorbed during reaction |

$$= +326.7 \text{ J/K} + (-1631 \text{ J/K})$$

$$= -1304 \text{ J/K}$$

These are heavy, but ultimately important concepts because they describe whether or not a chemical reaction will occur!

One way to force this decomposition reaction to occur is to raise the temperature to a point where the heat in the surroundings is able to drive the reactants to products. Interestingly, the entropy for the chemicals doesn't change significantly as the temperature is increased. The entropy that arises from the energy released or absorbed during the reaction, however, is substantially different. For example, if the temperature is raised to 1800 K, the entropy change from the energy absorbed during the reaction becomes −486,000 J divided by 1800 K = −270 J/K. The apparent change in the entropy of the universe thus becomes positive:

$$H_2O + H_2O \longrightarrow H_2 + H_2 + O_2$$

| change in S of universe | = | change in S from products − reactants | + | change in S from energy absorbed during reaction |

$$= +326.7 \text{ J/K} + (-270 \text{ J/K})$$

$$= +57 \text{ J/K}$$

So although this endothermic reaction is not favored at room temperature, it is favored at 1800 K.

We talk about the *apparent* change in the entropy of the universe because we must consider how we get to a temperature of 1800 K. To do so requires the heat from an exothermic reaction, say, from the burning of a fuel. The entropy increase resulting from such an exothermic reaction is much greater than the entropy decrease resulting from the water-decomposition reaction run at 298 K. In other words, an entropy-decreasing endothermic reaction

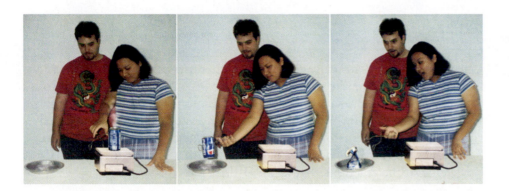

Figure 9.23
Chemists routinely force an otherwise nonspontaneous endothermic reaction to occur by increasing the temperature at which the reaction is run. In doing so, they rely on an accompanying entropy-producing reaction, such as the burning of natural gas. By making endothermic reactions favorable, chemists are able to create many compounds not found in nature, including such modern materials as plastics, computer chips, medicines, fertilizers, and metal alloys.

can be made to occur provided there is a supply of energy from a neighboring entropy-increasing exothermic reaction (Figure 9.23). When all the entropy changes are tallied, the net amount of entropy in the universe is always increasing—it's a natural law.

The Laws of Thermodynamics

In this and the previous section, we have focused on the role energy plays in chemical reactions. This is an area of science known as **thermodynamics**, which stems from Greek words meaning "movement of heat." The concepts we addressed, such as exothermic and endothermic reactions and entropy, fit neatly in the laws of thermodynamics, which are paraphrased as follows:

1. Energy is conserved. It may convert from one form to another, say from potential to kinetic energy, but the total amount of energy in the universe is constant. The energy an exothermic reaction releases always goes somewhere in the environment, usually in the form of heat.

2. Whenever anything happens on its own, energy becomes more dispersed. The degree of energy dispersal is measured by a quantity known as entropy, which is continuously increasing.

Concept Check ✓

1. Without looking at bond energies, deduce whether the following reaction is exothermic or endothermic. Should energy be written as a reactant or as a product?

$$O_2N-NO_2 \longrightarrow O_2N + NO_2$$

2. Is energy more dispersed in the reactants or products?

Was this your answer?

1. The nitrogen–nitrogen bond is broken during this reaction, but no new bonds are formed. Because energy is absorbed as a chemical bond breaks, this reaction is endothermic, and energy should be written as a reactant: energy + $N_2O_4 \longrightarrow$ 2 NO_2.

2. In this reaction, there are more product molecules than reactant molecules. This suggests that energy is more dispersed in the products than in the reactants, which tends to favor the reaction. In other words, because of this increase in entropy, the reaction will run at a lower temperature than it would otherwise. In fact, this endothermic reaction occurs spontaneously even at room temperature.

Figure 9.24
Magic is in the eye of the beholder.

In Perspective

Chemical reactions are truly the heart of chemistry, and their applications abound. For instance, the magician in Figure 9.24 has just ignited a sheet of nitrocellulose, also known as flash paper. In a moment, it will appear to have vanished. You know from the law of mass conservation, however, that materials don't simply vanish. Rather, they are transformed to new materials. Sometimes we can't see the new materials, but that doesn't mean they don't exist. One of the reactions that occur as flash paper burns is

$$4\ C_6H_7N_5O_{16}(s) + 19\ O_2(g) \longrightarrow 24\ CO_2(g) + 20\ NO_2(g) + 14\ H_2O(g)$$

| A component of nitrocellulose | Oxygen | Carbon dioxide | Nitrogen dioxide | Water |

The equation shows 24 carbon, 28 hydrogen, 20 nitrogen, and 102 oxygen atoms before and after the reaction. The difference is in how these atoms are grouped together. The products formed in this case are all gaseous materials that quickly mix into the atmosphere, escaping our notice.

To make the flash paper, the magician would have had to mix the starting materials cellulose and nitric acid. He could determine the proper proportions by knowing the formula masses of these two substances. And although the flash paper may be bathed in an atmosphere of oxygen, it will not react with the oxygen until an initial amount of energy (from the spark of the magician's lighter) is provided to overcome the energy barrier. We know the burning of flash paper is exothermic because the amount of energy released as product bonds form is greater than the amount absorbed as reactant bonds break. Also, because this reaction proceeds on its own, we know this reaction results in a dispersal of energy, which means an increase in entropy. The energy released is in the form of light and faster-moving molecules, which is why the air where the flash paper once was is now appreciably warmer. No true magic is involved, but it is enchanting all the same.

Key Terms and Matching Definitions

_____ activation energy
_____ Avogadro's number
_____ bond energy
_____ catalyst
_____ chemical equation
_____ coefficient
_____ endothermic
_____ entropy
_____ exothermic
_____ formula mass
_____ molar mass
_____ product
_____ reaction rate
_____ reactant
_____ thermodynamics

1. A representation of a chemical reaction.
2. A starting material in a chemical reaction, appearing before the arrow in a chemical equation.
3. A new material formed in a chemical reaction, appearing after the arrow in a chemical equation.
4. A number used in a chemical equation to indicate either the number of atoms/molecules or the number of moles of a reactant or product.
5. The sum of the atomic masses of the atoms in a chemical compound or element.
6. The number of particles—6.02×10^{23}—contained in 1 mole of anything.
7. The mass of 1 mole of a substance.
8. A measure of how quickly the concentration of products in a chemical reaction increases or the concentration of reactants decreases.
9. The minimum energy required in order for a chemical reaction to proceed.
10. Any substance that increases the rate of a chemical reaction without itself being consumed by the reaction.
11. A term that describes a chemical reaction in which there is a net release of energy.
12. A term that describes a chemical reaction in which there is a net absorption of energy.
13. The amount of energy either absorbed as a chemical bond breaks or released as a bond forms.
14. The total amount of energy in a given amount of substance divided by the substance's absolute temperature.
15. An area of science concerned with the role energy plays in chemical reactions.

Review Questions

Chemical Reactions Are Represented by Chemical Equations

1. What is the purpose of coefficients in a chemical equation?

2. How many chromium atoms and how many oxygen atoms are indicated on the right side of this balanced chemical equation:

$$4 \, Cr(s) + 3 \, O_2(g) \longrightarrow 2 \, Cr_2O_3(g)$$

3. What do the letters (s), (ℓ), (g), and (aq) stand for in a chemical equation?

4. Why is it important that a chemical equation be balanced?

5. Why is it important never to change a subscript in a chemical formula when balancing a chemical equation?

6. Which equations are balanced?
   a. $Mg(s) + 2 \, HCl(aq) \longrightarrow MgCl_2(aq) + H_2(g)$
   b. $3 \, Al(s) + 3 \, Br_2(\ell) \longrightarrow Al_2Br_3(s)$
   c. $2 \, HgO(s) \longrightarrow 2 \, Hg(\ell) + O_2(g)$

Chemists Use Relative Masses to Count Atoms and Molecules

7. Why don't equal masses of golf balls and Ping-Pong balls contain the same number of balls?

8. Why don't equal masses of carbon atoms and oxygen molecules contain the same number of particles?

9. How does formula mass differ from atomic mass?

10. What is the mass of a sodium atom in atomic mass units?

11. What is the formula mass of nitrogen monoxide, NO, in atomic mass units?

12. If you had 1 mole of marbles, how many marbles would you have?

13. If you had 2 moles of pennies, how many pennies would you have?

14. How many moles of water are there in 18 grams of water?

15. How many molecules of water are there in 18 grams of water?

16. Why is saying you have 1 mole of water molecules the same as saying you have 6.02×10^{23} water molecules?

Reaction Rate Is Influenced by Concentration and Temperature

17. Why don't all collisions between reactant molecules lead to product formation?

18. What two aspects of a collision between two reactant molecules determine whether or not the collision results in the formation of product molecules?

19. What generally happens to the rate of a chemical reaction with increasing temperature?

20. Why does food take longer to spoil when it is placed in the refrigerator?

21. Which reactant molecules are the first to pass over the energy barrier?

22. What term is used to describe the minimum amount of energy required in order for a reaction to proceed?

Catalysts Increase the Rate of Chemical Reactions

23. What catalyst is effective in the destruction of atmospheric ozone, O_3?

24. Can a catalyst react with a reactant?

25. What is the purpose of a catalytic converter?

26. What does a catalyst do to the energy barrier of a reaction?

27. What net effect does a chemical reaction have on a catalyst?

28. Why are catalysts so important to our economy?

Chemical Reactions Can Be Either Exothermic or Endothermic

29. If it takes 436 kilojoules to break a bond, how many kilojoules are released when the same bond is formed?

30. Is there any energy consumed at any time during an exothermic reaction?

31. What is released by an exothermic reaction?

32. What is absorbed by an endothermic reaction?

33. Which is higher in an endothermic reaction: the potential energy of the reactants or the potential energy of the products?

Entropy Is a Measure of Dispersed Energy

34. As energy disperses, where does it go?

35. What are the units of entropy?

36. True or false: chemists calculate the entropy change of a reaction by subtracting the entropies of the reactants from the entropies of the products.

37. What role does entropy play in chemical reactions?

38. Why do exothermic reactions typically favor the formation of products?

Hands-On Chemistry Insights

Warming and Cooling Water Mixtures

1. The mixing of rubbing alcohol and water is an exothermic process, as evidenced by the warmth you feel upon combining the two. At the molecular level, hydrogen bonds are being formed between alcohol molecules and water molecules. Recall from Section 7.1 that the hydrogen "bond" is a molecule-to-molecule attraction. It is the formation of these intermolecular attractions between alcohol and water molecules that results in the release of heat.

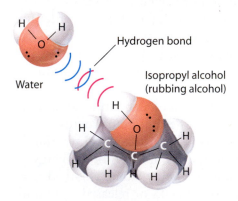

Water

Hydrogen bond

Isopropyl alcohol (rubbing alcohol)

2. You should have been able to feel that the salted water was cooler than the unsalted water, meaning the mixing of sodium chloride and water is an endothermic process. At the molecular level, two things are going on. First the ionic bonds

between Na^+ and Cl^- in the solid salt break, a process that absorbs energy. Then the ions form ion–dipole attractions with water molecules, a process that releases energy. The amount of energy absorbed in the first step is greater than the amount released in the second step.

Commercial "cold packs" work by the same principle. Instead of sodium chloride, however, these packs are made with ammonium nitrate, which absorbs much more energy as it dissolves in water. In order for the pack to be activated, it must be punched. This breaks an inner seal and allows the ammonium nitrate to mix with water. As the ammonium nitrate dissolves, heat is absorbed and the temperature of anything in contact with the pack—including a sprained ankle—decreases.

Exercises

1. Balance these equations:
   a. _____ $Fe(s)$ + _____ $O_2(g)$ $\longrightarrow$ _____ $Fe_2O_3(s)$

   b. _____ $H_2(g)$ + _____ $N_2(g)$ $\longrightarrow$ _____ $NH_3(g)$

2. Balance these equations:
   a. _____ $Fe(s)$ + _____ $S(s)$ $\longrightarrow$ _____ $Fe_2S_3(s)$

   b. _____ $P_4(s)$ + _____ $H_2(g)$ $\longrightarrow$ _____ $PH_3(g)$

3. What are the formula masses of water, H_2O; propene, C_3H_6; and 2-propanol, C_3H_8O?

4. What is the formula mass of sulfur dioxide, SO_2?

5. Which has more atoms: 17.031 grams of ammonia, NH_3, or 72.922 grams of hydrogen chloride, HCl?

6. Which has more atoms: 64.058 grams of sulfur dioxide, SO_2, or 72.922 grams of hydrogen chloride, HCl?

7. Which has the greatest number of molecules:
   a. 28 grams of nitrogen, N_2
   b. 32 grams of oxygen, O_2
   c. 32 grams of methane, CH_4
   d. 38 grams of fluorine, F_2

8. Which has the greatest number of atoms:
   a. 28 grams of nitrogen, N_2
   b. 32 grams of oxygen, O_2
   c. 16 grams of methane, CH_4
   d. 38 grams of fluorine, F_2

9. Hydrogen and oxygen always react in a 1:8 ratio by mass to form water. Early investigators took this to mean that oxygen was eight times more massive than hydrogen. What did these investigators assume about water's chemical formula?

10. Two atomic mass units equal how many grams?

11. What is the mass of an oxygen atom in atomic mass units?

12. What is the mass of a water molecule in atomic mass units?

13. What is the mass of an oxygen atom in grams?

14. What is the mass of a water molecule in grams?

15. Is it possible to have a sample of oxygen that has a mass of 14 atomic mass units? Explain.

16. Which is greater: 1.01 atomic mass units of hydrogen or 1.01 grams of hydrogen?

17. Which has the greater mass, 1.204×10^{24} molecules of molecular hydrogen or 1.204×10^{24} molecules of water?

18. You are given two samples of elements, and each sample has a mass of 10 grams. If the two samples contain the same number of atoms, what must be true of the two samples?

19. Does a refrigerator prevent or delay the spoilage of food? Explain.

20. The yeast used in bread dough feeds on sugar to produce carbon dioxide gas, which causes the dough to rise. Why is bread dough commonly left to rise in a warm area rather than in the refrigerator?

21. Why does a glowing splint of wood burn only slowly in air but burst into flames when placed in pure oxygen?

22. Why is heat often added to chemical reactions performed in the laboratory?

23. An Alka-Seltzer antacid tablet bubbles vigorously in room-temperature water but only slowly in a 50:50 mix of alcohol and water also

at room temperature. Propose an explanation involving the relationship between reaction speed and frequency of molecular collisions.

24. What can you deduce about the activation energy of a reaction that takes billions of years to go to completion? How about a reaction that takes only fractions of a second?

25. In the following reaction sequence for the catalytic formation of ozone from molecular oxygen, which chemical compound is the catalyst: nitrogen monoxide or nitrogen dioxide:

$$O_2 + 2\,NO \longrightarrow 2\,NO_2$$
$$2\,NO_2 \longrightarrow 2\,NO + 2\,O$$
$$2\,O + 2\,O_2 \longrightarrow 2\,O_3$$

26. What role do chlorofluorocarbons play in the catalytic destruction of ozone?

27. Many people hear about atmospheric ozone depletion and wonder why we don't simply replace that which has been destroyed. Knowing about chlorofluorocarbons and knowing how catalysts work, explain how this would not be a lasting solution.

28. Use the bond energies in Table 9.1 and the accounting format shown in Section 9.5 to determine whether these reactions are exothermic or endothermic:

$$H_2 + Cl_2 \longrightarrow 2\,HCl$$
$$2\,HC{\equiv}CH + 5\,O_2 \longrightarrow 4\,CO_2 + 2\,H_2O$$

29. Use the bond energies in Table 9.1 and the accounting format shown in Section 9.5 to determine whether these reactions are exothermic or endothermic:

$$\begin{array}{c}
H \qquad\qquad H \\
\diagdown \qquad\quad \diagup \\
N{-}N \quad \longrightarrow \quad H{-}H + H{-}H + N_2 \\
\diagup \qquad\quad \diagdown \\
H \qquad\qquad H
\end{array}$$

$$\begin{array}{c}
H \qquad\qquad\quad H \\
\diagdown \qquad\qquad\quad \diagdown \\
O{-}O \quad + \quad O{-}O \quad \longrightarrow \\
\diagdown \qquad\qquad\quad \diagdown \\
\;H \qquad\qquad\quad\; H \\[6pt]
O{=}O + H{-}O + H{-}O \\
\diagdown \qquad\quad \diagdown \\
H \qquad\qquad H
\end{array}$$

30. Note in Table 9.1 that bond energy increases going from H–N to H–O to H–F. Explain this trend based on the sizes of these atoms as deduced from their positions in the periodic table.

31. Are the chemical reactions that take place in a disposable battery exothermic or endothermic? What evidence supports your answer? Is the reaction going on in a rechargeable battery while it is recharging exothermic or endothermic?

32. Is the synthesis of ozone from oxygen exothermic or endothermic? How about the synthesis of oxygen from ozone?

33. Two people are looking at a brick. One person says that the energy in the brick is dispersed. The other says that the energy of the brick is contained. Who is right and why?

34. Why is the entropy of water vapor so much higher than that of liquid water?

35. Do combustible fuels tend to have high or low entropies.

36. How is it possible to cause an endothermic reaction to proceed when the reaction causes energy to become less dispersed?

37. For some endothermic reactions, the energy dispersal resulting from the change in the arrangement of atoms as products form is of greater significance than the energy lost in breaking reactant bonds. What is so special about these reactions is that the energy required to push them forward is drawn from the surroundings. Cite an example.

38. Our bodies synthesize tens of thousands of chemicals, such as proteins, that would not otherwise form on their own. What are two ways that our bodies accomplish this fantastic feat?

Problems

1. How many molecules of aspirin (formula mass 180 atomic mass units) are there in a 0.250-gram sample?

2. Small samples of oxygen gas needed in the laboratory can be generated by any number of simple chemical reactions, such as

$$2\,KClO_3(s) \longrightarrow 2\,KCl(s) + 3\,O_2(g)$$

What mass of oxygen (in grams) is produced when 122.55 grams of $KClO_3$ (formula mass 122.55 atomic mass units) takes part in this reaction?

3. How many grams of water, H_2O, and propene, C_3H_6, can be formed from the reaction of 6.0 grams of 2-propanol, C_3H_8O?

2-Propanol

Propene   Water

4. How many moles of water, H_2O, can be produced from the reaction of 16 grams of methane, CH_4, with an unlimited supply of oxygen, O_2? How many grams of water is this? The reaction is

$$CH_4 + 2\,O_2 \longrightarrow CO_2 + 2\,H_2O$$

5. How much energy, in kilojoules, is released or absorbed during the reaction of 1 mole of nitrogen, N_2, with 3 moles of molecular hydrogen, H_2, to form 2 moles of ammonia, NH_3? Consult Table 9.1 for bond energies.

6. The reaction of 1 mole of nitrogen, N_2, with 3 moles of molecular hydrogen, H_2, to form 2 moles of ammonia, NH_3, is exothermic, resulting in the release of 80,000 joules. Use Table 9.2 to calculate whether or not this reaction should proceed on its own at 298 K (25°C).

7. Assume that the entropy values given in Table 9.2 are good not just for 298 K but for any temperature. Does the reaction of 1 mole of nitrogen, N_2, with 3 moles of molecular hydro-

gen, H_2, to form 2 moles of ammonia, NH_3, become more or less favored when the temperature is increased from 298 K (25°C) to 723 K (450°C)?

Answers to Calculation Corner

Figuring Masses of Reactants and Products

1. According to the law of mass conservation, the amount of mass in the products must equal the amount of mass in the reactants. Given that this reaction involves only one reactant and one product, you should not be surprised to learn that 64 grams of reactant produces 64 grams of product:

Step 1. Convert grams of O_2 to moles of O_2:

$$(64\ g\,O_2)\left(\frac{1\ \text{mole}\ O_2}{32\ g\,O_2}\right) = 2\ \text{moles}\ O_2$$

Step 2. Convert moles of O_2 to moles of O_3:

$$(2\ \text{moles}\ O_2)\left(\frac{2\ \text{moles}\ O_3}{3\ \text{moles}\ O_2}\right) = 1.33\ \text{moles}\ O_3$$

Step 3. Convert moles of O_3 to grams of O_3:

$$(1.33\ \text{moles}\ O_3)\left(\frac{48\ g\ O_3}{1\ \text{mole}\ O_3}\right) = 64\ g\ O_3$$

2. There are several ways to answer this problem. One way would be to recognize that 28 grams of N_2 is 1 mole of N_2 and 32 grams of O_2 is 1 mole of O_2. According to the balanced equation, combining 1 mole of N_2 with 1 mole of O_2 yields 2 moles of NO. The mass of 2 moles of NO is

$$(2\ \text{moles}\ NO)\left(\frac{30\ g\ NO}{1\ \text{mole}\ NO}\right) = 60\ g\ NO$$

which is the sum of the masses of the reactants, as it must be because of the law of mass conservation.

Exploring Further

http://www.thecatalyst.org/wwwchem.html

This site has been developed as a resource for high school chemistry teachers, but anyone studying chemistry should find the links helpful. You might follow the link to the history of chemistry, for example, to learn more about Amadeo Avogadro and that huge number named after him.

http://www.wxumac.demon.co.uk

Nitrogen monoxide, also known as nitric oxide, NO, is a precursor to nitrate fertilizers and a common atmospheric pollutant, but it also plays a multitude of vital roles in our human biology. Use *nitric oxide* as a keyword in your Internet search engine to find a plethora of Web sites devoted to the many roles this small but important molecule plays in our physiology and in various diseases, such as Alzheimer's, Parkinson's, asthma, heart disease, and infections.

http://www.secondlaw.com
http://www.entropysimple.com

These sites emphasize the "big picture" of how the second law of thermodynamics applies to our everyday experiences, including our sense of time. Many practical and down-to-earth applications are provided. A great follow-up to Section 9.6, these sites will help you to see this law as one of the simplest yet most profound laws of nature.

An Overview of Chemical Reactions

Visit The Chemistry Place at:

www.aw.com/chemplace

10

Acids and Bases

Transferring Protons

As rainwater falls, it absorbs atmospheric carbon dioxide. Once in the rainwater, the carbon dioxide reacts with water to form an acid known as carbonic acid, H_2CO_3, which, as we discuss in this chapter, makes rainwater naturally acidic. As the rainwater passes through the ground, the carbonic acid reacts with various basic minerals, such as limestone, to form products that are water-soluble and thus carried away by the underground flow of water. This washing-away action over the course of millions of years creates caves. The world's most extensive cave system is in western Kentucky in Mammoth Cave National Park, where more than 300 miles of networked caves have been mapped.

Although Mammoth Cave National Park has the most extensive network of caves, its cave chambers are much smaller than those in Carlsbad Caverns National Park in southeastern New Mexico. This chapter's opening photograph shows the largest chamber at Carlsbad, which measures 25 stories high and half a kilometer wide. The great size of the chambers at Carlsbad is due to the "limestone-eating" action of an acid known as sulfuric acid, H_2SO_4, which is much stronger than carbonic acid. This sulfuric acid forms from gaseous hydrogen sulfide, H_2S, and gaseous sulfur dioxide, SO_2, both of which rise up from oil and gas deposits buried deep in the Earth.

In this chapter, we explore acids and bases and the chemical reactions they undergo. We begin with a definition of these two important substances

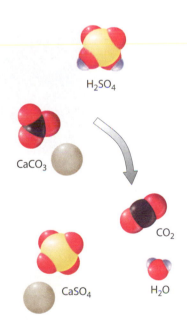

H_2SO_4

$CaCO_3$

CO_2

$CaSO_4$

H_2O

and then explore how some acids and bases are stronger than others. After learning about the pH scale, we close by looking at some environmental and physiological applications of acid–base concepts.

10.1 Acids Donate Protons, Bases Accept Them

The term *acid* comes from the Latin *acidus*, which means "sour." The sour taste of vinegar and citrus fruits is due to the presence of acids. Food is digested in the stomach with the help of acids, and acids are also essential in the chemical industry. Today, for instance, more than 85 billion pounds of sulfuric acid is produced annually in the United States, making this the number-one manufactured chemical. Sulfuric acid is used in fertilizers, detergents, paint dyes, plastics, pharmaceuticals, storage batteries, iron, and steel. It is so important in the manufacturing of goods that its production is considered a standard measure of a nation's industrial strength. Figure 10.1 shows only a very few of the acids we commonly encounter.

(a)

(c)

(d)

(b)

Figure 10.1
Examples of acids. (a) Citrus fruits contain many types of acids, including ascorbic acid, $C_6H_8O_8$, which is vitamin C. (b) Vinegar contains acetic acid, $C_2H_4O_2$, and can be used to preserve foods. (c) Many toilet bowl cleaners are formulated with hydrochloric acid, HCl. (d) All carbonated beverages contain carbonic acid, H_2CO_3, while many also contain phosphoric acid, H_3PO_4.

Bases are characterized by their bitter taste and slippery feel. Interestingly, bases themselves are not slippery. Rather, they cause skin oils to transform into slippery solutions of soap. Most commercial preparations for unclogging drains are composed of sodium hydroxide, NaOH (also known as lye), which is extremely basic and hazardous when concentrated. Bases are also heavily used in industry. Each year in the United States about 25 billion pounds of sodium hydroxide is manufactured for use in the production of various chemicals and in the pulp and paper industry. Solutions containing bases are often called *alkaline*, a term derived from the Arabic word for ashes (*al-qali*), a term we met in Section 2.6. Ashes are slippery when wet because of the presence of the base potassium carbonate, K_2CO_3. Figure 10.2 shows some common bases with which you are probably familiar.

(b)

(c)

(d)

(a)

Figure 10.2
Examples of bases. (a) Reactions involving sodium bicarbonate, $NaHCO_3$, make baked goods rise. (b) Ashes contain potassium carbonate, K_2CO_3. (c) Soap is made by reacting bases with animal or vegetable oils. The soap itself, then, is slightly alkaline. (d) Powerful bases, such as sodium hydroxide, NaOH, are used in drain cleaners.

Acids and bases may be defined in several ways. For our purposes, an appropriate definition is the one suggested in 1923 by the Danish chemist Johannes Brønsted (1879–1947) and the English chemist Thomas Lowry (1874–1936). In the Brønsted–Lowry definition, an **acid** is any chemical that donates a hydrogen ion, H^+, and a **base** is any chemical that accepts a hydrogen ion. Recall from Chapter 2 that, because a hydrogen atom consists of one electron surrounding a one-proton nucleus, a hydrogen ion formed from the loss of an electron is nothing more than a lone proton.

Thus, it is also sometimes said that an acid is a chemical that donates a proton and a base is a chemical that accepts a proton.

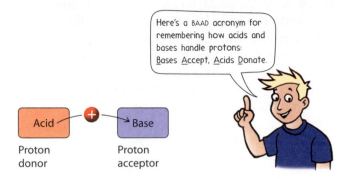

Consider what happens when hydrogen chloride is mixed into water:

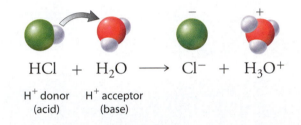

$$HCl \ + \ H_2O \ \longrightarrow \ Cl^- \ + \ H_3O^+$$

H$^+$ donor   H$^+$ acceptor
(acid)   (base)

Hydrogen chloride donates a hydrogen ion to one of the nonbonding electron pairs on a water molecule, resulting in a third hydrogen bonded to the oxygen. In this case, hydrogen chloride behaves as an acid (proton donor) and water behaves as a base (proton acceptor). The products of this reaction are a chloride ion and a **hydronium ion**, H_3O^+, which, as Figure 10.3 shows, is a water molecule with an extra proton.

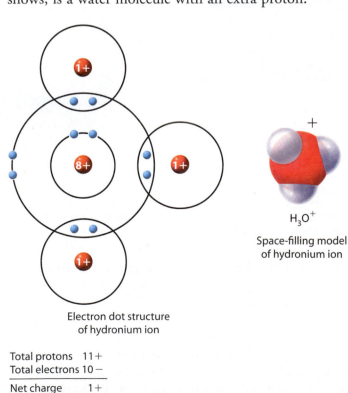

Figure 10.3
The hydronium ion's positive charge is a consequence of the extra proton this molecule has acquired. Hydronium ions, which play a part in many acid–base reactions, are *polyatomic ions*, which, as mentioned in Chapter 6, are molecules that carry a net electric charge.

Electron dot structure
of hydronium ion

Total protons   11+
Total electrons 10 −

Net charge   1+

When added to water, ammonia behaves as a base as its nonbonding electrons (see Figure 6.16) accept a hydrogen ion from water, which, in this case, behaves as an acid:

$$H_2O \; + \; NH_3 \; \longrightarrow \; OH^- \; + \; NH_4^+$$

H$^+$ donor   H$^+$ acceptor
(acid)   (base)

This reaction results in the formation of an ammonium ion and a **hydroxide ion**, which, as shown in Figure 10.4, is a water molecule without the nucleus of one of the hydrogen atoms.

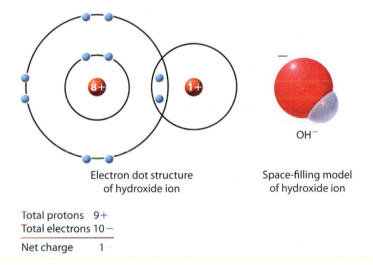

Electron dot structure
of hydroxide ion

Space-filling model
of hydroxide ion

OH$^-$

Total protons 9+
Total electrons 10−

Net charge   1−

Figure 10.4
Hydroxide ions have a net negative charge, which is a consequence of having lost a proton. Like hydronium ions, they play a part in many acid–base reactions.

An important aspect of the Brønsted–Lowry definition is that it recognizes acid–base as a *behavior*. We say, for example, that hydrogen chloride *behaves* as an acid when mixed with water, which *behaves* as a base. Similarly, ammonia *behaves* as a base when mixed with water, which under this circumstance *behaves* as an acid. Because acid–base is seen as a behavior, there is really no contradiction when a chemical like water behaves as a base in one instance but as an acid in another instance. By analogy, consider yourself. You are who you are, but your behavior changes depending on whom you are with. Likewise, it is a chemical property of water to behave as a base (accept H$^+$) when mixed with hydrogen chloride and as an acid (donate H$^+$) when mixed with ammonia.

The products of an acid–base reaction can also behave as acids or bases. An ammonium ion, for example, may donate a hydrogen ion back to a hydroxide ion to re-form ammonia and water:

$$H_2O \; + \; NH_3 \; \longleftarrow \; OH^- \; + \; NH_4^+$$

H$^+$ acceptor   H$^+$ donor
(base)   (acid)

Forward and reverse acid–base reactions proceed simultaneously and can therefore be represented as occurring at the same time by using two oppositely facing arrows:

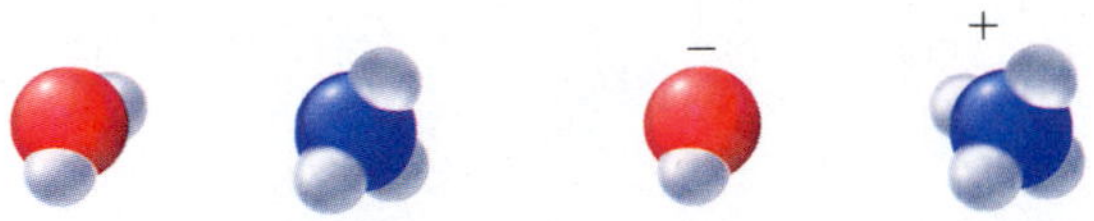

$$H_2O \;+\; NH_3 \;\rightleftharpoons\; OH^- \;+\; NH_4^+$$

| H$^+$ donor (acid) | H$^+$ acceptor (base) | H$^+$ acceptor (base) | H$^+$ donor (acid) |

When the equation is viewed from left to right, the ammonia behaves as a base because it accepts a hydrogen ion from the water, which therefore acts as an acid. Viewed in the reverse direction, the equation shows that the ammonium ion behaves as an acid because it donates a hydrogen ion to the hydroxide ion, which therefore behaves as a base.

Concept Check ✔

Identify the acid or base behavior of each participant in the reaction

$$H_2PO_4^- + H_3O^+ \;\rightleftharpoons\; H_3PO_4 + H_2O$$

Was this your answer? In the forward reaction (left to right), $H_2PO_4^-$ gains a hydrogen ion to become H_3PO_4. In accepting the hydrogen ion, $H_2PO_4^-$ is behaving as a base. It gets the hydrogen ion from the H_3O^+, which is behaving as an acid. In the reverse direction, H_3PO_4 loses a hydrogen ion to become $H_2PO_4^-$ and is thus behaving as an acid. The recipient of the hydrogen ion is the H_2O, which is behaving as a base as it transforms to H_3O^+.

A Salt Is the Ionic Product of an Acid–Base Reaction

In everyday language, the word *salt* implies sodium chloride, NaCl, table salt. In the language of chemistry, however, **salt** is a general term meaning any ionic compound formed from the reaction between an acid and a base. Hydrogen chloride and sodium hydroxide, for example, react to produce the salt sodium chloride and water:

$$HCl \;+\; NaOH \;\longrightarrow\; NaCl \;+\; H_2O$$

| Hydrogen chloride (acid) | Sodium hydroxide (base) | Sodium chloride (salt) | Water |

Similarly, the reaction between hydrogen chloride and potassium hydroxide yields the salt potassium chloride and water:

$$HCl \;+\; KOH \;\longrightarrow\; KCl \;+\; H_2O$$

| Hydrogen chloride (acid) | Potassium hydroxide (base) | Potassium chloride (salt) | Water |

Potassium chloride is the main ingredient in "salt-free" table salt, as noted in Figure 10.5.

Salts are generally far less corrosive than the acids and bases from which they are formed. A corrosive chemical has the power to disintegrate a material or wear away its surface. Hydrogen chloride is a remarkably corrosive acid, which makes it useful for cleaning toilet bowls and etching metal surfaces. Sodium hydroxide is a very corrosive base used for unclogging drains. Mixing hydrogen chloride and sodium hydroxide together in equal portions, however, produces an aqueous solution of sodium chloride—salt water, which is nowhere near as destructive as either starting material.

There are as many salts as there are acids and bases. Sodium cyanide, NaCN, is a deadly poison. "Salt peter," which is potassium nitrate, KNO_3, is useful as a fertilizer and in the formulation of gun powder. Calcium chloride, $CaCl_2$, is commonly used to de-ice roads, and sodium fluoride, NaF, prevents tooth decay. The acid–base reactions forming these salts are shown in Table 10.1.

The reaction between an acid and a base is called a **neutralization** reaction. As can be seen in the color-coding of the neutralization reactions in Table 10.1, the positive ion of a salt comes from the base and the negative ion comes from the acid. The remaining hydrogen and hydroxide ions join to form water.

Not all neutralization reactions result in the formation of water. In the presence of hydrogen chloride, for example, the drug cocaine behaves as a base by accepting H^+ from a hydrogen chloride. The negative Cl^- then joins the cocaine–H^+ ion to form the salt cocaine hydrochloride, shown in Figure 10.6 on page 314. This salt of cocaine is soluble in water and can be absorbed through the moist membranes of the nasal passages or mouth. The nonsalt form of cocaine, also known as "free-base cocaine" or "crack

Figure 10.5

"Salt-free" table-salt substitutes contain potassium chloride in place of sodium chloride. Caution is advised in using these products, however, because excessive quantities of potassium salts can lead to serious illness. Furthermore, sodium ions are a vital component of our diet and should never be totally excluded. For a good balance of these two important ions, you might inquire about commercially available half-and-half mixtures of sodium chloride and potassium chloride, such as the one shown here.

Table 10.1

Acid–Base Reactions and the Salts Formed

Acid		Base		Salt		Water
HCN	+	NaOH	$\longrightarrow$	NaCN	+	H_2O
Hydrogen cyanide		Sodium hydroxide		Sodium cyanide		
HNO_3	+	KOH	$\longrightarrow$	KNO_3	+	H_2O
Nitric acid		Potassium hydroxide		Potassium nitrate		
2 HCl	+	$Ca(OH)_2$	$\longrightarrow$	$CaCl_2$	+	2 H_2O
Hydrogen chloride		Calcium hydroxide		Calcium chloride		
HF	+	NaOH	$\longrightarrow$	NaF	+	H_2O
Hydrogen fluoride		Sodium hydroxide		Sodium flouride		

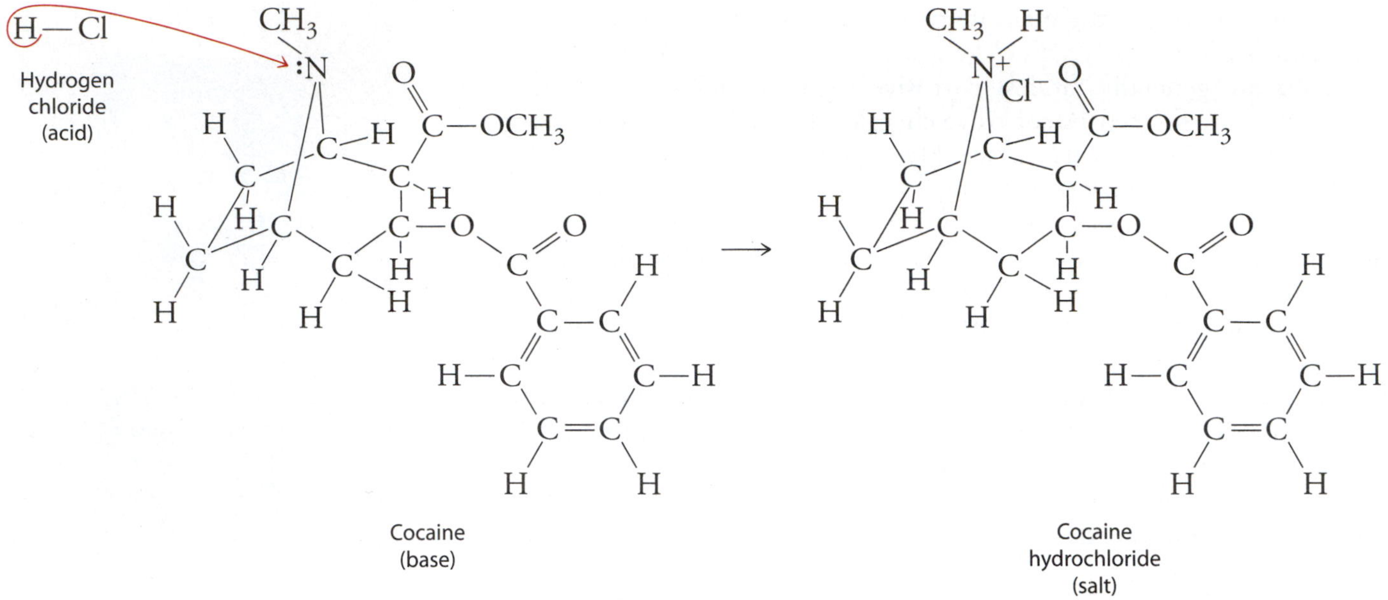

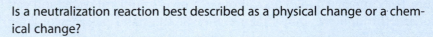

Cocaine
(base)

Cocaine
hydrochloride
(salt)

Figure 10.6
Hydrogen chloride and cocaine react to form the salt cocaine hydrochloride, which, because of its solubility in water, is readily absorbed into the body through moist membranes.

cocaine," is a nonpolar material that vaporizes easily when heated. Its vapors are inhaled directly into the lungs, resulting in dangerously high concentrations of cocaine in the bloodstream. We shall return to the actions of various drugs in Chapter 14.

Concept Check ✔

Is a neutralization reaction best described as a physical change or a chemical change?

Was this your answer? New chemicals are formed during a neutralization reaction, meaning the reaction is a chemical change.

10.2 Some Acids and Bases Are Stronger Than Others

In general, the stronger an acid, the more readily it donates hydrogen ions. Likewise, the stronger a base, the more readily it accepts hydrogen ions. An example of a strong acid is hydrogen chloride, HCl, and an example of a strong base is sodium hydroxide, NaOH. The corrosiveness of these materials is a result of their strength.

One way to assess the strength of an acid or base is to measure how much of it remains after it has been added to water. If little remains, the acid or base is strong. If a lot remains, the acid or base is weak. To illustrate this concept, consider what happens when the strong acid hydrogen chloride is added to water and what happens when the weak acid acetic acid, $C_2H_4O_2$ (the active ingredient of vinegar), is added to water.

Being an acid, hydrogen chloride donates hydrogen ions to water, forming chloride ions and hydronium ions. Because HCl is such a strong acid, nearly all of it is converted to these ions, as is shown in Figure 10.7.

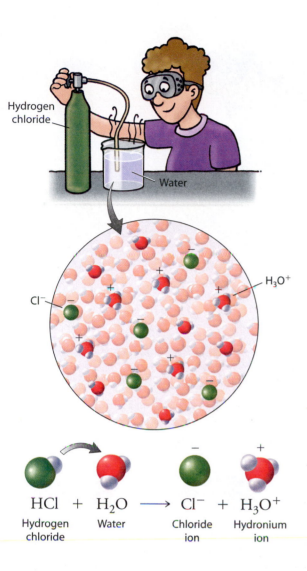

Figure 10.7
Immediately after hydrogen chloride, which is a gaseous substance, is added to water, it reacts with the water to form hydronium ions and chloride ions. That very little HCl remains (none shown here) tells us that HCl is a strong acid.

$$HCl \quad + \quad H_2O \quad \longrightarrow \quad Cl^- \quad + \quad H_3O^+$$

Hydrogen chloride · Water · Chloride ion · Hydronium ion

Because acetic acid is a weak acid, it has much less tendency to donate hydrogen ions to water. When this acid is dissolved in water, only a small portion of the acetic acid molecules are converted to ions, which occurs as the polar O–H bonds are broken (the C–H bonds of acidic acid are unaffected by the water because of their nonpolarity). The majority of acetic acid molecules remain intact in their original nonionized form, as shown in Figure 10.8.

Figures 10.7 and 10.8 (on page 316) show the submicroscopic behavior of strong and weak acids in water. However, molecules and ions are too small to see. How then does a chemist measure the strength of an acid? One way is by measuring a solution's ability to conduct an electric current, as Figure 10.9 illustrates on page 316. In pure water there are practically no ions to conduct electricity. When a strong acid is dissolved in water many ions are generated, as indicated in Figure 10.7. The presence of these ions allows for the flow of a large electric current. A weak acid dissolved in water generates only a few ions, as indicated in Figure 10.8. The presence of fewer ions means there can be only a small electric current.

Figure 10.8
When liquid acetic acid is added to water, only a few acetic acid molecules react with water to form ions. The majority of the acetic acid molecules remain in their nonionized form, which tells us that acetic acid is a weak acid.

$$C_2H_4O_2 \quad + \quad H_2O \quad \longrightarrow \quad C_2H_3O_2^- \quad + \quad H_3O^+$$

Acetic acid   Water   Acetate ion   Hydronium ion

Figure 10.9
(a) The pure water in this circuit is unable to conduct electricity because it contains practically no ions. The light bulb in the circuit therefore remains unlit. (b) Because HCl is a strong acid, nearly all of its molecules break apart in water, giving a high concentration of ions, which are able to conduct an electric current that lights the bulb. (c) Acetic acid, $C_2H_4O_2$, is a weak acid, and in water only a small portion of its molecules break up into ions. Because fewer ions are generated, only a weak current exists and the bulb is dimmer.

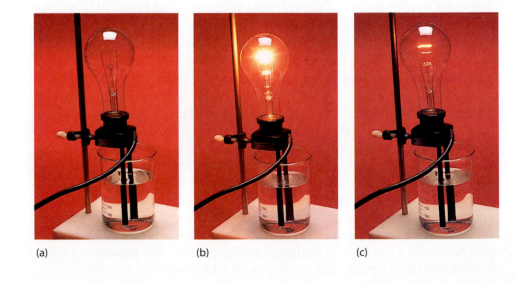

(a)   (b)   (c)

This same trend is seen with strong and weak bases. Strong bases, for example, tend to accept hydrogen ions more readily than weak bases. In solution, a strong base allows the flow of a large electric current and a weak base allows the flow of a small electric current.

Concept Check ✓

According to the aqueous solutions illustrated here, which is the stronger base, NH_3 or NaOH:

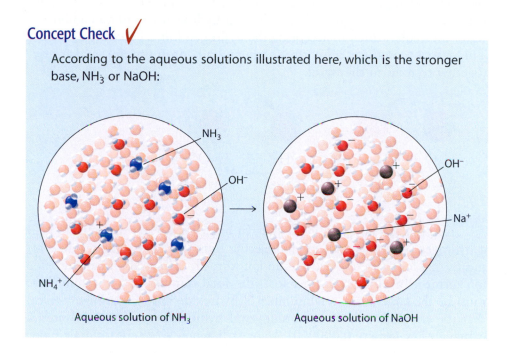

Aqueous solution of NH_3 Aqueous solution of NaOH

Was this your answer? The solution on the right contains the greater number of ions, meaning sodium hydroxide, NaOH, is the stronger base. Ammonia, NH_3, is the weaker base, indicated by the relatively few ions in the solution on the left.

Just because an acid or base is strong doesn't mean a solution of that acid or base is corrosive. The corrosive action of an acidic solution is caused by the hydronium ions rather than by the acid that generated those hydronium ions. Similarly, the corrosive action of a basic solution results from the hydroxide ions it contains, regardless of the base that generated those hydroxide ions. A *very* dilute solution of a strong acid or a strong base may have little corrosive action because in such solutions there are only a few hydronium or hydroxide ions. (Almost all the molecules of the strong acid or base break up into ions, but, because the solution is dilute, there are only a few acid or base molecules to begin with. As a result, there are only a few hydronium or hydroxide ions.) You shouldn't be too alarmed, therefore, when you discover that some toothpastes are formulated with small amounts of sodium hydroxide, one of the strongest bases known.

On the other hand, a concentrated solution of a weak acid, such as acetic acid, may be just as corrosive as or even more corrosive than a dilute solution of a strong acid, such as hydrogen chloride. The relative strengths of two acids in solution or two bases in solution, therefore, can be compared only when the two solutions have the same concentration.

10.3   Solutions Can Be Acidic, Basic, or Neutral

A substance whose ability to behave as an acid is about the same as its ability to behave as a base is said to be **amphoteric**. Water is a good example. Because it is amphoteric, water has the ability to react with itself. In behaving as an acid, a water molecule donates a hydrogen ion to a neighboring water molecule, which in accepting the hydrogen ion is behaving as a base. This reaction produces a hydroxide ion and a hydronium ion, which react together to re-form the water:

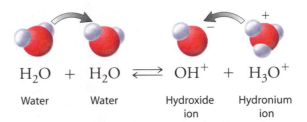

$$H_2O \ + \ H_2O \ \rightleftharpoons \ OH^+ \ + \ H_3O^+$$

Water   Water   Hydroxide   Hydronium
                              ion   ion

From this reaction we can see that, in order for a water molecule to gain a hydrogen ion, a second water molecule must lose a hydrogen ion. This means that for every one hydronium ion formed, there is also one hydroxide ion formed. In pure water, therefore, the total number of hydronium ions must be the same as the total number of hydroxide ions. Experiments reveal that the concentration of hydronium and hydroxide ions in pure water is extremely low—about 0.0000001 M for each, where M stands for molarity or moles per liter (Section 7.2). Water by itself, therefore, is a very weak acid as well as a very weak base, as evidenced by the unlit light bulb in Figure 10.9a.

Concept Check ✔

Do water molecules react with one another?

Was this your answer?   Yes, but not to any large extent. When they do react, they form hydronium and hydroxide ions. (Note: Make sure you understand this point because it serves as a basis for most of the rest of the chapter.)

Further experiments reveal an interesting rule pertaining to the concentrations of hydronium and hydroxide ions in any solution that contains water. The concentration of hydronium ions in any aqueous solution multiplied by the concentration of the hydroxide ions in the solution always equals the constant K_w, which is a very, very small number:

concentration H_3O^+ × concentration $OH^- = K_w$
$$= 0.00000000000001$$

Concentration is usually given as molarity, which is indicated by abbreviating this equation using brackets:

$$[H_3O^+] \times [OH^-] = K_w = 0.00000000000001$$

The brackets mean this equation is read "the molarity of H_3O^+ times the molarity of OH^- equals K_w." Writing in scientific notation (see Appendix A), we have

$$[H_3O^+][OH^-] = K_w = 1.0 \times 10^{-14}$$

For pure water, the value of K_w is the concentration of hydronium ions, 0.0000001 M, multiplied by the concentration of hydroxide ions, 0.0000001 M, which can be written in scientific notation as

$$[1.0 \times 10^{-7}][1.0 \times 10^{-7}] = K_w = 1.0 \times 10^{-14}$$

The constant value of K_w is quite significant because it means that, *no matter what is dissolved in the water*, the product of the hydronium ion and hydroxide ion concentrations always equals 1.0×10^{-14}. This means that if the concentration of H_3O^+ goes up, the concentration of OH^- must go down so that the product of the two remains 1.0×10^{-14}.

Suppose, for example, that a small amount of HCl is added to pure water to increase the concentration of hydronium ions to 1.0×10^{-5} M. (Be sure to see Appendix A if you're confused as to how 10^{-5} is larger than 10^{-7}.) The hydroxide ion concentration decreases to 1.0×10^{-9} M so that the product of the two remains equal to $K_w = 1.0 \times 10^{-14}$:

$$[H_3O^+][OH^-] = K_w = 1.0 \times 10^{-14}$$

pure water $[1.0 \times 10^{-7}][1.0 \times 10^{-7}] = K_w = 1.0 \times 10^{-14}$

HCl added $[1.0 \times 10^{-5}][1.0 \times 10^{-9}] = K_w = 1.0 \times 10^{-14}$

The hydroxide ion concentration goes down because some of the hydroxide ions from the water are neutralized by the added hydronium ions from the HCl, as shown in Figure 10.10b on page 320. In a similar manner, adding a base to water increases the hydroxide ion concentration. The response is a decrease in the hydronium ion concentration as hydronium ions from the water become neutralized by the added hydroxide ions from the base, as shown in Figure 10.10c. The net result is that the product of the hydronium and hydroxide ion concentrations is always equal to the constant $K_w = 1.0 \times 10^{-14}$.

Concept Check ✔

1. In pure water, the hydroxide ion concentration is 1.0×10^{-7} M. What is the hydronium ion concentration?
2. What is the concentration of hydronium ions in a solution if the concentration of hydroxide ions is 1.0×10^{-3} M?

Were these your answers?

1. 1.0×10^{-7} M, because in pure water $[H_3O^+] = [OH^-]$.
2. 1.0×10^{-11} M, because $[H_3O^+][OH^-]$ must equal $1.0 \times 10^{-14} = K_w$.

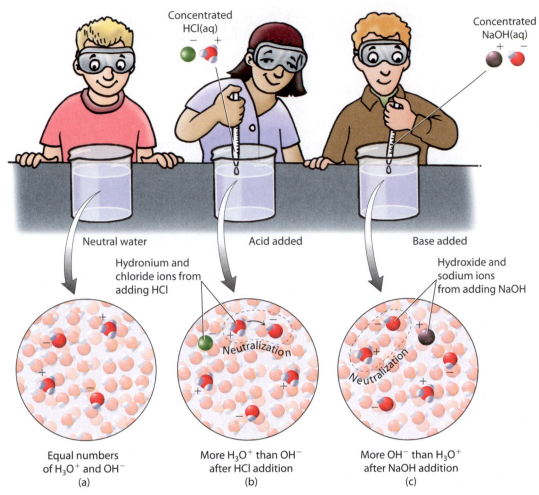

Figure 10.10
(a) Neutral water contains as many hydronium ions as hydroxide ions. (b) When the acid HCl is added to water, hydronium ions from the added HCl neutralize hydroxide ions from the water, thereby decreasing the hydroxide ion concentration. (c) When the base NaOH is added to water, the added hydroxide ions neutralize hydronium ions from the water, thereby decreasing the hydronium ion concentration.

In an **acidic** solution,
$[H_3O^+] > [OH^-]$.

In a **basic** solution,
$[H_3O^+] < [OH^-]$.

In a **neutral** solution,
$[H_3O^+] = [OH^-]$.

Figure 10.11
The relative concentrations of hydronium and hydroxide ions determine whether a solution is acidic, basic, or neutral.

An aqueous solution can be described as acidic, basic, or neutral, as Figure 10.11 summarizes. An **acidic solution** is one in which the hydronium ion concentration is higher than the hydroxide ion concentration. An acidic solution is made by adding an acid to water. The effect of this addition is to increase the concentration of hydronium ions, which necessarily decreases the concentration of hydroxide ions. A **basic solution** is one in which the hydroxide ion concentration is higher than the hydronium ion concentration. A basic solution is made by adding a base to water. This addition increases the concentration of hydroxide ions, which necessarily decreases the concentration of hydronium ions. A **neutral solution** is one in which the hydronium ion concentration equals the hydroxide ion concentration. Pure water is an example of a neutral solution—not because it contains so few hydronium and hydroxide ions but because it contains equal numbers of them. A neutral solution is also obtained when equal quantities of acid and base are combined, which is why acids and bases are said to *neutralize* each other.

Concept Check

How does adding ammonia, NH_3, to water make a basic solution when there are no hydroxide ions in the formula for ammonia?

Was this your answer? Ammonia indirectly increases the hydroxide ion concentration by reacting with water:

$$NH_3 + H_2O \longrightarrow NH_4^+ + OH^-$$

This reaction raises the hydroxide ion concentration, which has the effect of lowering the hydronium ion concentration. With the hydroxide ion concentration now higher than the hydronium ion concentration, the solution is basic.

The pH Scale Is Used to Describe Acidity

The *pH scale* is a numeric scale used to express the acidity of a solution. Mathematically, **pH** is equal to the negative of the base-10 logarithm of the hydronium ion concentration:

$$pH = -\log[H_3O^+]$$

Note again that brackets are used to represent molar concentrations, meaning $[H_3O^+]$ is read "the molar concentration of hydronium ions." For understanding the logarithm function, see the Calculation Corner on page 322.

Consider a neutral solution that has a hydronium ion concentration of 1.0×10^{-7} *M*. To find the pH of this solution, we first take the logarithm of this value, which is -7 (see the Calculation Corner on logarithms). The pH is by definition the negative of this value, which means $-(-7) = +7$. Hence, in a neutral solution, where the hydronium ion concentration equals 1.0×10^{-7} *M*, the pH is 7.

Acidic solutions have pH values less than 7. For an acidic solution in which the hydronium ion concentration is 1.0×10^{-4} *M*, for example, $pH = -\log(1.0 \times 10^{-4}) = 4$. The more acidic a solution is, the greater its hydronium ion concentration and the lower its pH.

Basic solutions have pH values greater than 7. For a basic solution in which the hydronium ion concentration is 1.0×10^{-8} *M*, for example, $pH = -\log(1.0 \times 10^{-8}) = 8$. The more basic a solution is, the smaller its hydronium ion concentration and the higher its pH.

Figure 10.12 shows typical pH values of some familiar solutions, and Figure 10.13 on page 322 shows two common ways of determining pH values.

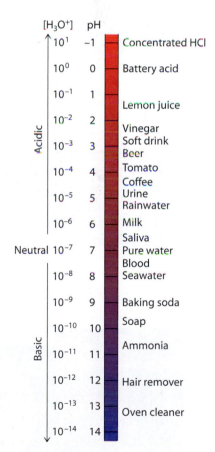

Figure 10.12
The pH values of some common solutions.

Calculation Corner: Logarithms and pH

The logarithm of a number can be found on any scientific calculator by typing in the number and pressing the [log] button. What the calculator does is find the power to which 10 is raised to give the number.

The logarithm of 10^2, for example, is 2 because that is the power to which 10 is raised to give the number 10^2. If you know that 10^2 is equal to 100, then you'll understand that the logarithm of 100 also is 2. Check this out on your calculator. Similarly, the logarithm of 1000 is 3 because 10 raised to the third power, 10^3, equals 1000.

Any positive number, including a very small one, has a logarithm. The logarithm of $0.0001 = 10^{-4}$, for example, is −4 (the power to which 10 is raised to equal this number).

Example
What is the logarithm of 0.01?

Answer
The number 0.01 is 10^{-2} (see Appendix A), the logarithm of which is −2 (the power to which 10 is raised).

The concentration of hydronium ions in most solutions is typically much less than 1 M. Recall, for example, that in neutral water the hydronium ion concentration is 0.0000001 M (10^{-7} M). The logarithm of any number smaller than 1 (but greater than zero) is a negative number. The definition of pH includes the minus sign so as to transform the logarithm of the hydronium ion concentration to a positive number.

When a solution has a hydronium ion concentration of 1 M, the pH is 0 because $1\ M = 10^0\ M$. A 10 M solution has a pH of −1 because $10\ M = 10^1\ M$.

Example
What is the pH of a solution that has a hydronium ion concentration of 0.001 M?

Answer
The number 0.001 is 10^{-3}, and so

$$\begin{aligned} \text{pH} &= -\log[\text{H}_3\text{O}^+] \\ &= -\log 10^{-3} \\ &= -(-3) = 3 \end{aligned}$$

Your Turn
1. What is the logarithm of 10^5?

2. What is the logarithm of 100,000?

3. What is the pH of a solution having a hydronium ion concentration of 10^{-9} M? Is this solution acidic, basic, or neutral?

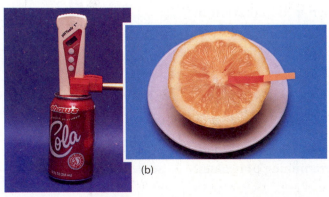

(a)

(b)

Figure 10.13
(a) The pH of a solution can be measured electronically using a pH meter. (b) A rough estimate of the pH of a solution can be obtained with litmus paper, which is coated with a dye that changes color with pH.

Hands-On Chemistry: Rainbow Cabbage

The pH of a solution can be approximated with a *pH indicator*, which is any chemical whose color changes with pH. Many pH indicators are found in plants; the pigment of red cabbage is a good example. This pigment is red at low pH values (1 to 5), light purple around neutral pH values (6 to 7), light green at moderately alkaline pH values (8 to 11), and dark green at very alkaline pH values (12 to 14).

Safety Note

Wear safety glasses. Do not use bleach products because they will oxidize the pigment, rendering it insensitive to any changes in pH. You also do not want to run the risk of accidentally mixing a bleach solution with the toilet-bowl cleaner because the resulting solution would generate harmful chlorine gas.

What You Need

Head of red cabbage, small pot, water, four colorless plastic cups or drinking glasses, toilet-bowl cleaner, vinegar, baking soda, ammonia cleanser.

Procedure

① Shred about a quarter of the head of red cabbage and boil the shredded cabbage in 2 cups of water for about 5 minutes. Strain and collect the broth, which contains the pH-indicating pigment.

② Pour one-fourth of the broth into each cup. (If the cups are plastic, either allow the broth to cool before pouring or dilute with cold water.)

③ Add a small amount of toilet-bowl cleaner to the first cup, a small amount of vinegar to the second cup, baking soda to the third, and ammonia solution to the fourth.

④ Use the different colors to estimate the pH of each solution.

⑤ Mix some of the acidic and basic solutions together and note the rapid change in pH (indicated by the change in color).

10.4 Rainwater Is Acidic and Ocean Water Is Basic

Rainwater is naturally acidic. One source of this acidity is carbon dioxide, the same gas that gives fizz to soda drinks. There is 670 billion tons of CO_2 in the atmosphere, most of it from such natural sources as volcanoes and decaying organic matter but a growing amount from human activities.

Water in the atmosphere reacts with carbon dioxide to form *carbonic acid*:

$$CO_2(g) + H_2O(\ell) \longrightarrow H_2CO_3(aq)$$

Carbon   Water   Carbonic
dioxide   acid

Carbonic acid, as its name implies, behaves as an acid and lowers the pH of water. The CO_2 in the atmosphere brings the pH of rainwater to about

5.6—noticeably below the neutral pH value of 7. Because of local fluctuations, the normal pH of rainwater varies between 5 and 7. This natural acidity of rainwater may accelerate the erosion of land and, under the right circumstances, can lead to the formation of underground caves, as was discussed in this chapter's introduction.

By convention, *acid rain* is a term used for rain having a pH lower than 5. Acid rain is created when airborne pollutants such as sulfur dioxide are absorbed by atmospheric moisture. Sulfur dioxide is readily converted to sulfur trioxide, which reacts with water to form *sulfuric acid*:

$$2\ SO_2(g)\ +\ O_2(g)\ \longrightarrow\ 2\ SO_3(g)$$

Sulfur   Oxygen   Sulfur
dioxide   trioxide

$$SO_3(g)\ +\ H_2O(\ell)\ \longrightarrow\ H_2SO_4(aq)$$

Sulfur   Oxygen   Sulfuric
trioxide   acid

As noted at the beginning of the chapter, the sulfuric acid that helped create the great chambers of Carlsbad Caverns was generated from sulfur dioxide (and hydrogen sulfide) from subterranean fossil-fuel deposits. When we burn these fossil fuels, the reactants that produce sulfuric acid are emitted into the atmosphere. Each year, for example, about 20 million tons of SO_2 is released into the atmosphere by the combustion of sulfur-containing coal and oil. Sulfuric acid is much stronger than carbonic acid, and as a result rain laced with sulfuric acid eventually corrodes metal, paint, and other exposed substances. Each year the damage costs billions of dollars. The cost to the environment is also high (Figure 10.14). Many rivers and lakes receiving acid rain become less capable of sustaining life. Much vege-

(b)

(a)

Figure 10.14
The two photographs in (a) show the same obelisk before and after the effects of acid rain. (b) Many forests downwind from heavily industrialized areas, such as in the northeastern United States and in Europe, have been noticeably hard hit by acid rain.

tation that receives acid rain doesn't survive. This is particularly evident in heavily industrialized regions.

Concept Check ✓

When sulfuric acid, H_2SO_4, is added to water, what makes the resulting aqueous solution corrosive?

Was this your answer? Because H_2SO_4 is a strong acid, it readily forms hydronium ions when dissolved in water. Hydronium ions are responsible for the corrosive action.

The environmental impact of acid rain depends on local geology, as Figure 10.15 illustrates. In certain regions, such as the midwestern United States, the ground contains significant quantities of the alkaline compound calcium carbonate (limestone), deposited when these lands were submerged

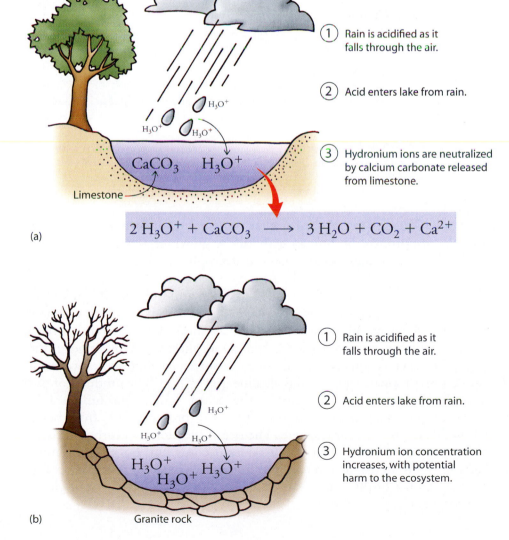

(a)

1. Rain is acidified as it falls through the air.

2. Acid enters lake from rain.

3. Hydronium ions are neutralized by calcium carbonate released from limestone.

$$2\,H_3O^+ + CaCO_3 \longrightarrow 3\,H_2O + CO_2 + Ca^{2+}$$

1. Rain is acidified as it falls through the air.

2. Acid enters lake from rain.

3. Hydronium ion concentration increases, with potential harm to the ecosystem.

(b)

Figure 10.15
(a) The damaging effects of acid rain do not appear in bodies of fresh water lined with calcium carbonate, which neutralizes any acidity. (b) Lakes and rivers lined with inert materials are not protected.

Figure 10.16
Most chalks are made from calcium carbonate, which is the same chemical found in limestone. The addition of even a weak acid, such as the acetic acid of vinegar, produces hydronium ions that react with the calcium carbonate to form several products, the most notable being carbon dioxide, which rapidly bubbles out of solution. Try this for yourself! If the bubbling is not as vigorous as shown here, then the chalk is made of other mineral components.

under oceans 200 million years ago. Acid rain pouring into these regions is often neutralized by the calcium carbonate before any damage is done. (Figure 10.16 shows calcium carbonate neutralizing an acid.) In the northeastern United States and many other regions, however, the ground contains very little calcium carbonate and is composed primarily of chemically less reactive materials, such as granite. In these regions, the effect of acid rain on lakes and rivers accumulates.

One demonstrated solution to this problem is to raise the pH of acidified lakes and rivers by adding calcium carbonate—a process known as *liming*. The cost of transporting the calcium carbonate coupled with the need to monitor treated water systems closely limits liming to only a small fraction of the vast number of water systems already affected. Furthermore, as acid rain continues to pour into these regions, the need to lime also continues.

A longer-term solution to acid rain is to prevent most of the generated sulfur dioxide and other pollutants from entering the atmosphere in the first place. Toward this end, smokestacks have been designed or retrofitted to minimize the quantities of pollutants released. Though costly, the positive effects of these adjustments have been demonstrated, as we discuss in Section 17.2. An ultimate long-term solution, however, would be a shift from fossil fuels to cleaner energy sources, such as nuclear and solar energy, as we discuss in Chapter 19.

Concept Check ✔

What kind of lakes are protected against the negative effects of acid rain?

Was this your answer?   Lakes that have a floor consisting of basic minerals, such as limestone, are more resistant to acid rain because the chemicals of the limestone (mostly calcium carbonate, $CaCO_3$) neutralize any incoming acid.

It should come as no surprise that the amount of carbon dioxide put into the atmosphere by human activities is growing. What is surprising, however, is that studies indicate that the atmospheric concentration of CO_2 is not increasing proportionately. A likely explanation has to do with the oceans and is illustrated in Figure 10.17. When atmospheric CO_2 dissolves in any body of water—a raindrop, a lake, or the ocean—it forms carbonic acid. In fresh water, this carbonic acid transforms back to water and carbon dioxide, which is released back into the atmosphere. Carbonic acid in the ocean, however, is quickly neutralized by dissolved alkaline substances such as calcium carbonate (the ocean is alkaline, pH ≈ 8.2). The products of this neutralization eventually end up on the ocean floor as insoluble solids. Thus carbonic acid neutralization in the ocean prevents CO_2 from being released back into the atmosphere. The ocean therefore is a carbon dioxide *sink*—most of the CO_2 that goes in doesn't come out. So, pushing more CO_2 into our atmosphere means pushing more of it into our vast oceans. This is another of the many ways in which the oceans regulate our global environment.

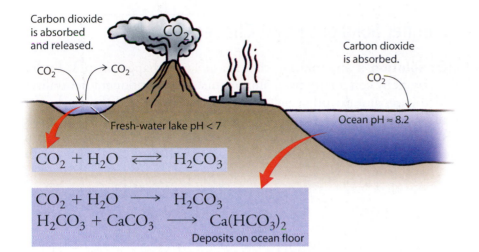

Fresh-water lake pH < 7

$$CO_2 + H_2O \rightleftharpoons H_2CO_3$$

$$CO_2 + H_2O \longrightarrow H_2CO_3$$
$$H_2CO_3 + CaCO_3 \longrightarrow Ca(HCO_3)_2$$

Deposits on ocean floor

Ocean pH ≈ 8.2

Figure 10.17
Carbon dioxide forms carbonic acid upon entering any body of water. In fresh water, this reaction is reversible, and the carbon dioxide is released back into the atmosphere. In the alkaline ocean, the carbonic acid is neutralized to compounds such as calcium bicarbonate, $Ca(HCO_3)_2$, which precipitate to the ocean floor. As a result, most of the atmospheric carbon dioxide that enters our oceans stays there.

Nevertheless, as Figure 10.18 shows, the concentration of atmospheric CO_2 *is* increasing. Carbon dioxide is being produced faster than the ocean can absorb it, and this may alter the Earth's environment. Carbon dioxide is a *greenhouse gas*, which means it helps keep the surface of the Earth warm by preventing infrared radiation from escaping into outer space. Without greenhouse gases in the atmosphere, the Earth's surface would average a frigid −18°C. However, with increasing concentration of CO_2 in the atmosphere, we might experience higher average temperatures. Higher temperatures may significantly alter global weather patterns as well as raise the average sea level as the polar ice caps melt and the volume of seawater increases because of thermal expansion. Global warming is explored in more detail in Section 17.4.

So we find that the pH of rain depends, in great part, on the concentration of atmospheric CO_2, which depends on the pH of the oceans. These systems are interconnected with global temperatures, which naturally connect to the countless living systems on the Earth. How true it is—all the parts are intricately connected, down to the level of atoms and molecules!

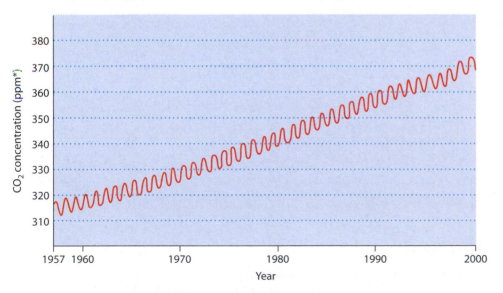

* ppm = parts per million, which tells us the number of carbon dioxide molecules for every million molecules of air.

Figure 10.18
Researchers at the Mauna Loa Weather Observatory in Hawaii have recorded increasing concentrations of atmospheric carbon dioxide since they began collecting data in the 1950s. The oscillations of this graph reflect seasonal changes in CO_2 levels.

10.5   Buffer Solutions Resist Changes in pH

A **buffer solution** is any solution that resists large changes in pH. Buffer solutions work by containing two components. One component neutralizes any added base, and the other neutralizes any added acid. Effective buffer solutions can be prepared by mixing a weak acid with a salt of the weak acid. An example would be a mixture of acetic acid, $C_2H_4O_2$, and sodium acetate, $NaC_2H_4O_2$. This salt can be made by reacting acetic acid with sodium hydroxide.

Acetic acid
(weak acid)

Sodium acetate
(salt of weak acid)

To make the buffer solution, a solution of acetic acid and a solution of sodium acetate are combined. To understand how this buffer solution resists changes in pH, first recall what happens when a strong acid is added to plain water, as in Figure 10.10b. The pH of the solution quickly *decreases* because the concentration of hydronium ions increases. Add a strong base to plain water, and you quickly *increase* the pH by decreasing the relative concentration of hydronium ions, as in Figure 10.10c.

Add the strong acid HCl to an acetic acid–sodium acetate buffer solution, however, and the H^+ ions produced by the HCl do not stay in solution to lower the pH because they react with the acetate ions, $C_2H_3O_2{}^-$, of sodium acetate to form acetic acid, as shown in Figure 10.19. (Remember that acetic acid, being a weak acid, stays mostly in its molecular form,

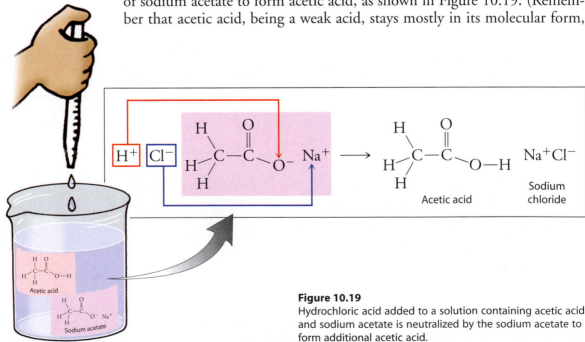

Figure 10.19
Hydrochloric acid added to a solution containing acetic acid and sodium acetate is neutralized by the sodium acetate to form additional acetic acid.

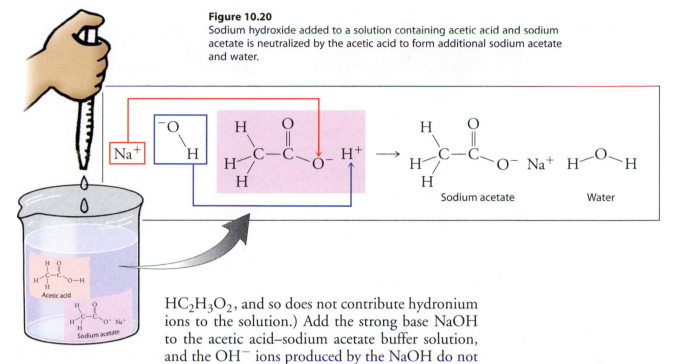

Figure 10.20
Sodium hydroxide added to a solution containing acetic acid and sodium acetate is neutralized by the acetic acid to form additional sodium acetate and water.

$HC_2H_3O_2$, and so does not contribute hydronium ions to the solution.) Add the strong base NaOH to the acetic acid–sodium acetate buffer solution, and the OH^- ions produced by the NaOH do not stay in solution to raise the pH because they combine with H^+ ions from the acetic acid to form water, as shown in Figure 10.20.

So, strong bases and acids are neutralized by the components of a buffer solution. This does not mean that the pH remains unchanged, however. When NaOH is added to the buffer system we are using as our example, sodium acetate is produced. Because sodium acetate behaves as a weak base (it accepts hydrogen ions but not very well), there is a slight increase in pH. When HCl is added, acetic acid is produced. Because acetic acid behaves as a weak acid, there is a slight decrease in pH. Buffer solutions therefore resist only *large* changes in pH.

Concept Check ✔

Why must a buffer solution consist of at least two dissolved components?

Was this your answer? One component is needed to neutralize any incoming acid, and the second component is needed to neutralize any incoming base.

There are many different buffer systems useful for maintaining particular pH values. The acetic acid–sodium acetate system is good for maintaining a pH around 4.8. Buffer solutions containing equal mixtures of a weak base and a salt of that weak base maintain alkaline pH values. For example, a buffer solution of the weak base ammonia, NH_3, and ammonium chloride, NH_4Cl, is useful for maintaining a pH about 9.3.

Blood has several buffer systems that work together to maintain a narrow pH range between 7.35 and 7.45. A pH value above or below these levels can be lethal, primarily because cellular proteins become *denatured*, which is what happens to milk when vinegar is added to it.

Figure 10.21
Carbonic acid and sodium bicarbonate.

Carbonic acid
(weak acid)

Sodium bicarbonate
(salt)

The primary buffer system of the blood is a combination of carbonic acid and its salt sodium bicarbonate, shown in Figure 10.21. Any acid that builds up in the bloodstream is neutralized by the basic action of sodium bicarbonate, and any base that builds up is neutralized by the carbonic acid.

The carbonic acid in your blood is formed as the carbon dioxide produced by your cells enters the bloodstream and reacts with water—this is the same reaction that occurs in a raindrop, as we discussed earlier. You fine-tune the levels of blood carbonic acid, and hence your blood pH, by your breathing rate, as Figure 10.22 illustrates. Breathe too slowly or hold your breath and the amount of carbon dioxide (and hence carbonic acid) builds up, causing a slight but significant drop in pH. Hyperventilate and the carbonic acid level decreases, causing a slight but significant increase in pH. Your body uses this mechanism to protect itself from changes in blood pH. One of the symptoms of a severe overdose of aspirin, for example, is hyperventilation. Aspirin, also known as acetylsalicylic acid, is an acidic chemical that when taken in large amounts can overwhelm the blood buffering system, causing a dangerous drop in blood pH. As you hyperventilate, however, your body loses carbonic acid, which helps to maintain the proper blood pH despite the overabundance of the acidic aspirin.

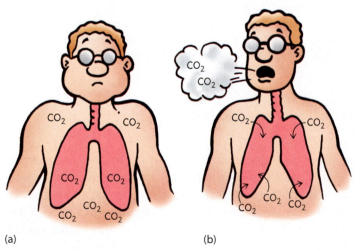

(a)   (b)

Figure 10.22
(a) Hold your breath, and CO_2 builds up in your bloodstream. This increases the amount of carbonic acid, which lowers your blood pH. (b) Hyperventilate and the amount of CO_2 in your bloodstream decreases. This decreases the amount of carbonic acid, which raises your blood pH.

In Perspective

To summarize the concepts of this chapter, consider the gardener shown in Figure 10.23. Using a pH-measuring kit, the gardener has found that the soil pH is unacceptably low, perhaps because of local atmospheric pollutants, which may arise from natural or human-made sources. At this low pH, the soil contains an overabundance of hydronium ions, which react with many of the basic nutrients of the soil, such as ammonia, to form water-soluble salts. Because of their water solubility, these nutrients in their salt form are readily washed away with the rainwater, and as a result the soil becomes nutrient-poor. The mechanism by which plants absorb whatever nutrients do remain in the soil is also disturbed by the soil's low pH.

As a result of all this, most plants do not grow well in acidic soil. To remedy this, the gardener spreads powdered limestone, a form of calcium carbonate, $CaCO_3$, which neutralizes the hydronium ions, thus raising the pH toward neutral.

Interestingly, the calcium carbonate reacts with the acidic soil to form carbon dioxide gas, which in the atmosphere helps to keep rainwater slightly acidic. This is the same gas that is generated by the cells of our bodies and tends to acidify our blood. The blood pH, however, is kept fairly constant at around 7.4 because it is buffered.

Figure 10.23
Raising the pH of garden soil by the addition of an alkaline mineral is known as liming.

Key Terms and Matching Definitions

_____ acid
_____ acidic solution
_____ amphoteric
_____ base
_____ basic solution
_____ buffer solution
_____ hydronium ion
_____ hydroxide ion
_____ neutral solution
_____ neutralization
_____ pH
_____ salt

1. A substance that donates hydrogen ions.
2. A substance that accepts hydrogen ions.
3. A water molecule after accepting a hydrogen ion.
4. A water molecule after losing a hydrogen ion.
5. An ionic compound formed from the reaction between an acid and a base.
6. A reaction in which an acid and base combine to form a salt.
7. A description of a substance that can behave as either an acid or a base.
8. A solution in which the hydronium ion concentration is higher than the hydroxide ion concentration.
9. A solution in which the hydroxide ion concentration is higher than the hydronium ion concentration.
10. A solution in which the hydronium ion concentration is equal to the hydroxide ion concentration.
11. A measure of the acidity of a solution, equal to the negative of the base-10 logarithm of the hydronium ion concentration.
12. A solution that resists large changes in pH, made from either a weak acid and one of its salts or a weak base and one of its salts.

Review Questions

Acids Donate Protons, Bases Accept Them

1. What are the Brønsted–Lowry definitions of acid and base?

2. When an acid is dissolved in water, what ion does the water form?

3. When a chemical loses a hydrogen ion, is it behaving as an acid or a base?

4. Does a salt always contain sodium ions?

5. What two classes of chemicals are involved in a neutralization reaction?

Some Acids and Bases Are Stronger Than Others

6. What does it mean to say that an acid is strong in aqueous solution?

7. What happens to most of the molecules of a strong acid when the acid is mixed with water?

8. Why does a solution of a strong acid conduct electricity better than a solution of a weak acid having the same concentration?

9. Which has a greater ability to accept hydrogen ions: a strong base or a weak base?

10. When can a solution of a weak base be more corrosive than a solution of a strong base?

Solutions Can Be Acidic, Basic, or Neutral

11. Is it possible for a chemical to behave as an acid in one instance and as a base in another instance?

12. Is water a strong acid or a weak acid?

13. Is K_w a very large or a very small number?

14. As the concentration of H_3O^+ ions in an aqueous solution increases, what happens to the concentration of OH^- ions?

15. What is true about the relative concentrations of hydronium and hydroxide ions in an acidic solution? How about a neutral solution? A basic solution?

16. What does the pH of a solution indicate?

17. As the hydronium ion concentration of a solution increases, does the pH of the solution increase or decrease?

Rainwater Is Acidic and Ocean Water Is Basic

18. What is the product of the reaction between carbon dioxide and water?

19. How can rain be acidic and yet not qualify as acid rain?

20. What does sulfur dioxide have to do with acid rain?

21. How do humans generate the air pollutant sulfur dioxide?

22. How does one lime a lake?

23. Why aren't atmospheric levels of carbon dioxide rising as rapidly as might be expected based on the increased output of carbon dioxide resulting from human activities?

Buffer Solutions Resist Changes in pH

24. What is a buffer solution?

25. A strong acid quickly drops the pH when added to water. Not so when added to a buffer solution. Why?

26. Do buffer solutions *prevent* or *inhibit* changes in pH?

27. Why is it so important that the pH of our blood be maintained within a narrow range of values?

28. Holding your breath causes the pH of your blood to decrease. Why?

Hands-On Chemistry Insights

Rainbow Cabbage

The change in color of a pH indicator is not permanent. Red cabbage juice brought to a pH of 4 turns red. This same solution brought to a pH of 8 turns green and then red again as it is brought back to a pH of 4. To demonstrate this, add a teaspoon of baking soda to the glass/cup to which you originally added the vinegar. The solution should turn green. (Why does this addition of baking soda also result in bubbling?) Add vinegar again to bring the color back to red.

Here is another interesting experiment. Boil a whole head of red cabbage for about 20 minutes to obtain a concentrated solution. Place several tablespoons of the concentrated broth in a large colorless glass container. Note the color of the broth and esti-

mate the pH (the extract is acidic). Then quickly pour water into the container, carefully watching for any change in color. What does the changing color tell you about change in pH as you add the water? Does the pH go up or down? How can adding pure water change the pH of a solution?

Exercises

1. Suggest an explanation for why people once washed their hands with ashes.

2. What is the relationship between a hydroxide ion and a water molecule?

3. An acid and a base react to form a salt, which consists of positive and negative ions. Which forms the positive ions: the acid or the base? Which forms the negative ions?

4. Water is formed from the reaction between an acid and a base. Why is water not classified as a salt?

5. Identify the acid or base behavior of each substance in these reactions:
   a. $H_3O^+ + Cl^- \leftrightarrows H_2O + HCl$
   b. $H_2PO_4^- + H_2O \leftrightarrows H_3O^+ + HPO_4^-$
   c. $HSO_4^- + H_2O \leftrightarrows H_3O^+ + SO_4^{2-}$

6. Identify the acid or base behavior of each substance in these reactions:
   a. $HSO_4^- + H_2O \leftrightarrows OH^- + H_2SO_4$
   b. $O^{2-} + H_2O \leftrightarrows OH^- + OH^-$

7. Sodium hydroxide, NaOH, is a strong base, which means it readily accepts hydrogen ions. What products are formed when sodium hydroxide accepts a hydrogen ion from a water molecule?

8. What happens to the corrosive properties of an acid and a base after they neutralize each other? Why?

9. What does the value of K_w say about the extent to which water molecules react with one another?

10. Why do we use the pH scale to indicate the acidity of a solution rather than simply stating the concentration of hydronium ions?

11. The amphoteric reaction between two water molecules is endothermic, which means the reaction requires the input of heat energy in order to proceed:

    $$\text{energy} + H_2O + H_2O \longrightarrow H_3O^+ + OH^-$$

    The warmer the water, the more heat energy is available for this reaction, and the more hydronium and hydroxide ions are formed.
    a. Does the value of K_w increase, decrease, or stay the same with increasing temperature?
    b. Which has a lower pH: pure water that is hot or pure water that is cold?
    c. Is it possible for water to be neutral but have a pH less than or greater than 7.0?

12. The pOH scale indicates the "basicity" of a solution, where $pOH = -\log[OH^-]$. For any solution, what is the sum $pH + pOH$ always equal to?

13. When the hydronium ion concentration of a solution equals 1 mole per liter, what is the pH of the solution? Is the solution acidic or basic?

14. When the hydronium ion concentration of a solution equals 10 moles per liter, what is the pH of the solution? Is the solution acidic or basic?

15. What is the concentration of hydronium ions in a solution that has a pH of -3? Why is such a solution impossible to prepare?

16. What happens to the pH of an acidic solution as pure water is added?

17. A weak acid is added to a concentrated solution of hydrochloric acid. Does the solution become more or less acidic?

18. What happens to the pH of soda water as it loses its carbonation?

19. Why might a small piece of chalk be useful for alleviating acid indigestion?

20. How might you tell whether or not your toothpaste contained calcium carbonate, $CaCO_3$, or perhaps baking soda, $NaHCO_3$, without looking at the ingredients label?

21. Why do lakes lying in granite basins tend to become acidified by acid rain more readily than lakes lying in limestone basins?

22. Cutting back on the pollutants that cause acid rain is one solution to the problem of acidified lakes. Suggest another.

23. How might warmer oceans accelerate global warming?

24. Sodium bicarbonate, $NaHCO_3$,

    Sodium bicarbonate
    (salt)

    is the active ingredient of baking soda. Compare this structure with those of the weak acids and weak bases presented in this chapter and explain how this compound by itself in solution moderates changes in pH.

25. Hydrogen chloride is added to a buffer solution of ammonia, NH_3, and ammonium chloride, NH_4Cl. What is the effect on the concentration of ammonia? On the concentration of ammonium chloride?

26. Sodium hydroxide is added to a buffer solution of ammonia, NH_3, and ammonium chloride, NH_4Cl. What is the effect on the concentration of ammonia? On the concentration of ammonium chloride?

27. At what point will a buffer solution cease to resist changes in pH?

28. Sometimes an individual going through a traumatic experience cannot stop hyperventilating. In such a circumstance, it is recommended that the individual breathe into a paper bag or cupped hands as a useful way to avoid an increase in blood pH, which can cause the person to pass out. Explain how this works.

Problems

1. What is the hydroxide ion concentration in an aqueous solution when the hydronium ion concentration is 1×10^{-10} mole per liter?

2. When the hydronium ion concentration of a solution is 1×10^{-10} mole per liter, what is the pH of the solution? Is the solution acidic or basic?

3. When the hydronium ion concentration of a solution is 1×10^{-4} mole per liter, what is the pH of the solution? Is the solution acidic or basic?

4. What is the hydroxide ion concentration in an aqueous solution having a pH of 5?

5. When the pH of a solution is 1, the concentration of hydronium ions is 1×10^{-1} $M =$ 0.1 M. Assume that the volume of this solution is 500 mL and that the solution is not buffered. What is the pH after 500 mL of pure water is added? You will need a calculator with a logarithm function to answer this question.

Answers to Calculation Corner

Logarithms and pH

1. "What is the logarithm of 10^5?" can be rephrased as "To what power is 10 raised to give the number 10^5?" The answer is 5.

2. You should know that 100,000 is the same as 10^5. Thus the logarithm of 100,000 is 5.

3. The pH is 9, which means this is a basic solution:

$$pH = -\log[H_3O^+]$$
$$= -\log 10^{-9}$$
$$= -(-9)$$
$$= 9$$

Exploring Further

http://www.nps.gov/cave
http://www.carlsbad.caverns.national-park.com/info.htm
http://www.nps.gov/maca
http://www.mammoth.cave.national-park.com/info.htm

Check these official and unofficial sites for Carlsbad Caverns National Park and Mammoth Cave National Park for details on how these underground landmarks formed. Ample travel information is included.

http://www.epa.gov

Go to this home page for the Environmental Protection Agency and use *acid rain* as a keyword in the agency's search engine to find numerous articles on this subject.

http://mlso.hao.ucar.edu/cgi-bin/mlso_homepage.cgi

This address itemizes the atmospheric projects of the Climate Monitoring and Diagnostic Laboratory of the Mauna Loa Weather Observatory. Links to the Network for the Detection of Stratospheric Changes are included.

Acids and Bases

Visit The Chemistry Place at:
www.aw.com/chemplace

Oxidation and Reduction

Transferring Electrons

What do our bodies have in common with the burning of a campfire or the rusting of old farm equipment? Why does silver tarnish? How can aluminum restore tarnished silver? Why is it unwise for people with fillings in their teeth to bite down on aluminum foil? How do batteries work and what is the source of their energy? Why is hydrogen the ultimate fuel of the future? The answers to all these questions involve the transfer of electrons from one substance to another. These kinds of chemical reactions are the main focus of this chapter.

As we learned in Chapter 9, chemicals that react with one another are called reactants. In the process of reacting, the reactants form new chemicals known as products. In an acid–base reaction, a proton is transferred from one reactant to another. In this chapter we look at a class of reactions in which one or more electrons are transferred from one reactant to another. These types of reactions are called **oxidation–reduction reactions**.

337

11.1   Oxidation Is the Loss of Electrons and Reduction Is the Gain of Electrons

Oxidation is the process whereby a reactant loses one or more electrons. **Reduction** is the opposite process whereby a reactant gains one or more electrons. Oxidation and reduction are complementary processes that occur at the same time. They always occur together; you cannot have one without the other. The electrons lost by one chemical in an oxidation reaction don't simply disappear; they are gained by another chemical in a reduction reaction.

An oxidation–reduction reaction occurs when sodium and chlorine react to form sodium chloride, as shown in Figure 11.1. The equation for this reaction is

$$2 \, Na + Cl_2 \longrightarrow 2 \, NaCl$$

To see how electrons are transferred in this reaction, we can look at each reactant individually. Each electrically neutral sodium atom changes to a positively charged ion. We can also say that each atom loses an electron and is therefore oxidized:

$$2 \, Na \longrightarrow 2 \, Na^+ + 2 \, e^- \qquad \text{Oxidation}$$

Each electrically neutral chlorine molecule changes to two negatively charged ions. Each of these atoms gains an electron and is therefore reduced:

$$Cl_2 + 2 \, e^- \longrightarrow 2 \, Cl^- \qquad \text{Reduction}$$

The net result is that the two electrons lost by the sodium atoms are transferred to the chlorine atoms. Therefore, each of the two equations shown above actually represents one half of an entire process, which is why they are each called a **half reaction**. In other words, an electron won't be lost from a sodium atom without there being a chlorine atom available to pick up that electron. Both half reactions are required to represent the *whole* oxidation–reduction process. Half reactions are useful for showing which reactant loses electrons and which reactant gains them, which is why half reactions are used throughout this chapter.

Because the sodium causes reduction of the chlorine, the sodium is acting as a *reducing agent*. A reducing agent is any reactant that causes another reactant to be reduced. Note that sodium is oxidized when it behaves as a reducing agent—it loses elec-

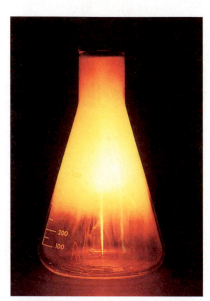

Figure 11.1
In the formation of sodium chloride, sodium metal is oxidized by chlorine gas and chlorine gas is reduced by sodium metal.

trons. Conversely, the chlorine causes oxidation of the sodium and so is act-ing as an *oxidizing agent.* Because it gains electrons in the process, an oxi-dizing agent is reduced. Just remember that **l**oss of **e**lectrons is **o**xidation, and **g**ain of **e**lectrons is **r**eduction. Here is a helpful mnemonic adapted from a once-popular children's story: **Leo** the lion went "**ger.**"

Different elements have different oxidation and reduction tendencies—some lose electrons more readily, while others gain electrons more readily, as Figure 11.2 illustrates. The tendency to do one or the other is a function of how strongly the atom's nucleus holds electrons. The greater the effective nuclear charge (Section 5.8), the greater the tendency of the atom to *gain* electrons. Because the atoms of elements at the upper right of the periodic table have the strongest effective nuclear charges (with the noble gases excluded), these atoms have the greatest tendency to gain electrons and hence behave as oxidizing agents. The atoms of elements at the lower left of the periodic table have the weakest effective nuclear charges and there-fore the greatest tendency to *lose* electrons and behave as reducing agents.

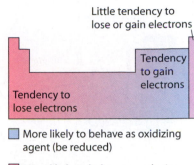

Little tendency to lose or gain electrons

Tendency to gain electrons

Tendency to lose electrons

■ More likely to behave as oxidizing agent (be reduced)

■ More likely to behave as reducing agent (be oxidized)

Figure 11.2
The ability of an atom to gain or lose electrons is a function of its position in the periodic table. Those at the upper right tend to gain electrons, and those at the lower left tend to lose them.

Concept Check ✓

True or false:
1. Reducing agents are oxidized in oxidation–reduction reactions.
2. Oxidizing agents are reduced in oxidation–reduction reactions.

Were these your answers? Both statements are true.

11.2 Photography Works by Selective Oxidation and Reduction

Lay some wax paper on the back of an open unloaded camera as shown in Figure 11.3. Hold the shutter open, then focus. Voila! You have an image. Let the shutter close, however, and the image is gone. This is the same image that forms on the photographic film inside a loaded camera. The dif-ference between the film and the wax paper is that the film is able to retain the image after the shutter has closed. How does it do that? The answer has to do with oxidation–reduction chemistry.

Follow the steps in Figure 11.4 on page 340 as you read this simplified explanation of how a black-and-white photograph is produced.

1. Unexposed black-and-white photographic film is a transparent strip of plastic coated with a gel containing microcrystals of silver bromide, AgBr. Light reflected from the subject being photographed passes through the camera lens and is focused on these microcrystals. The light causes many of the bromide ions in the microcrystals to oxidize. The electrons set loose by this oxidation are transferred to the silver ions, which are thereby reduced to opaque silver atoms. The more light received by a microcrystal, the greater the number of opaque silver

Figure 11.3
A camera can be used to focus an image on wax paper as well as it does on photographic film.

① The film is exposed.

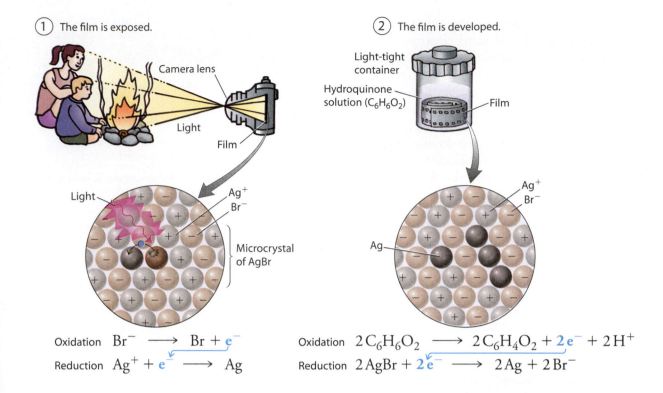

Camera lens

Light

Film

Light

Ag⁺
Br⁻

Microcrystal
of AgBr

② The film is developed.

Light-tight
container

Hydroquinone
solution ($C_6H_6O_2$)

Film

Ag⁺
Br⁻

Ag

Oxidation $Br^- \longrightarrow Br + e^-$

Reduction $Ag^+ + e^- \longrightarrow Ag$

Oxidation $2\,C_6H_6O_2 \longrightarrow 2\,C_6H_4O_2 + 2\,e^- + 2\,H^+$

Reduction $2\,AgBr + 2\,e^- \longrightarrow 2\,Ag + 2\,Br^-$

③ The film is fixed and washed.

Hypo solution
($Na_2S_2O_3$) followed
by water wash.

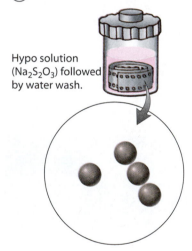

④ The negative is dark where Ag⁺ ions have been reduced to metallic silver.

⑤ Light projected through the negative is captured on photographic paper as a positive image.

Figure 11.4
Black-and-white photography involves a series of oxidation–reduction reactions.

atoms formed. In this way, the photographic image is encoded, and the film is said to have been *exposed*.

2. The light reflected from the subject does not typically result in the formation of enough silver atoms to make a visible image. The more silver atoms a microcrystal contains, however, the more susceptible it is to further oxidation–reduction reactions. To make a visible image, the photographer puts the film in a light-tight container to prevent further exposure. Then the film is treated with a reducing agent, such as hydroquinone, $C_6H_6O_2$, which reveals the encoded image by causing the formation of many more opaque silver atoms. Through this step the image *develops*.

Hands-On Chemistry: Silver Lining

Tarnish on silverware is a coating of silver sulfide, Ag_2S, an ionic compound consisting of two silver ions, Ag^+, and one sulfide ion, S^{2-}. Tarnishing begins when silver atoms in the silverware come into contact with airborne hydrogen sulfide, H_2S, a smelly gas produced by the digestion of food in mammals and other organisms. The half reaction for the silver and hydrogen sulfide is

$$4\,Ag + 2\,H_2S \longrightarrow 4\,Ag^+ + 4\,H^+ + 2\,S^{2-} + 4\,e^- \quad \text{Oxidation}$$

The silver ions and sulfide ions combine to form blackish silver sulfide, while at the same time the hydrogen ions and electrons combine with atmospheric oxygen to form water:

$$4\,H^+ + 4\,e^- + O_2 \longrightarrow 2\,H_2O \quad \text{Reduction}$$

The balanced chemical equation for the tarnishing of silver is the combination of these two half reactions:

$$4\,Ag + 2\,H_2S + O_2 \longrightarrow 2\,Ag_2S + 2\,H_2O$$

From these equations we see that the hydrogen sulfide causes the silver to lose electrons to oxygen. To restore the silver to its shiny elemental state, we need to return the electrons it lost. The oxygen won't relinquish electrons back to silver, but with the proper connection, aluminum atoms will.

What You Need

Very clean aluminum pot (or non-aluminum pot and aluminum foil), water, baking soda, piece of tarnished silver

Procedure

1. Put about a liter of water and several heaping tablespoons of baking soda in the aluminum pot or the non-aluminum pot containing aluminum foil.

2. Bring the water to boiling and then remove the pot from the heat source.

3. Slowly immerse the tarnished silver; you'll see an immediate effect as the silver and aluminum make contact. (Add more baking soda if you don't.) Also, as the silver ions accept electrons from the aluminum and are thereby reduced to shiny silver atoms, the sulfide ions are free to re-form hydrogen sulfide gas, which is released back into the air. You may smell it!

The baking soda serves as a conductive ionic solution that permits electrons to move from the aluminum atoms to the silver ions. What is the advantage of this approach over polishing the silver with an abrasive paste?

3. The reduction of silver ions by the hydroquinone developing solution is stopped by treating the film with a solution of sodium thiosulfate, $Na_2S_2O_3$, also called either *hypo* or *fixing solution*. The thiosulfate ion, $S_2O_3^{2-}$, binds with any unreduced silver ions to form a water-soluble salt. Subsequent washing with water removes everything except the silver atoms adhering to the film, which are most abundant where the greatest amount of light hit the film when the photograph was taken. The film is now *fixed*.

4. Because the silver atoms are opaque, the film appears as a *negative*, which is dark where the subject was light and light where the subject was dark.

5. Light is projected through the negative onto photographic paper, which is developed using the same reactions that produced the negative. The resulting developed image is a negative of the negative—in other words, a positive print.

Color photographic film is coated with a variety of chemicals that respond to light of different frequencies (colors). There are more oxidation–reduction reactions involved in the developing of a color photograph, but the basic principle is the same—the selective reduction of only those chemicals exposed to light. Digital photography, by contrast, is an outgrowth of photovoltaic cells, which are made of metalloids, such as silicon, that lose electrons upon exposure to light. We explore photovoltaic cells in Chapter 19 in our discussion of energy sources.

Concept Check ✔

Would a photographic negative be mostly transparent or mostly opaque if the camera shutter remained open too long and too much light fell on the film? What would the positive print from this negative look like?

Were these your answers? The more light that hits the film, the greater the number of silver ions reduced by the bromide ions or hydroquinone. The reduction of the silver ions results in opaque silver atoms that adhere to the film. Such an overexposed negative, therefore, would be mostly opaque because of all the opaque silver atoms.

The positive print would be very light because there would be very little light passing through the negative to sensitize the silver ions in the photographic paper.

11.3 The Energy of Flowing Electrons Can Be Harnessed

Electrochemistry is the study of the relationship between electrical energy and chemical change. It involves either the use of an oxidation–reduction reaction to produce an electric current or the use of an electric current to produce an oxidation–reduction reaction.

To understand how an oxidation–reduction reaction can generate an electric current, consider what happens when a reducing agent is placed in direct contact with an oxidizing agent: electrons flow from the reducing agent to the oxidizing agent. This flow of electrons is an electric current, which is a form of kinetic energy that can be harnessed for useful purposes.

Iron atoms, Fe, for example, are better reducing agents than copper ions, Cu^{2+}. So when a piece of iron metal and a solution containing copper ions are placed in contact with each other, electrons flow from the iron atoms to the copper ions, as Figure 11.5 illustrates. The result is the oxidation of iron atoms and the reduction of copper ions.

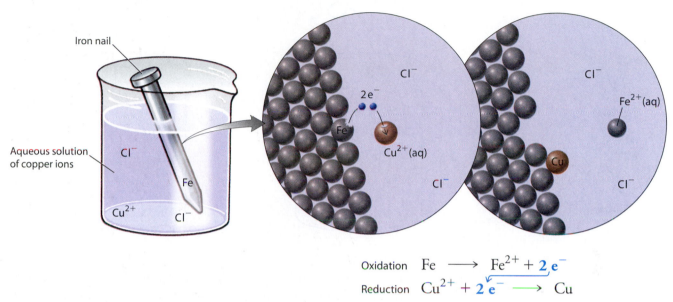

$$\text{Oxidation} \quad Fe \longrightarrow Fe^{2+} + 2\,e^-$$
$$\text{Reduction} \quad Cu^{2+} + 2\,e^- \longrightarrow Cu$$

Figure 11.5
A nail made of iron placed in a solution of Cu^{2+} ions oxidizes to Fe^{2+} ions, which dissolve in the water. At the same time, copper ions are reduced to metallic copper, which coats the nail. (Negatively charged ions, such as chloride ions, Cl^-, must also be present to balance these positively charged ions in solution.)

The elemental iron and copper ions need not be in physical contact in order for electrons to flow between them. If they are in separate containers but bridged by a conducting wire, the electrons can flow from the iron to the copper ions through the wire. The resulting electric current in the wire could be attached to some useful device, such as a light bulb. But alas, an electric current is not sustained by this arrangement.

The reason the electric current is not sustained is shown in Figure 11.6 on page 344. An initial flow of electrons through the wire immediately results in a buildup of electric charge in both containers. The container on the left builds up positive charge as it accumulates Fe^{2+} ions from the nail. The container on the right builds up negative charge as electrons accumulate on this side. This situation prevents any further migration of electrons through the wire. Recall that electrons are negative, and so they are repelled by the negative charge in the right container and attracted to the positive charge in the left container. The net result is that the electrons do not flow through the wire, and the bulb remains unlit.

The solution to this problem is to allow ions to migrate into either container so that neither builds up any positive or negative charge. This is accomplished with a *salt bridge*, which may be a U-shaped tube filled with a salt, such as sodium nitrate, $NaNO_3$, and closed with semiporous plugs. Figure 11.7 on page 345 shows how a salt bridge allows the ions it holds to

Figure 11.6
An iron nail is placed in water and connected by a conducting wire to a solution of copper ions. Nothing happens because this arrangement results in a buildup of charge that prevents the further flow of electrons.

This side immediately builds up a positive charge that attracts electrons, preventing them from migrating.

This side immediately builds up a negative charge that repels electrons, preventing them from entering.

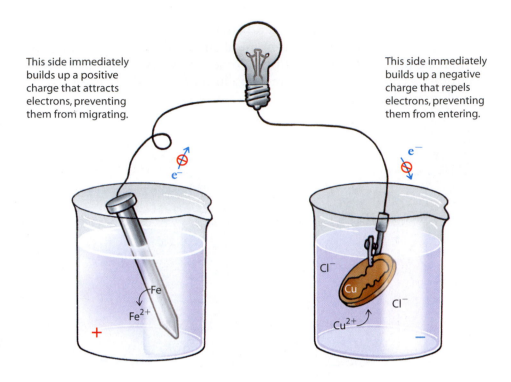

enter either container, permitting the flow of electrons through the conducting wire and creating a complete electric circuit.

The Electricity of a Battery Comes from Oxidation–Reduction Reactions

So we see that with the proper setup it is possible to harness electrical energy from an oxidation–reduction reaction. The apparatus shown in Figure 11.7 is one example. Such devices are called *voltaic cells*. Instead of two containers, a voltaic cell can be an all-in-one, self-contained unit, in which case it is called a *battery*. Batteries are either disposable or rechargeable, and here we explore some examples of each. Although the two types differ in design and composition, they function by the same principle: two materials that oxidize and reduce each other are connected by a medium through which ions travel to balance an external flow of electrons.

Let's look at disposable batteries first. The common *dry-cell battery* was invented in the 1860s and is still used today as probably the cheapest disposable energy source for flashlights, toys, and the like. The basic design consists of a zinc cup filled with a thick paste of ammonium chloride, NH_4Cl, zinc chloride, $ZnCl_2$, and manganese dioxide, MnO_2. Immersed in this paste is a porous stick of graphite that projects to the top of the battery, as shown in Figure 11.8 on page 346.

Graphite is a good conductor of electricity, and it is at the graphite stick that the chemicals in the paste receive electrons and so are reduced. The reaction for the ammonium ions, for instance, is

$$2\,NH_4^+(aq) + 2\,e^- \longrightarrow 2\,NH_3(g) + H_2(g) \qquad \text{Reduction}$$

An **electrode** is any material that conducts electrons into or out of a medium in which electrochemical reactions are occurring. The electrode

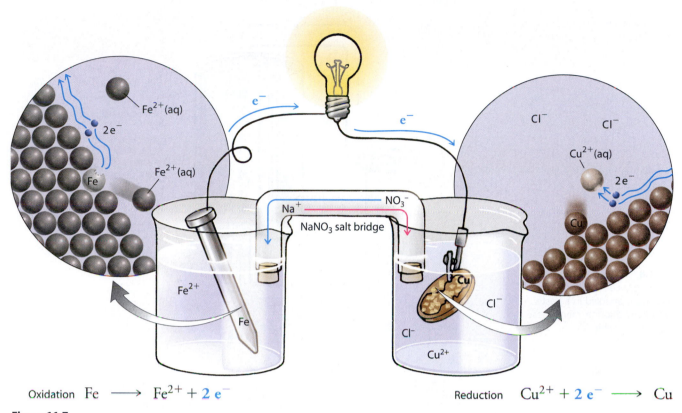

Oxidation Fe $\longrightarrow$ Fe^{2+} + **2 e**$^-$ Reduction Cu^{2+} + **2 e**$^-$ $\longrightarrow$ Cu

Figure 11.7
The salt bridge completes the electric circuit. Electrons freed as the iron is oxidized pass through the wire to the container on the right. Nitrate ions, NO$_3^-$, from the salt bridge flow into the left container to balance the positive charges of the Fe^{2+} ions that form, thereby preventing any buildup of positive charge. Meanwhile, Na$^+$ ions from the salt bridge enter the right container to balance the Cl$^-$ ions "abandoned" by the Cu^{2+} ions as the Cu^{2+} ions pick up electrons to become metallic copper.

where chemicals are reduced is called a **cathode**. For any battery, such as the one shown in Figure 11.8, the cathode is always positive (+), which indicates that electrons are naturally attracted to this location. The electrons gained by chemicals at the cathode originate at the **anode**, which is the electrode where chemicals are oxidized. For any battery, the anode is always negative (−), which indicates that electrons are streaming away from this location. The anode in Figure 11.8 is the zinc cup, where zinc atoms lose electrons to form zinc ions:

Zn(s) $\longrightarrow$ Zn^{2+}(aq) + 2 e$^-$ Oxidation

The reduction of ammonium ions in a dry-cell battery produces two gases—ammonia, NH$_3$, and hydrogen, H$_2$—that need to be removed to avoid a pressure buildup and a potential explosion. Removal is accomplished by having the ammonia and hydrogen react with the zinc chloride and manganese dioxide:

ZnCl$_2$(aq) + 2 NH$_3$(g) $\longrightarrow$ Zn(NH$_3$)$_2$Cl$_2$(s)
2 MnO$_2$(s) + H$_2$(g) $\longrightarrow$ Mn$_2$O$_3$(s) + H$_2$O(ℓ)

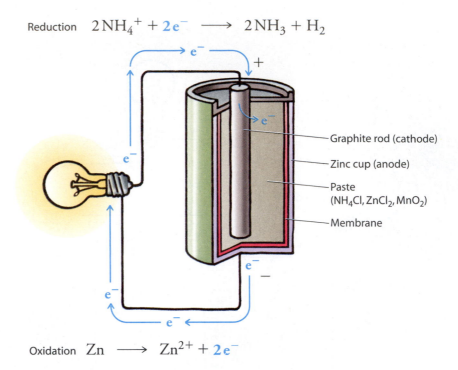

Reduction $2\,NH_4^+ + 2\,e^- \longrightarrow 2\,NH_3 + H_2$

Graphite rod (cathode)

Zinc cup (anode)

Paste
(NH_4Cl, $ZnCl_2$, MnO_2)

Membrane

Oxidation $Zn \longrightarrow Zn^{2+} + 2\,e^-$

Figure 11.8
A common dry-cell battery with a graphite rod immersed in a paste of ammonium chloride, manganese dioxide, and zinc chloride.

The life of a dry-cell battery is relatively short. Oxidation causes the zinc cup to deteriorate, and eventually the contents leak out. Even while the battery is not operating, the zinc corrodes as it reacts with ammonium ions. This zinc corrosion can be inhibited by storing the battery in a refrigerator. As discussed in Chapter 9, chemical reactions slow down with decreasing temperature. Chilling a battery therefore slows down the rate at which the zinc corrodes, which increases the life of the battery.

Another type of disposable battery, the more expensive *alkaline battery*, shown in Figure 11.9, avoids many of the problems of dry-cell batteries by operating in a strongly alkaline paste. In the presence of hydroxide ions, the zinc oxidizes to insoluble zinc oxide:

$$Zn(s) + 2\,OH^-(aq) \longrightarrow ZnO(s) + H_2O(\ell) + 2\,e^- \qquad \text{Oxidation}$$

while at the same time manganese dioxide is reduced:

$$2\,MnO_2(s) + H_2O(\ell) + 2\,e^- \longrightarrow Mn_2O_3(s) + 2\,OH^-(aq) \qquad \text{Reduction}$$

Note how these two reactions avoid the use of the zinc-corroding ammonium ion (which means alkaline batteries last a lot longer than dry-cell batteries) and the formation of any gaseous products. Furthermore, these reactions are better-suited to maintaining a given voltage during longer periods of operation.

The small mercury and lithium disposable batteries used for calculators and cameras are variations of the alkaline battery. In the mercury battery, mercuric oxide, HgO, is reduced rather than manganese dioxide. Manufacturers are phasing out these batteries because of the environmental hazard posed by mercury, which is poisonous. In the lithium battery, lithium

Figure 11.9
Alkaline batteries last a lot longer than dry-cell batteries and give a steadier voltage, but they are expensive.

metal is used as the source of electrons rather than zinc. Not only is lithium able to maintain a higher voltage than zinc, it is about 13 times less dense, which allows for a lighter battery.

Disposable batteries have relatively short lives because electron-producing chemicals are consumed. The main feature of rechargeable batteries is the reversibility of the oxidation and reduction reactions. In your car's rechargeable lead storage battery, for example, electrical energy is produced as lead dioxide, lead, and sulfuric acid are consumed to form lead sulfate and water. The elemental lead is oxidized to Pb^{2+}, and the lead in the lead dioxide is reduced from the Pb^{4+} state to the Pb^{2+} state. Combining the two half reactions gives the complete oxidation–reduction reaction:

$$PbO_2 + Pb + 2\,H_2SO_4 \longrightarrow 2\,PbSO_4 + 2\,H_2O + \text{electrical energy}$$

This reaction can be reversed by supplying electrical energy, as Figure 11.10 on page 348 shows. This is the task of the car's alternator, which is powered by the engine:

$$\text{electrical energy} + 2\,PbSO_4 + 2\,H_2O \longrightarrow PbO_2 + Pb + 2\,H_2SO_4$$

So running the engine maintains concentrations of lead dioxide, lead, and sulfuric acid in the battery. With the engine turned off, these reactants stand ready to supply electric power as needed to start the engine, operate the emergency blinkers, or play the radio.

Concept Check ✓

What is recharged in a car battery?

Was this your answer? When the battery is being recharged, electrical energy from a source (the alternator) outside the battery is used to regenerate reactants that were earlier transformed to products during the oxidation–reduction reaction that produced the electrical energy needed to start the engine. The reactants being regenerated are lead dioxide, elemental lead, and sulfuric acid.

Many rechargeable batteries smaller than car batteries are made of compounds of nickel and cadmium (ni–cad batteries). As with the lead storage battery, ni–cad reactants are replenished by supplying electrical energy from some external source, such as an electrical wall outlet. Like mercury batteries, ni–cad batteries pose an environmental hazard because cadmium is toxic to humans and other organisms. For this reason, alkaline batteries designed to be rechargeable are rapidly gaining a place in the market.

Fuel Cells Are Highly Efficient Sources of Electrical Energy

A *fuel cell* is a device that changes the chemical energy of a fuel to electrical energy. Fuel cells are by far the most efficient means of generating electricity. A hydrogen–oxygen fuel cell is shown in Figure 11.11 on page 349. It has two compartments, one for entering hydrogen fuel and the other for

Figure 11.10
(a) Electrical energy from the battery forces the starter motor to start the engine. (b) The combustion of fuel keeps the engine running and provides energy to spin the alternator, which recharges the battery. Note that the battery has a reversed cathode–anode orientation during recharging.

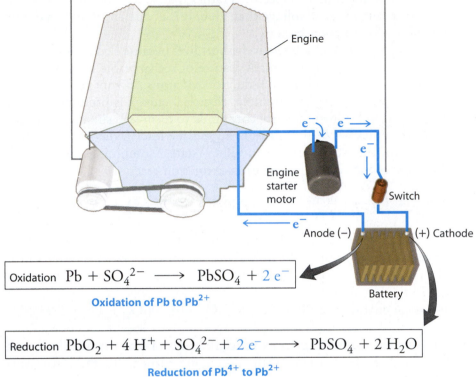

Oxidation $\quad Pb + SO_4^{2-} \longrightarrow PbSO_4 + 2\,e^-$

Oxidation of Pb to Pb^{2+}

Reduction $\quad PbO_2 + 4\,H^+ + SO_4^{2-} + 2\,e^- \longrightarrow PbSO_4 + 2\,H_2O$

Reduction of Pb^{4+} to Pb^{2+}

(a)

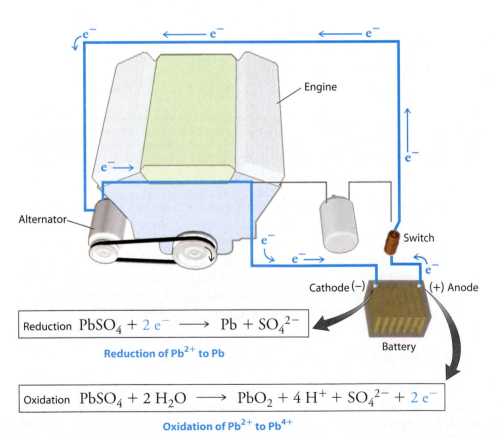

Reduction $\quad PbSO_4 + 2\,e^- \longrightarrow Pb + SO_4^{2-}$

Reduction of Pb^{2+} to Pb

Oxidation $\quad PbSO_4 + 2\,H_2O \longrightarrow PbO_2 + 4\,H^+ + SO_4^{2-} + 2\,e^-$

Oxidation of Pb^{2+} to Pb^{4+}

(b)

Oxidation

$$2\,H_2(g) + 4\,OH^-(aq) \longrightarrow 4\,H_2O(g) + 4\,e^-$$

Reduction

$$4\,e^- + O_2(g) + 2\,H_2O(g) \longrightarrow 4\,OH^-(aq)$$

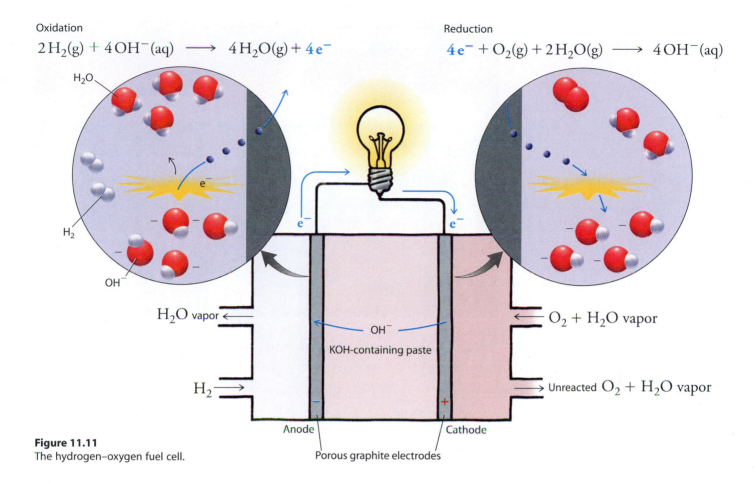

Figure 11.11
The hydrogen–oxygen fuel cell.

entering oxygen fuel, separated by a set of porous electrodes. Hydrogen is oxidized upon contact with hydroxide ions at the hydrogen-facing electrode (the anode). The electrons from this oxidation flow through an external circuit and provide electric power before meeting up with oxygen at the oxygen-facing electrode (the cathode). The oxygen readily picks up the electrons (in other words, the oxygen is reduced) and reacts with water to form hydroxide ions. To complete the circuit, these hydroxide ions migrate across the porous electrodes and through an ionic paste of potassium hydroxide, KOH, to meet up with hydrogen at the hydrogen-facing electrode.

As the oxidation equation shown at the top of Figure 11.11 demonstrates, the hydrogen and hydroxide ions react to produce energetic water molecules that arise in the form of steam. This steam may be used for heating or to generate electricity in a steam turbine. Furthermore, the water that condenses from the steam is pure water, suitable for drinking!

Fuel cells are similar to dry-cell batteries, but fuel cells don't run down as long as fuel is supplied. The space shuttle uses hydrogen–oxygen fuel cells to meet its electrical needs. The cells also produce more than 100 gallons of drinking water for the astronauts during a typical week-long mission. Back on the Earth, researchers are developing fuel cells for buses and automobiles. As shown in Figure 11.12 on page 350, experimental fuel-cell buses are already operating in several cities, such as Vancouver, British Columbia, and Chicago, Illinois. These vehicles produce very few pollutants and can run much more efficiently than vehicles that run on fossil fuels.

Figure 11.12
Because this bus is powered by a fuel cell, its tail pipe emits mostly water vapor.

In the future, commercial buildings as well as individual homes may be outfitted with fuel cells as an alternative to receiving electricity (and heat) from regional power stations. Researchers are also working on miniature fuel cells that could replace the batteries used for portable electronic devices, such as cellular phones and laptop computers. Such devices could operate for extended periods of time on a single "ampule" of fuel available at your local supermarket.

Amazingly, a car powered by a hydrogen–oxygen fuel cell requires only about 3 kilograms of hydrogen to travel 500 kilometers. However, this much hydrogen gas at room temperature and atmospheric pressure would occupy a volume of about 36,000 liters, the volume of about four midsize cars! Thus the major hurdle to the development of fuel-cell technology lies not with the cell but with the fuel. This volume of gas could be compressed to a much smaller volume, as it is on the experimental buses in Vancouver.

Compressing a gas takes energy, however, and as a consequence the inherent efficiency of the fuel cell is lost. Chilling hydrogen to its liquid phase, which occupies much less volume, poses similar problems. Instead, researchers are looking for novel ways of providing fuel cells with hydrogen. In one design, hydrogen is generated within the fuel cell from chemical reactions involving liquid hydrocarbons, such as methanol, CH_3OH. Alternatively, certain porous materials, including the recently developed carbon nanofibers shown in Figure 11.13, can hold large volumes of hydrogen on their surfaces, behaving in effect like hydrogen "sponges." The hydrogen is "squeezed" out of these materials on demand by controlling the temperature—the warmer the material, the more hydrogen released. We explore hydrogen as a fuel source in Chapter 19, in our discussions of sustainable energy sources.

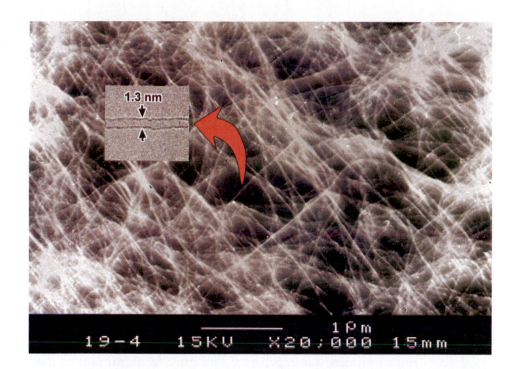

Figure 11.13
Carbon nanofibers consist of near-submicroscopic tubes of carbon atoms. They outclass almost all other known materials in their ability to absorb hydrogen molecules. With carbon nanofibers, for example, a volume of 36,000 liters of hydrogen can be reduced to a mere 35 liters. Carbon nanofibers are a recent discovery, however, and much research is still required to confirm their applicability to hydrogen storage and to develop the technology.

Concept Check ✔

As long as fuel is available to it, a given fuel cell can supply electrical energy indefinitely. Why can't batteries do the same?

Was this your answer? Batteries generate electricity as the chemical reactants they contain are reduced and oxidized. Once these reactants are consumed, the battery can no longer generate electricity. A rechargeable battery can be made to operate again, but only after the energy flow is interrupted so that the reactants can be replenished.

Electrical Energy Can Produce Chemical Change

Electrolysis is the use of electrical energy to produce chemical change. The recharging of a car battery is an example of electrolysis. Another, shown in Figure 11.14, is passing an electric current through water, a process that breaks the water down into its elemental components:

$$\text{electrical energy} + 2\ H_2O(\ell) \longrightarrow 2\ H_2(g) + O_2(g)$$

Electrolysis is used to purify metals from metal ores. An example is aluminum, the third most abundant element in the Earth's crust. Aluminum occurs naturally bonded to oxygen in an ore called *bauxite*. Aluminum metal wasn't known until about 1827, when it was prepared by reacting bauxite with hydrochloric acid. This reaction gave the aluminum ion, Al^{3+}, which was reduced to aluminum metal with sodium metal acting as the reducing agent:

$$Al^{3+} + 3\ Na \longrightarrow Al + 3\ Na^+$$

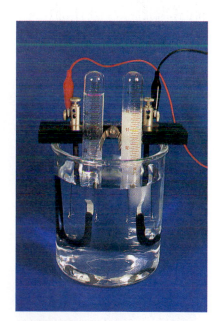

Figure 11.14
The electrolysis of water produces hydrogen gas and oxygen gas in a 2:1 ratio by volume, which is in accordance with the chemical formula for water: H_2O. For this process to work, ions must be dissolved in the water so that the electricity can be conducted between the electrodes.

Figure 11.15
The melting point of aluminum oxide (2030°C) is too high for it to be efficiently electrolyzed to aluminum metal. When the oxide is mixed with the mineral cryolite, the melting point of the oxide drops to a more reasonable 980°C. A strong electric current passed through the molten aluminum oxide–cryolite mixture generates aluminum metal at the cathode, where aluminum ions pick up electrons and so are reduced to elemental aluminum.

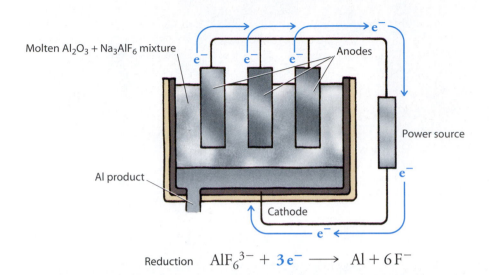

Oxidation $\quad 2\,AlOF_3{}^{2-} + 6\,F^- + C \longrightarrow 2\,AlF_6{}^{3-} + CO_2 + 4\,e^-$

Reduction $\quad AlF_6{}^{3-} + 3\,e^- \longrightarrow Al + 6\,F^-$

This chemical process was expensive. The price of aluminum at that time was about $100,000 per pound, and it was considered a rare and precious metal. In 1855, aluminum dinnerware and other items were exhibited in Paris with the crown jewels of France. Then, in 1886, two men working independently, Charles Hall (1863–1914) in the United States and Paul Heroult (1863–1914) in France, almost simultaneously discovered a process whereby aluminum could be produced from aluminum oxide, Al_2O_3, a main component of bauxite. In what is now known as the Hall–Heroult process, shown in Figure 11.15, a strong electric current is passed through a molten mixture of aluminum oxide and cryolite, Na_3AlF_6, a naturally occurring mineral. The fluoride ions of the cryolite react with the aluminum oxide to form various aluminum fluoride ions, such as $AlOF_3{}^{2-}$, which are then oxidized to the aluminum hexafluoride ion, $AlF_6{}^{3-}$. The Al^{3+} in this ion is then reduced to elemental aluminum, which collects at the bottom of the reaction chamber. This process, which is still in use today, greatly facilitated mass production of aluminum metal, and by 1890 the price of aluminum had dropped to about $2 per pound.

Today, worldwide production of aluminum is about 16 million tons annually. For each ton produced from ore, about 16,000 kilowatt-hours of electrical energy is required, as much as a typical American household consumes in 18 months. Processing recycled aluminum, on the other hand, consumes only about 700 kilowatt-hours for every ton. Thus recycling aluminum not only reduces litter but also helps reduce the load on power companies, which in turn reduces air pollution.

For a nerve-wracking experience involving the oxidation of elemental aluminum, bite a piece of aluminum foil with a tooth filled with dental amalgam. (If you don't have any dental fillings, hooray for you! You'll need to ask a less fortunate friend what this activity feels like.) The aluminum behaves as an anode and releases electrons to the amalgam (a mix of silver, tin, and mercury). The amalgam behaves as a cathode by transferring these

Hands-On Chemistry: Splitting Water

You can see the electrolysis of water by immersing the top of a disposable 9-volt battery in salt water. The bubbles that form contain hydrogen gas produced as the water decomposes. Why does this activity work better with salt water than with tap water? Why does this activity quickly ruin the battery (which should therefore not be used again)?

electrons to oxygen, which then combines with hydrogen ions to form water. The slight current that results produces a jolt of . . . ouch . . . pain.

Concept Check

Is the reaction that goes on in a hydrogen–oxygen fuel cell an example of electrolysis?

Was this your answer? During electrolysis, electrical energy is used to produce chemical change. In the hydrogen–oxygen fuel cell, chemical change is used to produce electrical energy. Therefore, the answer to the question is no.

11.4   Oxygen Is Responsible for Corrosion and Combustion

Look to the upper right of the periodic table, and you will find one of the most common oxidizing agents—oxygen. In fact, if you haven't guessed already, the term *oxidation* is derived from this element. Oxygen is able to pluck electrons from many other elements, especially those that lie at the lower left of the periodic table. Two common oxidation–reduction reactions involving oxygen as the oxidizing agent are *corrosion* and *combustion*.

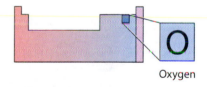

Oxygen

Concept Check

Oxygen is a good oxidizing agent, but so is chlorine. What does this tell you about their relative positions in the periodic table?

Was this your answer? Chlorine and oxygen must lie in the same area of the periodic table. Both have strong effective nuclear charges, and both are strong oxidizing agents.

Corrosion is the process whereby a metal deteriorates. Corrosion caused by atmospheric oxygen is a widespread and costly problem. About one-quarter of the steel produced in the United States, for example, goes into replacing corroded iron at a cost of billions of dollars annually. Iron

Figure 11.16
Rust itself is not harmful to the iron structures on which it forms. It is the loss of metallic iron that ruins the structural integrity.

corrodes when it reacts with atmospheric oxygen and water to form iron oxide trihydrate, which is the naturally occurring reddish-brown substance you know as *rust*, shown in Figure 11.16:

$$4\,Fe \;+\; 3\,O_2 \;+\; 3\,H_2O \;\longrightarrow\; 2\,Fe_2O_3 \cdot 3\,H_2O$$

| Iron | Oxygen | Water | Rust |

We can better understand rusting by considering this equation in steps, as shown in Figure 11.17. ① Iron loses electrons to form the Fe^{2+} ion. ② Oxygen accepts these electrons and then reacts with water to form hydroxide ions, OH^-. ③ Iron ions and hydroxide ions combine to form iron hydroxide, $Fe(OH)_2$, which is further oxidized by oxygen to form rust, $Fe_2O_3 \cdot 3\,H_2O$.

Another common metal oxidized by oxygen is aluminum. The product of aluminum oxidation is aluminum oxide, Al_2O_3, which is water-insoluble. Because of its water insolubility, aluminum oxide forms a protective coat that shields aluminum from further oxidation. This coat is so thin that it's transparent, which is why aluminum maintains its metallic shine.

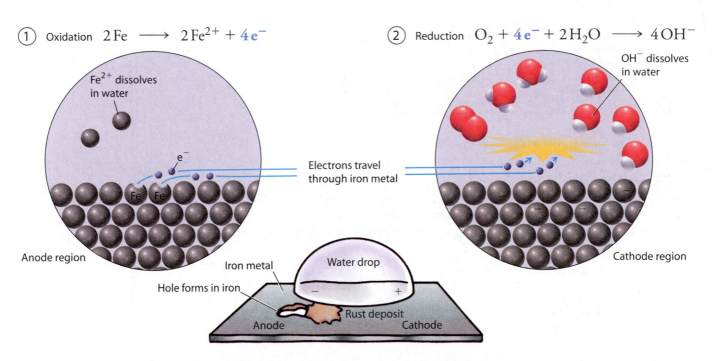

① Oxidation $2\,Fe \longrightarrow 2\,Fe^{2+} + 4\,e^-$

Fe²⁺ dissolves in water

e⁻

Fe Fe

Electrons travel through iron metal

Anode region

② Reduction $O_2 + 4\,e^- + 2\,H_2O \longrightarrow 4\,OH^-$

OH⁻ dissolves in water

Cathode region

Iron metal

Hole forms in iron

Water drop

Anode   Rust deposit   Cathode

③ Fe^{2+} and OH^- react in aqueous solution to form iron hydroxide, $Fe(OH)_2$, which reacts with H_2O and O_2 to form rust, $Fe_2O_3 \cdot 3\,H_2O$.

Figure 11.17
A piece of iron metal begins to rust when iron atoms lose electrons to form Fe^{2+} ions. These electrons are lost to oxygen atoms, which are thereby reduced to hydroxide ions, OH^-. One region of the piece of iron behaves as the anode while another region behaves as the cathode. Rust forms only in the region of the anode, where iron atoms lose electrons. The loss of elemental iron in this region causes a hole to form in the metal. The formation of rust, however, is not as much of a problem as is the loss of iron atoms, a loss that results in a decrease in structural integrity.

A protective water-insoluble oxidized coat is the principle underlying a process called *galvanization.* Zinc has a slightly greater tendency to oxidize than does iron. For this reason, many iron articles, such as the nails pictured in Figure 11.18, are *galvanized* by coating them with a thin layer of zinc. The zinc oxidizes to zinc oxide, an inert, insoluble substance that protects the inner iron from rusting.

In a technique called *cathodic protection,* iron structures can be protected from oxidation by placing them in contact with metals, such as zinc or magnesium, that have a greater tendency to oxidize. This forces the iron to accept electrons, which means it is behaving as a cathode—recall from Figure 11.17 that rusting occurs only where iron behaves as an anode. Ocean tankers, for example, are protected from corrosion by strips of zinc affixed to their hulls, as shown in Figure 11.19. Similarly, outdoor steel pipes are protected by being connected to magnesium rods inserted into the ground.

Yet another way to protect iron and other metals from oxidation is to coat them with a corrosion-resistant metal, such as chromium, platinum, or gold. *Electroplating* is the operation of coating one metal with another by electrolysis, and it is illustrated in Figure 11.20. The object to be electroplated is connected to a negative battery terminal and then submerged in a solution containing ions of the metal to be used as the coating. The positive terminal of the battery is connected to an electrode made of the coating metal. The circuit is completed when this electrode is submerged in the solution. Dissolved metal ions are attracted to the negatively charged object, where they pick up electrons and are deposited as metal atoms. The ions in solution are replenished by the forced oxidation of the coating metal at the positive electrode.

Figure 11.18
The galvanized nail (bottom) is protected from rusting by the sacrificial oxidation of zinc.

Figure 11.19
Zinc strips help protect the iron hull of an oil tanker from oxidizing. The zinc strip shown here is attached to the hull's interior surface.

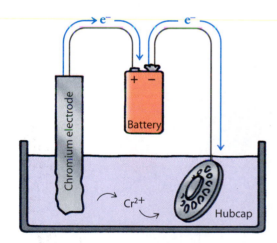

Figure 11.20
As electrons flow into the hubcap and give it a negative charge, positively charged chromium ions move from the solution to the hubcap and are reduced to chromium metal, which deposits as a coating on the hubcap. The solution is kept supplied with ions as chromium atoms in the cathode are oxidized to Cr^{2+} ions.

Combustion is an oxidation–reduction reaction between a nonmetallic material and molecular oxygen. Combustion reactions are characteristically exothermic (energy-releasing). A violent combustion reaction is the formation of water from hydrogen and oxygen. As discussed in Section 9.5, the energy from this reaction is used to power rockets into space. More common examples of combustion include the burning of wood and fossil fuels. The combustion of these and other carbon-based chemicals forms carbon dioxide and water. Consider, for example, the combustion of methane, the major component of natural gas:

$$CH_4 \;+\; 2\,O_2 \;\longrightarrow\; CO_2 \;+\; 2\,H_2O \;+\; \text{energy}$$

Methane Oxygen   Carbon   Water
           dioxide

In combustion, electrons are transferred as polar covalent bonds are formed in place of nonpolar covalent bonds, or vice versa. (This is in contrast to the other examples of oxidation–reduction reactions presented in this chapter, which involve the formation of ions from atoms or, conversely, atoms from ions.) This concept is illustrated in Figure 11.21, which compares the electronic structures of the combustion starting material molecular oxygen and the combustion product water. Molecular oxygen is a nonpolar covalent compound. Although each oxygen atom in the molecule has a fairly strong electronegativity, the four bonding electrons are pulled equally by both atoms and thus are unable to congregate on one side or the other. After combustion, however, the electrons are shared between the oxygen and hydrogen atoms in a water molecule and are pulled to the oxygen. This gives the oxygen a slight negative charge, which is another way of saying it has gained electrons and has thus been reduced. At the same time, the hydrogen atoms in the water molecule develop a slight positive charge, which is another way of saying they have lost electrons and have thus been oxidized. This gain of electrons by oxygen and loss of electrons by hydrogen is an energy-releasing process. Typically, the energy is released either as molecular kinetic energy (heat) or as light (the flame).

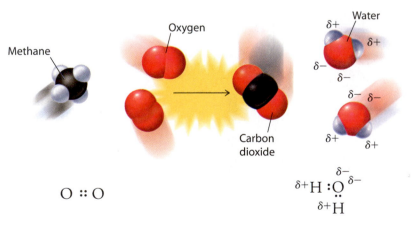

(a) Reactant oxygen atoms share electrons equally in O_2 molecules.

(b) Product oxygen atoms pull electrons away from H atoms in H_2O molecules and are reduced.

Figure 11.21
(a) Neither atom in an oxygen molecule is able to preferentially attract the bonding electrons. (b) The oxygen atom of a water molecule pulls the bonding electrons away from the hydrogen atoms on the water molecule, making the oxygen slightly negative and the two hydrogens slightly positive.

In Perspective

In the previous chapter we discussed acid–base reactions, which are chemical reactions involving the transfer of *protons* from one reactant to another. In this chapter, we explored oxidation–reduction reactions, which involve the transfer of one or more *electrons* from one reactant to another. Oxidation–reduction reactions have many applications, such as in photography, batteries, fuel cells, the manufacture and corrosion of metals, and the combustion of nonmetallic materials such as wood.

Interestingly, combustion oxidation–reduction reactions occur throughout your body. You can visualize a simplified model of your metabolism by reviewing Figure 11.21 and substituting a food molecule for the methane. Food molecules relinquish their electrons to the oxygen molecules you inhale. The products are carbon dioxide, water vapor, and energy. You exhale the carbon dioxide and water vapor, but much of the energy from the reaction is used to keep your body warm and to drive the many other biochemical reactions necessary for living.

Key Terms and Matching Definitions

_____ anode
_____ cathode
_____ combustion
_____ corrosion
_____ electrochemistry
_____ electrode
_____ electrolysis
_____ half reaction
_____ oxidation
_____ oxidation–reduction reaction
_____ reduction

1. A reaction involving the transfer of electrons from one reactant to another.
2. The process whereby a reactant loses one or more electrons.
3. The process whereby a reactant gains one or more electrons.
4. One portion of an oxidation–reduction reaction, represented by an equation showing electrons as either reactants or products.
5. The branch of chemistry concerned with the relationship between electrical energy and chemical change.
6. Any material that conducts electrons into or out of a medium in which electrochemical reactions are occurring.
7. The electrode where reduction occurs.
8. The electrode where oxidation occurs.
9. The use of electrical energy to produce chemical change.
10. The deterioration of a metal, typically caused by atmospheric oxygen.
11. An exothermic oxidation–reduction reaction between a nonmetallic material and molecular oxygen.

Review Questions

Oxidation Is the Loss of Electrons and Reduction Is the Gain of Electrons

1. Which elements have the greatest tendency to behave as oxidizing agents?

2. Write an equation for the half reaction in which a potassium atom, K, is oxidized.

3. Write an equation for the half reaction in which a bromine atom, Br, is reduced.

4. What is the difference between an oxidizing agent and a reducing agent?

5. What happens to a reducing agent as it reduces?

Photography Works by Selective Oxidation and Reduction

6. What special property of silver bromide makes it so useful for photography?

7. What gets reduced as bromine ions, Br^-, on photographic film are oxidized by light?

8. What role does hydroquinone play in the development of a black-and-white photograph?

The Energy of Flowing Electrons Can Be Harnessed

9. What is electrochemistry?

10. What is the purpose of the manganese dioxide in a dry-cell battery?

11. What chemical reaction is forced to occur while a car battery is being recharged?

12. Why don't the electrodes of a fuel cell deteriorate the way the electrodes of a battery do?

13. What is electrolysis, and how does it differ from what goes on inside a battery?

Oxygen Is Responsible for Corrosion and Combustion

14. Why is oxygen such a good oxidizing agent?

15. What do the oxidation of zinc and the oxidation of aluminum have in common?

16. What is electroplating, and how is it accomplished?

17. What are some differences between corrosion and combustion?

18. What are some similarities between corrosion and combustion?

Hands-On Chemistry Insights

Silver Lining

This is one of the better party tricks you can perform for any willing dinner host burdened with a cabinet full of tarnished silver pieces. Forewarn, however, that many pieces coming out of the treatment are still in need of some buffing with silver polish. Lively conversation is guaranteed, especially concerning the source of the tarnishing hydrogen sulfide gas.

Polishing with an abrasive paste removes both the thin layer of tarnish and some silver atoms. Silver-plated pieces are therefore susceptible to losing their thin coating of silver. The aluminum method, by contrast, restores the silver lost to the tarnishing.

For pieces too large to fit in the pot, try rubbing lightly with a paste of baking soda and water, using aluminum foil as your rubbing cloth.

Splitting Water

Try this activity with tap water instead of salt water to see the difference dissolved ions can make—the ions are needed to conduct electricity between the two electrodes.

The primary reaction occurs at the negative electrode (anode), where water molecules accept electrons to form hydrogen gas and hydroxide ions. Recall from Chapter 10 that an increase in hydroxide ion concentration causes the pH of the solution to rise. You can track the production of hydroxide ions by adding a pH indicator to the solution. The indicator of choice is phenolphthalein, which you might obtain from your instructor. Alternatively, you might use the red cabbage extract discussed in Chapter 10. Whichever indicator you use, note the swirls of color forming at the anode as hydroxide ions are generated.

The battery is quickly ruined because placing it in the conducting liquid short-circuits the terminals, which results in a large drain on the battery.

You may be wondering why oxygen gas is not generated along with the hydrogen gas. For reasons beyond the scope of this text, oxygen gas is generated only when the positive electrode (cathode) is made of certain metals, such as gold or platinum. The steel electrode of the 9-volt battery does not suffice.

Exercises

1. Which atom is oxidized, 🔵 or 🔴 :

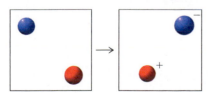

2. In the previous exercise, which atom behaves as the oxidizing agent, 🔵 or 🔴 ?

3. What correlation might you expect between an element's electronegativity (Section 6.6) and its ability to behave as an oxidizing agent? How about its ability to behave as a reducing agent?

4. What correlation might you expect between an element's ionization energy (Section 5.8) and its ability to behave as an oxidizing agent? How about its ability to behave as a reducing agent?

5. Based on their relative positions in the periodic table, which might you expect to be a stronger oxidizing agent, chlorine or fluorine? Why?

6. How does an atom's electronegativity relate to its ability to become oxidized?

7. Iron atoms, Fe, are better reducing agents than copper ions, Cu^{2+}. In which direction do electrons flow when an iron nail is submerged in a solution of Cu^{2+} ions?

8. What is the purpose of the salt bridge in Figure 11.7?

9. Why is the anode of a battery indicated with a minus sign?

10. Is sodium metal oxidized or reduced in the production of aluminum?

11. Why is the formation of iron hydroxide, $Fe(OH)_2$, from Fe^{2+} and OH^- not considered an oxidation–reduction reaction?

12. Your car lights were left on while you were shopping, and now your car battery is dead. Has the pH of the battery fluid increased or decreased?

13. Sketch a voltaic cell that uses the oxidation–reduction reaction

$$Mg(s) + Cu^{2+}(aq) \longrightarrow Mg^{2+}(aq) + Cu(s)$$

Which atom or ion is reduced? Which atom or ion is oxidized?

14. Jewelry is often manufactured by electroplating an expensive metal such as gold over a cheaper metal. Sketch a setup for this process.

15. Some car batteries require the periodic addition of water. Does adding the water increase or decrease the battery's ability to provide electric power to start the car? Explain.

16. Why does a battery that has thick zinc walls last longer than one that has thin zinc walls?

17. The oxidation of iron to rust is a problem structural engineers need to be concerned about, but the oxidation of aluminum to aluminum oxide is not. Why?

18. How many electrons are transferred from iron atoms to oxygen atoms in the formation of two molecules of iron hydroxide, $Fe(OH)_2$? See Figure 11.17.

19. Why are combustion reactions generally exothermic?

20. Which element is closer to the upper right corner of the periodic table, 🔵 or 🔴:

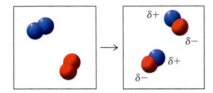

21. Water is 88.88 percent oxygen by mass. Oxygen is exactly what a fire needs to grow brighter and stronger. So why doesn't a fire grow brighter and stronger when water is added to it?

22. Clorox is a laundry bleaching agent used to remove stains from white clothes. Suggest why the name begins with *Clor-* and ends with *-ox*.

23. Iron atoms have a greater tendency to oxidize than do copper atoms. Is this good news or bad news for a home in which much of the plumbing consists of iron and copper pipes connected together? Explain.

24. Copper atoms have a greater tendency to be reduced than iron atoms do. Was this good news or bad news for the Statue of Liberty, whose copper exterior was originally held together by steel rivets?

25. One of the products of combustion is water. Why doesn't this water extinguish the combustion?

Exploring Further

http://www.aluminum.org

The Web site of the Aluminum Association, Inc., where you will find basic facts about the aluminum industry, recycling efforts, and the impact of our aluminum use on the environment.

http://www.kodak.com/us/en/corp/aboutKodak/kodakHistory/kodakHistory.shtml

The Eastman Kodak Company was founded in the late 1800s and was the first company to offer easy-to-use photography services to the general public. Explore this site for some of this history, and be sure to check out the link to "About Film and Imaging" to read about the chemistry and engineering required for the manufacture of photographic film.

http://www.internationalfuelcells.com
http://www.fuelcellworld.org

Use *fuel cells* as a search keyword and you will find a number of private companies and organizations, such as the two named here, that are dedicated to improving the efficiency of fuel cells and publicizing their use. Fuel cells are certainly a wave of the future.

Oxidation and Reduction
Visit The Chemistry Place at:
www.aw.com/chemplace

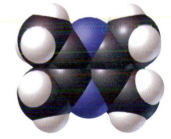

Vanillin

Tetramethylpyrazine

12

Organic Compounds

A Survey of Carbon-Based Molecules

Carbon atoms have the ability to link together and thereby form molecules made up of many carbon atoms. Add to this the fact that any of the carbon atoms in such a chain can also bond with atoms of other elements, and you see the possibility of an endless number of different carbon-based molecules. Each molecule has its own unique set of physical, chemical, and biological properties. The flavor of vanilla, for example, is perceived when the compound *vanillin* is absorbed by the sensory organs in the mouth and nose. This compound consists of a ring of carbon atoms with oxygen atoms attached in a particular fashion. Vanillin is the essential ingredient in anything having the flavor of vanilla—without vanillin, there is no vanilla flavor. The flavor of chocolate, on the other hand, is generated when not just one but a wide assortment of carbon-based molecules are absorbed in the mouth and nose. One of the more significant of these molecules is *tetramethylpyrazine*, which has a ring of nitrogen and carbon atoms attached in a particular fashion.

Life is based on carbon's ability to bond with other carbon atoms to form diverse structures. Reflecting this fact, the branch of chemistry that is the study of carbon-containing compounds has come to be known as **organic chemistry**. The term *organic* is derived from *organism* and is not necessarily related to the environment-friendly form of farming discussed in Chapter 15. Today, more than 13 million organic compounds are known, and about

361

100,000 new ones are added to the list each year. This includes those discovered in nature and those synthesized in the laboratory. (By contrast, there are only 200,000 to 300,000 known *inorganic* compounds, those based on elements other than carbon.)

Because organic compounds are so closely tied to living organisms and because they have many applications—from flavorings to fuels, polymers, medicines, agriculture, and more—it is important to have a basic understanding of them. We begin with the simplest organic compounds—those consisting of only carbon and hydrogen.

12.1 Hydrocarbons Contain Only Carbon and Hydrogen

Organic compounds that contain only carbon and hydrogen are **hydrocarbons**, which differ from one another by the number of carbon and hydrogen atoms they contain. The simplest hydrocarbon is methane, CH_4, with only one carbon per molecule. Methane is the main component of natural gas. The hydrocarbon octane, C_8H_{18}, has eight carbons per molecule and is a component of gasoline. The hydrocarbon polyethylene contains hundreds of carbon and hydrogen atoms per molecule. Polyethylene is a plastic used to make many items, including milk containers and plastic bags.

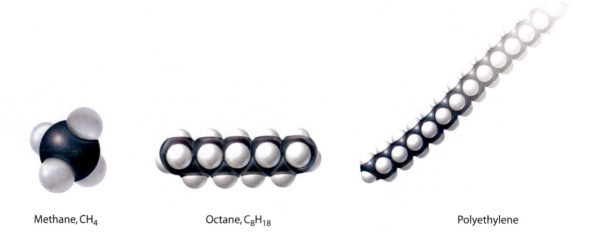

Methane, CH_4 Octane, C_8H_{18} Polyethylene

Hydrocarbons also differ from one another in the way the carbon atoms connect to each other. Figure 12.1 shows the three hydrocarbons *n*-pentane, *iso*-pentane, and *neo*-pentane. These hydrocarbons all have the same molecular formula, C_5H_{12}, but are structurally different from one another. The carbon framework of *n*-pentane is a chain of five carbon atoms. In *iso*-pentane, the carbon chain branches, so that the framework is a *four*-carbon chain branched at the second carbon. In *neo*-pentane, a central carbon atom is bonded to four surrounding carbon atoms.

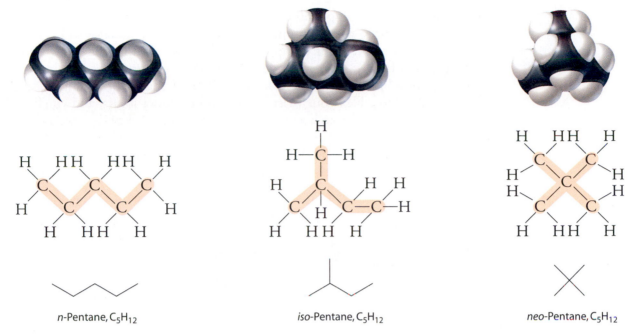

Figure 12.1
These three hydrocarbons all have the same molecular formula. We can see their different structural features by highlighting the carbon framework in two dimensions. Easy-to-draw stick structures that use lines for all carbon–carbon covalent bonds can also be used.

We can see the different structural features of *n*-pentane, *iso*-pentane, and *neo*-pentane more clearly by drawing the molecules in two dimensions, as shown in the middle row of Figure 12.1. Alternatively, we can represent them by the *stick structures* shown in the bottom row. A stick structure is a commonly used, shorthand notation for representing an organic molecule. Each line (stick) represents a covalent bond, and carbon atoms are understood to be wherever two or more straight lines meet and at the end of any line (unless another type of atom is drawn at the end of the line). Any hydrogen atoms bonded to the carbons are also typically not shown. Instead, their presence is implied so that the focus can remain on the skeletal structure formed by the carbon atoms.

When every carbon atom in a hydrocarbon except the two terminal ones is bonded to only two other carbon atoms, the molecule is called a *straight-chain hydrocarbon*. (Do not take this name literally, for, as the *n*-pentane structures in Figure 12.1 show, this is a straight-chain hydrocarbon despite the zigzag nature of the drawings representing it.) When at least one carbon atom in a hydrocarbon is bonded to either three or four carbon atoms, the molecule is a *branched hydrocarbon*. Both *iso*-pentane and *neo*-pentane are branched hydrocarbons.

Molecules such as *n*-pentane, *iso*-pentane, and *neo*-pentane, which have the same molecular formula but different structures, are known as **structural isomers**. Structural isomers have different physical and chemical properties. For example, *n*-pentane has a boiling point of 36°C, *iso*-pentane's boiling point is 30°C, and *neo*-pentane's is 10°C.

The number of possible structural isomers for a chemical formula increases rapidly as the number of carbon atoms increases. There are 3 structural isomers for compounds having the formula C_5H_{12}, 18 for C_8H_{18}, 75 for $C_{10}H_{22}$, and a whopping 366,319 for $C_{20}H_{42}$! Carbon-based molecules can have different spatial orientations called **conformations**. Flex your wrist, elbow, and shoulder joints, and you'll find your arm passing through a range of conformations. Likewise, organic molecules can twist and turn about their carbon–carbon single bonds and thus have a range of conformations. The structures in Figure 12.2, for example, are different conformations of *n*-pentane.

Figure 12.2
Three conformations for a molecule of *n*-pentane. The molecule looks different in each conformation, but the five-carbon framework is the same in all three conformations. In a sample of liquid *n*-pentane, the molecules are found in all conformations—not unlike a bucket of worms.

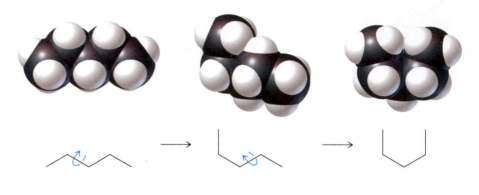

Concept Check ✔

Which carbon–carbon bond was rotated to go from the "before" conformation of *iso*-pentane to the "after" conformation:

Before   After

Was this your answer? The best way to answer any question about the conformation of a molecule is to play around with molecular models that you can hold in your hand. In this case, bond c rotates in such a way that the carbon at the right end of bond d comes up out of the plane of the page, momentarily points straight at you, and then plops back into the plane of the page below bond c. This rotation is similar to that of the arm of an arm wrestler who, her arm just above the table as she is on the brink of losing, suddenly gets a surge of strength and swings her opponent's arm (and her own) through a half-circle arc and wins.

Before   After

Hydrocarbons are obtained primarily from coal and petroleum, both formed when plant and animal matter decays in the absence of oxygen. Most of the coal and petroleum that exist today were formed between 280 and 395 million years ago. At that time, the Earth was covered with

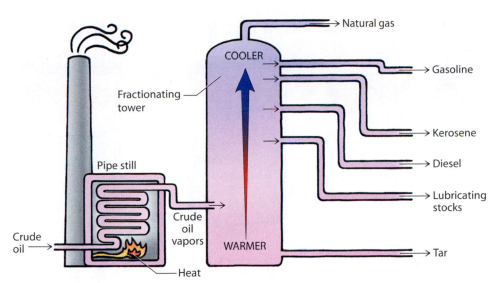

Figure 12.3
A schematic for the fractional distillation of petroleum into its useful hydrocarbon components.

extensive swamps that, because they were close to sea level, periodically became submerged. The organic matter of the swamps was buried beneath layers of marine sediments and was eventually transformed to either coal or petroleum.

Coal is a solid mineral containing many large, complex hydrocarbon molecules. Most of the coal mined today is used for the production of steel and for generating electricity at coal-burning power plants.

Petroleum, also called crude oil, is a liquid readily separated into its hydrocarbon components through a process known as *fractional distillation*, shown in Figure 12.3. The crude oil is heated in a pipe still to a temperature high enough to vaporize most of the components. The hot vapor flows into the bottom of a fractionating tower, which is warmer at the bottom than at the top. As the vapor rises in the tower and cools, the various components begin to condense. Hydrocarbons that have high boiling points, such as tar and lubricating stocks, condense first at warmer temperatures. Hydrocarbons that have low boiling points, such as gasoline, travel to the cooler regions at the top of the tower before condensing. Pipes drain the various liquid hydrocarbon fractions from the tower. Natural gas, which is primarily methane, does not condense. It remains a gas and is collected at the top of the tower.

Differences in the strength of molecular attractions explain why different hydrocarbons condense at different temperatures. As discussed in Section 7.1, in our comparison of induced dipole–induced dipole attractions in methane and octane, larger hydrocarbons experience many more of these attractions than smaller hydrocarbons do. For this reason, the larger hydrocarbons condense readily at high temperatures and so are found at the bottom of the tower. Smaller molecules, because they experience fewer attractions to neighbors, condense only at the cooler temperatures found at the top of the tower.

The gasoline obtained from the fractional distillation of petroleum consists of a wide variety of hydrocarbons having similar boiling points. Some of these components burn more efficiently than others in a car engine. The straight-chain hydrocarbons, such as *n*-hexane, tend to burn too quickly, causing what is called *engine knock*, as illustrated in Figure 12.4 on page 366. Gasoline hydrocarbons that have more branching, such as *iso*-octane,

Figure 12.4
(a) A straight-chain hydrocarbon, such as *n*-hexane, can be ignited from the heat generated as gasoline is compressed by the piston—before the spark plug fires. This upsets the timing of the engine cycle, giving rise to a knocking sound. (b) Branched hydrocarbons, such as *iso*-octane, burn less readily and are ignited not by compression alone but only when the spark plug fires.

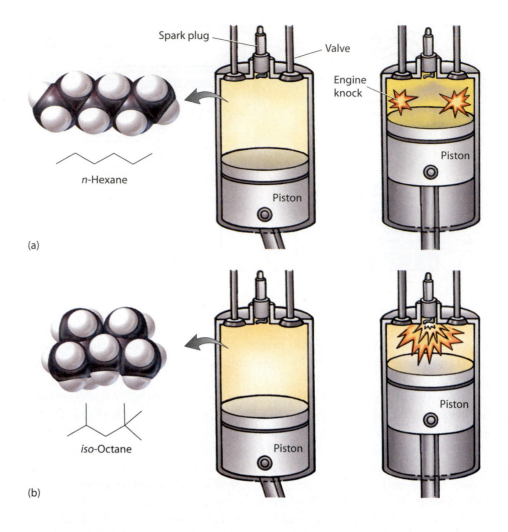

n-Hexane

(a)

iso-Octane

(b)

Figure 12.5
Octane ratings are posted on gasoline pumps.

burn slowly, and as a result the engine runs more smoothly. These two compounds, *n*-hexane and *iso*-octane, are used as standards in assigning *octane ratings* to gasoline. An octane number of 100 is arbitrarily assigned to *iso*-octane, and *n*-hexane is assigned an octane number of 0. The antiknock performance of a particular gasoline is compared with that of various mixtures of *iso*-octane and *n*-hexane, and an octane number is assigned. Figure 12.5 shows octane information on a typical gasoline pump.

Concept Check ✔

Which structural isomer in Figure 12.1 should have the highest octane rating?

Was this your answer? The structural isomer with the greatest amount of branching in the carbon framework will likely have the highest octane rating, making *neo*-pentane the clear winner. Just for the record, the ratings are

Compound	Octane rating
n-Pentane	61.7
iso-Pentane	92.3
neo-Pentane	116

12.2   Unsaturated Hydrocarbons Contain Multiple Bonds

Recall from Section 6.1 that carbon has four unpaired valence electrons. As shown in Figure 12.6, each of these electrons is available for pairing with an electron from another atom, such as hydrogen, to form a covalent bond.

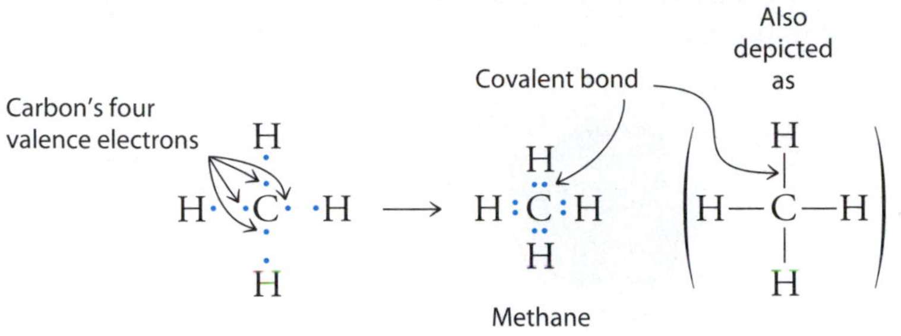

Figure 12.6
Carbon has four valence electrons. Each electron pairs with an electron from a hydrogen atom in the four covalent bonds of methane.

In all the hydrocarbons discussed so far, including the methane shown in Figure 12.6, each carbon atom is bonded to four neighboring atoms by four single covalent bonds. Such hydrocarbons are known as **saturated hydrocarbons**. The term *saturated* means that each carbon has as many atoms bonded to it as possible. We now explore cases where one or more carbon atoms in a hydrocarbon are bonded to fewer than four neighboring atoms. This occurs when at least one of the bonds between a carbon and a neighboring atom is a multiple bond. (See page 186 for a review of multiple bonds.)

A hydrocarbon containing a multiple bond—either double or triple— is known as an **unsaturated hydrocarbon**. Because of the multiple bond, two of the carbons are bonded to fewer than four other atoms. These carbons are thus said to be *unsaturated*.

Figure 12.7 compares the saturated hydrocarbon *n*-butane with the unsaturated hydrocarbon 2-butene. The number of atoms bonded to each of the two middle carbons of *n*-butane is four, whereas each of the two middle carbons of 2-butene is bonded to only three other atoms—a hydrogen and two carbons.

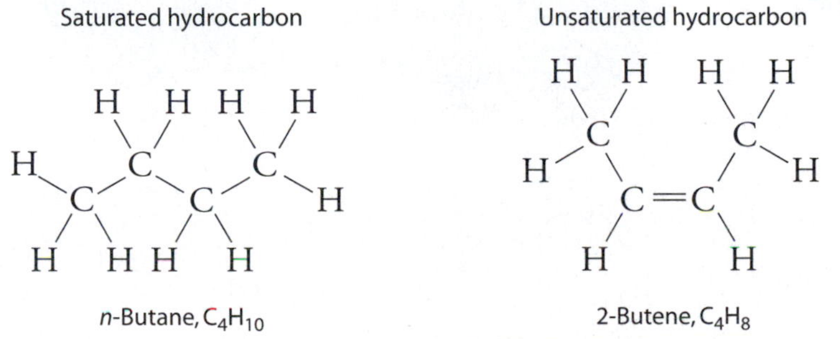

Figure 12.7
The carbons of the hydrocarbon *n*-butane are *saturated*, each being bonded to four other atoms. Because of the double bond, two of the carbons of the unsaturated hydrocarbon 2-butene are bonded to only three other atoms, which makes the molecule an unsaturated hydrocarbon.

An important unsaturated hydrocarbon is benzene, C_6H_6, which may be drawn as three double bonds contained within a flat hexagonal ring, as is shown in Figure 12.8a. Unlike the double-bond electrons in most other unsaturated hydrocarbons, however, the electrons of the double bonds in benzene are not fixed between any two carbon atoms. Instead, these electrons are able to move freely around the ring. This is commonly represented by drawing a circle within the ring, as shown in Figure 12.8b, rather than the individual double bonds.

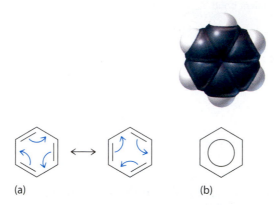

(a)   (b)

Figure 12.8
(a) The double bonds of benzene, C_6H_6, are able to migrate around the ring. (b) For this reason, they are often represented by a circle within the ring.

Many organic compounds contain one or more benzene rings in their structure. Because many of these compounds are fragrant, any organic molecule containing a benzene ring is classified as an **aromatic compound** (even if it is not particularly fragrant). Figure 12.9 shows a few examples. Toluene, a common solvent used as paint thinner, is toxic and gives airplane glue its distinctive odor. Some aromatic compounds, such as naphthalene, contain two or more benzene rings fused together. At one time, mothballs were made of naphthalene. Most mothballs sold today, however, are made of the less toxic 1,4-dichlorobenzene.

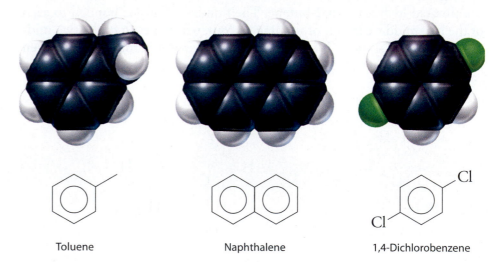

Toluene   Naphthalene   1,4-Dichlorobenzene

Figure 12.9
The structures for three odoriferous organic compounds containing one or more benzene rings: toluene, naphthalene, and 1,4-dichlorobenzene.

An example of an unsaturated hydrocarbon containing a triple bond is acetylene, C_2H_2. A confined flame of acetylene burning in oxygen is hot enough to melt iron, which makes acetylene a choice fuel for the welding shown in Figure 12.10.

H—C≡C—H

Acetylene

Figure 12.10
The unsaturated hydrocarbon acetylene, C_2H_2, burned in this torch produces a flame hot enough to melt iron.

Hands-On Chemistry: Twisting Jellybeans

Two carbon atoms connected by a single bond can rotate relative to each other. As we discussed in Section 12.1, this ability to rotate can give rise to numerous conformations (spatial orientations) of an organic molecule. Is it also possible for carbon atoms connected by a double bond to rotate relative to each other? Perform this quick activity to see for yourself.

What You Need

Jellybeans (or gumdrops), round toothpicks

Procedure

① Attach one jellybean to each end of a single toothpick. Hold one of the jellybeans firmly with one hand while rotating the second jellybean with your other hand. Observe how there is no restriction on the different orientations of the two jellybeans relative to each other.

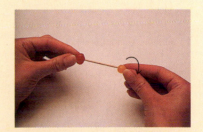

② Hold two toothpicks side by side and attach one jellybean to each end such that each jellybean has both toothpicks poked into it. As before, hold one jellybean while rotating the other. What kind of rotations are possible now?

Relate what you observe to the carbon–carbon double bond. Which structure of Figure 12.7 do you suppose has more possible conformations: n-butane or 2-butene? What do you suppose is true about the ability of atoms connected by a carbon–carbon triple bond to twist relative to each other?

Concept Check ✓

Prolonged exposure to benzene has been found to increase the risk of developing certain cancers. The structure of aspirin contains a benzene ring. Does this necessarily mean that prolonged exposure to aspirin will increase a person's risk of developing cancer?

Benzene ring

Aspirin

Was this your answer? No. Although benzene and aspirin both contain a benzene ring, these two molecules have different overall structures, which means the properties of one are quite different from the properties of the other. Each carbon-containing organic compound has its own set of unique physical, chemical, and biological properties. While benzene may cause cancer, aspirin is a safe remedy for headaches.

12.3 Organic Molecules Are Classified by Functional Group

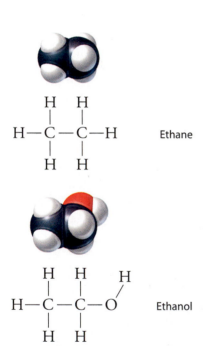

Ethane

Ethanol

Ethylamine

Carbon atoms can bond to one another and to hydrogen atoms in many ways, which results in an incredibly large number of hydrocarbons. But carbon atoms can bond to atoms of other elements as well, further increasing the number of possible organic molecules. In organic chemistry, any atom other than carbon or hydrogen in an organic molecule is called a **heteroatom**, where *hetero-* means "different from either carbon or hydrogen."

A hydrocarbon structure can serve as a framework to which various heteroatoms can be attached. This is analogous to a Christmas tree serving as the scaffolding on which ornaments are hung. Just as the ornaments give character to the tree, so do heteroatoms give character to an organic molecule. In other words, heteroatoms can have profound effects on the properties of an organic molecule.

Consider ethane, C_2H_6, and ethanol, C_2H_6O, which differ from each other by only a single oxygen atom. Ethane has a boiling point of −88°C, making it a gas at room temperature, and it does not dissolve in water very well. Ethanol, by contrast, has a boiling point of +78°C, making it a liquid at room temperature. It is infinitely soluble in water and is the active ingredient of alcoholic beverages. Consider further ethylamine, C_2H_7N, which has a nitrogen atom on the same basic two-carbon framework. This compound is a corrosive, pungent, highly toxic gas—most unlike either ethane or ethanol.

Organic molecules are classified according to the functional groups they contain, where a **functional group** is defined as a combination of atoms that behave as a unit. Most functional groups are distinguished by the heteroatoms they contain, and some common groups are listed in Table 12.1.

Table 12.1

Functional Groups in Organic Molecules

General Structure	Name	Class
—C—OH	Hydroxyl group	Alcohols
(ring structure) C—OH	Phenolic group	Phenols
—C—O—C—	Ether group	Ethers
—C—N	Amine group	Amines
O (ketone) —C—C—C—	Ketone group	Ketones
O (aldehyde) C—H	Aldehyde group	Aldehydes
O (amide) C—N	Amide group	Amides
O (carboxyl) C—OH	Carboxyl group	Carboxylic acids
O (ester) C—O—C—	Ester group	Esters

The remainder of this section introduces the classes of organic molecules shown in Table 12.1. The role heteroatoms play in determining the properties of each class is the underlying theme. As you study this material, focus on understanding the chemical and physical properties of the various classes of compounds, for doing so will give you a greater appreciation of the remarkable diversity of organic molecules and their many applications.

Concept Check

What is the significance of heteroatoms in an organic molecule?

Alcohols Contain the Hydroxyl Group

Alcohols are organic molecules in which a *hydroxyl group* is bonded to a saturated carbon. The hydroxyl group consists of an oxygen bonded to a hydrogen. Because of the polarity of the oxygen–hydrogen bond, low-formula-mass alcohols are often soluble in water, which is itself very polar. Some common alcohols and their melting and boiling points are listed in Table 12.2.

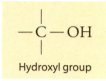

Hydroxyl group

Table 12.2

Some Simple Alcohols

Structure	Scientific Name	Common Name	Melting Point (°C)	Boiling Point (°C)
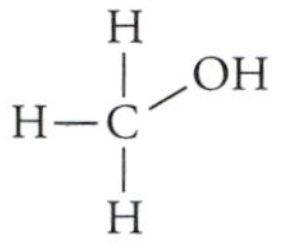 Methanol structure	Methanol	Methyl alcohol	−97	65
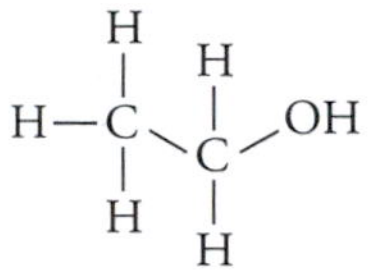 Ethanol structure	Ethanol	Ethyl alcohol	−115	78
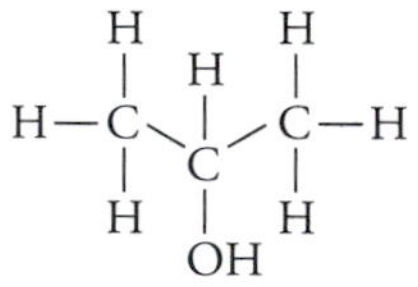 2-Propanol structure	2-Propanol	Isopropyl alcohol	−126	97

More than 11 billion pounds of methanol, CH_3OH, is produced annually in the United States. Most of it is used for making formaldehyde and acetic acid, important starting materials in the production of plastics. In addition, methanol is used as a solvent, an octane booster, and an anti-icing agent in gasoline. Sometimes called wood alcohol because it can be obtained from wood, methanol should never be ingested because in the body it is metabolized to formaldehyde and formic acid. Formaldehyde is harmful to the eyes, can lead to blindness, and was once used to preserve dead biological specimens. Formic acid, the active ingredient in an ant bite, can lower the pH of the blood to dangerous levels. Ingesting only about 15 milliliters (about 3 tablespoons) of methanol may lead to blindness, and about 30 milliliters can cause death.

Ethanol, C_2H_5OH, is one of the oldest chemicals manufactured by humans. The "alcohol" of alcoholic beverages, ethanol is prepared by feeding the sugars of various plants to certain yeasts, which produce ethanol through a biological process known as *fermentation*. Ethanol is widely used as an industrial solvent. For many years, ethanol intended for this purpose

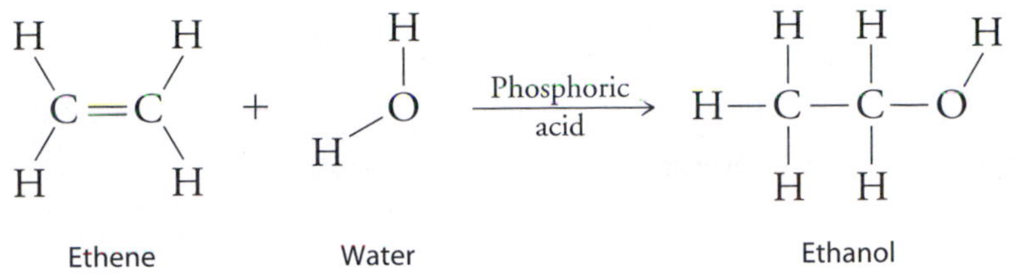

Ethene   Water   Ethanol

Figure 12.11
Ethanol can be synthesized from the unsaturated hydrocarbon ethene, with phosphoric acid as a catalyst.

was made by fermentation, but today industrial-grade ethanol is more cheaply manufactured from petroleum by-products, such as ethene, as Figure 12.11 illustrates.

The liquid produced by fermentation has an ethanol concentration no greater than about 12 percent because at this concentration the yeast begin to die. This is why most wines have an alcohol content of 11 or 12 percent—they are produced solely by fermentation. To attain the higher ethanol concentrations found in such "hard" alcoholic beverages as gin and vodka, the fermented liquid must be distilled. In the United States, the ethanol content of alcoholic beverages is measured as *proof,* which is twice the percent ethanol. An 86-proof whiskey, for example, is 43 percent ethanol by volume. The term *proof* evolved from a crude method once employed to test alcohol content. Gunpowder was wetted with a beverage of suspect alcohol content. If the beverage was primarily water, the powder would not ignite. If the beverage contained a significant amount of ethanol, the powder would burn, thus providing "proof" of the beverage's worth.

A third well-known alcohol is isopropyl alcohol, also called 2-propanol. This is the rubbing alcohol you buy at the drugstore. Although 2-propanol has a relatively high boiling point, it readily evaporates, leading to a pronounced cooling effect when applied to skin—an effect once used to reduce fevers. (Isopropyl alcohol is very toxic if ingested. Washcloths wetted with cold water are nearly as effective in reducing fever and far safer.) You are probably most familiar with the use of isopropyl alcohol as a topical disinfectant.

Phenols Contain an Acidic Hydroxyl Group

Phenols contain a phenolic group, which consists of a hydroxyl group attached to a benzene ring. Because of the presence of the benzene ring, the hydrogen of the hydroxyl group is readily lost in an acid–base reaction, which makes the phenolic group mildly acidic.

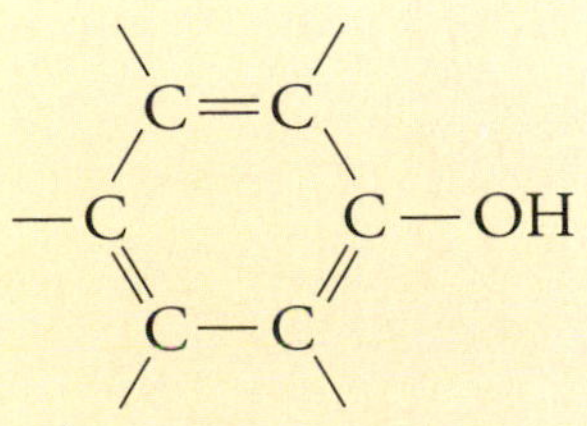

The reason for this acidity is illustrated in Figure 12.12 on page 374. How readily an acid donates a hydrogen ion is a function of how well the acid is able to accommodate the resulting negative charge it gains after donating the hydrogen ion. After phenol donates the hydrogen ion, it becomes a negatively charged phenoxide ion. The negative charge of the phenoxide ion, however, is not restricted to the oxygen atom. Recall that the electrons of the benzene ring are able to migrate around the ring. In a similar manner, the electrons responsible for the negative charge of the phenoxide ion are also able to migrate around the ring, as shown in Figure 12.12. Just as it is easy for

Phenolic group

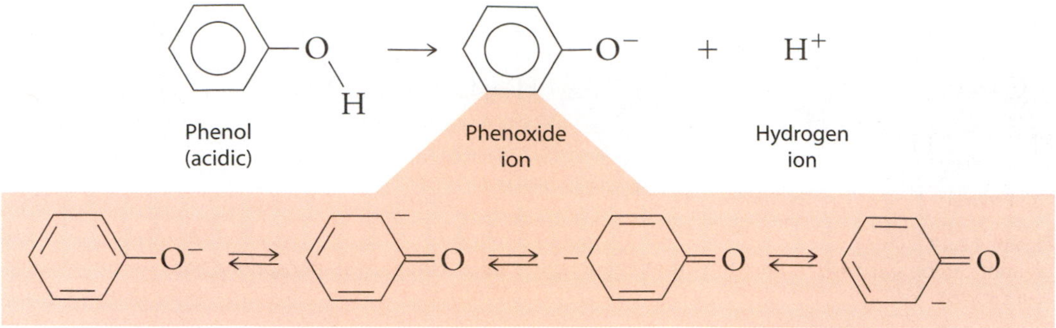

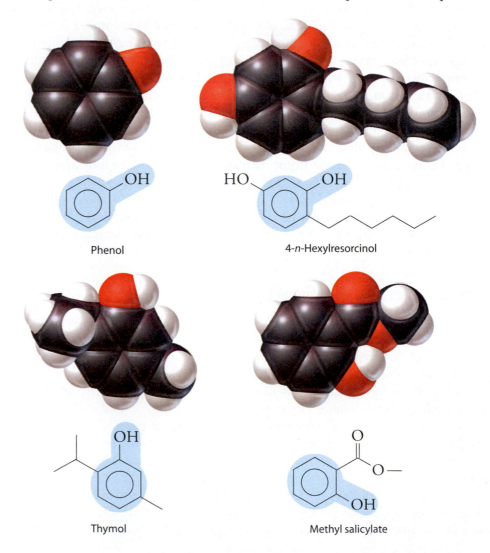

Figure 12.12
The negative charge of the phenoxide ion is able to migrate to select positions on the benzene ring. This mobility helps to accommodate the negative charge, which is why the phenolic group readily donates a hydrogen ion.

several people to hold a hot potato by quickly passing it around, it is easy for the phenoxide ion to hold the negative charge because the charge gets passed around. Because the negative charge of the ion is so nicely accommodated, the phenolic group is more acidic than it would be otherwise.

The simplest phenol, shown in Figure 12.13, is called phenol. In 1867, Joseph Lister (1827–1912) discovered the antiseptic value of phenol,

Phenol

4-*n*-Hexylresorcinol

Thymol

Methyl salicylate

Figure 12.13
Every phenol contains a phenolic group (highlighted in blue).

which, when applied to surgical instruments and incisions, greatly increased surgery survival rates. Phenol was the first purposefully used anti-bacterial solution, or *antiseptic*. Phenol damages healthy tissue, however, and so a number of milder phenols have since been introduced. The phenol 4-*n*-hexylresorcinol, for example, is commonly used in throat lozenges and mouthwashes. This compound has even greater antiseptic properties than phenol, and yet it does not damage tissue. Listerine brand mouthwash (named after Joseph Lister) contains the antiseptic phenols thymol and methyl salicylate.

Concept Check

Why are alcohols less acidic than phenols?

Was this your answer? An alcohol does not contain a benzene ring adjacent to the hydroxyl group. If the alcohol were to donate the hydroxyl hydrogen, the result would be a negative charge on the oxygen. Without an adjacent benzene ring, this negative charge has nowhere to go. As a result, an alcohol behaves only as a very weak acid, much the way water does.

The Oxygen of an Ether Group Is Bonded to Two Carbon Atoms

Ethers are organic compounds structurally related to alcohols. The oxygen atom in an ether group, however, is bonded not to a carbon and a hydrogen but rather to two carbons. As we see in Figure 12.14, ethanol and dimethyl ether have the same chemical formula, C_2H_6O, but their physical properties are vastly different. Whereas ethanol is a liquid at room temperature (boiling point 78°C) and mixes quite well with water, dimethyl ether is a gas at room temperature (boiling point −25°C) and is much less soluble in water.

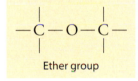

Ether group

Ethanol: Soluble in water, boiling point 78°C

Dimethyl ether: Insoluble in water, boiling point −25°C

Figure 12.14
The oxygen in an alcohol, such as ethanol, is bonded to one carbon atom and one hydrogen atom. The oxygen in an ether, such as dimethyl ether, is bonded to two carbon atoms. Because of this difference, alcohols and ethers of similar molecular mass have vastly different physical properties.

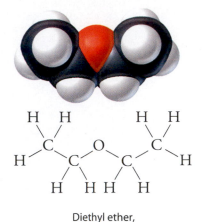

Diethyl ether,
boiling point 35°C

Figure 12.15
Diethyl ether is the systematic name for the "ether" historically used as an anesthetic.

Amine group

Ethers are not very soluble in water because, without the hydroxyl group, they are unable to form strong hydrogen bonds with water (Section 7.1). Furthermore, without the polar hydroxyl group, the molecular attractions among ether molecules are relatively weak. As a result, it does not take much energy to separate ether molecules from one another. This is why ethers have relatively low boiling points and evaporate so readily.

Diethyl ether, shown in Figure 12.15, was one of the first anesthetics. The anesthetic properties of this compound were discovered in the early 1800s and revolutionized the practice of surgery. Because of its high volatility at room temperature, inhaled diethyl ether rapidly enters the bloodstream. Because this ether has low solubility in water and high volatility, it quickly leaves the bloodstream once introduced. Because of these physical properties, a surgical patient can be brought in and out of anesthesia (a state of unconsciousness) simply by regulating the gases breathed. Modern-day gaseous anesthetics have fewer side effects than diethyl ether but work on the same principle.

Amines Form Alkaline Solutions

Amines are organic compounds that contain the amine group—a nitrogen atom bonded to one, two, or three saturated carbons. Amines are typically less soluble in water than are alcohols because the nitrogen–hydrogen bond is not quite as polar as the oxygen–hydrogen bond. The lower polarity of amines also means their boiling points are typically somewhat lower than those of alcohols of similar formula mass. Table 12.3 lists three simple amines.

Table 12.3

Three Simple Amines

Structure	Name	Melting Point (°C)	Boiling Point (°C)
	Ethylamine	−81	17
	Diethylamine	−50	55
	Triethylamine	−7	89

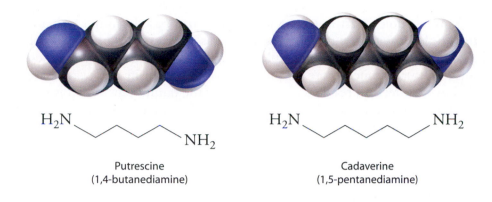

Figure 12.16
Low-formula-mass amines like these tend to have an offensive odor.

Putrescine
(1,4-butanediamine)

Cadaverine
(1,5-pentanediamine)

One of the most notable physical properties of many low-formula-mass amines is their offensive odor. Figure 12.16 shows two appropriately named amines, putrescine and cadaverine, responsible for the odor of decaying flesh.

Amines are typically alkaline because the nitrogen atom readily accepts a hydrogen ion from water, as Figure 12.17 illustrates.

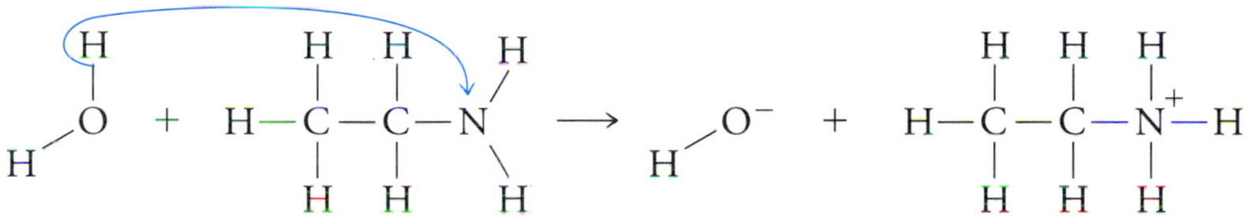

Water
(acid)

Ethylamine
(base)

Hydroxide
ion

Ethylammonium
ion

Figure 12.17
Ethylamine acts as a base and accepts a hydrogen ion from water to become the ethylammonium ion. This reaction generates a hydroxide ion, which increases the pH of the solution.

A group of naturally occurring complex molecules that are alkaline because they contain nitrogen atoms are often called *alkaloids*. Because many alkaloids have medicinal value, there is great interest in isolating these compounds from plants or marine organisms containing them. As shown in Figure 12.18, an alkaloid reacts with an acid to form a salt that is usually quite soluble in water. This is in contrast to the nonionized form of the alkaloid, known as a *free base* and typically insoluble in water.

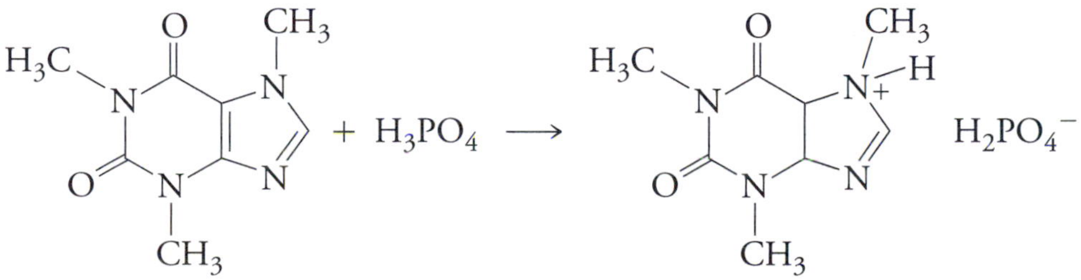

Caffeine, free-base form
(water-insoluble)

Phosphoric
acid

Caffeine–phosphoric acid salt
(water-soluble)

Figure 12.18
All alkaloids are bases that react with acids to form salts. An example is the alkaloid caffeine, shown here reacting with phosphoric acid.

Figure 12.19
Tannins are responsible for the brown stains in coffee mugs or on a coffee drinker's teeth. Because tannins are acidic, they can be readily removed with an alkaline cleanser. Use a little laundry bleach on the mug, and brush your teeth with baking soda.

Most alkaloids exist in nature not in their free-base form but rather as the salt of naturally occurring acids known as *tannins*, a group of phenol-based organic acids that have complex structures. The alkaloid salts of these acids are usually much more soluble in hot water than in cold water. The caffeine in coffee and tea exists in the form of the tannin salt, which is why coffee and tea are more effectively brewed in hot water. As Figure 12.19 relates, tannins are also responsible for the stains caused by these beverages.

Concept Check ✔

Why do most caffeinated soft drinks also contain phosphoric acid?

Was this your answer? The phosphoric acid, as shown in Figure 12.18, reacts with the caffeine to form the caffeine–phosphoric acid salt, which is much more soluble in cold water than the naturally occurring tannin salt.

Ketones, Aldehydes, Amides, Carboxylic Acids, and Esters All Contain a Carbonyl Group

The **carbonyl group** consists of a carbon atom double-bonded to an oxygen atom. It occurs in the organic compounds known as ketones, aldehydes, amides, carboxylic acids, and esters.

A **ketone** is a carbonyl-containing organic molecule in which the carbonyl carbon is bonded to two carbon atoms. A familiar example of a ketone is *acetone*, which is often used in fingernail polish remover and is shown in Figure 12.20a. In an **aldehyde**, the carbonyl carbon is bonded either to one carbon atom and one hydrogen atom, as in Figure 12.20b, or, in the special case of formaldehyde, to two hydrogen atoms.

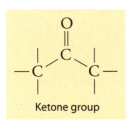

Ketone group

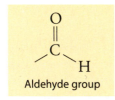

Aldehyde group

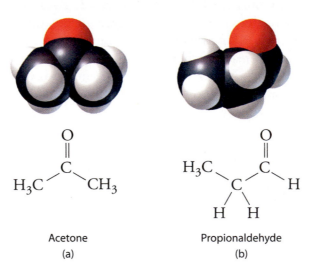

Acetone
(a)

Propionaldehyde
(b)

Figure 12.20
(a) When the carbon of a carbonyl group is bonded to two carbon atoms, the result is a ketone. An example is acetone. (b) When the carbon of a carbonyl group is bonded to at least one hydrogen atom, the result is an aldehyde. An example is propionaldehyde.

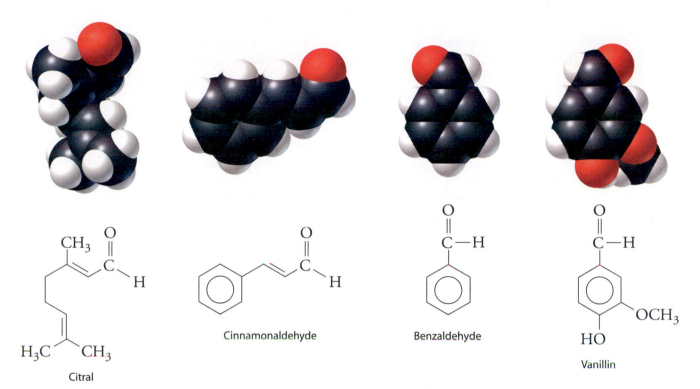

Figure 12.21
Aldehydes are responsible for many familiar fragrances.

Many aldehydes are particularly fragrant. A number of flowers, for example, owe their pleasant odor to the presence of simple aldehydes. The smells of lemons, cinnamon, and almonds are due to the aldehydes citral, cinnamaldehyde, and benzaldehyde, respectively. The structures of these three aldehydes are shown in Figure 12.21. The aldehyde vanillin, introduced at the beginning of this chapter, is the key flavoring molecule derived from the vanilla orchid. You may have noticed that vanilla seed pods and vanilla extract are fairly expensive. Imitation vanilla flavoring is less expensive because it is merely a solution of the compound vanillin, which is economically synthesized from the waste chemicals of the wood pulp industry. Imitation vanilla does not taste the same as natural vanilla extract, however, because in addition to vanillin many other flavorful molecules contribute to the complex taste of natural vanilla. Many books made in the days before "acid-free" paper smell of vanilla because of the vanillin formed and released as the paper ages, a process that is accelerated by the acids the paper contains.

An **amide** is a carbonyl-containing organic molecule in which the carbonyl carbon is bonded to a nitrogen atom. The active ingredient of most mosquito repellents is an amide whose chemical name is *N,N*-diethyl-*m*-toluamide but is commercially known as DEET, shown in Figure 12.22. This compound is actually not an insecticide. Rather, it causes certain insects, especially mosquitoes, to lose their sense of direction, which effectively protects DEET wearers from being bitten.

Amide group

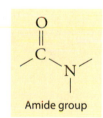

N,N-Diethyl-*m*-toluamide
(DEET)

Figure 12.22
N,N-diethyl-*m*-toluamide is an example of an amide. Amides contain the amide group, shown highlighted in blue.

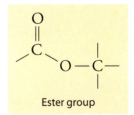

Carboxyl group

A **carboxylic acid** is a carbonyl-containing organic molecule in which the carbonyl carbon is bonded to a hydroxyl group. As its name implies, this functional group is able to donate hydrogen ions, and as a result organic molecules containing it are acidic. An example is acetic acid, $C_2H_4O_2$, the main ingredient of vinegar. You may recall that this organic compound was used as an example of a weak acid back in Chapter 10.

As with phenols, the acidity of a carboxylic acid results in part from the ability of the functional group to accommodate the negative charge of the ion that forms after the hydrogen ion has been donated. As shown in Figure 12.23, a carboxylic acid transforms to a carboxylate ion as it loses the hydrogen ion. The negative charge of the carboxylate ion is able to pass back and forth between the two oxygens. This spreading out helps to accommodate the negative charge.

Carboxyl group
in acetic acid

Carboxylate ion
in acetate ion

Hydrogen ion

Figure 12.23
The negative charge of the carboxylate ion is able to pass back and forth between the two oxygen atoms of the carboxyl group.

An interesting example of an organic compound that contains both a carboxylic acid and a phenol is salicylic acid, found in the bark of the willow tree and illustrated in Figure 12.24a. At one time brewed for its antipyretic (fever-reducing) effect, salicylic acid is an important analgesic (painkiller), but it causes nausea and stomach upset because of its relatively high acidity, a result of the presence of two acidic functional groups. In 1899, Friederich Bayer and Company, in Germany, introduced a chemically modified version of this compound in which the phenolic group was transformed to an ester functional group. Because both the carboxyl group and the phenolic group contribute to the high acidity of salicylic acid, getting rid of the phenolic group reduced the acidity of the molecule considerably. The result was the less acidic and more tolerable acetylsalicylic acid, the chemical name for aspirin, shown in Figure 12.24b.

An **ester** is an organic molecule similar to a carboxylic acid except that in the ester the hydroxyl hydrogen is replaced by a carbon. Unlike carboxylic acids, esters are not acidic because they lack the hydrogen of the hydroxyl group. Like aldehydes, many simple esters have notable fragrances and are used as flavorings. Some familiar ones are listed in Table 12.4 on page 382.

Ester group

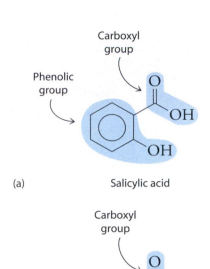

Carboxyl group

Phenolic group

(a) Salicylic acid

Carboxyl group

Ester

(b) Aspirin (acetylsalicylic acid)

Figure 12.24
(a) Salicylic acid, found in the bark of the willow tree, is an example of a molecule containing both a carboxyl group and a phenolic group. (b) Aspirin, acetylsalicylic acid, is less acidic than salicylic acid because it no longer contains the acidic phenolic group, which has been converted to an ester.

Concept Check ✓

Identify all the functional groups in these four molecules (ignore the sulfur group in penicillin G):

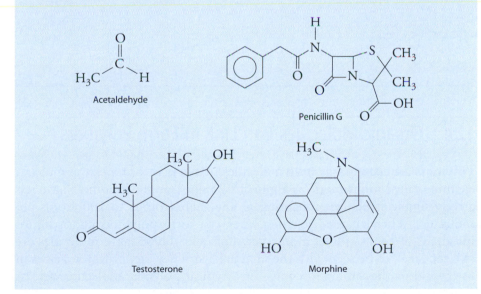

Acetaldehyde

Penicillin G

Testosterone

Morphine

Was this your answer? Acetaldehyde: aldehyde; penicillin G: amide (two amide groups), carboxylic acid; testosterone: alcohol and ketone; morphine: alcohol, phenol, ether, and amine.

Table 12.4

Some Esters and Their Flavors and Odors

Structure	Name	Flavor/Odor
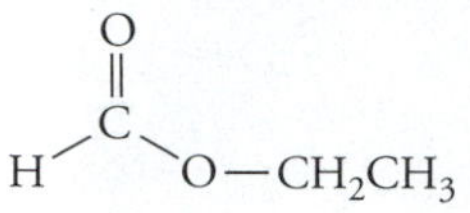	Ethyl formate	Rum
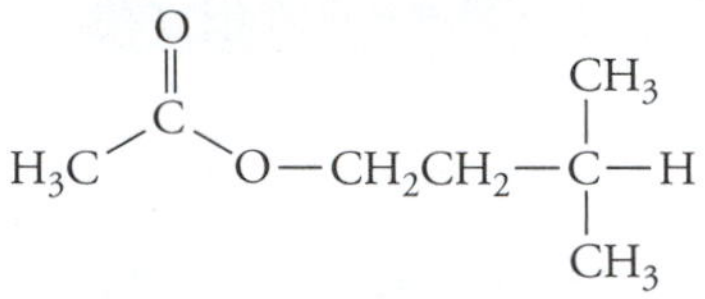	Isopentyl acetate	Banana
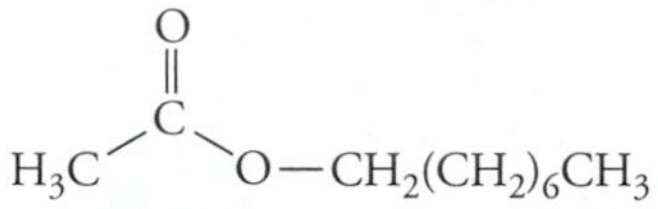	Octyl acetate	Orange
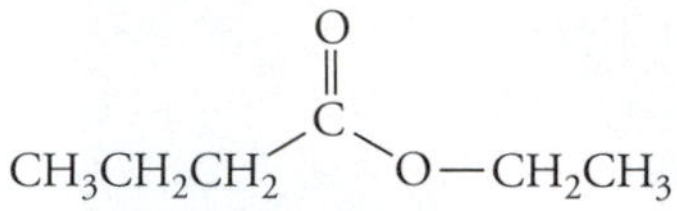	Ethyl butyrate	Pineapple
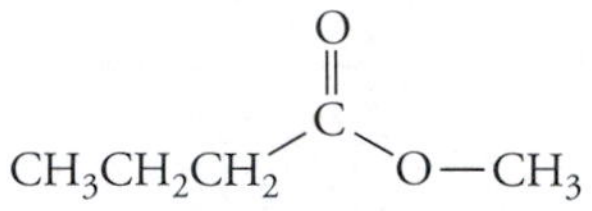	Methyl butyrate	Apple
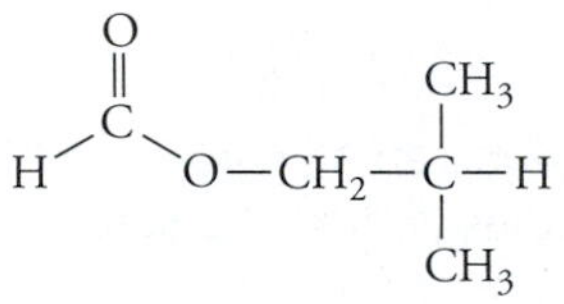	Isobutyl formate	Raspberry
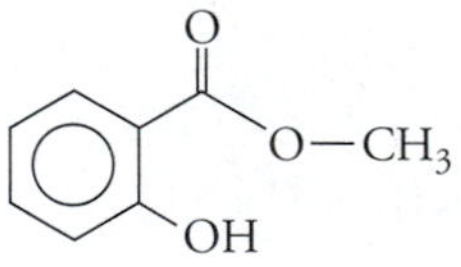	Methyl salicylate	Wintergreen

12.4   Organic Molecules Can Link to Form Polymers

Polymers are exceedingly long molecules that consist of repeating molecular units called **monomers**, as Figure 12.25 illustrates. Monomers have relatively simple structures consisting of anywhere from 4 to 100 atoms per molecule. When chained together, they can form polymers consisting of hundreds of thousands of atoms per molecule. These large molecules are still too small to be seen with the unaided eye. They are, however, giants in the world of the submicroscopic—if a typical polymer molecule were as thick as a kite string, it would be 1 kilometer long.

Many of the molecules that make up living organisms are polymers, including DNA, proteins, the cellulose of plants, and the complex carbohydrates of starchy foods. We leave a discussion of these important biological molecules to Chapter 13. For now, we focus on the human-made polymers,

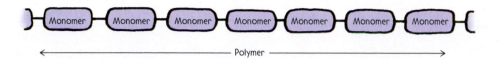

Figure 12.25
A polymer is a long molecule consisting of many smaller monomer molecules linked together.

also known as synthetic polymers, that make up the class of materials commonly known as plastics.

We begin by exploring the two major types of synthetic polymers used today—*addition polymers* and *condensation polymers*. This provides a good background for the discussion of plastics in Chapter 18.

As shown in Table 12.5 on pages 384 and 385, addition and condensation polymers have a wide variety of uses. Solely the product of human design, these polymers pervade modern living. In the United States, for example, synthetic polymers have surpassed steel as the most widely used material.

Addition Polymers Result from the Joining Together of Monomers

Addition polymers form simply by the joining together of monomer units. For this to happen, each monomer must contain at least one double bond. As shown in Figure 12.26, polymerization occurs when two of the electrons from each double bond split away from each other to form new covalent bonds with neighboring monomer molecules. During this process, no atoms are lost, meaning that the total mass of the polymer is equal to the sum of the masses of all the monomers.

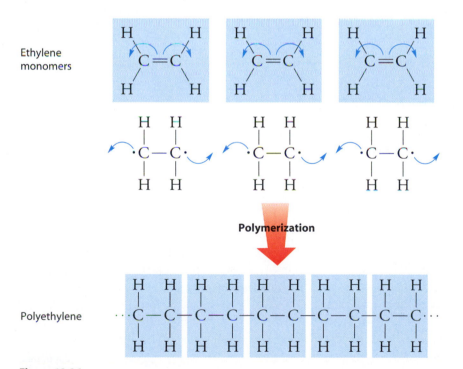

Figure 12.26
The addition polymer polyethylene is formed as electrons from the double bonds of ethylene monomer molecules split away and become unpaired valence electrons. Each unpaired electron then joins with an unpaired electron of a neighboring carbon atom to form a new covalent bond that links two monomer units together.

Table 12.5

Addition and Condensation Polymers

Addition Polymers	Repeating Unit	Common Uses	Recycling Code
Polyethylene (PE)	H H | | ···C—C··· | | H H	Plastic bags, bottles	2 HDPE 4 LDPE
Polypropylene (PP)	H H | | ···C—C··· | | H CH$_3$	Indoor–outdoor carpets	5 PP
Polystyrene (PS)	H H | | ···C—C··· | | H ⬡	Plastic utensils, insulation	6 PS
Polyvinyl chloride (PVC)	H H | | ···C—C··· | | H Cl	Shower curtains, tubing	3 V
Polyvinylidene chloride (Saran)	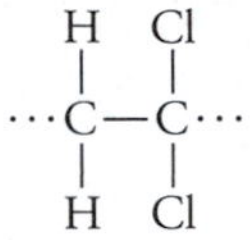 H Cl | | ···C—C··· | | H Cl	Plastic wrap	—
Polytetrafluoroethylene (Teflon)	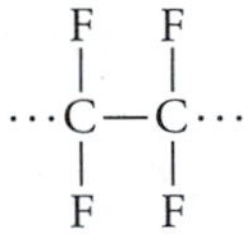 F F | | ···C—C··· | | F F	Nonstick coating	—
Polyacrylonitrile (Orlon)	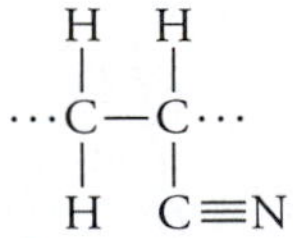 H H | | ···C—C··· | | H C≡N	Yarn, paints	—
Polymethyl methacrylate (Lucite, Plexiglas)	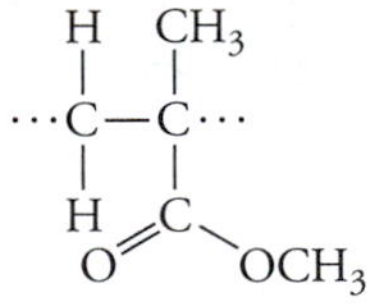 H CH$_3$ | | ···C—C··· | | H C O⫽ ＼OCH$_3$	Windows, bowling balls	—
Polyvinyl acetate (PVA)	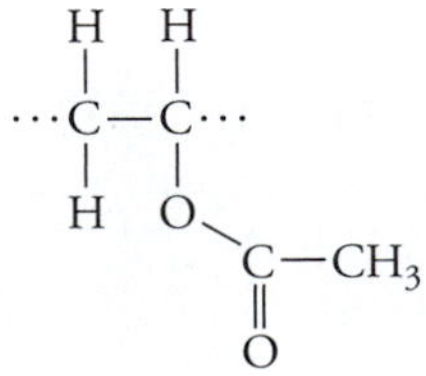 H H | | ···C—C··· | | H O | C—CH$_3$ ⫽ O	Adhesives, chewing gum	—

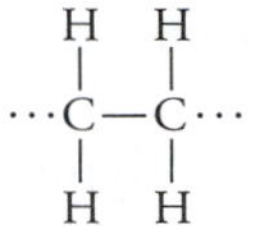

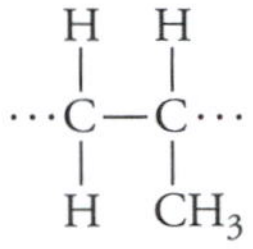

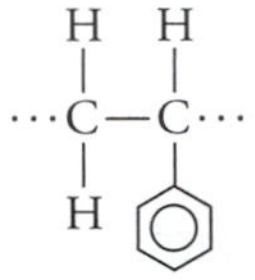

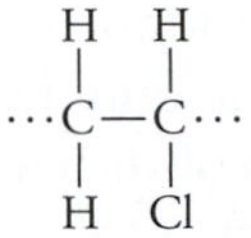

(continued)

Table 12.5 *(continued)*

Addition and Condensation Polymers

Condensation Polymers

Nylon	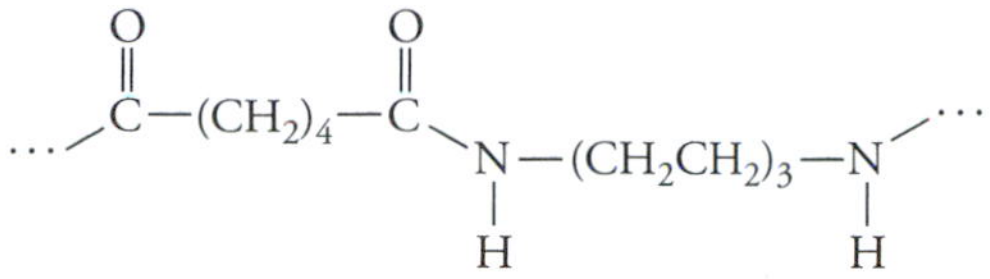	Carpeting, clothing	—
Polyethylene terephthalate (Dacron, Mylar)	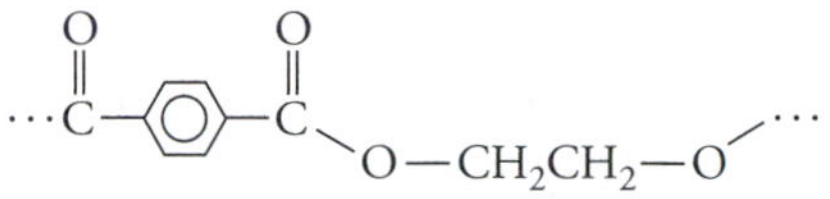	Clothing, plastic bottles	♻ 1 PET
Melamine–formaldehyde resin (Melmac, Formica)	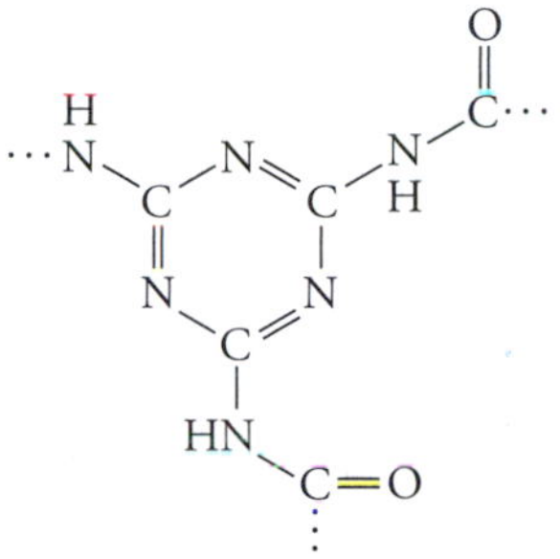	Dishes, countertops	—

Nearly 12 million tons of polyethylene is produced annually in the United States; that's about 90 pounds per U.S. citizen. The monomer from which it is synthesized, ethylene, is an unsaturated hydrocarbon produced in large quantities from petroleum.

Two principal forms of polyethylene are produced by using different catalysts and reaction conditions. High-density polyethylene (HDPE), shown schematically in Figure 12.27a, consists of long strands of straight-chain molecules packed closely together. The tight alignment of neighboring strands makes HDPE a relatively rigid, tough plastic useful for such things as bottles and milk jugs. Low-density polyethylene (LDPE), shown in Figure 12.27b, is made of strands of highly branched chains, an architecture that prevents the strands from packing closely together. This makes LDPE more bendable than HDPE and gives it a lower melting point. While HDPE holds its shape in boiling water, LDPE deforms. It is most useful for such items as plastic bags, photographic film, and electrical-wire insulation.

(a) Molecular strands of HDPE   (b) Molecular strands of LDPE

Figure 12.27
(a) The polyethylene strands of HDPE are able to pack closely together, much like strands of uncooked spaghetti. (b) The polyethylene strands of LDPE are branched, which prevents the strands from packing well.

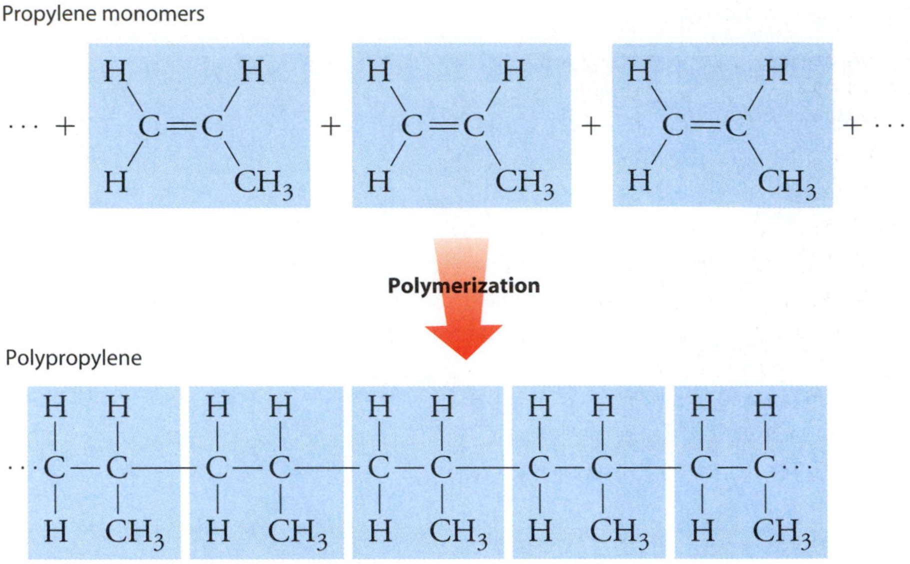

Figure 12.28
Propylene monomers polymerize to
form polypropylene.

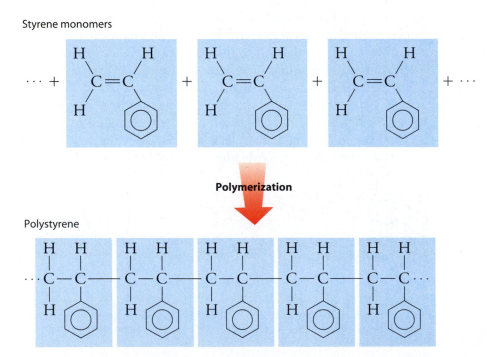

Other addition polymers are created by using different monomers. The only requirement is that the monomer must contain a double bond. The monomer propylene, for example, yields polypropylene, as shown in Figure 12.28. Polypropylene is a tough plastic material useful for pipes, hard-shell suitcases, and appliance parts. Fibers of polypropylene are used for upholstery, indoor–outdoor carpets, and even thermal underwear.

Figure 12.29 shows that using styrene as the monomer yields polystyrene. Transparent plastic cups are made of polystyrene, as are thousands of other household items. Blowing gas into liquid polystyrene generates Styrofoam, widely used for coffee cups, packing material, and insulation.

Figure 12.29
Styrene monomers polymerize to form
polystyrene.

Polyvinyl chloride (PVC)

Another important addition polymer is polyvinylchloride (PVC), which is tough and easily molded. Floor tiles, shower curtains, and pipes are most often made of PVC, shown in Figure 12.30.

The addition polymer polyvinylidene chloride (trade name Saran), shown in Figure 12.31, is used as plastic wrap for food. The large chlorine atoms in this polymer help it stick to surfaces such as glass by dipole–induced dipole attractions, as we saw in Section 7.1.

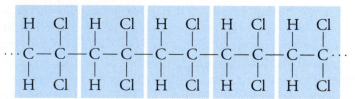

Polyvinylidene chloride (Saran)

Figure 12.31
The large chlorine atoms in polyvinylidene chloride make this addition polymer sticky.

Figure 12.32
The fluorine atoms in polytetrafluoroethylene tend not to experience
molecular attractions, which is why
this addition polymer is used as a
nonstick coating and lubricant.

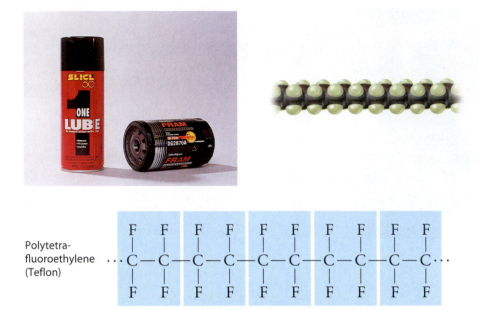

Polytetra-
fluoroethylene
(Teflon)

The addition polymer polytetrafluoroethylene, shown in Figure 12.32, is what you know as Teflon. In contrast to the chlorine-containing Saran, fluorine-containing Teflon has a nonstick surface because the fluorine atoms tend not to experience any molecular attractions. In addition, because carbon–fluorine bonds are unusually strong, Teflon can be heated to high temperatures before decomposing. These properties make Teflon an ideal coating for cooking surfaces. It is also relatively inert, which is why many corrosive chemicals are shipped or stored in Teflon containers.

Concept Check ✓

What do all monomers of addition polymers have in common?

Was this your answer? A double covalent bond between two carbon atoms.

Condensation Polymers Form with the Loss of Small Molecules

A **condensation polymer** is one formed when the joining of monomer units is accompanied by the loss of a small molecule, such as water or hydrochloric acid. Any monomer capable of becoming part of a condensation polymer must have a functional group on each end. When two such monomers come together to form a condensation polymer, one functional group of the first monomer links up with one functional group of the other monomer. The result is a two-monomer unit that has two terminal functional groups, one from each of the two original monomers. Each of these terminal functional groups in the two-monomer unit is now free to link up with one of the functional groups of a third monomer, and then a fourth, and so on. In this way a polymer chain is built. Figure 12.33 shows this process for the condensation polymer called nylon, created in 1937 by DuPont chemist Wallace

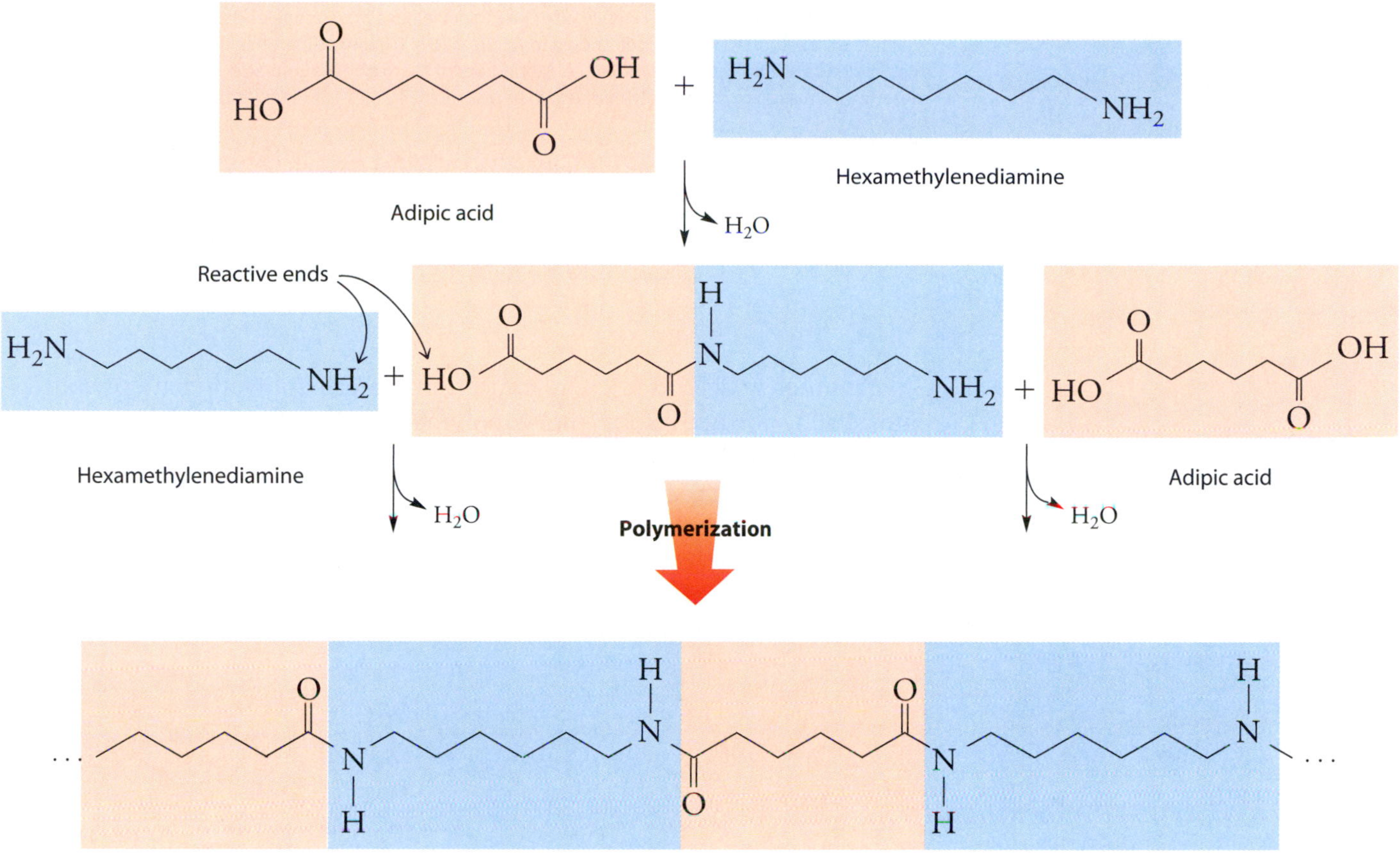

Figure 12.33
Adipic acid and hexamethylenediamine polymerize to form the condensation copolymer nylon.

Carothers (1896–1937). This polymer is composed of two different monomers, as shown in Figure 12.33, which classifies it as a *copolymer*. One monomer is adipic acid, which contains two reactive end groups, both carboxyl groups. The second monomer is hexamethylenediamine, in which two amine groups are the reactive end groups. One end of an adipic acid molecule and one end of a hexamethylamine molecule can be made to react with each other, splitting off a water molecule in the process. After two monomers have joined, reactive ends still remain for further reactions, which leads to a growing polymer chain. Aside from its use in hosiery, nylon also finds great use in the manufacture of ropes, parachutes, clothing, and carpets.

Concept Check ✔

The structure of 6-aminohexanoic acid is

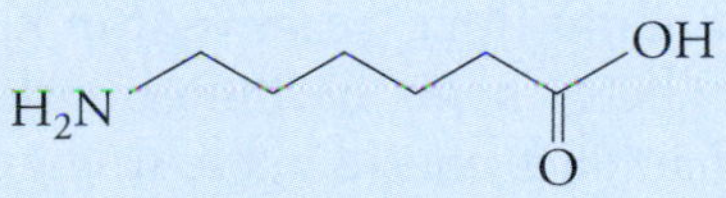

Is this compound a suitable monomer for forming a condensation polymer? If so, what is the structure of the polymer formed, and what small molecule is split off during the condensation?

Was this your answer? Yes, because the molecule has two reactive ends. You know both ends are reactive because they are the ends shown in Figure 12.33. The only difference here is that both types of reactive ends are on the same molecule. Monomers of 6-aminohexanoic acid combine by splitting off water molecules to form the polymer known as nylon-6:

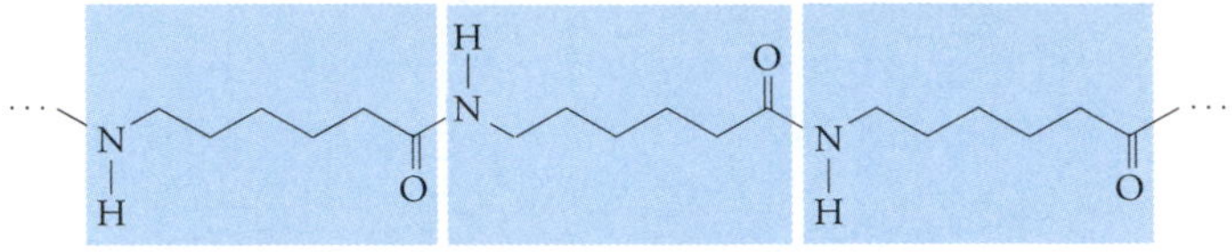

Another widely used condensation polymer is polyethylene terephthalate (PET), formed from the copolymerization of ethylene glycol and terephthalic acid, as shown in Figure 12.34. Plastic soda bottles are made from this polymer. Also, PET fibers are sold as Dacron polyester, used in clothing and stuffing for pillows and sleeping bags. Thin films of PET are called Mylar and can be coated with metal particles to make magnetic recording tape or those metallic-looking balloons you see for sale at most grocery store check-out counters.

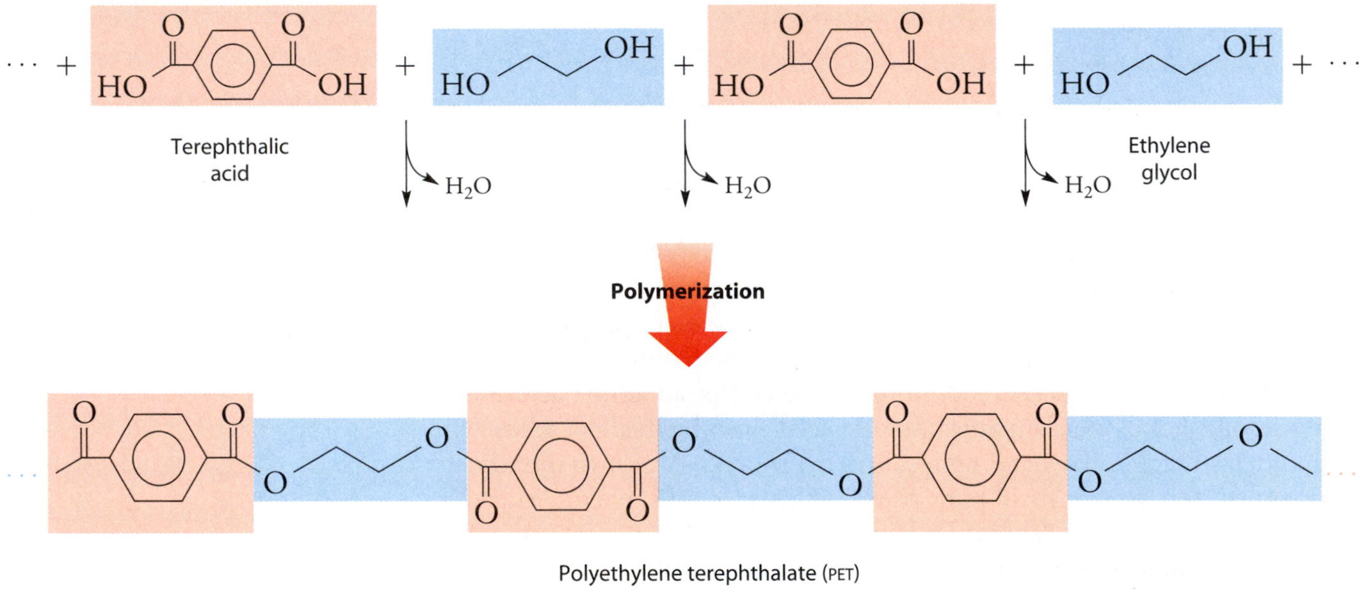

Figure 12.34
Terephthalic acid and ethylene glycol polymerize to form the condensation copolymer polyethylene terephthalate.

Monomers that contain three reactive functional groups can also form polymer chains. These chains become interlocked in a rigid three-dimensional network that lends considerable strength and durability to the polymer. Once formed, these condensation polymers cannot be remelted or reshaped, which makes them hard-set, or *thermoset*, polymers. A good example is the thermoset polymer shown in Figure 12.35, formed from the reaction of formaldehyde with melamine. Hard plastic dishes (Melmac) and countertops (Formica) are made of this material. A similar polymer,

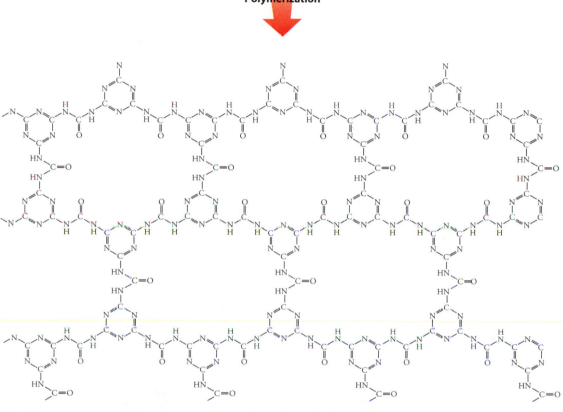

Figure 12.35
The three reactive groups of melamine allow it to polymerize with formaldehyde to form a three-dimensional network.

Bakelite, made from formaldehyde and phenols containing multiple oxygen atoms, is used to bind plywood and particle board. Bakelite was synthesized in the early 1900s, and it was the first widely used polymer.

In Perspective

The synthetic-polymers industry has grown remarkably over the past 50 years. Annual production of polymers in the United States alone has grown from 3 billion pounds in 1950 to 100 billion pounds in 2000. Today, it is a

Hands-On Chemistry: Racing Water Drops

The chemical composition of a polymer has a significant effect on its macroscopic properties. To see this for yourself, place a drop of water on a new plastic sandwich bag, and then tilt the bag vertically so that the drop races off. Observe the behavior of the water carefully. Now race a drop of water off a freshly pulled strip of plastic food wrap. How does the behavior of the drop on the wrap compare with the behavior of the drop on the sandwich bag?

Most brands of sandwich bags are made of polyethylene terephthalate, and most brands of food wrap are made of polyvinylidene chloride. Look carefully at the chemical composition of these polymers, shown in Table 12.5. Which contains larger atoms? Which might be involved in stronger dipole–induced dipole interactions with water? Need help with these questions? Refer back to Sections 6.7 and 7.1.

challenge to find any consumer item that does *not* contain a plastic of one sort or another. Try this yourself.

In the future, watch for new kinds of polymers having a wide range of remarkable properties. One interesting application is shown in Figure 12.36. We already have polymers that conduct electricity, others that emit light, others that replace body parts, and still others that are stronger but much lighter than steel. Imagine synthetic polymers that mimic photosynthesis by transforming solar energy to chemical energy or efficiently separate fresh water from the oceans. These are not dreams. They are realities chemists have already demonstrated in the laboratory. Polymers hold a clear promise for the future.

The plastics industry is but one outgrowth of our knowledge of organic chemistry. As we explore in the next chapter, our understanding of life itself is based on our understanding of the properties of carbohydrates, fats, proteins, and nucleic acids, all of which are polymers containing the functional groups introduced in this chapter.

Figure 12.36
Flexible and flat video displays can now be fabricated from polymers.

Key Terms and Matching Definitions

_____ addition polymer
_____ alcohol
_____ aldehyde
_____ amide
_____ amine
_____ aromatic compound
_____ carbonyl group
_____ carboxylic acid
_____ condensation polymer
_____ conformation
_____ ester
_____ ether
_____ functional group
_____ heteroatom
_____ hydrocarbon
_____ ketone
_____ monomer
_____ organic chemistry
_____ phenol
_____ polymer
_____ saturated hydrocarbon
_____ structural isomers
_____ unsaturated hydrocarbon

1. The study of carbon-containing compounds.
2. A chemical compound containing only carbon and hydrogen atoms.
3. Molecules that have the same molecular formula but different chemical structures.
4. One of the possible spatial orientations of a molecule.
5. A hydrocarbon containing no multiple covalent bonds, with each carbon atom bonded to four other atoms.
6. A hydrocarbon containing at least one multiple covalent bond.
7. Any organic molecule containing a benzene ring.
8. Any atom other than carbon or hydrogen in an organic molecule.
9. A specific combination of atoms that behave as a unit in an organic molecule.
10. An organic molecule that contains a hydroxyl group bonded to a saturated carbon.
11. An organic molecule in which a hydroxyl group is bonded to a benzene ring.
12. An organic molecule containing an oxygen atom bonded to two carbon atoms.

13. An organic molecule containing a nitrogen atom bonded to one or more saturated carbon atoms.
14. A carbon atom double-bonded to an oxygen atom, found in ketones, aldehydes, amides, carboxylic acids, and esters.
15. An organic molecule containing a carbonyl group the carbon of which is bonded to two carbon atoms.
16. An organic molecule containing a carbonyl group the carbon of which is bonded either to one carbon atom and one hydrogen atom or to two hydrogen atoms.
17. An organic molecule containing a carbonyl group the carbon of which is bonded to a nitrogen atom.
18. An organic molecule containing a carbonyl group the carbon of which is bonded to a hydroxyl group.
19. An organic molecule containing a carbonyl group the carbon of which is bonded to one carbon atom and one oxygen atom bonded to another carbon atom.
20. A long organic molecule made of many repeating units.
21. The small molecular unit from which a polymer is formed.
22. A polymer formed by the joining together of monomer units with no atoms lost as the polymer forms.
23. A polymer formed by the joining together of monomer units accompanied by the loss of a small molecule, such as water.

Review Questions

Hydrocarbons Contain Only Carbon and Hydrogen

1. What are some examples of hydrocarbons?
2. What are some uses of hydrocarbons?
3. How do two structural isomers differ from each other?
4. How are two structural isomers similar to each other?
5. What physical property of hydrocarbons is used in fractional distillation?
6. What types of hydrocarbons are more abundant in higher-octane gasoline?

7. To how many atoms is a saturated carbon atom bonded?

Unsaturated Hydrocarbons Contain Multiple Bonds

8. What is the difference between a saturated hydrocarbon and an unsaturated hydrocarbon?

9. How many multiple bonds must a hydrocarbon have in order to be classified as unsaturated?

10. Aromatic compounds contain what kind of ring?

Organic Molecules Are Classified by Functional Group

11. What is a heteroatom?

12. Why do heteroatoms make such a difference in the physical and chemical properties of an organic molecule?

13. Which molecule should have the higher boiling point and why:

$$CH_3CH_2CH_2CH_3$$
$$CH_3CH_2CH_2CH_2\text{—}OH$$

14. Why are low-formula-mass alcohols soluble in water?

15. What distinguishes an alcohol from a phenol?

16. What distinguishes an alcohol from an ether?

17. Why do ethers typically have lower boiling points than alcohols?

18. Which heteroatom is characteristic of an amine?

19. Do amines tend to be acidic, neutral, or basic?

20. Are alkaloids found in nature?

21. What are some examples of alkaloids?

22. Which elements make up the carbonyl group?

23. How are ketones and aldehydes related to each other? How are they different from each other?

24. What is one commercially useful property of aldehydes?

25. How are amides and carboxylic acids related to each other? How are they different from each other?

26. From what naturally occurring compound is aspirin prepared?

27. Identify each molecule as hydrocarbon, alcohol, or carboxylic acid:

$$CH_3CH_2CH_2CH_3$$
$$CH_3CH_2CH_2CH_2\text{—}OH$$

Organic Molecules Can Link to Form Polymers

28. What happens to the double bond of a monomer participating in the formation of an addition polymer?

29. What is released in the formation of a condensation polymer?

30. Why is plastic food wrap a stickier plastic than polyethylene?

31. What is a copolymer?

Hands-On Chemistry Insights

Twisting Jellybeans

What you should discover in this activity is that the carbon–carbon double bond greatly restricts the number of possible conformations for an organic molecule. While *n*-butane, for instance, can twist like a snake, 2-butene is restricted to one of two possible conformations. (Check back to Figure 12.7 for the structures of these two molecules.) In one conformation, the two end carbons are on the same side of the double bond—this is called the *cis* conformation. In the second conformation, the two end carbons are on opposite sides of the double bond—the *trans* conformation:

cis-2-Butene   trans-2-Butene

Because the double bond cannot rotate, the *cis* and *trans* conformations are not interconvertible. They therefore represent two different molecules (structural isomers), each having its own unique set of properties. The melting point of *cis*-2-butene, for example, is −139°C, while that of *trans*-2-butene is a warmer −106°C.

Racing Water Drops

You may need to play around with the drops for a while in order to see the differing affinities that the bag and wrap have for water. One way to do this is to tape the polymers side by side stretched out on a sturdy piece of cardboard. Tilt the cardboard to various angles, testing for the speed with which water drops roll down the incline on the two surfaces. Ultimately, you should find that the drops roll more slowly on the wrap (polyvinylidene chloride) than on the bag (polyethylene terephthalate). The source of this greater "stickiness" in the wrap is the fairly large chlorine atoms of the polyvinylidene chloride. Recall from Section 7.1 that the larger the atom, the greater its potential for forming induced dipole molecular interactions.

The greater stickiness of the wrap is also apparent when you try to glide one sheet of wrap over another.

Exercises

1. Which contains more hydrogen atoms: a five-carbon saturated hydrocarbon molecule or a five-carbon unsaturated hydrocarbon molecule?

2. Why does the melting point of hydrocarbons increase as the number of carbon atoms per molecule increases?

3. Draw all the structural isomers for hydrocarbons having the molecular formula C_4H_{10}.

4. Draw all the structural isomers for hydrocarbons having the molecular formula C_6H_{14}.

5. How many structural isomers are shown here:

6. Which two of these four structures are of the same structural isomer:

7. The temperatures in a fractionating tower at an oil refinery are important, but so are the pressures. Where might the pressure in a fractionating tower be greatest, at the bottom or at the top? Defend your answer.

8. Heteroatoms make a difference in the physical and chemical properties of an organic molecule because
   a. they add extra mass to the hydrocarbon structure.
   b. each heteroatom has its own characteristic chemistry.
   c. they can enhance the polarity of the organic molecule.
   d. all of the above.

9. Why might a high-formula-mass alcohol be insoluble in water?

10. What is the percent volume of water in 80-proof vodka?

11. How does ingested methanol lead to the damaging of a person's eyes?

12. One of the skin-irritating components of poison oak is tetrahydrourushiol:

The long, nonpolar hydrocarbon tail embeds itself in a person's oily skin, where the molecule initiates an allergic response. Scratching the itch spreads tetrahydrourushiol molecules over a greater surface area, causing the zone of irritation to grow. Is this compound an alcohol or a phenol? Defend your answer.

13. The phosphoric acid salt of caffeine has the structure

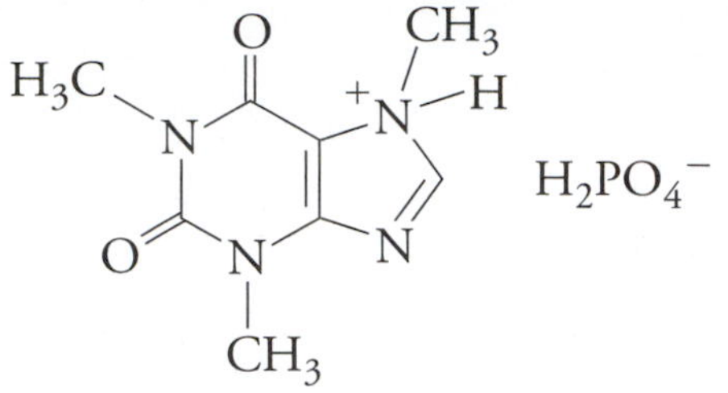

Caffeine–phosphoric acid salt

This molecule behaves as an acid in that it can donate a hydrogen ion, created from the hydrogen atom bonded to the positively charged nitrogen atom. What are all the products formed when 1 mole of this salt reacts with 1 mole of sodium hydroxide, NaOH, a strong base?

14. The solvent diethyl ether can be mixed with water but only by shaking the two liquids together. After the shaking is stopped, the liquids separate into two layers, much like oil and vinegar. The free-base form of the alkaloid caffeine is readily soluble in diethyl ether but not in water. Suggest what might happen to the caffeine of a caffeinated beverage if the beverage was first made alkaline with sodium hydroxide and then shaken with some diethyl ether.

15. Alkaloid salts are not very soluble in the organic solvent diethyl ether. What might happen to the free-base form of caffeine dissolved in diethyl ether if gaseous hydrogen chloride, HCl, were bubbled into the solution?

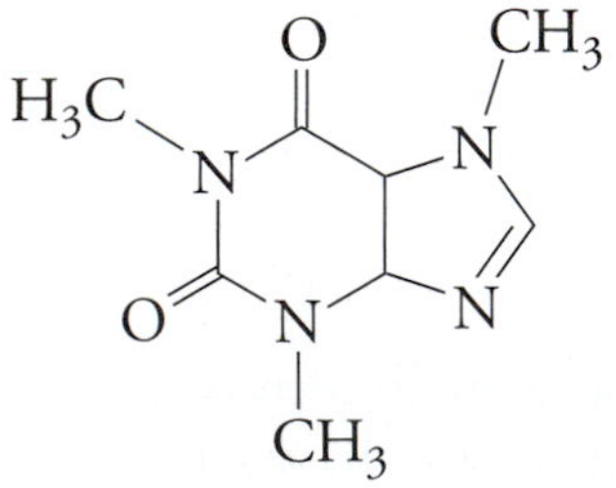

Caffeine
(free base)

16. Draw all the structural isomers for amines having the molecular formula C_3H_9N.

17. Explain why caprylic acid, $CH_3(CH_2)_6COOH$, dissolves in a 5 percent aqueous solution of sodium hydroxide but caprylaldehyde, $CH_3(CH_2)_6CHO$, does not.

18. In water, does the molecule

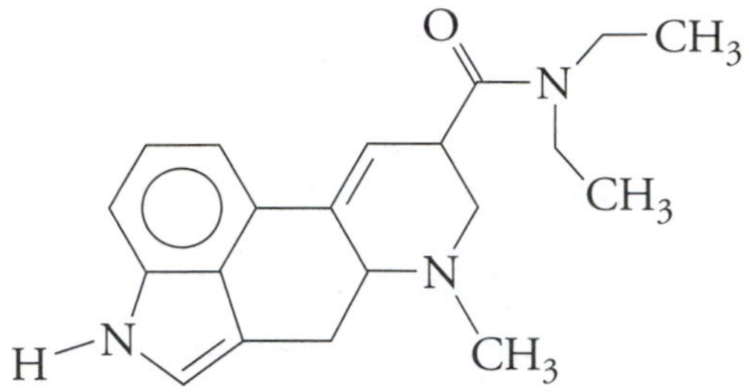

Lysergic acid diethylamide

act as an acid, a base, neither, or both?

19. If you saw the label phenylephrine · HCl on a decongestant, would you worry that consuming it would expose you to the strong acid hydrochloric acid? Explain.

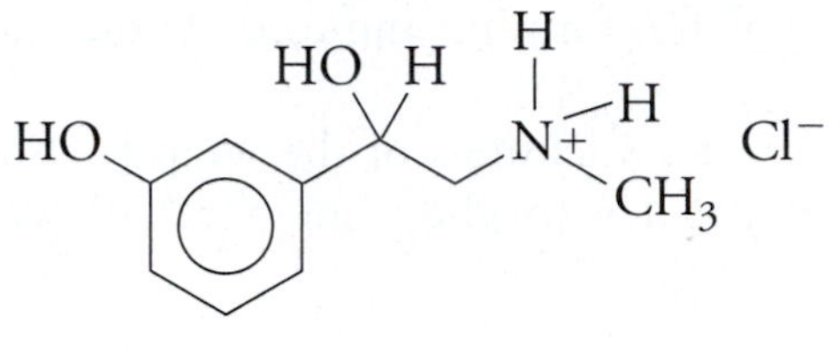

Phenylephrine–hydrochloric acid salt

20. Suggest an explanation for why aspirin has a sour taste.

21. An amino acid is an organic molecule that contains both an amine group and a carboxyl group. At neutral pH, which structure is more likely:

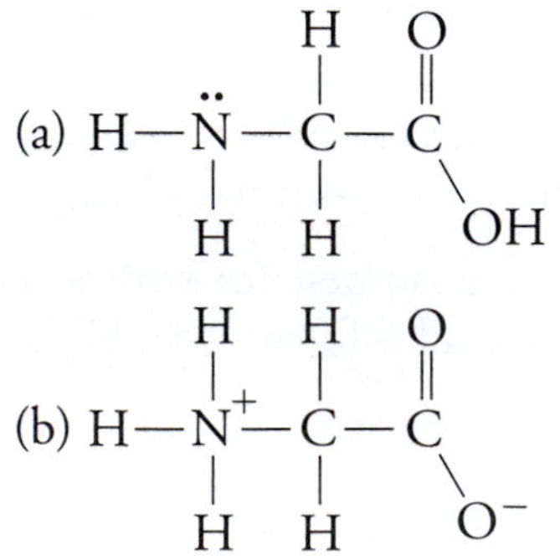

Explain your answer.

22. An amino acid is an organic molecule that contains both an amine group and a carboxyl group. At an acidic pH, which structure is most likely:

(a)
$$H-\overset{\overset{\displaystyle H}{|}}{\underset{\underset{\displaystyle H}{|}}{\ddot{N}}}-\overset{\overset{\displaystyle H}{|}}{\underset{\underset{\displaystyle H}{|}}{C}}-\overset{\overset{\displaystyle O}{\parallel}}{C}\diagdown_{O^-}$$

(b)
$$H-\overset{\overset{\displaystyle H}{|}}{\underset{\underset{\displaystyle H}{|}}{\ddot{N}}}-\overset{\overset{\displaystyle H}{|}}{\underset{\underset{\displaystyle H}{|}}{C}}-\overset{\overset{\displaystyle O}{\parallel}}{C}\diagdown_{OH}$$

(c)
$$H-\overset{\overset{\displaystyle H}{|}}{\underset{\underset{\displaystyle H}{|}}{\overset{+}{N}}}-\overset{\overset{\displaystyle H}{|}}{\underset{\underset{\displaystyle H}{|}}{C}}-\overset{\overset{\displaystyle O}{\parallel}}{C}\diagdown_{OH}$$

Explain your answer.

23. Identify the following functional groups in this organic molecule—amide, ester, ketone, ether, alcohol, aldehyde, amine:

24. Would you expect polypropylene to be denser or less dense than low-density polyethylene? Why?

25. Many polymers emit toxic fumes when burning. Which polymer in Table 12.5 produces hydrogen cyanide, HCN? Which two produce toxic hydrogen chloride gas, HCl?

26. One solution to the problem of our overflowing landfills is to burn plastic objects instead of burying them. What would be some of the advantages and disadvantages of this practice?

27. Which would you expect to be more viscous, a polymer made of long molecular strands or one made of short molecular stands? Why?

28. Hydrocarbons release a lot of energy when ignited. Where does this energy come from?

29. What type of polymer would be best to use in the manufacture of stain-resistant carpets?

30. As noted in the Concept Check on page 389, the compound 6-aminohexanoic acid is used to make the condensation polymer nylon-6. Polymerization is not always successful, however, because of a competing side reaction. What is this side reaction? Would polymerization be more likely in a dilute solution of this monomer or in a concentrated solution? Why?

Exploring Further

P. W. Atkins, *Molecules.* New York: W. H. Freeman, 1987.

An enchanting account of some of the more important organic molecules of nature as well as those produced by chemists. Written for the general public, the dialogue is warm, intriguing, and accompanied by spectacular photographs.

http://www.icco.org

The home page of the International Cocoa Organization. Through this site, you can find answers to many of the questions you may have regarding the chemistry of chocolate and its path from the cocoa tree to your mouth.

http://www.chevron.com/about/learning_center

Web address for the Learning Zone of the Chevron Corporation, where you can find information about crude oil and the refining process.

Organic Compounds
Visit The Chemistry Place at:
www.aw.com/chemplace

13

Chemicals of Life

The Nutrients That Make Up Our Bodies

We absorb molecules from the food we eat and either use them for their energy content or else incorporate them into the various structures that give our bodies both form and function. No molecule in a living organism is a permanent resident, however. Rather, there is perpetual change as molecules ingested in the latest batch of food are transformed to replace older molecules in the organism. Within seven years, for example, most of the molecules in a human body have been replaced by new ones—the body you have today is not the one you had seven years ago!

What then is an individual? If not the molecules themselves, then how about the patterns in which they are assembled? A glance at a set of identical twins tells you the answer to this second question is no. The molecular patterns in any organism are determined by the organism's genetic code, and identical twins have identical genetic codes. Each member in a set of identical twins has its own unique personality, however, despite the fact that the two persons have identical molecular patterns.

Interestingly, the genetic code you have today is the same as the one you had in your yesteryears, but you are now made of a different set of molecules—not unlike two identical twins. Aside from memories, perhaps you are as different from your past self as two identical twins are from each other. Perhaps an individual's identity is continually re-established each and every moment.

Although this chapter cannot promise any insights into the intriguing questions of existence, it will give you a basic understanding of *biomolecules*—the molecules that make up living organisms—and the remarkable roles they play in your body.

13.1 Biomolecules Are Produced and Utilized in Cells

The fundamental unit of almost all organisms is the *cell*. Cells are typically so small that you need a microscope to see them individually. About ten average-sized human cells, for example, could fit within the period at the end of this sentence. Figure 13.1 shows a typical animal cell and a typical plant cell.

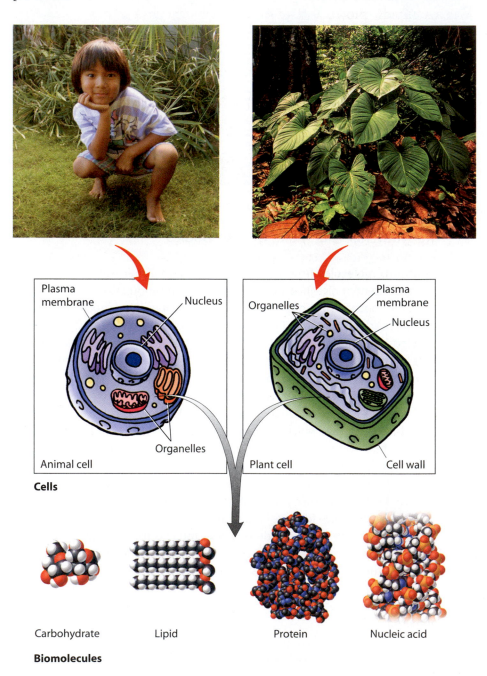

Figure 13.1
Macroscopic, microscopic, and sub-microscopic views of an animal and a plant.

Every cell has a *plasma membrane*. More than just a boundary, the plasma membrane allows molecules to pass into and out of the cell and provides sites where important chemical reactions occur. In animal cells, the plasma membrane is the outermost part of the cell, but the plasma membranes of plant cells are bounded by a rigid *cell wall* that protects the cells and gives them structure.

Housed within each cell is the *cell nucleus*, which contains the genetic code. Everything between the plasma membrane and the cell nucleus is the *cytoplasm*, which consists of a variety of microstructures suspended in a viscous liquid. These microstructures, called *organelles*, work together in the synthesis, storage, and export of important biomolecules.

The great majority of the biomolecules used by cells are either carbohydrates, lipids, proteins, or nucleic acids. In addition, most cells need small amounts of vitamins and minerals in order to function properly. We now discuss all these categories of biomolecules.

13.2   Carbohydrates Give Structure and Energy

Carbohydrates are molecules of carbon, hydrogen, and oxygen produced by plants through photosynthesis. The term *carbohydrate* is derived from the fact that plants make these molecules from carbon (from atmospheric carbon dioxide) and water. The term **saccharide** is a synonym for *carbohydrate*, and a *monosaccharide* ("one saccharide") is the fundamental carbohydrate unit. In most monosaccharides, each carbon atom is bonded to at least one oxygen atom, most often in the form of a hydroxyl group. There are many kinds of monosaccharides. The structures of the two most common ones, glucose and fructose, are shown in Figure 13.2.

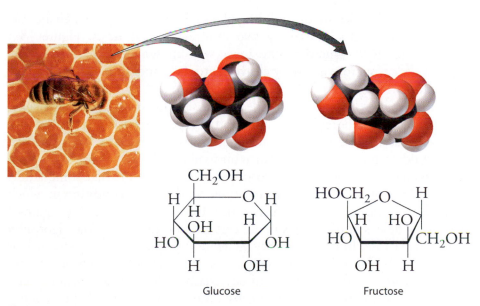

Figure 13.2
Honey is a mixture of the monosaccharides glucose and fructose. Glucose is a six-membered ring, and fructose is a five-membered ring. For simplicity, the stick structures introduced in Chapter 12 are shown below each molecular model.

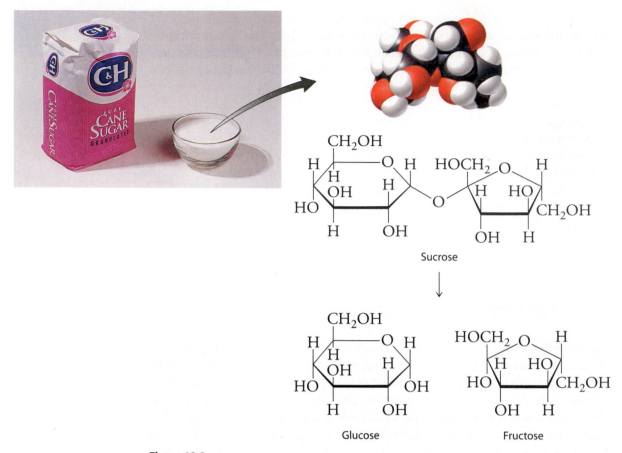

Figure 13.3
Disaccharides, such as sucrose, consist of two chemically bonded monosaccharide units, which are cleaved during digestion.

Monosaccharides are the building blocks of *disaccharides*, which are carbohydrate molecules containing two monosaccharide units. Figure 13.3 shows table sugar—sucrose—the most well-known disaccharide. In the digestive tract, sucrose is readily cleaved into its monosaccharide units, glucose and fructose.

Another important disaccharide is lactose, shown in Figure 13.4. Lactose is the main carbohydrate in milk. In the digestive tract, it is cleaved into the monosaccharides galactose and glucose by the enzyme lactase, which most children produce in abundance up to about the age of six. Thereafter, the production of this enzyme decreases, with the result that some adults produce little or none. This leads to *lactose intolerance*, a condition in which ingestion of milk or milk products causes bloating, flatulence, and painful cramps. These symptoms result as certain intestinal bacteria vigorously digest the lactose. In doing so, they generate large amounts of gases, such as hydrogen, H_2. To relieve these symptoms, some milk products are treated with *Acidophilus bifudus*, a strain of bacteria that destroy gas-causing bacteria in our digestive tract. Aside from abstaining from milk products, an alternative solution is to add a commercially available lactase enzyme to milk or milk products before consuming them.

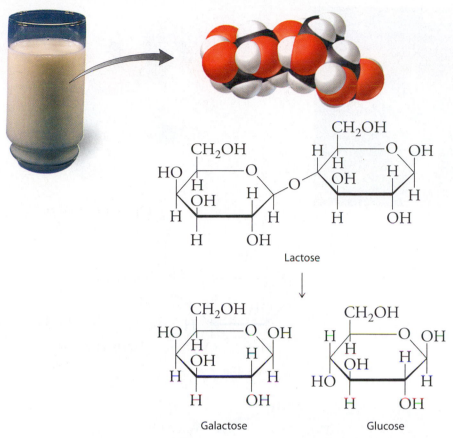

Figure 13.4
Milk and milk products contain the disaccharide lactose, which is digested to galactose and glucose.

Monosaccharides and disaccharides are classified as *simple carbohydrates*, where the word *simple* is used because these food molecules consist of only one or two monosaccharide units. Most simple carbohydrates have some degree of sweetness and are also known as sugars.

Polysaccharides Are Complex Carbohydrates

Recall from Chapter 12 that polymers are large molecules made of repeating monomer units. Monosaccharides are the monomers that link to form the biomolecular polymers called *polysaccharides*, which contain hundreds to thousands of monosaccharide units. Polysaccharides can be built from any type of monosaccharide units. Our bone joints are lubricated by the polysaccharide hyaluronic acid, for example, which consists of alternating glucuronic acid and *N*-acetylglucosamine, as shown in Figure 13.5a on page 404.

The exoskeletons (protective shells) of insects and some marine organisms, such as crabs and shrimp, are made of chitin, a hard, resilient polysaccharide made of the monosaccharide *N*-acetylglucosamine, as shown in Figure 13.5b. Wood varnish once contained chitin from the exoskeletons of insects. In powdered form, chitin is now finding use as a dietary fiber supplement.

Carbohydrates	
Simple	**Complex**
Monosaccharides	*Polysaccharides*
Glucose	Hyaluronic
Fructose	acid
	Chitin
Disaccharides	Starch
Sucrose	Glycogen
Lactose	Cellulose

Figure 13.5
(a) Hyaluronic acid, a lubricant in bone joints, is a polysaccharide consisting of the monosaccharides glucuronic acid and *N*-acetylglucosamine. (b) The exoskeletons of insects, crabs, shrimp, and lobsters are made of chitin, a polysaccharide containing only *N*-acetylglucosamine units.

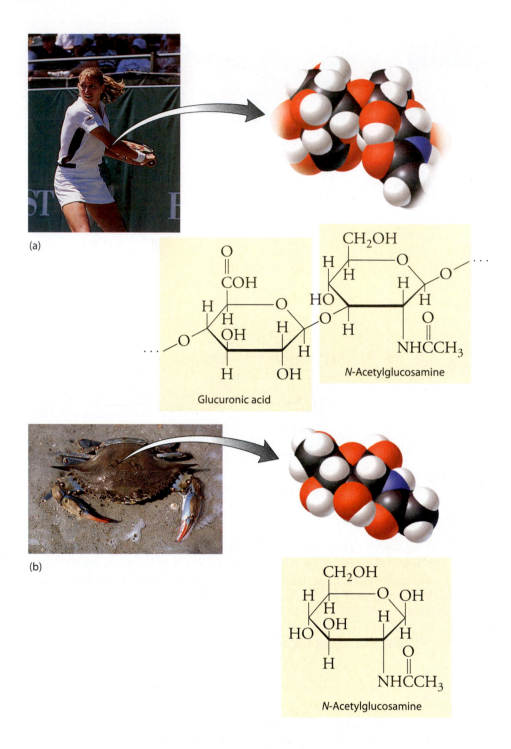

(a)

Glucuronic acid

N-Acetylglucosamine

(b)

N-Acetylglucosamine

Although a polysaccharide can be made of any type of monosaccharide units, the polysaccharides of the human diet are made only of glucose. These polysaccharides include *starch*, *glycogen*, and *cellulose*, which differ from one another only in how the glucose units are chained together. All polysaccharides, but especially the ones in our diet, are known as *complex carbohydrates*, where *complex* refers to the multitude of monosaccharide units linked together.

Concept Check ✓

What makes a simple carbohydrate simple and a complex carbohydrate complex?

Was this your answer? Simple carbohydrates are simple in the sense that they consist of only one or two monosaccharide units per molecule. Complex carbohydrates are complex in the sense of consisting of high numbers of monosaccharide units per molecule.

Starch is a polysaccharide produced by plants to store the abundance of glucose formed during photosynthesis, which is the process whereby plants absorb solar energy and convert it to the chemical energy of sugar molecules. (We shall return to a discussion of photosynthesis in Chapter 15.) On cloudy days or at night, the breakdown of starch polymers to glucose gives the plant a constant energy supply. Animals can also obtain glucose from plant starch, which makes plant starch an all-important food source. Most plants store the starch they produce either in their seeds or in their roots.

A starch molecule may contain up to 6000 glucose units, although this number is highly variable and can be as low as 200. Plants produce two forms of starch, *amylose* and *amylopectin*, as illustrated in Figure 13.6.

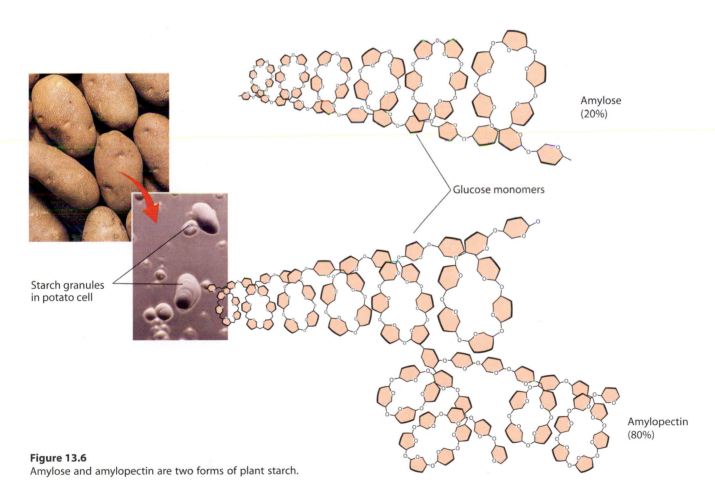

Amylose
(20%)

Glucose monomers

Starch granules
in potato cell

Amylopectin
(80%)

Figure 13.6
Amylose and amylopectin are two forms of plant starch.

In amylose, the glucose units are linked together in chains that coil. In amylopectin, the glucose units are also linked together in coiling chains, but in addition there is periodic branching in the chains. The starch of most starchy foods, such as bread and potatoes, is about 20 percent amylose and 80 percent amylopectin. As these foods are digested, glucose units are broken off the ends of the chains. In Figure 13.6, you can see that, because of its branching, amylopectin has more ends than amylose does. Therefore, amylopectin releases glucose units at a faster rate than amylose does.

If you hold a piece of bread in your mouth for a few minutes, it begins to taste sweet, which is a signal that digestion has begun and glucose is being released.

Hands-On Chemistry: Spit in Blue

When iodine is mixed with starch, the amylose strands coil around the iodine molecules, forming a starch–iodine complex that has a characteristic dark blue color. The more starch present in a solution, the deeper the blue. The formation of this color is used to identify the presence of starch. In this activity, you will use an iodine solution to test for the presence of amylose and the amylose-digesting action of your saliva.

What You Need

Potato, saucepan, water, two drinking glasses, iodine solution (from drugstore)

Safety Note

While moderate topical application of disinfecting iodine solution is regarded as safe, ingestion is not. So keep your iodine solution away from children, and be sure not to ingest any yourself.

Procedure

① Prepare a solution of starch by boiling several potato slices in about two cups of water. Add a teaspoon of the solution to each glass and dilute with a tablespoon of water.

② Collect a good wad of saliva in your mouth, and then gracefully spit into one of the glasses. Swirl to mix.

③ Wait for a few minutes, and then add a drop of iodine solution to each glass. Swirl to mix while looking for a difference in color between the two samples. If no difference is noted, repeat the experiment several times, changing one parameter each time, such as the number of potato slices boiled.

You should see a noticeable difference between the intensity of blue in the glass containing your saliva and the other glass. The intensity being directly proportional to the amount of starch present, which glass should contain the darker solution? Why?

Concept Check ✓

When amylose and amylopectin are digested, which one is a more immediate supply of glucose?

Was this your answer? There is more branching in amylopectin and therefore a greater number of ends per molecule. The digestion of amylopectin produces more glucose molecules per unit of time and therefore is a quicker and more immediate supply of glucose.

Animals store some of their excess glucose by converting it to **glycogen**, a polymer made of hundreds of glucose monomers and sometimes referred to as *animal starch*. Glycogen has a structure much like that of amylopectin, but there is more branching in glycogen, as shown in Figure 13.7. Between meals, when glucose levels drop, the body metabolizes glycogen to glucose. Glycogen therefore serves as our glucose reserve. The areas of the body most abundant in glycogen are the liver and muscle tissue.

Cellulose, a structural component of plant cell walls, is also a polysaccharide of glucose. The glucose in cellulose, however, is slightly different from the glucose in starch and glycogen. In the glucose of starch and glycogen, called α-glucose, the hydroxyl group highlighted in Figure 13.8 on page 408 faces in one direction. In the glucose of cellulose, called β-glucose, this hydroxyl group faces in the opposite direction. Figure 13.8 illustrates how the linking between α-glucose units results in a coiling of the polysaccharide strands, while the link-

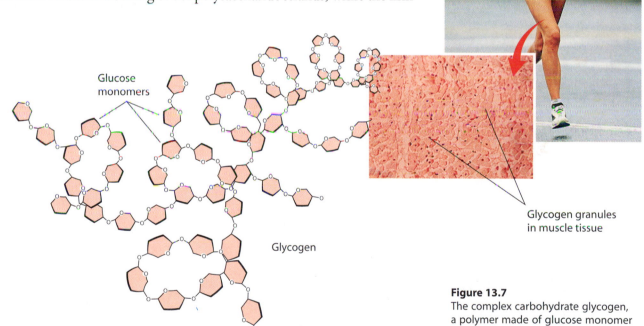

Glucose monomers

Glycogen

Glycogen granules in muscle tissue

Figure 13.7
The complex carbohydrate glycogen, a polymer made of glucose monomer units, is found in animal tissue.

ing between β-glucose units results in straight chains, with no coiling possible. In addition to no coiling, there is also no branching in the chains. These two attributes of cellulose molecules allow the polysaccharide strands to align much like strands of uncooked spaghetti. This alignment maximizes the number of hydrogen bonds between strands, which makes cellulose a tough material. Added strength is given as the plant lays down microscopic fibers of cellulose in a criss-cross fashion, as shown in Figure 13.9 on page 408.

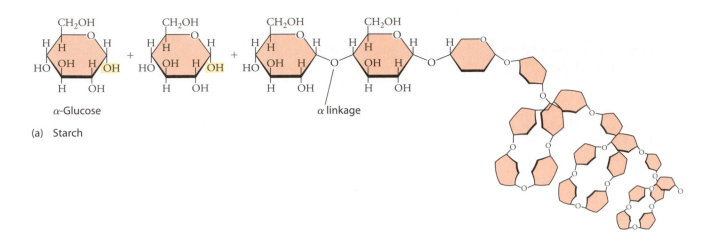

(a) Starch

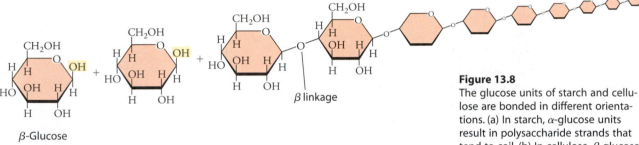

(b) Cellulose

Figure 13.8
The glucose units of starch and cellulose are bonded in different orientations. (a) In starch, α-glucose units result in polysaccharide strands that tend to coil. (b) In cellulose, β-glucose units result in polysaccharide strands that do not coil and as a result can align with one another.

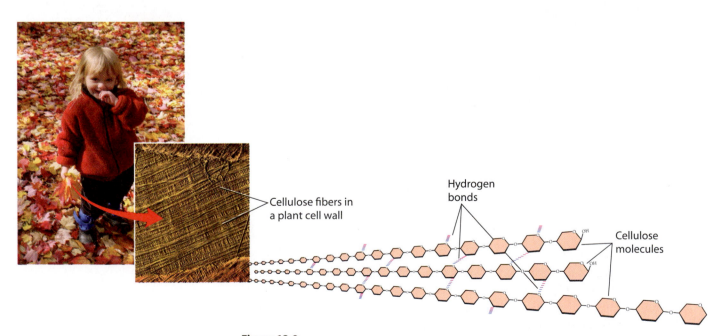

Figure 13.9
Strands of cellulose in a plant, including the plant leaves held by Maitreya, are joined by hydrogen bonds. These microscopic fibers are laid down in a criss-cross pattern to give strength in many directions.

Cellulose serves as the primary structural component of all plants. Cotton is nearly pure cellulose. Wood, made largely of cellulose, can support trees that are as much as 30 meters tall. Cellulose is by far the most abundant organic compound on the Earth.

Most animals, including humans, are not able to break cellulose down to glucose. Instead, the cellulose in the food we eat serves as dietary fiber that helps in regulating bowel movements. In the large intestine, cellulose-based fiber absorbs water and has a laxative effect. Waste products are therefore moved along faster, and with them harmful bacteria and carcinogens. Microorganisms that live in the digestive tracts of wood-eating termites and grass-eating ruminants (cows, sheep, and goats) can break cellulose down to glucose. Strictly speaking, termites and ruminants do not digest cellulose. Rather, they digest the glucose produced by the cellulose-digesting microorganisms that live inside them, as shown in Figure 13.10.

Figure 13.10
Termites digest glucose produced by cellulose-digesting microorganisms living in the termites' digestive tract.

13.3 Lipids Are Insoluble in Water

Lipids are a broad class of biomolecules. Although structurally diverse, all lipids are insoluble in water because each lipid molecule contains a large number of nonpolar hydrocarbon units. In this section we discuss two important types of lipids: *fats* and *steroids*.

Fats Are Used for Energy and Insulation

A **fat** is any biomolecule formed from the reaction between a glycerol molecule, $C_3H_8O_3$, and three fatty acid molecules, as shown in Figure 13.11 on page 410. A *fatty acid* is a long-chain hydrocarbon terminating in a carboxylic acid group. The chain is typically between 12 and 18 carbon atoms long and may be either saturated or unsaturated. Recall from Chapter 12 that a *saturated* chain contains no carbon–carbon double bonds; an *unsaturated* chain contains one, two, or three carbon–carbon double bonds. Note that, like carbohydrates, fats contain only carbon, hydrogen, and

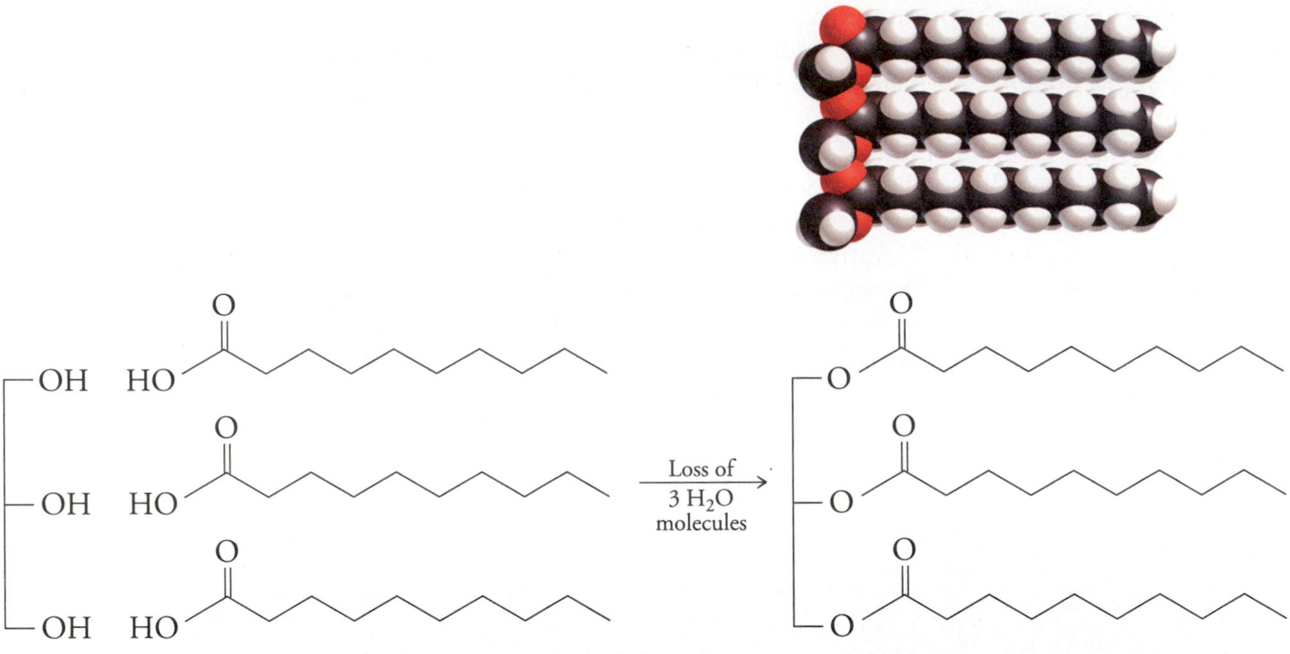

Figure 13.11
A typical fat molecule, also known as a triglyceride, is the combination of one glycerol unit and three fatty acid molecules. Note that this reaction involves the formation of three ester functional groups.

Figure 13.12
The walrus and other polar species are insulated from the cold by a thick layer of fat beneath their skin.

oxygen. Fats and carbohydrates have similar compositions because both plants and animals synthesize fats from carbohydrates. There is a significant difference in the structures of these two types of biomolecules, however, as you can readily see by comparing Figures 13.8 and 13.11. Because fats are made from three fatty acids and glycerol, they are commonly referred to as *triglycerides*.

Fats are stored in the body in localized regions known as fat deposits. These deposits serve as important reservoirs of energy. Fat deposits directly under the skin help to insulate us from the cold, good news for the walrus shown in Figure 13.12. In addition, vital organs, such as the heart and kidneys, are cushioned against injury by fat deposits.

The digestion of fat is accompanied by the release of considerably more energy than is produced by the digestion of an equivalent amount of either carbohydrate or protein. There is about 38 kilojoules (9 Calories) of energy in 1 gram of fat but only about 17 kilojoules (4 Calories) of energy in 1 gram of carbohydrate or protein. (Recall that the energy content of food is often reported in Calories, with an uppercase C, where 1 Calorie = 1 kilocalorie = 1000 calories.)

Concept Check ✔

Give two reasons animals living in cold climates tend to form a thick layer of fat just prior to the onset of winter.

Was this your answer? Fat provides a source of energy during the winter, when food is generally scarce, and it provides insulation from cold winter temperatures.

Fats made from saturated fatty acids are referred to as saturated fats. As you can see in Figure 13.13a, saturated fat molecules are able to pack tightly together because their fatty acids point straight out and align with one another. Induced dipole–induced dipole attractions hold the aligned chains together. This gives saturated fats, such as lard, relatively high melting points, and as a result they tend to be solid at room temperature. The fatty acid chains of unsaturated fats—those made from unsaturated fatty acids—are "kinked" wherever double bonds occur, as shown in Figure 13.13b. The kinks inhibit alignment, and as a result unsaturated fats tend to have relatively low melting points. These fats are liquid at room temperature and are commonly referred to as *oils*. Most vegetable oils are liquid at room temperature because of the high proportion of unsaturated fats they contain.

The fat from an animal or plant is a mixture of different fat molecules having various degrees of unsaturation. Fat molecules containing only one carbon–carbon double bond per fatty acid chain are *monounsaturated*.

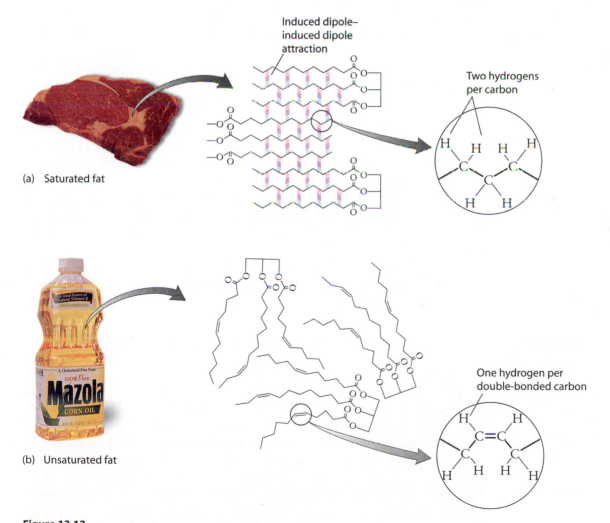

(a) Saturated fat

(b) Unsaturated fat

Figure 13.13
(a) Saturated fats are typically solid at room temperature because of molecular attractions between fatty acid chains. (b) Unsaturated fats are typically liquid at room temperature because molecular attractions are inhibited by the kinked nature of the fatty acid chains.

Those containing more than one double bond per fatty acid chain are *polyunsaturated*. Table 13.1 shows the percentage of saturated, monounsaturated, and polyunsaturated fats in a number of widely used dietary fats.

Table 13.1

Degree of Unsaturation in Some Common Fats

Fat	Percentage of Total Fatty Acid Content		
	Saturated	**Monounsaturated**	**Polyunsaturated**
Coconut	93	6	1
Palm	57	36	7
Lard	44	46	10
Cottonseed	26	22	52
Peanut	21	49	30
Olive	15	73	12
Corn	14	29	57
Soybean	14	24	62
Sunflower	11	19	70
Safflower	10	14	76
Canola oil	6	58	36

Steroids Contain Four Carbon Rings

Steroids are a class of lipids that have in common a system of four linked carbon rings. Cholesterol, shown in Figure 13.14, is the most abundant steroid by far, and it serves as the starting material for the biosynthesis of most all other steroids, including the sex hormones estradiol and testosterone, also shown in Figure 13.14. *Hormones* are chemicals produced by one part of the body to influence other parts of the body. For example, estradiol, produced by the ovaries, and testosterone, produced by the testes, are responsible for the development of secondary sex characteristics in other parts of the body.

Cholesterol is found throughout the body. In fact, the human brain is about 10 percent cholesterol by weight. Our bodies synthesize cholesterol in the liver. In addition, we obtain cholesterol through a diet of animal products.

Many synthetic steroids having a wide variety of biological effects have been prepared. Prednisone, for instance, is often prescribed as an anti-inflammatory in the treatment of arthritis. Also available are synthetic steroids that mimic the muscle-building properties of testosterone. These are the *anabolic steroids*, which physicians prescribe to assist patients suffering from hormone imbalances and those recovering from severe starvation. Athletes have found that these steroids improve performance, but they have many negative side effects, including impotence, changes in sexual desires and characteristics, and liver toxicity. Most sports organizations have banned the use of anabolic steroids because of their deleterious effects. Any athlete should carefully consider whether the short-term gains provided by these agents are worth the long-term losses. For further information, start by exploring the steroid Web site of the National Institute on Drug Abuse at http://www.steroidabuse.org.

Lipids	
Fats	**Steroids**
Saturated	Cholesterol
Lard	Testosterone
	Estradiol
Unsaturated	
Vegetable oils	

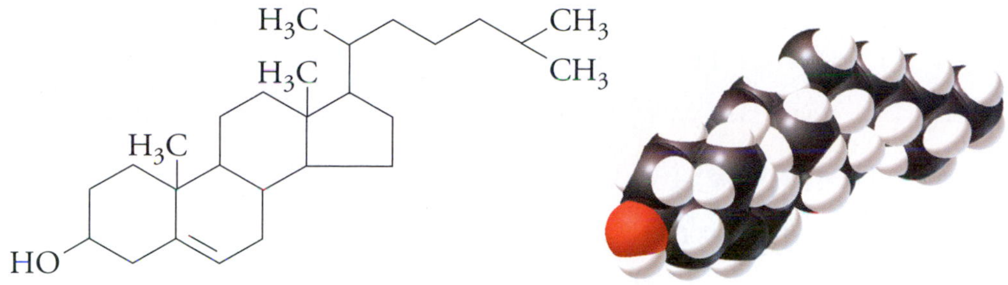

Cholesterol

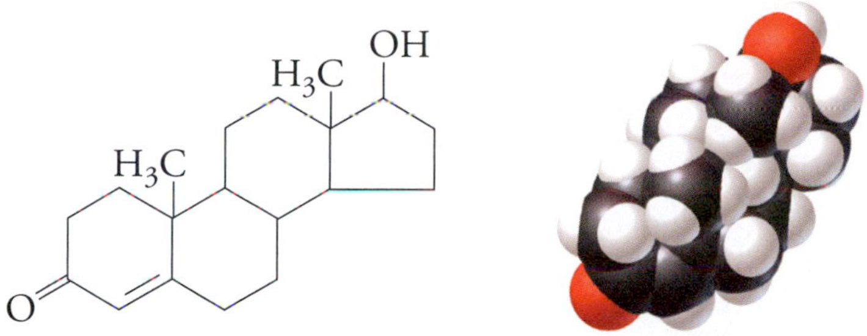

Testosterone

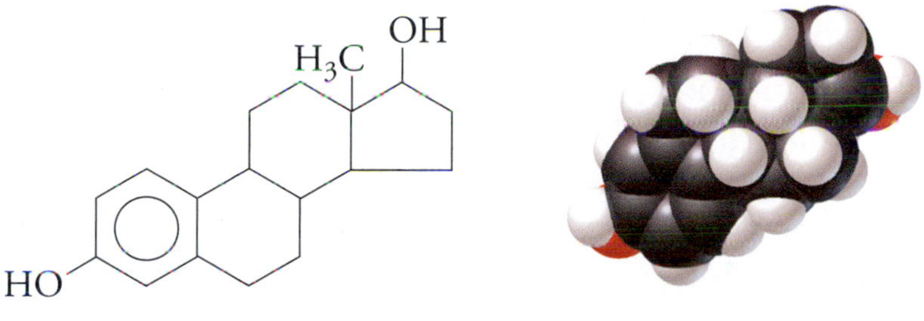

Estradiol

Figure 13.14
The sex hormones estradiol and testosterone are steroids that the body produces from the most abundant steroid of all—cholesterol.

Concept Check ✔

How are fats and steroids similar to each other?

Was this your answer? They are both lipids and, hence, insoluble in water.

13.4 Proteins Are Polymers of Amino Acids

Proteins are large polymeric biomolecules made of monomer units called amino acids. An **amino acid** consists of an amine group and a carboxylic acid group bonded to the same carbon atom, as shown in Figure 13.15. A side group is also attached to the same carbon. All proteins are made from 20 amino-acid building blocks, and these 20 amino acids differ from one another by the chemical identity of their side groups, as shown in Figure 13.16.

Amino acids are linked together by *peptide bonds*, which are formed in a condensation reaction in which the carboxylic acid end of one amino

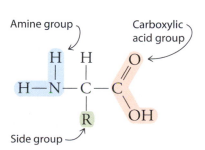

Figure 13.15
The general structure of an amino acid, with R representing the side group that makes each amino acid different from all others.

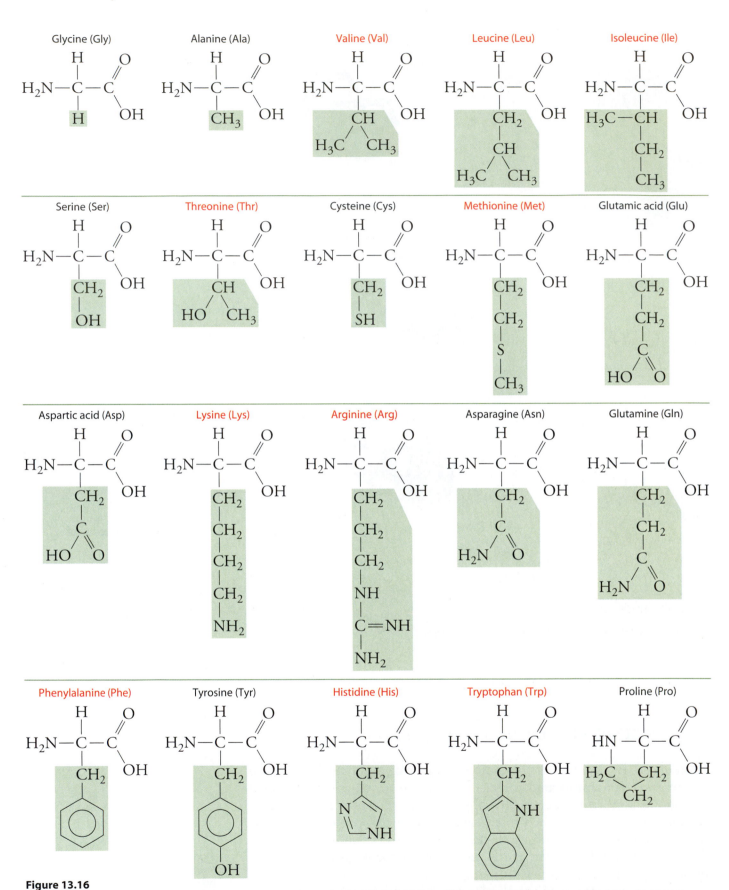

Figure 13.16
These 20 amino acids are the building blocks of proteins. The side groups are highlighted in green, and the names printed in red indicate *essential* amino acids, discussed in Section 13.8.

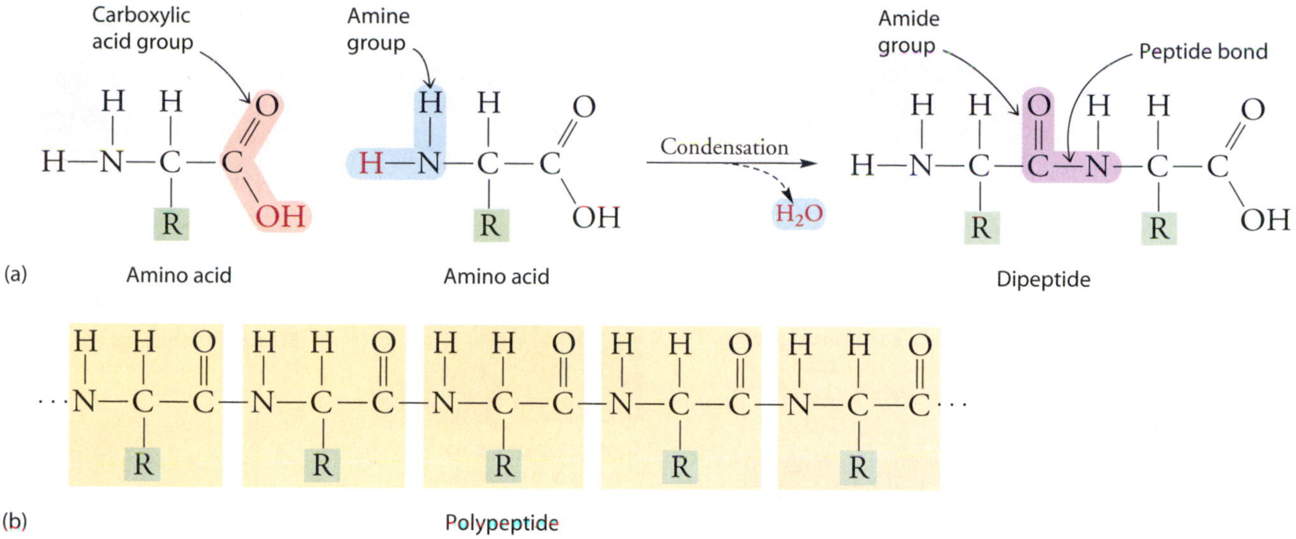

Figure 13.17
(a) Formation of a peptide bond from the condensation of two amino acids. The resulting dipeptide contains an amide group. (b) A polypeptide is many amino acids linked together by peptide bonds.

acid is joined to the amine end of a second amino acid, as illustrated in Figure 13.17. (Recall from Section 12.4 that a condensation reaction is marked by the loss of a small molecule, such as water.) A group of amino acids linked together in this fashion is called a *peptide*. The number of amino acids in a peptide is indicated by a prefix. A dipeptide is made from two amino acids, a tripeptide from three, a tetrapeptide from four, and so on. Peptides of ten or more amino acids are generally called *polypeptides*. Proteins are naturally occurring polypeptides that have some biological function, and they consist of large numbers of amino acids, up to several hundred. For example, the molecular formula of one of the proteins found in milk is $C_{1864}H_{3012}O_{576}N_{468}S_{21}$, which gives you an idea of the size and complexity of some protein molecules.

Plant and animal tissues contain proteins both in solution and in solid form. Dissolved proteins reside in the liquid found inside cells and in other liquids in the body, such as blood. Solid proteins form skin, muscles, hair, nails, and horns. The human body contains many thousands of different proteins, such as those in Figure 13.18 on page 416.

Amino acid: a small biomolecule

Polypeptide: a polymer of amino acids

Protein: a polypeptide having biological function

Protein Structure Is Determined by Attractions Between Neighboring Amino Acids

The structure of proteins determines their function and can be described on four levels, illustrated on page 417. The *primary structure* is the sequence of amino acids in the polypeptide chain. The *secondary structure* describes how various short portions of a chain are either wrapped into a coil called an *alpha helix* or folded into a thin *pleated sheet*. The *tertiary structure* is the way in which an entire polypeptide chain may either twist into a long fiber or bend into a globular clump. The *quaternary structure* describes how separate proteins may join to form one larger complex. Each level of structure is determined by the level before it, which means that ultimately it is the

Some proteins are **hormones** that regulate body metabolism and growth.

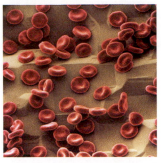

Hemoglobin, a **transport protein**, is the component of blood that takes the oxygen we breathe to our cells.

Contractile proteins in muscles allow us to move.

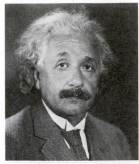

Structural proteins are found in skin, hair, and bones.

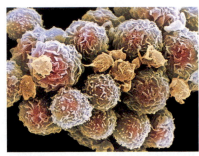

Storage proteins serve as a source of amino acids in milk.

Storage proteins serve as a source of amino acids in milk.

White blood cells produce **antibodies**, which are proteins that fight infections.

Enzymes are proteins that act as catalysts for reactions in the body, including the digestion of food.

Figure 13.18
The variety of proteins in the human body.

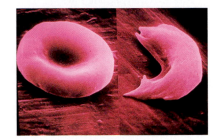

Figure 13.19
On the left is a red blood cell containing the normal hemoglobin protein. On the right, a sickled red blood cell containing hemoglobin that has one out-of-place amino acid. The curved shape is reminiscent of the curved-blade harvesting tool known as a sickle.

sequence of amino acids that creates the overall protein shape. This final shape is maintained both by chemical bonds and by weaker molecular attractions between amino acid side groups.

In long polypeptides, the number of possible variations in primary structure is astronomical. For example, the number of 20-unit polypeptides possible from a single set of 20 different amino acids is a whopping 2.43×10^{18}! The number of possible polypeptides when more than 100 units are being combined is seemingly infinite. This diversity is exactly what is needed for building a living organism.

Although functioning proteins have very specific amino acid sequences, slight variations can often be tolerated. In some cases, however, a slight variation can be disastrous. For example, some people have a version of hemoglobin—a protein found in red blood cells—that has one incorrect amino acid in about 300. That "minor" error is responsible for sickle-cell anemia, an inherited condition with painful and often lethal effects. The sickle shape characteristic of this disease is shown in Figure 13.19.

Attractions between neighboring amino acids in a polypeptide chain are what cause the local contortions that constitute the secondary structure of the polypeptide. This secondary structure takes its alpha-helix form when simpler amino acids, such as glycine and alanine, are grouped together along the polypeptide chain. As shown in Figure 13.20, the shape

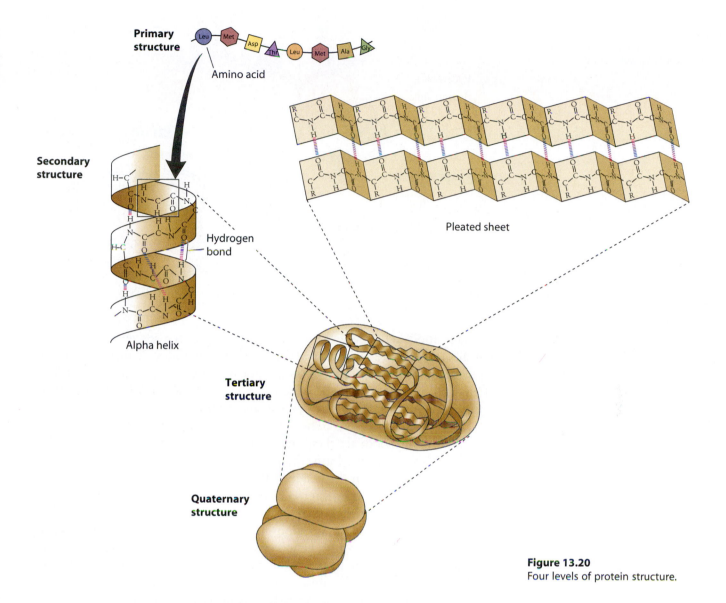

Figure 13.20
Four levels of protein structure.

of the alpha helix is maintained by hydrogen bonds between successive turns of the spiral. Proteins containing many alpha helices, like wool, can be stretched because of the springlike properties of the alpha helix. The pleated-sheet version of secondary structure forms when predominately nonpolar amino acids, such as phenylalanine and valine, are grouped together. Proteins containing many pleated sheets, like silk, are strong and flexible but not easily stretched.

Secondary structure can vary from one portion of a polypeptide chain to another. (For example, one part of a given chain may have a helical secondary structure while another part of the same chain has a pleated-sheet secondary structure).

Tertiary structure refers to the way an entire chain is shaped. As with secondary structure, tertiary structure is also maintained by various chemical attractions between amino acid side groups. For some proteins, such as the hypothetical one shown in Figure 13.21 on page 418, these attractions include disulfide bonds between opposing cysteine amino acids. Also important are the ionic bonds (also known as *salt bridges*) that occur

Figure 13.21
Chemical forces that maintain the tertiary structure of a polypeptide.

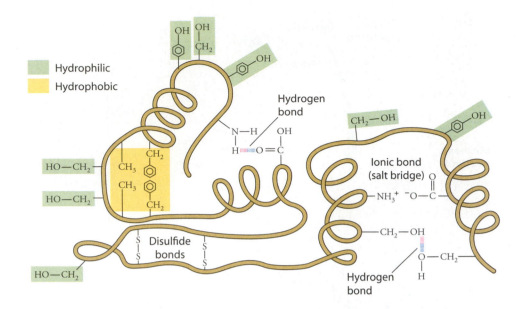

between oppositely charged ions. Hydrogen bonding between side groups can also contribute to tertiary structure, as can induced dipole–induced dipole attractions between nonpolar side groups, such as those found in phenylalanine and valine. Because these latter attractions tend to exclude water, they are also known as *hydrophobic attractions*.

Hydro*philic* attractions between a protein and an aqueous medium, such as cytoplasm or blood, also help maintain tertiary structure. In a protein dissolved in an aqueous medium, the polypeptide chain is folded so that nonpolar side groups are on the inside of the molecule and polar side groups are on the outside, where they interact with the water.

In keratin, the protein that is the main component of hair and fingernails, an important force shaping tertiary structure is disulfide bonds cross-linking adjacent alpha helices, as shown in Figure 13.22. The more disulfide cross-links in keratin, the tougher this protein is. In general, thicker hair has more disulfide cross-links than does fine hair. Also, fingernails are hard because of extensive disulfide cross-linking.

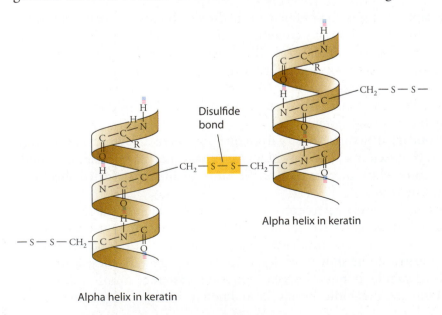

Figure 13.22
Parallel polypeptide chains may be cross-linked by a disulfide bond between two cysteines.

Disulfide cross-links enable hair to hold a particular shape, such as a curl. Figure 13.23 illustrates a permanent wave, which can modify the degree of curl in hair. The hair is first treated with a reducing agent that cleaves some of the disulfide cross-links. This is usually a smelly step because some of the sulfur is reduced to odorous hydrogen sulfide. Cleaving disulfide cross-links, however, allows the keratin to become more flexible. The hair is then set into the desired shape, using rollers if the desired shape is curls or boards if the desired shape is straight. An oxidizing agent is then applied to restore the disulfide cross-links, which hold the hair in its new orientation.

Hydrogen bonds between adjacent alpha helices also play an important role in making keratin a tough material. When keratin gets wet, these hydrogen bonds are disrupted, which is why fingernails soften in water. Hair also softens in water. As water molecules slip between the alpha helices, the polypeptide strands slide past one another to the extent permitted by the disulfide cross-links. As water molecules evaporate, hydrogen bonds between alpha helices are re-established, and the hair toughens to its original shape—unless, of course, the hair is mechanically held in a different shape by, for example, rollers. Hair fashioned by wetting maintains its new shape only temporarily, however, because disulfide cross-links ultimately pull the hair back to its natural look. Interestingly, a small amount of water in hair enhances the molecular attractions between alpha helices. Curls therefore hold longer in humid climates than in dry climates.

Many proteins consist of two or more polypeptide chains. Such proteins have a quaternary structure resulting from bonding and interactions among these chains. A good example is hemoglobin (Figure 13.24), the

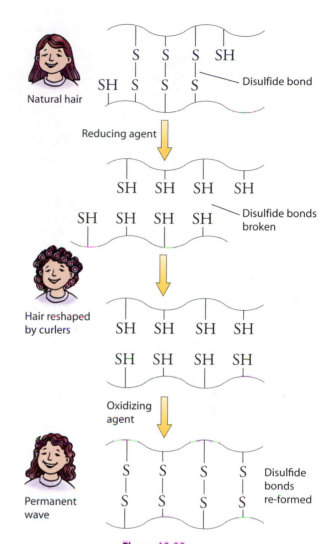

Figure 13.23
A permanent wave breaks and re-forms disulfide bonds in hair.

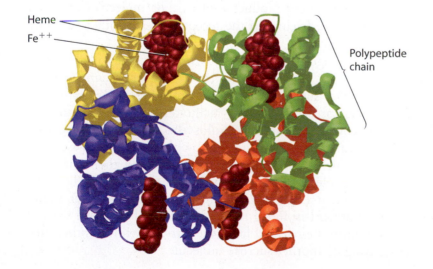

Figure 13.24
Computer-generated model of the quaternary structure of hemoglobin, a protein consisting of four interlinked polypeptide chains, each shown in a different color. Computers are important tools for the study of biomolecules, as they help scientists visualize complex three-dimensional structures.

oxygen-holding component of red blood cells. It is a complex of four polypeptides inside of which four iron-bearing heme groups are tightly nestled.

Concept Check ✔

Distinguish among hemoglobin's primary, secondary, tertiary, and quaternary structures.

Was this your answer?   Hemoglobin's primary structure is its sequence of amino acids along each polypeptide. The twisting of each polypeptide into an alpha helix is its secondary structure. The folding up of the full length of each alpha helix into a globular shape is its tertiary structure. The combined four polypeptides is the quaternary structure.

Proteins are viable only under very specific conditions, such as a particular pH and temperature. Changes in these conditions can break the chemical attractions within a protein and thereby cause a loss of structure, which necessarily means a loss of biological function. A protein with lost structure is said to be *denatured*. A hard-boiled egg, for example, consists of denatured proteins that can in no way support the development of a chick. The same atoms are there, but, as usual, it is the arrangement of atoms and their spatial orientations that make all the difference.

Enzymes Are Biological Catalysts

Enzymes are a class of proteins that catalyze (speed up) biochemical reactions. Their function has everything to do with their structure. Wad up a piece of paper, and you will find many nooks and crannies. In a similar fashion, there are nooks and crannies on the surface of an enzyme. Some of these sites, called *receptor sites*, are special in that reactant molecules, called *substrates*, are able to fit inside them. Like a hand in a glove, a substrate molecule must have the right shape in order to fit a receptor site. Molecular attractions, such as hydrogen bonding, then hold the substrate molecule in the receptor site, where the substrate is activated for reaction. The resulting product molecule or molecules are then released, freeing up the receptor site for other substrate molecules.

Figure 13.25 shows how an enzyme called sucrase breaks sucrose down to its monosaccharide units. Once sucrose binds to the empty receptor site on the sucrase, the enzyme facilitates the breaking of the covalent bond that holds the glucose and fructose units together. It does so by holding the sucrose molecule in a certain conformation and then changing the electronic characteristics of this covalent bond so that it easily breaks when attacked by water molecules, which are ever-present in living tissue. In the final step, the severed glucose and fructose units are released from the enzyme, which is then free to catalyze the splitting of another sucrose molecule.

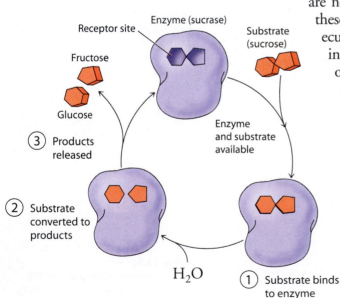

Figure 13.25
Upon binding to the receptor site on the enzyme sucrase, the substrate sucrose is split into its two monosaccharide units, glucose and fructose.

While some enzymes, such as sucrase, split substrate molecules apart, others join substrate molecules together. In all cases, enzymes are so efficient that a single enzyme molecule may act on thousands or even millions of substrate molecules per second. Without enzymes, most biochemical reactions would not occur at rates fast enough to support life.

A chemical that interferes with an enzyme's activity is called an *inhibitor*. Inhibitors work by binding to an enzyme and thereby preventing a substrate from binding. They are important regulators of cell metabolism. In many instances, an inhibitor is the very product created in an enzyme-catalyzed reaction. Once the enzyme produces a given concentration of product, it begins to shut down because the product molecules kick in as inhibitors. As we shall see in the next chapter, many drugs act either by inhibiting enzymes or by mimicking an enzyme's natural substrate.

Proteins (polypeptides with biological functions)	
Structure	**Description**
Primary	Sequence of amino acids
Secondary	Alpha helix or pleated sheet
Tertiary	Overall shape of single polypeptide
Quaternary	Complex of two or more polypeptides

13.5   Nucleic Acids Code for Proteins

Our bodies are built of proteins—from microscopic cell constituents, such as enzymes and organelles, to such macroscopic structures as bones, hair, skin, and teeth. The number of different ways amino acids can link together to make proteins is fantastic. Yet, somehow our bodies are able to assemble amino acids in just the right order to build proteins that have highly functional structures.

In order to understand how the myriad specific proteins in our bodies are built, you need to know about nucleotides and nucleic acids. A **nucleotide**, represented in Figure 13.26, is a structural unit consisting of a phosphate group, a ribose sugar, and a nitrogenous base. (Don't worry about these unfamiliar words. *Ribose* simply means a sugar containing five carbon atoms, and *nitrogenous* means "containing nitrogen atoms.") A **nucleic acid** is a polymer made up of nucleotide monomers.

Proteins are built by nucleic acids, which are found in almost every cell of an organism and hold the instructions for building every sort of protein.

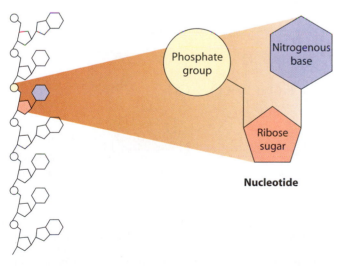

Nucleotide

Nucleic acid

Figure 13.26
A nucleic acid is a long polymeric chain of nucleotides, each nucleotide consisting of a nitrogenous base, a ribose sugar, and a phosphate group.

An organism's nucleic acids are inherited from its parents, which is why the organism resembles its parents. Thus, like begets like—bears have baby bears, seals have baby seals, and humans have baby humans.

Figure 13.27 illustrates the two major classes of nucleic acids, distinguished from each other by the type of ribose sugar in the nucleotide monomers. Those without an oxygen atom on one of the carbon atoms in the ring of the ribose sugar are **deoxyribonucleic acids**, or simply DNA. These polymers, which are the primary source of genetic information in plants and animals, are found in the cell nucleus as well as in certain organelles known as mitochondria. Nucleic acids that have an oxygen atom on the carbon of the ribose sugar, as shown in Figure 13.27b, are **ribonucleic acids** (RNA). These polymers occur mostly outside the cell nucleus in the cytoplasm, where they piece together amino acids to make proteins.

There are four types of nucleotides in a DNA polymer (Figure 13.27a). These four nucleotides differ from one another in the nitrogenous base they contain, which may be adenine (A), guanine (G), cytosine (C), or thymine (T). There are also four types of RNA nucleotides used to build the RNA polymer (Figure 13.27b). The RNA nucleotides contain the same nitrogenous bases as found in DNA nucleotides except for thymine. Instead of thymine, RNA nucleotides contain the nitrogenous base uracil (U), whose structure is only slightly different from that of thymine.

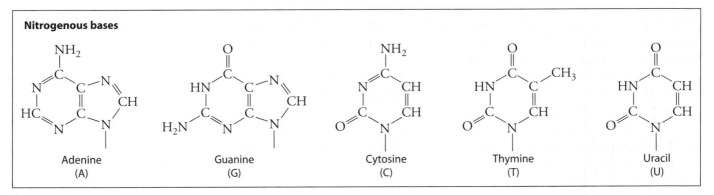

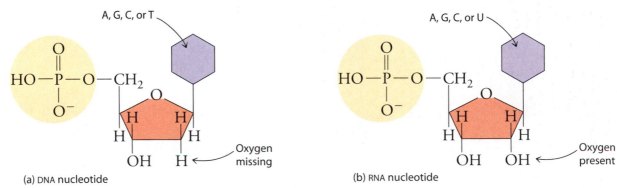

Figure 13.27

(a) The nucleotides of DNA have ribose sugars that are missing an oxygen on one of the carbons. The nitrogenous bases of DNA are adenine, guanine, cytosine, and thymine. (b) The nucleotides of RNA have fully oxygenated ribose sugars. The nitrogenous bases of RNA are adenine, guanine, cytosine, and uracil.

Concept Check ✔

What are the two ways in which DNA nucleotides differ structurally from RNA nucleotides?

Was this your answer? All DNA nucleotides lack an oxygen atom on the ribose sugar. Also, the nitrogenous base in a DNA nucleotide may be adenine, guanine, cytosine, or thymine. The nitrogenous base in an RNA nucleotide may be adenine, guanine, cytosine, or uracil.

DNA Is the Template of Life

The story of our modern-day understanding of genetics began in the 1850s in an abbey garden, where a monk named Gregor Mendel (1822–1884) documented how varieties of sweet peas could pass traits, such as flower color, from one generation to the next. From Mendel's work arose the idea that heritable traits are passed from parents to offspring in discrete units called *genes*. In the early 1900s, researchers correlated Mendel's heritable genes to cellular microstructures known as **chromosomes**, which are elongated bundles of DNA and protein that form whenever a cell is getting ready to divide (Figure 13.28). Each gene, it was found, resides at a specific location on a particular chromosome. Offspring inherit genes by receiving replicates of their parents' chromosomes.

Up until the 1940s, it was not known whether the DNA portion or the protein portion of the chromosomes was the carrier of genetic information. Most investigators were inclined to believe that proteins, with their great diversity, were the carriers. In the 1940s, however, DNA was found to be a polymer containing adenine, guanine, cytosine, and thymine. Along the DNA chain, these bases could be sorted in any order. This potential variability in base sequencing opened up the possibility that DNA was the carrier of genetic information.

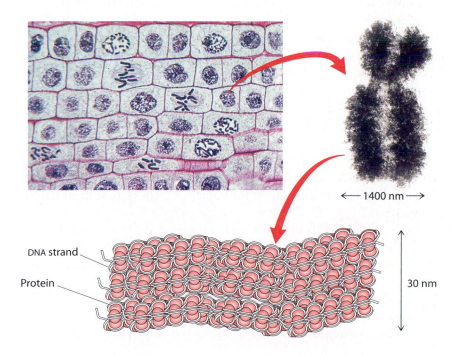

←— 1400 nm —→

DNA strand

Protein

30 nm

Figure 13.28
Onion cells in the process of dividing show chromosomes that consist of DNA and protein molecules clumped together. During division, the chromosomes duplicate themselves such that each new cell receives a full set of chromosomes identical to the set in the parent cell.

Figure 13.29
A ladder made of rope sides and rigid wooden rungs can be twisted to create a double helix. The ropes are the equivalent of DNA's two sugar–phosphate backbones, which are shown as the yellow and orange atoms in the computer rendition shown on the right. The rungs represent pairs of nitrogenous bases, which are seen in the computer rendition as shades of lavender and blue.

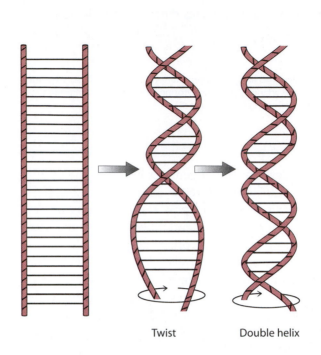
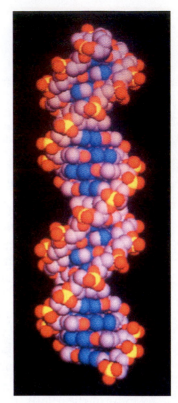

Twist   Double helix

Hydrogen bond

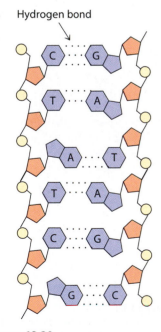

Figure 13.30
The two strands of nucleotides in a DNA molecule are held together by hydrogen bonding between complementary nitrogenous bases: adenine with thymine and guanine with cytosine.

In 1953, James Watson (b. 1928), an American biologist, and Francis Crick (b. 1916), a British biophysicist, together deduced that DNA occurs as two separate strands of nucleotides coiled around each other in a double helix, as illustrated in Figure 13.29. The strands are held together by hydrogen bonding between opposing nitrogenous bases.

What is most critical about Watson and Crick's model is that hydrogen bonding occurs only between specific bases—guanine pairs only with cytosine, and adenine pairs only with thymine, as shown in Figure 13.30. This means that if you know the sequence of one strand, you can automatically deduce the sequence of the second strand, also known as the *complementary strand*. For example, if the first strand contains the sequence CTGA, the complementary strand must contain the sequence GACT.

In living tissue, cells divide to create duplicates of themselves. In a maturing organism, cells divide frequently to provide for growth. In a mature organism, cells divide at a rate sufficient for the replacement of those that die. Each time a cell divides, genetic information must be preserved. In a process called **replication**, DNA strands are duplicated so that each newly formed cell receives a copy. Replication also allows for copies of DNA to pass from parent to offspring.

With their model, Watson and Crick proposed that DNA replication begins with the unraveling of the double helix. Each single strand then serves as a template for the synthesis of its complementary strand. Free nucleotides are coupled to the single strand of DNA according to the rules of base pairing: guanine + cytosine, adenine + thymine. One double helix thus turns into two, as shown in Figure 13.31. As a cell divides, one of the two new strands is segregated in each of the two newly formed cells.

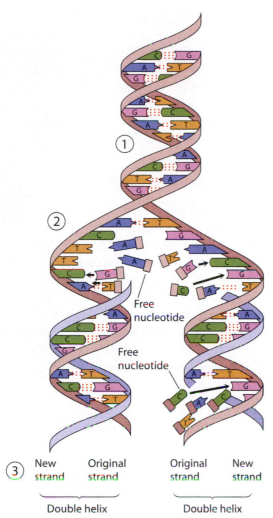

(a)

(b)

Figure 13.31
The replication of DNA. ① Double helix of DNA unwinds. ② Each single strand serves as a template for the formation of a new DNA strand containing the complementary sequence. ③ Two daughter double helices are formed, each containing one of the parent strands.

For their elucidation of DNA's secondary structure and function, Watson and Crick, along with biophysicist Maurice Wilkens (b. 1916), received the 1962 Nobel Prize in Physiology or Medicine. Subsequent research soon led to an understanding of how the nucleotide sequences in DNA translate into the synthesis of proteins. The Nobel laureates are shown in Figure 13.32.

One Gene Codes for One Polypeptide

As we saw in Section 13.4, the shape of a protein molecule is determined by the protein's primary structure, which is the sequence of amino acids. So what, then, controls a protein's amino acid sequence? The answer to this question, as we are now ready to explore, is the gene.

In modern terms, a **gene** is a particular sequence of DNA nucleotides along the DNA strand in a chromosome. Each gene codes for the synthesis of one or more proteins in an organism. Organisms of course need enormous numbers of genes to produce all the proteins they need. The number

(c)

Figure 13.32
(a) Watson and Crick in 1953 with their model of the DNA double helix. In discovering this model, Watson and Crick relied heavily on the experimental evidence gathered by other researchers, most notably the research team of (b) Maurice Wilkens and (c) Rosalind Franklin (1921–1958). Franklin did not share in the 1962 Nobel Prize because it is never awarded posthumously.

of genes contained in the 46 chromosomes in a single human cell, for instance, is estimated to be on the order of 40,000. To accommodate this many genes, each DNA molecule is very long, containing on the order of 3.1 billion base pairs. Interestingly, genes make up only about 20 percent of a DNA molecule. The other 80 percent of the nucleotides in a DNA strand appear to serve merely as spacers, their main job being to separate genes on the same DNA molecule. Other functions these spacer nucleotides might have are not yet fully understood.

RNA Is Largely Responsible for Protein Synthesis

The translating of a sequence of DNA nucleotides into a sequence of amino acids—in other words, into the manufacture of a protein—involves many intricate cellular mechanisms that are still the subject of much research. The overall process, however, is conceptually straightforward, involving two steps—*transcription* followed by *translation*. These steps are mediated by the other major nucleic acid, RNA, of which there are three forms—messenger RNA (mRNA), ribosomal RNA (rRNA), and transfer RNA (tRNA).

Protein synthesis begins in the cell nucleus with **transcription**, illustrated in Figure 13.33. In this step, the cell manufactures a single strand of messenger RNA. The genetic information encoded in a gene in the cell's DNA

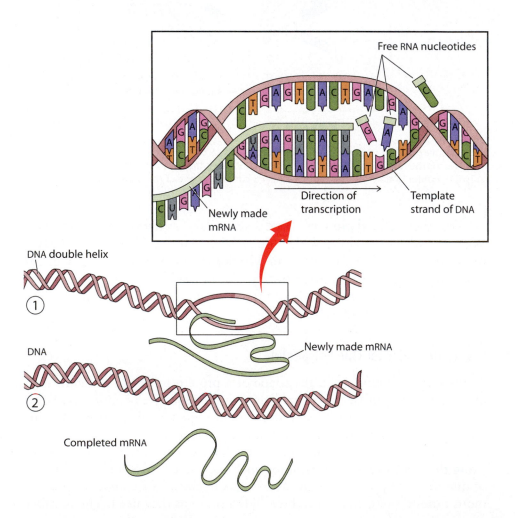

Figure 13.33
Transcription resembles DNA replication except that only one of the two DNA strands is copied and the newly synthesized mRNA single strand does not remain associated with its template DNA strand.

is used to specify the nucleotide sequence in this new strand of messenger RNA. During transcription, the section of DNA that is the gene unwinds, and then an enzyme binds free RNA nucleotides to one of the single DNA strands according to a base-pairing scheme similar to that for DNA replication, with one difference. Every cytosine on the DNA pulls in a guanine, every DNA guanine pulls in a cytosine, every DNA thymine pulls in an adenine, but—here's the difference—every DNA adenine now pulls in not a thymine but rather a uracil. Thus the mRNA synthesized during transcription contains the four RNA bases, as it must. The developing mRNA single strand is conformationally different from DNA because of the additional oxygen on the ribose sugar and as a result does not bind well to the DNA. Instead, the information-rich mRNA strand migrates away from the DNA, exiting the cell nucleus and entering the cytoplasm, where the protein will be built.

In the next step, **translation**, the sequence of nucleotides in the newly synthesized mRNA strand is used to determine the sequence of amino acids in the protein to be synthesized. This is done by way of a *genetic code*, which was fully deciphered by 1966 and is shown in Figure 13.34. According to the genetic code, it takes three mRNA nucleotides—each three-nucleotide unit is called a *codon*—to code for a single amino acid. The mRNA nucleotide sequence AGU, for example, codes for the amino acid serine, and AAG codes for lysine. (Note from Figure 13.34 that more than one codon can call for the same amino acid.) A few codons, such as AUG and UGA, are the signals for protein synthesis to either start or stop.

When using Figure 13.34 to determine which amino acid a particular codon codes for, you can quickly locate the codon by using the three colored strips. First run your finger down the red strip until you come to the first letter in the codon, then run your finger to the right until it is under

	U		C		A		G		
U	UUU	Phe	UCU	Ser	UAU	Tyr	UGU	Cys	U
	UUC		UCC		UAC		UGC		C
	UUA	Leu	UCA		UAA	Stop	UGA	Stop	A
	UUG		UCG		UAG	Stop	UGG	Trp	G
C	CUU	Leu	CCU	Pro	CAU	His	CGU	Arg	U
	CUC		CCC		CAC		CGC		C
	CUA		CCA		CAA	Gln	CGA		A
	CUG		CCG		CAG		CGG		G
A	AUU	Ile	ACU	Thr	AAU	Asn	AGU	Ser	U
	AUC		ACC		AAC		AGC		C
	AUA		ACA		AAA	Lys	AGA	Arg	A
	AUG	Met Start	ACG		AAG		AGG		G
G	GUU	Val	GCU	Ala	GAU	Asp	GGU	Gly	U
	GUC		GCC		GAC		GGC		C
	GUA		GCA		GAA	Glu	GGA		A
	GUG		GCG		GAG		GGG		G

Figure 13.34
RNA codons are three-nucleotide units that either code for specific amino acids in a protein being synthesized or else signal the synthesis to either start or stop. For example, the codon CUA codes for the amino acid leucine.

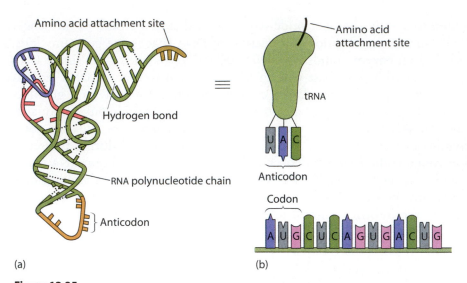

Figure 13.35

Translation involves transfer RNA. (a) The structure of a transfer RNA molecule, with an anticodon at one end and an amino acid attachment site at the other end. (b) A highly simplified symbol for tRNA. The anticodon is a series of three nucleotides that complement the mRNA codon and code for a specific amino acid at the amino acid attachment site.

the letter in the yellow strip that matches the second letter of the codon, then run your finger either up or down this column until your finger is aligned with the letter in the blue strip that matches the third letter of the codon. There you can identify which amino acid the codon codes for.

The key player in translation is transfer RNA, which has an angular conformation and is illustrated in Figure 13.35. Cells contain large numbers of tRNA molecules, as well as large numbers of free amino acids (which came from some protein food the organism ate). On the lower tip of each tRNA molecule as drawn in Figure 13.35 lie three nucleotides that complement a codon on mRNA. These three tRNA nucleotides are called the *anticodon*. On the opposite end of the tRNA molecule is an attachment site for an amino acid. This site is unique for a particular amino acid. One tRNA containing a particular anticodon, for example, might attach only to the amino acid glycine, while another tRNA containing a different anticodon might attach only to the amino acid alanine.

As the mRNA leaves the cell nucleus in which it was created and enters the cytoplasm, it binds with specialized structures called *ribosomes*, as shown in Figure 13.36. Ribosomes are microscopic complexes of rRNA and proteins, and they are the site where proteins are built. As the mRNA is scrolled sequentially over the ribosome, the anticodon end of a free tRNA molecule binds to an mRNA codon. In this manner, tRNA molecules and their tagalong amino acids are placed adjacent to one another along the mRNA strand. The amino acids then chemically bond with one another, forming a long polypeptide chain that breaks away from the tRNA as it forms. This process continues until a stop mRNA codon, for which there are no tRNA anticodons, is encountered. At this point, the primary structure of a new protein has been built.

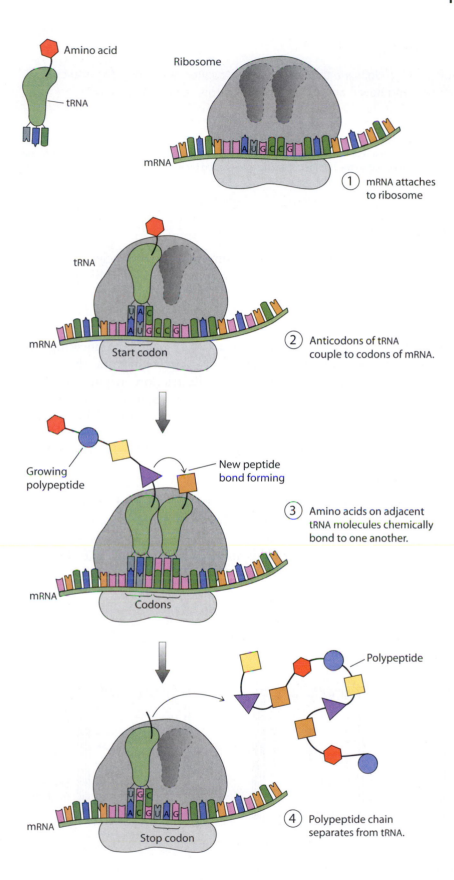

Amino acid

tRNA

Ribosome

mRNA

① mRNA attaches to ribosome

tRNA

mRNA

Start codon

② Anticodons of tRNA couple to codons of mRNA.

Growing polypeptide

New peptide bond forming

mRNA

Codons

③ Amino acids on adjacent tRNA molecules chemically bond to one another.

Polypeptide

mRNA

Stop codon

④ Polypeptide chain separates from tRNA.

Figure 13.36
Ribosomes are where messenger RNA is translated into a polypeptide chain by way of the action of transfer RNA.

Concept Check ✔

How many codons are there in the mRNA sequence AUG CUU AAA AGU CAA GCA UAA, and how many amino acids does this sequence code for?

Was this your answer? A codon is a triplet of mRNA nucleotides holding instructions for the building of a polypeptide strand, and so this sequence contains seven codons. The first and last are start/stop instructions, however, and so the number of amino acids coded for is only five. The sequence is Leu-Lys-Ser-Gln-Ala.

Genetic Engineering

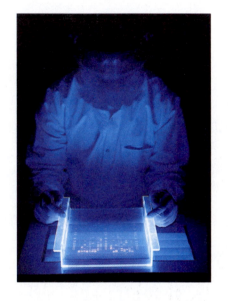

As they learned how proteins are made from nucleic acids, scientists developed tools to alter and map this process in order to do such things as treat human disease, uncover submicroscopic archaeologic evidence, and create new agricultural crops. All these activities come under the title of *genetic engineering*.

One of the most important early advances in genetic engineering was the discovery of a class of enzymes, called restriction enzymes, that cleave long strands of DNA into smaller fragments. Restriction enzymes reside in bacteria and viruses as a means of self-defense—they protect the host organism by degrading any invading foreign DNA molecules. Most restriction enzymes recognize a specific nucleotide sequence in the foreign DNA. By random chance, the sequence occurs many times throughout any DNA strand. The enzyme latches onto these sites and then catalyzes a cut across the double helix.

Figure 13.37 shows how fragments of DNA created by restriction enzymes can be separated from one another by a process based on differences in molecular size. The technique, called *gel electrophoresis*, takes advantage of the fact that nucleic acids are negatively charged and therefore

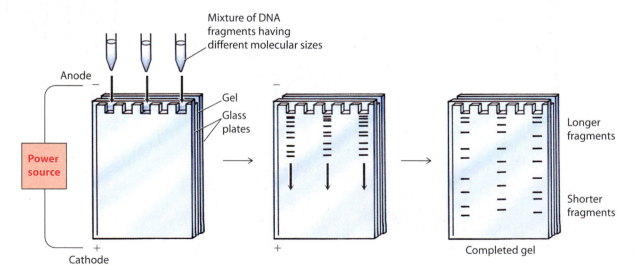

Figure 13.37
The profile of a series of DNA fragments separated by gel electrophoresis.

attracted to a positive electrode. As they move through a gel toward the positive electrode, the shorter fragments travel faster than the longer ones. The result is a series of bands, each containing fragments having the same size. The pattern these bands form is characteristic of both the type of restriction enzyme used and the identity of the DNA. Treating DNA with a series of different restriction enzymes can result in a gel electrophoresis pattern that is quite intricate and extremely characteristic not only of the species but of the individual. What forensic experts call a DNA fingerprint, used to identify a crime suspect, is an example of a gel electrophoresis pattern.

Nucleotide sequences recognized by restriction enzymes are usually symmetrical such that the two strands of the double helix have the same base sequences but in opposite directions. The base sequences shown here in blue, for example, are symmetrical:

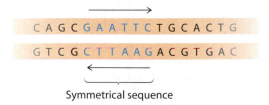

Symmetrical sequence

Some restriction enzymes cleave a symmetrical sequence in a staggered manner such that the nucleotide chain is cleaved at one place on one of the strands of the double helix and on a different place on the other strand, as shown at the top of Figure 13.38. The DNA fragments resulting from

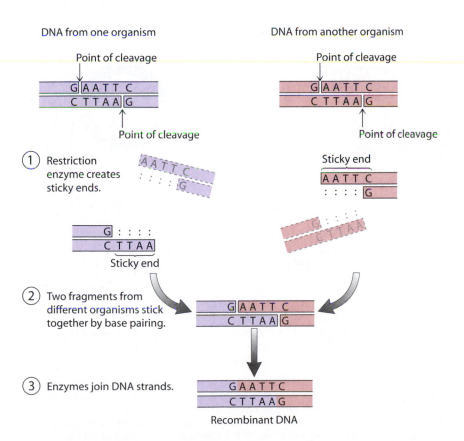

Recombinant DNA

Figure 13.38
The formation of recombinant DNA.

staggered cleaving thus have one strand of the double helix longer than the other. These single-strand ends are called *sticky ends* because they have complementary base sequences and so tend to reassociate by base pairing.

Sticky ends allow the formation of **recombinant DNA** (pronounced re-COM-bin-ant), which is formed by combining DNA from different organisms, as shown in Figure 13.38. DNA molecules from two sources are cleaved with the same restriction enzyme, yielding two sets of fragments having identical sticky ends. When these fragments are mixed together, the complementary base pairs of the sticky ends combine to form modified new DNA.

Concept Check ✓

What are restriction enzymes? What is their function, and how do scientists exploit this function?

Were these your answers? They are enzymes that catalyze the cleavage of the DNA double helix. They occur naturally in bacteria and viruses and are used as a means of self-defense. Scientists use restriction enzymes to splice together DNA fragments from different organisms.

A great significance of recombinant DNA technology lies in the ease with which large amounts of selected DNA sequences, such as human genes, can be synthesized. This is done by inserting a desired sequence into bacterial cells and allowing the sequence to be replicated as the bacterial cells multiply. This technique, called **gene cloning**, is illustrated in Figure 13.39

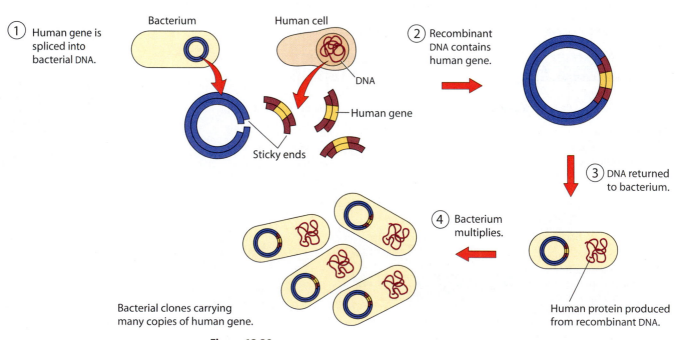

Figure 13.39
A gene is spliced into bacterial DNA. Bacteria containing the recombinant DNA multiply, and the many copies of the human gene now present in the bacteria are available to produce large quantities of protein. Note that the cloning of a gene, which is what is shown here, is markedly different from cloning a whole organism, which is done by transplanting the entire DNA sequence of an organism into an egg that then gets implanted into a surrogate uterus.

and is now a routine procedure in biological laboratories. Growing large batches of the bacteria containing the inserted gene yields large quantities of the gene, meaning all these genes are now available to produce large quantities of the protein for which they code. Currently, various medicinally important proteins that are otherwise difficult to obtain are manufactured in this manner. These include human insulin for diabetes; human growth hormone for osteoporosis and growth disorders in children; epidermal growth factor for skin regeneration in burn victims; interferon, which is a promising anticancer agent; and bovine growth hormone, which increases milk production in cows by up to 20 percent.

There are many potential uses for recombinant DNA technology. Agricultural products, as is discussed in Chapter 15, can be made bug-resistant and spoil-resistant. The technology may also be used to treat genetic disorders, such as sickle-cell anemia, muscular dystrophy, cystic fibrosis, and lupus, and may even address the question of why our bodies deteriorate with age.

Toward these and other goals, scientists from around the world are working on the Human Genome Project, which aims to identify all the genes in human DNA and the proteins they code for. (A *genome* is the collection of all the genes in an organism.) In June 2000, the Human Genome Project announced that it had sequenced nearly all of the 3.1 billion nucleotide base pairs of human DNA—a historical milestone on par with our landing on the moon. It will likely take many more years, perhaps decades, before all our genes and the proteins they code for are identified. This will be an accomplishment like none other ever achieved by humans, allowing for truly revolutionary advances in medicine. The impact of this new knowledge will also be far-reaching, raising numerous ethical, philosophical, and even religious issues.

13.6 Vitamins Are Organic, Minerals Are Inorganic

In addition to carbohydrates, lipids, proteins, and nucleic acids, our bodies require vitamins and minerals in order to survive. **Vitamins** are organic chemicals that, by assisting in various biochemical reactions, help us maintain good health. **Minerals** are inorganic chemicals that play a variety of roles in the body. Some, such as iron in hemoglobin, are vital components of biomolecules. Others, such as calcium in bone, are integral parts of structures. Deficiencies in vitamins or minerals are the cause of certain diseases. Lack of vitamin C, for example, leads to scurvy, a disease marked by a deterioration of the gums. Lack of iron leads to anemia, which results in general fatigue and an irregular heartbeat.

Vitamins are classified as either lipid-soluble or water-soluble, as shown in Table 13.2 on page 434. The lipid-soluble vitamins tend to accumulate in fatty tissue, where they may be stored for years. Adults can remain free of a deficiency disease for quite some time because of these vitamin reserves. Children, on the other hand, because they have yet to build up these reserves, are particularly vulnerable to these diseases. In developing nations, for example, many children suffer permanent blindness because of a lack of vitamin A.

Table 13.2

Some Vitamins Needed by the Human Body

Vitamin	Function	Deficiency Syndrome
Lipid-Soluble		
Vitamin A (retinol)	Precursor to rhodopsin, a chemical used for vision; assists in inhibiting bacterial and viral infections	Night blindness
Vitamin D (calciferol)	Helps incorporate calcium into body	Weak bones
Vitamin E (tocopherol)	Inhibits oxidation of polyunsaturated fats; free radical scavenger; helps maintain circulatory and nervous systems	Diminished hemoglobin
Vitamin K (phylloquinone)	Helps maintain ability to form blood clots	Abnormal bleeding
Water-Soluble		
B vitamins	Coenzymes in biochemical reactions for growth and energy production	Various nerve and skin disorders, anemia
Vitamin C (ascorbic acid)	Antioxidant; assists in inhibiting bacterial and viral infections	Scurvy

A lack of lipid-soluble vitamins can be detrimental, but so can excessive amounts, particularly of vitamins A and D, which can accumulate to dangerous levels. Too much vitamin A causes dry skin, irritability, and headaches. Excessive amounts of vitamin D lead to diarrhea, nausea, and calcification of joints and other body parts. Vitamins E and K are less harmful in large quantities because they are readily metabolized.

The water-soluble vitamins are not retained by the body for long periods of time. Instead, because they are soluble in water, they are readily excreted in urine and must therefore be ingested frequently. It is difficult to harm yourself by taking in too much of the water-soluble vitamins. Your body simply absorbs what it immediately needs and excretes the rest. Foods boiled in water tend to lose vitamins B and C because of the solubility of these vitamins in water. The vitamins are then poured down the drain along with the water. For this reason, many people prefer steaming or microwaving their vegetables. Also, foods should not be overcooked, as both lipid-soluble and water-soluble vitamins are destroyed by heat.

All minerals are ionic compounds of various elements. They are classified according to the quantities we need. *Macrominerals*, the ones we need in greatest quantity, make up about 4 percent of our body weight. Macrominerals are listed in Table 13.3. The amounts of macrominerals we need are measured in grams. The amounts of *trace minerals* we ingest daily are measured in milligrams. Beyond trace minerals are *ultratrace minerals*, which we use on the microgram or even picogram scale.

Table 13.3

Some Macrominerals Needed by the Human Body

Macromineral (Ionic Form)	Some Functions	Deficiency Syndrome
Sodium (Na^+)	Transportation of molecules across cell membrane, nerve function	Muscle cramps, reduced appetite
Potassium (K^+)	Transportation of molecules across cell membrane, nerve function	Muscular weakness, paralysis, nausea, heart failure
Calcium (Ca^{2+})	Bone and tooth formation, nerve and muscle function	Retarded growth, possible loss of bone mass
Magnesium (Mg^{2+})	Enzyme function	Nervous system disturbances
Chlorine (Cl^-)	Transportation of molecules across cell membrane, digestive fluid, nerve function	Muscle cramps, reduced appetite
Phosphorus ($H_2PO_4^-$)	Bone and tooth formation, nucleotide synthesis	Weakness, calcium loss
Sulfur (SO_4^{2-})	Amino acid component	Protein deficiency

The body requires balanced amounts of minerals, meaning that too much is as harmful as too little. Ultratrace minerals are particularly toxic when taken in large quantities. Cadmium, chromium, and nickel, for example, are potent carcinogens, and arsenic is a well-known poison. Yet our bodies need microquantities of these minerals if we are to stay healthy. Eating a well-balanced diet is often the best way to obtain a good balance of minerals. Mineral supplements may also be taken, but doses should be monitored carefully.

Two of the most abundant minerals in our diet are the ions of potassium and sodium. Both these ions are involved in nerve-signal transmission and in the transport of molecules into and out of cells. For good health, we need more potassium ions than sodium ions. So do all other living organisms, including the plants and animals we eat. When we eat these plants or animals without additives and without excessive processing, our need for greater amounts of potassium is met because these organisms naturally contain more potassium than sodium. When food is boiled or deep-fried, however, both potassium ions and sodium ions are stripped away along with the liquids in which they are dissolved. Salting the food with sodium chloride then puts the sodium ion content way above the potassium ion content.

Another important dietary mineral is phosphorus, which comes to us as the phosphate ion, $H_2PO_4^-$. As you can see by checking back to Figure 13.26 on page 421, phosphate ions form the backbone of nucleic acids. In addition, they are components of the energy-packed compound adenosine triphosphate (ATP), shown in Figure 13.40 on page 436, which is produced in the oxidation of carbohydrates, lipids, and proteins.

ATP is the direct source of energy for most of the energy-requiring processes in the body, such as tissue building, muscle contraction, transmission of nerve impulses, heat production, and movement of molecules into

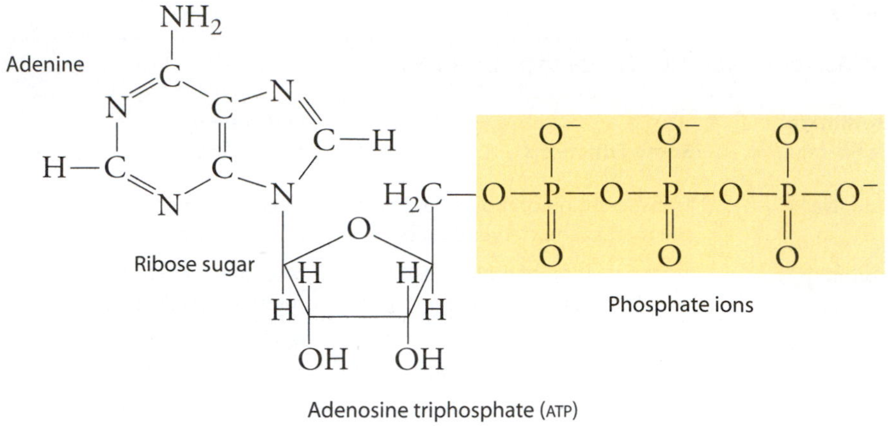

Figure 13.40
Phosphate ions are an important part of the ATP molecule.

and out of cells. The human body goes through lots of ATP—about 8 grams per minute during strenuous exercise. It is a short-lived molecule and so must be produced continuously. The many chemical pathways by which foods are oxidized to yield ATP have been mapped extensively by biochemists.

Various poisons act by blocking the synthesis of ATP. Carbon monoxide, for example, binds to the iron of hemoglobin, thereby preventing hemoglobin from carrying oxygen. The reason the body needs oxygen, however, is so that it can be used to oxidize carbohydrates, lipids, and proteins to form ATP. So without oxygen, the body becomes starved of energy-yielding ATP and quickly dies. Cyanide also blocks the synthesis of ATP but does so by incapacitating enzymes that play an important role in ATP synthesis. Interestingly, ATP is also used by the body to allow muscles to relax after contraction. When the body dies, no matter what the cause, ATP synthesis comes to a halt, and all body muscles become stiff—a condition known as rigor mortis.

13.7 Metabolism Is the Cycling of Biomolecules Through the Body

Your body takes in biomolecules in the food you eat and breaks them down to their molecular components. Then one of two things happens: either your body "burns" these molecular components for their energy content through a process known as *cellular respiration,* or these components are used as the building blocks for your body's own versions of carbohydrates, lipids, proteins, and nucleic acids. The sum total of all these biochemical activities is what we call **metabolism**. Two forms of metabolism are *catabolism* and *anabolism*, and Figure 13.41 shows the major catabolic and anabolic pathways of living organisms.

All metabolic reactions that involve the tearing down of biomolecules are grouped under the heading of **catabolism**. Digestion and cellular respiration

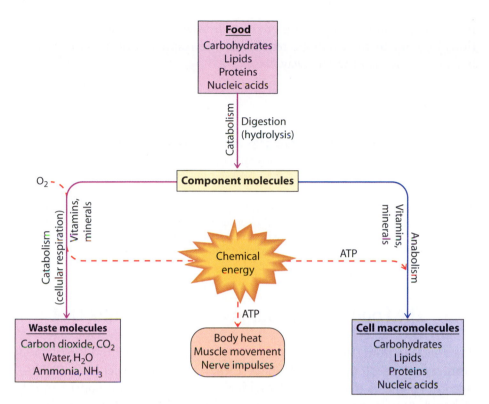

Figure 13.41
Metabolic pathways for the food we ingest. Catabolic pathways are indicated by the purple arrows, anabolic pathways by the blue arrow.

are examples of catabolic reactions. Digestion begins with the hydrolysis of food molecules, a reaction in which water is used to sever bonds in the molecules. The small molecules formed in digestion, such as the glucose units of complex carbohydrates, then migrate to all the various cells of the body and take part in cellular respiration. There the small food molecules lose electrons to the oxygen that was inhaled through our lungs, and as a result break down to even smaller molecules as carbon dioxide, water, and ammonia, which are excreted. Through this process, high-energy molecules, such as ATP, are created. These high-energy molecules are able to drive reactions that produce body heat, muscle movement, and nerve impulses. They also are responsible for allowing **anabolism**, which is the general term for all the chemical reactions that produce large biomolecules from smaller molecules.

The types of biomolecules produced by anabolism are the same as the types found in food—carbohydrates, lipids, proteins, and nucleic acids. These products of anabolism are, if you will, the host's own version of what the food once was. And if the host ever becomes food, anabolic reactions in the subsequent host will result in different versions of the molecules. Thus, organisms in a food chain live off one another by absorbing one another's energy via catabolic reactions and then rearranging the remaining atoms and molecules via anabolic reactions into the biomolecules they need to survive.

Catabolism and anabolism work together. In healthy muscle tissue, for example, the rate of muscle degradation (catabolism) is matched by the rate of muscle building (anabolism). If you increase your food supply and exercise vigorously, it is possible to favor the muscle-building anabolic reactions over the muscle-destroying catabolic reactions. The result is an increase in muscle mass. Stop eating and exercising, however, and these anabolic reactions lose out to the catabolic reactions. The result is a decrease in muscle mass—you begin to waste away.

Concept Check ✔

> Anabolic steroids help people gain muscle mass. If there were such a thing as a catabolic steroid, what would be its effect?

Was this your answer? Anabolic? Catabolic? Which is which? Many students recognize the term *anabolic steroids* from the sports news media, which are quick to report on famous athletes caught using these steroids for improved performance. Anabolism therefore is muscle-building, and so catabolism must be muscle-degrading. A catabolic steroid would cause a loss of muscle mass.

13.8   The Food Pyramid Summarizes a Healthful Diet

The food pyramid, shown in Figure 13.42, summarizes the food intake recommendations of the United States Department of Agriculture (USDA). According to this pyramid, an individual's daily diet should consist mostly of bread, cereals, grains, pastas, fruits, and vegetables, with the amount of dairy products and meats fairly limited and foods high in sugars or fats consumed only sparingly.

We can get some insight into the reasons behind these recommendations by looking at how the body handles the biomolecules contained in these foods.

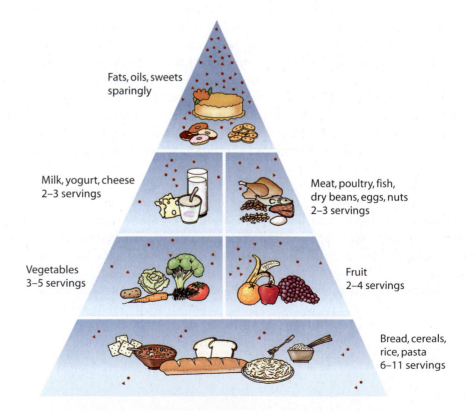

Fats, oils, sweets
sparingly

Milk, yogurt, cheese
2–3 servings

Meat, poultry, fish,
dry beans, eggs, nuts
2–3 servings

Vegetables
3–5 servings

Fruit
2–4 servings

Bread, cereals,
rice, pasta
6–11 servings

Figure 13.42
The food pyramid.

Carbohydrates Predominate in Most Foods

The breads, cereals, grains, pastas, fruits, and vegetables of the lower two tiers of the pyramid are important sources of food primarily because they contain a good balance of all nutrients—carbohydrates, fats, proteins, nucleic acids, vitamins, and minerals. The predominant component of these foods is carbohydrates, however.

There are two types of carbohydrates—nondigestible, called dietary *fiber*, and digestible, mainly starches and sugars.

As discussed in Section 13.2, dietary fiber helps keep things moving in the bowels, especially in the large intestine. There are two kinds of fiber—water-insoluble and water-soluble. Insoluble fiber consists mainly of cellulose, which is found in all food derived from plants. In general, the less processed the food, the higher the insoluble-fiber content. Brown rice, for example, has a greater proportion of insoluble fiber than does white rice, which is made by milling away the rice seed's outer coating (along with numerous vitamins and minerals).

Soluble fiber is made of certain types of starches that are resistant to digestion in the small intestine. An example is pectin, which is added to jams and jellies because it acts as a thickening agent, becoming a gel when dissolved in a limited amount of water. Soluble fiber tends to lower cholesterol levels in the blood because of how it interacts with *bile salts*, which are cholesterol-derived substances produced in the liver and then secreted into the intestine. As shown in Figure 13.43, one of the functions of bile salts is to carry ingested lipids through the membranes of the intestine and into the bloodstream. The bile salts are then reabsorbed by the liver and cycled back to the intestine. Soluble fiber in the intestine binds to bile salts, which are then efficiently passed out of the body rather than being reabsorbed. The liver responds by producing more bile salts, but to do so it must utilize

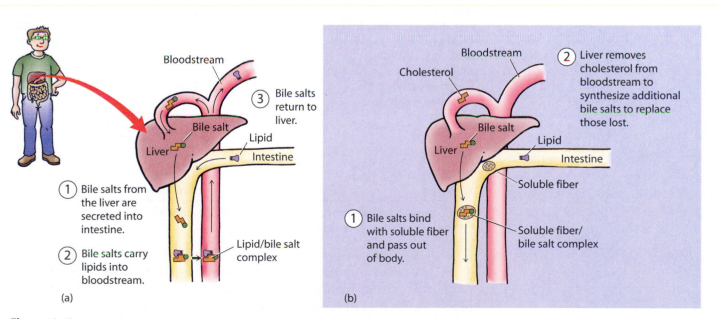

Figure 13.43
(a) With no soluble fiber present, bile salts recycle to the liver and no new ones need to be made. (b) In the presence of soluble fiber, bile salts are removed from the body. The liver then must use cholesterol from the blood to make new supplies of bile salts. Thus by binding with bile salts, soluble fiber indirectly decreases the amount of cholesterol in the blood.

cholesterol, which it collects from the bloodstream. By this indirect route of binding with bile salts, soluble fiber tends to lower a person's cholesterol level. Foods rich in soluble fiber include fruits and certain grains, such as oats and barley.

During digestion, the digestible carbohydrates—both starches and sugars—are transformed to glucose, which is absorbed into the bloodstream through the walls of the small intestine. The body then utilizes this glucose to build energy molecules, such as ATP.

Carbohydrate-containing foods are rated for how quickly they cause an increase in blood glucose levels. This rating is done with what is known as the *glycemic index*. The index compares how much a given food increases a person's blood glucose level relative to the increase seen when pure glucose is ingested, with the latter increase assigned a standard value of 100. In general, foods that are high in starch or sugar but low in dietary fiber are high on the glycemic index, a baked potato being a prime example.

The glycemic index for a particular food can vary greatly from one person to the next. How the food was prepared can also make a big difference. Thus, index values, such as the ones shown in Table 13.4, are to be taken lightly—merely as ballpark figures. Given this qualification, however, the index provides valuable information for people, such as those with diabetes, who need to pay close attention to their blood sugar levels.

Table 13.4

Glycemic Index for Select Foods

Glucose	100	Honey	58
Baked potato	85	Sweet corn	55
Cornflakes	83	Brown rice	55
Microwaved potato	82	Popcorn	55
Jelly beans	80	Oatmeal cookies	55
Vanilla Wafers	77	Sweet potato	54
French fries	75	Banana	54
Cheerios	74	Milk chocolate	49
White bread	71	Orange	44
Mashed potato	70	Snickers candy bar	40
Life-Savers candy	70	Pinto beans	39
Shredded Wheat	69	Apple	38
Wheat bread	68	Spaghetti, boiled 5 minutes	36
Sucrose	64	Skim milk	32
Raisins	64	Whole milk	27
Mars candy bar	64	Grapefruit	25
High-fructose corn syrup	62	Soy beans	18
White rice	58	Peanuts	15

Source: Jennie Brand Miller *et al., The Glucose Revolution: The Authoritative Guide to the Glycemic Index.* Sydney: Marlowe & Company, 1999.

There are a number of problems associated with eating carbohydrate foods that have a high glycemic index. For example, the rapid spike in blood glucose level causes the body to produce extra *insulin*, a blood-soluble protein that causes glucose to be moved out of the blood and into cells to be metabolized. Insulin is very effective at what it does, however, and soon the extra insulin in the blood leads to a depletion of blood glucose. The body responds by releasing glucose-yielding glycogen but also by triggering a sense of hunger, even if the person just ate. A meal rich in foods high on the glycemic index therefore promotes overeating, which usually leads to obesity.

Many professional organizations, such as the American Diabetes Association, caution that priority should be given to the quantities of carbohydrates ingested rather than to the glycemic index of the food containing those carbohydrates. What really counts is the total number of calories absorbed, not whether these calories came from foods high or low on the index. For most people, however, ingesting foods low on the index makes maintaining a healthful caloric intake more manageable.

Another advantage of eating carbohydrates from foods that are low on the index is that these foods provide energy to the body over an extended period of time. They do this because the glucose molecules they contain are released slowly. Furthermore, maintaining moderate glucose levels in the blood allows the body to continue using fats for its energy needs. As was discussed in Section 13.3, fats provide much more energy per gram (and thus more ATP) than do carbohydrates.

For athletes, a diet rich in foods low on the index, such as spaghetti, translates to greater endurance. Interestingly, this greater endurance is just as useful for body-builders as it is for marathon runners. The energy required for building muscles is far more critical than the supply of raw materials needed. Furthermore, the body's metabolism is versatile enough to generate proteins out of glucose (just as it is able to generate glucose out of proteins). Thus, the body-builder's supply of proteins is assured. A diet rich in carbohydrates is therefore more effective at allowing a body-builder to build muscles than is a diet rich in proteins.

Despite the many advantages of eating carbohydrates low on the glycemic index, foods rich in carbohydrates high on the index, such as sucrose, are now more popular than ever. Many of these foods are highly processed and are found at the apex of the food pyramid. Although they are good at providing energy, the USDA recommends that they be consumed only sparingly because they lack many of the essential nutrients present in the foods of the lower two tiers.

A candy bar is good for a quick energy fix, but chow down on a spaghetti feast the night before a strenuous workout for long-run energy.

Unsaturated Fats Are Generally More Healthful than Saturated Fats

Because your body uses saturated fats to synthesize cholesterol, the more saturated fats you ingest, the more cholesterol your body is able to synthesize. Unsaturated fats, by contrast, are not ideal starting materials for cholesterol synthesis.

Another reason unsaturated fats are more healthful has to do with how fats associate with cholesterol. Fats and cholesterol are both nonpolar lipids, which, on their own, are insoluble in blood. In order to move through the bloodstream, these compounds are packaged with bile salts, as was discussed earlier. Most lipids, however, are made water-soluble by being packaged with water-soluble proteins in complexes called *lipoproteins*. Lipoproteins are classified according to density, as noted in Table 13.5. Very-low-density lipoproteins (VLDL) serve primarily in the transport of fats throughout the body. Low-density lipoproteins (LDL) transport cholesterol to the cells, where it is used to build cell walls. High-density lipoproteins (HDL) bring cholesterol to the liver, where it is transformed to a variety of useful biomolecules.

Table 13.5

The Classification of Lipoproteins

Lipoprotein	Percent Protein	Density (g/mL)	Primary Function
Very-low-density (VLDL)	5	1.006–1.019	Fat transport
Low-density (LDL)	25	1.019–1.063	Cholesterol transport (to cells to build cell walls)
High-density (HDL)	50	1.063–1.210	Cholesterol transport (to liver for processing)

A diet high in saturated fats leads to elevated VLDL and LDL levels in the bloodstream. This is undesirable because these lipoproteins tend to form fatty deposits called *plaque* in the artery walls. Plaque deposits can become inflamed to the point where they rupture, releasing blood-clotting factors into the bloodstream. A blood clot formed around the rupture site is let loose into the bloodstream, where it can become lodged and block the flow of blood to a particular region of the body. When that region is in the heart, the result is a heart attack. When that region is in the brain, the result is a stroke.

In contrast to saturated fats, unsaturated fats tend to increase blood HDL levels, which is desirable because these lipoproteins are effective at *removing* plaque from artery walls.

Concept Check ✔

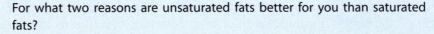

For what two reasons are unsaturated fats better for you than saturated fats?

Was this your answer? Unsaturated fats are not so readily used by your body to synthesize cholesterol. They also tend to increase the proportion of high-density lipoproteins, which lower the level of cholesterol in your blood and help relieve the buildup of arterial plaque.

Unsaturated fats, as noted in Section 13.3, tend to be liquids at room temperature. They can be transformed to a more solid consistency, however, by *hydrogenation*, a chemical process in which hydrogen atoms are added to carbon–carbon double bonds. Mix a partially hydrogenated vegetable oil with yellow food coloring, a little salt, and the organic compound butyric acid for flavor, and you have margarine, which was first prepared around the time of World War II as an alternative to butter. Many food products, such as chocolate bars, contain partially hydrogenated vegetable oils so that they are of a consistency that sells well in the marketplace. Hydrogenation increases the percentage of saturated fats, however, and therefore makes these fats less healthful. Furthermore, as Figure 13.44 shows, some of the double bonds that remain are transformed to the *trans* structural isomer (see *Insights: Twisting Jellybeans* on page 369). Because carbon chains containing *trans* bonds tend to be less kinked than chains containing *cis* bonds, the partially hydrogenated fat has straighter chains. This means the fat is more likely to mimic the action of saturated fats in the body.

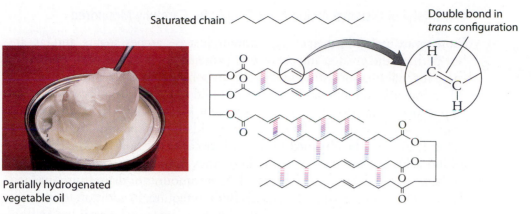

Saturated chain

Double bond in *trans* configuration

Partially hydrogenated vegetable oil

Figure 13.44
Hydrogenation can lead to *trans* double bonds in the fatty acid chain, which as a result points straight out, much as the chain of a saturated fatty acid does.

Hands-On Chemistry: Sizzle Sources

Did you ever notice that butter and margarine sizzle on a hot griddle but vegetable oil does not? The sizzling is the sound of water boiling away rapidly as the butter or margarine hits the hotter-than-100°C griddle. The sizzling subsides once the water is gone. Vegetable oil contains no appreciable quantities of water, and so it does not sizzle. Different brands of margarine contain different proportions of water, which is the subject of this investigation.

What You Need

Several brands of margarine (be sure to include a number of "light" spreads), series of same-sized drinking glasses, microwave oven, kitchen baster or eye dropper

Procedure

① Place each margarine sample in a separate glass. Add enough margarine so that it is at least 0.5 inch deep.

② Label each glass with the brand it contains.

③ Melt all the samples in the microwave oven. (Watch carefully because this doesn't take long.) As the margarine melts, the water and lipid layers separate.

④ Note the relative water content of the various brands by comparing the depths of the water layers.

⑤ Use the eye dropper or kitchen baster to pull off only the water layer, which will be beneath the lipid layer. Cool the lipid layers in the refrigerator, and then look for differences in consistencies.

Based on the consistencies you noted in step 5, which sample do you suppose contains the greatest proportion of saturated fats?

Our Intake of Essential Amino Acids Should Be Carefully Monitored

Proteins are useful for their energy content, just as starches, sugars, and fats are, but perhaps the greatest importance of proteins lies with how our bodies use them for building such structures as enzymes, bones, muscles, and skin. Of the 20 amino acids the human body uses to build proteins, the adult body is able to produce 12 of them in amounts sufficient for its needs—it produces these amino acids from carbohydrates and fatty acids. The remaining eight, listed in Table 13.6, must be obtained from food. Because the body needs these eight amino acids but cannot synthesize them, they are called *essential amino acids*, in the sense that it is essential we get adequate amounts of them from our food. To support rapid growth, infants and children require, in addition to the eight amino acids listed for adults in Table 13.6, large amounts of arginine and histidine, which can be obtained only from the diet. Infants and juveniles therefore have a total of ten essential amino acids. (The term *essential* is unfortunate because, in truth, all 20 amino acids are vital to our good health.)

Table 13.6

The Essential Amino Acids

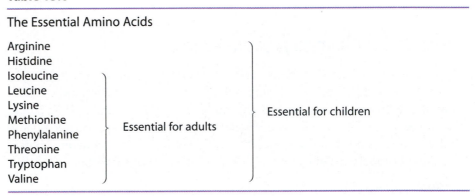

Arginine
Histidine
Isoleucine
Leucine
Lysine
Methionine
Phenylalanine
Threonine
Tryptophan
Valine

Essential for adults

Essential for children

Why our bodies produce ample amounts of some amino acids and not others can be explained by looking at the chemical structures of the amino acid side groups, shown in Figure 13.16 back on page 414. The nonessential amino acids have side groups that tend to be simple and therefore can be produced by the body without much effort. The essential amino acids, however, tend to be biochemically more difficult to make. The body therefore can save energy by obtaining these amino acids from outside sources. Over the course of evolution, our capacity to build these amino acids diminished. In a similar manner, we lost the capacity to build vitamins, which are also complex molecules more efficiently obtained through our diet. In other words, we let other living organisms go through the metabolic expense of building these biomolecules, and then we eat those organisms.

In general, the more closely the amino acid composition of ingested protein resembles the amino acid composition of the animal eating the protein, the higher the nutritional quality of that protein. For humans, mammalian protein is of the highest nutritional quality, followed by fish and poultry, then by fruits and vegetables. Plant proteins in particular are often deficient in lysine, methionine, or tryptophan. A vegetarian diet provides adequate protein only if it contains a variety of protein sources, with a deficiency in one source being compensated for by an excess in another source, as shown in Figure 13.45.

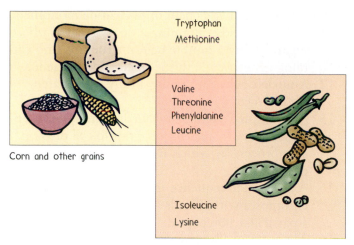

Corn and other grains

Tryptophan
Methionine

Valine
Threonine
Phenylalanine
Leucine

Isoleucine
Lysine

Beans and other legumes

Figure 13.45
Sufficient protein can generally be obtained in a vegetarian diet by combining a legume, such as peas or beans, with a grain, such as wheat or corn. Familiar meals containing such a combination include a peanut butter sandwich, corn tortillas and refried beans, and rice and tofu.

In Perspective

The old adage "you are what you eat" has a literal foundation. With the exception of the oxygen you obtain through your lungs, nearly every atom or molecule in your body got there by first passing through your mouth and into your stomach. All the biomolecules needed for the energy and growth of a fetus growing in the womb, like Maitreya Suchocki in Figure 13.46, must first pass through the lungs and mouth of her mother, which is why it is so vital that her mother eat right and maintain a healthful lifestyle while pregnant. And then, a mere 40 weeks later, her mother's food has been transformed by the actions of Maitreya's DNA into a whole new body ripe for exploring the world around her.

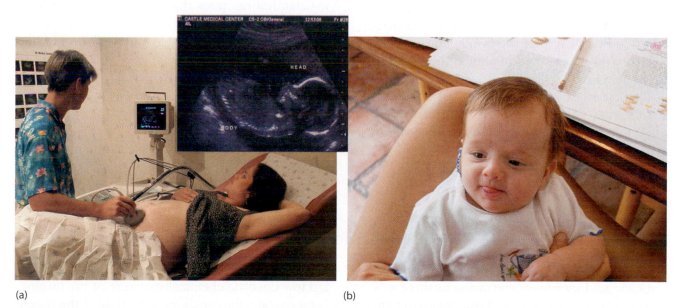

(a)　　　　　(b)

Figure 13.46
(a) As a fetus, Maitreya is undergoing the most rapid growth rate of her life, and thus her dependence on a healthful diet is as great as it will ever be. (b) As a baby, Maitreya's nutritional needs are still great, as is her ability to imitate her mother during a copyedit session.

Key Terms and Matching Definitions

_____ amino acid
_____ anabolism
_____ carbohydrate
_____ catabolism
_____ chromosome
_____ deoxyribonucleic acid
_____ enzyme
_____ fat
_____ gene
_____ gene cloning
_____ glycogen
_____ lipid
_____ metabolism
_____ mineral
_____ nucleic acid
_____ nucleotide
_____ protein
_____ recombinant DNA
_____ replication
_____ ribonucleic acid
_____ saccharide
_____ transcription
_____ translation
_____ vitamin

1. A biomolecule that contains only carbon, hydrogen, and oxygen atoms and is produced by plants through photosynthesis.
2. Another term for carbohydrate. The prefixes *mono-*, *di-*, and *poly-* are used before this term to indicate the length of the carbohydrate.
3. A glucose polymer stored in animal tissue and also known as animal starch.
4. A broad class of biomolecules that are not soluble in water.
5. A biomolecule that packs a lot of energy per gram and consists of a glycerol unit attached to three fatty acid molecules.
6. A polymer of amino acids, also known as a polypeptide.
7. The monomers of polypeptides, each monomer consisting of an amine group and a carboxylic acid group bonded to the same carbon atom.
8. A protein that catalyzes biochemical reactions.
9. A nucleic acid monomer consisting of three parts: a nitrogenous base, a ribose sugar, and an ionic phosphate group.
10. A long polymeric chain of nucleotide monomers.
11. A nucleic acid containing a deoxygenated ribose sugar, having a double helical structure, and carrying genetic code in the nucleotide sequence.
12. A nucleic acid containing a fully oxygenated ribose sugar.
13. An elongated bundle of DNA and protein that appears in a cell's nucleus just prior to cell division.
14. The process by which DNA strands are duplicated.
15. A nucleotide sequence in the DNA strand in a chromosome that leads a cell to manufacture a particular polypeptide.
16. The process whereby the genetic information of DNA is used to specify the nucleotide sequence of a complementary single strand of messenger RNA.
17. The process of bringing amino acids together according to the codon sequence on mRNA.
18. A hybrid DNA composed of DNA strands from different organisms.
19. The technique of incorporating a gene from one organism into the DNA of another organism.
20. Organic chemicals that assist in various biochemical reactions in the body and can be obtained only from food.
21. Inorganic chemicals that play a wide variety of roles in the body.
22. The general term describing all chemical reactions in the body.
23. Chemical reactions that break down biomolecules in the body.
24. Chemical reactions that synthesize biomolecules in the body.

Review Questions

Biomolecules Are Produced and Utilized in Cells

1. Do plant cells have a plasma membrane?

2. What are the four major categories of biomolecules discussed in this chapter?

Carbohydrates Give Structure and Energy

3. Are all carbohydrates digestible by humans?

4. How does the chemical structure of the monosaccharide glucose differ from that of the monosaccharide fructose?

5. Why do plants produce starch?

6. How does amylose differ from amylopectin?

7. Which monosaccharide do starches and cellulose have in common?

8. What is the most abundant organic compound on the Earth?

Lipids Are Insoluble in Water

9. What are the structural components of a triglyceride?

10. What makes a saturated fat saturated?

11. What do all steroids have in common?

Proteins Are Polymers of Amino Acids

12. What are the building blocks of a protein molecule?

13. How do various amino acids differ from one another?

14. What do a peptide, polypeptide, and protein all have in common?

15. Name the four structural levels possible in a protein, and describe the details of each.

16. What are the two most common secondary structures in a protein? The two most common tertiary structures?

17. What does disulfide cross-linking do for a protein?

18. What is the role of enzymes in the body?

19. What holds a substrate to its receptor site?

Nucleic Acids Code for Proteins

20. What is the difference between a nucleic acid and a nucleotide?

21. Where in the cell are ribonucleic acids found?

22. Where in the cell are deoxyribonucleic acids found?

23. Which four nitrogenous bases are found in DNA? In RNA?

24. Are codons found in DNA or in RNA?

25. On what form of RNA are anticodons found?

26. What is the difference between transcription and translation?

27. Which amino acid is coded for by the nitrogenous-base sequence AGG?

28. What is a restriction enzyme, and what does it do?

29. What is a sticky end in a DNA fragment, and how are sticky ends useful in the formation of recombinant DNA?

Vitamins Are Organic, Minerals Are Inorganic

30. What are two classes of vitamins?

31. Why is it often more healthful to eat vegetables that have been steamed rather than boiled?

Metabolism Is the Cycling of Biomolecules Through the Body

32. What is the general outcome of catabolism?

33. What is the general outcome of anabolism?

The Food Pyramid Summarizes a Healthful Diet

34. Which type of biomolecule does the food pyramid recommend we eat the most of?

35. Are all dietary fibers made of cellulose?

36. Is it possible to eat a food low on the glycemic index and still experience a significant increase in blood glucose?

37. Which type of lipoproteins have a greater association with the formation of plaque on artery wall: LDLs or HDLs?

38. Why doesn't the human body synthesize the essential amino acids?

Hands-On Chemistry Insights

Spit in Blue

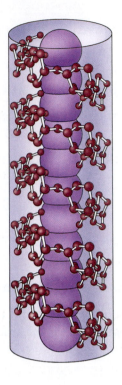

The amylose–iodine complex is shown above. Amylase, an enzyme in your saliva, broke down the amylose in the solution you spit into, meaning there is less starch present to react with the iodine and consequently a lighter blue solution.

The reason you had to wait a few minutes before adding the iodine has to do with how the amylase attacks each starch molecule. The enzyme does not attack in the middle of an amylose strand so that the strand is broken first in half, then in quarters, and so on. Instead, the amylase attacks only at the two ends of each strand, cleaving only one glucose unit at a time from each end and so destroying the starch molecule only very slowly.

Here are some questions relating to this activity. Enzymes such as amylase are destroyed by heat—how could you confirm this experimentally? If you boil one starch solution for only a few minutes and a second starch solution for an excessively long time, which will be light blue and which will be dark blue when you add iodine? Many instant Cream of Wheat cereals contain papain, an enzyme related to amylase. Can it be said that these instant cereals are being digested *before* reaching your mouth?

Sizzle Sources

With the chilled lipid layers, you can assume that, in general, the more solid the sample, the higher its proportion of saturated fats.

As you should have discovered from this activity, the "light" brands of spread contain fewer calories simply because they contain a greater proportion of water. Rather than water, some brands whip air into the spread. Either way, the net result is fewer lipid molecules per serving, which for saturated fats is not too bad a deal. Note that many of the "light" brands are labeled "for spread purposes only, not for cooking." Why do you suppose this is so?

Exercises

1. Does a carbohydrate contain water?

2. What is another biological use for carbohydrates besides energy?

3. In what ways are cellulose and starch similar to each other? In what ways are they different from each other?

4. Why does starch begin to taste sweet after it has been in your mouth for a few minutes?

5. Why are lipids insoluble in water?

6. Why is it important to have cholesterol in your body?

7. Could a food product containing glycerol and fatty acids but no triglycerides be advertised as being fat-free? If so, how might such advertising be misleading?

8. Silk is more waterproof than cotton. Why?

9. You are a beautician about to apply a reducing agent to a customer with fine hair who wants to have his hair curly. Should the reducing agent be regular strength, concentrated, or diluted?

10. Why is a permanent wave not really permanent?

11. When an unknown peptide containing five amino acids is treated with an enzyme that hydrolyzes only the serine–leucine peptide bond, (but not the leucine–serine bond) the fragments Leu–Cys, Ser, and Leu–Ser are formed. What was the original amino acid sequence in the peptide?

12. Why can't your body produce proteins from carbohydrates and fats alone?

13. Identify the molecular attractions occurring in this large protein at the locations a, b, and c:

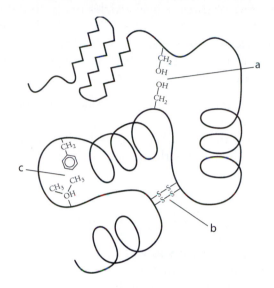

Distinguish among the primary, secondary, and tertiary structures of this protein.

14. Why do changes in pH interfere with the function of proteins? For your answer, consider the acid–base character of many amino acid side chains.

15. A common source of DNA damage is the spontaneous loss of the amine group on cytosine. This occurs at a rate of about 100 times a day. Fortunately, the body produces enzymes able to detect and repair such deaminated cytosines. Given this information, suggest why DNA differs from RNA in possessing the nucleotide thymine rather than uracil.

16. List codon, gene, nucleic acid, and nucleotide in order of increasing size.

17. Which amino acid does the DNA sequence ATG code for?

18. Why is the number of adenines in a DNA molecule always the same as the number of thymines?

19. What polypeptide is coded for by the mRNA sequence AUGGACCCAGCGUGAUGUA?

20. What polypeptide would be coded for by the mRNA sequence in Exercise 19 if the second G

from the left were somehow deleted? What problems might this change cause in a gene?

21. Why does mRNA not remain associated with DNA after being generated through transcription?

22. How many symmetrical sequences can you find in the DNA segment

> G T A G T T A A C C A G T C C G G A A G
>
> C A T C A A T T G G T C A G G C C T T C

23. Both water-soluble and water-insoluble vitamins can be toxic in large quantities. Our bodies are much more tolerant of the water-soluble ones, however. Why?

24. The dietary minerals must be in ionic form in order for the body to make use of them. Why?

25. A friend of yours loads up on vitamin C once a week instead of spacing it out over time. She argues the convenience of not having to take pills every day. What advice do you have for her?

26. Which statement is more accurate:

   a. Vitamins are needed by the body to avoid vitamin-deficiency diseases, such as scurvy.

   b. Vitamins are needed by the body so that many of its catabolic and anabolic reactions can proceed efficiently.

27. Suggest why the glycemic index for sucrose is only about 64 percent that of glucose.

28. Suggest why, in the intestine, breaking starch down to glucose takes longer than breaking sucrose down to glucose.

29. Mammals cannot produce polyunsaturated fatty acids. How is it then that the lard obtained from beef fat contains up to 10 percent polyunsaturated fatty acids?

30. Peanut butter has more protein per gram than a hard-boiled egg, and yet the egg represents a better source of protein. Why?

31. The human body stores excess glucose as glycogen and excess fat as fatty tissue that can accumulate beneath the skin. How does the body store any excess amino acids?

32. Cold cereal is often fortified with all sorts of vitamins and minerals but is deficient in the amino acid lysine. How might this deficiency be compensated for in a breakfast meal?

Discussion Topics

1. The Human Genome Project is being conducted by both private groups and government agencies. This raises the issue of who should profit from the information gathered and to what degree. What do you think? Should all the information gathered be in the public domain so that anyone has the right to capitalize on this information? Alternatively, should some of the information be owned privately so that it can be sold to help recover expenses and/or to make a profit?

2. Some diets, most notably the Atkins diet, call for large amounts of protein and fat and small amounts of carbohydrate. One of the claims of such diets is that for the same number of calories, a meal high in protein and fat leaves a person with less of an urge to eat later on. One of the arguments against such diets is that they are hard on the kidneys and liver. Find out more about the advantages at www.atkinscenter.com and the disadvantages at www.healthcentral.com (keyword search: high-protein diet), and discuss the issues with your classmates and friends.

Exploring Further

John Rennie, editor, "The Business of the Human Genome: Special Industry Report." *Scientific American* **283**(1), July 2000.

A series of three review articles on the progress of the Human Genome Project. Included are overviews of the science involved in mapping human DNA and how the knowledge gained may be put to practical use.

http://www.ornl.gov/TechResources/Human_Genome/home.html

Site for the Human Genome Project, which began in 1990 with the goal of identifying all the genes in human DNA.

http://vm.cfsan.fda.gov/list.html

Site for the Food and Drug Administration's Center for Food and Safety and Applied Nutrition. Here you will find links to the FDA's stance on such topics as dietary supplements, food labeling, and nutrition.

http://www.dietitian.com

The question-and-answer site of Joanne Larsen, a reputable expert in the field of dietetics. Peruse her well-formed answers, and you'll find ample use of the many terms and concepts introduced in this chapter.

the **Chemistry** place

Chemicals of Life

Visit The Chemistry Place at:
www.aw.com/chemplace

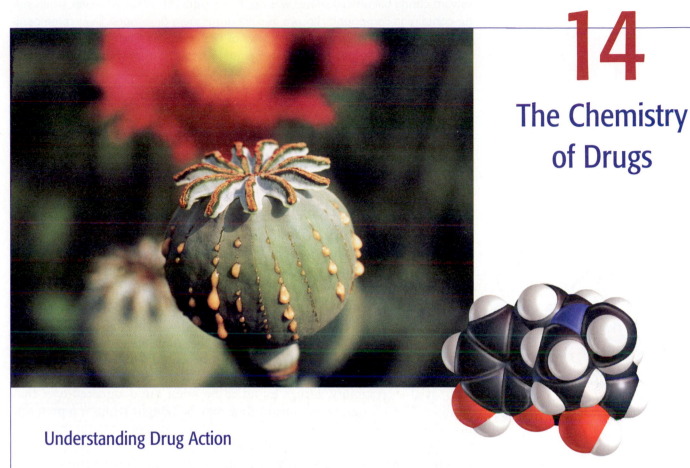

14

The Chemistry of Drugs

Understanding Drug Action

Archaeological evidence shows that early civilizations were keenly aware of the medicinal properties of certain plants. In A.D. 78, for example, the Greek physician Dioscorides wrote *Materia Medica*, a treatise in which he described about 600 plants known to have medicinal properties. Included in this list was the opium poppy, shown in this chapter's opening photograph. Incisions in the seed capsules of this plant yield a milky sap. When air-dried and kneaded, the sap forms a soft material known as opium, which contains *opioids*, a class of alkaloids known for their pain-killing and tranquilizing effects. The molecule shown is morphine, one of the more abundant and potent opioids.

With the development of chemistry in the early 1800s came the understanding that natural products owe their medicinal properties to certain substances they contain. In 1806, for example, morphine was isolated from opium, and in 1820 quinine, a drug useful in fighting malaria, was isolated from the bark of the cinchona tree. Soon, compounds produced in the laboratory were also found to have medicinal properties. In the 1840s, for example, anesthetic activity was found in the synthetic chemicals chloroform, nitrous oxide, and ethyl ether, making painless surgery and dentistry possible.

In the 1860s, Louis Pasteur (1822–1895) confirmed the germ theory of disease with his discovery of bacteria. This led to the discovery of the antiseptic properties of phenol and related compounds, which, as discussed in Chapter 12, could be used to prevent bacterial infection. The first major advance toward curing bacterial diseases was not made until the 1930s, however, when sulfur-containing compounds known as sulfa drugs were developed. Next came penicillin, an antibiotic derived from extracts of the mold *Penicillium notatum*. Subsequent research has led to an ever-expanding array of drugs—both natural and synthetic. Today, there are more than 25,000 prescription drugs and 300,000 nonprescription drugs available in the United States.

This chapter describes the major categories of drugs and the methods used in developing new drugs. It also addresses some of the social issues arising from our reliance on these chemicals.

14.1   Drugs Are Classified by Safety, Social Acceptability, Origin, and Biological Activity

Loosely defined, a *drug* is any substance other than food or water that affects how the body functions. A drug having therapeutic properties is also referred to as a *medicine*. Drugs, some legal and others illegal, are also used for nonmedical purposes. Legal nonmedical drugs include alcohol, caffeine, and nicotine. Illegal nonmedical drugs include heroin and cocaine.

There are a variety of ways to classify drugs. As Table 14.1 shows, the U.S. Drug Enforcement Agency (DEA) classifies them according to safety and social acceptability. Drugs found to be safest are designated over-the-counter (OTC) drugs, which means they may be bought without a prescription. Prescription drugs are those that should be taken only under the supervision of a physician because of their strong potency or their potential for misuse or abuse. The DEA further classifies drugs according to abuse potential, using the *schedule system* shown in Table 14.1.

Drugs may also be classified according to origin, as is done in Table 14.2. Drugs that are natural products come directly from terrestrial or marine plants or animals. Drugs that are chemical derivatives are natural products that have been chemically modified to increase potency or decrease side effects. Synthetic drugs are those that originate in the laboratory.

Table 14.1

U.S. Drug Enforcement Agency Classification of Drugs

Classification	Description	Examples
Over-the-counter (OTC) drugs	Available to anyone	Aspirin, cough medicines
Permitted nonmedical drugs	Available in food, beverages, and tobacco products	Alcohol, caffeine, nicotine
Prescription drugs	Require physician authorization	Antibiotics, birth control pills
Controlled substances		
Schedule 1	No medical use, high abuse potential	Heroin, LSD, mescaline, marijuana
Schedule 2	Some medical use, high abuse potential	Amphetamines, cocaine, morphine, codeine
Schedule 3	Prescription drugs, abuse potential	Barbiturates, tranquilizers

Table 14.2

The Origin of Some Common Drugs

Origin	Drug	Biological Effect
Natural product	Caffeine	Nerve stimulant
	Reserpine	Hypertension reducer
	Vincristine	Anticancer agent
	Penicillin	Antibiotic
	Morphine	Analgesic
Chemical derivative of natural product	Prednisone	Antirheumatic
	Ampicillin	Antibiotic
	LSD	Hallucinogenic
	Chloroquinine	Antimalarial
	Ethynodiol diacetate	Contraceptive
Synthetic	Valium	Antidepressant
	Benadryl	Antihistamine
	Allobarbital	Sedative–hypnotic
	Phencyclidine	Veterinary anesthetic
	Methadone	Analgesic

Perhaps the most common way to classify drugs is according to their primary biological effect, which is how they are presented in this chapter. It must be noted, however, that most drugs exhibit a broad spectrum of activity and therefore may fall under several classifications. Aspirin, for example, relieves pain, but it also reduces fever and inflammation, thins the blood, causes ringing in the ears, and may lead to Reye's syndrome in children. Morphine relieves pain, but it also constipates and suppresses the urge to cough.

At times, the multiple effects of a drug are desirable. Aspirin's pain-reducing and fever-reducing properties work well together in treating flu symptoms in adults, for instance, and its blood-thinning ability is useful in the prevention of heart disease. Morphine was widely used during the American Civil War both for relieving the pain of battle wounds and for controlling diarrhea. Often, however, the side effects of a drug are less desirable. Ringing in the ears, Reye's syndrome, and upset stomach are a few of the negative side effects of aspirin, and a major side effect of morphine is its addictiveness. A main goal of drug research, therefore, is to find drugs that are specific in their action and have minimal side effects.

Although two drugs being taken together may have different primary activities, they may share a common secondary activity that can be amplified when the two drugs are taken together. One drug enhancing the action of another is called the **synergistic effect**, and a synergistic effect is often more powerful than the sum of the activities of the two drugs taken separately. One of the great challenges of physicians and pharmacists is keeping track of all the possible combinations of drugs and potential synergistic effects.

The synergism that results from mixing drugs that have the same primary effect is particularly hazardous. For example, a moderate dose of a sedative combined with a moderate amount of alcohol may be lethal. In fact, most drug overdoses are the result of a combination of drugs rather than the abuse of a single drug.

Concept Check ✔

Distinguish between a drug and a medicine.

Was this your answer? A drug is any substance administered to affect body function. A medicine is any drug administered for its therapeutic effect. All medicines are drugs, but not all drugs are medicines.

14.2 The Lock-and-Key Model Guides Chemists in Synthesizing New Drugs

To find new and more effective medicines, chemists use various models that describe how drugs work. By far, one of the most useful models of drug action is the **lock-and-key model**. The basis of this model is that there is a connection between a drug's chemical structure and its biological effect. For example, morphine and all related pain-relieving opioids, such as codeine and heroin, have the T-shaped structure shown in Figure 14.1.

According to the lock-and-key model, illustrated in Figure 14.2, biologically active molecules function by fitting into *receptor sites* on proteins in the body, where they are held by molecular attractions, such as hydrogen bonding. When a drug molecule fits into a receptor site the way a key fits into a lock, a particular biological event is triggered, such as a nerve

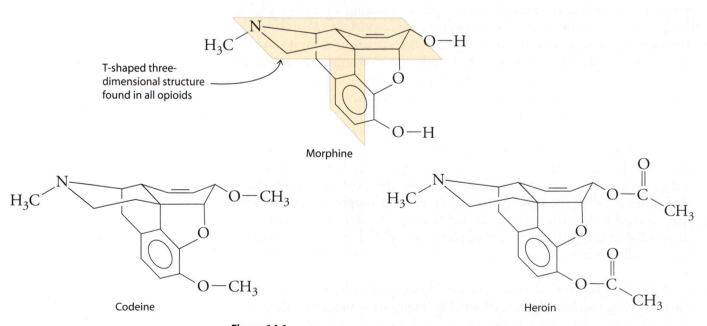

Figure 14.1
All drugs that act like morphine have the same basic three-dimensional shape as morphine.

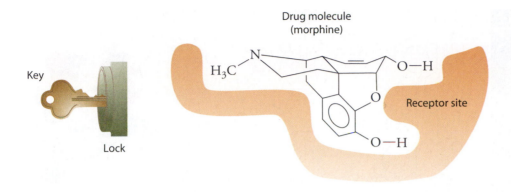

Key

Lock

Drug molecule
(morphine)

Receptor site

Figure 14.2
Many drugs act by fitting into receptor sites on molecules found in the body, much as a key fits in a lock.

impulse, a change in the shape of a protein, or even a chemical reaction. In order to fit into a particular receptor site, however, a molecule must have the proper shape, just as a key must have properly shaped notches in order to fit a lock.

Another facet of this model is that the molecular attractions holding a drug to a receptor site are easily broken. (Recall from Chapter 7 that most molecular attractions are many times weaker than chemical bonds.) A drug is therefore held to a receptor site only temporarily. Once the drug is removed from the receptor site, body metabolism destroys the drug's chemical structure and the effects of the drug are said to have worn off.

Using this model, we can understand why some drugs are more potent than others. Heroin, for example, is a more potent pain killer than is morphine because the chemical structure of heroin allows for tighter and longer binding to its receptor sites.

The lock-and-key model has developed into one of the central tenets of pharmaceutical study. Knowing the precise shape of a target receptor site allows chemists to design molecules that have an optimum fit and a specific biological effect.

Biochemical systems are so complex, however, that our knowledge is still limited, as is our capacity to design effective medicinal drugs. For this reason, most new medicinal drugs are still discovered instead of designed. One important avenue for drug discovery is ethnobotany. An *ethnobotanist* is a researcher who learns about the medicinal plants used in indigenous cultures, such as the root of the Bobgunnua tree, shown in Figure 14.3.

Figure 14.3
Ethnobotanists directed natural-products chemists to the yellow coating on the root of the African Bobgunnua tree. Indigenous people have known for many generations that this coating has medicinal properties. From extracts of the coating, the chemists isolated a compound that is highly effective in treating fungal infections. This compound, produced by the tree to protect itself from root rot, shows much promise in the treatment of the opportunistic fungal infections that plague those suffering from AIDS.

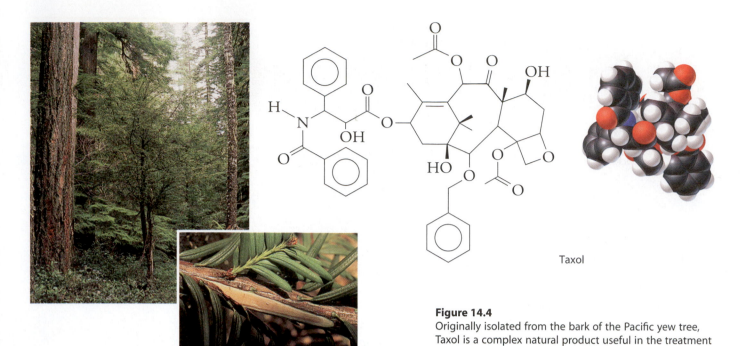

Taxol

Figure 14.4
Originally isolated from the bark of the Pacific yew tree, Taxol is a complex natural product useful in the treatment of various forms of cancer.

Today there are hundreds of clinically useful prescription drugs derived from plants. About three quarters of these came to the attention of the pharmaceutical industry as a result of their use in folk medicine.

Another important method of drug discovery is the random screening of vast numbers of compounds. Each year, for example, the National Cancer Institute screens some 20,000 compounds for anticancer activity. One successful hit was the compound Taxol, shown in Figure 14.4. This compound has significant activity against several forms of cancer, especially ovarian cancer.

A drug isolated from a natural source is not necessarily better or more benign than one produced in the laboratory. Aspirin, for example, is a human-made chemical derivative, and it is certainly more benign than cocaine, which is 100 percent natural. The main advantage of natural products is their great *diversity*. Each year, more than 3000 new chemical compounds are discovered from plants. Many of these compounds are biologically active, serving the plant as a chemical defense against disease or predators. Nicotine, for example, is a naturally occurring insecticide produced by the tobacco plant to protect itself from insects.

It has been estimated that only 5000 plant species have been studied exhaustively for possible medical applications. This is a minor fraction of the estimated 250,000 to 300,000 plant species on our planet, most of which are located in tropical rainforests. That we know little or nothing about much of the plant kingdom has raised justified and well-publicized concern, for as these forests are being destroyed, also being destroyed are species that might yield useful medicines.

A recent laboratory approach intended to mimic nature's chemical diversity is known as **combinatorial chemistry** and is a method of generating a large "library" of related compounds. Combinatorial chemistry

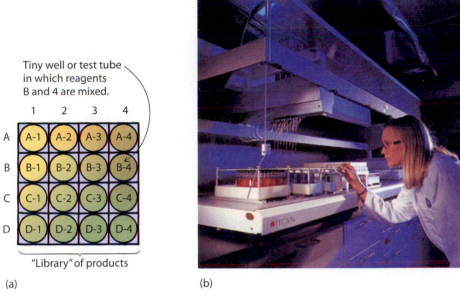

Tiny well or test tube in which reagents B and 4 are mixed.

"Library" of products

(a)

(b)

Figure 14.5
(a) Eight hypothetical starting materials A through D and 1 through 4 can be combined in various ways to yield 16 products, each of which may have some biological activity not found in any of the starting materials. (b) A multitude of products are thus immediately available to be screened for medicinal activity.

takes advantage of the many different ways in which a series of chemicals may be combined. Microquantities of reagents are combined in a grid so as to maximize the number of possible products, as is illustrated in Figure 14.5a. The result is a great number of closely related compounds that can be screened for biological activity. The most active derivatives are analyzed for chemical structure and then synthesized on a larger scale for further testing or clinical trials. A typical array is shown in Figure 14.5b.

Concept Check ✓

Why are organic chemicals so suitable for making drugs?

Was this your answer? Because their vast diversity permits the manufacture of the many different types of medicines needed to combat the many different types of illnesses humans are subject to.

14.3 Chemotherapy Cures the Host by Killing the Disease

The use of drugs that destroy disease-causing agents without destroying the animal host is known as **chemotherapy**. This approach is effective in the treatment of many diseases, including bacterial infections, viral infections, and cancer. It works by taking advantage of the ways a disease-causing agent, also known as a *pathogen*, is different from a host.

Sulfa Drugs and Antibiotics Treat Bacterial Infections

Sulfa drugs are synthetic drugs first used to treat bacterial infections in the 1930s. They work by taking advantage of a striking difference between humans and bacteria—even though both must have the nutrient folic acid in order to remain healthy, we humans can absorb folic acid from what we eat, but bacteria cannot. Instead, bacteria must make their own supply of folic acid. For this, they possess enzymes that help make folic acid from a simpler molecule found in all bacteria, para-aminobenzoic acid (PABA). The PABA attaches to its specific receptor site on the bacterial enzyme and is converted to folic acid, as shown in Figure 14.6.

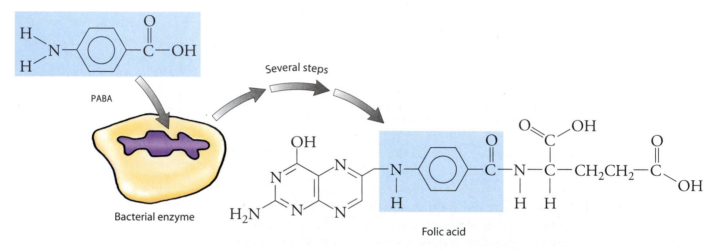

Figure 14.6
Bacterial enzymes use para-aminobenzoic acid (PABA) to synthesize folic acid.

Sulfa drugs have a close structural resemblance to PABA. When taken by a person suffering from a bacterial infection, a sulfa drug is transformed by the body to the compound *sulfanilamide*, which attaches to the bacterial receptor sites designed for PABA, as shown in Figure 14.7, thereby preventing the synthesis of folic acid. Without folic acid, the bacteria soon die. The patient, however, because he or she receives folic acid from the diet, lives on.

Antibiotics are chemicals that prevent the growth of bacteria. They are produced by such microorganisms as molds, other fungi, and even bacteria. The first antibiotic discovered was penicillin. Many derivatives of penicillin, such as the penicillin G shown in Figure 14.8, have since been isolated from microorganisms as well as prepared in the laboratory. Penicillins and the closely related compounds known as cephalosporins, also shown in Figure 14.8, kill bacteria by inactivating an enzyme responsible for strengthening the bacterial cell wall. Without this enzyme, bacterial cell walls grow weak and the cells eventually burst.

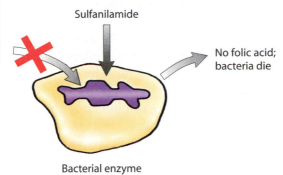

Figure 14.7
In the body all sulfa drugs are metabolized to sulfanilamide, which binds to the bacterial enzymes and keeps them from doing their job.

Sulfanilamide

PABA

No folic acid; bacteria die

Bacterial enzyme

Figure 14.8
Penicillins, such as penicillin G, and cephalosporins, such as cephalexin, as well as most other antibiotics, are produced by microorganisms that can be mass-produced in large vats. The antibiotics are then harvested and purified.

Penicillin G

Cephalexin

Concept Check ✔

How is sulfanilamide poisonous to bacteria but not to humans?

Was this your answer? Sulfanilamide is poisonous to bacteria because it prevents them from synthesizing the folic acid they need to survive. Humans utilize folic acid from their diet, and so they are not bothered by sulfanilamide's ability to disrupt the synthesis of folic acid.

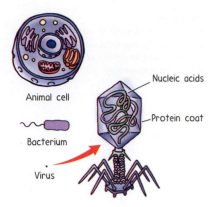

Figure 14.9
Viruses are much smaller than bacteria and many times smaller than animal cells. (Notice the small dot representing the virus.) The smallest of all pathogens, viruses consist mostly of nucleic acids enclosed in a protein coat.

Chemotherapy Can Inhibit the Ability of Viruses to Replicate

So far, chemotherapy has been more successful in treating bacterial infections than in treating viral infections. Perhaps the greatest obstacle to effective viral treatment lies in the nature of viruses. When not attached to a host, a virus is an inert, lifeless bundle of biomolecules—and it's difficult to kill something that's not alive! A typical virus, shown in Figure 14.9, consists of only one or several strands of either RNA or DNA encapsulated by a protein coat. Some viruses infect by attaching to a cell and then injecting their genetic contents into the cell. Once inside the cell, the virus's genetic information is incorporated into the host DNA and replicated by the host cell. Eventually, the cell bursts because it is overstuffed with a multitude of viral replicates, which then spread to infect other host cells.

The most common antiviral drugs are derivatives of *nucleosides*, which are similar to nucleotides (Section 13.5) but do not have a phosphate group. Nucleosides roam freely in all cells and are used by the cells to create RNA or DNA. Before being used, however, the nucleosides must first be primed with three phosphate groups, as shown in Figure 14.10. Various synthetic derivatives of nucleosides are readily primed by virus-infected cells but not by uninfected cells. Two synthetic nucleoside derivatives, both shown in Figure 14.11, are acyclovir, sold under the trade name Zovirax, and zidovudine, sold under the trade name AZT. Once incorporated in the RNA or DNA of a virus-infected host cell, these nucleoside derivatives disrupt protein synthesis, and the infected cell dies before replicating the virus. Proliferation of the virus, though not halted, is thus brought under control.

Acyclovir is useful in the treatment of herpes. Oral herpes is caused by the herpes simplex virus 1 (HSV-1), and genital herpes is caused by the herpes simplex virus 2 (HSV-2). More than 90 percent of the world's population is infected with the oral herpes virus, though there are many infected people who do not exhibit symptoms. Genital herpes is the most prevalent noncurable sexually transmitted disease. In the United States, there are about 30 million people infected with HSV-2 and an estimated 200,000 to 500,000 new cases each year.

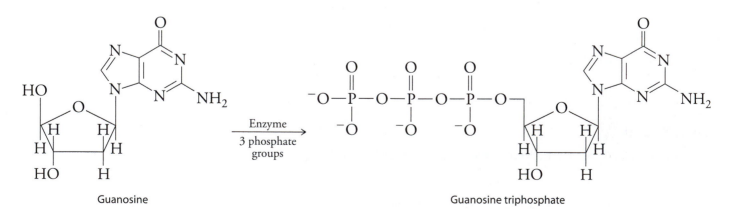

Figure 14.10
Before a nucleoside, such as guanosine, can be incorporated into RNA or DNA, it must be activated by having three phosphate groups attached to it.

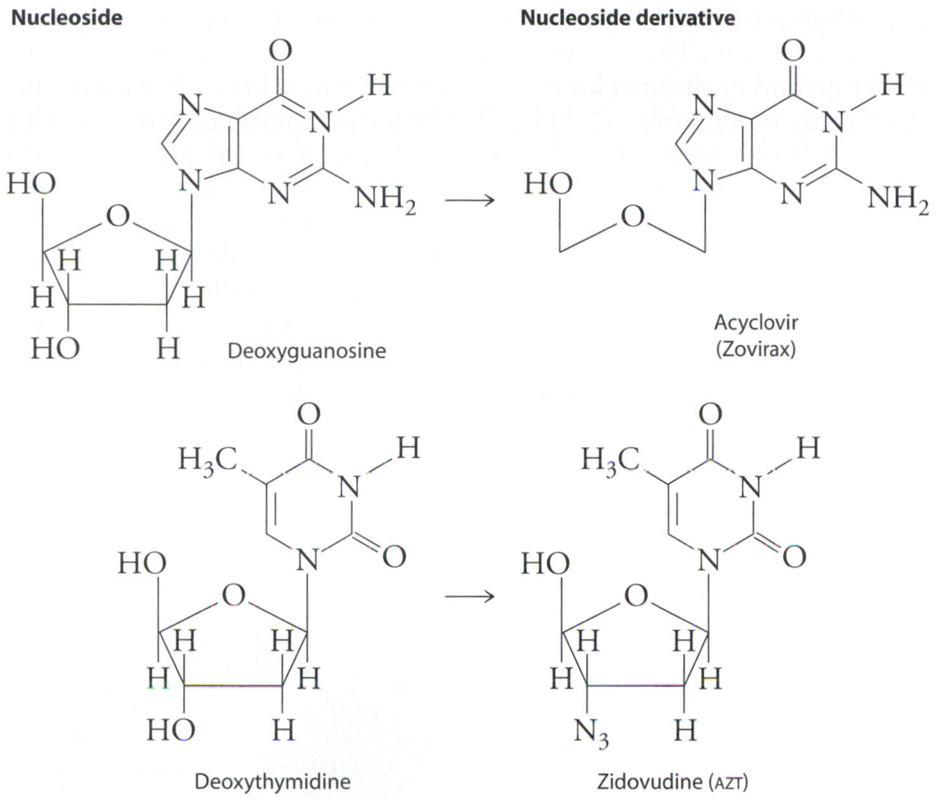

Nucleoside   **Nucleoside derivative**

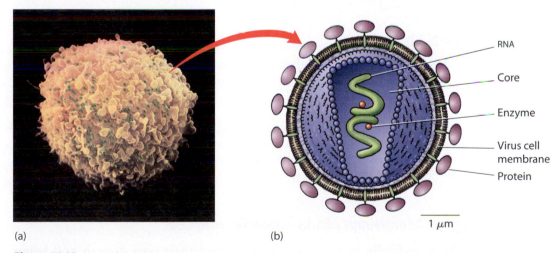

Deoxyguanosine

Acyclovir
(Zovirax)

Deoxythymidine

Zidovudine (AZT)

Figure 14.11
Acyclovir (Zovirax) is a derivative
of the nucleoside deoxyguanosine,
and zidovudine (AZT) is a derivative
of the nucleoside deoxythymidine.

Zidovudine is used to suppress the replication of the human immu-nodeficiency virus (HIV), which is responsible for aquired immune defi-ciency syndrome, AIDS (Figure 14.12). According to the World Health Organization, by the beginning of 2000 more than 34 million people were infected with HIV and about 19 million people around the world had already lost their lives to AIDS.

RNA

Core

Enzyme

Virus cell
membrane

Protein

1 μm

(a)   (b)

Figure 14.12
(a) The small green bodies covering this white blood cell are human immunodeficiency viruses.
(b) The anatomy of HIV. After the initial infection, the infected person's immune response elimi-nates most of the virus. Some of the virus, however, remains dormant in infected cells and evades the immune response. Over a period of years, HIV reactivates itself, the immune system collapses, and the person succumbs to opportunistic diseases, such as cancer or pneumonia.

HIV research has led to a new class of antiviral agents known as *protease inhibitors*. The life cycle of many viruses, including HIV, depends on the actions of enzymes known as proteases, which break down proteins. A protease, for example, might be used by a virus to penetrate the proteins on the cell membrane of a host cell or to break down the host's polypeptides to create a supply of amino acids necessary for viral replication. Drugs that block the action of proteases control viral proliferation. An example of an effective protease inhibitor is nelfinavir, sold under the trade name Viracept and shown in Figure 14.13. Patients receiving a "cocktail" of a protease inhibitor and nucleoside antiviral agents may have their HIV count brought below detectable levels. Though it is unlikely that this regimen can totally eliminate HIV from the body, the highly reduced viral counts tend to significantly delay the onset of AIDS and reduce the patient's infectiousness to others.

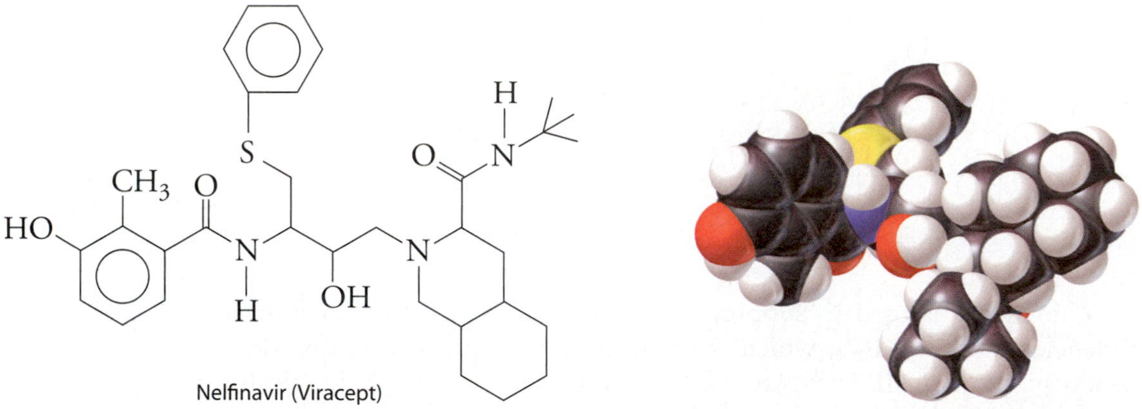

Nelfinavir (Viracept)

Figure 14.13
The protease inhibitor nelfinavir.

Concept Check ✔

A virus is so much simpler than a bacterium. Why then are viruses so much more difficult to target with chemotherapeutics?

Was this your answer? Chemotherapeutics act by interfering with one or more of the chemical reactions a pathogen needs in order to exist. The more complex a pathogen, the more ways there are to interfere with its life cycle. That viruses are so simple means we have few avenues for a chemotherapeutic approach.

Cancer Chemotherapy Attacks Rapidly Growing Cells

Cells periodically lose the ability to control their own growth and begin multiplying rapidly. Normally, these renegade cells are recognized by the immune system and destroyed. Occasionally, however, they escape this line of defense and continue to multiply unchecked. The result is a hard mass of tissue, called a *tumor*, that deprives healthy cells of oxygen and nutrients. Cells from a tumor may break away and be carried to other sites in the body, where they lodge and continue to multiply, forming additional tumors. As tumors multiply, more and more healthy cells suffer and eventually die. Ultimately, the whole body may die. This is cancer, the second leading cause of death in most developed nations. At present mortality rates, one in six of us will die of this disease.

Cancer is actually not a single disease but rather a group of many different kinds of disease, each having its own behavior and its own susceptibility to treatment. While some cancers respond to chemotherapy alone, most require a combination of chemotherapy, radiation therapy, and surgery.

Chemotherapy is most effective at the early stages of cancer because drugs work best on cells that are in the process of dividing, a process called *cellular mitosis*, shown in Figure 14.14. When a tumor is young, most of its cells are undergoing mitosis. As the tumor ages, however, the fraction of cells in this growth phase decreases, and so drug sensitivity is reduced. Drugs also have a difficult time destroying all the cells in a large tumor. A 100-gram tumor, for example, may contain about 100 billion cells. Killing 99.9 percent of these cells would still leave 100 million cells—too many for the patient's immune response to control. The same treatment against a 50-milligram tumor containing about a million cells would leave only about 1000 cells, which can be controlled by the immune system. Survival rates from cancer are therefore greatly increased by early diagnosis. Hence, you are advised to keep close watch over your body for unusual signs and schedule regular checkups with your physician.

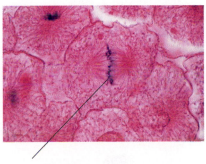

Parent cell with single set of chromosomes

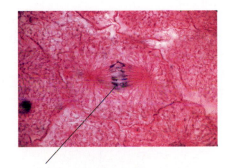

Duplicated chromosomes

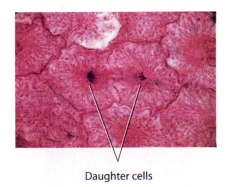

Daughter cells

Figure 14.14
During cellular mitosis, DNA and certain cellular proteins bundle together into chromosomes, which are visible under a microscope. These chromosomes duplicate themselves and then divide evenly into two separate cells called daughter cells.

Unfortunately, cancerous cells are not the only cells in the body that divide. Normal cells divide periodically, and some types of cells, such as those in the gastrointestinal tract and in hair follicles, are always in a state of cellular division. As a consequence, cancer chemotherapeutics are noted for their toxicity, with patients undergoing treatments often experiencing gastrointestinal problems and hair loss.

DNA is the target of many anticancer compounds because during cellular mitosis strands of DNA are unwound and therefore susceptible to chemical attack. A variety of chemicals may be used to selectively kill cells

Figure 14.15
These anticancer agents all kill dividing cells by targeting the cells' DNA.

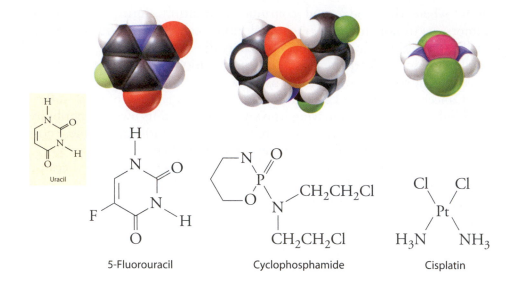

Uracil

5-Fluorouracil

Cyclophosphamide

Cisplatin

that are in the process of dividing. The compound 5-fluorouracil, for example, shown in Figure 14.15, is mistaken by a cell for the nucleotide base uracil. Once incorporated into the DNA of the cancerous cell, 5-fluorouracil's non-nucleotide structure interferes with the normal DNA workings, and the cell dies. Harsher agents, such as cyclophosphamide and cisplatin, also shown in Figure 14.15, destroy DNA's ability to function by chemically bonding to the DNA or by cross-linking the two strands of the double helix.

Some anticancer drugs kill cancerous cells without acting on DNA. Certain alkaloids, such as vincristine, shown in Figure 14.16, and Taxol, kill dividing cells by preventing the formation of cellular microstructures needed for division.

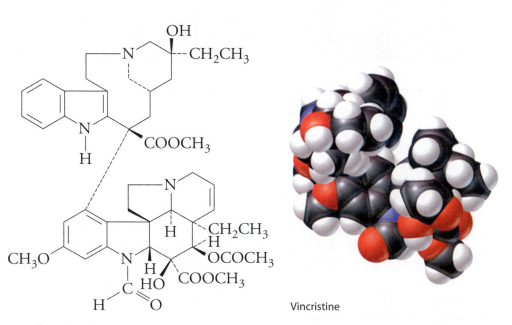

Vincristine

Figure 14.16
Vincristine is a naturally occurring alkaloid that has significant anticancer activity. It is isolated from a plant closely related to the periwinkle, a common ornamental plant of tropical and temperate regions.

As another point of attack, cancerous cells have high metabolic rates, which means they rely heavily on biochemical nutrients, such as the dihydrofolic acid shown in Figure 14.17. The anticancer agent methotrexate is structurally very similar to dihydrofolic acid and works by binding to dihydrofolic acid receptor sites in the cancerous cells, thereby interfering with metabolic reactions in the cells.

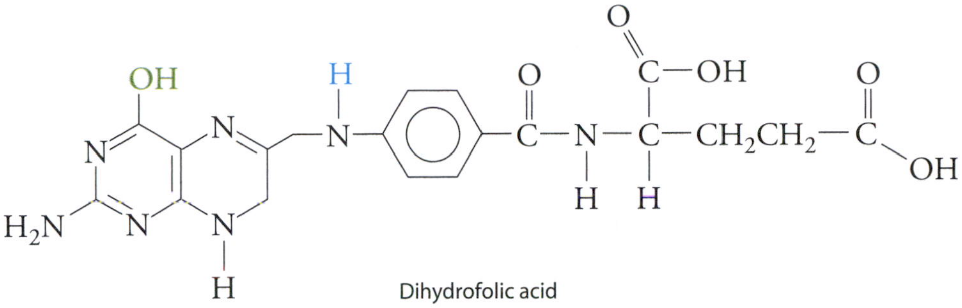

Dihydrofolic acid

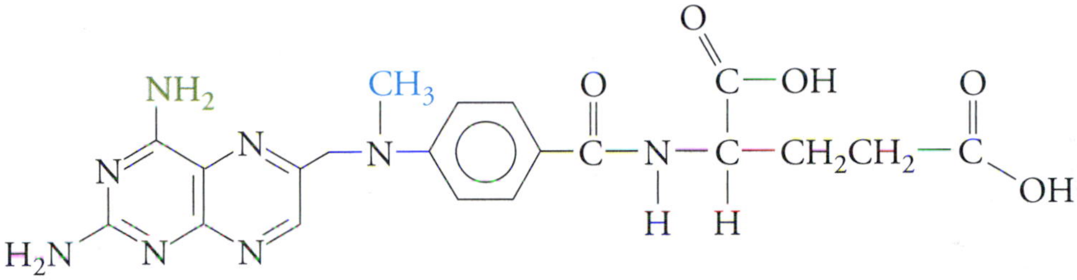

Methotrexate

Figure 14.17
Methotrexate disturbs the metabolism of cancerous cells by substituting for dihydrofolic acid at dihydrofolic acid receptor sites.

Cancer chemotherapy combined with radiation therapy and/or surgery can be effective in restraining or even curing many forms of cancer. As our understanding of cellular mechanics continues to grow, so will our ability to increase the overall survival rates of cancer patients.

14.4   Some Drugs Either Block or Mimic Pregnancy

In the 1930s, it was found that injected doses of the hormone progesterone maintain a state of false pregnancy during which a woman does not ovulate and therefore cannot conceive. Oral administration of progesterone does not have the same effect because the progesterone is quickly broken down by the digestive system. In the early 1950s, chemists developed compounds that were very similar to progesterone but retained their birth control properties when taken orally. The first birth control pill was marketed in 1960 and contained the progesterone mimicker norethynodrel as well as an estrogen derivative, mestranol, to help regulate the menstrual cycle. Many other derivatives of progesterone and estrogen have since been formulated into birth control pills, which are about 99 percent effective at preventing pregnancy. Table 14.3 on page 466 illustrates some drugs used for birth control. Today, more than 60 million women worldwide use birth control pills.

Table 14.3

Drugs for Birth Control

Drug	Structure	Action
Norethynodrel	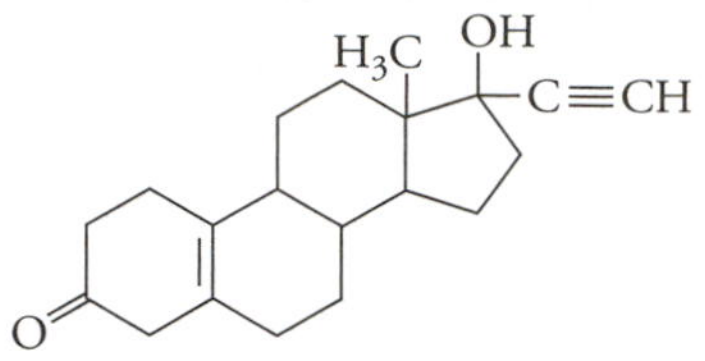	Mimics progesterone
Mifepristone	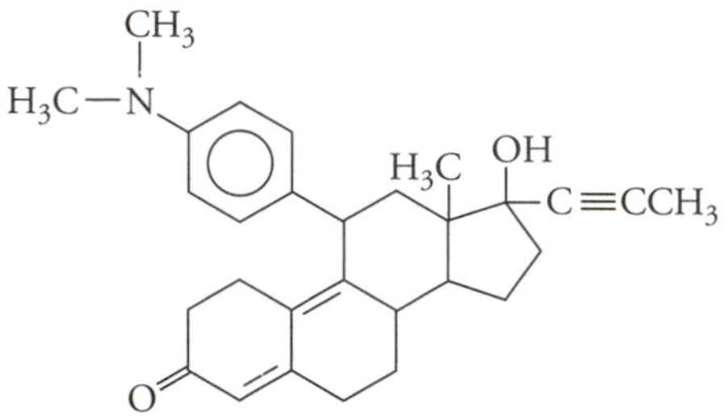	Blocks progesterone
Nonoxynol-9	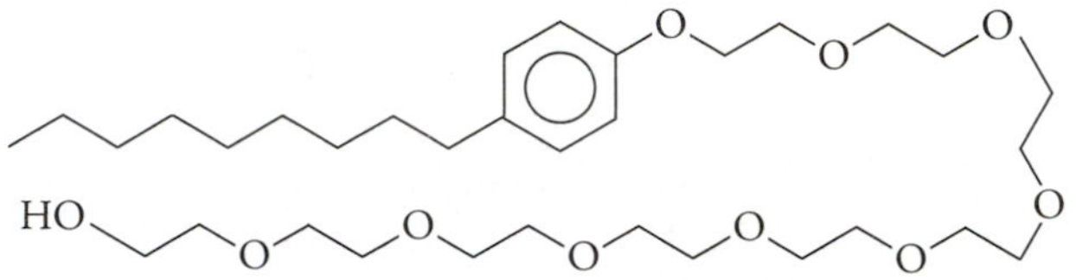	Kills sperm

Progesterone derivatives that block rather than mimic the action of progesterone are also effective as a form of birth control. Progesterone is vital to maintaining a pregnancy. Without it, or when its activity is blocked, the lining of the uterus is sloughed away, along with any fertilized ovum bound to the lining. A most effective progesterone blocker currently available is mifepristone, also known as RU-486. A controversy surrounding mifepristone is that, rather than preventing fertilization, it prevents a fertilized ovum from establishing itself in the uterus. Those who view a fertilized ovum as a human life tend to be opposed to the use of mifepristone. Those who differentiate between a fertilized ovum and a developing fetus, on the other hand, are more likely to approve of its use.

Birth control may also involve a spermicide, such as the nonoxynol-9 shown in Table 14.3. When used in conjunction with a barrier device, such as a condom or a cervical diaphragm, spermicides can be close to 95 percent effective at preventing pregnancy.

Birth control can also be achieved by causing a drop in a man's sperm count. This can be done by injecting testosterone, which at high blood levels inhibits the production of sperm. Recent advances have provided derivatives of testosterone that may be taken orally. Such male birth control pills have been shown to lower the sperm count in semen from normal values of about 100 million sperm per milliliter to less than 3 million sperm per milliliter, which is considered a very low concentration. Taken correctly, these pills are about as effective at preventing pregnancy as female birth control pills. Their long-term effects are still being studied.

14.5   The Nervous System Is a Network of Neurons

Many drugs function by affecting the nervous system. To understand how these drugs work, it is important to know the basic structure and functions of the nervous system.

Thoughts, physical actions, and sensory input all involve the transmission of electrical signals through the body. The conduit for these signals is a network of **neurons**, which are specialized cells capable of sending electrical impulses. First, in what is called the *resting phase*, a neuron primes itself for an impulse by ejecting sodium ions, as shown in Figure 14.18a. More sodium ions outside the neuron than inside creates a separation of charge. And a separation of charge gives rise to an electric potential of around −70 millivolts across the cell membrane. As shown in Figure 14.18b, a nerve impulse is a reversal in this electric potential that travels down the length of the neuron to the *synaptic terminals*. The reversal of the electric potential in an impulse occurs as sodium ions flush back into the neuron.

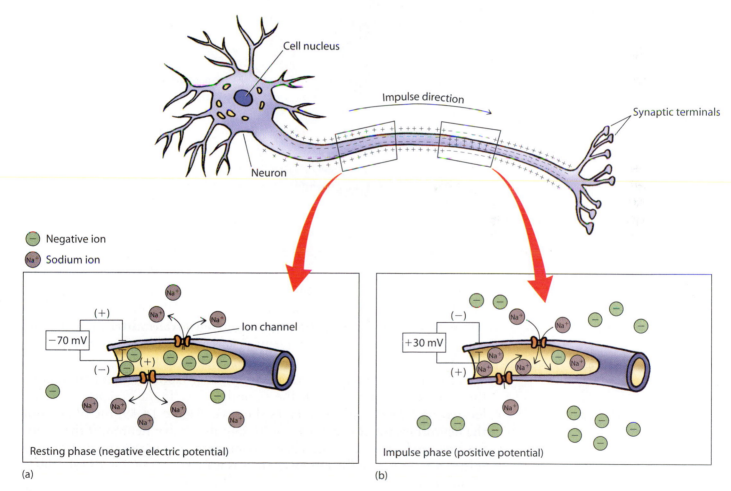

Figure 14.18
(a) The resting phase of a neuron maintains a greater concentration of sodium ions outside the cell. This results in a voltage of about −70 millivolts across the cell membrane. (b) In the impulse phase, sodium ions flush back into the cell to give a voltage of about +30 millivolts across the cell membrane.

After the impulse passes a given point along the neuron, the cell again ejects sodium ions at that point to re-establish the original distribution of ions and the potential of −70 millivolts.

Unlike the wires in an electric circuit, most neurons are not physically connected to one another. Nor are they connected to the muscles or glands on which they act. Rather, as Figure 14.19 shows, they are separated from one another or from a muscle or gland by a narrow gap known as the **synaptic cleft**.

Figure 14.19
The passage of neurotransmitters across a synaptic cleft.

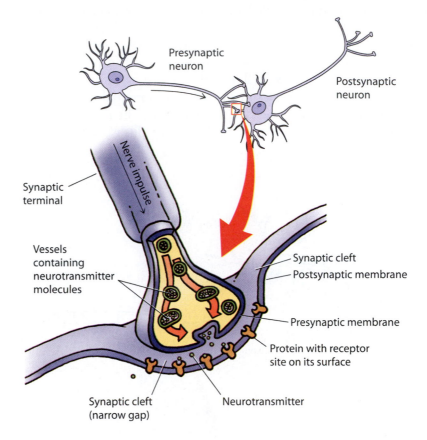

In the synaptic terminal of every neuron are bubble-like compartments called *vessels*. A nerve impulse reaching a synaptic cleft causes these vessels to release neurotransmitters into the cleft. **Neurotransmitters** are organic compounds that are released by a neuron and are capable of activating receptor sites.

A neurotransmitter, once released into the synaptic cleft, migrates across the cleft to receptor sites on the opposite side. If the receptor sites are located on a *postsynaptic neuron*, as shown in Figure 14.19, the binding of the neurotransmitter may start a nerve impulse in that neuron. If the receptor sites are located on a muscle or organ, then binding of the neurotransmitter may start a bodily response, such as muscle contraction or the release of hormones.

Two important classes of neurons are *stress neurons* and . Both types are always firing, but in times of stress, as when facing an angry bear or giving a speech, the stress neurons are more active than the maintenance neurons.

This condition is the *fight-or-flight* response, during which fear causes stress neurons to trigger rapid physiological changes to help defend against impending danger: the mind becomes alert, air passages in the nose and lungs open to bring in more oxygen, the heart beats faster to spread the oxygenated blood throughout the body, and nonessential activities such as digestion are temporarily stopped. In times of relaxation, such as sitting down in front of the television with a bowl of potato chips, the maintenance neurons are more active than the stress neurons. Under these conditions, digestive juices are secreted, intestinal muscles push food through the gut, the pupils constrict to sharpen vision, and the heart pulses at a minimal rate.

Concept Check ✔

What is a neurotransmitter?

Was this your answer? A neurotransmitter is a small organic molecule released by a neuron into a synaptic cleft. It influences neighboring tissue, such as the postsynaptic membrane of a neuron on the opposite side of the cleft, by binding to receptor sites.

Hands-On Chemistry: Diffusing Neurons

Nerve impulses can travel in a neuron at speeds of up to 100 meters per second (250 miles per hour!), but neurotransmitters travel across the synaptic cleft at only about 10^{-5} meter per second. One of the reasons for this comparative slowness is the process that moves the neurotransmitters across the synaptic cleft. Once the neurotransmitters are released into the cleft, they are prodded to the other side merely by the random bumping of jiggling molecules in the cleft—a process known as *diffusion*.

Recall from Section 1.6 that molecules slow down with decreasing temperature. The effect of temperature on diffusion can be readily seen by adding food coloring to water.

What You Need

Food coloring, three drinking glasses, ice-cold water, warm water, hot water

Procedure

① Fill one glass with the ice-cold water, one with the warm water, and one with the hot water. Allow the glasses of water to stand for a couple of minutes so that the water is perfectly still.

② Add a drop of food coloring to each glass. The drop will sink to the bottom and then begin to diffuse. Observe how long it takes until the water is uniformly colored.

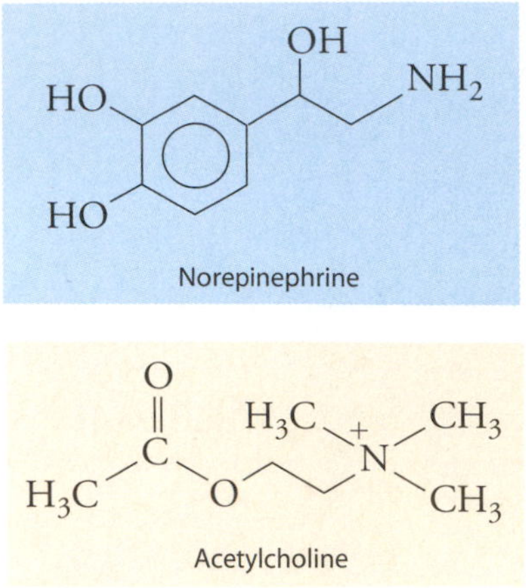

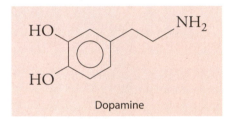

Figure 14.20
The chemical structures of the stress neurotransmitter norepinephrine and the maintenance neurotransmitter acetylcholine.

Neurotransmitters Include Norepinephrine, Acetylcholine, Dopamine, Serotonin, and GABA

On the chemical level, stress and maintenance neurons can be distinguished by the types of neurotransmitters they use. The primary neurotransmitter for stress neurons is *norepinephrine*, and the primary neurotransmitter for maintenance neurons is *acetylcholine*, both shown in Figure 14.20. As we shall see in the following sections, many drugs function by altering the balance of stress and maintenance neuron activity.

In addition to norepinephrine and acetylcholine, a host of other neurotransmitters contribute to a broad range of effects. Three examples are the neurotransmitters dopamine, gamma aminobutyric acid, and serotonin, all shown in Figure 14.21.

Dopamine plays a significant role in activating the brain's reward center, which is located in the hypothalamus, an area at the lower middle of the brain, as illustrated in Figure 14.22. The hypothalamus is the main control center for emotional response and behavior. Stimulation of the reward center by dopamine results in a pleasurable sense of *euphoria*, which is an exaggerated sense of well-being.

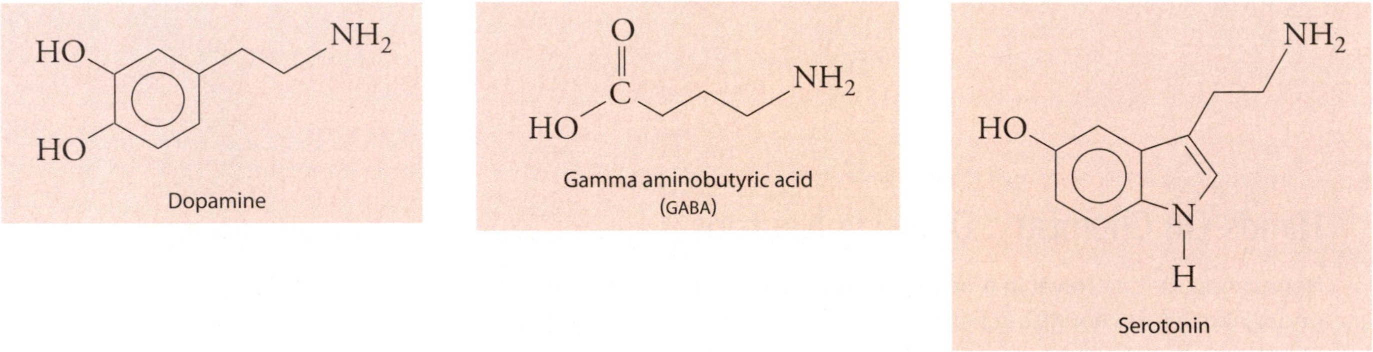

Figure 14.21
The chemical structures of three important neurotransmitters.

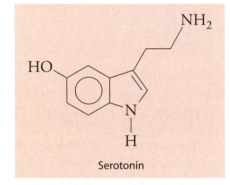

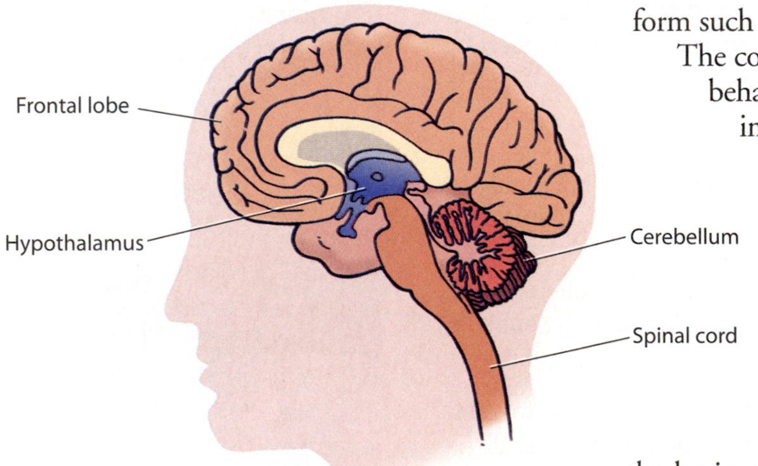

Figure 14.22
The human brain.

The control of physical responses ultimately allows us to perform such complex tasks as driving a car or playing the piano. The control of emotional responses allows us to refine our behavior, such as overcoming anxiety in tense social interactions or remaining calm in an emergency. The brain controls both physical and emotional responses by inhibiting the transmission of nerve impulses. The neurotransmitter responsible for this inhibition—*gamma aminobutyric acid* (GABA)—is *the* major inhibitory neurotransmitter of the brain. Without it, coordinated movements and emotional skills would not be possible.

Serotonin is another neurotransmitter used by the brain to block unneeded nerve impulses. To make sense of the world, the frontal lobes of the brain selectively block out a multitude of signals coming from the lower brain and from parts of the nervous system. We are not born with this ability to selectively block out information. In order to have an appropriate focus on the world, newborns must learn from experience which lights, sounds, smells, and feelings outside and inside their bodies must be dampened. A healthy, mature brain is one in which serotonin successfully suppresses lower-brain nerve signals. Information that does make it to the higher brain can then be sorted efficiently.

Drugs that modify the action of serotonin alter the brain's ability to sort information, and this alters perception. LSD is one such drug. While hallucinating, an LSD user rarely sees something that isn't there. Rather, the user has an altered perception of something that does exist.

Concept Check

Match the neurotransmitter to its primary function:

_____ norepinephrine   a. inhibits nerve transmission

_____ acetylcholine   b. stimulates reward center

_____ dopamine   c. selectively blocks nerve impulses

_____ serotonin   d. maintains stressed state

_____ GABA   e. maintains relaxed state

Were these your answers? d, e, b, c, a.

14.6 Psychoactive Drugs Alter the Mind or Behavior

Any drug affecting the mind or behavior is classified as **psychoactive**. In this section we focus on three classes of psychoactive drugs: stimulants, hallucinogens, and depressants.

Stimulants Activate the Stress Neurons

By enhancing the intensity of our reactions to stimuli, *stimulants* cause brief periods of heightened awareness, quick thinking, and elevated mood. Four widely recognized stimulants are amphetamines, cocaine, caffeine, and nicotine.

Amphetamines are a family of stimulants that include the parent compound *amphetamine* (also known as "speed") and such derivatives as methamphetamine and pseudoephedrine. As you can see by comparing Figure 14.23 on page 472 with Figures 14.20 and 14.21, these drugs are structurally similar to the neurotransmitters norepinephrine and dopamine. As might be expected, amphetamines bind to receptor sites for these neurotransmitters, thereby mimicking many of their effects, including the fight-or-flight response and the ability to give a person a sense of euphoria.

Amphetamines not only mimic the action of norepinephrine and dopamine; they also boost the levels of these neurotransmitters in a synaptic cleft by blocking their removal. Normally, neurotransmitters are reabsorbed by presynaptic neurons after they have exerted their effect on postsynaptic receptor sites. This process, commonly called **neurotransmitter re-uptake** and illustrated in Figure 14.24 on page 472, is the body's way of

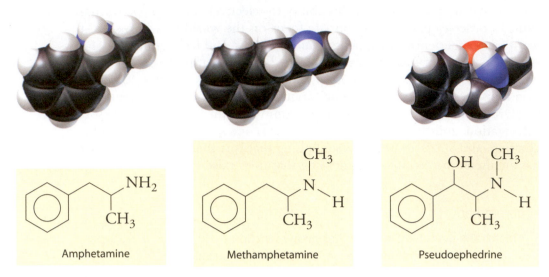

Figure 14.23
Amphetamines are a family of compounds structurally related to the neurotransmitters norepinephrine and dopamine.

recycling neurotransmitters, molecules that are difficult to synthesize. Special membrane-embedded proteins are required to pull once-used neurotransmitter molecules back into a presynaptic neuron. Amphetamines inactivate norepinephrine and dopamine re-uptake proteins by binding to them. As a consequence, the concentration of these stimulating neurotransmitters in the synaptic cleft is maintained at a higher-than-normal level.

The stimulating and mood-altering effects of amphetamines give them a high abuse potential. Side effects include insomnia, irritability, loss of appetite, and paranoia. Amphetamines take a particularly hard toll on the

Figure 14.24
① Neurotransmitters bind to their postsynaptic receptors. ② Neurotransmitters are reabsorbed by the presynaptic neuron that released them through proteins embedded in the presynaptic membrane. ③ A drug that interferes with re-uptake causes a buildup of neurotransmitters in the synaptic cleft.

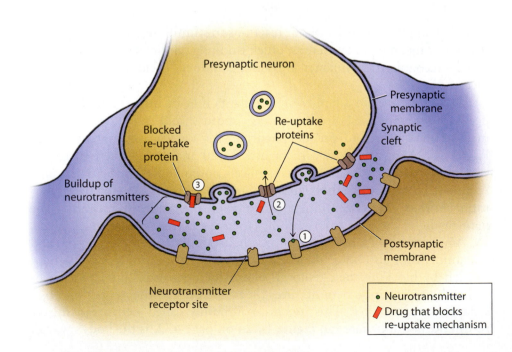

heart. Hyperactive heart muscles are prone to tearing. Subsequent scarring of tissue ultimately leads to a weaker heart. Furthermore, amphetamines cause blood vessels to constrict and blood pressure to rise, conditions that increase the likelihood of heart attack or stroke.

Drug addiction is not completely understood, but scientists do know that it involves both physical and psychological dependence. **Physical dependence** is the need to continue taking the drug to avoid withdrawal symptoms, which for amphetamines include depression, fatigue, and a strong desire to eat. **Psychological dependence** is the *craving* to continue drug use. This craving may be the most serious and deep-rooted aspect of addiction in that it can persist even after withdrawal from physical dependence, frequently leading to renewed drug-seeking behavior.

One of the more notorious and abused stimulants is *cocaine*, a natural product isolated from the South American coca plant, shown in Figure 14.25. Once in the bloodstream, cocaine produces a sense of euphoria and increased stamina. It is also a powerful local anesthetic when applied topically. Within a few decades of its first isolation from plant material in 1860, cocaine was used as a local anesthetic for eye surgery and dentistry—a practice that stopped once safer local anesthetics were discovered in the early 1900s.

Figure 14.25
The South American coca plant has been used by indigenous cultures for many years in religious ceremonies and as an aid to staying awake on long hunting trips. Leaves are either chewed or ground to a powder that is inhaled nasally.

Cocaine

Cocaine and amphetamines share a similar profile of addictiveness, though cocaine's addictive properties are more intense. The cocaine that is inhaled nasally is the hydrochloride salt. The free-base form of cocaine, called crack cocaine, is also abused. As with the street drug "ice," which is the free-base form of methamphetamine, crack cocaine is volatile and may be smoked for what is an intense but profoundly dangerous and addictive high.

Amphetamines and cocaine work the same way in the body, but cocaine is much more vigorous at blocking the re-uptake of dopamine. How cocaine

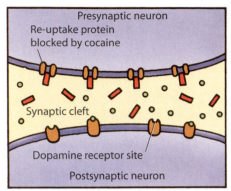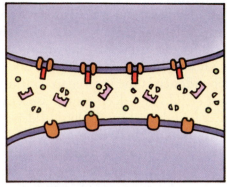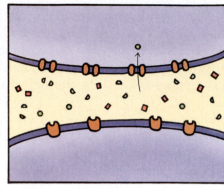

① High levels of dopamine remain active in the synaptic cleft as cocaine blocks dopamine re-uptake sites. This causes cocaine's euphoric effect.

② Dopamine is metabolized and deactivated as it loiters in the synaptic cleft awaiting re-uptake.

③ After cocaine is metabolized and deactivated, dopamine re-uptake is no longer blocked, but there is very little dopamine in the synaptic cleft or in the neurons, and the cocaine user experiences extreme depression.

Figure 14.26
Cocaine affects dopamine levels in the synaptic clefts of the brain's reward center.

works is illustrated in Figure 14.26. The great buildup of dopamine in synaptic clefts in the brain's reward center is the source of cocaine's euphoric effect. As cocaine keeps dopamine from being reabsorbed by the presynaptic neuron, the dopamine remains active in the synaptic cleft, and as a result the reward center stays stimulated. This euphoric state is only temporary, however, because enzymes in the cleft metabolize, and hence deactivate, the dopamine. Once the cocaine is metabolized by enzymes, dopamine re-uptake is again permitted. By this time, however, there is very little dopamine in the cleft to be reabsorbed. Nor is there an adequate supply of dopamine in the presynaptic neuron, which is unable to make sufficient quantities of dopamine without the recycling process. The net result is a depletion of dopamine that causes severe depression.

Long-term cocaine or amphetamine abuse leads to a deterioration of the nervous system. The body recognizes the excessive stimulatory actions produced by these drugs. To deal with the overstimulation, the body creates more depressant receptor sites for neurotransmitters that inhibit nerve transmission. A tolerance for the drugs therefore develops. Then, to receive the same stimulatory effect, the abuser is forced to increase the dose, which induces the body to create even more depressant receptor sites. The end result over the long term is that the abuser's natural levels of dopamine and norepinephrine are insufficient to compensate for the excessive number of depressant sites. Lasting personality changes are thus often observed.

Concept Check ✔

What are two ways in which amphetamines and cocaine exert their effects?

Was this your answer? Amphetamines and cocaine in the synaptic cleft both mimic the action of neurotransmitters. They also block the re-uptake of neurotransmitters, which results in an increase in the concentration of neurotransmitters in the cleft. The main action of amphetamines is their mimicking of neurotransmitters, whereas the main action of cocaine is the blocking of neurotransmitter re-uptake.

A much milder and legal stimulant is *caffeine*, depicted in Figure 14.27. A number of mechanisms have been proposed for caffeine's stimulatory effects. Perhaps the most straightforward mechanism is that caffeine facilitates the release of norepinephrine into synaptic clefts. Caffeine also exerts many other effects on the body, such as dilation of arteries, relaxation of bronchial and gastrointestinal muscles, diuretic action on the kidneys, and stimulation of stomach-acid secretion.

The caffeine we ingest comes from various natural sources, including coffee beans, teas, kolanuts, and cocoa beans. Kolanut extracts are used for making cola drinks, and cocoa beans (not to be confused with the cocaine-producing coca plant) are roasted and then ground to a paste used for making chocolate. Caffeine is relatively easy to remove from these natural products using high-pressure carbon dioxide, which selectively dissolves the caffeine. This allows for the economical production of "decaffeinated" beverages, many of which, however, still contain small amounts of caffeine. Interestingly, cola drink manufacturers use decaffeinated kolanut extract in their beverages. The caffeine is added in a separate step to guarantee a particular caffeine concentration. In the United States, about two million pounds of caffeine is added to soft drinks each year. Table 14.4 shows the caffeine content of various commercial products. For comparison, the maximum daily dose of caffeine tolerable by most adults is about 1500 milligrams.

Caffeine

Figure 14.27
A coffee plant with its ripening caffeine-containing beans.

Table 14.4

Approximate Caffeine Content of Various Products

Product	Caffeine Content
Brewed coffee	100–150 mg/cup
Instant coffee	50–100 mg/cup
Decaffeinated coffee	2–10 mg/cup
Black tea	50–150 mg/cup
Cola drink	35–55 mg/12 oz
Chocolate bar	1–2 mg/oz
Over-the-counter stimulant	100 mg/dose
Over-the-counter analgesic	30–60 mg/dose

Another legal, but far more toxic, stimulant is *nicotine*. As noted earlier, tobacco plants produce nicotine as a chemical defense against insects. This compound is so potent that a lethal dose in humans is only about 60 milligrams. A single cigarette may contain up to 5 milligrams of nicotine. Most of it is destroyed by the heat of the burning embers, however, so that less than 1 milligram is typically inhaled by the smoker.

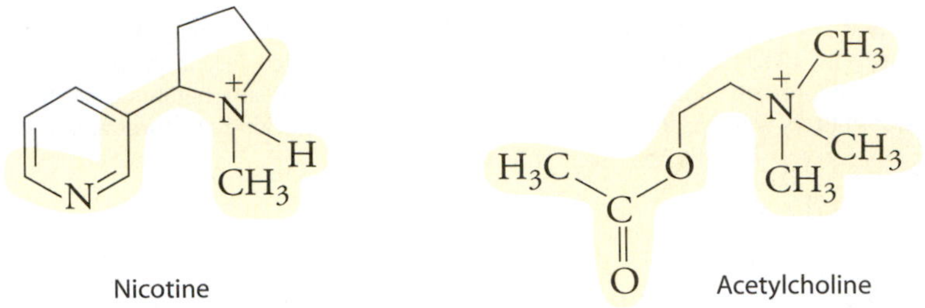

Figure 14.28
Nicotine is able to bind to receptor sites for acetylcholine because of structural similarities.

Nicotine

Acetylcholine

Nicotine and the neurotransmitter acetylcholine, which acts on maintenance neurons, have similar structures, as Figure 14.28 illustrates. Nicotine molecules are therefore able to bind to acetylcholine receptor sites and trigger many of acetylcholine's effects, including relaxation and increased digestion, which explains the tendency of smokers to smoke after eating meals. In addition, acetylcholine is used for muscle contraction, and so the smoker may also experience some muscle stimulation immediately after smoking. After these initial responses, however, nicotine molecules remain inertly bound to the acetylcholine receptor sites, thereby blocking acetylcholine molecules from binding. The result is that the activity of these neurons is depressed.

Recall that maintenance neurons and stress neurons are both always working. Thus inhibiting the activity of one type increases the activity of the other type. So, as nicotine depresses the maintenance neurons, it favors the stress neurons, thereby raising the smoker's blood pressure and stressing the heart.

Up in the brain, nicotine effects the stress system directly by enhancing the release of stress neurotransmitters, such as norepinephrine. Nicotine also increases the levels of dopamine in the reward center. Furthermore, when inhaled, nicotine is a fast-acting drug. All of these factors give nicotine a high level of addictiveness. Animal studies show inhaled nicotine to be about six times more addictive than injected heroin. Because nicotine leaves the body quickly, withdrawal symptoms begin about 1 hour after a cigarette is smoked, which means the smoker is inclined to light up frequently.

Figure 14.29 shows what a smoker's lungs look like after years of smoking. In the United States, about 450,000 individuals die each year from such tobacco-related health problems as emphysema, heart failure, and various forms of cancer, especially lung cancer, which is brought on primarily by tobacco's tar component. Some relief from the addiction can be obtained with nicotine chewing gum and nicotine skin patches. In order for any method to be effective, however, the smoker must first genuinely desire to quit smoking.

Concept Check ✔

Caffeine and nicotine both add stress to the nervous system, but they do so by different means. Briefly describe the difference.

Was this your answer? Caffeine stimulates the release of the stress neurotransmitter norepinephrine, and nicotine both depresses the action of the maintenance neurotransmitter acetylcholine and enhances the release of norepinephrine.

Tobacco field

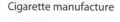

Tobacco curing on racks

Cigarette manufacture

User

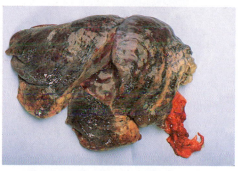

Blackened lungs

Figure 14.29
The path of tobacco from the field to a smoker's lungs. About 46 million Americans smoke despite an awareness of the dangers of this habit.

Hallucinogens and Cannabinoids Alter Perceptions

A *hallucinogen*, also known as a *psychedelic*, is any drug that can alter visual perceptions and skew the user's sense of time. Hallucinogens have a pronounced effect on moods, thought patterns, and behavior. Two main categories of hallucinogens are represented by the compounds lysergic acid diethylamide (LSD) and mescaline. A closely related category of drugs is the *cannabinoids*, the psychoactive components of marijuana. Cannabinoids do not alter visual perceptions and so are not true hallucinogens. They are similar to hallucinogens in other regards, however, such as in their ability to alter our sense of time.

LSD is the prototypic hallucinogen. Its molecular structure, shown in Figure 14.30 on page 478, is very similar to that of serotonin, and this similarity permits LSD to activate serotonin receptors even better than serotonin does. This means that unusually high numbers of nerve impulses are blocked, increasing our normal dampening powers. It is LSD's ability to interfere with serotonin's work that causes LSD users to experience an altered sense of reality. Because LSD also stimulates the reward center, the

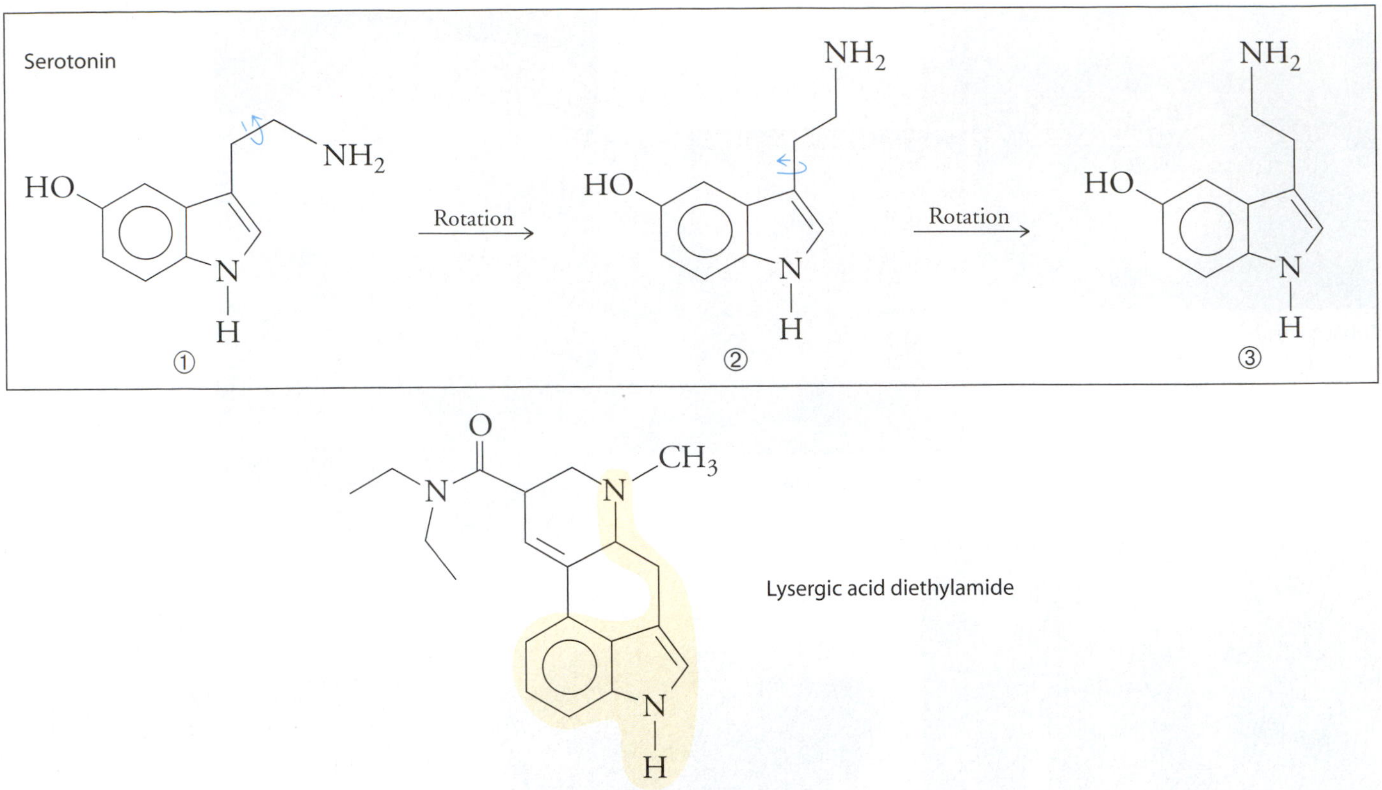

Serotonin

Lysergic acid diethylamide

Figure 14.30
The side chain of serotonin can rotate into a number of conformations. Upon binding to a receptor site, however, the side chain is likely to be held in conformation 3. Note how the LSD molecule can be superimposed on structure 3. LSD may therefore be thought of as a modified serotonin molecule in which the side chain is held in the ideal conformation for receptor binding.

change in sensory organization is usually, but not always, characterized as a favorable experience. LSD also triggers the stress neurons, resulting in enlarged pupils, elevated blood pressure and heart rate, nausea, and tremors. These stress effects can shift the mood of an affected person to panic and anxiety. Because the LSD molecule is nonpolar, quantities may be trapped and hidden away in nonpolar fatty tissue, only to be released months later and result in a mild recurrence of the experience.

In the early 1970s, there was a significant rise in the street use of hallucinogenic derivatives of the compound phenylethylamine, shown in Figure 14.31. Initial interest was given to *mescaline*, the hallucinogenic component of several species of cacti used in religious ceremonies by Native American tribes in the western United States. One plant yielding this compound is pictured in Figure 14.32. Synthetic derivatives, such as methylenedioxyamphetamine (MDA), also became popular. Unlike LSD, these hallucinogens do not exert their effects by binding to serotonin receptor sites. Rather, they stimulate the release of excess quantities of serotonin. This pathway is not as effective as LSD's direct approach, and as a result these drugs are 200 to 4000 times less potent than LSD. Because larger doses of mescaline and MDA are required, a multitude of other effects are seen, such as a marked stimulation of the stress neurons. In addition, the regular use of these compounds causes withdrawal symptoms.

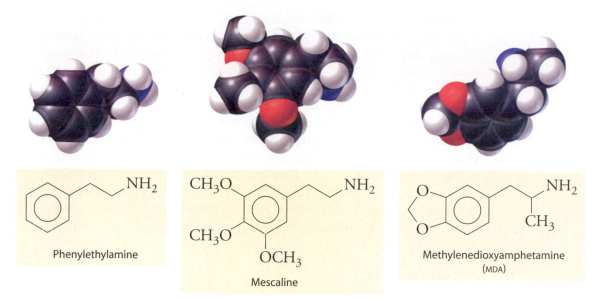

Phenylethylamine

Mescaline

Methylenedioxyamphetamine (MDA)

Figure 14.31
Hallucinogenic derivatives of phenylethylamine.

Cannabinoids are the psychoactive components of marijuana, which has the species name *Cannabis sativa*. Concentrations of cannabinoids vary greatly from plant to plant. The original strains of this plant species contain very little of these psychoactive components and have been used for many centuries for their great fiber qualities. Strains of *Cannabis* that may be smoked for psychoactive effects on average contain about 4 percent cannabinoid derivatives. The most active of these derivatives is the compound Δ^9-tetrahydrocannabinol (THC), shown in Figure 14.33.

How THC exerts its psychoactive effect is not completely understood. In 1990, a specific receptor site for the THC molecule was discovered. A few years later a peptide occurring naturally in the body was found to bind to this receptor and initiate marijuana-like responses. These results suggest that THC functions by mimicking this naturally occurring peptide.

Figure 14.32
The peyote cactus is a source of the hallucinogen mescaline.

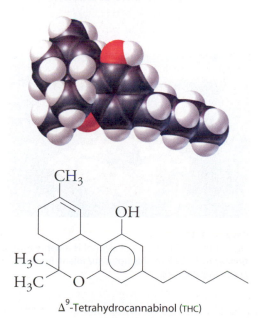

Δ^9-Tetrahydrocannabinol (THC)

Figure 14.33
The major psychoactive component of marijuana is Δ^9-tetrahydrocannabinol.

Most notably, cannabinoids accumulate in the area of the brain where short-term memories are sorted. Every experience we ever have goes through this center. Some things—the image of a sidewalk crack you saw on a morning walk, say—get thrown out. Other experiences, like your first date, get filed away in long-term memory storage. Cannabinoids disrupt this filing system so that memories are not sorted appropriately. In addition, people under the influence of cannabinoids may have a distorted sense of time and unclear thoughts. Another effect of cannabinoids is unrestful sleep. The brain sorts through memories during a phase of sleep marked by rapid eye movement (REM). People who smoke marijuana lose REM sleep time, which results in irritability the following day. Once the memory filing center is cleared of the drug, which may take days or even weeks, the brain makes up for lost time by having extra long periods of REM.

Concept Check ✓

Why is marijuana not considered a true hallucinogen?

Was this your answer? The chemicals in marijuana do not significantly alter visual perceptions.

Depressants Inhibit the Ability of Neurons to Conduct Impulses

Depressants are a class of drugs that inhibit the ability of neurons to conduct impulses. Two examples of depressants are ethanol and benzodiazepines.

Ethanol, also known simply as *alcohol*, is by far the most widely used depressant. Its structure is shown in Figure 14.34. In the United States, about a third of the population, or about 100 million people, drink alcohol. It is well established that alcohol consumption leads to about 150,000 deaths each year in the United States. The causes of these deaths are overdoses of alcohol alone, overdoses of alcohol combined with other depressants, alcohol-induced violent crime, cirrhosis of the liver, and alcohol-related traffic accidents.

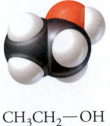

CH_3CH_2-OH

Ethanol

Figure 14.34
One of the initial effects of alcohol is a depression of social inhibitions, which can serve to bolster mood. Alcohol is not a stimulant, however. From the first sip to the last, body systems are being depressed.

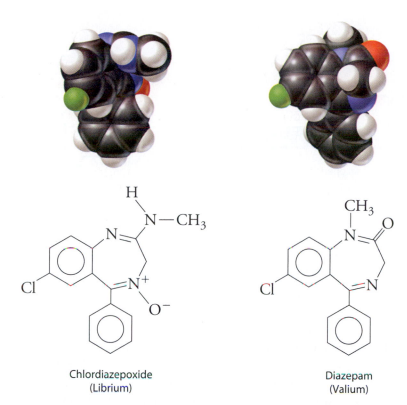

Figure 14.35
The benzodiazepines Librium and Valium.

Chlordiazepoxide
(Librium)

Diazepam
(Valium)

Benzodiazepines are a potent class of antianxiety agents. Compared with many other types of depressants, benzodiazepines are relatively safe and rarely produce cardiovascular and respiratory depression. Their antianxiety effects were identified in 1957 by chance. During a routine laboratory cleanup, a compound that had been sitting on the shelf for two years was submitted for routine testing despite the fact that compounds thought to have similar structures had shown no promising pharmacologic activity. This particular compound, however, shown in Figure 14.35 and now known as chlordiazepoxide, contained an unusual seven-membered ring. Chlordiazepoxide showed a significant calming effect in humans and by 1960 was marketed under the trade name Librium as an antianxiety agent. Shortly thereafter, a derivative, diazepam, was found to be five to ten times more potent than chlordiazepoxide. In 1963 diazepam hit the market under the trade name Valium.

A primary way in which alcohol and benzodiazepines exert their depressant effect is by enhancing the action of GABA. As shown in Figure 14.36 on page 482, GABA keeps electrical impulses from passing through a neuron by binding to a receptor site on a channel that penetrates the cell membrane of the neuron. Figure 14.36a shows that when GABA binds to the receptor site, the channel opens, allowing chloride ions to migrate into the neuron. The resulting negative charge buildup in the neuron maintains the negative electric potential across the cell membrane, thereby inhibiting a reversal to a positive potential and preventing an impulse from traveling along the neuron. (If you are having trouble understanding this, go back and review Figure 14.18 and the text describing it.)

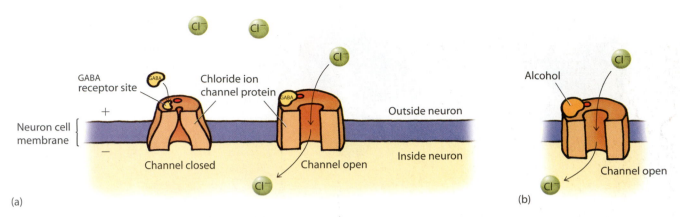

Figure 14.36
(a) When GABA binds to its receptor site, a channel opens to allow negatively charged chloride ions into the neuron. The high concentration of negative ions inside the neuron prevents the electric potential from reversing from negative to positive. Because that reversal is necessary if an impulse is to travel through a neuron, no impulse can move through the neuron. (b) Alcohol mimics GABA by binding to GABA receptor sites.

Alcohol mimics the effect of GABA by binding to GABA receptor sites, allowing chloride ions to enter the neuron, as shown in Figure 14.36b. The effect of alcohol is dose-dependent—the greater the amount drunk, the greater the effect. At small concentrations, few chloride ions are permitted into the neuron; these low concentrations of ions decrease inhibitions, alter judgment, and impair muscle control. As the person continues to drink and the chloride ion concentration inside the neuron rises, both reflexes and consciousness diminish, eventually to the point of coma and then death.

Recall from our discussion of cocaine and amphetamines that the body responds to the long-term abuse of these stimulants by creating more depressant receptor sites. Likewise, the body recognizes the excessive inhibitory actions produced by alcohol and tries to recover by increasing the number of synaptic receptor sites that lead to nerve excitation. A tolerance for alcohol therefore develops. To receive the same inhibitory effect, the drinker is forced to drink more, which induces the body to create even more excitable synaptic receptor sites. Eventually, an excess of these excitatory receptor sites leads to perpetual body tremors, which can be subdued either by more drinking or, with greater difficulty, by a long-term cessation of alcohol consumption.

Figure 14.37 illustrates how benzodiazepines exert their depressant effects by binding to receptor sites located adjacent to GABA receptor sites.

Figure 14.37
The receptor site for a benzodiazepine is adjacent to the GABA receptor site. (a) A benzodiazepine cannot open up the chloride channel on its own. (b) Rather, the benzodiazepine helps GABA in its channel-opening task.

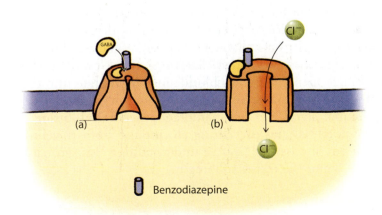

Benzodiazepine binding helps GABA bind. Because benzodiazepines don't directly open chloride-ion channels, overdoses of these compounds are less hazardous than overdoses of alcohol. For this reason, benzodiazepines have become a drug of choice for treating symptoms of anxiety.

Concept Check ✔

Does the activity of a neuron increase or decrease as chloride ions are allowed to pass into it?

Was this your answer? Chloride ions inside the neuron help maintain the negative electric potential. This inhibits the neuron from conducting an impulse (see Figure 14.18). Chloride ions therefore decrease the activity of a neuron.

14.7   Pain Relievers Inhibit the Transmission or Perception of Pain

Physical pain is a complex body response to injury. On the cellular level, pain-inducing biochemicals are rapidly synthesized at the site of injury, where they initiate swelling, inflammation, and other responses that get your body's attention. These pain signals are sent through the nervous system to the brain, where the pain is perceived. To alleviate pain, drugs act at various stages of this process, as shown in Figure 14.38.

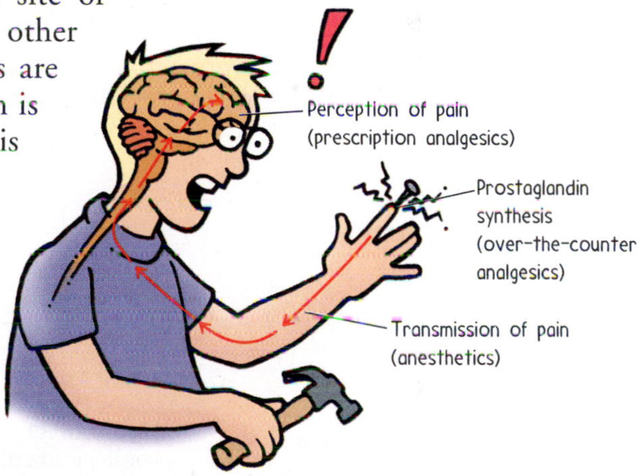

Anesthetics prevent neurons from transmitting sensations to the brain. *Local anesthetics* are applied either topically to numb the skin or by injection to numb deeper tissues. These mild anesthetics are useful for minor surgical or dental procedures. As described earlier, cocaine was the first medically used local anesthetic. Others having fewer side effects soon followed, such as the ones shown in Figure 14.39 on page 484. In general, molecules that have strong local anesthetic properties have an aromatic ring linked to an amine group by an intermediate chain of a particular length. Benzocaine lacks an amine group and so has low activity, which permits its use as an over-the-counter topical anesthetic good for mouth sores and sunburn.

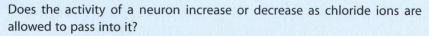

Perception of pain
(prescription analgesics)

Prostaglandin synthesis
(over-the-counter analgesics)

Transmission of pain
(anesthetics)

Figure 14.38
Injury to tissue causes the transmission of pain signals to the brain. Pain relievers prevent this transmission, inhibit the inflammation response, or dampen the brain's ability to perceive the pain.

A *general anesthetic* blocks out pain by rendering the patient unconscious. Gaseous general anesthetics are commonly used because with them the anesthesiologist has excellent control over how much anesthetic is given. As discussed in Section 12.3, diethyl ether was one of the first general anesthetics. Two of the more popular gaseous general anesthetics used by anesthesiologists today are those shown in Figure 14.40 on page 484, sevoflurane and nitrous oxide. When inhaled, these compounds enter the bloodstream and are distributed throughout the body. At certain blood concentrations, general anesthetics render the individual unconscious, which is useful for invasive surgery. General anesthesia must be monitored very carefully, however, so as to avoid a major shutdown of the nervous system and subsequent death.

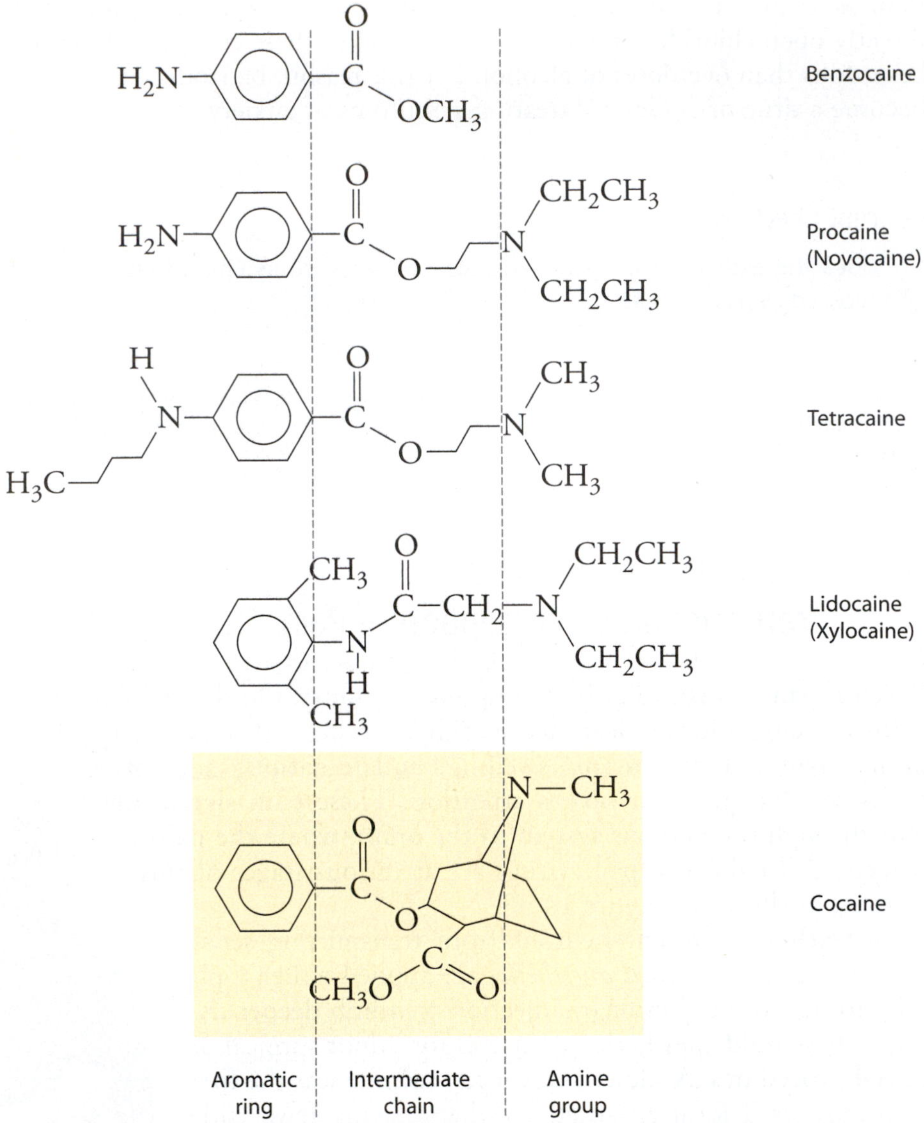

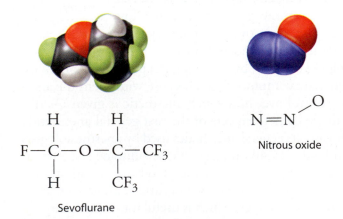

Benzocaine

Procaine
(Novocaine)

Tetracaine

Lidocaine
(Xylocaine)

Cocaine

| Aromatic ring | Intermediate chain | Amine group |

Figure 14.39
Local anesthetics have similar structural features, including an aromatic ring, an intermediate chain, and an amine group. Ask your dentist which ones he or she uses for your treatment.

Sevoflurane

Nitrous oxide

Figure 14.40
The chemical structures of sevoflurane and nitrous oxide.

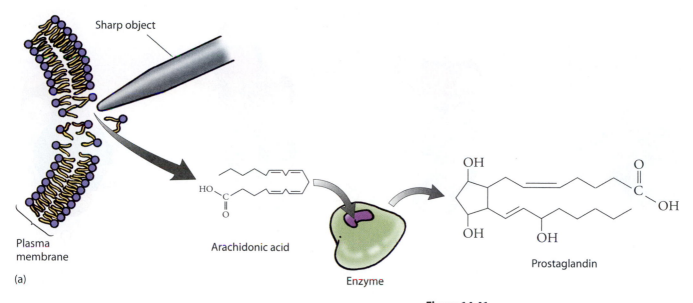

Sharp object

Plasma membrane

Arachidonic acid

Enzyme

Prostaglandin

(a)

(b)

Over-the-counter analgesic bound to enzyme

No prostaglandin

Enzyme

Figure 14.41
(a) Prostaglandins, which cause pain signals to be sent to the brain, are synthesized by the body in response to injury. The starting material for all prostaglandins is arachidonic acid, which is found in the plasma membrane of all cells. Arachidonic acid is transformed to prostaglandins with the help of an enzyme. There are a variety of prostaglandins, each having its own effect, but all have a chemical structure resembling the one shown here. (b) Analgesics inhibit the synthesis of prostaglandins by binding to the arachidonic acid receptor site on the enzyme. With no prostaglandins, no pain signals are generated.

Analgesics are a class of drugs that enhance our ability to tolerate pain without abolishing nerve sensations. Over-the-counter analgesics, such as aspirin, ibuprofen, naproxen, and acetaminophen, inhibit the formation of *prostaglandins*, which, as Figure 14.41 illustrates, are biochemicals the body quickly synthesizes to generate pain signals. Analgesics also reduce fever because of the role prostaglandins play in raising body temperature. In addition to reducing pain and fever, aspirin, ibuprofen, and naproxen act as anti-inflammatory agents because they block the formation of a certain type of prostaglandin responsible for inflammation. Acetaminophen does not act on inflammation. These four analgesics are shown in Figure 14.42 on page 486.

The more potent opioid analgesics—morphine, codeine, and heroin (see Figure 14.1)—moderate the brain's perception of pain by binding to receptor sites on neurons. Initial discovery of these receptor sites raised the question of why they exist. Perhaps, it was hypothesized, opioids mimic the action of a naturally occurring brain chemical. *Endorphins*, a group of polypeptides that have strong opioid activity, were subsequently isolated from brain tissue. It has been suggested that endorphins, an example of which is shown in Figure 14.43 on page 487, evolved as a means of suppressing awareness of pain that would otherwise be incapacitating in life-threatening situations. The "runner's high" experienced by many athletes after a vigorous workout is caused by endorphins.

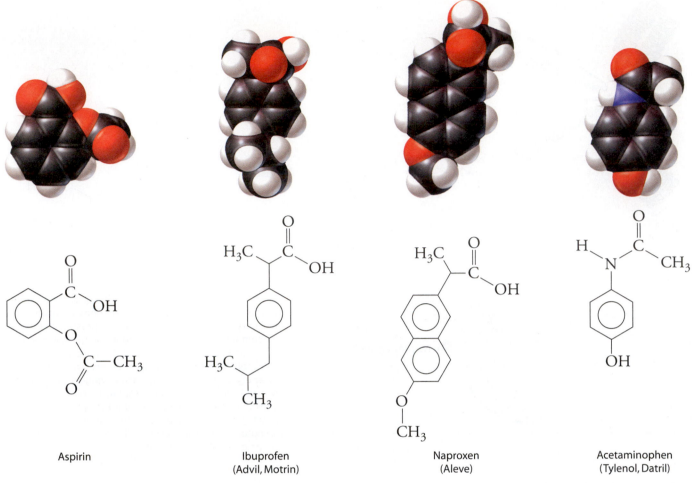

Aspirin

Ibuprofen
(Advil, Motrin)

Naproxen
(Aleve)

Acetaminophen
(Tylenol, Datril)

Figure 14.42
Aspirin, ibuprofen, and naproxen block the formation of prostaglandins responsible for pain, fever, and inflammation. Acetaminophen blocks the formation only of prostaglandins responsible for pain and fever.

Endorphins are also implicated in the *placebo effect*, in which patients experience a reduction in pain after taking what they believe is a drug but is actually a sugar pill. (A *placebo* is any inactive substance used as a control in a scientific experiment.) Through the placebo effect, it is the patients' belief in the effectiveness of a medicine rather than the medicine itself that leads to pain relief. The involvement of endorphins in the placebo effect has been demonstrated by replacing the sugar pills with drugs that block opioids or endorphins from binding to their receptor sites. Under these circumstances, the placebo effect vanishes.

In addition to acting as analgesics, opioids can induce euphoria, which is why they are so frequently abused. With repeated use, the body develops a tolerance to these drugs: more and more must be administered to achieve the same effect. Abusers also become physically dependent on opioids, which means they must continue to take the opioids to avoid severe withdrawal symptoms, such as chills, sweating, stiffness, abdominal cramps, vomiting, weight loss, and anxiety. Interestingly, when opioids are used pri-

OH

Tyrosine

H_2C O

$H_2N—C—C—N$ Glycine

H   H

Glycine

Phenylalanine

Methionine

Met-enkephalin

Figure 14.43
Met-enkephalin is just one of many endorphins, which are pain-relieving polypeptides produced by the body. Studies show a similarity between the structure of opioids and the structure of the tyrosine–glycine–glycine portion of this macromolecule, a similarity that supports the notion that endorphins and opioids fit the same receptor sites.

marily for pain relief rather than for pleasure, the withdrawal symptoms are much less dramatic—especially when the patient does not know he or she has been taking these drugs.

The most widely used approach to treating opioid addiction is methadone maintenance. *Methadone*, shown in Figure 14.44, is a synthetic opioid derivative that has most of the effects of other opioids, including euphoria, but differs in that it retains much of its activity when taken orally. This means that doses are very easy to control and monitor. The withdrawal symptoms of methadone are also far less severe, and the addict may be slowly weaned off the opioid without excessive stress. An addict may be freed of physical dependence in a matter of months. The psychological dependence, however, usually persists throughout the individual's life, which is why the relapse rate is so high.

Concept Check ✔

Distinguish between an anesthetic and an analgesic.

Was this your answer? An anesthetic blocks pain signals from reaching the brain, and an analgesic facilitates the ability to manage pain signals once they are received by the brain.

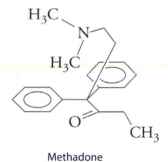

Methadone

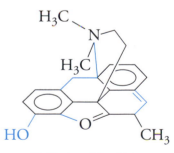

Methadone/Morphine

Figure 14.44
The structure of methadone (black) superimposed on that of morphine (blue and black).

14.8 Drugs for the Heart Open Blood Vessels or Alter Heart Rate

Heart disease is any condition that diminishes the heart's ability to pump blood. A common heart disease is *arteriosclerosis*, a buildup of plaque on the inside walls of arteries. As discussed in Section 13.8, plaque deposits are mostly an accumulation of low-density lipoproteins, which are high in cholesterol and saturated fats. Plaque-filled arteries are less elastic and have a decreased volume. Both these effects make pumping blood more difficult, and the heart becomes overworked and weakens. Accumulated damage to heart muscle from arteriosclerosis or other stresses can result in abnormal heart rhythms, known as *arrhythmia*. Chest pains, known as *angina*, result from an insufficient oxygen supply to heart muscles. Ultimately, the weakened heart does not adequately circulate blood to the body. People with heart disease have decreased stamina and frequently need to catch their breath.

As discussed in Section 13.8, another great danger of arteriosclerosis is the potential for a blood clot around the site of plaque formation. Such a clot is carried through the bloodstream until it clogs a blood vessel, effectively cutting off the blood supply to tissue, which then begins to die. A heart attack occurs when the dying tissue is heart muscle. Some heart attacks progress slowly, allowing the victim time to seek medical assistance, which may involve the administration of a quick-acting clot-dissolving enzyme. Other heart attacks are more rapid, killing the victim within minutes. Surviving a heart attack means living with a heart weakened by dead tissue.

Vasodilators are a class of drugs that increase the blood supply to the heart by expanding blood vessels. They are useful for treating angina. They also reduce the work load of the heart because opening up the blood vessels makes pumping blood easier. Traditional vasodilators include nitroglycerin and amyl nitrite, both shown in Figure 14.45. They can be administered by a number of routes, including orally, sublingually (under the tongue), or as a transdermal patch. A benefit of the latter two approaches is that they allow the drug

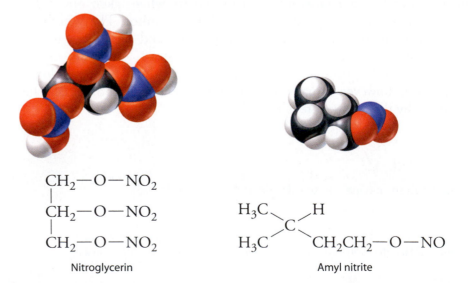

Figure 14.45
The vasodilators nitroglycerine and amyl nitrite.

Nitroglycerin

Amyl nitrite

to enter the body slowly, in contrast to what happens when the drug is taken orally or by injection. These organic nitrates are metabolized to nitric oxide, NO, which has been shown to relax muscles in blood vessels.

Drugs that relax the pumping action of the heart have also been developed. When bound to receptor sites called beta-adrenoceptors in heart muscle, the neurotransmitters norepinephrine and epinephrine stimulate the heart to beat faster. A series of drugs called *beta blockers* slow down and relax an overworked heart by blocking norepinephrine and epinephrine from binding to the beta-adrenoceptors. The first beta blocker developed, propranolol (Inderal), shown in Figure 14.46, is useful for treating angina, arrhythmias, and high blood pressure.

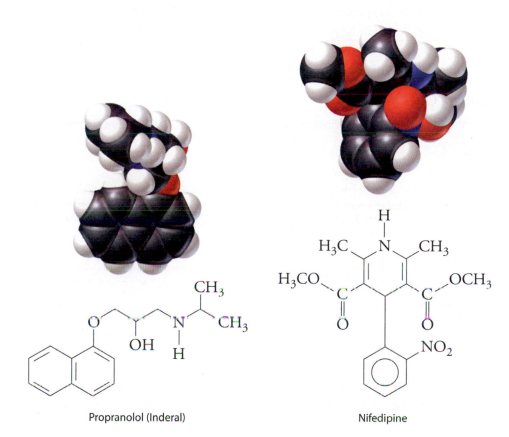

Propranolol (Inderal)   Nifedipine

Figure 14.46
Propranolol is a beta blocker, and nifedipine is a calcium-channel blocker.

Another group of drugs that relax heart muscle are the *calcium-channel blockers*. One example is nifedipine, shown in Figure 14.46. Muscle contraction is initiated as a nerve impulse signals calcium ions to enter muscle cells. As their name implies, calcium-channel blockers inhibit the flow of calcium ions into muscles, thereby inhibiting muscle contraction. The heart rate slows down, and muscles of blood vessels relax and dilate, lowering blood pressure.

In the United States and in most other developed nations, heart disease is the number one cause of death for individuals over the age of 65. Because most people in these nations live past this age, heart disease is actually the leading cause of death for all age groups combined, as noted in Table 14.5.

Table 14.5

Causes of Death in the United States

Number One Cause by Age Group		Top Ten Causes, All Ages Combined
Age Group (Years)	Cause	
15–24	Accident	1. Heart disease
25–44	HIV infection	2. Cancer
45–64	Cancer	3. Stroke
> 65	Heart disease	4. Lung disease
		5. Accident
		6. Pneumonia
		7. Diabetes
		8. Suicide
		9. HIV infection
		10. Homicide

Concept Check ✓

Why do long-time alcoholics require relatively greater doses of a beta blocker in order to relax cardiac muscle?

Was this your answer? As discussed in Section 14.6, long-time excessive drinking leads to an increase in the number of receptor sites for stress neurotransmitters. With more of these receptor sites to block, the alcoholics require a correspondingly greater dose of the beta blocker to achieve the desired degree of cardiac relaxation.

In Perspective

Perhaps nowhere is the impact of chemistry on society more evident than in the development of drugs. On the whole, drugs have increased our lifespan and improved our quality of living. They have also presented us with a number of ethical and social questions. Should an abortion pill such as mifepristone be allowed? How do we care for an expanding elderly population? What drugs, if any, should be permissible for recreational use? How do we deal with drug addiction—as a crime, as a disease, or both? As we continue to learn more about ourselves and our ills, we can be sure that more powerful drugs will become available. All drugs, however, carry certain risks that we should be aware of. As most physicians would point out, drugs offer many benefits, but they are no substitute for a healthy lifestyle and preventative approaches to medicine.

Key Terms and Matching Definitions

_____ analgesic
_____ anesthetic
_____ chemotherapy
_____ combinatorial chemistry
_____ lock-and-key model
_____ neuron
_____ neurotransmitter
_____ neurotransmitter re-uptake
_____ physical dependence
_____ psychoactive
_____ psychological dependence
_____ synaptic cleft
_____ synergistic effect

1. One drug enhancing the effect of another.
2. A model that explains how drugs interact with receptor sites.
3. The production of a large number of compounds in order to increase the chances of finding a new drug having medicinal value.
4. The use of drugs to destroy pathogens without destroying the animal host.
5. A specialized cell capable of receiving and sending electrical impulses.
6. A narrow gap across which neurotransmitters pass either from one neuron to the next or from a neuron to a muscle or gland.
7. An organic compound capable of activating receptor sites on proteins embedded in the membrane of a neuron.
8. Said of a drug that affects the mind or behavior.
9. A mechanism whereby a presynaptic neuron absorbs neurotransmitters from the synaptic cleft for reuse.
10. A dependence characterized by the need to continue taking a drug to avoid withdrawal symptoms.
11. A deep-rooted craving for a drug.
12. A drug that prevents neurons from transmitting sensations to the brain.
13. A drug that enhances the ability to tolerate pain without abolishing nerve sensations.

Review Questions

Drugs Are Classified by Safety, Social Acceptability, Origin, and Biological Activity

1. What are the three origins of drugs?

2. Are a drug's side effects necessarily bad?
3. What is the synergistic effect?

The Lock-and-Key Model Guides Chemists in Synthesizing New Drugs

4. In the lock-and-key model, is a drug viewed as the lock or the key?
5. What holds a drug to its receptor site?
6. What are most receptor sites made of?
7. What advantages are there to synthesizing a naturally occuring medicine, such as Taxol, in the laboratory rather than isolating it from nature?
8. How is the laboratory method called combinatorial chemistry similar to the search for drugs in nature?

Chemotherapy Cures the Host by Killing the Disease

9. Why do bacteria need PABA but humans can do without it?
10. When is chemotherapy most effective against cancer?
11. What is cancer?
12. How does methotrexate work?

Some Drugs Either Block or Mimic Pregnancy

13. How effective are birth control pills at preventing pregnancy?
14. How does the action of mifepristone differ from that of the progesterone-based birth control pill?

The Nervous System Is a Network of Neurons

15. How does a neuron maintain an electric potential difference across its membrane?
16. What are the symptoms that a person's stress neurons have been activated?
17. What are some of the things going on in the body when maintenance neurons are more active than stress neurons?

18. Which neurotransmitter functions most in the brain's reward center?

19. What is the role of GABA in the nervous system?

Psychoactive Drugs Alter the Mind or Behavior

20. What is neurotransmitter re-uptake?

21. Of the different psychoactive drugs presented in this chapter, which act by blocking the re-uptake of neurotransmitters?

22. How is psychological dependence distinguished from physical dependence?

23. What is one mechanism for how caffeine stimulates the nervous system?

24. Which neurotransmitter does nicotine mimic?

25. Which neurotransmitter does LSD mimic?

26. Is marijuana better described as a drug that has few side effects or a drug that has many?

27. What drugs enhance the action of GABA?

Pain Relievers Inhibit the Transmission or Perception of Pain

28. What is an anesthetic?

29 What is an analgesic?

30. Where are the major opioid receptor sites located?

31. What biochemical is thought to be responsible for the placebo effect?

32. How does methadone work in treating opioid addiction?

Drugs for the Heart Open Blood Vessels or Alter Heart Rate

33. What is angina, and what is its cause?

34. What role does nitrogen oxide play in the treatment of angina?

35. How does a vasodilator reduce the workload on the heart?

36. Distinguish between a beta blocker and a calcium-channel blocker.

Hands-On Chemistry Insights

Diffusing Neurons

Because molecules slow down with decreasing temperature, the rate at which neurotransmitters are able to diffuse across the synaptic cleft decreases with temperature. This is one of the reasons cold-blooded animals become sluggish at colder temperatures. As neurotransmitters take longer to diffuse across the synaptic cleft, the rate at which nerve signals can reach target muscles slows down.

When stuck outside in the cold without adequate clothing, you may find your extremities becoming numb and your muscles sluggish. This isn't just because of a decrease in the rate of diffusion in your synaptic clefts. Your body responds to cold by diverting blood from your extremities to your internal organs. The speed of neuron transmission, however, depends on blood supply. As the blood supply diminishes, neurons lose out on needed oxygen and nutrients and thus begin to shut down. The result is a numbing effect or loss of muscle control. The same thing happens to your foot after you've been sitting on it for too long and it has "gone to sleep." Ice packs work by the same mechanism.

Most but not all neuron connections rely on the passage of neurotransmitters across a synaptic cleft. The neurons that connect to muscles controlling eye movement, however, are different in that they have what are called *electrical synapses*, which are direct connections between the neurons and the muscles (there's no gap). The high speeds of these direct connections provide for eye motion that is quick and jerky—a useful trait. Some fish have electrical synapses in their tails, an arrangement that provides for rapid escape from predators.

As an interesting aside, you might find that the descending drop of food coloring in this activity resembles a neuron, complete with knoblike synaptic terminals.

Exercises

1. Why are organic chemicals so suitable for making drugs?

2. Aspirin can cure a headache, but when you take an aspirin tablet, how does the aspirin know to go to your head rather than your big toe?

3. Which is better for you: a drug that is a natural product or one that is synthetic?

4. When might two drugs taken together not exert a synergistic effect?

5. Why is cancer treated most successfully in its earliest stages?

6. Would formulating a sulfa drug with PABA be likely to increase or decrease its antibacterial properties?

7. Why do protease inhibitors work so well when used in conjunction with antiviral nucleosides?

8. Why do some antiviral agents exhibit anticancer activity?

9. About 60 million women use birth control pills, which when taken correctly are about 99 percent effective in preventing pregnancy. Build an argument for whether such pills have had a major or a minor impact on the growth of the human population, which now stands at more than 6 billion.

10. What is an advantage of synaptic clefts between neurons rather than direct connections?

11. Why do so many stimulant drugs result in a depressed state after an initial high?

12. How is a drug addict's addiction similar to our need for food? How is it different?

13. Nicotine solutions are available from lawn and garden stores as an insecticide. Why must gardeners handle this product with extreme care?

14. Seeds of the morning glory plant contain the natural product lysergic acid, and yet they are only marginally hallucinogenic. Why?

15. Suggest why withdrawal symptoms are observed after repeated use of MDA but not after repeated use of LSD.

16. Why do heavy drinkers have a greater tolerance for alcohol?

17. One of the active components of marijuana, Δ^9-tetrahydrocannabinol, is available as a prescription drug under the trade name Marinol, which is taken orally. What advantage and what disadvantage does Marinol hold for a person suffering from nausea?

18. A variety of gaseous compounds behave as general anesthetics even though their structures have very little in common. Does this support the role of a receptor site for their mode of action?

19. How might the structure of benzocaine be modified to create a compound having greater anesthetic properties?

20. At one time, halothane, $CF_3CHBrCl$, was widely used as a general anesthetic. Suggest why its use is now banned.

21. Which is the more appropriate statement: opioids have endorphin activity, or endorphins have opioid activity? Explain your answer.

22. Why is methadone not very useful in the treatment of cocaine addiction?

23. What are the problems associated with a buildup of plaque on the inner walls of blood vessels?

24. A person may feel more relaxed after smoking a cigarette, but his or her heart is actually stressed. Why?

Discussion Topics

1. Alcohol-free and caffeine-free beverages have been quite successful in the marketplace, while nicotine-free tobacco products have yet to be introduced. Speculate about possible reasons.

2. What would be the advantages and disadvantages of making the production, sale, and consumption of tobacco products illegal?

3. Historically, drug abuse has been viewed as criminal behavior. Advances in medicine, however, show the complexity of drug abuse, leading some people to believe it should be viewed and treated as a disease. This newer view suggests that education and medical treatment, not fines and jail sentences, should be the major weapons combating drug abuse. Assume this newer view is the predominant legal one, and prioritize where you think resources should be spent to help alleviate drug abuse. Under what circumstances do you think a drug user should be incarcerated?

4. Worldwide, tuberculosis and malaria are far greater killers than AIDS. Why then does there seem to be more of a focus on AIDS and AIDS prevention?

5. Research suggests that skin loses its flexiblity as we get older because of how sugar molecules are able to bind to the skin protein called collagen. New classes of drugs that prevent this from happening and are even able to reverse the process may hit the market within a decade, promising a long-lasting youthful look for everyone. Should such drugs be sold as prescription or over-the-counter medicine? What if there is a minor side effect, such as increased susceptability to infections, or a major side effect, such as a weakening of heart tissue? What if there are no significant side effects except for a doubling of the average life-span of humans?

Exploring Further

http://www.usdoj.gov/dea

Home page for the Drug Enforcement Administration, the lead federal agency for the enforcement of narcotics and controlled-substance laws and regulations. The agency's priority is the long-term immobilization of major drug trafficking organizations through the jailing of their leaders, termination of their trafficking networks, and seizure of their assets.

http://www.nida.nih.gov

A primary mission of The National Institute on Drug Abuse is to ensure that science, not ideology or anecdote, forms the foundation for our nation's drug abuse policies. This Web site is designed to ensure the rapid and effective transfer of scientific data to policy makers, health care practitioners, and the general public.

http://www.norml.org/home.shtml

Since its founding in 1970, the National Organization to Reform Marijuana Laws has been the principal national advocate for legalizing marijuana. Today NORML serves as the umbrella group for a national network of activists committed to ending marijuana prohibition.

http://www.cancer.org

The American Cancer Society has been the lead organization in the fight against cancer since its inception in the early 1900s. At this site, you will find information on the history of our understanding of cancer as well as information regarding the latest treatments for the different forms of cancer.

http://www.americanheart.org

Explore the site of the American Heart Association to learn about early warning signs of heart disease and about what you can do to decrease your risk. This site is a good place to start researching the statistics of heart disease and latest trends in heart research.

http://www.lungusa.org

This site for the American Lung Association can lead you to information about all kinds of lung disease, the air you breathe, and events and services in your area. Since 1904, the Association has worked to fight lung disease by helping people quit smoking, funding research, improving indoor and outdoor air quality, and educating millions about asthma.

The Chemistry of Drugs

Visit The Chemistry Place at:
www.aw.com/chemplace

From the Good Earth

Each year, about 500,000 children in developing countries become irreversibly blind because of a deficiency of vitamin A. In an effort to stop this tragedy, two European scientists recently created a new strain of rice enriched in the orange pigment β-carotene, which the body uses to make vitamin A. The amount of β-carotene in this rice gives the rice a golden hue. Current plans are to provide this golden rice free of charge to farmers in developing nations, such as Vietnam and Bangladesh, where the need is greatest.

Developing a new strain of a crop to meet our nutritional needs is nothing new. Most of our major crops are the result of centuries, if not millennia, of *selective breeding*, a process whereby plants that offer more nutritional value are selectively bred over ones of less nutritional value. Today's potato plant, for example, was once a small, wild shrub with tiny but nutritious root nodules growing in the Andes of South America. What is different about golden rice is that it was created over a single generation, by inserting genes from a daffodil and a bacterium into the DNA of some other strain of rice. Golden rice is therefore an example of a *transgenic plant*—one in which genes of different species have been artificially combined.

This chapter is a showcase of the chemistry involved in the production of food, primarily the food derived from plants, which feeds both us and our livestock. Along the way, you will be introduced to many of the fundamental

Optimizing Food Production

concepts of **agriculture**, which may be loosely defined as the organized use of resources for the production of food. Special attention is paid to chemistry-related aspects of agriculture, such as soil composition, fertilizers, and pesticides. Also discussed are many of the new approaches to agriculture, including the creation of transgenic crops, which offer many advantages but also require our careful consideration.

15.1   Humans Eat at All Trophic Levels

The formation of food begins with *photosynthesis*, the biochemical process used by plants to create carbohydrates and oxygen from solar energy, water, and atmospheric carbon dioxide:

$$6\,CO_2 + 6\,H_2O \xrightarrow{\text{Sunlight}} C_6H_{12}O_6 + 6\,O_2$$

Carbon   Water   Carbohydrate   Oxygen
dioxide

Each day, only about 1 percent of the solar energy reaching the Earth's surface is used in photosynthesis. On a global scale, this is enough to produce 170 billion tons of organic material per year. The energy content in this amount of organic matter is the total annual energy budget for virtually all living organisms.

The route food energy takes through a community of organisms is determined by the community's **trophic structure**, which is the pattern of feeding relationships in the community. Trophic structures, also known as *food chains*, consist of a number of hierarchical levels, shown in Figure 15.1. The first trophic level is **producers**, most of which are photosynthetic organisms that use light energy to power the synthesis of organic compounds. Plants are the main producers on land. In water, the main producers are photosynthetic organisms known as *phytoplankton* (*phyto-* means "plants").

Producers support all other trophic levels, collectively called **consumers**. Organisms that consume producers are *primary* consumers. These are herbivores ("grass-eaters"), such as grazing mammals, most insects, and most birds. The primary consumers in aquatic environments are the many microscopic organisms collectively known as *zooplankton*. Above the primary consumers, the trophic levels are made of *carnivores* ("meat-eaters"), each level eating consumers from lower levels. Secondary consumers eat primary consumers, tertiary consumers eat secondary consumers, and quaternary consumers eat tertiary consumers. Any organism that dies before being eaten becomes subject to the action of **decomposers**, organisms that break down organic material into simpler substances that then act as soil nutrients. Common decomposers are earthworms, insects, fungi, and microorganisms.

With each transfer of energy from one trophic level to the next, there is a significant loss of energy. Typically, not more than 10 percent of the

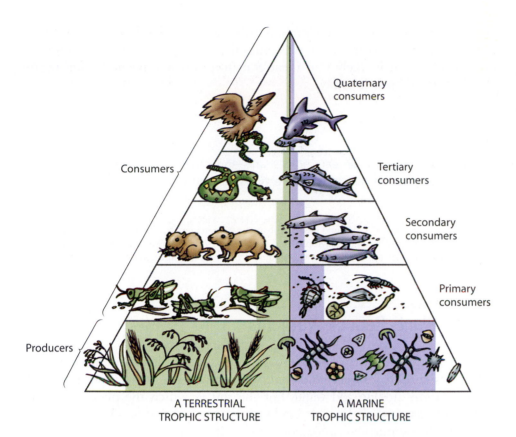

Figure 15.1
Terrestrial and marine trophic structures. Energy and nutrients pass through the trophic levels when one organism feeds on another. The shaded blocks represent the amount of energy transferred from one trophic level to the next.

energy contained in the organic material of one trophic level is incorporated into the next higher level. The availability of food energy is therefore greatest for the organisms lowest on the food chain. A grasshopper, for example, will find many more blades of grass to feed on than a field mouse will find grasshoppers. And the field mouse will find more grasshoppers than a snake will find field mice. This dwindling supply of food resources quickly limits the number of trophic levels, which rarely exceeds the quaternary level. Accordingly, the higher the trophic level, the smaller the possible population of organisms.

We humans eat at all trophic levels. When we eat such things as fruits, vegetables, or the grains shown in Figure 15.2, we are primary consumers; when we eat beef or other meat from herbivores, we are secondary consumers. When we eat fish like trout or salmon, which eat insects and other small animals, we are tertiary or quaternary consumers. Our great and growing numbers, however, are possible only because of our ability to eat as primary consumers.

Eating meat is a luxury. For the people who will eat the chickens shown in Figure 15.3, for instance, the amount of biochemical energy they will obtain from eating the chickens is minuscule compared with the amount of biochemical energy used in raising the chickens. In the United States, more than 70 percent of grain production is fed to livestock. Producing meat therefore requires that more land be cultivated, more water be used for irrigation, and more fertilizers and pesticides be applied to

Figure 15.2
Most people are primary consumers, with a diet consisting primarily of grains.

Figure 15.3
In affluent countries, eating meat is quite common. In the United States, for example, chickens outnumber people 44 to 1.

croplands. If people in the United States ate 10 percent less meat, the savings in resources could feed 100 million people! As the human population expands, it is likely that meat consumption will become even more of a luxury than it is today.

Concept Check

> The orca killer whale eats both sharks and phytoplankton-feeding gray whales. In which case is the orca eating at a higher trophic level?

Was this your answer? Sharks feed on fish and marine mammals, which makes them secondary, tertiary, or quaternary consumers. Phytoplankton-feeding gray whales, however, are primary consumers. When an orca feeds on a shark, it is eating at a higher trophic level than when it feeds on a gray whale.

15.2 Plants Require Nutrients

Plant material consists primarily of carbohydrates, which are made of the elements carbon, oxygen, and hydrogen. Plants get these three elements from carbon dioxide and water, but the soil in which the plants live also provides many other elements vital to their survival and good health. Table 15.1 lists these nutrients as *macronutrients*, those needed in large quantities, and *micronutrients*, those needed only in trace quantities. Some micronutrients are needed in such trace quantities that a plant's lifetime supply is provided by the seed from which the plant grew.

Table 15.1

Essential Elements for Most Plants

Element	Form Available to Plants	Relative Number of Ions in Dry Plant Material*
Macronutrients		
Nitrogen, N	NO_3^-, NH_4^+	1,000,000
Potassium, K	K^+	250,000
Calcium, Ca	Ca^{2+}	125,000
Magnesium, Mg	Mg^{2+}	80,000
Phosphorus, P	$H_2PO_4^-$, HPO_4^{2-}	60,000
Sulfur, S	SO_4^{2-}	30,000
Micronutrients		
Chlorine, Cl	Cl^-	3000
Iron, Fe	Fe^{3+}, Fe^{2+}	2000
Boron, B	$H_2BO_3^-$	2000
Manganese, Mn	Mn^{2+}	1000
Zinc, Zn	Zn^{2+}	300
Copper, Cu	Cu^+, Cu^{2+}	100
Molybdenum, Mo	MoO_4^{2-}	1

*Measured relative to molybdenum = 1.
Source: Salisbury and Ross, *Plant Physiology*. Belmont, CA: Wadsworth, 1985.

Plants Utilize Nitrogen, Phosphorus, and Potassium

Plants need nitrogen to build proteins and a variety of other biomolecules, such as *chlorophyll*, the green pigment responsible for photosynthesis. As Table 15.1 shows, plants are able to absorb nitrogen from the soil in the form of ammonium ions, NH_4^+, and nitrate ions, NO_3^-. Figure 15.4 shows the natural sources of nitrogen for a plant. The two reactions shown there are examples of **nitrogen fixation**, which is defined as any chemical reaction that converts atmospheric nitrogen to a form of nitrogen that plants can use. The two most common forms are ammonium ions and nitrate ions.

Most of the ammonium ions in soil come from nitrogen fixation carried out either by bacteria living in the soil or by microorganisms living in root nodules of certain plants, especially those of the legume family, including clover, alfalfa, beans, and peas (plants generally called *nitrogen fixers*). These microorganisms possess the enzyme nitrogenase, which catalyzes the formation of ammonium ions from atmospheric nitrogen and soil-bound hydrogen ions, as Figure 15.4a shows.

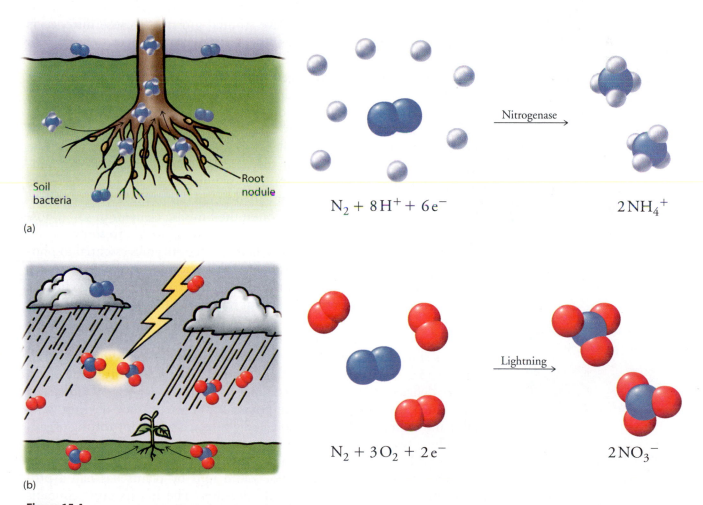

(a)

$$N_2 + 8H^+ + 6e^- \xrightarrow{\text{Nitrogenase}} 2NH_4^+$$

(b)

$$N_2 + 3O_2 + 2e^- \xrightarrow{\text{Lightning}} 2NO_3^-$$

Figure 15.4
Two pathways for nitrogen fixation, a source of nitrogen for plants. (a) Both free-living bacteria in the soil and microorganisms in root nodules produce ammonium ions. (b) Lightning provides the energy needed to form nitrate ions from atmospheric nitrogen.

Nitrogen fixation also results, to a lesser extent, from the action of lightning on atmospheric nitrogen, as Figure 15.4b shows. The high energy of the lightning is sufficient to initiate the oxidization of atmospheric nitrogen to nitrate ions, which are then washed into the soil by rain.

In a natural setting, nitrogen fixation is the original source of ammonium and nitrate ions in the soil. Most of this fixed nitrogen, however, is recycled from one organism to the next. After a plant dies, for example, bacterial decomposition of the plant releases ammonium and nitrate ions back into the soil, where these ions become available to plants that are still living.

Concept Check ✔

Why are the seeds of nitrogen-fixing plants, such as soybeans and peanuts, unusually high in protein?

Was this your answer? Plants use nitrogen to build proteins. Nitrogen-fixing plants assimilate a lot of nitrogen and so produce proteins in large quantities.

Plants deficient in nitrogen have stunted growth. Because nitrogen is needed for making chlorophyll, another symptom of nitrogen deficiency in plants is yellow leaves, as Figure 15.5a shows. The yellowing is most pronounced in older leaves—younger ones remain green longer because soluble forms of nitrogen are transported to them from older, dying leaves.

Plants need phosphorus to build nucleic acids, phospholipids, and a variety of energy-carrying biomolecules, such as ATP. All phosphorus comes to plants in the form of phosphate ions. The major natural source of these ions is eroded phosphate-containing rock. Significant amounts of phosphates are also recycled as organisms die and become incorporated into the soil. After nitrogen, phosphorus is most often the limiting element in soil. Phosphorus-deficient plants are stunted, as Figure 15.5b shows.

Potassium ions activate many of the enzymes essential to photosynthesis and respiration. As with phosphates, the major natural sources of potassium ions are eroded rock and recycling of potassium ions from decomposing plant material. After nitrogen and phosphorus, soils are usually most deficient in potassium. Symptoms of potassium deficiency include small areas of dead tissue, usually along leaf tips or edges, as shown in Figure 15.5c. As with nitrogen and phosphorus, potassium ions are easily redistributed from mature to younger parts of the plant, and so deficiency symptoms first appear on older leaves. When cereal grains, such as wheat or corn, are potassium-deficient, they develop weak stalks, and their roots become more easily infected with root-rotting organisms. These two factors cause potassium-deficient plants to be easily bent to the ground by wind, rain, or snow.

Interestingly, the uptake of potassium ions by plants has had a profound effect on the composition of sea water. The Earth's crust contains nearly equal portions of sodium ions and potassium ions. Both readily leach from the soil, traveling into streams, rivers, and then ultimately the ocean, where they are the cause of much of the ocean's salinity. Plants absorb

(a)

(b)

(c)

Figure 15.5
(a) The leaves of nitrogen-deficient plants turn yellow prematurely.
(b) Phosphorus deficiencies are marked by stunted growth. (c) The leaves of potassium-deficient plants develop dead areas here seen as the pale yellow region along the leaf perimeter.

potassium ions, but in most instances the route for sodium ions is in the other direction—plants give off sodium ions to the soil and so ultimately to stream and river water. So, while sodium and potassium ions are about equally abundant in the Earth's crust, seawater contains about 2.8 percent NaCl but only about 0.8 percent KCl. (As another interesting aside, all living cells, not just plant cells, have a low tolerance for sodium ions and therefore have evolved mechanisms for pumping out these ions, as noted in Section 14.5.)

Plants Also Utilize Calcium, Magnesium, and Sulfur

Both calcium and magnesium are absorbed by plants as the positively charged calcium and magnesium ions, Ca^{2+} and Mg^{2+}, and sulfur is absorbed as the negatively charged sulfate ion, SO_4^{2-}. Most topsoils contain enough of these ions for adequate plant growth.

Calcium ions are essential for building cell walls. Once absorbed by the plant, calcium ions are relatively immobile; that is, they do not travel well from one part of the plant to another. The plant therefore is not so capable of reallocating calcium supplies in times of need. This is why new-growth zones, such as the tips of roots and stems, are most susceptible to calcium deficiencies. The results are twisted and deformed growth patterns.

Magnesium ions are essential for the formation of chlorophyll, which is the green-pigmented molecule essential for photosynthesis. Chlorophyll houses a magnesium ion at the center of a structure called a porphyrin ring, as shown in Figure 15.6. Besides its presence in chlorophyll, magnesium is essential because it activates many metabolic enzymes. Deficiencies in magnesium, which are rare, result in yellow leaves because the plant is unable to generate chlorophyll.

Most of the sulfur in plants occurs in proteins, especially in the amino acids cysteine and methionine. Other essential compounds that contain

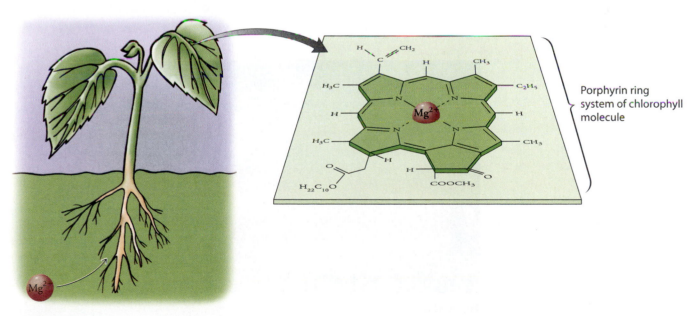

Figure 15.6
Magnesium ions in the green pigment chlorophyll are vital to photosynthesis.

sulfur are coenzyme A, a compound essential for cellular respiration and for the synthesis and breakdown of fatty acids, and the vitamins thiamine and biotin. Sulfur can be absorbed by leaves as gaseous sulfur dioxide, SO_2, an environmental pollutant released from active volcanoes and from the burning of wood or fossil fuels.

Concept Check ✓

Plants require more calcium and magnesium than phosphorus and potassium, and yet deficiencies of calcium and magnesium are much more infrequent than deficiencies of phosphorus and potassium. How can this be so?

Was this your answer? The potential for nutrient deficiency depends not only on the amount of nutrient needed but on the nutrient's availability. Calcium and magnesium are abundant in most soils, but phosphorus and potassium are not. Deficiencies of phosphorus and potassium therefore tend to be more frequent.

15.3   Soil Fertility Is Determined by Soil Structure and Nutrient Retention

Soil is a mixture of sand, silt, and clay. All three of these components are ground-up rock, and the differences are in how finely the particles are ground. Sand particles are the largest, and clay particles are the smallest.

Soil often occurs as a series of horizontal layers called **soil horizons**, shown in Figure 15.7. The deepest horizon, which lies just above solid rock, is the *substratum*, which is rock just beginning to disintegrate into soil by the action of water that has seeped down to this level. No growing plant material is found in the substratum. Above the substratum is the *subsoil*,

Topsoil

Subsoil

Substratum

Figure 15.7
The vertical structure of soil is a series of layers called soil horizons.

which consists mostly of clay. Only the deepest roots penetrate into the subsoil, which may be up to 1 meter thick. Above the subsoil is the *topsoil*, which lies on the surface and varies in thickness from a few centimeters to up to 2 meters. The topsoil usually contains sand, silt, and clay in about equal amounts. This is the horizon where the roots of plants absorb most of their nutrients.

Fertile topsoil is a mixture of at least four components—mineral particles, water, air, and organic matter. The mineral particles are the particles of sand, silt, and clay. Many of the nutrients plants need are released as these particles are formed from the erosion of rock. The size of the particles greatly affects soil fertility. Large particles result in porous soil that has many pockets of space that collect water and air—up to 25 percent of the volume of fertile topsoil consists of pockets. Roots absorb water and the oxygen from the air in these pockets. Small particles pack tightly together, so that none or only very few air pockets are present, and as a result there is little or no oxygen or water available to roots. This is why plants do not grow well in clay soils. Figure 15.8 compares these two extremes of soil.

The organic matter in topsoil is a mixture of fallen plant material, the remains of dead animals, and such decomposers as bacteria and fungi, as Figure 15.9 illustrates. This organic matter is called **humus**, and it is rich in a variety of plant nutrients. Humus tends to be porous, giving roots access to subterranean water and oxygen. It also binds the soil, helping to prevent erosion.

The flow of water through soil is called *percolation*. The more porous the soil, the greater the rate of percolation. With excessive percolation, flowing water removes many water-soluble nutrients needed to make the soil productive. This process is known as *leaching*. With too little percolation, topsoil becomes waterlogged, choking off a plant's supply of oxygen. Soils with optimal percolation drain water from all but the smallest air pockets.

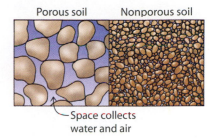

Porous soil   Nonporous soil

Space collects water and air

Figure 15.8
Large soil particles create larger pockets of space than do smaller soil particles.

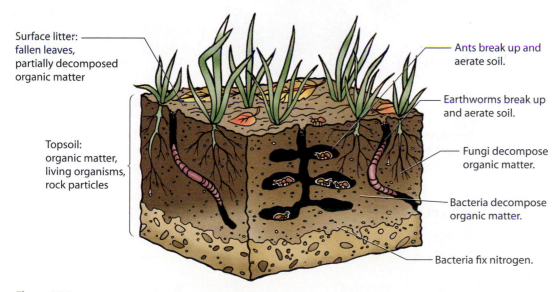

Surface litter: fallen leaves, partially decomposed organic matter

Topsoil: organic matter, living organisms, rock particles

Ants break up and aerate soil.

Earthworms break up and aerate soil.

Fungi decompose organic matter.

Bacteria decompose organic matter.

Bacteria fix nitrogen.

Figure 15.9
Organic matter and living organisms are important components of topsoil.

Soil Readily Retains Positively Charged Ions

Mineral particles play an important role in keeping nutrients in soil. As Table 15.1 shows, many plant nutrients are positively charged ions. The surface of most mineral particles, however, is negatively charged. Figure 15.10 illustrates how the resulting ionic attractions help keep nutrients from being washed away. The degree of nutrient retention is most pronounced in clay soils because these mineral particles are the smallest ones and thus have the largest surface area relative to volume.

Decaying matter in humus contains many carboxylic acid and phenolic groups, which at a typical soil pH are ionized to negatively charged carboxylate and phenolate ions. So, like mineral particles, humus helps retain positively charged nutrients.

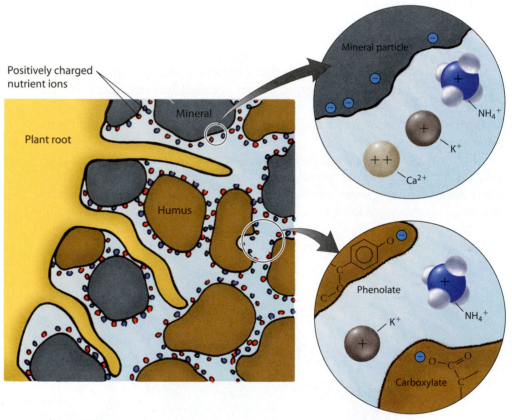

Figure 15.10
The negatively charged surfaces of soil mineral particles and humus help retain positively charged nutrient ions.

Concept Check ✔

Why is soil able to retain ammonium ions, NH_4^+, better than nitrate ions, NO_3^-?

Was this your answer? Mineral particles and bits of humus in soil are negatively charged and therefore hold on to positively charged ammonium ions but repel negatively charged nitrate ions.

The pH of soil is largely a function of the amount of carbon dioxide present. Recall from Section 10.4 that carbon dioxide reacts with water to form carbonic acid, which in turn forms hydronium ions:

$$O=C=O + H_2O \longrightarrow HO-C(=O)-OH + H_2O \longrightarrow HO-C(=O)-O^- + H_3O^+$$

| Carbon dioxide | Water | Carbonic acid | Water | Bicarbonate ion | Hydronium ion |

The greater the amount of carbon dioxide in soil, the more hydronium ions and so the lower the pH. Soil that has a low pH is referred to as *sour*. (Recall from Chapter 10 that many acidic foods, such as lemon, are characteristically sour.) Two main sources of soil carbon dioxide are humus and plant roots. The humus releases carbon dioxide as it decays, and plant roots release carbon dioxide as a product of cellular respiration. A healthy soil may have enough carbon dioxide released from these processes to give a pH range from about 4 to 7. If the soil becomes too acidic, a weak base, such as calcium carbonate (known as lime or limestone), can be added.

Hydronium ions are able to displace nutrient ions held to mineral particles and humus. Plants use this fact to great advantage. Figure 15.11 illustrates how a plant releases carbon dioxide through its root system and in

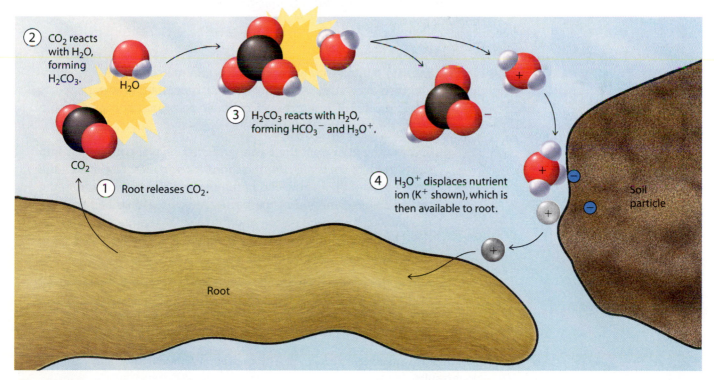

Figure 15.11
By releasing carbon dioxide, a plant guarantees a steady flow of nutrients from the soil to its roots.

doing so generates nutrient-displacing hydronium ions. The displaced nutrients, no longer attached to soil particles, thus become available to the plant.

Concept Check ✔

How might a below-normal pH in soil lead to nutrient deficiencies in plants?

Was this your answer? When the soil pH is below normal, the water in the soil pockets contains an abundance of hydronium ions, which displace large numbers of nutrient ions from soil and humus particles. Most of these nutrients wash away before the plant is able to absorb them. In a short time, the soils are depleted of nutrients, and the plants become nutrient-deficient.

Hands-On Chemistry:
Your Soil's pH—A Qualitative Measure

You can measure the pH of soil samples using the red-cabbage pH indicator introduced in Chapter 10.

What You Need

1 cup of soil, water, concentrated red-cabbage pH indicator (see page 323), oven baster with suction bulb, cotton balls, chopstick or metal skewer, two drinking glasses.

Procedure

① Saturate the soil with water, making a thin mud slurry. Stir the slurry, and then allow it to settle until about 1 centimeter of water appears above the soil. If this top layer of water does not appear, add more water, stir, and allow for further settling.

② Remove the bulb from the baster and stuff several cotton balls into the top end of the baster. Use the chopstick or skewer to compact the cotton toward the narrow end of the baster. Pour the water from the settled soil into the top of the baster. Attach the bulb and squeeze the liquid through the cotton and into one of the glasses. If the filtered water is still brown or cloudy, repeat the filtering, using clean cotton balls each time.

③ Add to the second glass as much fresh water as there is filtered water in the first glass. Add equal amounts of the pH indicator to the two glasses. Compare colors, recalling from Chapter 10 that a deeper red color means greater acidity and that green indicates alkaline.

You might want to test soil samples from several different sites and compare your results.

15.4   Natural and Synthetic Fertilizers Help Restore Soil Fertility

As a soil loses its plant nutrients to harvested crops and to leaching, it loses its fertility. Farmers amend soil by adding *fertilizers*, which are replacement sources for these lost nutrients. Naturally occurring fertilizers are compost and minerals. **Compost** is decayed organic matter, which can be animal manure, food scraps, or plant material. Mineral fertilizers are mined. Saltpeter, $NaNO_3$, for example, was once used extensively as a source of nitrogen, but by the late 1800s the supply of this nitrogen-containing mineral was almost exhausted. A new source of nitrogen for fertilizers came along in 1913, when Fritz Haber (1868–1934), a German chemist, developed a process for producing ammonia from hydrogen and atmospheric nitrogen:

$$N_2 \quad + \quad 3\,H_2 \quad \longrightarrow \quad 2\,NH_3$$

Nitrogen   Hydrogen   Ammonia

This technique is now the primary means of producing ammonia, which can be stored in high-pressure tanks as a liquid and injected into the soil. Alternatively, the ammonia can be converted to a water-soluble salt, such as ammonium nitrate, NH_4NO_3, that is then applied to the soil either as a solid or in solution. The mining of other nutrients, such as phosphorus and potassium, still remains an important endeavor.

In times past, mineral fertilizers were used just as they came from the ground. Today, however, chemists have learned how to mix and match minerals to obtain many different formulations, each suitable for a different soil problem or the specific requirements of a particular plant. All these formulated mineral fertilizers are referred to as either *chemically manufactured fertilizers* or, more frequently, *synthetic fertilizers*. Don't take the word *synthetic* literally, though, because except for what is produced by the Haber reaction, all the minerals in synthetic fertilizers originally came from the ground.

A fertilizer that contains only one nutrient is called a **straight fertilizer**. Ammonium nitrate, NH_4NO_3, is an example of a straight fertilizer, yielding only nitrogen. Any fertilizer containing a mixture of the three most essential nutrients (nitrogen, phosphorus, and potassium) is called either a *complete fertilizer* or a **mixed fertilizer**. All mixed fertilizers are graded by the N-P-K system, which lists the percent of nitrogen (N), phosphorus (P), and potassium (K) they contain, as Figure 15.12 shows. A typical mixed fertilizer might be graded 6-12-12. A typical compost, by contrast, might be rated anywhere from 0.5-0.5-0.5 to 4-4-4. Compost N-P-K ratings are much lower because of their high percentage of organic bulk. This organic bulk helps to keep the soil loose for aeration, however, and also serves as food for beneficial organisms that live in the soil. Because of the negative electric charges it carries, the organic bulk also attracts positively charged nutrient ions, which are then not so readily leached away.

Figure 15.12
Fertilizers are rated by the percentages of nitrogen, phosphorus, and potassium they contain.

The effect that nitrogen-containing synthetic fertilizers can have on yields is illustrated in Figure 15.13. It requires a lot of energy to mine and refine synthetic fertilizers, however, and so they are expensive. For example, of the total energy required to produce corn in the United States, at least one-third is needed to produce, transport, and apply the fertilizer. Nevertheless, synthetic fertilizers are widely used, and our present food supply depends on them.

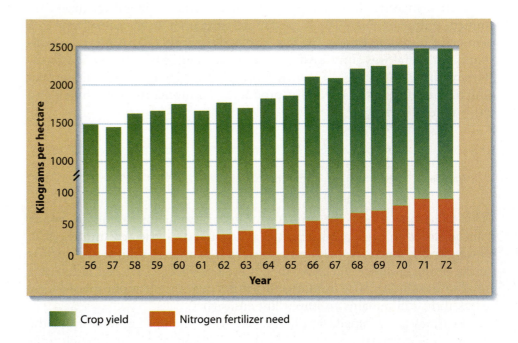

Figure 15.13
Between 1956 and 1972, world crop yields grew in tandem with increases in the use of nitrogen fertilizers.

Concept Check ✔

Which N-P-K rating would you expect for coffee grounds, which contain significant quantities of the alkaloid caffeine: 2-0.3-0.2, 0.3-2-0.2, or 0.3-0.2-2?

Caffeine

Was this your answer? That caffeine is a nitrogen-containing compound means that coffee grounds must contain a relatively high proportion of nitrogen, as is indicated by their N-P-K rating of 2-0.3-0.2.

15.5 Pesticides Kill Insects, Weeds, and Fungi

A high-yield crop needs more than adequate nutrition. It also needs defense against a host of natural enemies, a few of which are shown in Figure 15.14. To control these pests, farmers can apply substances known as pesticides. There are several kinds of pesticides, including insect-killing insecticides, weed-killing herbicides, and fungus-killing fungicides.

Figure 15.14
Crop pests such as these threaten crop yields.

Insecticides Kill Insects

Most species of insects are beneficial or even essential to agriculture. Honeybees, for example, are responsible for the pollination of $10 billion worth of produce in the United States. Countless other species take part in nutrient recycling and help maintain soil quality. A small minority of insect species, however, have continually threatened our capacity to grow, harvest, and store crops, and it is against these species that insecticides are used. The most widely used insecticides are chlorinated hydrocarbons, organophosphorus compounds, and carbamates.

The *chlorinated hydrocarbons* have a remarkable persistence, killing insects for months and years on treated surfaces. There are at least two reasons for this persistence. First, chlorinated hydrocarbons tend to be nonbiodegradable, which means there are no natural pathways to break them down chemically. Second, they are nonpolar compounds, which means they are insoluble in water and so are not washed away by rainwater.

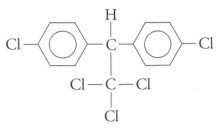

Figure 15.15
The chemical name for DDT is dichlorodiphenyltrichloroethane.

Figure 15.16
Rachel Carson was a pioneer in the fight against excessive use of pesticides.

In 1939 there was a breakthrough in the fight against insect pests with the chemical synthesis of the chlorinated hydrocarbon DDT, shown in Figure 15.15. During the 1940s and 1950s, DDT was applied liberally to crops, resulting in markedly greater yields. In addition to protecting plants, DDT protected people from disease. It was applied to rivers, streams, and villages to help control the proliferation of mosquitoes, lice, and tsetse flies, which spread malaria, typhus, and sleeping sickness, respectively. According to the World Health Organization, by protecting against these diseases, DDT has saved an estimated 25 million human lives.

Insect populations began to develop a resistance to DDT within a few years of its first application. Furthermore, DDT was found to be toxic to wildlife, including the natural predators of insects, such as birds. With fewer natural predators, DDT-resistant insects were able to thrive. The early increased crop yields resulting from DDT use were therefore not sustainable.

In the 1950s and 1960s, these and other negative aspects of DDT and other pesticides were brought to the public's attention by a number of publications, including the biologist Rachel Carson's book *Silent Spring*. Through the use of poetry, Carson, shown in Figure 15.16, described the importance of understanding the dynamics of ecosystems, most of which are highly sensitive to human activities. She also described a phenomenon known as **bioaccumulation**, whereby a toxic chemical that enters a food chain at a low trophic level becomes more concentrated in organisms higher up the chain, as illustrated in Figure 15.17. In bodies of water sprayed with DDT, for example, small amounts of the pesticide were ingested and stored

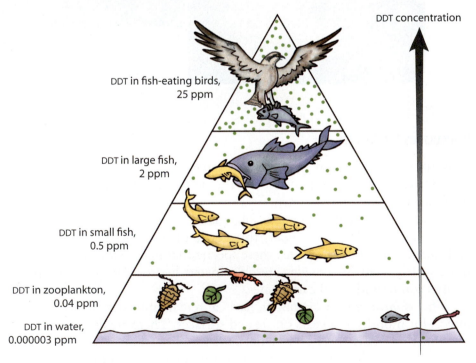

Figure 15.17
The DDT concentration in a food chain can be magnified from 0.000003 parts per million (ppm) as a pollutant in the water to 25 ppm in a bird at the top of the chain.

in the nonpolar lipids of aquatic microorganisms. Because these microorganisms serve as food for animals at higher trophic levels, DDT became more concentrated in the body fat of these larger creatures. Predatory birds at the top of the chain accumulated the greatest amounts of DDT. Eventually, the elevated DDT levels affected avian population numbers because the eggshells of affected birds were too thin and fragile to support the growing chick embryos. DDT contributed to the decline of many bird populations and the near extinction of some species of osprey, hawks, eagles, and falcons. In the early 1970s, the United States and many other countries banned the use of DDT. Within a matter of years, many wildlife species in these countries were able to recover.

Not all nations have banned DDT, however. Many nations still rely on it as an economical method of controlling insects that carry human disease. The use of this insecticide remains controversial.

Many chlorinated hydrocarbon alternatives to DDT have been developed. One of the earliest substitutes was methoxychlor, shown in Figure 15.18. This compound has a much lower toxicity in most animals and, unlike DDT, is not readily stored in animal fat. Look carefully at the structures of methoxychlor and DDT, and you'll see that they are identical except that methoxychlor has two ether groups where DDT has two chlorine atoms. Because the structures are nearly identical, they have nearly the same level of toxicity in insects. In higher animals, however, the oxygen atoms facilitate detoxification. Specifically, enzymes in the liver cleave the ether groups to synthesize polar products that are readily excreted through the kidneys.

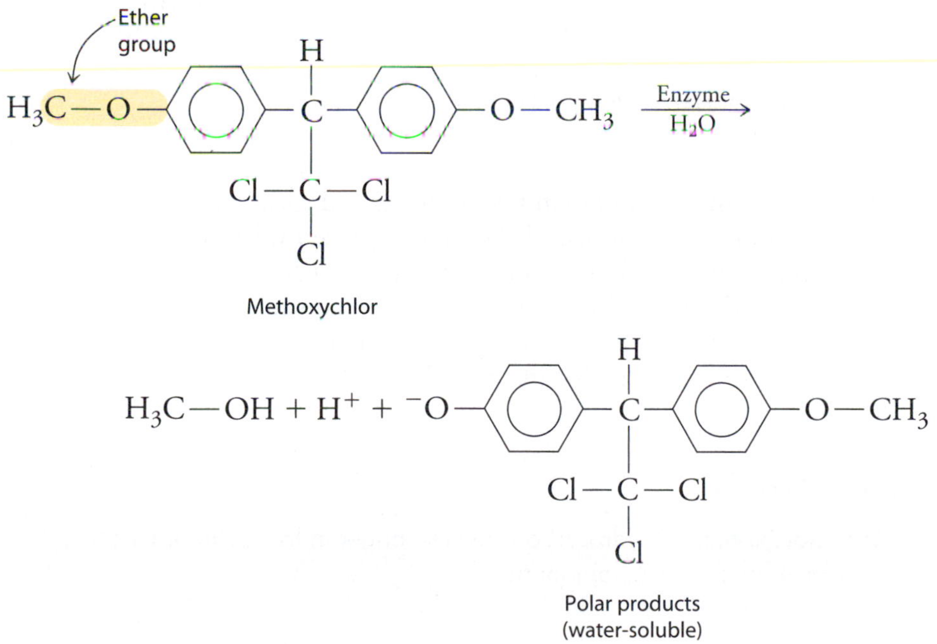

Figure 15.18
Methoxychlor is one of many alternatives to DDT. Enzymes in the liver can cleave the ether groups to produce polar products. Look back to Figure 15.15, and you will see that DDT lacks ether groups.

Figure 15.19
Honeybees do not forage at night. Quick-acting pesticides, such as organophosphorus compounds and carbamates, are therefore best applied in the evening. By the time the bees return the next day, these pesticides have lost much of their toxicity.

Organophosphorus compounds and carbamates, in contrast to chlorinated hydrocarbons, readily decompose to water-soluble components and so do not act over extended periods of time. Their immediate toxicity to both insects and animals is much greater than that of chlorinated hydrocarbons, however. Added safety precautions are required during the application of both organophosphates and carbamates, especially because of their toxicity to honeybees (Figure 15.19).

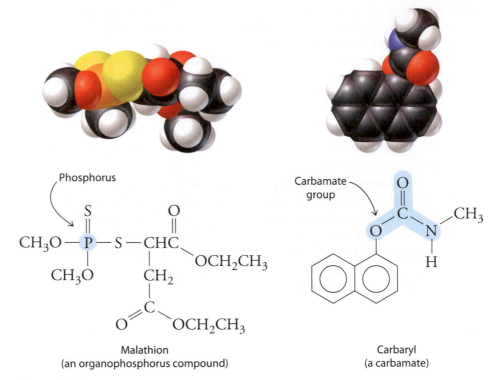

Malathion
(an organophosphorus compound)

Carbaryl
(a carbamate)

Figure 15.20
The widely used pesticides malathion and carbaryl.

There are hundreds of organophosphorus and carbamate insecticides in agricultural and household use. Two important examples are malathion, an organophosphorus compound, and carbaryl, a carbamate, both shown in Figure 15.20. Malathion kills a variety of insects, such as aphids, leafhoppers, beetles, and spider mites. Carbaryl, like many other carbamates, is relatively selective in the types of insects killed.

Concept Check ✔

Why do chlorinated hydrocarbons persist longer in the environment than do organophosphorus compounds?

Was this your answer?   There are no natural pathways by which chlorinated hydrocarbons are readily broken down to less harmful compounds. Also, because they are nonpolar, chlorinated hydrocarbons are not readily washed away by rainwater. Organophosphorus compounds readily decompose to water-soluble products that can be carried away by rainwater.

Hands-On Chemistry: Cleaning Your Insects

Perhaps the most environmentally friendly insecticides are solutions of soap or detergent. Insects are perforated with tiny holes, called *spiracles*, through which atmospheric oxygen migrates directly into cells. These holes are easily penetrated by liquid soap or detergent, which then blocks the exchange of atmospheric gases and so suffocates the insect. In general, the larger the insect, the more concentrated a soap solution needs to be in order to kill efficiently. A dilute soap solution, for example, will quickly annihilate aphids, but only a relatively concentrated one will kill a cockroach.

What You Need

Liquid soap or detergent, measuring spoons, pump-spray bottles, plants infested with household or garden insect pests such as houseflies or aphids.

Procedure

Use the measuring spoons to create various concentrations of soap solution. Stir your solutions gently to avoid excessive foaming. Pour each solution into a pump-spray bottle, and write the concentration (in teaspoons per cup, say) on each bottle. Test the effectiveness of each solution on the infested plants. Follow with a spray of fresh water to remove any residual soap from the plants.

Herbicides Kill Weeds

Weeds compete with crop plants for valuable nutrients. The traditional method for controlling weeds is to plow them under the soil, where in decomposing they release the nutrients they absorbed while they were alive. Plowing also aerates the soil, but it is either labor-intensive or energy-intensive and can lead to topsoil erosion. In the early 1900s, farmers noted that certain fertilizers, such as calcium cyanamide, CaNCN, selectively kill weeds while causing little harm to crops. This prompted a broad search for chemicals that act as herbicides. Today, a farmer can choose from hundreds of herbicides, many tailored for a specific weed. Farmers in the United States apply almost 600 million pounds of herbicides annually, which is about three times more than the amount of insecticides they apply.

Two selective herbicides are the carboxylic acids 2,4-dichlorophenoxyacetic acid (2,4-D) and 2,4,5-trichlorophenoxyacetic acid (2,4,5-T), shown in Figure 15.21 on page 514. Both mimic the action of plant growth hormones and are selective in killing broad-leafed plants but not grasslike crops such as corn and wheat. A herbicide known as *agent orange* is a blend of 2,4-D and 2,4,5-T. During the Vietnam War, U.S. military forces applied more than 15 million gallons of agent orange and related herbicides in an effort to defoliate jungle areas that could harbor enemy troops. Health problems in Vietnamese troops and civilians, U.S. troops, and others exposed to agent orange have since been linked to a minor contaminant of the agent orange—the highly toxic compound 2,3,7,8-tetrachlorodibenzo-*p*-dioxin (TCDD). This contaminant was generated as a side product in the

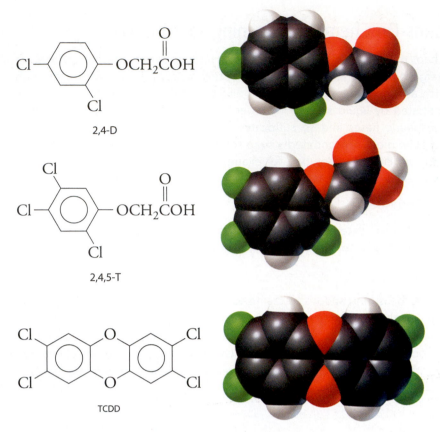

Figure 15.21
The herbicides 2,4-D and 2,4,5-T and the dioxin contaminant TCDD.

manufacture of 2,4,5-T. In 1985, because of this contamination, the use of 2,4,5-T was prohibited by the U.S. Environmental Protection Agency. TCDD-free methods of 2,4,5-T production, however, have since been developed, which raises the possibility that 2,4,5-T may once again be introduced as an effective herbicide.

Three other commonly used herbicides are atrazine, paraquat, and glyphosate, all shown in Figure 15.22. Atrazine is toxic to common weeds but not to many grasslike crops, which can rapidly detoxify this herbicide through metabolism.

Paraquat kills weeds in their sprouting phase. During the 1970s and 1980s, this herbicide was sprayed aerially to destroy drug-producing poppy and marijuana fields in the United States, Mexico, and much of Central America and South America. Paraquat residues made their way into the illicit drug products, however, causing lung damage in users. So, for ethical reasons, the spraying of paraquat on drug-producing plants is no longer common practice.

Glyphosate is a nonselective herbicide that affects a biochemical process common to all plants—the biosynthesis of the amino acids tyrosine and phenylalanine. Glyphosate has low toxicity in animals because most animals do not synthesize these amino acids, obtaining them from food instead. Glyphosate is the active ingredient of the herbicide Round-up.

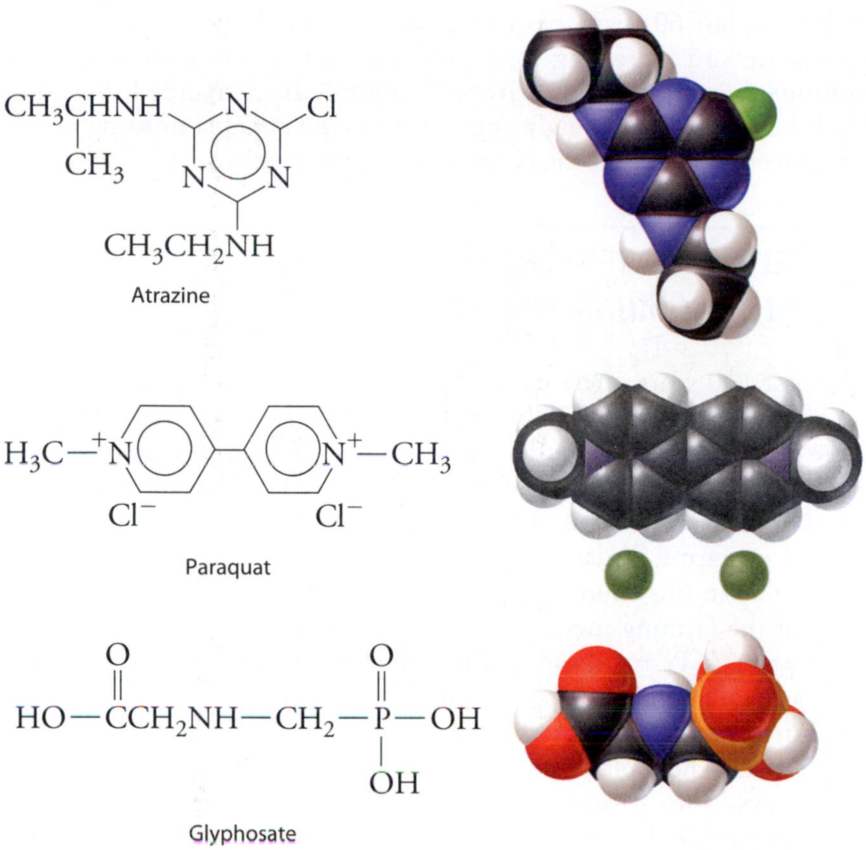

Figure 15.22
The herbicides atrazine, paraquat, and glyphosate.

Atrazine

Paraquat

Glyphosate

Fungicides Kill Fungi

As decomposers, fungi play an important role in soil formation, but they can also harm crops. Most of the harm they cause occurs during a plant's early growth stages. Fungi can also spoil stored food and are particularly devastating to the world's fruit harvest.

In the United States, farmers use about 100 million pounds of fungicides annually, meaning fungicides rank third after herbicides and insecticides in the amounts used. An example of a fungicide is thiram, widely used on fruits and vegetables and shown in Figure 15.23.

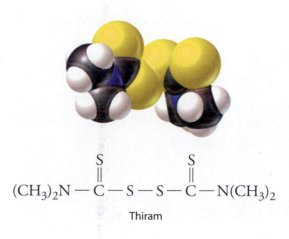

Thiram

Figure 15.23
The fungicide thiram.

During the last 60 years, pesticides have benefited our society by preventing disease and increasing food production. Our need for pesticides will continue, but greater specificity will certainly be demanded. Furthermore, it is becoming increasingly apparent that the benefits of using pesticides must be considered in the context of potential risks.

15.6 There Is Much to Learn from Past Agricultural Practices

Over the past 100 years, there have been dramatic increases in crop yields. An acre of U.S. farmland in 1900 yielded about 30 bushels of corn. Today, that same acre yields on the order of 130 bushels of corn. This increased efficiency has meant a significant drop in the number of people needed to farm. In the early 1900s, about 33 million people in the United States lived and worked on farms. Today, only 2 million people are engaged in commercial farming in this country, producing crops and livestock.

Many of the farming methods used to obtain high yields have significant disadvantages. Pesticides and fertilizers, for example, pose certain risks. Pesticides are inherently toxic, and each year thousands of people working in agriculture are poisoned by the mishandling of these dangerous compounds. Fertilizers help plants grow, but major portions of applied fertilizer are washed away into streams, rivers, ponds, and lakes, where they upset ecosystems, especially by promoting an excessive growth of algae (see Section 16.6). Fertilizer runoff from fields, illustrated in Figure 15.24, can also contaminate drinking water supplies and thus affect human health. An ailment known as blue-baby syndrome, for example, results from drinking water containing high concentrations of nitrate ions, a main ingredient of most fertilizers. Nitrate ions in the bloodstream compete with oxygen for

Figure 15.24
The water running off this farm field contains many pesticides and fertilizers that can be harmful to ecosystems and human health.

Figure 15.25
Poor soil-conservation practices in the early 1900s contributed to the loss of much topsoil to wind storms thick with dust.

the positively charged iron ions of hemoglobin molecules. This leads to a form of anemia known as methemoglobinemia, to which babies are particularly sensitive. Aside from a shortness of breath, one of the major symptoms is a bluish color to the skin.

Poor maintenance of topsoil is also a major concern. Synthetic fertilizers have no organic bulk and do not provide a food source for soil microorganisms and earthworms. Over time, a soil treated with only these fertilizers loses biological activity, which diminishes the soil's fertility. Soils void of organic bulk become chalky and susceptible to wind erosion. Chalky soils lose their capacity to hold water, which means that more applied fertilizer is leached away. Ever-increasing amounts of fertilizer are thus needed.

Over the past 100 years, damaging farming practices have decreased the amount of topsoil in parts of the United States by as much as 50 percent. During the 1930s, farming practices and drought conditions created giant dust storms, such as the one shown in Figure 15.25, that removed major portions of the topsoil in Kansas, Oklahoma, Colorado, and Texas. In one storm, large dust clouds were carried all the way from the Midwest to Washington, D.C., and then into the Atlantic Ocean. Politicians in that city observing the effects of poor soil management right outside their windows quickly passed legislation that created the Soil Erosion Service, which became the National Resources Conservation Service and continues to this day in its efforts to help protect the nation's topsoil for future generations.

Another limited resource required for farming is fresh water. In regions where rainfall is insufficient to support large crops, water is either channeled into fields from lakes, rivers, and streams or else pumped up from the ground. In many areas, groundwater is the primary source of fresh water, but excessive use of groundwater can lead to land subsidence, illustrated in Figure 15.26.

Any source of water other than rainwater requires an irrigation method to deliver water to the fields. Flooding is a common method, but it is not

Figure 15.26
The San Joaquin Valley of California has subsided by more than 35 feet since the pumping of groundwater began in the 1920s.

Figure 15.27
As a river flows along, runoff from agricultural fields can add to the river's natural salinity. By the time the Colorado River reaches the Gulf of California, for example, it is too salty for productive farming. A typical safe drinking water standard for salt content is 500 milligrams/liter. Agricultural damage occurs when soil salinity reaches a concentration of about 800 milligrams/liter.

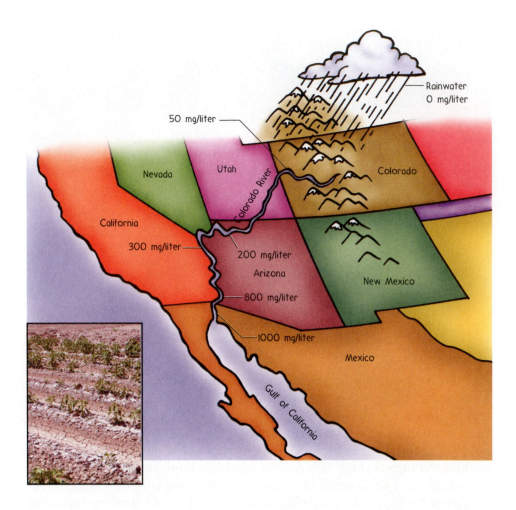

efficient because most of the water is lost in runoff and evaporation. Sprinkler systems are an improvement over flooding because they do not cause soil erosion. Such systems also lose large amounts of water, however, because a significant portion of the airborne water evaporates before reaching the ground.

All liquid water on the Earth's surface, no matter how fresh, contains some salts. After irrigation water evaporates from farmland, these salts are left behind, and so over time repeated irrigation causes the salinity of the soil to increase. This process is known as **salinization**, and it leads to a rapid decrease in productivity. To counteract growing soil salinity, farmers flood the land with huge quantities of water. As the water drains into a river, it washes the unwanted salts—along with significant amounts of topsoil—into the river. Thus a river passing through farmlands gets saltier and saltier as it runs to the sea, as depicted in Figure 15.27.

Concept Check ✔

Are there any dissolved salts in a mountain stream?

Was this your answer? Land contains a variety of salts. As water runs over and through the land, these salts dissolve in the water. In general, the farther the water travels, the saltier it becomes. Thus even water that has traveled only a short distance, which is the case with a mountain stream, contains some salts. So the answer to this question is a qualified "Yes, but not enough to make the water undrinkable."

15.7 High Agricultural Yields Can Be Sustained with Proper Practices

At the beginning of this chapter, agriculture was defined as the organized use of resources for the production of food. Whether or not these resources—mainly topsoil and fresh water—will be available for future generations depends on how well we manage them now. We already know from experience that pesticides and fertilizers cannot be applied liberally without threatening both topsoil quality and our supplies of clean groundwater (not to mention the health of ourselves and the planet).

Over the past several decades, there have been strong movements toward developing methods and technologies that will sustain agricultural resources over the long term. Problems associated with irrigation, for example, can be solved by **microirrigation**, which is any method of delivering water directly to plant roots. Microirrigation not only prevents topsoil erosion but also minimizes the loss of water through evaporation, which in turn minimizes the salinization of farmland. One method of microirrigation is shown in Figure 15.28.

Figure 15.28
The microirrigation method known as drip irrigation delivers water through long, narrow strips of punctured plastic tubing. These farmers are placing tubing in a soon-to-be-planted field. Once installed, the system will provide only as much water as the plants need.

Organic Farming Is Environmentally Friendly

For controlling pests and maintaining fertile soil, the conventional agricultural industry is now looking at the efforts of many small-scale farmers who have demonstrated that significant crop yields can be obtained without pesticides and synthetic fertilizers. This method of farming is known as **organic farming**, where the term *organic* is used to indicate a concern for the environment and a commitment to using only chemicals that occur in nature.

Figure 15.29
Odor-free backyard compost bins are easy to build and maintain.

To protect against pests, organic farmers alternate the crops planted on a particular plot of land. Such *crop rotation* works fairly well because different crops are damaged by different pests. A pest that thrives on one season's crop of corn, for example, will do poorly on the next season's crop of alfalfa. For fertilizer, organic farmers rely on compost, shown in Figure 15.29. They also include nitrogen-fixing plants in their crop-rotation schedules.

Many claims are made that food produced organically is better for human consumption. Chemically, however, the atoms that plants absorb from synthetic fertilizers are the same as those they absorb from natural fertilizers. If organically grown produce tastes any better or is more nutritious than conventionally grown produce, the reason likely has to do with the genetic strain of the produce or with the greater attention paid to growing conditions, such as the amount of water made available to plants or the careful control of soil pH.

Organic farming tends to be benevolent to the environment. In addition to avoiding the potential runoff of pesticides and fertilizers, organic farming is energy-efficient, using only about 40 percent as much energy as conventional farming. A large portion of the energy savings arises because the manufacture of pesticides and fertilizers is energy-intensive. For instance, each year in the United States, about 300 million barrels of oil is consumed for the production of nitrogen fertilizers.

Because much organically grown food, such as that shown in Figure 15.30, is grown locally, by purchasing it you help support local farmers. Ultimately, though, your purchase of organically grown food is a vote in favor of environmentally friendly methods of farming.

Figure 15.30
(a) Pound for pound, organically grown food is cheaper to produce than conventionally grown food. Market forces, however, often result in organically grown foods costing more. (b) A product can bear this USDA organic seal if at least 95% of its ingredients are certified organic. Interestingly, the annual U.S. sales of organic foods is projected to double from about $10 billion in 2002 to about $20 billion in 2005.

(a)

(b)

Concept Check ✔

Which are made of organic chemicals, organically grown foods or conventionally grown foods?

Was this your answer? Regardless of whether they are grown with natural fertilizer or synthetic fertilizer, all foods are made of organic chemicals—carbohydrates, lipids, proteins, nucleic acids, and vitamins. A thoroughly rinsed conventionally grown carrot may be just as good for you—or even better—as one grown without the use of synthetic fertilizers and pesticides. The *organic* in organic farming is a term used to designate a natural method of farming.

Integrated Crop Management Is a Strategy for Sustainable Agriculture

To meet concerns about sustaining agricultural resources over the long term, groups from industry, government, and academia have identified a whole-farm strategy called **integrated crop management** (ICM). This method of farming involves managing crops profitably and with respect for the environment in ways that suit local soil, climatic, and economic conditions. Its aim is to safeguard a farm's natural assets over the long term through the use of practices that avoid waste, enhance energy efficiency, and minimize pollution. ICM is not a rigidly defined form of crop production but rather a dynamic system that adapts and makes sensible use of the latest research, technology, advice, and experience.

One of the more significant aspects of ICM is its emphasis on multicropping, which means growing different crops on the same area of land either simultaneously as shown in Figure 15.31 or in rotation from season to season. As with organic farming, multicropping achieves significant pest control, and it also can be used to improve soil fertility. For example, nitrogen-generating crops, such as legumes, are a good complement to nitrogen-depleting crops, such as corn.

An important component of ICM is **integrated pest management** (IPM), one of the aims of which is to reduce the initial severity of pest infestation. This can be accomplished through a number of routes. Upon starting a farm, for example, only crops that fit the local climate, soil, and topography should be grown. This selectivity makes for crops that are hardy and pest-resistant. Crops should also be rotated as much as possible to reduce pest and weed problems. Another IPM strategy is growing tree crops or hedges either around the perimeter of a farm or interspersed throughout the farm. These trees and hedges provide habitat, cover, and refuge for beneficial insects and such pest predators as spiders, snakes, and birds. As an added benefit, the trees and hedges also protect the land from wind erosion. Yet another IPM strategy is to breed and cultivate plants that have a natural resistance to pests. For centuries this was accomplished by selectively mating plants that showed the greatest resistance. Today, this age-old method is quickly being supplanted by the techniques of genetic engineering.

Another aim of IPM is to minimize the use of pesticides. For example, many farmers now use the global positioning satellite system (GPS) to target precise pesticide applications. Using infrared satellite photography, illustrated in Figure 15.32, and careful walk-through assessment of field conditions, farmers can match pesticide blends with crop needs. Computers link application equipment with the GPS satellites, which "beam" pesticide application adjustments every few seconds as a farmer moves across a field. This same technology also works well with selective delivery of synthetic fertilizers.

There are many other methods of pest control that can be used in place of chemical pesticides or in combination with them to minimize the need for these agents. Depending on the availability of labor, egg masses or larvae can be hand-picked off plants. Instead of using herbicides, weeds can be tilled under. An insect population can also be controlled by various biological approaches, such as by introducing large numbers of sterile insects into a population or by introducing natural predators, as shown in Figure 15.33 on page 522.

Figure 15.31
Complementary crops such as legumes and corn are grown in alternating strips to enhance soil fertility. The strips follow the contour of the land to minimize erosion from rainwater or irrigation.

Figure 15.32
Satellite images can reveal information about crop growth and potential pest infestation. The darker areas of this infrared satellite image show where corn growth has been stunted by some form of infestation.

Figure 15.33
The almond trees on the right side of the road were decimated by spider mites. Those on the left were protected by the introduction of spider mite predators.

Another way to control the proliferation of insects is to modify their behavior through the use of **pheromones**, which are volatile organic molecules that insects release to communicate with one another. Each insect species produces its own set of pheromones, some as warning signals and others as sexual attractants. Sexual pheromones synthesized in the laboratory can be used to lure harmful insects to localized insecticide deposits, thereby reducing the need for spraying an entire field, as depicted in Figure 15.34.

Figure 15.34
Female gypsy moths emit the pheromone disparlure (top) to entice male gypsy moths (bottom left) into mating. The males are so sensitive to this compound that they can detect one molecule in 10^{17} molecules of air. This astounding sensitivity enables them to respond to a female who may be more than 1 kilometer away. However, they can also be tricked into responding to insecticide traps laced with synthetic disparlure (bottom right).

Nature is sophisticated, and if we are to work with nature in a sustainable way, our methods must also be sophisticated. New and improved techniques provide the farmer with a menu of possible actions in response to nature's ever-changing parameters. Each action, however, must be taken with an awareness of its potential environmental impact. In this sense, the human who farms sustainably is not dominating nature but rather working with it.

15.8   A Crop Can Be Improved by Inserting a Gene from Another Species

Over the past couple of decades, advances in genetic engineering have led to profound developments in agriculture. (See Section 13.5 for a review of genetic engineering techniques.) For centuries, farmers have improved crops and domestic animals by breeding for desirable traits. This uncertain and often lengthy process can now be performed relatively quickly and with great certainty by using the tools of modern molecular biology to introduce genes for desired traits into plants and animals. The resulting organisms are called **transgenic organisms** because they contain one or more genes from another species.

Transgenic bacteria have been developed to mass-produce a variety of valuable proteins, including bovine growth hormone (BGH). When this hormone is injected into dairy or beef cattle, it raises milk production or improves weight gain. It has passed all safety tests so far and is now being used extensively in dairy herds.

Most of the progress in transgenic agriculture has been with plants. Several major crops have been engineered with genes that create proteins having insecticidal properties. The insect pest is killed only when it feeds on the crop. Figure 15.35 illustrates this technique for corn. With such a mechanism,

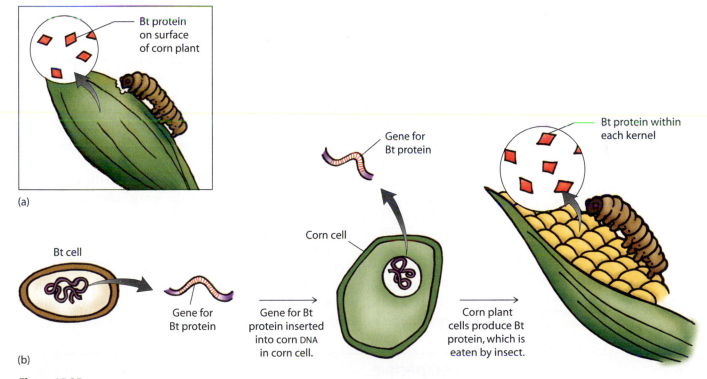

(a)

(b)

Figure 15.35
(a) The bacterium *Bacillus thuringiensis* (Bt) produces proteins that are toxic to insects, such as the corn borer, a devastating corn pest. The external application of Bt proteins on corn, however, cannot control the corn borer once it is inside the stalk. (b) Corn is made resistant to the corn borer by splicing the gene for the Bt protein into corn DNA. The resulting corn plant produces the Bt protein in its cells and is thus fully resistant to the corn borer. Recent studies, however, have indicated that pollen from Bt corn may have adverse effects on Monarch butterflies, which are beneficial insects.

Figure 15.36
Gene transfer has made sweet potatoes a better protein source.

most—though not necessarily all—nontarget benevolent organisms are left unharmed and the need for pesticide application is reduced. Other major crops have been engineered with genes that make them resistant to the herbicide glyphosate, meaning that the herbicide kills weeds in a field but doesn't threaten the crop planted there. Researchers have also inserted a gene coding for a dietary protein into sweet potato plants. This protein contains significant amounts of the eight amino acids essential to adult humans (see Table 13.6). Figure 15.36 shows these protein-rich sweet potatoes, which are easy to cultivate and hold special value for developing nations, where high-quality protein foods are hard to come by.

The examples just described all involve the transposing of only one or a couple of genes into the transgenic organism. There are many desirable traits, however, that involve clusters of genes. An important example is the ability to fix nitrogen. Intense research is currently under way to transpose all the genes necessary for nitrogen fixation into plants that do not naturally fix nitrogen. With such a transgenic species, the expensive production and application of nitrogen fertilizers becomes unnecessary. Because many genes are involved, the system is complicated and currently beyond science's biotechnical capabilities, but perhaps that won't be true for long.

There is heated debate about genetically engineered agricultural products. Some scientists argue that producing transgenic organisms is only an extension of traditional cross-breeding, the procedure that has given us such new and interesting products as the tangelo (a tangerine–grapefruit hybrid). In general, the Food and Drug Administration has held that if the result of genetic engineering is not significantly different from a product already on the market, testing is not required. On the other side of the argument are scientists who believe that creating transgenic organisms is radically different from hybridizing closely related species of plants or animals. There is concern, for example, that genetically engineered crops might grow too well, ultimately reseeding themselves in areas where they are not desired and thus becoming "superweeds." Transgenic crops might also pass their new genes to close relatives in neighboring wild areas, creating offspring that would be difficult to control.

Worldwide, more than 75 million acres is cultivated with transgenic crops each year. As a result, about one-third of the world corn harvest and more than one-half of the world soybean harvest now come from genetically engineered plants. Stay tuned for developments in the area of transgenic agriculture, such as the development of the golden rice discussed at the beginning of this chapter. This form of agriculture holds much promise. The power of genetic engineering, however, demands that we move cautiously, with all necessary safeguards in place. One of the more important safeguards, no doubt, will be a well-informed general public.

In Perspective

Within a century, the Earth may have 12 billion inhabitants, about twice as many as it has now. If projections made by demographers are correct, the human population will be approaching a stable level by then, just as the populations of many developed nations already have. Will we be able to feed ourselves when this steady state is reached? The answer is probably yes, but potentially only for a while. Even if world food production were to grow more slowly than at the current rate, there would still be enough food for 12 billion mouths by the time they arrive.

However, not only must the food supply expand, but it must expand in a way that does not destroy the natural environment. For agriculture to be sustainable, a steady stream of new technologies that minimize environmental damage must be developed.

The most critical problems faced by those seeking to counteract world hunger are more likely to be social rather than technical. Above all, efforts toward stabilizing the world population must continue in earnest. The size of our planet is limited, and practically all the world's farmable land is now under cultivation. As the population grows, more food will be required, while at the same time more farmable land will be lost to residential and business development. In tropical areas, economic pressures to slash and burn rainforests for the formation of additional farmland will probably continue.

Even with a stable world population, it cannot be assumed that a large-enough food supply will lead to the end of world hunger. Today, the abundance of food is at an all-time high, and yet an estimated 8.7 million individuals, most of them young children, die each year from a lack of adequate nutrition. Amartya Sen, a leader in the fight against world hunger and the 1998 Nobel laureate in economics, points out that in most circumstances, malnutrition arises not from a lack of food but from a lack of appropriate social infrastructure, as the graph of Figure 15.37 shows. Backed by strong evidence, Sen argues that "public action can eradicate the terrible and resilient problems of starvation and hunger in the world in which we live." Efforts toward optimizing the food yields from agriculture must therefore be matched by efforts to build social, political, and economic systems that give those facing starvation the means for survival. World hunger is not inevitable.

Amartya Sen, professor of economics at Trinity College, Cambridge, England, is well known for his work in the economics of poverty and famine.

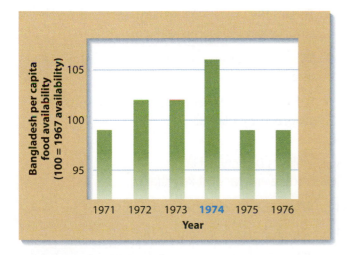

Figure 15.37
The Bangladesh famine of 1974 occurred during a period when the amount of food available per person in that country was at a peak. It was unemployment, hoarding, and inflated food prices that drove millions to their death.

Key Terms and Matching Definitions

_____ agriculture
_____ bioaccumulation
_____ compost
_____ consumer
_____ decomposer
_____ humus
_____ integrated crop management
_____ integrated pest management
_____ microirrigation
_____ mixed fertilizer
_____ nitrogen fixation
_____ organic farming
_____ pheromone
_____ producer
_____ salinization
_____ soil horizon
_____ straight fertilizer
_____ transgenic organism
_____ trophic structure

1. The organized use of resources for the production of food.
2. The pattern of feeding relationships in a community of organisms.
3. An organism at the bottom of a trophic structure.
4. An organism that takes in the matter and energy of other organisms.
5. An organism in the soil that transforms once-living matter to nutrients.
6. A chemical reaction that converts atmospheric nitrogen to some form of nitrogen usable by plants.
7. A layer of soil.
8. The organic matter of topsoil.
9. Fertilizer formed by the decay of organic matter.
10. A fertilizer that contains only one nutrient.
11. A fertilizer that contains the plant nutrients nitrogen, phosphorus, and potassium.
12. The process whereby a toxic chemical that enters a food chain at a low trophic level becomes more concentrated in organisms higher up the chain.
13. The process whereby irrigated land becomes saltier as the irrigation water evaporates.
14. A method of delivering water directly to plant roots.
15. Farming without the use of pesticides or synthetic fertilizers.

16. A whole-farm strategy that involves managing crops in ways that suit local soil, climatic, and economic conditions.
17. A pest-control strategy that emphasizes prevention, planning, and the use of a variety of pest-control resources.
18. An organic molecule secreted by insects to communicate with one another.
19. An organism that contains one or more genes from another species.

Review Questions

Humans Eat at All Trophic Levels

1. What are the two major chemical products of photosynthesis?

2. In a trophic structure, what distinguishes a producer from a consumer?

3. Why are the number of trophic levels limited?

4. What happens to most biochemical energy as it passes from one trophic level to the next?

Plants Require Nutrients

5. What is the major natural source of the nitrogen used by plants?

6. Do plants require oxygen?

7. What is the major natural source of phosphorus in soil?

8. What effect have terrestrial plants had on the composition of ocean water?

9. Why are calcium and magnesium deficiencies rare in plants?

10. What is the most common form of sulfur absorbed by plants?

Soil Fertility Is Determined by Soil Structure and Nutrient Retention

11. What are the three soil horizons?

12. Which horizon is void of life?

13. What are four important components of fertile topsoil?

14. What is one of the great advantages to having humus in soil?

Natural and Synthetic Fertilizers Help Restore Soil Fertility

15. What process is used to generate most synthetic nitrogen fertilizer?

16. What is the difference between a straight fertilizer and a mixed fertilizer?

17. What advantages do mixed synthetic fertilizers have over compost?

18. What advantages does compost have over mixed synthetic fertilizers?

Pesticides Kill Insects, Weeds, and Fungi

19. Name three kinds of pesticides.

20. What are three classes of insecticides?

21. Is DDT still being used today?

22. Which herbicide tends to kill weeds in their sprouting phase?

23. Glyphosate interferes with the biosynthesis of which two amino acids?

There Is Much to Learn from Past Agricultural Practices

24. How do pesticides and fertilizers end up in our drinking water?

25. What is missing from synthetic fertilizers that makes them harmful to soil?

26. How is irrigation damaging to topsoil?

High Agricultural Yields Can Be Sustained with Proper Practices

27. What are the advantages of microirrigation?

28. What is organic farming?

29. What role did organic farming play in the development of integrated crop management?

30. How is space technology used to reduce the amount of pesticides applied to farmlands?

31. How are pheromones used to diminish insect populations?

A Crop Can Be Improved by Inserting a Gene from Another Species

32. What type of biomolecule is generated by a gene?

33. How might a transgenic plant reduce the amount of pesticide needed for crop protection?

34. Why is it so difficult to develop a transgenic corn variety that fixes its own nitrogen?

35. What harm might a transgenic plant bring to the environment?

In Perspective

36. What is the present world population?

37. About how many people die each year from hunger-related causes?

Hands-On Chemistry Insights

Soil pH–A Qualitative Measure

The soil particles adsorb the red pigment, which is why it is important to remove as much suspended solid matter as possible from the water before you add your indicator. Any soil particles suspended in the water will cause the indicator to turn blackish.

This procedure gives only a rough approximation of the soil pH. More quantitative do-it-yourself soil pH measuring kits are available at your local gardening store.

Cleaning Your Insects

The advantages of using soap or detergent to kill insects are that these materials are inexpensive, easily washed away, and environmentally friendly. The disadvantages are that they are not selective for harmful insect pests over beneficial ones. Also, they work only by direct contact, and the plant must be wiped clean once the pests are destroyed; getting rid of all the soap can be difficult if you are using a concentrated solution. Lastly, they have no lasting insecticidal effect.

Interestingly, the fine network of spiracles is what limits the size of insects. If an insect were any heavier, its weight would collapse all the tiny channels. In prehistoric times, higher atmospheric concentrations of oxygen permitted the evolution of much larger insects, such as the 1-meter-long dragonflies often depicted in dinosaur books.

Exercises

1. Thoroughly dried dead plant material is 44.4 percent oxygen by weight. How is this oxygen

held in the plant so that it doesn't escape into the atmosphere?

2. What would happen to a forest without the action of decomposers?

3. Why is the number of *Tyrannosaurus rex* fossils found so much lower than the number of fossils found for other dinosaurs?

4. Why would a predominately meat-based diet be a severe restriction on the possible size of the human population?

5. How does humus contribute to the acidity of soil?

6. Why do groundskeepers poke holes in football fields?

7. Pockets of open space in soil are necessary for plant health. If these pockets are too large, however, plant health suffers. Why?

8. The use of tractors and other heavy farm machinery compacts soil. How might this compacting affect soil fertility?

9. Why do soils that contain a high percentage of clay retain plant nutrients so well?

10. Why don't plants grow well in nutrient-rich soils that contain a high percentage of clay?

11. What happens to ammonium ions in alkaline soil, and how might this reaction favor the loss of nitrogen from the soil?

12. Why do the leaves of plants deficient in nitrogen turn yellow? Consider the structure of the porphyrin ring shown in Figure 15.6.

13. How does the organic bulk of compost help to maintain fertile soil conditions?

14. Why does DDT have such a strong affinity for fat tissue?

15. When only synthetic fertilizers are used on a crop, the quantity used needs to be increased over time. Why?

16. Earthworms are repelled by the high concentrations of nutrients in soils treated with synthetic fertilizers. How might this affect the soil structure?

17. Shown below are the N-P-K ratings for three fertilizer additives: sawdust, fish meal, and wood

ashes. Using what you know about the chemical composition of these substances, assign each to its most likely N-P-K rating.

5-3-3	0-1.5-8	0.2-0-0.2

18. How might periodic floods from rainstorms benefit irrigated cropland?

19. Which is better for a plant: an ammonium ion from compost or an ammonium ion from synthetic fertilizer?

20. Distinguish between organic farming and integrated crop management.

21. Why does organic farming exclude the use of pesticides when pesticides themselves are organic?

22. Why is it beneficial to grow two or three crops simultaneously in the same field?

23. Many farmers in the United States are paid by the government not to farm their land. Based on your understanding of the concepts presented in this chapter, cite one reason why this might be so.

24. Besides luring insects to an insecticide how else might pheromones be used to decrease insect populations?

25. Why are advances in genetic engineering so significant to agriculture?

26. What are the chances that you have already ingested a product made from a transgenic organism?

27. How might genetic engineering be used to counteract the negative effects of salinization?

28. Why might a sweet potato plant genetically engineered for a higher protein content have a greater intolerance for nitrogen-poor soils?

Discussion Topics

1. What are the benefits of eating organisms low on the food chain? What are the benefits of eating higher up on the food chain? Identify these benefits as being for the entire human population or for an individual person.

2. Should organically grown food that has been irradiated to kill pathogens be permitted

to carry the label "organic"? How about a crop grown organically but from transgenic seeds? Why or why not?

3. Should transgenic foods be labeled as such in the stores where they are sold? Why or why not?

4. Should there be an international ban on the production and use of DDT? Why or why not?

5. The caption to Figure 15.30 notes that market forces often result in higher prices for organically grown foods. Identify some of these market forces.

6. You are a Brazilian farmer looking to clear-cut a rainforest to make room for cattle pasture. An environmental activist from the predominately beef-eating United States comes knocking at your door and tries to convince you not to cut. What are some of the arguments the activist might present, and what counterarguments can you think of for proceeding with your plan?

7. Are you willing to drink milk from a cow whose milk production has been increased by injections of bovine growth hormone? Does it matter to you that this bovine growth hormone was created by transgenic bacteria? Why or why not?

8. Why is growing more food not necessarily the solution to world hunger?

Exploring Further

http://www.croplifeamerica.org

The home page of Crop Life America, formerly the American Crop Protection Association, which was established to represent companies that produce, sell, and distribute chemicals used for crop protection.

http://www.epa.gov

Use the search engine at this site for the Environmental Protection Agency to find numerous articles related to pesticides, including articles on the recovery of contaminated ecosystems.

http://www.nrcs.usda.gov

Home page for the Natural Resources Conservation Service of the U.S. Department of Agriculture. The mission of this department is to help people conserve, improve, and sustain our natural resources and environment.

http://www.thp.org

Home page for The Hunger Project, an international agency dedicated to ending world hunger.

http://www.cspinet.org

The Center for Science in the Public Interest is a nonprofit educational and advocacy organization that focuses on improving the safety and nutritional quality of our food supply.

http://cipotato.org

The International Potato Center, known worldwide by its Spanish acronym CIP, sees the potato and other Andean root and tuber crops as underexploited resources for agricultural development and hunger relief in developing countries. Founded in 1971, CIP has worked to enhance the cultivation, yield, processing, and consumption of potatoes. Its original mandate was expanded to include sweet potatoes and, more recently, other Andean roots and tubers in danger of extinction.

http://www.ams.usda.gov/nop

Check out the details of the USDA's program for certifying organic foods. They claim that this program will help safeguard against creatively worded packages that identify a product as organic when only a few ingredients are.

the Chemistry place

Optimizing Food Production
Visit The Chemistry Place at:
www.aw.com/chemplace

Fresh Water Resources

Saline water in oceans: 97.2%

Ice caps and glaciers: 2.14%

Available fresh water: 0.66%

Our Roles and Responsibilities

About 97.2 percent of the Earth's water is saline (salty) ocean water. Another 2.14 percent is fresh water frozen in polar ice caps and glaciers. All the remaining water, less than 1 percent of the Earth's total, comprises water vapor in the atmosphere, water in the ground, and water in rivers and lakes—the fresh water we rely on in our daily lives.

If you have ever stood on the shore of one of the Great Lakes or experienced the power of a waterfall or been caught in a rainstorm, it may seem to you that the supply of fresh water on the Earth is inexhaustible. From the perspective of one person, it is. Our population, however, has grown to more than 6 billion people. If we were to spread ourselves evenly across all habitable lands, there would be about 50 of us in every square kilometer. Thus it should come as no surprise that collectively we have a significant impact on the Earth's relatively limited resources, such as fresh water.

We are reminded daily that fresh water is a limited resource. In the United States, we see the signs of water shortage when farmers fight for the privilege to irrigate, when our water utility bills rise, or when the water supply of downstream municipalities is endangered as upstream municipalities release sewage into the water. Globally, there are many nations in which the primary supply of fresh water is rivers that originate in some other nation. As the upstream nation diverts fresh water for its own expanding population, political tensions rise. Over the next decade, for example, it is projected that agricultural development in Ethiopia and Sudan will reduce the Nile River flow to Egypt by 15 percent. Similarly, Turkey is currently escalating its damming and irrigation projects along the headwaters of the Euphrates River. Once fully implemented, these projects could result in a 40 percent reduction of river flow into Syria and an 80 percent reduction of river flow into Iraq. Not surprisingly, this issue has been a major source of political tension among these nations.

In this chapter, we explore some of the fundamental dynamics of fresh water resources, the chemistry used to treat and protect these resources, and the impact human activity has on them.

16.1 Water Circulates Through the Hydrologic Cycle

The Earth's water is constantly circulating, powered by the heat of the sun and the force of gravity. The sun's heat causes water from the Earth's oceans, lakes, rivers, and glaciers to evaporate into the atmosphere. As the atmosphere becomes saturated with moisture, the water precipitates in the form or either rain or snow. This constant water movement and phase changing is called the **hydrologic cycle**. As Figure 16.1 shows, the route of water through the cycle can be either from ocean directly back to ocean or a more circuitous route over the ground and even underground.

In the direct route, water molecules in the ocean evaporate into the atmosphere, condense to form clouds, and then precipitate into the ocean as either rain or snow, to begin the cycle anew.

The cycle is more complex when precipitation falls on land. As with the direct route, the cycle "begins" with ocean water evaporating into the atmosphere. Instead of forming clouds over the water, however, the moist air is blown by winds until it is over land. Now there are four possibilities

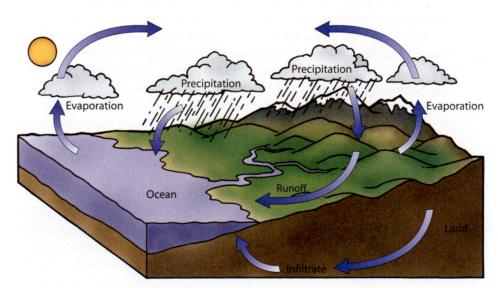

Figure 16.1
The hydrologic cycle. Water evaporated at the Earth's surface enters the atmosphere as water vapor, condenses into clouds, precipitates as rain or snow, and falls back to the surface, only to go through the cycle yet another time.

for what happens to the water once it precipitates. It may (1) evaporate from the land back into the atmosphere, (2) infiltrate the ground, (3) become part of a snow pack or glacier, or (4) drain to a river and then back to the ocean.

Water that seeps below the Earth's surface fills the spaces between soil particles until the soil reaches *saturation*, at which point every space is filled with water. The upper boundary of this saturated zone is called the **water table**. The depth of the water table varies with precipitation and with climate. It ranges from zero depth in marshes and swamps (meaning the water table is at ground level at these locations) to hundreds of meters deep in some desert regions. The water table also tends to follow the contours of the land and lowers during times of drought, as shown in Figure 16.2. Many lakes and streams are simply regions where the water table lies above the land surface.

All water that is below the Earth's surface is called *groundwater*. (Liquid water that is on the surface—in streams, rivers, and lakes—is called, naturally enough, *surface water*.) Any water-bearing soil layer is called an **aquifer**, which can be thought of as an underground water reservoir. Aquifers underlie the land in many places and collectively contain an enormous amount of fresh water—approximately 35 times the total volume of water in freshwater lakes, rivers, and streams combined. More than half the land area of the United States is underlain by aquifers, such as the Ogallala Aquifer stretching from South Dakota to Texas and from Colorado to Arkansas!

As populations grow, the demand for fresh water grows. Precipitation is the Earth's only natural source of groundwater recharge. Although the reservoir of groundwater is great, when the pumping rate exceeds the recharge rate, there can be a problem. In wet climates, such as in the Pacific Northwest, extraction is often balanced by recharge. In dry climates, however, extraction can easily exceed recharge. To support large populations, these areas must import their water from distant sources, typically through aqueducts. In southern California, for example, most of the fresh water comes from the Colorado River through aqueducts that stretch hundreds of miles.

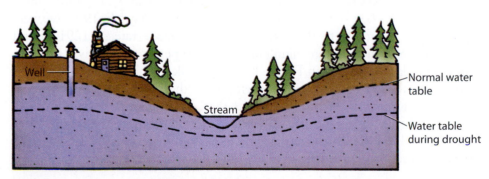

Figure 16.2
The water table in any location roughly parallels surface contouring. In times of drought, the water table falls, reducing stream flow and drying up wells. It also falls when the amount of water pumped out of a well exceeds the amount replaced as precipitated water infiltrates the ground and recharges the supply.

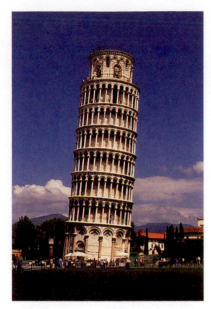

Figure 16.3
The Leaning Tower of Pisa, built centuries ago, slowly acquired a deviation from the vertical of about 4.6 meters as a result of groundwater withdrawal. The tower's foundation has been stabilized by groundwater withdrawal management, and the tower should remain stable for years to come.

The Ogallala Aquifer, which is below the dry High Plains, has supplied water to this thirsty agricultural region for more than 100 years. Most of the aquifer is water that has been locked underground for thousands of years, however, with few sources of replenishment. If pumping were to cease, it would take thousands of years for the water table to return to its original level. In this respect, the Ogallala, unlike most other aquifers, is a limited and nonrenewable resource. Despite its vast size, withdrawal has been so great in the past 20 years that the area of farmland serviceable by the Ogallala has decreased by about 20 percent.

As water is removed from the spaces between soil particles, the sediments compact and the ground surface is lowered—it *subsides*. In areas where groundwater withdrawal has been extreme, the ground surface has subsided significantly. In the United States, extensive groundwater withdrawal for irrigation of the San Joaquin Valley of California has caused the water table to drop 75 meters in 20 years, and the resulting land subsidence has been significant.

Probably the most well-known example of land subsidence is the Leaning Tower of Pisa in Italy, shown in Figure 16.3. Over the years, as groundwater has been withdrawn to supply the growing city, the tilt of the tower has increased.

Concept Check ✓

An aquifer is a body of underground fresh water. Where does this water come from?

Was this your answer? The source of all natural underground (and aboveground) fresh water is the atmosphere, which gets most of its moisture from the evaporation of ocean water.

16.2 Collectively, We Consume Huge Amounts of Water

The U.S. Geological Survey (USGS) began compiling national water-use data in 1950. Since then, this federal agency has been conducting surveys at five-year intervals. According to their 1995 report, the rate at which water enters all U.S. aquifers combined is about 6790 billion liters/day.* In 1995, we were withdrawing water from these aquifers at an average rate of 1291 billion liters/day, meaning we were taking out about 20 percent of the available volume each day. As shown in Figure 16.4, the bulk of this water was used for irrigation and as a coolant in the generation of thermoelectric power.

The numbers in Figure 16.4 tell us that, based on a population of 267 million, the 1995 per-person water usage in the United States was 4835 liters/day (1291 billion liters/day ÷ 267,000,000 persons). Each person's personal use was 8 percent of that amount, which comes to about 390

---

* These reports are made every five years, but it takes several years to compile and cross-check the data. Thus, at the time of the printing of this textbook, the data for 2000 was not yet available. To check on the status of the year 2000 report, go to http://water.usgs.gov/watuse.

liters/day. However, only about one-fourth of that 390 liters is water we either drink or use to water our lawns or gardens. The other three-fourths—about 290 liters!—becomes household wastewater from our bathtubs, toilets, sinks, and washing machines.

It may seem that because we consume only about 20 percent of the fresh water available to us that there is little need to conserve. This percentage, however, is only an average. In many drier regions of the western United States, water usage already exceeds the rate at which aquifers in the region are recharged. In Albuquerque, New Mexico, for example, escalating water consumption has caused the underlying aquifer to drop by about 50 meters over the past 40 years. In response, the local government is seeking a 30 percent reduction in water usage in the next decade. Such a reduction would reduce the water demand by about 140 million liters/day and bring water usage to sustainable rates.

The quality of fresh water varies from one region to another. Deep water deposits, for example, are often high in dissolved solids. So, even in regions where fresh water is plentiful, it's important to conserve water to help protect that smaller portion of the water supply that is of greatest purity. Furthermore, we are not the only species that relies on fresh water. Many ecosystems, such as lakes and wetlands, are already stressed by our increasing water demands. Water conservation can go far to alleviate this stress even in the face of our growing population.

According to the USGS, there is good news regarding water conservation efforts in the United States. As shown in Figure 16.5, total water withdrawal peaked in 1980 at around 1420 billion liters/day. By 1995, however, the withdrawal rate had declined to 1300 billion liters/day even though the

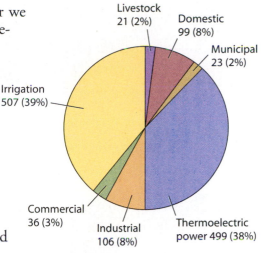

Figure 16.4
Water usage in the United States (1995) in billions of liters per day.

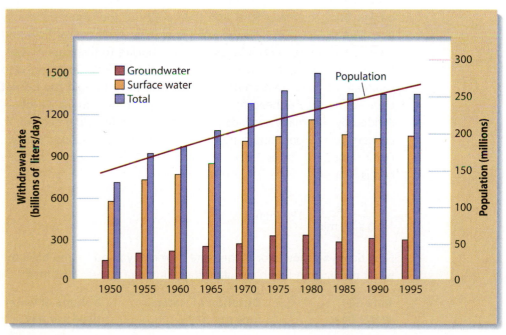

Figure 16.5
Total withdrawals of fresh water increased from 1950 to 1980 largely because of expanded irrigation systems and urban developments. After 1980, however, conservation measures reduced water usage even in the face of a growing population.

Hands-On Chemistry: Water Wiser

In this activity, you will measure the water flow rate in your bathroom and use this rate to estimate the amount of water you use for personal hygiene per day.

What You Need

Ruler calibrated in centimeters, bucket, 500-milliliter measuring cup, clock or watch with second hand

Procedure

① *Flushing*: Calculate the volume of the water in your toilet's tank by multiplying three dimensions: distance across front of tank times front-to-back distance times height of water in tank, using units of centimeters. (Determine the water height by removing the lid from the tank, estimating how many centimeters below the top of the tank the water level is, and subtracting this distance from the total tank height.) Divide the number you calculate by 1000 to convert cubic centimeters to liters. Alternatively, shut off the water valve to the toilet, flush to empty the tank, and then fill to the normal fill-line using the measuring cup and keeping track of how much water you add. This is the amount of water used each time you flush. The number of liters you flush each day is therefore

Volume of water used per flush		Average number of flushes per day		Volume of water used per day
_____ L/flush	×	_____ flushes/day	=	_____ L/day

② *Shower/bath*: Use the measuring cup to add 1 liter of water to the bucket. Mark the water level and then pour out the water, preferably over some plants. Turn on the shower or bath to a typical flow and time how many seconds it takes to fill the bucket to the marked level. (If you are running the shower, make sure the bucket is positioned to catch all the water that comes out. You may have to hold the bucket close to the shower head to accomplish this.) Your volume—1 liter—divided by this many seconds is the flow rate in liters per second. To convert to liters per minute, multiply this value by 60 seconds/minute. For instance, if it takes 5 seconds to collect 1 liter of water, the flow rate is

$$\frac{1\,L}{5\,s} \times \frac{60\,s}{1\,min} = 12\,L/min$$

Next time you shower or bathe, note how many minutes you have the water running, and then calculate the volume of water consumed:

Rate of water flow		Duration of water flow per day		Volume of water used per day
$\dfrac{\_\_\_\_\ L}{\_\_\_\_\ min}$	×	_____ min/day	=	_____ L/day

③ *Bathroom sink activities*: Turn on your bathroom sink faucet to a typical flow rate and measure the number of seconds it takes to fill the measuring cup. (Recall that 500 milliliters equals 0.5 liter.) Log the number of seconds you have the faucet running at this rate over the course of the day as you wash, brush your teeth, or shave, and then calculate the volume you've used:

Rate of water flow		Duration of water flow per day		Volume of water used per day
$\dfrac{\_\_\_\_ \text{ L}}{\_\_\_\_ \text{ s}}$	$\times$	_____ s/day	$=$	_____ L/day

④ Add up your total bathroom water usage for the day: _____ L/day

population grew from 230 million to 267 million over the same time period. This remarkable savings in fresh water was largely the result of improved irrigation techniques (Section 15.7), but enhanced public awareness of water resources and conservation programs was also a contributing factor.

16.3 Water Treatment Facilities Make Water Safe for Drinking

Water that is safe for drinking is said to be *potable*. In the United States, potable water is currently used for everything from cooking to flushing toilets. The first step in producing potable water from natural sources is to remove any dirt particles and pathogens, such as bacteria. Figure 16.6 shows how this is done by mixing the water with slaked lime and aluminum sulfate, which coagulate into gelatinous aluminum hydroxide. The aluminum hydroxide starts out dispersed all through the water, but slow stirring causes it to clump and settle to the bottom of the basin, carrying with

$$3\,Ca(OH)_2 + Al_2(SO_4)_3 \longrightarrow 2\,Al(OH)_3 + 3\,CaSO_4$$

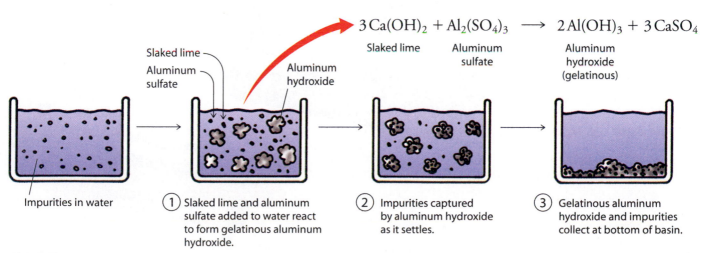

Figure 16.6
Slaked lime, $Ca(OH)_2$, and aluminum sulfate, $Al_2(SO_4)_3$, react to form aluminum hydroxide, $Al(OH)_3$, which forms a gelatinous material used to capture impurities.

Figure 16.7
Volatile impurities are removed from drinking water by cascading it through the columns of air within each of these stacks.

it many of the dirt particles and bacteria. The water is then filtered through sand and gravel.

To improve odor and flavor, many treatment facilities *aerate* the water by cascading it through a column of air, as shown in Figure 16.7. Aeration removes many unpleasant-smelling volatile chemicals, such as sulfur compounds. Aeration also removes the radioactive gaseous element radon, Rn, which occurs naturally in many water sources, especially groundwater. At the same time, air dissolves in the water, giving it a better taste (without dissolved air, water tastes flat). As a final step, the water is treated with a disinfectant, usually chlorine gas, Cl_2, but sometimes ozone, O_3, and then stored in a holding tank that feeds into distribution pipelines.

Developed nations have the technology and infrastructure to produce vast quantities of potable water, and as a result, many citizens take their drinking water for granted. The number of water treatment facilities in developing nations is relatively small, however. In these locations, many people drink their water in the form of a hot beverage, such as tea, which is disinfected through boiling. Alternatively, disinfecting iodine tablets can be used. However, fuel for boiling and tablets for disinfecting are not always available. As a result, more than 400 people (mostly children) die every hour from such preventable diseases as cholera, typhoid fever, dysentery, and hepatitis, which they contract by drinking contaminated water. In response, several U.S. manufacturers have developed tabletop systems that bathe water with pathogen-killing ultraviolet light. One model, shown in Figure 16.8, disinfects 60 liters per minute, and weighs about 7 kilograms.

Aside from pathogens, untreated water from wells or rivers may contain toxic metals that seep into the water from natural geologic formations. Many of the wells in Bangladesh, for example, are dug deep so as to be free of the pathogens that run rampant in the surface waters of this region. The

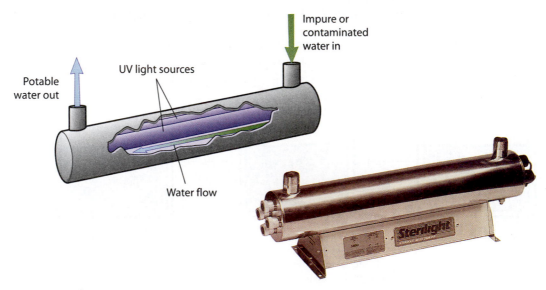

Figure 16.8
Small-scale water disinfecting units, such as the one shown here, hold great value in regions of the world where potable water is scarce.

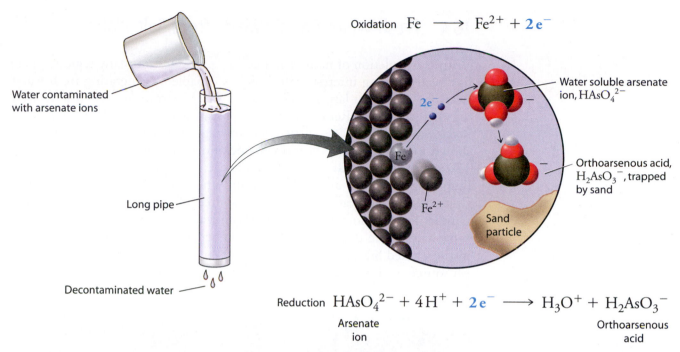

Figure 16.9
A method for removing arsenic from well water. The water is filtered through a long pipe filled with iron filings and sand. Water-soluble arsenate ions gain electrons from metallic iron to form less-soluble orthoarsenous acid, which gets trapped by the sand and does not exit the pipe.

water obtained from these deep wells, however, is highly contaminated with arsenic. The source of this arsenic is the underlying rock, which formed from river sediments carried down from the Himalayas. Because this region is so densely populated, as many as 70 million people may be subject to some level of arsenic poisoning, which manifests itself as skin lesions and a higher-than-normal susceptibility to cancer. Low-cost methods for removing arsenic from well water are being developed. Figure 16.9 shows a method that involves passing contaminated water through a long pipe filled with a mixture of iron filings and sand. As the water passes through the pipe, the iron reduces arsenic from water-soluble arsenate ions to orthoarsenous acid, which binds with sand particles and is thus removed from solution.

Concept Check ✔

At a water treatment facility, chemicals are added to water to purify it. As you learned in Chapter 2, however, adding anything to water makes the water less pure. This being true, how can *adding* chemicals *increase* the purity of the treated water?

Was this your answer? The water coming into a treatment plant is usually a heterogeneous mixture containing suspended solids. The chemicals added capture these suspended solids, which then sink to the bottom and are easily removed. The water therefore contains fewer materials, and so its purity has been increased.

16.4 Fresh Water Can Be Made from Salt Water

With the depletion of natural sources of fresh water in many regions, there has been growing interest in techniques capable of generating fresh water from the Earth's far larger reserves of seawater or from *brackish* (moderately salty) groundwater. Removing salts from seawater or brackish water is called *desalinization*, a process carried out in large installations such as the one shown in Figure 16.10. Worldwide, desalinization plants operate in about 120 countries and have a combined capacity to produce about 16 billion liters a day. In many areas of the Caribbean, North Africa, and the Middle East, desalinized water is the main source of municipal supply. In the United States, more than 1000 desalinization plants have a combined capacity of more than 400 million liters per day. Most of the treated water in the United States is used for industrial purposes and comes from brackish sources or from water high in dissolved minerals.

Figure 16.10
Saudi Arabia is the world's leading producer of desalinized water. Its desalinization plants, such as the one shown here, together have a generating capacity of about 4 billion liters a day.

The two primary methods of removing salts from seawater or brackish water are distillation and reverse osmosis. These techniques are also highly effective in removing a host of other contaminants, such as hard-water ions, pathogens, fertilizers, and pesticides, and so are also used to purify fresh water. Many brands of bottled water, for example, are fresh water that has been treated by either distillation or reverse osmosis.

As noted in Section 2.4, distillation involves vaporizing a liquid and then condensing the vapors to purified liquid. More than 60 percent of the world's desalinized water is produced in this way. Because water has such a high heat of vaporization, however, this technique is energy-intensive. Today, most distilling plants heat the water by burning large quantities of fossil fuels and so generate levels of pollution that are excessive relative to the volume of fresh water produced. A solar distiller, shown in Figure 16.11, avoids the burning of fuels but requires about 1 square meter of surface area to produce 4 liters of fresh water per day. For a single home or small village, this surface area requirement may be easily accommodated.

Figure 16.11
These solar distillers are popular in remote communities along the Texas–Mexico border where the waters from the Rio Grande basin are saline and tainted by agricultural chemicals.

For larger urban areas where land is scarce, solar distillation is less practical, especially when the maintenance costs of vast fields of solar distillers are taken into account.

In many regions of the world, reverse osmosis is the preferred method of desalinization. In order to understand reverse osmosis, you must first understand **osmosis**, which is the net flow of water molecules from a region where the concentration of some solute (or solutes) is lower (or zero) to a region where the solute concentration is higher. As Figure 16.12 shows, this flow is across a **semipermeable membrane**, defined as one containing submicroscopic pores that allow the passage of water molecules but not of any solute particles larger than a water molecule.

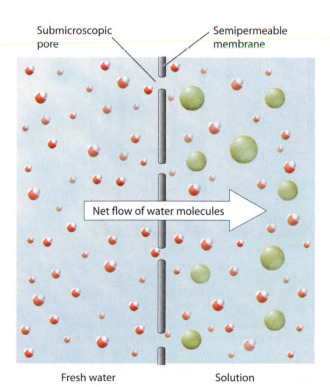

Figure 16.12
Osmosis. The submicroscopic pores of a semipermeable membrane allow only water molecules to pass. Because there are more water molecules along the freshwater face of the membrane than along the solution face, more water molecules are available to migrate into the solution than are available to migrate into the fresh water.

When fresh water is separated from salt water by a semipermeable membrane, the rate at which water molecules pass from the fresh water into the salt water is higher than the rate at which they pass from the salt water into the fresh water. The reason is that there are more water molecules along the freshwater face of the membrane than along the saltwater face.

The result of osmosis is an increase in the volume of the salt water and a decrease in the volume of the fresh water. These changes in volume cause a buildup in pressure, called *osmotic pressure*. For the system in Figure 16.13a, osmotic pressure is a consequence of the salt water's greater height. This pressure builds as water molecules move across the membrane from fresh to salt, and the volume of the salt water continues to increase. The greater pressure on this side forces some water molecules to move across the membrane from salt to fresh, against osmosis. Eventually, the rate at which water molecules pass from salt to fresh is the same as the rate at which they pass from fresh to salt, and the system reaches equilibrium, as shown in Figure 16.13b. If an external pressure is now applied to the salt water, even more water molecules are squeezed across the membrane from the salt side to the fresh side, as shown in Figure 16.13c. Contrary to what happens in osmosis, water molecules are now being forced across the semipermeable membrane from the side having the higher concentration of solute molecules to the side having the lower concentration of solute molecules, and this is the process called **reverse osmosis**. The utility of reverse osmosis is that it is a mechanism for generating fresh water from salt water.

The osmotic pressure generated when seawater and fresh water are separated by a semipermeable membrane is an astounding 24.8 atmospheres (365 pounds per square inch). If reverse osmosis is to take place with seawater, therefore, an external pressure higher than this must be exerted on

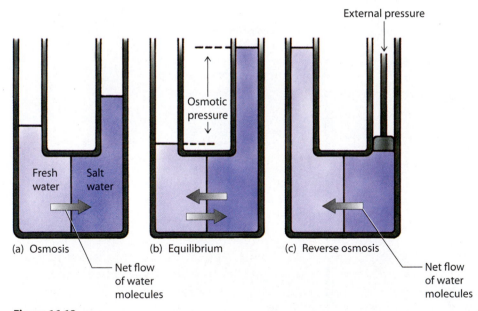

Figure 16.13
(a) Osmosis results in a greater volume of salt water, which causes the pressure to increase on the salt side of the membrane. (b) When the pressure on the salt side gets high enough, equal numbers of water molecules pass in both directions. (c) The application of external pressure forces water molecules to pass from the salt water to the fresh water, so that now the salt-to-fresh flow rate is higher than the fresh-to-salt flow rate.

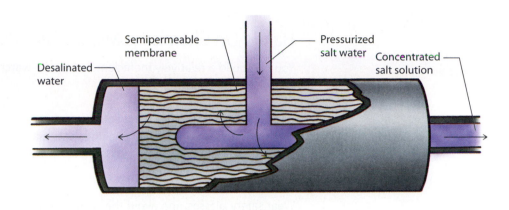

Figure 16.14
An industrial reverse osmosis unit consists of many semipermeable membranes packed around highly pressurized salt water. As desalinated water is pushed out one side, the remaining salt water, which is now even more concentrated, exits on the other side. A network of reverse osmosis units operating parallel to one another can produce enormous volumes of fresh water from salt water.

the water. Generating such high pressures has its share of technical difficulties and is an energy-intensive process. Nonetheless, engineers have succeeded in building durable reverse osmosis units, shown in Figure 16.14, that can be networked to generate fresh water from seawater at rates of millions of liters per day. Reverse osmosis desalinization facilities treating brackish waters, which require much lower external pressures, are proportionately more economical.

Concept Check ✓

The cell membranes in cucumber cells are semipermeable, meaning water molecules pass back and forth through the membranes but solute molecules do not. A cucumber left in a concentrated salt solution shrivels up because water molecules leave the cells and enter the salt solution. Is this an example of osmosis or reverse osmosis?

Was this your answer? That no external pressure is involved rules out reverse osmosis. Instead, the shriveling tells you that the cells are losing water molecules, which must be entering the salt water. This is osmosis. Add spices and the right kinds of microorganisms to the solution, and you've made a pickle.

Desalinated seawater and desalinated brackish water are important new sources of fresh water. Although this fresh water is more costly than fresh water from natural sources, one could argue that the higher cost reflects fresh water's true value. In the United States, natural sources of fresh water are relatively plentiful, allowing companies to sell fresh water at rates of a fraction of a penny per liter. Nonetheless, consumers are still willing to buy bottled water at up to $2 per liter! Each year Americans spend about $400 million dollars on bottled water, and the market continues to grow rapidly. Unless we conserve fresh water, it is easy to project a growing reliance on distillation and reverse osmosis.

Hands-On Chemistry: Micro Water Purifier

You can build a relatively inefficient but fun-to-watch distiller at home.

What You Need

Deep cooking pot, water, food coloring, table salt, heavy ceramic mug at least 3 centimeters shorter than cooking pot, plastic food wrap, strong rubber band that can fit around pot, scissors, ice cubes, small sponge

Safety Note

Wear safety glasses, and avoid the steam produced in this experiment because steam burns can be particularly harmful.

Procedure

1. Fill the cooking pot with water to a depth of about 1 centimeter. Add several drops of food coloring and 1 teaspoon of salt to the water and stir.

2. Place the mug in the center of the pot.

3. Lay plastic wrap loosely across the top of the pot and secure with the rubber band. The seal should NOT be airtight. Instead, leave two opposite edges of the wrap outside the rubber band to prevent a buildup of pressure in the pot as the water is brought to a boil. Use scissors to trim away any wrap hanging below the rubber band. Place an ice cube at the center of the wrap, which should then sag above the mug.

4. Put on your safety glasses, place your "distiller" on the stove, and turn the burner on low to bring the water to a low boil. Look for signs of cloud formation below the ice cube. Liquid water condensing here will drop into the mug. Once boiling begins, the mug may jostle. Turn the heat down or off if the jostling becomes too pronounced.

5. Boil the water only until the ice cube melts. As it melts, remove the meltwater with the sponge.

Examine the water in the mug, both visually and by tasting a few drops. Why isn't the food coloring or the salt carried over into the mug? How might you modify your distiller to use sunlight to drive the distillation?

16.5 Human Activities Can Pollute Water

Water pollution can arise from either point sources or nonpoint sources. A **point source** is one, specific, well-defined location where a pollutant enters a body of water. One example of a point source would be the wastewater pipes of a factory or sewage treatment plant, as shown in Figure 16.15a. Point sources are relatively easy to monitor and regulate. A **nonpoint source** is one in which pollutants originate at diverse locations, oil residue on streets being one example. Water becomes polluted as rain washes the

(a)

(b)

Figure 16.15
(a) This technician is assessing the clarity of the effluent coming from a wastewater treatment facility, a common point source of pollution. (b) Nonpoint sources are not so easily regulated and depend more on public awareness.

oil residue into streams, rivers, and lakes. Agricultural runoff and house-hold chemicals making their way into the storm drains of Figure 16.15b are two other common examples of nonpoint sources of water pollution. Because it is difficult to monitor and regulate nonpoint sources, the most effective solutions are often public awareness campaigns emphasizing responsible disposal practices. As Figure 16.16 illustrates, lawn care in the United States is a major nonpoint source of water pollution.

The rate of water contamination from many point sources has decreased markedly since the passing of the Clean Water Act of 1972 and its subsequent amendments. Prior to 1972, the user of a water supply, such as a municipality, was responsible for protecting the supply. Because it is far more efficient to control water pollutants before they are released into the environment, the Clean Water Act shifted the burden of protecting a water supply to anyone discharging wastes into the water, such as a local industry.

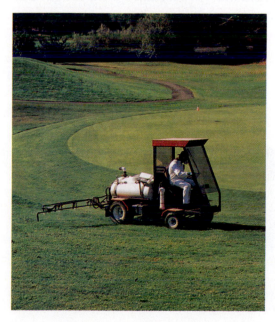

Figure 16.16
Lawns cover 25 to 30 million acres in the United States, an area larger than Virginia. People taking care of these lawns use up to two and a half times more pesticides per acre than farmers use on croplands.

Figure 16.17
The arrows indicate some major sources of groundwater contamination.

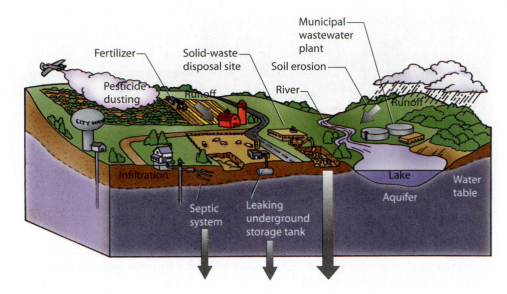

When rivers and lakes become polluted, they can be cleaned because they are accessible. When groundwater becomes polluted, however, it's a different story. Even after the sources of the pollution have been removed, it can be a lifetime before the contaminants are removed, not only because the groundwater is so inaccessible but also because the flow rate of many aquifers is extremely slow—on the order of only a few centimeters per day! As shown in Figure 16.17, groundwater is susceptible to a wide variety of point and nonpoint pollution sources.

Municipal solid-waste disposal sites are a common source of groundwater pollution. Rainwater infiltrating a disposal site may dissolve a variety of chemicals from the solid waste. The resulting solution, known as **leachate**, can move into the groundwater, forming a contamination plume that spreads in the direction of groundwater flow, as shown in Figure 16.18. To reduce the chances of groundwater contamination, the site can be underlain and capped with layers of compacted clay or plastic sheeting that prevent leachate from entering the ground. A collection system designed to catch any draining leachate may also be used.

Figure 16.18
A contaminant plume of leachate spreads in the direction of groundwater flow.

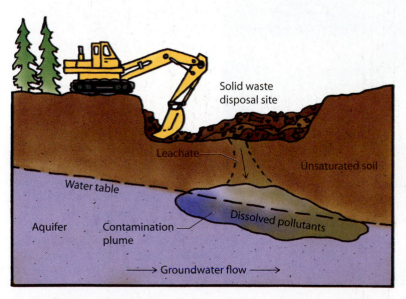

Another common source of groundwater pollution is sewage, which includes drainage from septic tanks and inadequate or broken sewer lines. Animal sewage, especially from factory-style animal farms, is also a source of groundwater (and river water) pollution. Sewage water contains bacteria, which if untreated can cause waterborne diseases such as typhoid, cholera, and infectious hepatitis. If the contaminated groundwater travels relatively quickly through underground sediments in which there are large air pockets, bacteria and viruses can be carried considerable distances. However, if the contaminated groundwater flows through underground sediments in which the air pockets are very small, as is the case with sand, pathogens are filtered out of the water.

Concept Check

What is the difference between a point source of pollution and a nonpoint source?

Was this your answer? A point source arises from a specific location—you can pinpoint it on a map. A nonpoint source represents a collection of many sources, each difficult to trace. To specify a nonpoint source on a map, you need to draw a circle.

16.6   Microorganisms in Water Alter Levels of Dissolved Oxygen

Naturally occurring water is alive with organisms. At the microscopic level, there are microorganisms, some of them disease-causing and others benign. One of the natural functions of these microorganisms is breaking down organic matter. The body of a dead fish, for example, does not remain at the bottom of a pond forever. Instead, microorganisms such as bacteria digest the organic matter into small compounds of carbon, hydrogen, oxygen, nitrogen, and sulfur.

We identify bacteria as either aerobic or anaerobic. **Aerobic bacteria** decompose organic matter only in the presence of oxygen, O_2. **Anaerobic bacteria** can decompose organic matter in the absence of oxygen. The products of aerobic decomposition are entirely different from the products of anaerobic decomposition. Aerobic bacteria in water utilize oxygen dissolved in the water to transform organic matter to such compounds as carbon dioxide, CO_2, water, H_2O, nitrates, NO_3^-, and sulfates, SO_4^{2-}. All of these products are odorless and, in the quantities produced, cause little harm to the ecosystem. Anaerobic bacteria in water use different chemical mechanisms to decompose organic matter to such products as methane, CH_4 (which is flammable), foul-smelling amines such as putrescine, $NH_2C_4H_8NH_2$, and foul-smelling sulfur compounds such as hydrogen sulfide, H_2S. Cesspools owe their wretched stink to a lack of dissolved oxygen and the resulting anaerobic decomposition.

When organic matter is introduced into a body of water, aerobic bacteria need (or "demand") dissolved oxygen to decompose the organic matter. The phrase used to describe this demand is **biochemical oxygen demand**

Figure 16.19
Sewage entering a river dramatically decreases the level of dissolved oxygen in the water. Because it takes time for aerobic bacteria to decompose organic wastes and because rivers flow, the drop in dissolved oxygen is often most pronounced far downstream. Fish start to die when the concentration of dissolved oxygen dips below 3 milligrams/liter.

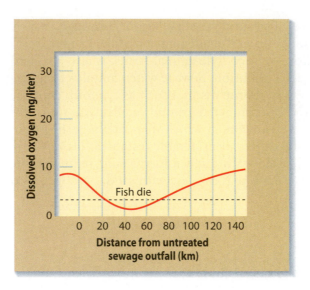

(BOD). As more organic matter is introduced, the BOD increases, resulting in a drop in the amount of dissolved oxygen as the bacteria use more and more of it to do their work. If too much organic matter is introduced, say from the outfall of a sewage treatment plant, dissolved oxygen levels can get so low that aquatic organisms start to die, as shown in Figure 16.19. Aerobic bacteria start to work on the bodies of these dead organisms, which lowers the oxygen level even further, killing off even the hardiest aquatic organisms. Ultimately, the dissolved oxygen level reaches zero. At this point, the noxious anaerobic bacteria take over.

Concept Check ✓

Which should have a greater capacity to decompose organic matter aerobically: a still pond or a babbling brook?

Was this your answer? The capacity for aerobic decomposition is limited by the amount of dissolved oxygen. In a babbling brook, aeration guarantees that any dissolved oxygen lost to aerobic decomposition is quickly replaced. This is not so with a still pond. Cubic meter for cubic meter, a babbling brook has a greater capacity to decompose organic matter aerobically.

Figure 16.20
This algal bloom consumes oxygen dissolved in the water and prevents atmospheric oxygen from mixing into the pond, thereby choking off aquatic life.

In addition to organic wastes, inorganic wastes, such as nitrate and phosphate ions from fertilizers, can also cause the level of dissolved oxygen to drop. These ions are nutrients for algae and aquatic plants, which grow rapidly in their presence, an event called an *algal bloom*. Significantly, the plants and algae in a bloom consume more oxygen at night than they produce through photosynthesis during the day. Also, in some instances, a bloom can cover the surface of a body of water, as shown in Figure 16.20, effectively choking off the supply of atmospheric oxygen. As a result, aquatic organisms suffocate and fall to the bottom, along with large portions of dead algae. Aerobic microorganisms decompose this organic matter to the point that the water loses all its dissolved oxygen, and anaerobic microorganisms start functioning. This process, whereby inorganic wastes fertilize algae and plants and the resulting overgrowth reduces the concentration of dissolved oxygen in the water, is called **eutrophication**, from the Greek word for "well nourished."

16.7 Wastewater Is Processed by Treatment Facilities

The contents of the sewer systems that underlie most municipalities must be treated before being released to a body of water. The level of treatment depends in great part on whether the treated water is released to a river or the ocean. Wastewater destined for a river requires the highest level of treatment for the benefit of communities downstream. However, in a facility located in a region surrounded by very deep ocean water, as is the facility shown in Figure 16.21 on page 550, treatment requirements are less stringent.

Figure 16.21
In the City of Honolulu, about 280 million liters of wastewater passes through the largest of several wastewater facilities each day. This water can be piped to depths of hundreds of meters below sea level, from where it continues to flow toward the bottom of the ocean. Water treatment requirements are therefore much less stringent than those at mainland facilities, where the effluent is not so easily discarded.

Human waste loses form by the time it reaches the wastewater facility, and as a result raw sewage is a murky stream. In this stream, however, are many nonsoluble products, including small plastic items, such as tampon applicators, relatively large pieces of gritty material, such as coffee grinds and tiny rocks, and hardened balls of oil from cooking grease. As Figure 16.22 shows, the first step to all wastewater treatment involves removing these materials by passing the raw sewage first through a screening device to remove plastic items and grease, and then through a tank called a grit chamber to allow any grit to settle out. (You should know that wastewater treatment managers point out that these insolubles—even cooking grease—should be disposed of as solid waste and not thrown down the drain or toilet.)

In the next step, called *primary treatment*, the screened wastewater enters a large settling basin where solid particles too fine to have been caught by the screen settle out as sludge. After a period of time, the sludge is removed from the bottom of the basin and is often sent to a disposal site as solid waste. Some facilities, however, are equipped with large furnaces in which dried sludge is burned, sometimes along with other municipal waste, such as paper products. The resulting ash is more compact and so takes up less space in a disposal site.

Figure 16.22
A schematic for the screening and primary treatment steps in a municipal treatment facility. The rotating skimmer on the settling basin removes any buoyant materials not captured in the screening step.

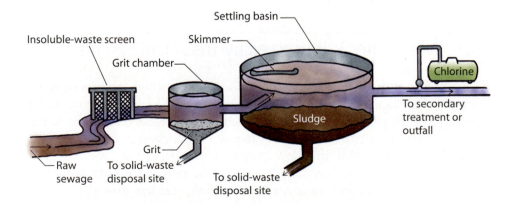

Wastewater effluent from primary treatment is commonly disinfected with either chlorine gas or ozone gas prior to release into the environment. A great advantage of using chlorine gas is that it remains in the water for an extended time after leaving the facility, thereby providing residual protection against diseases. However, the chlorine reacts with organic compounds in the water to form chlorinated hydrocarbons, many of which are known carcinogens (cancer-causing agents). Also, chlorine kills only bacteria, leaving viruses unharmed. Ozone gas is more advantageous in that it kills both bacteria and viruses. Also, there are no carcinogenic by-products that result from treating wastewater effluent with ozone. A disadvantage of ozone is that it provides no residual protection. Most facilities in the United States use chlorine for disinfecting. European facilities tend to favor ozone. In a few locations, both chlorine and ozone have been replaced by strong ultraviolet lamps, which, like ozone, kill both bacteria and viruses but offer no long-term residual protection.

The BOD of primary effluent is extremely high, and, by virtue of the Clean Water Act, in most places its release is not permitted. A frequently used *secondary treatment*, shown in Figure 16.23, involves passing the primary effluent first through an aeration tank, which supplies the oxygen necessary for continued aerobic decomposition, and then into a tank where any fine particles not removed in primary treatment can settle out. Because the sludge from this settling step is high in aerobic microorganisms, some of it is recycled back to the aeration tank to increase efficiency. The remainder of the sludge is hauled off to a disposal site or incinerator. The main advantage of secondary treatment is a marked decrease in effluent BOD.

Many municipalities also require *tertiary treatment*. There are a number of tertiary processes, most involving filtration of some sort. A common method is to pass secondary effluent through a bed of finely powdered carbon, which captures any remaining particulate matter and many of the organic molecules not removed in earlier stages. The advantage of tertiary treatment is greater protection of our water resources. Unfortunately, tertiary treatment is costly and ordinarily used only in situations where the need is deemed vital.

Primary treatment and secondary treatment are also costly. Where appropriate, therefore, the millions of dollars spent on wastewater treatment might be shifted to alternate methods of waste management, such as advanced integrated pond systems, which are described next.

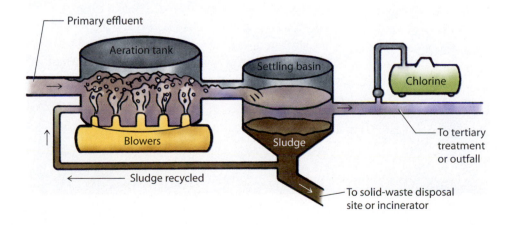

Figure 16.23
A schematic for secondary treatment of wastewater from a municipal sewer system.

Concept Check ✔

> Distinguish between the main functions of primary, secondary, and tertiary wastewater treatment.

Was this your answer? Primary treatment uses settling basins to remove the bulk of solid waste and sludge from sewage effluent. Secondary treatment uses aeration to decrease effluent BOD. Tertiary treatment removes pathogens and wastes not removed by earlier treatments by filtering effluent through powdered carbon or other fine particles.

Advanced Integrated Pond Systems Treat Wastewater

The advanced integrated pond (AIP) system, shown in Figure 16.24, is a method of wastewater treatment that makes sense for many communities of 2000 to 10,000 people in both developed and developing nations. Wastewater is channeled into an extensive pond system where plants use nutrients from the sewage as fertilizer. A system of paddlewheels guarantees constant aeration of the effluent, which is naturally disinfected by ultraviolet light from the sun. The effluent coming out of an AIP system is as clean as—and sometimes even cleaner than—effluent that has gone through secondary treatment in a conventional facility. Researchers at the University of California at Berkeley designed an AIP prototype that costs one third to one half as much to build as a conventional facility of equal capacity. A more significant source of savings lies in using solar energy rather than electrical energy to run aeration pumps. Conventional secondary plants use electrical energy to blow or mix air bubbles into the wastewater. This aeration step consumes 60 percent or more of the total electrical energy used in wastewater treatment. In an AIP system, algae and other plants use solar energy and photosynthesis to supersaturate the wastewater with the oxygen that

Figure 16.24
The pilot advanced integrated pond system in St. Helena, California, which has been in operation since the 1960s. There are now more than 85 AIP systems in operation in various spots around the world.

aerobic microbes need to break down waste. AIP systems are particularly applicable in sunbelt communities, where solar energy is plentiful, and in developing nations, where the supply of electrical energy is minimal or nonexistent.

Another important advantage of AIP systems is the small amount of sludge they produce. In these ponds, sludge ferments until only a small volume of it remains, a substantial benefit in meeting environmental regulations for sludge disposal. Furthermore, harvested plants are a significant source of biomass, which may be fermented to produce methane fuel (natural gas) or incinerated in gas turbines to produce electrical energy.

Perhaps one of the biggest obstacles to improving waste management lies not in available technology but in our attitude and willingness to accept the large one-time setup costs. Human waste is not a waste—it is a resource waiting to be utilized.

In Perspective

Fresh water is one of the most valuable resources our planet has to offer. Clearly, each of us must be mindful of fresh water's true value and practice conservation measures whenever possible.

Toward this goal, there is much you can do to conserve water and thus help ensure the water supply for future generations. A typical shower, for example, flows at a rate of about 40 liters per minute. You could reduce this rate by up to 70 percent by installing an inexpensive water-saving shower head, like the one shown in Figure 16.25. If every home in the United States were equipped with these shower heads, about 2400 billion liters of potable water would be conserved each year—about 5 percent of our total annual use. For more water conservation ideas worthy of your good citizenship, please consider the following guidelines adapted from the American Water and Energy Savers Association.

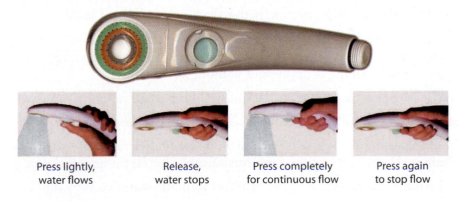

| Press lightly, | Release, | Press completely | Press again |
| water flows | water stops | for continuous flow | to stop flow |

Figure 16.25
This water-saving shower head has a reduced flow rate and a quick-action stop button that allows the user to turn the water off while lathering up.

20 Ways To Save Water

1. Read your water meter before and after a 2-hour period when no water is being used. If the reading has changed, there is a leak somewhere in your water system. Get this leak fixed.

2. Replace washers in dripping faucets. If a faucet is dripping at the rate of one drop per second, you can expect to waste 11,000 liters per year!

3. Check for toilet tank leaks by adding food coloring to the tank. If the toilet is leaking, color will appear in the bowl within 30 minutes.

4. Avoid flushing the toilet unnecessarily.

5. Install a composting toilet, which uses no water. Rather, it allows human wastes to decompose *aerobically* as air is vented over the waste, which is buried in peat moss. Dried odorfree compost is removed every few months and is useful as a garden fertilizer.

6. Take shorter showers.

7. When washing dishes by hand, do so in a sink or large pot filled with soapy water. Quickly rinse under a small stream from the faucet.

8. Operate automatic dishwashers and clothes washers only when they are fully loaded, or properly set the water level for the size of load you are washing.

9. Store drinking water in the refrigerator rather than letting the tap run every time you want a glass of cool water.

10. Start a compost pile as an alternative method of disposing of food waste instead of using a garbage disposal, which both consumes a lot of water and places food matter in water rather than in soil, where it belongs.

11. Don't overwater your lawn. Most lawns need watering only every 5 to 7 days in the summer and every 10 to 14 days in the winter.

12. Water lawns during the early morning hours when temperature and wind speed are lowest. This reduces losses from evaporation.

13. Raise the blade of your lawn mower to at least 8 centimeters. A lawn cut this high encourages grass roots to grow deeper, shades the root system, and holds soil moisture better than a close-clipped lawn.

14. Plant native and/or drought-tolerant grasses, ground covers, shrubs, and trees. Once established, they do not need to be watered frequently and usually survive a dry period without any watering.

15. Use irrigation techniques in place of sprinklers when watering your lawn.

16. Do not clean driveways or sidewalks with a hose. Use a broom.

17. Make your bed yourself when staying at a hotel for more than one night. Hotels use copious amounts of water for cleaning bed sheets.

18. Patronize businesses that practice and promote water conservation.

19. Always follow all water conservation and water shortage rules and restrictions in effect in your area.

20. Encourage family, friends, and neighbors to be part of a water-conscious community.

Key Terms and Matching Definitions

_____ aerobic bacteria
_____ anaerobic bacteria
_____ aquifer
_____ biochemical oxygen demand
_____ eutrophication
_____ hydrologic cycle
_____ leachate
_____ nonpoint source
_____ osmosis
_____ point source
_____ reverse osmosis
_____ semipermeable membrane
_____ water table

1. The natural circulation of water throughout our planet.
2. The upper boundary of a soil's zone of saturation, which is the area where every space between soil particles is filled with water.
3. A soil layer in which groundwater may flow.
4. The net flow of water across a semipermeable membrane from a region where the water concentration of some solute is lower to a region where the solute concentration is higher.
5. A membrane that allows water molecules to pass through its submicroscopic pores but not solute molecules.
6. A technique for purifying water by forcing it through a semipermeable membrane.
7. A specific, well-defined location where pollutants enter a body of water.
8. A pollution source in which the pollutants originate at different locations.
9. A solution formed by water that has percolated through a solid-waste disposal site and picked up water-soluble substances.
10. Bacteria able to decompose organic matter only in the presence of oxygen.
11. Bacteria able to decompose organic matter in the absence of oxygen.
12. A measure of the amount of oxygen consumed by aerobic bacteria in water.
13. The process whereby inorganic wastes in water fertilize algae and plants growing in the water and the resulting overgrowth reduces the dissolved oxygen concentration of the water.

Review Questions

Water Circulates Through the Hydrologic Cycle

1. In what form does most of the fresh water on our planet exist?
2. How much of the Earth's supply of water is fresh water?
3. What two forces power the water cycle?
4. Where is it possible to see the water table above ground?
5. Is most of the liquid fresh water on our planet located above or below ground?

Collectively, We Consume Huge Amounts of Water

6. Has annual water usage in the United States increased or decreased over the past couple of decades?
7. What human activity consumes most of our fresh water?
8. Has the amount of fresh water we collectively consume ever decreased from one five-year period to the next?

Water Treatment Facilities Make Water Safe for Drinking

9. Why is treated water sprayed into the air prior to being piped to users?
10. What do most treatment plants use to filter water?
11. What are two ways in which people disinfect water where municipal treatment facilities are not available?
12. What naturally occurring element has been contaminating the water supply of Bangladesh?

Fresh Water Can Be Made from Salt Water

13. What are the two main techniques for desalinizing water?
14. What is a disadvantage of solar distillation?
15. How does a semipermeable membrane allow for the passage of water molecules but not any solute ions or molecules?
16. What reverses in reverse osmosis?

Human Activities Can Pollute Water

17. Why does groundwater take so long to rid itself of contaminants?

18. How can solid-waste disposal sites be designed to minimize the spread of leachates?

19. What type of soil are pathogens not able to pass through?

20. What did the Clean Water Act of 1972 shift?

Microorganisms in Water Alter Levels of Dissolved Oxygen

21. What are some of the main products of aerobic decomposition?

22. What are some of the main products of anaerobic decomposition?

23. What effect does organic matter in water have on the amount of oxygen dissolved in the water?

24. Nitrates in groundwater tend to come from what source?

Wastewater Is Processed by Treatment Facilities

25. Why can the wastewater treatment requirements in Hawaii be less stringent than the requirements in most locations on the U.S. mainland?

26. What is the first step in treating raw sewage?

27. What is the source of the energy used to aerate wastewater in an advanced integrated pond system?

In Perspective

28. What are two advantages the composting toilet has over the flush toilet?

29. How does raising the height of a lawn mower blade to 8 centimeters help conserve fresh water?

Hands-On Chemistry Insights

Water Wiser

A leaky faucet or toilet can significantly increase the amount of water consumed in your bathroom. For a leaky faucet, time how many minutes it takes to collect a measurable quantity, such as 0.1 liter. Multiply this rate by the number of minutes in a day (1440) to arrive at the number of liters lost each day.

To measure any water loss from a toilet leak (see item 3 in "20 Ways To Save Water" on page 554 to see how to determine if a toilet is leaking), mark the level of water in the filled tank, then turn off the toilet's water source. Wait for several hours, recording your wait time in minutes. Estimate the volume of water lost by multiplying the drop in water height by the distance across the tank front and then by the tank's front-to-back distance, using units of centimeters for all three distances to get your answer in cubic centimeters. To convert to liters, divide this number by 1000 cubic centimeters/liter. Then divide this volume in liters by the number of minutes it took for this volume to leak out of the tank; this gives you the number of liters lost each minute. Multiply this rate by 1440 minutes/day to arrive at the number of liters lost each day.

Micro Water Purifier

Water readily evaporates in your purifier, but the dissolved solute, which has a much higher boiling temperature, does not. Thus, the solute—either food coloring or salt—is not carried over into the mug.

The warm air above the heated water is saturated with water vapor, and warm air can hold more water vapor than cool air. So, as the warm saturated air is cooled by the ice cube, tiny cloud droplets of liquid water begin to form, and eventually it "rains" into the mug. The cooling of rising warm, moist air is how many rain clouds form. Review Chapter 8 for more details about evaporation and condensation.

Sunlight can also be used to drive your purifier, but to maximize heat absorption, use a dark, nonreflective pot and make a complete seal with the plastic wrap.

Exercises

1. Look at a map of any part of the world, and you'll see that older cities are either next to rivers or next to where rivers used to be. Why?

2. The oceans are salt water, and yet evaporation over the ocean surface produces clouds that precipitate fresh water. Explain.

3. Where does most rainfall on the Earth finally end up before becoming rain again?

4. Withdrawal of groundwater for irrigation in the San Joaquin Valley of California caused the

water table to drop by 75 meters. Pumping has been greatly reduced, the aquifer is slowly recharging, and water for irrigation is now provided by canals that bring water from the adjacent Sierra Nevada. What else might be done to ensure an adequate water supply for the forseeable future?

5. Removal of groundwater can cause subsidence. If the water removal is stopped, will the land likely rise to its original level? Defend your answer.

6. Are our bodies apart from or a part of the water cycle?

7. Why is pollution of groundwater a greater environmental hazard than pollution of surface water?

8. Are polar or nonpolar chemical compounds more often found in a leachate? Explain.

9. Compare the advantages and disadvantages of chlorine gas and ozone gas as disinfectants for municipal water supplies.

10. Why is it important to conserve fresh water?

11. Might reverse osmosis be used to obtain fresh water from a sugar–water solution? Explain.

12. Cells at the top of a tree have a higher concentration of sugars than cells at the bottom. How might this fact assist a tree in moving water upward from its roots?

13. Why is flushing a toilet with clean water from a municipal supply about as wasteful as flushing it with bottled water? Make a rough sketch of a home plumbing system that uses water from an upstairs bathtub to flush a downstairs toilet.

14. Why is it significantly less costly to purify fresh water through reverse osmosis than to purify salt water through reverse osmosis?

15. Why do red blood cells, which contain an aqueous solution of dissolved ions and minerals, burst when placed in fresh water?

16. Some people fear drinking distilled water because they have heard it leaches minerals from the body. Using your knowledge of chemistry, explain how these fears have no basis and how distilled water is in fact very good for drinking.

17. How might water be desalinized by freezing? What would be a major advantage and a major disadvantage of such a method?

18. Which smells worse: a babbling brook or a stagnant pond? Why?

19. Phosphates were once a common component of laundry detergents because they soften water. Why has their use been restricted?

20. How might an air pump be used in the treatment of a small pond affected by eutrophication? What should be done to the pond before the pump is used?

21. Is the decomposition of food by bacteria in our digestive systems aerobic or anaerobic? What evidence supports your answer?

22. Groundwater can be contaminated by bacteria and other pathogens that seep from septic tanks, broken sewer lines, and barnyard wastes. In many instances, however, this is not a serious health problem even in the presence of a large population. Why?

23. Where does most of the solid mass of raw sewage end up after being collected at a treatment facility?

24. How are the disinfecting properties of ultraviolet light and ozone similar to each other?

25. What prevents an urban or suburban community from developing an advanced integrated pond system?

26. Why are there few odors emanating from a properly managed composting toilet?

Discussion Topics

1. Many brands of bottled water cost more per liter than gasoline. Why do you think people are willing to buy such expensive water?

2. It is possible to tow huge icebergs to coastal cities as a source of fresh water. What obstacles—technological, social, environmental, and political—do you foresee for such an endeavor?

3. The lowest point on our planet is the Dead Sea (elevation −413 meters), located in Israel about 80 kilometers from the Mediterranean Sea and about 175 kilometers from the Red Sea. Plans

are under consideration to build a canal connecting the Dead Sea to either the Mediterranean or the Red. The elevation difference along the canal would provide enough pressure to desalinate seawater by reverse osmosis, yielding as much as 800 million cubic meters of fresh water per year. Identify some of the pros and cons of such a plan.

4. After considering the plan to build a desalinization plant at the Dead Sea (see previous discussion topic), discuss how a long tube with reverse osmosis filters on one end might be used to extract fresh water from seawater. Sketch a design and plan for the technical hurdles and daily operations of your own private company.

5. Pretend you are the president of Egypt, a country whose sole source of fresh water is the Nile River. What actions would you take upon learning that Ethiopia, which is upstream from Egypt, had begun building water projects that would restrict the flow of the Nile?

6. In reference to human nature, Jerome Delli Priscoli, a social scientist with the U.S. Army Corps of Engineers, stated, "The thirst for water may be more persuasive than the impulse toward conflict." Do you agree or disagree with his statement? Might our universal need for water be our salvation or our demise?

Exploring Further

Michael A. Mallin, "Impacts of Industrial Animal Production on Rivers and Estuaries." *American Scientist*, January–February 2000.

Factory-style animal farms are a growing trend that began about 20 years ago. This article explores the impact that such farming has had on the rivers and estuaries of North Carolina, the state that is the second-largest pork producer in the United States.

http://water.usgs.gov

Explore this site of the U.S. Geological Survey for up-to-date news about water resources in

the United States and references to published articles about any U.S. water system. Be sure to explore http://water.usgs.gov/watuse.

http://www.w-ww.com

The mission of the Water–Wastewater Web site is to promote communication, education, and information exchange among water-management professionals. Follow the "plant pages" link to find the Web site of your local wastewater treatment facility.

http://www.ida.bm

Home page for the International Desalination Association, whose goals are the development and promotion of the appropriate use of desalination technology worldwide.

http://www.waterwiser.org

This site is a cooperative project of several environmental organizations, including the American Water Works Association and the U.S. Environmental Protection Agency. Books, brochures, and papers for purchase.

http://www.awwa.org

Home page for the American Water Works Association, an international nonprofit scientific and educational society dedicated to the improvement of drinking water quality and supply.

http://www.bicn.com/acic

To learn more about the arsenic crisis in Bangladesh, visit this home page for the Arsenic Crisis Information Center. A detailed description of the chemistry involved in the removal of arsenic can be found at www.eng2.uconn.edu/~nikos/asrt-brochure.html.

Fresh Water Resources
Visit The Chemistry Place at:
www.aw.com/chemplace

17

Air Resources

One Planet, One Atmosphere

Most of the Earth's atmosphere is contained within 30 kilometers of the planet's surface. Given the size of our planet—its diameter is 13,000 kilometers—this 30-kilometer thickness is ultrathin, so thin that from space the atmosphere appears only as a narrow band along the horizon. Indeed, if the Earth were the size of an apple, its atmosphere would be about as thick as the skin of the apple. We learned in the previous chapter that fresh water is a limited resource. Now we learn that the air around us is also a limited resource.

Air pollution is a well-known problem, one that migrates across international boundaries. Chemists sampling the air in North America, for example, can detect heavy metals released by smelters operating in China. Chlorofluorocarbons released in the Northern Hemisphere affect ozone levels in the air above the South Pole. Since the introduction of the internal-combustion engine, atmospheric carbon dioxide levels have been rising markedly, and global warming is a potential consequence.

Our planet is a gigantic terrarium, and collectively we are its caretakers. With this job comes the responsibility to learn how the Earth's resources can be properly managed for the benefit of all its inhabitants. In this chapter we explore some of the fundamental dynamics of the atmosphere and the impact of human activities.

17.1   The Earth's Atmosphere Is a Mixture of Gases

If the sun no longer provided heat, as represented in Figure 17.1a, the air molecules surrounding our planet would settle to the ground—much like popcorn at the bottom of an unplugged popcorn machine. Plug in the popcorn machine, and the exploding kernels bumble their way to higher altitudes. Likewise, add solar energy to the air molecules, and they, too, bumble their way to higher altitudes. Popcorn kernels attain speeds of 1 meter per second and can rise 1 or 2 meters, but solar-heated air molecules move at about 1600 kilometers per hour, and a few make their way up to more than 50 kilometers in altitude. Figure 17.1b shows that, if there were no gravity, air molecules would fly into outer space and be lost from our planet. Combine heat from the sun with the Earth's gravity, however, as in Figure 17.1c, and the result is a layer of air more than 50 kilometers thick that we call the *atmosphere*. This atmosphere provides oxygen, nitrogen, carbon dioxide, and other gases needed by living organisms. It protects the Earth's inhabitants by absorbing and scattering cosmic radiation. It also protects us from being rained on by cosmic debris because any headed toward the Earth burns up before reaching us. It is the heat generated by friction between the flying debris and our atmosphere that causes the debris to burn.

(a) Atmosphere with gravity but no solar heat: molecules lie on the Earth's surface.

(b) Atmosphere with solar heat but no gravity: molecules escape into outer space.

(c) Atmosphere with solar heat and gravity: molecules reach high altitudes but are prevented from escaping into outer space.

Figure 17.1
Our atmosphere is a result of the actions of both solar heat and gravity.

Table 17.1 shows that the Earth's present-day atmosphere is a mixture of gases—primarily nitrogen and oxygen, with small amounts of argon, carbon dioxide, and water vapor and traces of other elements and compounds. This has not always been the composition of the Earth's atmosphere. Oxygen, for example, was not a component until the evolution of photosynthesis in primitive life-forms 3 billion years ago. Carbon dioxide levels have also varied significantly over time.

Table 17.1

Composition of the Earth's Atmosphere

Gases Having Fairly Constant Concentrations	Percent by Volume	Gases Having Variable Concentrations	Percent by Volume
Nitrogen, N_2	78	Water vapor, H_2O	0 to 4
Oxygen, O_2	21	Carbon dioxide, CO_2	0.034
Argon, Ar	0.9	Ozone, O_3	0.000004*
Neon, Ne	0.0018	Carbon monoxide, CO	0.00002*
Helium, He	0.0005	Sulfur dioxide, SO_2	0.000001*
Methane, CH_4	0.0001	Nitrogen dioxide, NO_2	0.000001*
Hydrogen, H_2	0.00005	Particles (dust, pollen)	0.00001*

*Average value in polluted air.

We have adapted so completely to the invisible air around us that we sometimes forget it has mass. At sea level, 1 cubic meter of air has a mass of 1.25 kilograms. So the air in an average-sized room has a mass of about 60 kilograms—about the average mass of a human.

When you are under water, the weight of the water above you exerts a pressure that pushes against your body. The deeper you go, the more water there is above you and hence the greater the pressure exerted on you. The behavior of air is the same. Because air has mass, gravity acts upon the air, giving it weight. The weight of the air, in turn, exerts a pressure on any object submerged in the air. This pressure is known as **atmospheric pressure**, and the deeper you go in the atmosphere, the greater this pressure becomes. At sea level, you are at the bottom of an "ocean of air," and so the atmospheric pressure is greatest. Climb a mountain so that you are no longer so deep, and the atmospheric pressure is less. Venture above the atmosphere, and you have entered space, where there is no atmospheric pressure.

If you have ever gone mountain climbing, you have probably noticed that the air grows cooler with increasing elevation. At lower elevations, the air is generally warmer. This is because the Earth's surface radiates much of the heat it absorbs from the sun. As this heat radiates upward, it warms the air—an effect that decreases with increasing distance from the ground.

You have probably also noticed that the air grows less dense with increasing elevation; that is, there are fewer air molecules to breathe for a given volume. You can understand why this is so by considering a deep pile of feathers. At the bottom of the pile, the feathers are squished together by the weight of the feathers above. At the top of the pile, the feathers remain fluffy and are much less dense. For the same reasons, air molecules close to the Earth's surface are squeezed together by the greater atmospheric pressure. With increasing elevation, the density of the air gradually decreases because of decreasing atmospheric pressure. Unlike a pile of feathers, however, the atmosphere doesn't have a distinct top. Rather, it gradually thins to the near vacuum of outer space. More than half of the atmosphere's mass lies below an altitude of 5.6 kilometers, and about 99 percent lies below an altitude of 30 kilometers.

Scientists classify the atmosphere by dividing it into layers, each layer distinct in its characteristics. The lowest layer is the **troposphere**, which

Calculation Corner: Dense as Air

Knowing the density of air (1.25 kilograms/cubic meter), it's a straightforward calculation to find the mass of air for any given volume—simply multiply air's density by the volume. In the example given in the text, the volume of the average-sized room was assumed to be 4.00 meters × 4.00 meters × 3.00 meters = 48.0 cubic meters. Thus the mass of the air in the room is

$$\frac{1.25 \text{ kg}}{m^3} \times 48.0 \text{ } m^3 = 60.0 \text{ kg}$$

If you're curious to know how many pounds this is, multiply by the conversion factor 2.20 pounds/1 kilogram.

$$60.0 \text{ kg} \times \frac{2.20 \text{ lb}}{\text{kg}} = 132 \text{ lb}$$

Example

What is the mass in kilograms of the air in a classroom that has a volume of 796 cubic meters?

Answer

Each cubic meter of air has a mass of 1.25 kilograms, and so

$$796 \text{ } m^3 \times \frac{1.25 \text{ kg}}{m^3} = 995 \text{ kg}$$

which is as much as the combined mass of 17 students having a mass of about 58 kilograms (145 pounds) each.

Your Turn

1. What is the mass in kilograms of the air in an "empty" nonpressurized scuba tank that has an internal volume of 0.0100 cubic meter?

2. What is the mass in kilograms of the air in a scuba tank that has an internal volume of 0.0100 cubic meter and is pressurized so that the density of the air in the tank is 240 kilograms/cubic meter?

contains 90 percent of the atmospheric mass and essentially all of the atmosphere's water vapor and clouds, as Figure 17.2 shows. This is where weather occurs. Commercial jets generally fly at the top of the troposphere to minimize the buffeting and jostling caused by weather disturbances. The troposphere extends to a height of about 16 kilometers. Its temperature decreases steadily with increasing altitude. At the top of the troposphere, temperatures average about −50°C.

Above the troposphere is the **stratosphere**, which reaches a height of 50 kilometers. In the stratosphere at an altitude of 20 to 30 kilometers lies the *ozone layer*. Stratospheric ozone acts as a sunscreen, protecting the Earth's surface from harmful solar ultraviolet radiation. Stratospheric ozone also affects stratospheric temperatures. At the lowest altitudes, the temperature is coolest because of the solar screening effect of ozone; air at this altitude is literally in the shade of ozone. At higher altitudes, less ozone is available for shading and temperature increases all the way to a warm 0°C at the top of the stratosphere.

Hands-On Chemistry: Atmospheric Can-Crusher

When water vapor condenses in a closed container, very low pressure is created inside the container. The atmospheric pressure on the outside then has the capacity to crush the container. In this activity you will see how this works for water vapor condensing inside an aluminum soda can.

What You Need

Water, aluminum soda can, saucepan, tongs

Safety Note

Wear safety goggles, and avoid touching the steam produced in this experiment—steam burns can be severe.

Procedure

① Fill the saucepan with water and set aside.

② Put about a tablespoon of water in the can and heat on a stove until steam comes out. The steam you see indicates that air has been driven out of the can and replaced by water vapor.

③ Quickly grasp the can with the tongs, invert it, and dip into the water in the saucepan just enough to place the can opening under water. Crunch! The can is crushed by atmospheric pressure! Why?

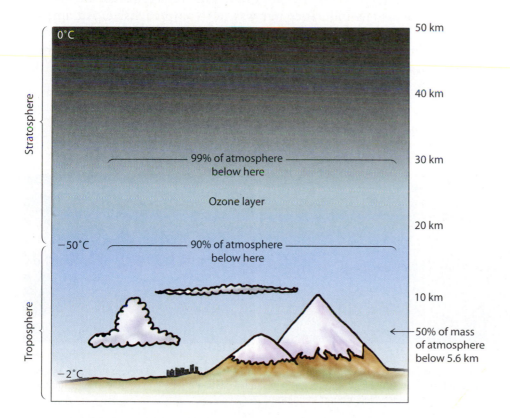

Figure 17.2
The two lowest atmospheric layers—troposphere and stratosphere.

Concept Check ✔

What effect does the Earth's gravity have on the atmosphere?

Was this your answer? The Earth's gravity pulls molecules in the atmosphere downward, preventing them from escaping into outer space.

17.2 Human Activities Have Increased Air Pollution

Any material in the atmosphere that is harmful to health is defined as an *air pollutant.* One major source of air pollutants is volcanoes. The largest volcanic blast of the 20th century, for example, was the 1991 eruption of Mount Pinatubo in the Philippines, an eruption that released 20 million tons of the noxious gas sulfur dioxide, SO_2. As Figure 17.3 shows, this sulfur dioxide managed to travel all the way to India in only four days.

In a number of ways, however, humans have surpassed volcanoes as sources of pollution. In the United States alone, for example, industrial and other human activities have been depositing about 20 million tons of sulfur dioxide in the air *every year* since around 1950! By one estimate, human activities account for about 70 percent of all sulfur that enters the global atmosphere.

To stem the human production of air pollutants, the U.S. government passed the Clean Air Act in 1970. This act regulated the gaseous emissions of various industries but was not comprehensive. An amendment in 1977 greatly restricted car emissions, and the most recent amendment, enacted in 1990, overhauled the act by regulating the emissions of nearly all air pollutants, including *aerosols*, *particulates*, and the components of *smog*.

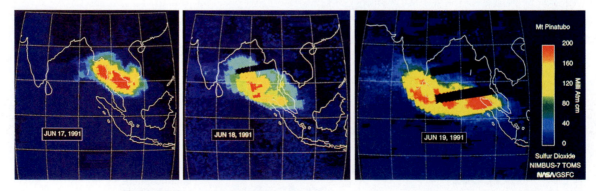

Figure 17.3
The cloud of sulfur dioxide generated by the June 15, 1991, eruption of Mount Pinatubo reached India in four days. (The black strips are where satellite data are missing.) By July 27, the sulfur dioxide cloud had traveled around the globe.

Aerosols and Particulates Facilitate Chemical Reactions Involving Pollutants

Airborne solid particles such as ash, soot, metal oxides, and even sea salts play a major role in air pollution. Particles up to 0.01 millimeter in diameter (too small to be seen with the naked eye) attract water droplets and thereby form **aerosols** that may be visible as fog or smoke. Aerosol particles remain suspended in the atmosphere for extended periods of time and, as Figure 17.4 shows, serve as sites for many chemical reactions involving pollutants.

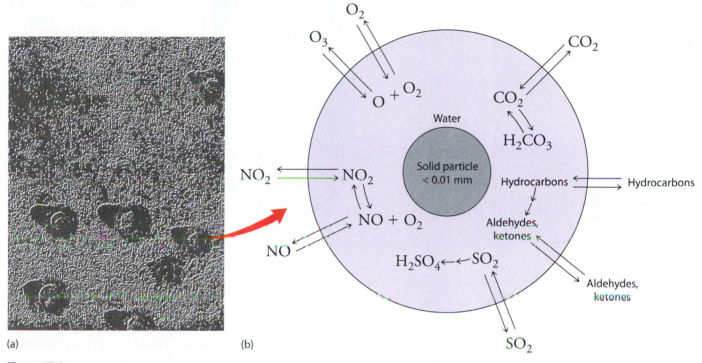

(a)   (b)

Figure 17.4
(a) Micrograph of aerosols in the atmosphere. (b) An aerosol is the site of many chemical reactions involving pollutants. Water surrounding the solid particle attracts airborne molecules that then readily react in aqueous solution before being released back into the atmosphere.

Larger solid particles, called **particulates**, tend to settle to the ground faster than the particles that form aerosols and hence do not play as big a role in facilitating atmospheric chemical reactions. While they are airborne, however, particulates obscure visibility. Atmospheric particulates (and aerosols) also have a global cooling effect because they reflect some sunlight back into space.

Industries use a variety of techniques to cut back on emissions of solid particles. Physical methods include filtration, centrifugal separation, and scrubbing, which, as Figure 17.5 shows, involves spraying

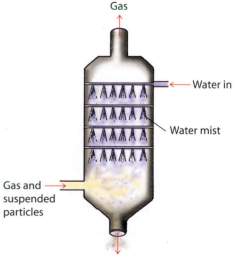

Figure 17.5
During scrubbing of industrial gaseous effluents, a fine mist of water captures and removes solid particles that have diameters as small as 0.001 millimeter.

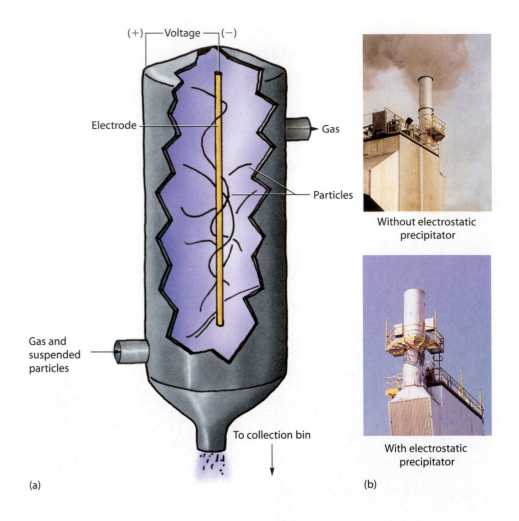

Figure 17.6
(a) Particles in industrial gaseous effluents become negatively charged by an electrode and are attracted to the positively charged wall of the electrostatic precipitator. Once it touches the wall, a particle loses its charge and falls into a collection bin. (b) Smokestacks with and without electrostatic precipitators.

gaseous effluents with water. Another method, electrostatic precipitation, shown in Figure 17.6, is energy-intensive but more than 98 percent effective at removing particles.

There Are Two Kinds of Smog

The term *smog* was coined in 1911 to describe a poisonous mixture of smoke, fog, and air that settled over the city of London and killed 1150 people. Smog has since grown to be a major problem, especially over urban areas where industrial and human activities abound.

Weather plays an important role in smog formation. Normally, air warmed by the Earth's surface rises to the upper troposphere, where pollutants are dispersed, as shown in Figure 17.7a. Parcels of dense, cold air, however, sometimes settle below warm air in a *temperature inversion*, shown in Figure 17.7b. Now the air tends to stagnate, which allows for a buildup of air pollutants. Temperature inversions may occur just about anywhere, but local geographies make some areas more prone than others. The smog of Los Angeles, for example, is trapped by an inversion created when low-level cold air moving eastward from the ocean is capped by a layer of hot air moving westward from the Mojave Desert. Temperature inversions tend to disperse at night because air at higher altitudes cools more quickly than

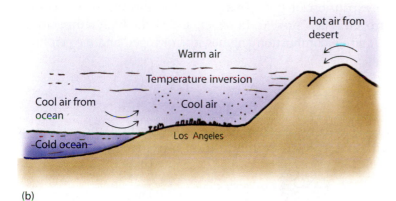

Figure 17.7
(a) Smog is removed by rising warm air. (b) In a temperature inversion, smog is trapped as cool air settles below warm air. (The normal scheme of cool above warm is *inverted*.)

does lower air, which is closer to the Earth's warm surface. This is one reason the skies in many urban areas tend to have less smog in the early morning than in the late afternoon.

Concept Check ✔

Why are temperature inversions more common during the day than at night?

Was this your answer? A temperature inversion occurs when a body of warm air sits above a body of denser cold air. The warm air is created by the heat of the sun, which is out only during the day.

There are two general types of smog: industrial and photochemical. **Industrial smog**, produced largely from the combustion of coal and oil, is high in particulates. Its main chemical ingredient, however, is sulfur dioxide, which accumulates in the water coating of aerosols and is transformed to sulfuric acid:

$$2\,SO_2 \;+\; 2\,O_2 \;\longrightarrow\; 2\,SO_3$$

Sulfur dioxide   Oxygen   Sulfur trioxide

$$SO_3 \;+\; H_2O \;\longrightarrow\; H_2SO_4$$

Sulfur trioxide   Water   Sulfuric acid

Breathing aerosols containing even very low concentrations of sulfuric acid can cause severe breathing distress. As Section 10.4 discussed, airborne sulfuric acid is also a leading cause of acid rain.

Although many industries still exceed federal standards regulating sulfur emissions, levels of industrial smog have dropped markedly since the passage of the 1970 Clean Air Act and its subsequent amendments. In the future, however, maintaining low levels of sulfur dioxide emissions will become more difficult as both national economies and world population continue to grow.

Photochemical smog consists of pollutants that participate either directly or indirectly in chemical reactions induced by sunlight. These pollutants are predominately nitrogen oxides, ozone, and hydrocarbons, and their prime source is the internal-combustion engine. In the combustion chamber, oxygen is mixed with vaporized hydrocarbons for the production of heat, which causes an expansion of gases that drives the piston's power stroke. Atmospheric nitrogen is also present, however, and at the high temperatures characteristic of internal-combustion engines, the nitrogen and oxygen form nitrogen monoxide:

$$\text{heat} + N_2 + O_2 \longrightarrow 2\,NO$$

Nitrogen monoxide is fairly reactive. Once released from the engine, it reacts rapidly with atmospheric oxygen to form nitrogen dioxide:

$$2\,NO + O_2 \longrightarrow 2\,NO_2$$

Nitrogen dioxide is a powerful corrosive agent that acts on metal, stone, and even human tissue. Its brown color is responsible for the brown haze typically seen over a polluted city.

Sunlight initiates the transformation of nitrogen dioxide to nitric acid, HNO_3, which, along with sulfuric acid, is a prime component of acid rain. In aerosols, sunlight splits nitrogen dioxide into nitrogen monoxide and atomic oxygen:

$$\text{sunlight} + NO_2 \longrightarrow NO + O$$

The nitrogen monoxide reacts with atmospheric oxygen to re-form nitrogen dioxide, and the atomic oxygen reacts with atmospheric oxygen to form ozone:

$$O + O_2 \longrightarrow O_3$$

Ozone is a pungent pollutant. It causes eye irritation and at high levels can be lethal. Plant life suffers when exposed to even relatively low concentrations of ozone, and it causes rubber to harden and turn brittle. To protect tires from ozone, manufacturers have incorporated paraffin wax, which reacts preferentially with the ozone, sparing the rubber. As we'll see in Section 17.3, ozone is also formed by natural processes in the Earth's stratosphere, where it filters out as much as 95 percent of the sun's ultraviolet rays. So, at the Earth's surface, ozone is a harmful pollutant, while 25 kilometers straight up it serves as a sunscreen and is vital for the good health of all living organisms.

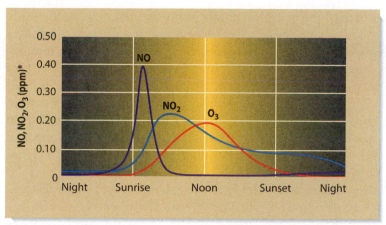

Figure 17.8
The average daily concentration of nitrogen monoxide, nitrogen dioxide, and ozone in Los Angeles.

*ppm = parts per million

A profile of average urban concentrations of nitrogen monoxide, nitrogen dioxide, and ozone is given in Figure 17.8, which neatly shows how the formation of one is related to the formation of the other two. Early morning rush hour causes a rapid increase in nitrogen monoxide, which by midmorning has largely been converted to nitrogen dioxide. On a sunny day, following nitrogen dioxide formation, ozone levels begin to peak. In the absence of a temperature inversion, late-afternoon winds clear the pollutants away. After a night of calm, the cycle begins again.

Another class of components in photochemical smog is hydrocarbons, such as those found in gasoline. In the presence of ozone, airborne hydrocarbons are transformed to aldehydes and ketones, many of which add to the foul odor of smog. Also, incomplete combustion of gasoline leads to the release of *polycyclic aromatic hydrocarbons*, which are known carcinogens. Significant amounts of hydrocarbons are also released each time a car is filled with gasoline. Because gasoline is a volatile liquid, any air in a closed tank of gasoline is loaded with gasoline vapors—even when the tank is near empty. Every time you fill up at the pump, these vapors, about 10 grams worth, are displaced and vented directly into the atmosphere. Newer gasoline pumps have nozzles designed to trap most of these vapors, as shown in Figure 17.9.

Figure 17.9
Some gasoline nozzles are equipped with a jacket that keeps gasoline vapors from escaping into the atmosphere. Instead, the vapors are directed back to the main tank of the gas station through a secondary hose hidden within the nozzle.

Concept Check ✔

How does the sun help disperse air pollutants?

Was this your answer? Sunlight warms the ground, which in turn warms the air, which then rises, carrying with it many air pollutants.

Catalytic Converters Reduce Automobile Emissions

To reduce the production of photochemical smog, the Clean Air Act amendments require that every automobile exhaust system be equipped with a *catalytic converter*. As we discussed in Section 9.4, a catalyst is an agent that increases the rate of a chemical reaction without being consumed

Figure 17.10
The catalyst of an automotive catalytic converter is typically a form of platinum, Pt; palladium, Pd; or rhodium, Rd. The production of catalytic converters is by far the largest market for these precious and semiprecious metals.

by the reaction. Ideally, the complete combustion of gasoline generates only carbon dioxide, CO_2, and water vapor, H_2O. Less-than-ideal conditions lead to the release of uncombusted hydrocarbons and toxic carbon monoxide, CO. Hot engine exhaust fumes containing these components pass through the catalytic converter, where catalysts catalyze the reactions that didn't take place in the engine, meaning that hydrocarbons and carbon monoxide are converted to carbon dioxide and water. As noted in Section 9.4, the catalysts of catalytic converters also reduce emissions of nitrogen monoxide, NO, by converting it to atmospheric nitrogen, N_2, and oxygen, O_2. A typical converter is shown in Figure 17.10.

The catalysis need not be restricted to the exhaust system. One innovative design, shown in Figure 17.11, coats an automobile's radiator with a base-metal catalyst that "eats up" ozone, one of the main components of smog.

Figure 17.11
The PremAir® catalyst-coated radiator assists in the destruction of airborne pollutants, potentially turning vehicles into net pollution absorbers.

17.3 Stratospheric Ozone Protects the Earth from Ultraviolet Radiation

Ozone is formed from automobile emissions and is an urban air pollutant, but it is also formed naturally in the stratosphere. At altitudes of 20 to 30 kilometers, high-energy ultraviolet (UV) radiation breaks diatomic oxygen down to atomic oxygen, which then reacts with additional O_2 to form ozone:

$$O_2 + \text{UV radiation} \longrightarrow 2\,O$$
$$2\,O + 2\,O_2 \longrightarrow 2\,O_3$$

Net reaction $\quad 3\,O_2 + \text{UV radiation} \longrightarrow 2\,O_3$

This synthesis of ozone is of great benefit to life on the Earth because if the ultraviolet radiation absorbed in the reaction were to reach the Earth's surface, it would cause immediate harm to living tissue.

When it absorbs ultraviolet radiation, ozone fragments into molecular oxygen and atomic oxygen. These molecules eventually re-form ozone. Because chemical bonds are created when ozone re-forms, heat energy is released:

$$O_3 + UV \text{ radiation} \longrightarrow O_2 + O$$
$$O_2 + O \longrightarrow O_3 + heat$$

Net reaction $O_3 + UV \text{ radiation} \longrightarrow O_3 + heat$

Thus harmful ultraviolet radiation is transformed by ozone into a not-so-harmful slight heating of the stratosphere. Note that ozone is not lost in this transformation, which means it can continue to shield the Earth's surface indefinitely from ultraviolet radiation.

The concentration of ozone in the stratosphere is quite small. If all of the ozone there were subjected to the amount of atmospheric pressure present at the Earth's surface, the ozone layer would be only 3 millimeters thick instead of 10 kilometers! Nevertheless, this ozone layer absorbs more than 95 percent of the ultraviolet radiation that comes to our planet from the sun. It is the safety blanket of Planet Earth.

Concept Check ✔

Is there any chemical difference between stratospheric ozone and the ozone found in air pollution?

Was this your answer? Absolutely not. Ozone, no matter where it is found, is a molecule made of three oxygen atoms.

In the early 1970s, Mario Molina (b. 1943) of the Massachusetts Institute of Technology, F. Sherwood Rowland (b. 1927) of the University of California Irvine, and Paul J. Crutzen (b. 1933) of the Max Planck Institute in Germany, all shown in Figure 17.12, recognized the potential threat to stratospheric ozone posed by chlorofluorocarbons (CFCs). Because CFCs are

Figure 17.12
Mario Molina, F. Sherwood Rowland, and Paul Crutzen at a press conference before receiving the 1995 Nobel Prize in Chemistry for their work in atmospheric chemistry, particularly concerning the formation and decomposition of stratospheric ozone.

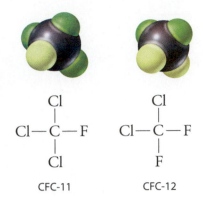

Figure 17.13
Two of the most common CFCs, also known as *freons,* were CFC-11, trichlorofluoromethane, and CFC-12, dichlorodifluoromethane. At the height of CFC production in 1988, some 1.13 million tons was produced worldwide. Because of their inertness, CFCs were once thought to pose little threat to the environment.

inert gases, they were once commonly used in air conditioners and aerosol propellants. Two of the most frequently used CFCs are shown in Figure 17.13.

Estimates are that CFCs are so stable that they remain in the atmosphere from 80 to 120 years, and they are now spread throughout the atmosphere. Even in the most remote location, there are no fewer than 25 trillion CFC molecules in every liter of air you breathe! Molina, Rowland, and Crutzen realized that CFC molecules reaching the stratosphere are fragmented when exposed to the harsh ultraviolet rays at this altitude, as illustrated in Figure 17.14. One of these fragments is atomic chlorine, which can catalyze the destruction of ozone. One chlorine atom, it is now estimated, can cause the destruction of at least 100,000 ozone molecules in the one or two years before the chlorine forms a hydrogen chloride molecule, HCl, and is carried away by atmospheric moisture.

The fragility of stratospheric ozone came to the world's attention in 1985 with the discovery of a seasonal depletion of stratospheric ozone over the Antarctic continent, a phenomenon known as the *ozone hole.* That atomic chlorine plays an active role in the destruction of Antarctic ozone can be shown by measuring chlorine monoxide concentrations over this region. As shown in Figure 17.14, chlorine monoxide is an intermediate molecule that forms during the chlorine-catalyzed destruction of ozone. The data shown in Figure 17.15, collected during flights over the South Pole, show the close relationship between stratospheric levels of chlorine monoxide and ozone. The satellite photographs of Figure 17.16 reveal how the shape of the ozone hole corresponds to the shape of a map showing chlorine monoxide distribution.

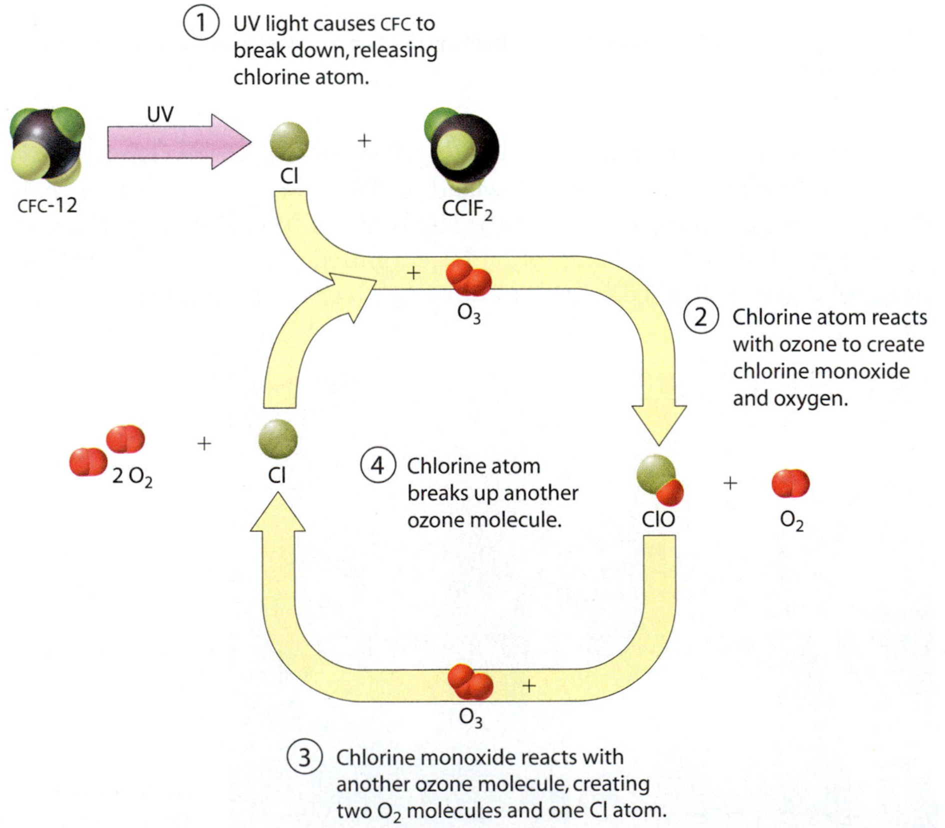

① UV light causes CFC to break down, releasing chlorine atom.

UV

CFC-12   Cl   +   CCIF$_2$

② Chlorine atom reacts with ozone to create chlorine monoxide and oxygen.

+   O$_3$

ClO   +   O$_2$

④ Chlorine atom breaks up another ozone molecule.

2 O$_2$ +   Cl

O$_3$ +

③ Chlorine monoxide reacts with another ozone molecule, creating two O$_2$ molecules and one Cl atom.

Figure 17.14
Pathway for the destruction of stratospheric ozone by chlorofluorocarbons.

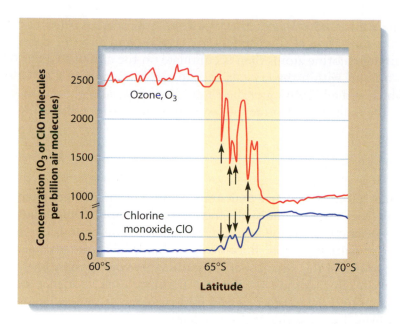

Figure 17.15
Concentrations of stratospheric ozone and chlorine monoxide in southern latitudes. As chlorine monoxide levels increase, ozone levels decrease. The yellow highlighting shows where small fluctuations in ClO concentrations result in large fluctuations in O_3 concentrations. This is consistent with catalytic behavior.

Chlorine Monoxide and the Ozone Hole

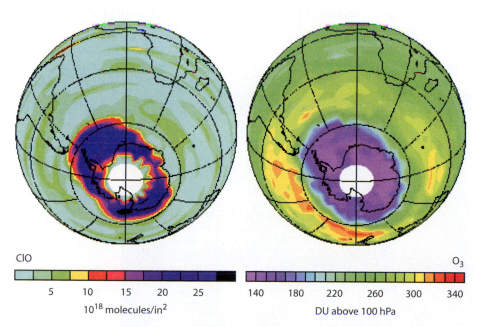

Figure 17.16
Satellite images of the Southern Hemisphere, showing concentrations of chlorine monoxide adjacent to concentrations of stratospheric ozone in September 1996.

Subsequent studies have shown that the cold, still, sunless Antarctic winter favors the formation of stratospheric ice crystals, and airborne compounds containing chlorine atoms then accumulate on the crystals. Chemical reactions on and within the crystals lead to the formation of Cl_2. In September, which is the beginning of the Antarctic spring, sunlight returns and fragments the molecular chlorine into vast amounts of ozone-depleting atomic chlorine:

$$Cl_2 + \text{sunlight} \longrightarrow 2\ Cl$$

An important piece of evidence that some of this chlorine comes from the breakdown of CFCs was the unusually high levels of fluorine compounds detected in the Antarctic stratosphere. Whereas chlorine compounds come from a number of natural sources, fluorine compounds in nature are relatively rare. The source of this stratospheric fluorine, therefore, is most likely chlorofluorocarbons. In addition to elevated fluorine levels, evidence of ozone depletion around the North Pole and mid-latitudes, shown in Figure 17.17, also confirmed the severity of the problem.

There has been an unprecedented level of international cooperation toward banning ozone-destroying chemicals. The first major step was the signing of the 1987 Montreal Protocol on Substances That Deplete the Ozone Layer, which called for the reduction of CFC production to one-half of 1986 levels by the year 1998. However, only a few years after the protocol was ratified, in 1990, scientists confirmed that the CFC problem was much more serious than what they believed in 1987 when the protocol was drafted. The protocol was soon amended to call for cessation of all CFC production by 1996. Even with the protocol in place and the continued cooperation of all signed nations, however, the ozone-destroying actions of CFCs will be with us for some time. Atmospheric CFC levels are not expected to drop back to the levels found before the ozone hole was formed until sometime in the twenty-second century.

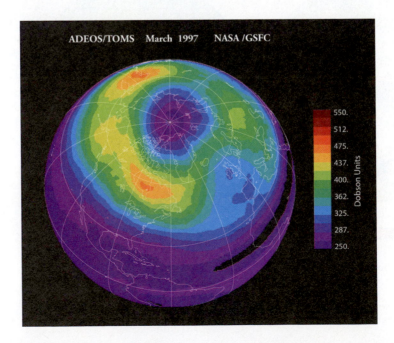

Figure 17.17
False-color image of ozone levels over the Northern Hemisphere, recorded by NASA's total-ozone mapping spectrometer (TOMS). Purple and blue areas are areas of ozone depletion; green through red areas are areas of higher-than-normal ozone levels.

17.4 Air Pollution May Result in Global Warming

Park your car with its windows closed in the bright sun, and its interior soon becomes quite toasty. The inside of a greenhouse is similarly toasty. This happens because glass is transparent to visible light but not to infrared, as illustrated in Figure 17.18. As you may recall from Figure 5.5, wavelengths of visible light are shorter than wavelengths of infrared. Visible light wavelengths range from 400 nanometers to 740 nanometers, while infrared wavelengths range from 740 nanometers to a million nanometers. Short-wavelength visible light from the sun enters your car or a greenhouse and is absorbed by various objects—car seats, plants, soil, whatever. The warmed objects then emit infrared energy, which cannot escape through the glass, and so the infrared energy builds up inside, increasing the temperature.

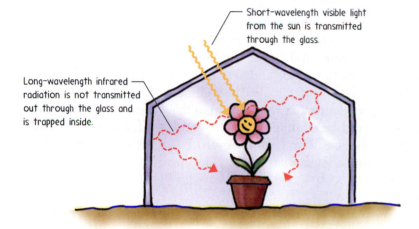

Short-wavelength visible light from the sun is transmitted through the glass.

Long-wavelength infrared radiation is not transmitted out through the glass and is trapped inside.

Figure 17.18
Glass acts as a one-way valve, letting visible light in and preventing infrared energy from exiting.

A similar effect occurs in the Earth's atmosphere, which, like glass, is transparent to visible light emitted by the sun. The ground absorbs this energy but radiates infrared waves. Atmospheric carbon dioxide, water vapor, and other select gases absorb and re-emit much of this infrared energy back to the ground, as Figure 17.19 illustrates. This process, called the **greenhouse effect**, helps keep the Earth warm. The greenhouse effect is quite desirable because the Earth's average temperature would be a frigid −18°C otherwise. Greenhouse warming also occurs on Venus but to a far greater extent. The atmosphere surrounding Venus is much thicker than the Earth's atmosphere, and its composition is 95 percent carbon dioxide, which brings surface temperatures to a scorching 450°C.

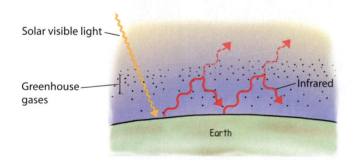

Solar visible light

Greenhouse gases

Infrared

Earth

Figure 17.19
The greenhouse effect in the Earth's atmosphere. Visible light from the sun is absorbed by the ground, which then emits infrared radiation. Carbon dioxide, water vapor, and other greenhouse gases in the atmosphere absorb and re-emit heat that would otherwise be radiated from the Earth into space.

Concept Check ✔

What does it mean to say that the greenhouse effect is like a one-way valve?

Was this your answer? The Earth's atmosphere and glass both allow incoming visible waves to pass but not outgoing infrared waves. As a result, radiant energy is trapped.

Atmospheric Carbon Dioxide Is a Greenhouse Gas

The role of carbon dioxide as a greenhouse gas is well documented. Core samples from polar ice sheets, for example, show a close relationship between atmospheric levels of carbon dioxide and global temperatures over the past 160,000 years. This relationship is graphed in Figure 17.20. Ancient air in bubbles trapped in the ice core, shown in Figure 17.21, can be sampled directly. The age of the air is a function of the depth of the core. Past global temperatures are determined by measuring the deuterium-to-hydrogen ratio in the trapped air. When global temperatures are relatively high, the ocean is warmer and larger fractions of water containing deuterium evaporate from the ocean and fall as snow. A high deuterium-to-hydrogen ratio therefore indicates a warmer climate.

There is strong evidence that recent human activities, such as the burning of fossil fuels and deforestation, are responsible for some dramatic increases in atmospheric carbon dioxide levels. Prior to the Industrial Revolution, carbon dioxide levels were fairly constant at about 280 parts per

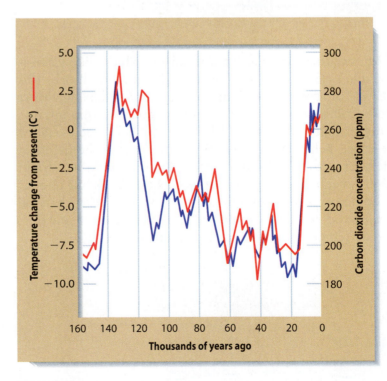

Figure 17.20
Levels of atmospheric carbon dioxide and global temperatures appear to be closely related to each other.

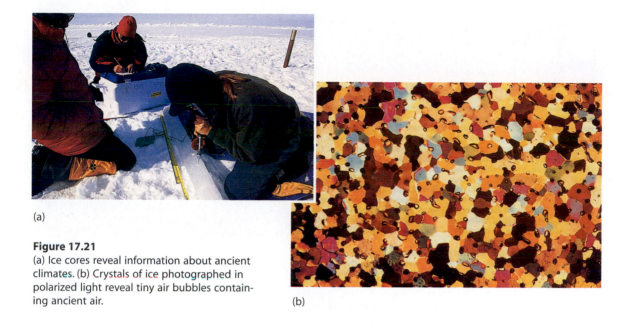

(a)

(b)

Figure 17.21
(a) Ice cores reveal information about ancient climates. (b) Crystals of ice photographed in polarized light reveal tiny air bubbles containing ancient air.

million, as shown in Figure 17.22 on page 578. During the 1800s, however, levels began to climb, as Figure 17.23 on page 579 shows, reaching a level of 300 parts per million in about 1910. Today's level is a worrisome 360 parts per million! Interestingly, as can be seen in Figure 17.20, ice samples dating as far back as 160,000 years ago do not show atmospheric carbon dioxide levels ever exceeding 300 parts per million. In step with these increases, average global temperatures since 1860 have increased by about 0.8 C°. Current estimates are that a doubling of today's atmospheric carbon dioxide levels will increase the average global temperature by an additional 1.5 to 4.5 C°.

Carbon dioxide ranks as the number-one gas emitted by human activities. When speaking of atmospheric pollutants such as sulfur dioxide, we talk in terms of millions of tons. The amount of carbon dioxide we pump into the atmosphere, however, is measured in *billions* of tons, as Figure 17.23 shows. A single tank of gasoline in an automobile produces up to 90 kilograms of carbon dioxide. A jet flying from New York to Los Angeles releases more than 200,000 kilograms (about 300 tons). Above all, our population increases by about 236,000 individuals every day, which is about 86 million individuals every year. In 1999, we passed the milestone of 6 billion humans, each of us responsible for activities that result in the output of carbon dioxide.

When direct monitoring of atmospheric carbon dioxide began in 1958, the global atmospheric reservoir of carbon dioxide was about 671 billion tons, a figure calculated from the observed concentration of 315 parts per million. By 1995, this amount had grown to 767 billion tons, which simple subtraction tells us is an increase of 96 billion tons:

1995 Global atmospheric reservoir of CO_2:   767 billion tons

1958 Global atmospheric reservoir of CO_2: $-$ 671 billion tons

Net increase: $+$ 96 billion tons

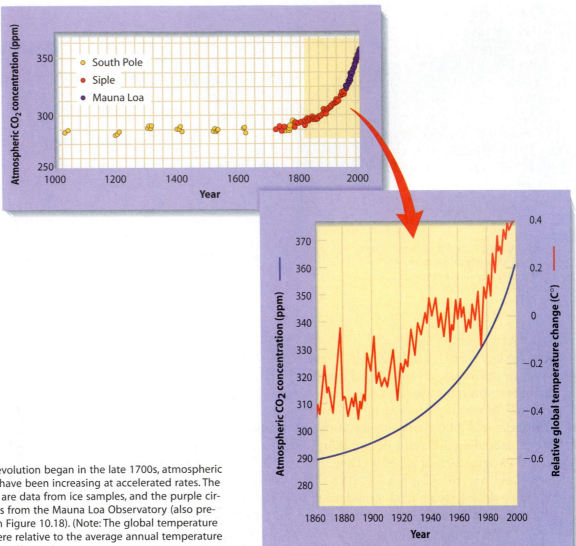

Figure 17.22
Since the Industrial Revolution began in the late 1700s, atmospheric carbon dioxide levels have been increasing at accelerated rates. The yellow and red circles are data from ice samples, and the purple circles are measurements from the Mauna Loa Observatory (also presented on page 327 in Figure 10.18). (Note: The global temperature changes are shown here relative to the average annual temperature for the years 1950 through 1979.)

Over the same period, humans released about 175 billion tons of atmospheric carbon dioxide from fossil-fuel emissions alone. From these data, we can get a feel for nature's ability to absorb carbon dioxide. Even though we pumped out 175 billion tons of carbon dioxide, the total quantity in the atmosphere went up by only 96 billion tons. Models suggest that most of the difference was absorbed by the oceans. As we saw in Section 10.4, the ocean, because its water is alkaline, can absorb carbon dioxide. Carbon dioxide can also be absorbed by vegetation during photosynthesis. It has been shown, for example, that trees grow more rapidly when exposed to higher concentrations of carbon dioxide.

That levels of atmospheric carbon dioxide have gone up by 96 billion tons tells us that we are exceeding nature's absorbing power. Bear in mind, 96 out of 767 is about 12 percent. Take a deep breath—at least 12 percent of the carbon dioxide you just inhaled came from the combustion of fossil fuels and deforestation.

With less than 5 percent of the world's population, the United States ranks first in carbon dioxide emissions and is responsible for about 25 per-

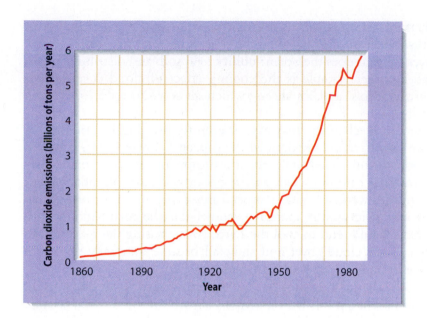

Figure 17.23
Carbon dioxide emissions from the burning of fossil fuels have grown dramatically since 1860.

cent of global carbon dioxide emissions. Industrial nations worldwide are responsible for about 58 percent of carbon dioxide emissions, primarily from the combustion of fossil fuels. Developing nations account for the remaining 42 percent, but their sources of carbon dioxide are split between fossil fuels (19 percent) and deforestation (23 percent).

Deforestation presents multiple threats to atmospheric resources. If the cut wood is used for fuel rather than lumber, burning the fuel releases carbon dioxide into the atmosphere. Whether the wood is used for fuel or for lumber, though, cutting down any forest destroys a net absorber of carbon dioxide. Furthermore, tropical forests have the capacity to evaporate vast volumes of water vapor, which assist in the formation of clouds. The clouds in turn keep regions cool by reflecting sunlight and moist by precipitating rain. Farmers who burn down rainforests for farmland, therefore, are simultaneously cutting off their future supply of rainwater. When their farms become desert, they are then spurred to burn even more of the rainforest. So far, about 65 percent of all rainforests have been destroyed. At present rates, within a few decades remaining rainforests will not be able to sustain regional climates, which will leave more than a billion citizens of the rapidly growing communities of South America, Africa, and Indonesia in the midst of arid land.

As their economies and populations continue to grow over the next several decades, developing nations will likely surpass industrial nations in the amounts of carbon dioxide and other pollutants they emit. As discussed in Chapter 19, however, new energy-efficient technologies that minimize emissions are now available. In a best-case scenario, developing nations will be able to utilize these new technologies while maintaining needed economic growth.

The Potential Effects of Global Warming Are Uncertain

There is general consensus that increased levels of atmospheric carbon dioxide and other greenhouse gases will result in global warming. How much temperatures may rise, however, is uncertain, as are the potential effects of the temperature increases. This uncertainty is due to the large number of variables that determine global weather. The sun's intensity, for example, changes over time, as does the ocean's ability to absorb and distribute greenhouse heat. Another variable is the cooling effect of cloud cover, atmospheric dust, aerosols, and ice sheets, which all serve to reflect incoming solar radiation.

A number of mechanisms may ease or even reverse global warming. For example, we may have underestimated the capacity of oceans and plants to absorb carbon dioxide. Greater levels of atmospheric carbon dioxide may simply mean more carbon dioxide in the ocean and more abundant plant life. Furthermore, warmer global temperatures could mean an increase in cloud cover worldwide and an increase in snowfall in the polar regions. Both of these effects would tend to cool the Earth by increasing the reflection of solar energy. If the cloud cover and snowfall became unusually extensive, continued reflection of solar radiation could even trigger an ice age.

On the other hand, mechanisms may enhance global warming. Warmer oceans might have a diminished capacity to absorb carbon dioxide because the solubility of carbon dioxide in water decreases with increasing temperature. Rapid climatic changes might destroy vast regions of forests and vegetation, meaning those reservoirs for carbon dioxide absorption would no longer exist. Alternatively, more abundant plant life might not be as beneficial for the atmosphere as we hope because, although plants absorb carbon dioxide, they also emit other greenhouse gases, such as methane. Warmer global temperatures might also enhance microbial activity in the soil. Microbes decaying organic matter are a significant source of carbon dioxide in dry soils and a source of methane in wet soils. Furthermore, large quantities of methane locked in Arctic permafrost may also be released as a consequence of warmer terrain. As represented in Figure 17.24, we just don't know.

Figure 17.24
Which weather extreme might become more prevalent as greenhouse gases continue to increase? Either one is possible.

An average global temperature increase of only a few degrees would not be felt uniformly around the world. Instead, some places would experience wider fluctuations than others. For example, the number of days temperatures reach above 32°C (90°F) might double in New York City but remain unchanged in Los Angeles. The number of days in polar regions when temperatures rise above 0°C might double or even triple, causing glaciers and polar ice sheets to melt. Melting ice combined with the thermal expansion of ocean waters would lead to an increase in sea level. Many climatologists project that a global temperature increase of a few degrees over the next 50 to 100 years may raise sea level by about 1 meter, enough to inundate many coastal regions and displace millions of people.

Small changes in average global temperatures would also change weather patterns. The warming of the equatorial eastern Pacific Ocean during an El Niño, for example, is already known to change local weather patterns throughout the world. If the whole planet were to warm by a few degrees, the impact would be far greater. What is now fertile agricultural land may turn barren while land now barren may turn fertile. Over the past several decades, for example, average global temperatures have edged up by about 0.2 C°. In step with this warming trend, the growing seasons of the Great Plains of Canada are now more than one week longer than they were just several decades ago. As weather patterns change, one nation's gain may well be another nation's loss. Developing nations lacking the resources to make adjustments, however, would be the hardest hit.

Concept Check ✔

Why are scientists uncertain about the potential effects of global warming?

Was this your answer? The uncertainty is due to the large number of variables that determine global weather. As the debates continue, bear in mind that the issue is not global warming itself but rather its potential effects. One thing that is certain is that if the levels of greenhouse gases in the atmosphere continue to rise, there will definitely be an increase in average global temperatures.

In Perspective

Science tells us that the potential for human-induced global warming is real, but the extent of global warming is uncertain. Actual detection will come only from a slow but steady accumulation of evidence. What should be done in the meantime is not a scientific issue but rather a societal one.

One societal response to global warming is to adapt to the changes as they occur. Economists argue that large uncertainties in climate projections make it unwise to spend large sums trying to avert disasters that may never materialize. Adjusting to immediate changes would be more directed and far less costly. Some measures, however, could be taken now to simplify future difficulties. Irrigation systems, for example, might be made more efficient because even without a major climatic change, such an improvement would make it easier to cope with normal extremes in weather.

A second societal response is to take preventive measures to minimize global warming. Emissions of greenhouse gases could be kept low by energy conservation and by shifting to fuels containing lower percentages of carbon, such as natural gas or, ultimately, hydrogen. Alternative energy sources such as biomass, solar thermal electric generation, wind power, and photovoltaics could also be exploited. Governments may also come to agree on a set of standards for emissions of carbon dioxide and other greenhouse gases. "Polluting rights" may be granted to each nation based on such factors as population and the need for economic growth.

The best public policies will be the ones that yield benefits even in the absence of global warming. A reduction in fossil-fuel use, for example, would curb air pollution, acid rain, and the dependence of many countries on foreign oil producers. Developing alternative energy sources, revising water laws, searching for drought-resistant crop strains, and negotiating international agreements are all steps offering widespread benefits.

Key Terms and Matching Definitions

_____ aerosol
_____ atmospheric pressure
_____ greenhouse effect
_____ industrial smog
_____ particulate
_____ photochemical smog
_____ stratosphere
_____ troposphere

1. The pressure exerted on any object immersed in the atmosphere.
2. The atmospheric layer closest to the Earth's surface, containing 90 percent of the atmosphere's mass and essentially all water vapor and clouds.
3. The atmospheric layer that lies just above the troposphere and contains the ozone layer.
4. A moisture-coated microscopic airborne particle up to 0.01 millimeter in diameter that is a site for many atmospheric chemical reactions.
5. An airborne particle having a diameter greater than 0.01 millimeter.
6. Visible airborne pollution containing large amounts of particulates and sulfur dioxide and produced largely from the combustion of coal and oil.
7. Airborne pollution consisting of pollutants that participate in chemical reactions induced by sunlight.
8. The process by which visible light from the sun is absorbed by the Earth, which then emits infrared energy that cannot escape and so warms the atmosphere.

Review Questions

The Earth's Atmosphere Is a Mixture of Gases

1. Why doesn't gravity flatten the atmosphere against the Earth's surface?

2. Which elements make up today's atmosphere?

3. Which chemical compounds make up today's atmosphere?

4. In which atmospheric layer does all our weather occur?

5. Does temperature increase or decrease as one moves upward in the troposphere? As one moves upward in the stratosphere?

Human Activities Have Increased Air Pollution

6. What is the difference between an aerosol and a particulate?

7. What is a temperature inversion?

8. What is the difference between industrial smog and photochemical smog?

9. How do unburned hydrocarbons contribute to air pollution?

10. How does a catalytic converter reduce the output of air pollutants from an automobile?

Stratospheric Ozone Protects the Earth from Ultraviolet Radiation

11. How is ozone produced in the stratosphere?

12. Why is stratospheric ozone so important?

13. Where else besides Antarctica are there signs of the depletion of stratospheric ozone?

14. When was the potential harm of chlorofluorocarbons first recognized?

15. During what time of the year is Antarctic stratospheric ozone depletion the greatest? Explain.

Air Pollution May Result in Global Warming

16. How do greenhouse gases keep the Earth's surface warm?

17. How do scientists estimate the age of ancient air in bubbles trapped in an ice core?

18. How is the burning of tropical rainforests a triple threat to weather patterns?

19. What is the most abundant air pollutant produced by human activities?

20. Why do scientists differ in their opinions about the potential effects of global warming?

21. Contrast the two types of societal responses to global warming discussed in the text.

Hands-On Chemistry Insights

Atmospheric Can-Crusher

When the molecules of water vapor come in contact with the room-temperature water in the saucepan, they condense, leaving a very low pressure inside the can. The much greater surrounding atmospheric pressure crushes the can. Here you see dramatically how pressure is reduced by condensation. This occurs because liquid water occupies much less volume than does the same mass of water vapor. As the vapor molecules come together to form the liquid, they leave a void (low pressure).

This activity also shows how the atmospheric pressure surrounding us is very real and significant.

Exercises

1. Gas fills all the space available to it. Why then doesn't the atmosphere go off into space?

2. How does the density of air in Death Valley, which is 86 meters below sea level, compare with the density of air at sea level? Explain.

3. Why do your ears pop when you go up in an airplane?

4. Before boarding an airplane, you buy an airtight foil package of peanuts to eat during your journey. While in flight you notice that the package is puffed up. Explain.

5. Should the atmospheric ratio of nitrogen molecules to oxygen molecules increase or decrease with increasing altitude?

6. We can understand how pressure in water depends on depth by considering a stack of bricks. The pressure at the bottom face of the bottom brick corresponds to the weight of the entire stack. Halfway up the stack, the pressure is half the bottom value because the weight of the bricks above is half the total weight. To explain atmospheric pressure, we

should consider a similar scenario but with compressible bricks made from a material like foam rubber. Why?

7. Burning coal produces sulfur oxides. Where did the sulfur originate?

8. In a still room, cigar smoke sometimes rises only partway to the ceiling. Explain.

9. Airborne sulfur dioxide doesn't remain airborne indefinitely. How is it removed from the atmosphere, and where does most of it end up?

10. Once formed, why is a temperature inversion such a stable weather system?

11. The atmosphere is primarily nitrogen, N_2, and oxygen, O_2. Under what conditions do these two materials react to form nitrogen monoxide, NO? Write a balanced equation for this reaction.

12. A catalytic converter increases the amount of carbon dioxide emitted by an automobile. Is this good news or bad news? Explain.

13. Explain the connection between photosynthetic life on the Earth and the ozone layer.

14. How close are you to a CFC molecule right now? Explain.

15. Does the ozone pollution from automobiles help alleviate the ozone hole over the South Pole? Defend your answer.

16. Chlorine is put into the atmosphere by volcanoes in the form of hydrogen chloride, HCl, but this form of chlorine does not remain in the atmosphere for very long. Why?

17. Why is whitewash sometimes applied to the glass of greenhouses in the summer?

18. Humans are pumping more and more carbon dioxide into the atmosphere, but the atmo-

spheric concentrations of this gas are not increasing proportionately. Suggest an explanation.

19. If the composition of the upper atmosphere were changed so that a greater amount of terrestrial radiation escaped, what effect would this have on the Earth's climate?

20. Geological records indicate that many ice ages were initiated by periods of unusually warm weather. How can warm weather precipitate an ice age?

Problems

1. As stated in the text, there are about 25 trillion $(25,000,000,000,000 = 2.5 \times 10^{13})$ chlorofluorocarbon molecules in every liter of air you breathe. Think critically about this statement. What information would you need to know if you wanted to calculate the percentage of CFCs in our air?

2. There are 1000 liters in 1 cubic meter and 1000 grams in 1 kilogram. How many grams of air are there in 1.00 liter of air? (Assume a density of 1.25 kg/m^3.)

3. Assume air has an average molar mass of 28 g/mole, and determine how many moles of air molecules there are in 1 liter of air. (See Section 9.2 for a review of this calculation.)

4. How many moles of CFC molecules are there in every liter of air you breathe, and what percentage of air is this? (Assume a CFC molecular mass of 120 g/mole.)

Discussion Topics

1. What kinds of air pollution have you been directly and indirectly responsible for in the past 24 hours? Rank these pollutants in order of the quantities you think you produced. Make a list of the sources of these pollutants, such as gasoline or electricity.

2. Which is a more pressing problem: ozone depletion or global warming? To what degree are the two interrelated?

3. Politicians in the United States have proposed taxing individuals and industries for the amount of CO_2 they emit by burning fossil fuels. Collected revenue would be placed in a trust fund to subsidize consumers who purchase energy-efficient devices, such as fluorescent light bulbs and high-mileage automobiles. Discuss the pros and cons of such a proposal. What modifications might you make to it?

Answers to Calculation Corner

Dense as Air

1. The density of the air in the tank is 1.25 kilograms/cubic meter. The mass of this air is found by multiplying by the tank volume:

$$\frac{1.25 \text{ kg}}{m^3} \times 0.0100 \ m^3 = 0.0125 \text{ kg}$$

2. The mass of the air is found by multiplying the density of the air in the tank by the tank volume:

$$\frac{240 \text{ kg}}{m^3} \times 0.0100 \ m^3 = 2.40 \text{ kg}$$

Exploring Further

http://www.epa.gov/oar/oarhome.html
The Office of Air and Radiation of the U.S. Environmental Protection Agency protects air quality. Explore links to information on such topics as urban air quality, automobile emissions, the ozone layer, acid rain, and indoor air.

http://www.ucsusa.org/resources/index.html
The global-resources site for the Union of Concerned Scientists. Follow the link to ozone depletion for a list of frequently asked questions and an overview of the science behind ozone depletion.

http://www.cmdl.noaa.gov
The home page for the Climate Monitoring and Diagnostics Laboratory (CMDL) of the National Oceanic and Atmospheric Administration. This

organization monitors greenhouse gases, aerosols, ozone, ozone-depleting gases, and solar and terrestrial radiation at sites located around the world.

http://www.engelhard.com

To learn more about the PremAir ozone to oxygen catalyst, type PremAir into the search engine of Engelhard.

http://toms.gsfc.nasa.gov/ozone/ozone.html

Since 1977, global ozone maps have been generated continuously by the TOMS satellite, and they are available to you here through this website. Also, be sure to check out their online textbook about stratospheric ozone.

http://toms.gsfc.nasa.gov/aerosols/aerosols.html

Global aerosol levels as measured by the Earth Probe, ADEOS, and Nimbus 7 satellites can be found at this Web address. Scientists use this data to observe a wide range of phenomena such as desert dust storms, forest fires, and biomass burning.

Air Resources

Visit The Chemistry Place at:
www.aw.com/chemplace

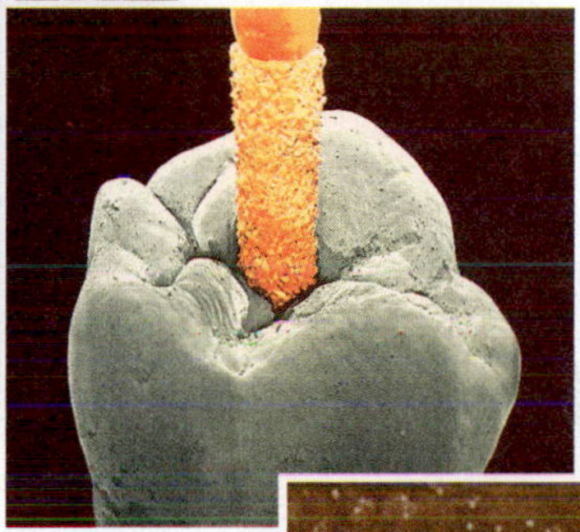

A Look at the Materials of Our Society

In addition to their stunning beauty, diamonds also have great usefulness because of their remarkable properties, which include extreme hardness, superlative ability to conduct heat, and resistance to corrosive chemicals. In the 1950s, researchers learned how to produce artificial diamonds. Although not of gem quality, laboratory-produced diamonds find many applications, particularly for abrasive surfaces, such as that of a dentist's drill bit. In the 1970s and 1980s, techniques were developed to produce thin films of diamond, and by the 1990s, scientists could fabricate diamond films into such devices as nanometer-scale calipers. Used for microscopic measurements, these calipers are made entirely of diamond, and their width is smaller than the diameter of a human hair. Other diamond nanostructures are currently under development, including a miniature robotic motor that may one day roam through your bloodstream removing plaque or delivering anticancer agents directly to tumors.

These applications of diamond films may seem fantastic now and will receive much attention once they are realized. Eventually our amazement will fade, however, and we will take this technology for granted. Our standard of living will have been raised, as will our expectations for what we can achieve by understanding the submicroscopic world.

The main goal of this chapter is to instill in you an excitement for those "common" materials that have already had an impact on society. Some advanced materials are introduced, but for the most part the chapter focuses on paper, plastic, metals, glass, and ceramics. The history of these materials is presented for your general interest and to help you gain a better perspective on the impact materials have had—and will surely continue to have—on the way we live.

18.1   Paper Is Made of Cellulose Fibers

No material has been more vital to the development of modern civilization than paper. With its advent, knowledge could be recorded, allowing us to learn from our mistakes as well as our successes. Furthermore, with paper, and the subsequent invention of the printing press, knowledge became accessible to a broad range of people. Technological, political, and social revolutions followed, and from these movements arose our complex modern societies.

Paper is reported to have been made in China as early as A.D. 100, when cellulose fibers from mulberry bark were pounded into thin sheets. Finer paper was eventually produced by lifting a silk screen up through a suspension of cellulose fibers in water, so that entangled fibers collected on the screen. After drying, the fibers remained intertwined, forming a sheet of paper, as shown in Figure 18.1.

Paper was introduced to the Western world by the Arabs, who in the eighth century learned how to make paper from Chinese papermakers captured in battle. Paper mills were soon established, and the technology spread swiftly to Europe. The first European paper mills were built in Spain as early as A.D. 1100.

Only handmade single sheets of paper were fabricated until, in 1798, a machine that could make continuous rolls of paper was invented in France. This machine consisted of a conveyor belt submerged at one end in a vat of suspended cellulose fibers. The conveyor belt was a screen so that water would drain from it as fibers were pulled out of the suspension. The entangled fibers were then squeezed through a series of rollers to create a long continuous sheet of paper. Within a few years, an improved version of this machine that used heated rollers was produced in England. The improved machine was called the Fourdrinier, after a wealthy industrialist who financed its creation. Automated Fourdrinier machines, such as the one shown in Figure 18.2, are still used today.

Most early paper was made from fibrous plant materials, such as bark, shrubs, and various grasses. Wood was not used because its fibers are embedded in a highly adhesive matrix of *lignin*, a naturally occurring polymer that does not dissolve in water. In 1867, researchers in the United States discovered that wood fibers could be isolated by soaking wood chips in sulfurous acid, which dissolves the lignin. The ability to make paper from wood was a great boon to entrepreneurs in North America because of that continent's once vast supply of trees. At about the same time, additives such as rosin and alum were added to strengthen paper and to make it accept ink well. Paper manufacturers whitened paper by bleaching it with chlorine and by incorporating opaque white minerals, such as titanium dioxide. Most large-scale papermaking processes today are basically unchanged from these earlier methods, except that chlorine is no longer

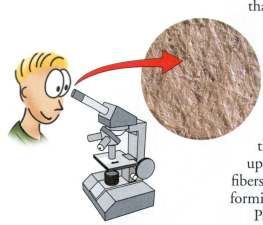

Figure 18.1
Take a close look at a piece of paper, and you'll see that paper is made of numerous tiny cellulose fibers.

Figure 18.2
A modern Fourdrinier machine for manufacturing paper.

Figure 18.3
In 1984, the Library of Congress esti-mated that 25 percent, or 3 million volumes, of its collection had become too brittle for circulation.

used because of the cancer-causing dioxins that form as paper is bleached. About 75 percent of the paper made in the United States comes from wood pulp, with the remainder coming from recycled wastepaper.

Paper made with rosin and alum tends to turn yellow and brittle within a matter of decades, as shown in Figure 18.3. This is because of the acidic nature of these additives, which catalyze the decomposition of the cellulose fibers. Noncorrosive alkaline alternatives were developed in the 1950s. Paper made with alkaline binders are designated *archival*, which means they last a minimum of 300 years. Until recently, alkaline binders were not used much because of their greater cost and because of a lack of demand from consumers. Paper companies were finally prompted to make the switch to "acid-free" paper in the late 1980s when the cost of wood pulp skyrocketed. Paper made with acidic rosin and alum binders must be at least 90 percent fiber in order to maintain strength, whereas paper made with alkaline binders maintains strength with as little as 75 percent fiber. Less fiber trans-lates to monetary savings.

A secondary benefit of alkaline binders is that, in a nonacidic environ-ment, cheap and abundant calcium carbonate can be used as a whitener in place of relatively expensive titanium dioxide. Acid-free paper has the added advantage that peroxide bleaches can be used on it instead of chlorine. Fur-thermore, consumers have come to recognize the value of archival-quality paper and are now demanding its production.

Concept Check

Was paper made before the mid-1800s acid-free?

Was this your answer? Yes, because acidic rosin and alum binders were not used in papermaking until the mid-1800s. (Books from the early 1800s can be found in pristine condition, whereas books from the 1960s are today often too brittle to read.)

Hands-On Chemistry: Papermaking

You can make paper at home by passing a screen through a soupy suspension of cellulose fibers, also known as pulp.

What You Need

Clean foam meat tray, window screen (available from a hardware store), duct tape, cellulose-containing material (scraps of paper, lint from clothes-dryer screen, cotton rags, dried leaves, flower petals, grass clippings, coffee grounds), measuring cup, blender, warm water, cooking pot having diameter greater than longer dimension of meat tray, 2 large towels, sponge, toothpicks

Procedure

1. Build a dipping screen by cutting out most of the bottom of the tray. Cut a piece of window screen to fit the hole, and tape it to the tray.

2. Cut the cellulose-containing material into very small pieces (so as not to clog the blender).

3. Fill the blender three-fourths full of warm water. Turn on low speed and *slowly* add 1 cup of the shredded material. There should always be much more water in the blender than solid matter. Blend the material to a fine pulp.

4. Pour the pulp into the pot. Add more water if necessary to give a depth of about 10 centimeters. Submerge the dipping screen and agitate the mixture to maintain homogeneity. Then hold the screen flat and lift it straight up out of the suspension to collect a thin layer of pulp.

5. Carefully invert the screen onto the towel, rub across the inverted screen with a damp sponge, and then carefully lift the screen, leaving the pulp behind. If necessary, use a toothpick to poke or pry the pulp free of the screen.

6. Press down on the pulp with a second towel to flatten it and remove as much water as possible. Lay the resulting paper out to dry on a flat surface.

Figure 18.4
Fiber-rich industrial hemp being harvested on a modern farm in Spain.

We use trees to meet our paper needs because they are abundant and renewable—for every tree cut down, a new one can be planted. However, growing and harvesting trees has some significant drawbacks. Timber takes decades to mature. This time frame is already not fast enough to keep up with our increasing demand. Also, replanting timber does not re-create a forest. Timber is typically replanted as a monoculture suited primarily for industrial needs. A true forest is a more evolved system of many species that thrive in one another's presence.

There are a number of possible alternatives to trees for the mass production of paper, including willow, kenaf, and industrial hemp, the latter shown being harvested in Figure 18.4. Each of these plants produces more than three times as much fiber per acre than trees do. They are also fast-growing plants, which means they replenish quickly. Whereas it takes 20 years before trees can be harvested, these plants, when grown in favorable climates, can be harvested three times in a single year. Also, the lignin content of willow, kenaf, and industrial hemp is quite low, which means their cellulose fibers are relatively easy to separate. If paper mills switched from wood to one of these alternatives, they could avoid the use of countless gallons of sulfurous acid, which is not only hazardous to the environment but largely responsible for the stink associated with paper production. The energy content of these rapidly growing plants is also of interest, especially to utility companies, who may one day burn plant material rather than oil for the production of electricity (Section 19.6).

Today, we use an enormous amount of paper—about 70 million tons a year in the United States alone. That's about 230 kilograms, or six full-sized trees, for each U.S. citizen. Currently, we recycle only about 25 percent of the paper we use. The rest we throw away. This is unfortunate because recycling paper does more than just save trees. A lot of energy is required to make paper from a tree. Less than half as much energy is needed to process old paper into new. Remarkably, about 50 percent of the solid waste produced in North America is paper. What have you done with your 230 kilograms this past year?

18.2   The Development of Plastics Involved Experimentation and Discovery

In Chapter 12, you learned about the great variety of synthetic polymers and their uses. This section covers the inventors who developed these polymers and the effect these new materials have had on society. The successes and failures in the story of plastics show how chemistry is a process of experimentation and discovery. As you read through this section, you may want to refer back to the structures and reactions shown in Chapter 12.

The search for a lightweight, nonbreakable, moldable material began with the invention of vulcanized rubber. This material is derived from natural rubber, which is a semisolid, elastic, natural polymer. The fundamental chemical unit of natural rubber is polyisoprene, which plants produce from isoprene molecules, as shown in Figure 18.5. In the 1700s, natural

Figure 18.5
Isoprene molecules react with one another to form polyisoprene, the fundamental chemical unit of natural rubber, which comes from rubber trees.

Isoprene   +   Isoprene   +   Isoprene

Polymerization

Polyisoprene

rubber was noted for its ability to rub off pencil marks, which is the origin of the term *rubber*. Natural rubber has few other uses, however, because it turns gooey at warm temperatures and brittle at cold temperatures.

In 1839, an American inventor, Charles Goodyear, discovered *rubber vulcanization*, a process in which natural rubber and sulfur are heated together. (His discovery occurred after he accidentally tipped an open jar of sulfur into a pot of heated natural rubber.) The product, vulcanized rubber, is harder than natural rubber and retains its elastic properties over a wide range of temperatures. This is the result of *disulfide cross-linking* between polymer chains, as illustrated in Figure 18.6.

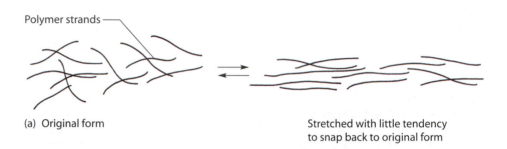

(a) Original form

Stretched with little tendency to snap back to original form

(b) Vulcanized form with disulfide cross-links

Stretched with great tendency to snap back because of cross-links

Figure 18.6
(a) When stretched, the individual poly-isoprene strands in natural rubber slip past one another and the rubber stays stretched. (b) When vulcanized rubber is stretched, the sulfur cross-links hold the strands together, allowing the rubber to return to its original shape.

Vulcanized rubber has found innumerable applications, from tires to rain gear, and has grown into a multibillion dollar industry. To help quench our ever-growing thirst for vulcanized rubber, natural rubber (poly-isoprene) is now synthesized from petroleum distillates.

Goodyear unfortunately reaped very few rewards from his discovery. He was a man of ill health who died in jail serving time for debts he was unable to pay. The present-day Goodyear Corporation was founded not by Goodyear but by others who sought to pay tribute to his name 15 years after he died.

In 1845, as vulcanized rubber was becoming popular, the Swiss chemistry professor Christian Schobein wiped up a spilled mixture of nitric and sulfuric acids with a cotton rag that he then hung up to dry. Within a few minutes, the rag burst into flames and then vanished, leaving only a tiny bit of ash. Schobein had discovered nitrocellulose, in which most of the hydroxyl groups in cellulose are bonded to nitrate groups, as Figure 18.7 illustrates. Schobein's attempts to market nitrocellulose as a smokeless gunpowder (*guncotton*) were unsuccessful, mainly because of a number of lethal explosions at plants producing the material.

Nitrate group

Nitrocellulose (cellulose nitrate)

Figure 18.7
Nitrocellulose, also known as cellulose nitrate, is highly combustible because of its many nitrate groups, which facilitate oxidation.

Collodion and Celluloid Begin with Nitrocellulose

Although Schobein failed at marketing guncotton, researchers in France discovered that solvents such as diethyl ether and alcohol transformed nitrocellulose to a gel that could be molded into various shapes. Furthermore, spread thin on a flat surface, the gel dried to a tough, clear, transparent film. This workable nitrocellulose material was dubbed *collodion*, and its first application was as a medical dressing for cuts.

In 1855, the moldable features of collodion were exploited by the British inventor and chemist Alexander Parkes, who marketed the material as Parkesine. Combs, earrings, buttons, bracelets, billiard balls, and even false teeth were manufactured in his factories. Parkes chose to focus more on quantity than on quality, however. Because he used low-grade cotton and cheap but unsuitable solvents, many of his products lacked durability, which led to commercial failure. In 1870, John Hyatt, a young inventor from Albany, New York, discovered that collodion's moldable properties were vastly improved by using camphor as a solvent. Hyatt's brother Isaiah named this camphor-based nitrocellulose material *celluloid*. Because of its greater workability, celluloid became the plastic of choice for the manufacture of many household items. In addition, thin transparent films of celluloid made excellent supports for photosensitive emulsions, a boon to the photography industry and a first step in the development of motion pictures.

As wonderful as celluloid was, it still had the major drawback of being highly flammable. Today, one of the few commercially available products made of celluloid is Ping-Pong balls, shown in Figure 18.8.

Bakelite Was the First Widely Used Plastic

About 1899, Leo Baekeland, a chemist who had immigrated to the United States from Belgium, developed an emulsion for photographic paper that was exceptional in its sensitivity to light. He sold his invention to George Eastman, who had made a fortune selling celluloid-based photographic film along with his portable Kodak camera. Expecting no more than $50,000 for his invention, Baekeland was shocked at Eastman's initial offer of $750,000 (in today's dollars, that would be about $25 million). Suddenly a very wealthy man, Baekeland was free to pursue his chemical interests.

He decided that the material the world needed most was a synthetic shellac to replace the natural shellac produced from the resinous secretions of the lac beetle native to southeastern Asia. At the time, shellac was the

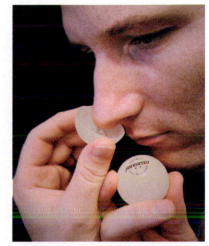

(a)

(b)

Figure 18.8
(a) Smell a freshly cut Ping-Pong ball, and you will note the distinct odor of camphor, which is the same smell that arises from heat cream for sore muscles. This camphor comes from the celluloid from which the ball is made. (b) Ping-Pong balls burn rapidly because they are made of nitrocellulose.

Figure 18.9
(a) The molecular network of Bakelite shown in two dimensions. (The actual structure projects in all three dimensions.) (b) The first handset telephones were made of Bakelite.

(a)

Formaldehyde

Polymerization

Phenol

Phenol–formaldehyde resin
(Bakelite)

(b)

optimal insulator for electrical wires. Ever since Edison's 1879 invention of the incandescent light bulb, miles of shellac-coated metal wire were being stretched across the land. The supply of shellac, however, was unable to keep up with demand.

Baekeland explored a tarlike solid once produced in the laboratories of Alfred von Baeyer, the German chemist who played a role in the development of aspirin. Whereas Baeyer had dismissed the solid as worthless, Baekeland saw it as a virtual gold mine. After several years, he produced a resin that, when poured into a mold and then heated under pressure, solidified into a transparent positive of the mold. Baekeland's resin was a mixture of formaldehyde and a phenol that polymerized into the complex network shown in Figure 18.9.

The solidified material, which he called *Bakelite*, was impervious to harsh acids or bases, wide temperature extremes, and just about any solvent. Bakelite quickly replaced celluloid as a molding medium, finding a wide variety of uses for several decades. It wasn't until the 1930s that alternative thermoset polymers (Section 12.4) began to challenge Bakelite's dominance in the evolving plastics industry.

The First Plastic Wrap Was Cellophane

Cellophane had its beginnings in 1892, when Charles Cross and Edward Beven of England found that treating cellulose with concentrated sodium hydroxide followed by carbon disulfide created a thick, syrupy, yellow liquid they called *viscose*. Extruding the viscose into an acidic solution generated a tough cellulose filament that could be used to make a synthetic silky cloth today called *rayon* (Figure 18.10).

Figure 18.10
Viscose is still used today in the manufacture of fibers used to make the synthetic fabric called rayon.

In 1904, Jacques Brandenberger, a Swiss textile chemist, observed restaurant workers discarding fine tablecloths that had only slight stains on them. Working with viscose at the time, he had the idea of extruding it not as a fiber but as a thin, transparent sheet that might be adhered to tablecloths and provide an easy-to-clean surface. By 1913, Brandenberger had perfected the manufacture of a viscose-derived, thin, transparent sheet of cellulose, which he named *cellophane*. After failing to form an adequate adhesion between cellophane and cloth, Brandenberger investigated cellophane's possible use as a film support for photography and motion pictures. This idea didn't work because of the cellophane's tendency to warp when heated. From these failures, Brandenberger began to realize that the most likely utility of his newly created cellophane was as a wrapping material.

Within several years, the DuPont Corporation bought the rights to cellophane. After producing several batches, investigators discovered that cellophane did not provide an effective barrier to water vapor and hence did little to keep foods from drying out. By 1926, DuPont chemists had solved this problem by incorporating small amounts of nitrocellulose and wax. Vaporproof cellophane gained wide popularity as a wrapping for such products as chocolates, cigarettes, cigars, and bakery goods. Hermetically sealed by cellophane, a product could be kept free of dust and germs. And unlike paper or tin foil—the alternatives of the day—cellophane was transparent and thus allowed the consumer to view the packaged contents, as seen in Figure 18.11. With properties like these, cellophane played a great role in the success of supermarkets, which first appeared in the 1930s. Perhaps cellophane's greatest appeal to the consumer, however, was its shine. As marketing people soon discovered, nearly any product—soaps, canned goods, or golf balls—would sell faster when wrapped in cellophane.

Figure 18.11
Cellophane transformed the way foods and other items were marketed.

Polymers Win in World War II

In the 1930s, more than 90 percent of the natural rubber used in the United States came from Malaysia. In the days after Pearl Harbor was attacked in December 1941 and the United States entered World War II, however, Japan captured Malaysia. As a result, the United States—the land with plenty of everything, except rubber—faced its first natural resource crisis. The military implications were devastating because without rubber for tires, military airplanes and jeeps were useless. Petroleum-based synthetic rubber had been developed in 1930 by DuPont chemist Wallace Carothers but was not widely used because it was much more expensive than natural rubber. With Malaysian rubber impossible to get and a war on, however, cost was no longer an issue. Synthetic rubber factories were constructed across the nation, and within a few years, the annual production of synthetic rubber rose from 2000 tons to about 800,000 tons.

Also in the 1930s, British scientists developed radar as a way to track thunderstorms. With war approaching, these scientists turned their attention to the idea that radar could be used to detect enemy aircraft. Their equipment was massive, however. A series of ground-based radar stations could be built, but placing massive radar equipment on aircraft was not feasible. The great mass of the equipment was due to the large coils of wire needed to generate the intense radio waves. The scientists knew that if they could coat the wires with a thin, flexible electrical insulator, they would be able to design a radar device that was much less massive. Fortunately, the recently developed polymer polyethylene turned out to be an ideal electrical insulator. This permitted British radar scientists to construct equipment light enough to be carried by airplanes. These planes were slow, but flying at night or in poor weather, they could detect, intercept, and destroy enemy aircraft. Midway through the war, the Germans developed radar themselves, but without polyethylene, their radar equipment was inferior, and the tactical advantage stayed with the Allied forces.

Four other polymers that had a significant impact on the outcome of World War II were Plexiglas, polyvinyl chloride, Saran, and Teflon. Plexiglas, shown in Figure 18.12, is a polymer known to chemists as poly-(methyl methacrylate). This glasslike but moldable and lightweight material made excellent domes for the gunner's nests on fighter planes and bombers. Although both Allied and German chemists had developed poly(methyl methacrylate), only the Allied chemists learned how small amounts of this polymer in solution could prevent oil or hydraulic fluid from becoming too thick at low temperatures. Equipped with only a few gallons of a poly(methyl methacrylate) solution, Soviet forces were able to keep their tanks operational in the Battle of Stalingrad during the winter of 1943. While Nazi equipment halted in the bitter cold, Soviet tanks and artillery functioned perfectly, resulting in victory and an important turning point in the war.

Figure 18.12
The bulky side groups in poly(methyl methacrylate) prevent the polymer chains from aligning with one another. This makes it easy for light to pass through the material, which is tough, transparent, lightweight, and moldable. (Plexiglas® is a registered trademark belonging to ATOFINA.)

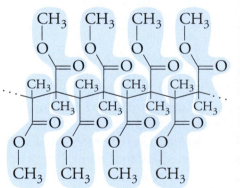

Poly(methyl methacrylate)

Polyvinyl chloride (PVC) had been developed by a number of chemical companies in the 1920s. The problem with this material, however, was that it lost resiliency when heated. In 1929, Waldo Semon, a chemist at BFGoodrich, found that PVC could be made into a workable material by the addition of a plasticizer. Semon got the idea of using plasticized PVC as a shower curtain when he observed his wife sewing together a shower curtain made of rubberized cotton. Other uses for PVC were slow to appear, however, and it wasn't until World War II that this material became recognized as an ideal waterproof material for tents and rain gear. After the war, PVC replaced Bakelite as the medium for making phonograph records.

Originally designed as a covering to protect theater seats from chewing gum, Saran found great use in World War II as a protective wrapping for artillery equipment during sea voyages. (Before Saran, the standard operating procedure had been to disassemble and grease the artillery to avoid corrosion.) After the war, the polymer was reformulated to eliminate the original formula's unpleasant odor and soon pushed cellophane aside to become the most popular food wrap of all time (Figure 18.13).

The discovery of Teflon was described in Chapter 1 to show how a scientist's sense of curiosity and analytical thinking can lead to success. Initially, the discoverers of Teflon were impressed by the long list of things this new material would *not* do. It would not burn, and it would not completely melt. Instead, at 620°F it congealed into a gel that could be conveniently molded. It would not conduct electricity, and it was impervious to attack by mold or fungus. No solvent, acid, or base could dissolve or corrode it. And most remarkably, nothing would stick to it, not even chewing gum.

Because of all the things Teflon would not do, DuPont was not quite sure what to do with it. Then, in 1944 the company was approached by governmental researchers in desperate need of a highly inert material to line the valves and ducts of an apparatus being built to isolate uranium-235 in the manufacture of the first nuclear bomb. Thus Teflon found its first application, and one year later, World War II came to a close with the nuclear bombing of Japan.

Figure 18.13
The now-familiar plastic food wrap carton with a cutting edge was introduced in 1953 by Dow Chemical for its brand of Saran wrap.

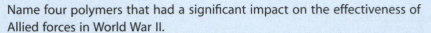

Concept Check ✓

> Name four polymers that had a significant impact on the effectiveness of Allied forces in World War II.

Was this your answer?   Plexiglas, polyvinyl chloride, Saran, and Teflon.

Attitudes About Plastics Have Changed

With a record of wartime successes, plastics were readily embraced in the postwar years. In the 1950s, Dacron polyester was introduced as a substitute for wool. Also, the 1950s were the decade during which the entrepreneur Earl Tupper created a line of polyethylene food containers known as Tupperware.

By the 1960s, a decade of environmental awakening, many people began to recognize the negative attributes of plastics. Being cheap, disposable, and nonbiodegradable, plastic readily accumulated as litter and as landfill. With petroleum so readily available and inexpensive, however, and with a growing population of plastic-dependent baby boomers, little stood in the way of an ever-expanding array of plastic consumer products. By 1977, plastics surpassed steel as the number-one material produced in the United States. Environmental concerns also continued to grow, and in the 1980s plastics-recycling programs began to appear. Although the efficiency of plastics recycling still holds room for improvement, we now live in a time when sports jackets made of recycled plastic bottles are a valued commodity.

In the past 50 years, there have been a number of significant technological advances in plastics. Polymers that emit light, for example, can be used to build display monitors that roll up like a newspaper or can be rolled onto walls like wallpaper. We have polymers that conduct electricity, replace body parts, and are stronger but much lighter than steel. Imagine synthetic polymers that mimic photosynthesis by transforming solar energy to chemical energy or synthetic polymers that efficiently separate fresh water from the oceans. These are not dreams. They are realities that chemists have already demonstrated in the laboratory. Polymers have played a significant role in our past, and they hold a clear promise for our future. Let's work to ensure that the petroleum starting materials from which we fabricate most polymers are not exhausted before this promise is realized.

Concept Check ✓

> Compressed plastics make up about 20 percent of the volume of a typical landfill. How does this compare with the amount of paper in landfills?

Was this your answer?   As noted at the end of Section 18.1, about 50 percent of the solid waste produced in North America is paper. So, significantly more paper ends up in our landfills than does plastic.

18.3 Metals Come from the Earth's Limited Supply of Ores

In Section 2.6 you learned about the properties of metals. They conduct electricity and heat, are opaque to light, and deform—rather than fracture—under pressure. Because of these properties, metals have found numerous applications. We use them to build homes, appliances, cars, bridges, airplanes, and skyscrapers. We stretch metal wire across poles to transmit communication signals and electricity. We wear metal jewelry, exchange metal currency, and drink from metal cans. Yet, what is it that gives a metal its metallic properties? We can answer this question by looking at the behavior of the atoms of the metallic elements.

The outer electrons of most metal atoms tend to be weakly held to the atomic nucleus. Consequently, these outer electrons are easily dislodged, leaving behind positively charged metal ions. The many electrons easily dislodged from a large group of metal atoms flow freely through the resulting metal ions, as is depicted in Figure 18.14. This "fluid" of electrons holds the positively charged metal ions together in the type of chemical bond known as a **metallic bond**.

The mobility of electrons in a metal accounts for the metal's significant ability to conduct electricity and heat. Also, metals are opaque and shiny because the free electrons easily vibrate to the oscillations of any light falling on them, reflecting most of it. Furthermore, the metal ions are not rigidly held to fixed positions, as ions are in an ionic crystal. Rather, because the metal ions are held together by a "fluid" of electrons, these ions can move into various orientations relative to one another, which is what happens when a metal is pounded, pulled, or molded into a different shape.

Two or more metals can be bonded to each other by metallic bonds. This occurs, for example, when molten gold and molten palladium are blended to form the homogeneous solution known as white gold. The quality of the white gold can be modified simply by changing the proportions of gold and palladium. White gold is an example of an **alloy**, which is any mixture composed of two or more metallic elements. By playing around with proportions, metal workers can readily modify the properties of an

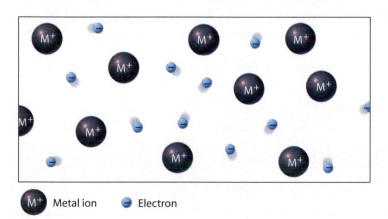

| M⁺ Metal ion | ⊖ Electron |

Figure 18.14
Metal ions are held together by freely flowing electrons. These loose electrons form a kind of "electronic fluid" that flows through the lattice of positively charged ions.

Figure 18.15
The gold color of the Sacagawea U.S. dollar coin is achieved by an outer surface made of an alloy of 77 percent copper, 12 percent zinc, 7 percent manganese, and 4 percent nickel. The interior of the coin is pure copper.

Figure 18.16
This open-pit copper mine at Bingham Canyon, Utah, is the world's biggest open-pit mine.

alloy. For example, in designing the Sacagawea dollar coin, shown in Figure 18.15, the U.S. Mint needed a metal that had a gold color—so that it would be popular—and also had the same electrical characteristics as the Susan B. Anthony dollar coin—so that the new coin could substitute for the Anthony coin in vending machines.

In this section we focus on the chemistry of naturally occurring metal-containing compounds and their large-scale transformation to metals. As you read, keep in mind that, whereas a metal consists of nothing but neutral metal atoms, a metal-containing compound is an ionic compound in which positively charged metal ions are bonded to negatively charged nonmetal ions. Aluminum oxide, Al_2O_3, and sodium chloride, $NaCl$, are two examples. These compounds may or may not be opaque to light, are poor conductors of heat and electricity, and often shatter under pressure.

Only a few metals—gold and platinum are two examples—appear in nature in metallic form. Deposits of these natural metals, also known as *native metals,* are quite rare. For the most part, metals are found in nature as chemical compounds. Iron, for example, is most frequently found as iron oxide, Fe_2O_3, and copper is found as chalcopyrite, $CuFeS_2$. Geologic deposits containing relatively high concentrations of metal-containing compounds are called **ores**. The metals industry mines these ores from the ground, as shown in Figure 18.16, and then processes them into metals. Although metal-containing compounds occur just about everywhere, only ores are concentrated enough to make the extraction of the metal economically feasible.

Metal ions bond with only five major types of negatively charged ions, shown in Figure 18.17. Consequently, metal-containing compounds are classified according to which type of negative ion they contain. Iron oxide is classified as an oxide, for instance, and chalcopyrite is classified as a sulfide.

Halides, such as sodium chloride and magnesium chloride, are commonly referred to as *salts*. They have good solubility in water and so are readily washed away by the action of either surface water or groundwater. Most of these and other water-soluble metal-containing compounds therefore end up in the ocean. These compounds are recovered by evaporating

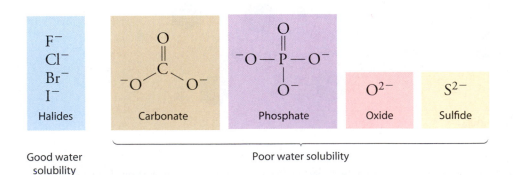

| Halides | Carbonate | Phosphate | Oxide | Sulfide |

Good water solubility   Poor water solubility

Figure 18.17
Five negatively charged ions to which positively charged metal ions bond.

seawater. Alternatively, water-soluble compounds may end up in land basins, such as the Bonneville salt flats of Utah, where they are readily mined. In some regions, such as along the Gulf of Mexico, vast deposits of halides remain undissolved hundreds of meters below the surface, where groundwater cannot reach. The compounds in these deposits tend to be very pure, which makes deep mining excavations like the one shown in Figure 18.18 worthwhile.

In contrast to halides, compounds containing carbonate, phosphate, oxide, or sulfide ions tend to have relatively low solubilities in water. Hence, their ores tend to stay put and are found in more diverse geologic locations.

The form in which a metal is most likely to be found in nature is a function of its position in the periodic table. Figure 18.19 shows that group 1 metals tend to be found mostly as halides, group 2 metals mostly as carbonates, and group 3 metals and lanthanides mostly as phosphates. Most metals from groups 4 to 8 along with aluminum, Al, and tin, Sn, tend to be found as oxides, and most metals from groups 9 to 15 along with molybdenum, Mo, tend to be found as sulfides.

Figure 18.18
This subterranean salt deposit contains relatively pure metal-containing compounds. After the deposits are mined, the resulting caverns are very dry and thus make excellent archival storage sites for moisture-sensitive equipment or documents.

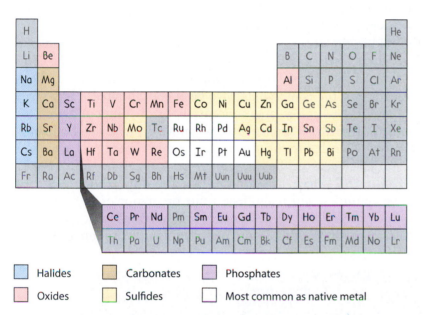

Halides Carbonates Phosphates
Oxides Sulfides Most common as native metal

Figure 18.19
Which compound of a metal is most prevalent in nature is related to the metal's position in the periodic table.

Concept Check ✔

According to Figure 18.19, which is more abundant in nature: iron oxide, Fe_2O_3, or iron sulfide, FeS?

Was this your answer? Figure 18.19 shows the most common forms found in nature, meaning iron is more abundant as iron oxide.

We Should Conserve and Recycle Metals

Because our planet is chock-full of metal-containing compounds, it is difficult to imagine how we could ever incur a shortage of metals. Experts suggest, however, that if we continue with our present rate of consumption, such shortages will occur within the next two centuries. The problem is not a shortage

Figure 18.20
Natural resources are unavailable to us when the energy required to collect them far exceeds the resource's inherent value. For example, most of the world's gold is found in the oceans, but this gold is too dilute for extraction to be worthwhile.

of metal-containing compounds but rather a shortage of ores from which these compounds can be extracted *at a reasonable cost.*

Consider the recovery of gold. All the gold in the world isolated from nature so far could be placed in a single cube 18 meters on a side, which would have a mass of about 130,000 tons. This includes all the naturally occurring elemental gold we have mined plus all the gold purified from gold-containing ores. Because the rate of gold production is steadily decreasing, one might think we have already isolated a significant portion of the Earth's total gold reserves. Our oceans, however, are laden with gold—as much as 2 milligrams per ton of seawater. Given that there is about 1.5×10^{18} tons of seawater on the planet, our oceans contain 3.4 billion tons of gold! As yet, however, no method has been found for recovering gold from seawater profitably—this gold is simply too dilute (Figure 18.20).

Like the gold in the ocean, most of the metal-containing compounds in the Earth's crust are finely mixed with other stuff, which is to say the compounds are diluted. Ores are, by definition, parts of the Earth's crust where, for geological reasons, the compounds have been concentrated. High-grade ores, those containing relatively large concentrations of compounds, are the first to be mined. After these are depleted, we move on to lower-grade ores, which have lower yields that translate into greater costs. Eventually, a nation's ore supplies are depleted, as are the aluminum oxide ores in the United States, as noted in Figure 18.21, and the nation is forced to import metals or their ores from other countries, which also have finite ore resources.

We should conserve and recycle metals whenever possible because it is far cheaper to produce metals from recycled products than from ore.

Figure 18.21
An open-pit aluminum mine in Australia. Aluminum ore is no longer mined in the United States because the reserves have dwindled to the point where it is less expensive to import high-grade aluminum ore from other countries, including Australia.

Environmentally sound exploration of new reserves is also required. Ore nodules discovered on the ocean floor, for example, contain as much as 24 percent manganese and 14 percent iron. Significant quantities of copper, nickel, and cobalt have also been found in this submarine terrain. Perhaps mining of the ocean floor may one day replace the mining we now do on land. And in the not too distant future, perhaps the mining of metal-rich asteroids in space will become a reality.

18.4   Metal-Containing Compounds Can Be Converted to Metals

To convert a metal-containing compound to a metal requires an oxidation–reduction reaction. Recall from Chapter 11 that oxidation is the loss of electrons and reduction is the gain of electrons. In the metal-containing compound, the metal exists as a positively charged ion because it has lost one or more of its electrons to its bonding partner. To convert metal ions to neutral metal atoms requires that they gain electrons; that is, they must be reduced:

$$M^+ \quad + \quad e^- \longrightarrow \quad M^0$$

Metal ion   Electron   Metal

The tendency of a metal ion to be reduced depends on its location in the periodic table, as summarized in Figure 18.22. As discussed in Chapters 5 and 11, metals on the left of the periodic table readily lose electrons. This means it is relatively difficult to give electrons back to these metal ions—in other words, they are difficult to reduce. A sodium atom, for example, being on the left of the periodic table, loses electrons easily. Any ionic compound it forms, such as sodium chloride, tends to be very stable. Reducing the sodium ion, Na^+, to sodium metal, Na^0, is difficult because doing so requires giving electrons to the sodium ion.

Metals on the left and especially the lower left of the periodic table therefore require the most energy-intensive methods of recovery, which includes *electrolysis*. As was shown in Section 11.3, during electrolysis an electric current supplies electrons to positively charged metal ions, thus reducing them. Metals commonly recovered by electrolysis include the metals of groups 1 through 3, which occur most frequently as halides, carbonates, and phosphates. In addition, aluminum is also commonly recovered by electrolysis, and other metals are also obtained using electrolysis when very high purity is needed. The reactions involved when copper is produced this way are shown in Figure 18.23 on page 604.

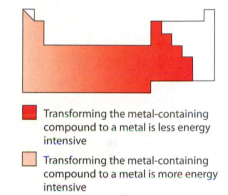

■ Transforming the metal-containing compound to a metal is less energy intensive

□ Transforming the metal-containing compound to a metal is more energy intensive

Figure 18.22
The ions of metallic elements at the lower left of the periodic table are most difficult to reduce. For this reason, obtaining these elements from the metal-containing compounds they form is energy-intensive. Metallic elements at the upper right of the periodic table tend to form compounds that require less energy to convert to metals.

Concept Check

Why is it so difficult to obtain a group 1 metal from a compound containing ions of that metal?

Was this your answer?   The metal ions do not readily accept electrons to form metal atoms.

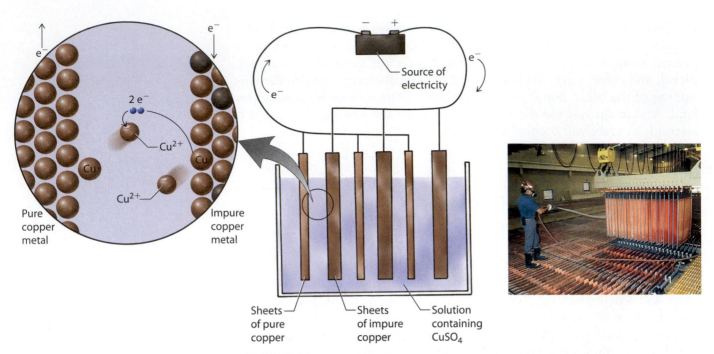

Figure 18.23
High-purity copper is recovered by electrolysis. Pure copper metal deposits on the negative electrode as copper ions in solution gain electrons. The source of these copper ions is a positively charged electrode made of impure copper.

Some Metals Are Most Commonly Obtained from Metal Oxides

Ores containing metal oxides can be converted to metals fairly efficiently in a *blast furnace*. First, the ore is mixed with limestone and coke, which is a concentrated form of carbon obtained from coal. The mixture is dropped into the furnace, where the coke is ignited and used as a fuel. At high temperatures, the coke also behaves as a reducing agent, yielding electrons to the positively charged metal ions in the oxide and reducing them to metal atoms. Figure 18.24 shows this method being used on iron oxide.

In the furnace, the limestone reacts with ore impurities—predominantly silicon compounds—to form *slag*, which is primarily calcium silicate:

$$SiO_2(s) \; + \; CaCO_3(s) \; \longrightarrow \; CaSiO_3(\ell) \; + \; CO_2(g)$$

| Silica sand (ore impurity) | Limestone | Molten slag (calcium silicate) | Carbon dioxide |

Because of the high temperatures, both the metal and the slag are molten. They drain to the bottom of the blast furnace, where they collect in two layers, the less-dense slag on top. The metal layer is then tapped off through an opening at the bottom of the blast furnace.

Once cooled, the metal from a blast furnace is known as a *cast metal.* (When the ore is an iron ore, the cast metal is known as *pig iron.*) A cast metal is brittle and soft because it still contains impurities, such as phosphorus, sulfur, and carbon. To remove these impurities, oxygen is blown through the molten cast metal in a *basic oxygen furnace*, shown in Figure

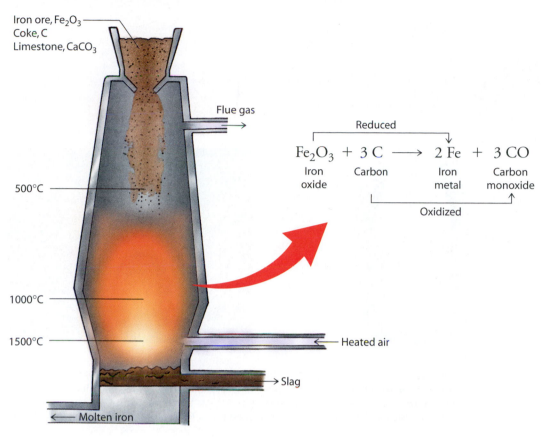

Figure 18.24
A mixture of iron oxide ore, coke, and limestone is dropped into a blast furnace, where the iron ions in the oxide are reduced to metal atoms.

18.25 on page 606. The oxygen oxidizes the impurities to form additional slag, which floats to the surface and is skimmed off.

Most phosphorus and sulfur impurities are removed in a basic oxygen furnace, but the purified metal still contains about 3 percent carbon. For the production of iron, this carbon is desirable. Iron atoms are relatively large, and when they pack together, small voids are created between atoms, as shown in Figure 18.26 on page 606. These voids tend to weaken the iron. Carbon atoms are small enough to fill the voids, and having the voids filled strengthens the iron substantially. Iron strengthened by small percentages of carbon is called **steel**. The tendency of steel to rust can be inhibited by alloying the steel with noncorroding metals, such as chromium or nickel. This yields the *stainless steel* used to manufacture eating utensils and countless other items.

Other Metals Are Most Commonly Obtained from Metal Sulfides

Metal sulfides can be purified by *flotation*, a technique that takes advantage of the fact that metal sulfides are relatively nonpolar and therefore attracted to oil. An ore containing a metal sulfide is first ground to a fine powder and then mixed vigorously with a lightweight oil and water. Compressed air is

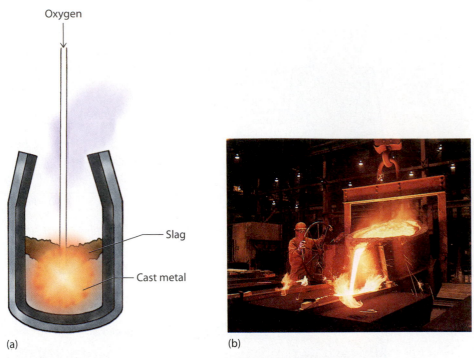

Oxygen

Slag

Cast metal

(a)   (b)

Figure 18.25
(a) A flow of oxygen through a basic oxygen furnace oxidizes most of the impurities in a cast metal to form slag that may be skimmed away as it floats to the surface. (b) The basic oxygen furnace is hoisted and its purified contents poured into a reservoir used for molding iron pieces.

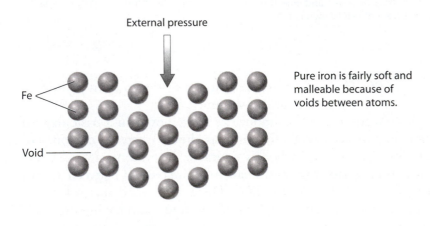

External pressure

Fe

Void

Pure iron is fairly soft and malleable because of voids between atoms.

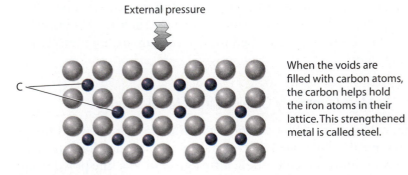

External pressure

C

When the voids are filled with carbon atoms, the carbon helps hold the iron atoms in their lattice. This strengthened metal is called steel.

Figure 18.26
Steel is stronger than iron because of the small amounts of carbon it contains.

then forced up through the mixture. As air bubbles rise, they become coated with oil and metal sulfide particles. At the surface of the liquid, the coated bubbles form a floating froth, as shown in Figure 18.27. This froth, which is very rich in the metal sulfide, is then skimmed off.

The metal sulfides recovered from the froth are then *roasted* in the presence of oxygen. The net reaction is the oxidation of S^{2-} in the sulfide to S^{4+} in sulfur dioxide and the reduction of the metal ion to its elemental state:

$$MS(s) \ + \ O_2(g) \ \longrightarrow \ M(\ell) \ + \ SO_2(g)$$

Metal sulfide   Oxygen   Metal   Sulfur dioxide

Figure 18.27
Air bubbles rising through a flotation container transport metal sulfide particles to the surface.

The isolation of copper from its most common ore, chalcopyrite, requires several additional steps because of the presence of iron. First the chalcopyrite is roasted in the presence of oxygen:

$$2\,CuFeS_2(s) + 3\,O_2(g) \ \longrightarrow \ 2\,CuS(s) + 2\,FeO(s) + 2\,SO_2(g)$$

Chalcopyrite   Oxygen   Copper   Iron   Sulfur
                                  sulfide   oxide   dioxide

The copper sulfide and iron oxide from this reaction are then mixed with limestone, $CaCO_3$, and sand, SiO_2, in a blast furnace, where CuS is converted to Cu_2S. The limestone and sand form molten slag, $CaSiO_3$, in which the iron oxide dissolves. The copper sulfide melts and sinks to the bottom of the furnace. The less-dense iron-containing slag floats above the molten copper sulfide and is drained off. The isolated copper sulfide is then roasted to copper metal:

$$Cu_2S(s) + O_2(g) \ \longrightarrow \ 2\,Cu(\ell) + SO_2(g)$$

Copper   Oxygen   Copper   Sulfur
sulfide   metal   dioxide

Roasting metal sulfides requires a fair amount of energy. Furthermore, sulfur dioxide is a toxic gas that contributes to acid rain, and so its emission must be minimized. Most companies comply with EPA emissions standards by converting the sulfur dioxide to marketable sulfuric acid, H_2SO_4.

Concept Check

Why isn't iron metal commonly obtained by roasting iron ores?

Was this your answer? Most iron ores are oxides, which are more suitably converted to metals using the blast furnace.

18.5   Glass Is Made Primarily of Silicates

As noted earlier, the prime component of slag is silicates, and as early as 500 B.C. metal workers noticed that solidified slag had properties not unlike those of the highly prized volcanic glass *obsidian*. This prompted

Figure 18.28
Eyeglasses, a telescope, and a distillation apparatus—pivotal glass inventions.

people to heat all sorts of rocks in various combinations. Slags of many colors were produced, including some that were transparent. When set to a useful purpose, such as in the making of beads, containers, or windows, the hardened slag became known as *glass*.

Over many centuries, glassmaking gradually improved. Clearer and stronger glass made possible many important inventions, such as those shown in Figure 18.28.

Eyeglasses were first made in northern Italy in the 13th century. Their use quickly spread, and the effect on society was significant, especially in that eyeglasses prompted people to remain intellectually active throughout their lives.

Another significant achievement made possible by the development of glass lenses was the telescope, which in 1610 was pointed skyward by Galileo, who observed moons orbiting the planet Jupiter and opened the door to a golden age of astronomy.

Improved glass also made possible the design of distillation equipment used to isolate alcohol from fermented broths. Distilled spirits, as they were called, became noted not only for their intoxicating effects but for their ability to disinfect and promote the healing of wounds.

Glass is an *amorphous material*. In such a substance, submicroscopic units are randomly oriented relative to one another. This is distinguished from a crystal, in which submicroscopic units are arranged in a periodic and orderly fashion (Section 6.3).

Glass can have a variety of chemical compositions. Common glass is a mixture of sodium silicate, Na_2SiO_3, and calcium silicate, $CaSiO_3$. It is formed by heating a blend of sodium carbonate, calcium carbonate, and silica:

$$Na_3CO_2 + SiO_2 \longrightarrow Na_2SiO_3 + CO_2$$

Sodium carbonate   Silica   Sodium silicate   Carbon dioxide

$$CaCO_3 + SiO_2 \longrightarrow CaSiO_3 + CO_2$$

Calcium carbonate   Silica   Calcium silicate   Carbon dioxide

Various additives provide glass with special properties. Adding potassium oxide, K_2O, for example, makes a very hard glass commonly used for optical applications. When lead oxides are added, glass becomes dense and has a high refractive index, which means it readily bends white light into a rainbow of colors. Such glass is called *crystal glass* because these properties resemble those of true crystals, such as quartz, which is a crystalline form of silica.

Adding boron oxide, B_2O_3, markedly lowers the rate at which glass expands when it is warmed and contracts when it is cooled. This enables the glass to withstand sudden changes in temperature without breaking. Such glass is known as *Pyrex glass* and is used in laboratory glassware and cooking utensils.

Glass can be colored by adding various chemicals. Cobalt oxide gives a blue glass used in specialty dinnerware. Adding the element selenium gives glass a red color, and the element iridium makes glass black. Glassmaking artists have experimented with chemical additives to give a variety of colored glasses, such as those shown in Figure 18.29.

The physical properties of glass are wholly remarkable. It is easily melted, can be poured or blown into almost any shape, and retains that shape when it cools. Moreover, pieces of glass can be melted together to become sealed without glue or cement. This allows for the manufacture of many intricate glassware designs. Glass is transparent and resistant to even the most corrosive chemicals, which is why chemical reactions in the laboratory are usually carried out in glass containers. Long, thin strands of glass are flexible enough to be encased in cables that can stretch for hundreds of kilometers. These are the *fiber optic cables* shown in Figure 18.30, through which pulses of light can travel, carrying information data with remarkable efficiency.

Fortunately, the starting materials for glass are abundant, and so, unlike the situation with paper, plastics, and metals, we are not in imminent danger of a glass shortage. Resources can still be saved, however, by recycling glass. The energy needed to produce glass from recycled products is only 70 percent of the energy needed to produce it from raw materials. Glass by any standard has been one of humankind's best bargains.

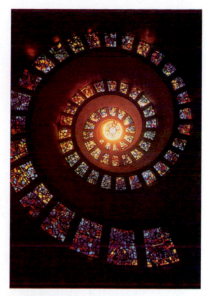

Figure 18.29
Glass as an art form.

18.6 Ceramics Are Hardened with Heat

Wet clay is a mixture of microcapsules of aluminum oxides and silicon oxides surrounded by water. It can be shaped easily because the water serves as a lubricant that allows the microcapsules to slip over one another. When dry, the microcapsules become locked in position, and the clay holds its shape. Heating the dried clay to high temperatures causes the silicon oxides to melt into a glass that on cooling bonds the microcapsules together. At this point, the clay is transformed to a hard, water-resistant ceramic useful for making the pottery shown in Figure 18.31 as well as a myriad of other products.

In general, a *ceramic* is any solid that has been hardened by heat. Modern ceramics contain nonmetallic elements such as oxygen, carbon, or silicon, but they may also contain metallic elements, such as aluminum or a transition metal. Unlike metals, ceramics cannot be pounded into thin sheets or drawn into wires. Instead, they tend to fracture, as anyone who has dropped a ceramic dinner plate knows. But ceramics are superior to metals for some applications. For example, ceramics are able to withstand extreme temperatures without melting or corroding, and they are lightweight and can be very hard. An example of a modern ceramic almost as hard as diamond is silicon carbide, SiC, also known as carborundum. This lightweight ceramic does not conduct heat and can withstand temperatures up to 2000°C. These properties make it ideal for coating spacecraft, which are subject to extreme conditions as they re-enter the Earth's atmosphere. Components for hot turbine engines are also being built of either silicon carbide or a similar ceramic, silicon nitride, Si_3N_4.

Figure 18.30
Information-bearing light pulses travel through the glass of fiber optic cables, a revolution in long-distance communication.

Figure 18.31
Ceramic pottery.

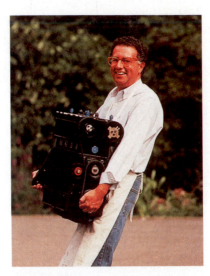

Figure 18.32
An engine made of the ceramic silicon nitride.

Modern ceramics are ideal for making automobile engine parts, such as those shown in Figure 18.32. The efficiency of an engine goes up with increasing operating temperatures. Today's metal automobile engines are relatively inefficient because their operating temperatures must be kept below the melting point of the metals. Radiators remove valuable heat that otherwise would raise efficiency. In fact, about 36 percent of fuel energy is lost through the radiator.

So why aren't today's engines made of ceramics? The short answer is that, unlike metals, ceramics cannot bend and deform to absorb impacts. Intense research is currently under way to solve the problem of ceramic brittleness, with some success. Improved resistance to fracturing, for example, can be attained by careful quality control of starting materials and processing. As we shall see in the next section, brittleness can also be combated by *compositing* ceramics with other materials.

Engines consisting of mostly ceramic parts are being developed in Japan, and Figure 18.33 shows a high-powered sedan featuring a small ceramic-based engine and no radiator. This car uses heat instead of rejecting it, as do the prototype turbine ceramic-based engines developed in the United States by the Department of Energy. In place of pistons, the U.S. versions feature two gas turbines made of durable silicon nitride, and the engine is designed to run at a hot 1300°C, with higher efficiency and cleaner emissions.

Ceramic Superconductors Have No Electrical Resistance

In ordinary electrical conductors, such as copper wire, moving electrons that flow as electric current often collide with the atoms of the conductor, transferring some of their kinetic energy to the conductor as heat, which is lost to the environment. This heat loss adds up when electrical energy is being carried over vast distances, meaning that some of the energy generated at a power plant never makes it to the consumer. In the late 1980s, researchers found that specially formulated ceramic compounds lost all electrical resistance when bathed in liquid nitrogen at −196°C. The electrons in these conductors travel in pathways that avoid atomic collisions, permitting them to flow indefinitely. Steady currents have been observed to persist for extended periods of time without apparent loss. These ceramics

Figure 18.33
This vehicle has no radiator because many of its engine parts are made of heat-resistant ceramic materials.

are called **superconductors** and have zero electrical resistance. The current through them does not decrease, and they generate no heat.

Transmission of electrical energy is one obvious application for ceramic superconductors. This energy might even be stored for later use in large circular loops of a superconducting material. Unfortunately, ceramics tend to be brittle and cannot be drawn into wires. A solution to these mechanical problems is to homogenize the superconducting ceramic starting materials with an organic thermoplastic. At warm temperatures, the mixture can be drawn into long fibers, which are then baked into a medium. When the fibers are cooled by liquid nitrogen, they become superconducting. Another technological hurdle is that present-day ceramics lose their superconductivity when large currents are applied.

The mechanisms of superconductivity in ceramics are not fully understood, and most progress is still being made by trial and error. Perhaps once the secrets are unlocked, the many obstacles posed by ceramic superconductors will be overcome and their great potential realized. It might even be possible to design materials that superconduct at room temperature, thereby avoiding the need for relatively inexpensive but cumbersome liquid nitrogen or other coolants.

18.7 Composites Combine Fibers and a Thermoset Medium

Thermoset plastics (Section 12.4) are pretty tough, as are various polymeric fibers. Combine the two, however, and you have a **composite** material that is more than twice as strong and just as lightweight. Composite materials are made by incorporating fibers into any thermoset medium, such as thermoset plastics, metals, or even ceramics. Examples of composite materials are shown in Figure 18.34 on page 612.

There are many examples of composites in nature. A tree can grow to great heights and support heavy branches because it is a composite of flexible cellulose fibers in a lignin matrix, as noted in Section 18.1. Seashells and limestone are both made of calcium carbonate, but sea shells are much harder because they are composites of crystalline calcium carbonate with embedded polypeptide fibers. Humans have been fabricating composites ever since they began mixing straw with clay. The straw not only makes for stronger pots and bricks but also keeps them from cracking as the clay is dried.

The composite industry was launched in the early 1960s with the development of fiberglass, which consists of short glass fibers in a matrix of some thermoset resin. Fiberglass composites are tough, lightweight, and inexpensive to make, and they have found many marine, housing, construction, sports, and industrial applications.

The strongest new composites are the *advanced composites*, in which fibers are aligned or interwoven before being set within the resin. Advanced composites have extraordinary strength in the direction of the aligned fibers and are relatively weak in the perpendicular direction. Weakness in one direction can be overcome by laminating layers together at different angles, as in plywood, a familiar composite. Strength in all directions can be

Rocket cone

Wood

Fiberglass

Graphite fiber composite

Figure 18.34
A few examples of composite materials.

Figure 18.35
The all-composite *Voyager* airplane.

achieved by weaving the fibers into a three-dimensional network. Besides strength, advanced composites are also known for their lightness, which makes them ideal for car parts, sporting goods, and artificial limbs. Advanced composites tend to be expensive, however, because much of their production is still done by hand.

Airplane parts, and even whole airplanes, are now being fabricated out of lightweight advanced composites in order to save fuel. In 1986, the all-composite *Voyager*, shown in Figure 18.35, was the first plane to fly around the world without refueling. Many private jets are now fabricated out of composites. The general public may one day have its chance to fly in fuel-efficient, stress-resistant composite airplanes. In the meantime, we fly in passenger airplanes made of metal, which are still more economical to build.

The aerospace industry is particularly interested in advanced composites that are lightweight and can withstand the extreme pressures and temperatures experienced by spacecraft. Efforts are now under way to build reusable launch vehicles that are much more efficient than the present space shuttle at getting payloads into orbit. One prototype is the X-33 *National Aerospace Plane*, shown in Figure 18.36.

The same materials developed for aerospace planes may one day be applied to commercial travel. Hypersonic aircraft traveling at many times the speed of sound would reduce the time of a transpacific flight from America to Australia from 16 hours to a mere 3 hours. What's more, the

Hands-On Chemistry:
A Composite of White Glue and Thread

It's easy to stretch a layer of dried white glue until it splits in half, and it's easy to yank a single strand of sewing thread into two pieces. Combine the glue with many strands of thread, however, and the resulting composite is most resistant to breakage—but only in the directions parallel to the threads. Make this composite and test its strength.

What You Need

White glue, sewing thread, aluminum foil, scissors

Procedure

① Pour the glue onto the aluminum foil to make three strips, each measuring about 4 centimeters by 8 centimeters.

② On one of the strips, lay strands of thread parallel to the longer side of the strip. Lay enough strands so that they are no more than 0.5 centimeter apart.

③ On a second strip, lay strands of thread both parallel and perpendicular to the longer side of the strip. Again lay enough strands so that they are no more than 0.5 centimeter apart.

④ After the glue has thoroughly dried, separate the three strips from the foil and cut off the ends of any loose threads. Pull on the three strips in various directions, and compare their strengths and weaknesses.

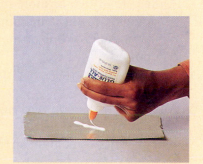

altitude of the aircraft would necessarily be so high that travelers would be able to see the curvature of planet Earth.

In Perspective

Technological feats are made possible only after the appropriate materials have been developed. Columbus made it to the Americas with rugged sails of canvas, firm ropes of hemp, and a sturdy wood hull bound by metal. Likewise, whether our dreams are to remain fiction or turn into fact depends on the materials available to us. Today, with a growing number of remarkable materials, we can send data through fiber optic cables, criss-cross the skies, or fly to Mars and beyond—a sign that the human spirit of exploration is more empowered than ever before.

Figure 18.36
Future generations of space vehicles, such as the X-33 prototype illustrated here, will depend greatly on the development of advanced composite materials.

Key Terms and Matching Definitions

_____ alloy
_____ composite
_____ metallic bond
_____ ore
_____ steel
_____ superconductor

1. A chemical bond in which the metal ions in a piece of solid metal are held together by their attraction to a "fluid" of electrons in the metal.
2. A mixture of two or more metallic elements.
3. A geologic deposit containing relatively high concentrations of one or more metal-containing compounds.
4. Iron strengthened by small percentages of carbon.
5. Any material having zero electrical resistance.
6. Any thermoset medium strengthened by the incorporation of fibers.

Review Questions

Paper Is Made of Cellulose Fibers

1. What component of plants is most useful in the fabrication of paper?

2. When was paper introduced in Europe?

3. What is a Fourdrinier machine, and what was the impact of its invention?

4. When did people start using trees to make paper?

The Development of Plastics Involved Experimentation and Discovery

5. How did Schobein discover nitrocellulose?

6. What chemical is used to make celluloid a workable material?

7. What is one of the major drawbacks of celluloid?

8. Who provided Baekeland with the financial resources to develop Bakelite?

9. What prompted Brandenberger to seek a way to make thin sheets of viscose?

10. How did chemists transform cellophane into a vaporproof wrap?

11. What was one of the prime motivations for the Japanese to invade Malaysia at the beginning of World War II?

12. What polymer proved useful in the development of radar equipment lightweight enough to be carried on airplanes?

13. What role did Teflon play in World War II?

Metals Come from the Earth's Limited Supply of Ores

14. What are the five types of negatively charged ions found in metal-containing compounds?

15. Which metal-containing compounds are most soluble in water?

Metal-Containing Compounds Can Be Converted to Metals

16. Which group of metals requires the most energy to be recovered from metal-containing compounds?

17. When iron ions are reduced to neutral iron atoms in a blast furnace, where do the electrons come from?

18. How are copper sulfide, CuS, and iron oxide, FeO, separated from each other in a blast furnace in the preparation of copper metal?

Glass Is Made Primarily of Silicates

19. In what ways did the development of glass affect human history?

20. How is glass different from crystal?

21. Is crystalware really made of crystal?

22. How is colored glass made?

23. What are the benefits of recycling glass?

Ceramics Are Hardened with Heat

24. What is a ceramic?

25. What advantages do ceramic automobile engines have over metal ones?

26. What is the primary drawback of a ceramic?

27. Is it possible for a ceramic to conduct electricity?

Composites Combine Fibers and a Thermoset Medium

28. What is a composite, and what are some examples found in nature?

29. Where are you most likely to find composites in the marketplace today?

30. Why are composites an ideal material for aircraft?

Hands-On Chemistry Insights

Papermaking

All plants contain cellulose, which is why paper can be made from all plants. However, all plants also contain cellulose-binding lignins. Those that have the highest lignin content, such as trees, require the strongest chemicals in order to free the cellulose fiber. Plants that have lower lignin content, such as grasses and nonwoody shrubs, can be treated with milder chemicals in order to free the cellulose fibers. Such pretreatment can be avoided altogether when working with recycled fibers, as you did in this activity using scraps of old paper or cloth. Thus, recycling paper not only saves trees but also makes the papermaking process more efficient.

A Composite of White Glue and Thread

The strength provided by the ten or so parallel strands you used for this activity is considerable. Consider, then, the strength provided by a fabric made of hundreds of strands laid on the glue.

The strength of a composite is maximum only in directions parallel to the fibers. Thus, criss-crossing fibers makes for a composite that has maximum strength in several directions. In some forms of fiberglass, the strands of reinforcing glass are strewn in random directions. This random arrangement decreases the strength in any one direction but optimizes the overall strength.

Exercises

1. How are paper and cooked spaghetti similar to each other?

2. What are the advantages and disadvantages of using trees to make paper?

3. What are the advantages and disadvantages of using industrial hemp to make paper?

4. Fallen trees rot as naturally occurring fungi digest the lignins that normally bind the wood fibers together. How might these fungi be used in the manufacture of paper?

5. What role did chance discovery play in the history of polymers? Cite some examples.

6. What is the difference between collodion and celluloid?

7. What is the chemical difference between celluloid and cellophane?

8. Why does a freshly cut Ping-Pong ball smell of camphor?

9. Why are Ping-Pong balls so highly flammable?

10. Melmac is a thermoset polymer discussed in Chapter 12. How are the chemical structures of Bakelite and Melmac similar to each other? How are they different?

11. List these plastics in order of the year in which they were developed: cellophane, celluloid, collodion, parkesine, PVC, Teflon, viscose.

12. Why are ores so valuable?

13. Can only group 1 elements form halides?

14. Distinguish between a metal and a metal-containing compound.

15. Metal ores are isolated from rock by taking advantage of differences in both physical and chemical properties. Cite examples given in the text where differences in physical properties are used. Cite examples given in the text where differences in chemical properties are used.

16. Iron is useful for reducing copper ions to copper metal. Might it also be used to reduce sodium ions to sodium metal? Why or why not?

17. Do we really have shortages of the metals we need to run our industrialized society? Isn't the planet made essentially of metals?

18. Is transparent glass a homogeneous or heterogenous mixture?

19. When is glass not fragile?

20. Commercial beverages are transported in both plastic and glass containers. How might it be that the glass containers are responsible for the release of more atmospheric pollution than the plastic containers?

21. In what sense is glass the glue that holds a ceramic together?

22. If the efficiency of an engine increases with increasing temperature, why are conventional automotive engines equipped with radiators to keep them cool?

23. In what direction is a sample of plywood weakest? Why?

24. Concrete by itself is not strong enough to build large structures. How do engineers overcome this inherent weakness?

Discussion Topics

1. Which of the materials described in this chapter are most important to recycle for economic reasons? For environmental reasons? For political reasons? Explain your answers.

2. Should the government require that certain materials be recycled? If so, how should this requirement be enforced?

3. What are some of the obstacles people face when trying to recycle materials? How might these obstacles be overcome in your community?

4. You are given the choice of shopping either from a modern catalog or from one from the 1930s. Which catalog offers more goods? Which catalog would you choose and why?

Exploring Further

Stephen Fenichell, *Plastic: The Making of a Synthetic Century.* New York: HarperCollins, 1997.
> This engaging historical account tells how the development of plastics has had a profound effect on our society.

http://www.invent.org
> Home page for the National Inventors Hall of Fame. Use this site to find additional information about many of the people referred to in this chapter. To search by name, follow the links to www.invent.org/book/index.html.

http://www.steelnet.org/index.html
> Web site for the Steel Manufacturers Association.

http://trc.dfrc.nasa.gov/gallery/photo/x-33/index.html
> A photograph gallery showing the development of the X-33 prototype. You might also be interested in exploring the Web site of the Marshall Space Flight Center at http://www1.msfc.nasa.gov.

Material Resources
Visit The Chemistry Place at:
www.aw.com/chemplace

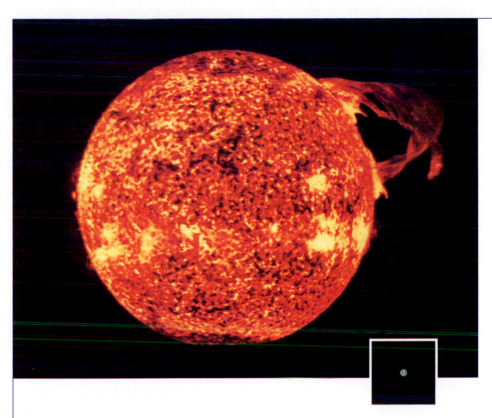

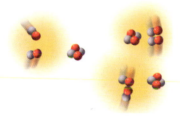

19

Energy Resources

Managing for the Present and the Future

In our search for energy sources, it is only natural to look to the sun. The warmth you feel from the sun, however, isn't so much because the sun is hot. Indeed, the sun's surface temperature of 6000°C is no hotter than the flame of some welding torches. Rather, the primary reason you are warmed by the sun is because it is so big. Look at the lower right corner of this photograph, and you'll see a small blue dot. This dot (painted on the photograph) is the approximate relative size of the Earth. Clearly, when we think about possible energy sources, the enormous energy wealth at the heart of our solar system demands our utmost attention.

Indeed, whenever we burn plant material, we are releasing solar energy that was captured through photosynthesis. Solar energy is also released when we burn fossil fuels, which are the decayed remains of plants and plant-eating animals. Electricity-producing hydroelectric dams depend on the water cycle, which is driven by solar radiation. Windmills harness wind to produce electricity and also to pump water, and winds exist because the sun heats different parts of the planet at different rates. Photovoltaic cells directly generate electric currents when exposed to solar radiation.

In addition, a number of energy sources that do not depend on the sun are now available, including nuclear, geothermal, and tidal energies.

All usable energy, whatever the source, is delivered to us either in the form of fuel or in the form of electricity. The wonder of electricity is the ease with which it can be transmitted to many sites. This property makes electricity one of our most convenient forms of energy. To produce electricity, however, requires the input of some other source of energy, such as the burning of a fuel. We therefore begin this chapter with a brief overview of how electricity is generated and how its consumption is measured.

19.1 Electricity Is a Convenient Form of Energy

Electricity is the flow of electric charge. It is generated in a metal wire when the wire is forced to move through a magnetic field and the field causes the electrons of the metallic bonds to flow. By coiling the wire into many loops and rotating the loops through powerful magnetic fields, power companies are able to generate enough electricity to light up cities. Figure 19.1 illustrates such an *electric generator*.

The many loops of wire wrapped around an iron core form what is known as an *armature*. The armature is connected to an assembly of paddle wheels called a *turbine*. Energy from wind or falling water can cause the turbine and thus the armature to rotate, but most commercial turbines are *steam turbines*, meaning they are driven by steam. To boil the water to create steam requires an energy source, which is usually a fossil fuel or a nuclear fuel.

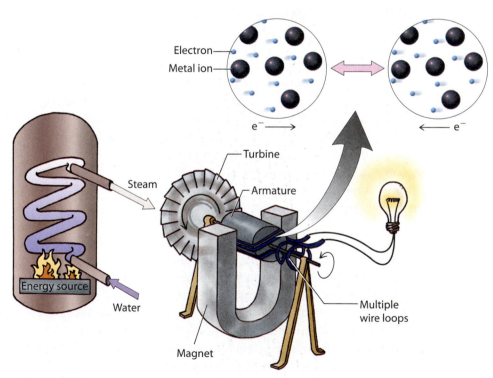

Figure 19.1
Basic anatomy of an electric generator. Electricity is generated in a looped wire as the wire rotates through a magnetic field. This motion causes electrons in the wire to slosh back and forth. Because the electrons are moving, they possess kinetic energy and so have the capacity to do work.

Still under development are more efficient *gas turbines*, which are driven not by steam but by the hot combustion products of vaporized alcohols and lightweight hydrocarbons.

Concept Check ✓

Is electricity more accurately thought of as a source of energy or as a carrier of energy?

Was this your answer? Electricity is energy that is readily transported through wires. In this sense, it can be thought of as a carrier of energy. The energy of electricity is used to run a light bulb, true, but the source of this energy is not the electricity. Rather, the electricity is merely delivering the energy that was generated by some electric generator, which received energy from some nonelectrical source, such as a fossil fuel or a waterfall.

What's a Watt?

Power is defined as the rate at which electrical energy (or any other form of energy) is expended. Power is measured in watts, where 1 **watt** is equal to 1 joule per second:

$$1 \text{ watt} = \frac{1 \text{ joule}}{1 \text{ second}}$$

A lot of watts means that a lot of energy is being consumed quickly. A 100-watt lightbulb, for example, consumes 100 joules of energy each second, and a 40-watt bulb consumes only 40 joules each second.

The typical U.S. household consumes electrical energy at an average rate of about 800 joules per second, or 800 watts. For a small city of 100,000 households, this adds up to a rate of 80,000,000 watts, or 80 megawatts (MW). This, however, is just the average rate of energy consumption. To meet peak demands, electric power plants must sometimes quadruple their average output. This is why small cities require electric power plants that can produce energy at a power rating of 300 megawatts or higher.

These needs are easily met by present-day power plants. A typical coal-fired plant produces on the order of 500 megawatts of electrical energy, a large nuclear plant can produce on the order of 1500 megawatts, and a large hydroelectric dam can produce more than 10,000 megawatts.

One factor affecting the cost of electricity is the source of the electrical energy. Fossil fuels and nuclear fuels produce hundreds of megawatts of power from a single power plant and are thus able to serve large areas, including cities (Figure 19.2). Therefore economies of scale make electricity from fossil fuels and nuclear fuels relatively inexpensive. Electricity from sources that are

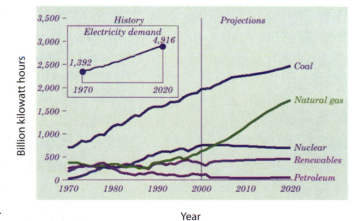

Figure 19.2
Coal and natural gas are expected to be the predominant energy sources for the production of electricity in the United States.

Calculation Corner: Kilowatt-Hours

Take a careful look at your next electric bill. Note that you pay for electrical energy in units of kilowatt-hours. A kilowatt is 1000 watts, and a **kilowatt-hour** (kWh) represents the amount of energy consumed in 1 hour at a rate of 1 kilowatt (1000 joules per second). Therefore, if electrical energy costs 15 cents per kilowatt-hour, a light bulb that has a power rating of 100 watts (0.1 kilowatt) can be run for 10 hours at a cost of 15 cents.

The calculation to arrive at this cost is done as follows:

Step 1 Calculate the total amount of energy consumed in kilowatt-hours:

power in kilowatts × time = energy consumed in kilowatt-hours

$$0.1 \text{ kW} \times 10 \text{ h} = 1 \text{ kWh}$$

Step 2 Calculate the cost of consuming this much energy:

kilowatt-hours consumed × price per kilowatt-hour = cost

$$1 \text{ kWh} \times \frac{15¢}{1 \text{ kWh}} = 15¢$$

Alternatively, ten 100-watt bulbs can also be run for 1 hour at a cost of 15 cents.

Step 1 Calculate the total amount of energy consumed in kilowatt-hours:

power in kilowatts × time = energy consumed in kilowatt-hours

$$10 \text{ bulbs} \times \frac{0.1 \text{ kW}}{1 \text{ bulb}} \times 1 \text{ h} = 1 \text{ kWh}$$

Step 2 Calculate the cost of consuming this much energy:

kilowatt-hours consumed × price per kilowatt-hour = cost

$$1 \text{ kWh} \times \frac{15¢}{1 \text{ kWh}} = 15¢$$

Your Turn

How much does it cost to operate ten 100-watt light bulbs for 10 hours at a cost of 15 cents per kilowatt-hour?

not so easily centralized, such as wind energy, have traditionally been more expensive. However, this gap has narrowed significantly as technology has improved.

19.2 Fossil Fuels Are a Widely Used but Limited Energy Source

Our fossil-fuel supplies were created hundreds of millions of years ago when ancient plants and animals died and became buried in swamps, lakes, and seabeds. These supplies cannot be replaced after we have used them up, which is why they are often referred to as *nonrenewable* energy sources. Estimates vary on the world's supply of fossil fuels, which include coal, petroleum, and natural gas. Even the most conservative estimates, however, show that, at present consumption rates, recoverable petroleum reserves will be depleted within 100 years and recoverable natural-gas reserves within 150 years. As depletion approaches, these valuable commodities will become too costly. Coal reserves, on the other hand, are more abundant and may last another 300 years. Worldwide, nearly all our present energy needs are met by fossil fuels—38 percent from petroleum, about 30 percent from coal, and about 20 percent from natural gas.

Why are fossil fuels so popular? First, they are readily available in many regions of the world, as shown in Figure 19.3. Second, gram for gram, they store much more chemical energy than other combustible fuels, such as wood. Third, they are portable and make excellent fuels for vehicles.

Gases emitted when fossil fuels are burned have negative environmental effects. As discussed in Section 17.2, sulfur and nitrogen oxides lead to

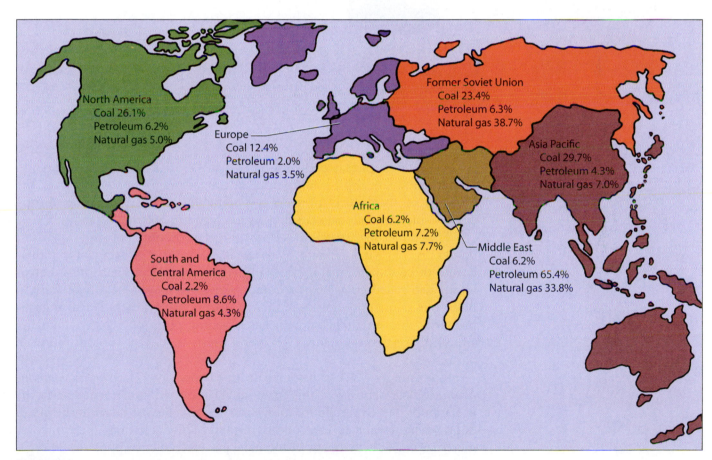

Figure 19.3

Fossil-fuel deposits are not distributed evenly throughout the world. For instance, 65 percent of the world's recoverable petroleum deposits are in the Middle East, along with 34 percent of recoverable natural-gas deposits. North America is relatively poor in petroleum and natural gas but has a bit more than one-fourth of the world's supply of coal.

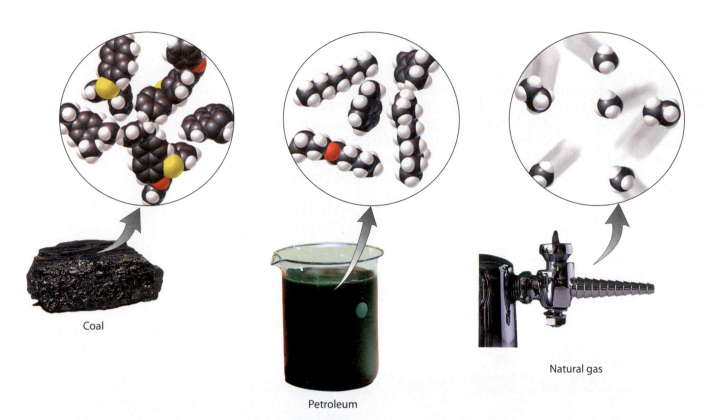

Coal

Petroleum

Natural gas

Figure 19.4
Typical molecular structures of coal, petroleum, and natural gas.

acid rain. These gases, along with particulates created from the combustion of fossil fuels, are also a leading cause of urban smog. On a global level, the burning of fossil fuels presents a potentially more devastating disturbance—increased global warming, as discussed in Section 17.4.

The molecular structure of fossil fuels accounts for their physical phases. As shown in Figure 19.4, **coal** is a solid consisting of a tightly bound three-dimensional network of hydrocarbon chains and rings. **Petroleum**, also called *crude oil*, is a liquid mixture of loosely held hydrocarbon molecules containing not more than 30 carbon atoms each. **Natural gas** is primarily methane, CH_4, which has a boiling point of $-163°C$. Smaller amounts of gaseous ethane, C_2H_6, and propane, C_3H_8, are also found in natural gas.

Interestingly, there is a fourth form of fossil fuel, known as *methane hydrate*. Most deposits of this material are located kilometers beneath the ocean floor, but in certain locations, the deposits lie just beneath the ocean floor, where researchers can collect samples. Methane hydrate is a white, icy material, and it is made up of methane gas molecules trapped inside cages of frozen water, as shown in Figure 19.5. Most experts agree that the supply of methane hydrate is at least twice that of coal, petroleum, and natural gas *combined*. Because methane hydrate is much more difficult to collect than the other forms of fossil fuel, however, this energy source will likely remain untapped for quite some time.

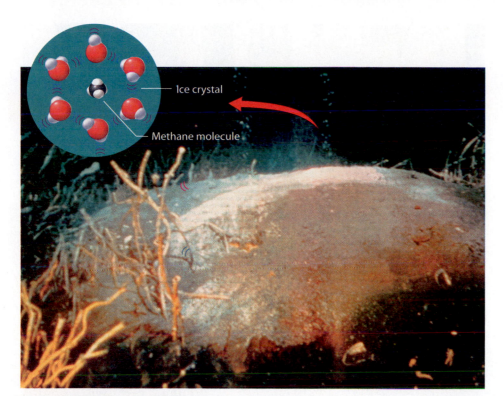

Ice crystal

Methane molecule

(a)

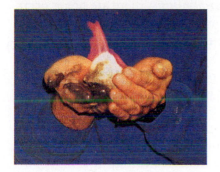

Figure 19.5
(a) Bubbles of methane gas escaping from a deposit of methane hydrate found on the ocean floor. (b) Once brought to the surface, methane hydrate crystals quickly decompose as the ice melts and the methane gas—seen here burning—is released.

Coal Is the Filthiest Fossil Fuel

Coal is the most abundant nonrenewable energy source on the planet. Worldwide, the amount of energy available from coal is estimated to be about ten times greater than the amount available from all petroleum and natural-gas reserves combined. Along with being the most abundant fossil fuel, coal is also the filthiest because it contains large amounts of such impurities as sulfur, toxic heavy metals, and radioactive isotopes. Burning coal is therefore one of the quickest ways to introduce a variety of pollutants into the air. More than half of the sulfur dioxide and about 30 percent of the nitrogen oxides released into the atmosphere by humans come from the combustion of coal. As with other fossil fuels, the combustion of coal also produces large amounts of carbon dioxide.

Extracting coal from the ground is also harmful to human health and to the environment. Mining coal from underground mines, as the workers in Figure 19.6 did, is a dangerous job with many health hazards. Local

Figure 19.6
Coal miners in Pennsylvania in the 1930s.

Figure 19.7
Pulverized coal floats on water, but impurities sink. This difference in densities allows for a simple and efficient means of purifying coal before it is burned.

waterways are contaminated as they receive effluents from the mines. These effluents tend to be very acidic because of the sulfuric acid that forms from the oxidation of such waste minerals as iron sulfide, FeS_2. When coal is mined from the surface, a process called *strip mining*, there are fewer occupational hazards, but the tradeoff is that whole ecosystems are destroyed. Although strip mining is initially cheaper than digging, the cost of restoring the ecosystem can be prohibitive.

Despite these drawbacks, coal reserves remain abundant, and so close to 60 percent of the electric power generated in the United States comes from coal-fired plants.

There are several ways to make burning coal a cleaner process. The coal can be purified before it is burned, pollutants can be filtered out after combustion, or the combustion process can be modified so that it is more efficient and fewer pollutants are produced.

Purifying coal prior to combustion usually involves pulverizing the coal and mixing it with detergents and water, as is demonstrated in Figure 19.7. The density of coal is lower than the density of any of its mineral impurities. A proper adjustment of the solution's density therefore allows the coal to float to the surface, where it is skimmed off, while the impurities sink to the bottom. This method of purification, called *flotation*, adds further cost to the coal, which is already expensive because of the high mining and shipping costs. Nonetheless, flotation is successful at removing most of the coal's mineral content, including up to 90 percent of the iron sulfide. Large quantities of sulfur still remain chemically locked in the coal, however. This sulfur can be removed only after combustion.

Most coal-fired utilities today remove any sulfur dioxide created when coal is burned by directing gaseous effluents into a *scrubber*, illustrated in Figure 19.8. Within the scrubber, the effluents come in contact with a slurry of limestone, $CaCO_3$. Up to 90 percent of the sulfur dioxide is removed as it reacts with the limestone to form solid calcium sulfate, $CaSO_4$, which is readily collected and sent to a solid-waste disposal site.

As a result of flotation and scrubbing technologies, sulfur dioxide emissions have decreased by 30 percent over the past 20 years, despite a 50 percent increase in the use of coal. This is promising, but given our dependence on coal, there is still plenty of room for improvement. For example, nitrogen oxide emissions have remained relatively constant. Also, equipping a coal-fired power plant with a scrubber reduces the efficiency at which the coal energy is converted to electrical energy from 37 percent to 34 percent.

Fewer pollutants and greater efficiencies are achieved by redesigning the combustion process. In conventional power plants, pulverized coal is burned in a combustion chamber, where the heat vaporizes water in steam tubes. Newer chambers send jets of compressed air into the pulverized coal, and as a result the coal becomes suspended as it burns. This allows the coal to burn more efficiently and also provides a better transfer of heat from the coal to the steam tubes. Because air-suspended coal burns more efficiently, lower temperatures can be maintained, resulting in a tenfold decrease in nitrogen oxide emissions. (Recall from Chapter 17 that nitrogen oxides form as atmospheric nitrogen and oxygen are subjected to extreme temperatures.) Air-suspended coal can be burned in the presence of limestone, which removes more than 90 percent of the sulfur dioxide as it forms, thus avoiding the need for a scrubber. Cleaned of sulfur and nitrogen oxides, the hot, pressurized effluent can be directed into a gas turbine that generates electricity in tandem with the steam turbine. Overall, this system converts coal energy to electrical energy with an efficiency of about 42 percent.

More efficient combustion is one possible future for coal. Another and more hopeful future involves treating coal with pressurized steam and oxygen, a process that produces clean-burning fuel gases such as hydrogen, H_2. These futures for coal, however, are only short-term. Like all other fossil fuels, coal is a nonrenewable energy source, and it will not be with us for the long haul if we continue to burn it.

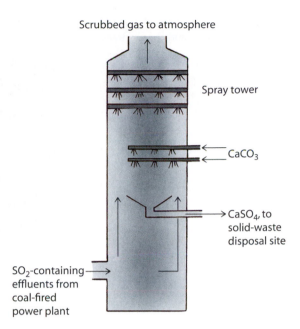

Figure 19.8
A scrubber is used to remove most of the sulfur dioxide created when coal is burned.

Concept Check

What is the major advantage of using coal as an energy source?

Was this your answer? Coal is relatively abundant, which means it can be used as an energy source for many years to come.

Petroleum Is the King of Fossil Fuels

The energy content of the coal reserves in the United States far exceeds that of the fossil-fuel reserves of all Middle East nations combined. Why, then, does the United States import so much petroleum from these nations? The immediate answer is because petroleum is a liquid, and liquids are far more convenient to handle in bulk quantities.

Figure 19.9
Petroleum is easily and cheaply transported because it is a liquid.

Figure 19.9
Petroleum is easily and cheaply transported because it is a liquid.

Consider, for example, that because petroleum is a liquid, it is easy to extract from the Earth. Punch a hole in the ground in the right place, and up it comes—no underground mining necessary. Being a liquid, petroleum is also easy to transport. Huge oil tankers easily load and unload liquid petroleum, as shown in Figure 19.9. On land, petroleum can be pumped vast distances via a network of pipelines. Coal, on the other hand, must be dug out of the ground with heavy machinery and shipped as solid cargo, usually aboard trucks or freight trains.

Petroleum is also versatile. It contains all the commercially important hydrocarbons, such as those that make up gasoline, diesel fuel, jet fuel, motor oil, heating oil, tar, and even natural gas. Using fractional distillation (Section 12.1), oil refineries can convert one type of petroleum hydrocarbon to another, thereby tailoring their output to fit consumer demand. Furthermore, petroleum contains much less sulfur than does coal and so produces less sulfur dioxide when burned. So, despite its vast coal reserves, the United States has a royal thirst for petroleum, the king of fossil fuels, consuming about 17 million barrels each day. This is about 10 liters per U.S. citizen per day.

Of the 17 million barrels of petroleum consumed each day in the United States, 16 million is burned for energy. The remaining 1 million is used to provide raw material for the production of organic chemicals and polymers. Thus only one-seventeenth of the hydrocarbons consumed daily goes into useful materials. The rest is burned for energy and ends up as heat and smoke.

Natural Gas Is the Purest Fossil Fuel

Natural gas is a component of petroleum, but there are also vast deposits of free natural gas in underground geologic formations. The natural gas in these deposits can be collected and stored in tanks like the ones shown in Figure 19.10.

Natural gas burns more cleanly than petroleum and much more cleanly than coal. This purest of fossil fuels contains negligible quantities of sulfur; hence, insignificant amounts of sulfur dioxide are produced. Also, because natural gas burns at lower temperatures, only small amounts of nitrogen oxides are released. Perhaps most important, however, is that generating energy from natural gas produces less carbon dioxide—about half as much

Figure 19.10
Natural gas is stored in large spherical tanks because this shape holds the greatest volume for a given amount of building material.

as is produced from burning coal. Because it is a gas, however, this fossil fuel is cumbersome to isolate and transport. Also, its natural abundance is not much greater than that of petroleum. Therefore, we cannot rely on natural gas for meeting our long-term energy needs. Experts suggest, however, that switching to natural gas as much as possible may buy time and protect the environment until technologies for nonfossil energy sources are perfected and made competitive.

Another advantage of natural gas is that it can be used to generate electricity with great efficiency. With a steam turbine, fossil fuel is burned in a boiler to produce steam, which runs the electricity-generating turbine, as shown in Figure 19.1. Such a system burning natural gas to boil the water produces electricity with an efficiency of about 36 percent, comparable to the 34 percent efficiency attained using coal. However, as mentioned in Section 19.1, the latest development in turbine technology is the *gas turbine*, in which the step of converting water to steam is eliminated. Instead, the hot combustion products of natural gas are what drive the paddle wheels of the turbine. In addition, the exhaust from the gas turbine is sufficiently hot to convert water to steam, which is then directed to an adjacent steam turbine to generate even more electricity. This system of using a gas turbine in tandem with a steam turbine can produce electricity with an efficiency as high as 47 percent. Even higher efficiencies can be attained by chemically converting natural gas to molecular hydrogen, H_2, which can be used to generate electricity in fuel cells, discussed in Section 11.3.

There are two types of natural gas supplied to consumers, one containing primarily methane, CH_4, and the other containing primarily propane, C_3H_8. Methane is lighter than air, which makes it relatively safe to deliver through a network of pipes extending throughout a municipality. If a leak occurs, the methane merely rises skyward, minimizing the fire hazard. Propane is heavier than air and readily liquefies under pressure. Because of these properties, propane is best stored as a liquid in pressurized tanks like the one shown in Figure 19.11. Propane tanks are used in areas not connected to municipal gas lines and require periodic filling.

Figure 19.11
If you use natural gas and it is stored outside your home in a pressurized tank, you are using propane. If there is no tank outside your home, you are using methane.

Concept Check ✔

> Of the three forms of fossil fuels, which is most abundant worldwide? Which burns most cleanly? Which is the easiest to transport?

Were these your answers? Coal is the most abundant. Because it contains few impurities, natural gas burns most cleanly. Because it is a liquid, petroleum is easiest to transport.

19.3   There Are Two Forms of Nuclear Energy

Figure 19.12 summarizes the two forms of nuclear energy. One form is nuclear fission, which involves the splitting apart of large atomic nuclei, such as uranium or plutonium. The other is nuclear fusion, which involves the combining of two small atomic nuclei, such as deuterium and tritium, into a single atomic nucleus, helium. All nuclear power plants to date use nuclear fission. These plants produce electrical energy without emitting any atmospheric pollutants. For a review of the concepts underlying nuclear fission and fusion, see Chapter 4. In this section we discuss some of the social and technological issues related to nuclear energy.

Figure 19.12
Nuclear fission involves the splitting apart of large atomic nuclei. Nuclear fusion involves the coming together of small nuclei.

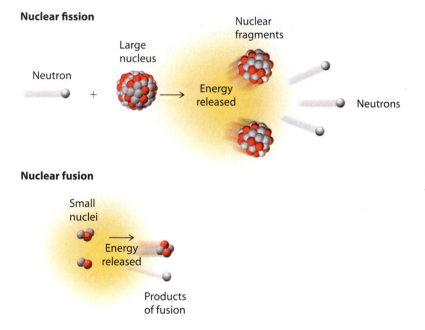

Nuclear Fission Generates Some of Our Electricity

Nuclear fission energy for the commercial production of electricity has been with us since the 1950s. In the United States, about 20 percent of all electrical energy now originates from 104 nuclear fission reactors situated throughout the country. Other countries also depend on nuclear fission energy, as is shown in Figure 19.13. Worldwide, there are about 435 nuclear reactors in operation and 36 currently under construction.

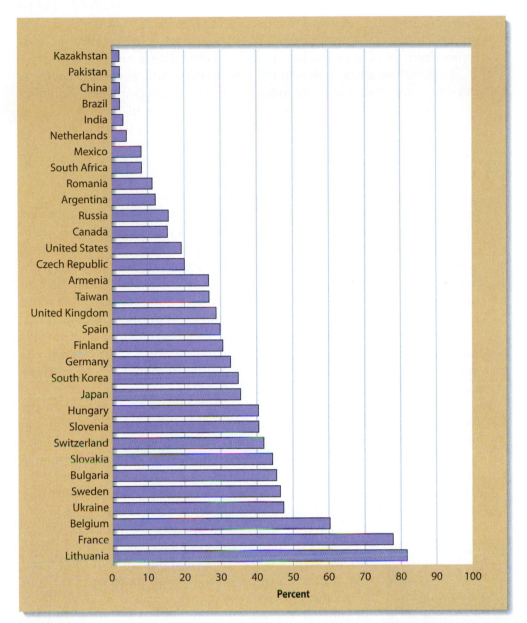

Figure 19.13
Percentage of electricity generated from nuclear fission reactors in selected countries.

Countries turning to nuclear fission energy have decreased their dependence on fossil fuels and have diminished their output of carbon dioxide, sulfur oxides, nitrogen oxides, heavy metals, airborne particulates, and other pollutants. Money that would have been spent on foreign oil payments has been saved. It is estimated, for example, that nuclear fission energy has saved the United States $150 billion in foreign oil payments.

Without an enormous and concerted conservation effort, the world energy demand is going to increase, especially in light of growing populations and the dire need for economic growth in developing countries. Should nuclear energy production come to a standstill, allowing fossil fuels to accommodate this increased energy demand? Or should we continue to operate existing nuclear power plants—and even build new ones—until the alternative sources of energy discussed later in this chapter become feasible on a large scale? Nuclear advocates suggest a fivefold increase in the number of nuclear power plants over the next 50 years. They argue that nuclear fission is an environmentally friendly alternative to increased dependence on fossil fuels.

In many countries, however, the public perception of nuclear energy is less than favorable. There are indeed formidable disadvantages, including the creation of radioactive wastes and the possibility of an accident that releases radioactive substances into the environment. In rebuttal, advocates point out that we cannot insist that nuclear fission energy be absolutely safe while at the same time accept tanker spills, global warming, acid rain, and coal-miner diseases.

So how much radioactive waste is there? According to the U.S. Department of Energy, there is on the order of 32,000 tons of spent nuclear fuel rods stored at reactor sites around the nation, and this amount continues to grow by about 2000 tons each year. The military is also a significant source of radioactive wastes. Holding tanks at the Hanford nuclear weapons plant in the State of Washington, for example, contain about 200 million liters of highly radioactive wastes. The general consensus among scientists is that our radioactive wastes are best handled by storing them in underground repositories located in geologically stable regions. Water seeping into such a repository, however, could greatly accelerate the corrosion of casks housing the radioactive wastes. Therefore an effective repository should also be relatively void of water and placed hundreds of meters above any water table.

To date, no long-term repositories are in operation anywhere in the world, primarily because few, if any, communities want such a repository in their "backyard." Furthermore, once a potential site gets chosen, extensive and time-consuming evaluations are necessary. For example, tests at a promising site beneath Yucca Mountain in Nevada, shown in Figure 19.14, have been going on since 1982 and are not expected to yield conclusive results until sometime after 2007. In 2002, however, President George W. Bush gave his approval to the site. If approval is upheld, a network of tunnels 150 kilometers in combined length would accommodate more than 80,000 tons of radioactive wastes. Signs of water seepage, however, raise concerns that this site is less than ideal. If so, what then? Because of the difficulty in finding locations and the costs involved, Yucca Mountain is currently the only site under consideration in the United States.

In addition to generating radioactive wastes, nuclear power plants pose the risk of having an accident in which radioactive material is released into the environment. The safety design of a nuclear power plant, however, has great bearing on the risks associated with generating nuclear fission energy. In 1979, a nuclear reactor at a facility on Three Mile Island, near Harrisburg, Pennsylvania, heated to the point that the reactor core began to melt. No appreciable radioactivity leaked into the environment because the core was housed in a containment building (shown in Figure 4.27). Seven years later, in 1986, a total meltdown occurred at the nuclear power plant shown in Figure 19.15, the Chernobyl plant in what is now Ukraine. Notably, the reactor core of the Chernobyl

Figure 19.14
Yucca Mountain in Nevada is a promising site for a permanent repository for nuclear wastes. Exploratory tunnels have already been drilled, and extensive tests are under way.

plant was not built and operated in accordance with internationally accepted nuclear safety principles. For example, the medium used to control the fission reactions was graphite, which loses its ability to control the reactions as the core temperature rises. Also, the reactor was not housed in a containment building. Because there was no containment building, large amounts of radiation escaped into the environment.

Risk analysis of all nuclear power plants operated according to internationally accepted safety standards indicates that one significant release of radioactive material from one of these plants can be expected every 200 years. Recent technological advances, however, hold the promise of lowering this rate considerably. New plant designs involve smaller reactors that generate between 155 and 600 megawatts of power rather than the 1500 megawatts that is the usual output of today's reactors. Smaller reactors are easier to manage and can be used in tandem to build a generating capacity suited to the community being served.

Significant advances have also been made in reactor safety. Early reactors rely on a series of active measures, such as water pumps, that come into play to keep the reactor core cool in the event of an accident. A major drawback is that these safety devices are subject to failure, thereby requiring backups and, in some cases, backups to the backups! The new reactor designs provide for what is called *passive stability*, in which natural processes, such as evaporation, are used to keep the reactor core cool. Furthermore, the core has a negative temperature coefficient, which means the reactor shuts itself down as its temperature rises owing to a number of physical effects, such as any swelling of the control rods.

The percentage of the world's electricity produced by fission reactors peaked in 1993 at 17 percent but slipped to 15 percent in 2000. The reason for this decline is that older plants have been shut down as repair costs have exceeded profit margins. Meanwhile, newer plants have not been

Figure 19.15
In 1986, a meltdown occurred at this nuclear power plant in Chernobyl, Ukraine. Because there was no containment building, large amounts of radioactive material were released into the environment. Three people died outright, and dozens more died from radiation sickness within a few weeks. Thousands who were exposed to high levels of radiation stand an increased risk of cancer. Today, 10,000 square kilometers of land remains contaminated with high levels of radiation.

Figure 19.16
Total worldwide electricity consumption projected through the year 2020.

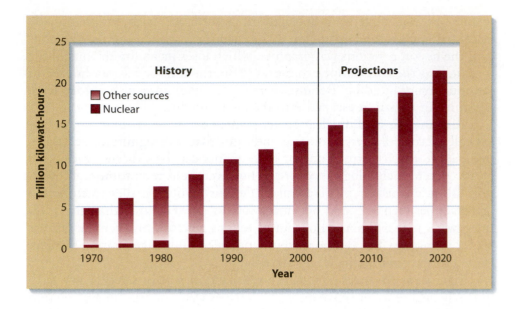

built, largely because of negative public sentiment about nuclear energy. In the United States, for example, from 1989 to 1999 the number of nuclear power plants in operation dropped from 114 to 104. The projection is that, without major efforts to replace aging plants, there will be only 52 U.S. plants in 2020. Worldwide, the International Atomic Energy Agency projects that by 2020, nuclear reactors will account for as little as 9 percent of the world's electricity generation. This poses a tough dilemma because projections are that by 2020 the world's electrical needs will have increased by about 75 percent, as indicated in Figure 19.16.

If nuclear fission energy is phased out, what will replace it? This question is of particular concern when viewed in the context of the Kyoto Protocol, an agreement among some 160 nations that by 2012 they will have reduced their emissions of greenhouse gases back to or below 1990 levels.

Concept Check ✓

What is the major disadvantage of nuclear fission as an energy source?

Was this your answer? Nuclear fission reactors generate large amounts of radioactive wastes that require permanent large-scale storage facilities.

Nuclear Fusion Is a Potential Source of Clean Energy

A potential major source of energy for the mid to late 21st century is nuclear fusion. In today's experimental fusion reactors, deuterium and tritium atoms (both isotopes of hydrogen) fuse to create helium and fast-flying neutrons. The neutrons escape from the reaction chamber, carrying with them vast amounts of kinetic energy.

Two positively charged hydrogen nuclei do not just willingly come together, however, because there is a powerful electric force of repulsion to be overcome. In Section 4.10, we discussed one possible method for fusing hydrogen nuclei: laser confinement, which involves dropping pellets of deuterium and tritium into the crossfire of powerful lasers that squeeze the fuel to a density 20 times that of lead. Another technique is magnetic confinement, which involves bringing nuclei to star-hot temperatures (about 350 million degrees Celsius) so that they are moving so fast that their inertia brings them into contact with one another and they fuse. Because the star-hot deuterium and tritium fuels are fully ionized, they can be contained by strong magnetic fields. For both systems, energetic neutrons escape into a surrounding "blanket" that absorbs heat that can be used to create either steam or hot ionized gases for generating electricity.

Fusion has already been achieved in several devices, but not beyond the break-even point, where the amount of energy produced is the same as the amount consumed. Much basic research is still required and is the focus of a number of international collaborative efforts. Research projects include the Tokamak Fusion Test Reactor (shown in Figure 19.17a) and the Large Helical Device, shown in Figure 19.17b. For the latest information on these and other fusion projects, explore the Web sites listed at the end of this chapter.

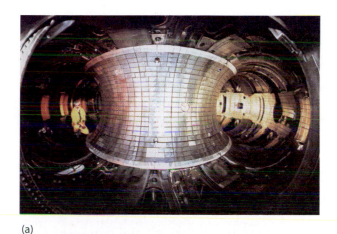

(a)

Small-scale tabletop nuclear fusion devices are routinely used as a source of neutron radiation. Because of their design, however, these devices consume more energy than they release.

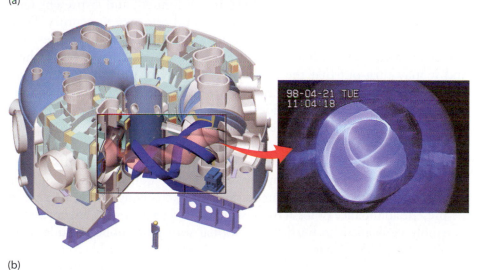

(b)

Figure 19.17
(a) An interior view of the Tokamak Fusion Test Reactor at the Princeton Plasma Physics Laboratory. Magnetic fields confine a fast-moving plasma to a circular path. At a high enough temperature, the atomic nuclei in the confined plasma fuse to produce energy. (b) A fusion reactor called the Large Helical Device is being built and tested by researchers in Japan. These coils create a magnetic field that contains the hot ionized gas within which nuclear fusion takes place.

Fusion energy offers a number of advantages over all other energy sources, including fission. Fusion reactors do not produce air pollutants that contribute to global warming or acid rain. The deuterium fuel they use is available in essentially unlimited supply from seawater, and tritium can be generated on site as part of the fusion process. Lastly, the amounts of radioactive wastes produced are much smaller than the amounts produced by fission reactors.

One important possible drawback for fusion power is that the cost of building fusion facilities may well exceed the value of returns. Scientists already know that a fission facility generally needs to be operating for about 20 years before investments are recovered. Because fusion power is necessarily more complex, the time span for financial recovery is likely to be much longer. Thus, only the most developed nations may be able to afford nuclear fusion power plants—and not so many at that. Meanwhile, demographers tell us that the greatest energy needs in the future will be in rapidly growing developing nations. So despite the potential technical advantages of fusion power, on a social level its development may widen the gap between the haves and have nots of this world.

Concept Check ✔

How many nuclear fusion plants are generating electricity for communities today?

Was this your answer? Because there are many technical hurdles yet to be overcome in building fusion plants, there are currently none in operation.

19.4   What Are Sustainable Energy Sources?

The amounts of fossil fuels available to us are limited, and at present rates of consumption they will not last us far beyond our present century. Fuels for nuclear fission turn into significant quantities of radioactive wastes, and sooner or later, society's holding capacity for these wastes will be reached. And it may turn out that the technical feasibility of nuclear fusion is not decades but rather centuries away. What we ultimately need, therefore, are *sustainable* energy sources. The ideal sustainable energy source is one that is not only inexhaustible but also environmentally benign.

Because no one sustainable source can meet the world's total energy needs, the development of a variety of technologies provides our greatest hope. Switching to sustainable energy sources will require commitment from the general public. Perhaps the largest obstacle to changing public attitudes is the present abundance of fossil fuels, which are so packed with energy and so incredibly convenient to burn. In a national survey of public attitudes conducted by the U.S. Department of Energy, however, sustainable sources were the most desirable form of energy by far. But can people put their money where their hearts are? Fortunately, technologies are progressing rapidly, and sustainable energy sources may soon actually cost the consumer less! This is a critical point because in a market economy, it is the dollar that speaks. Let's take a look at what some of the major alternative sustainable energy sources have to offer as well as some of their potential drawbacks.

Concept Check ✔

> What is a sustainable energy source?

Was this your answer? A sustainable energy source is one that has the potential for being available for many years to come and is also environmentally benign.

19.5   Water Can Be Used to Generate Electricity

Water power originates from one of three sources: the sun, the Earth's hot interior, or the moon. As you read the following descriptions of methods we use to harness energy from water, trace the energy to one of these three sources.

Hydroelectric Power Comes from the Kinetic Energy of Flowing Water

Water flowing through a hydroelectric dam rotates a turbine that spins an electric generator to produce electricity, as Figure 19.18 illustrates. In a modern facility, the efficiency with which kinetic energy is converted to electrical energy can be as high as 95 percent, which translates to low costs for the consumer. Hydroelectric power is clean, producing no pollutants such as carbon dioxide, sulfur dioxide, and other wastes. The source of this power is the sun, which transports water to mountain altitudes by way of the hydroglogic cycle. Hydroelectric power is the most widely used sustainable energy source in the United States, supplying about 10 percent of our electricity needs. In developing countries, hydroelectric power supplies about 30 percent of all electricity needs.

There is great potential for increasing the output of hydroelectric power, but not by building more dams. Only 2400 of the 80,000 existing dams in the United States are used to generate power. Many of these "untapped" dams could be fitted with turbines and generators. Furthermore, most hydroelectric dams in the United States were built in the 1940s,

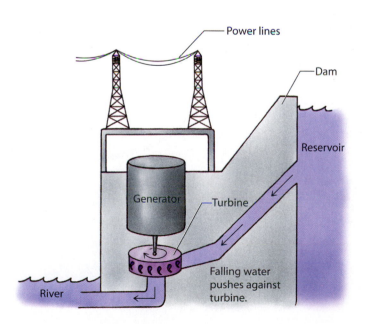

Figure 19.18
An overview of a hydroelectric dam.

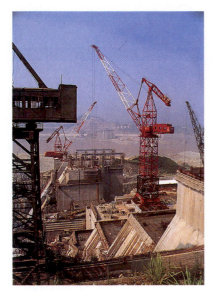

Figure 19.19
The Three Gorges Dam under construction in China. The Chinese government estimates it will spend about $25 billion to build what will be the world's largest hydroelectric project. When completed in 2009, the dam will extend 2.2 kilometers across, creating a reservoir that will be 600 kilometers long and will have displaced close to 2 million people from their homes. The capacity of the dam will be on the order of 18,000 megawatts. This combined with another 82,000 megawatts from other current hydroelectric projects will provide an estimated 30 percent of the country's electricity.

when equipment was not as efficient as it is today. New technologies can allow these older plants to operate more efficiently, producing more electricity. The U.S. Department of Energy estimates that a 1 percent improvement in the efficiency of existing U.S. hydropower plants would produce enough power to supply 283,000 households.

Hydroelectric power plants may be pollutant-free, but for the local environment around the dam, there are surely consequences (Figure 19.19). Fish and wildlife habitats are significantly affected. One tragic consequence of some dams is that they prevent fish from reaching their spawning grounds; therefore, fish populations decline. To address this concern, many dams have been retrofitted with fish "ladders" that are supposed to encourage upstream migration to spawning grounds. The success of these ladders has been limited, however. Reservoirs created by dams tend to fill with silt, affecting water quality and limiting the life-span of the dam. Furthermore, dams detract from the natural beauty of a river, and in many cases, valuable downstream cropland is lost or disrupted. Dams also need to be well maintained and inspected regularly to minimize the potential of a break, which would lead to catastrophic flooding.

Temperature Differences in the Ocean Can Generate Electricity

There is always a difference between the temperature of surface seawater (warmer) and the temperature of deep seawater (colder). A process known as *ocean thermal energy conversion* (OTEC) exploits this difference to produce electricity. As shown in Figure 19.20, warm surface water is used to boil a liquid that has a low boiling point, such as ammonia. The resulting high-pressure vapor pushes against the turbine, and the movement of the

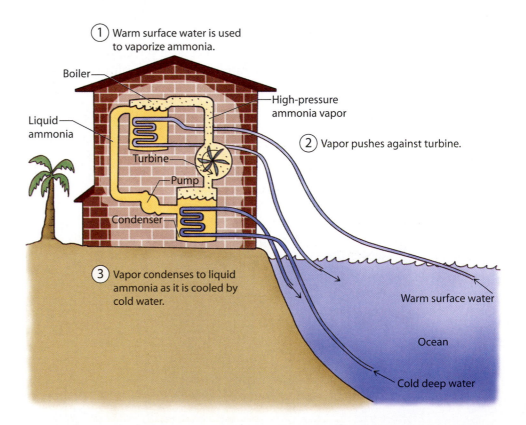

Figure 19.20
An ocean thermal energy conversion operation.

(a)

(b)

Figure 19.21
(a) An aerial view of The Natural Energy Laboratory of Hawaii—the world's first OTEC facility—located oceanside of the Kona-Kailua airport. (b) The vertical axis turbine of the OTEC facility. The vapor-to-liquid heat exchanger is seen to the right.

turbine generates electricity. After passing through the turbine, the vapor enters a condenser, where it is exposed to pipes containing cold water pumped up from great ocean depths. As its temperature drops, the vapor condenses to a liquid, which is recycled through the system.

OTEC is limited to regions where differences between ocean surface temperatures and deep-water temperatures are greatest. Fleets of floating offshore rigs have been proposed, with the electricity generated used to produce transportable hydrogen fuel made from seawater. As Section 19.8 discusses, hydrogen is the ultimate clean fuel—the only product of its combustion is water. Onshore OTEC plants are best suited for islands, such as Hawaii, Guam, and Puerto Rico, where deep waters are relatively close to shore. The world's first OTEC plant, shown in Figure 19.21, has been running in Hawaii since 1990, producing about 210 kilowatts of electricity. Most of this electricity is used by a local aquaculture industry, which uses the nutrient-rich deep ocean water piped by the OTEC unit to breed specialty fishes and crustaceans for the U.S. and Japanese markets. The piped cold water is also used to air-condition the OTEC offices and laboratories!

Geothermal Energy Comes from the Earth's Interior

The Earth's interior is quite warm because of radioactive decay and gravitational pressures. In some areas, the heat comes relatively close to the Earth's surface. When this heat pokes through, we see it as lava from a volcano or steam from a geyser. This is *geothermal energy*, and it can be tapped for our benefit. Figure 19.22 on page 638 shows some areas in the United States that have geothermal activity.

Hydrothermal energy is produced by pumping naturally occurring hot water or steam from the ground. It is the predominate form of geothermal energy now being used commercially for electricity generation. Californians currently receive about 1000 megawatts from hydrothermal power plants operating in a region known as the Geysers, a large steam reservoir north of San Francisco. There are 45 hydrothermal power plants in the United States, and many more in Italy, New Zealand, and Iceland. The

Figure 19.22
Regions of the United States where the prospects for geothermal energy are greatest.

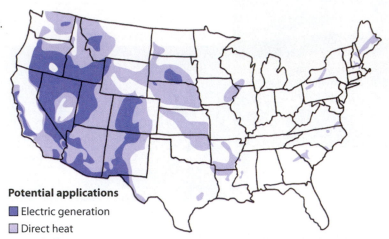

Potential applications
■ Electric generation
□ Direct heat

present generating capacity of all hydrothermal power plants worldwide is about 7000 megawatts, and one installation is shown in Figure 19.23.

Hydrothermal plants produce electric power at a cost competitive with the cost of power from fossil fuels. Besides generating electricity, hydrothermal energy is used directly to heat buildings. Across the United States, geothermal hot-water reservoirs are much more common than geothermal steam reservoirs. Most of the untapped hot-water reservoirs are in California, Nevada, Utah, and New Mexico. The temperatures of these reservoirs are not hot enough to drive steam turbines efficiently, but the water is used to boil a secondary fluid, such as butane, whose vapors then drive gas turbines.

Hot dry rock energy, another form of geothermal energy, involves using hydraulic pressure to open up a large reservoir deep underground. Liquid water is injected into the reservoir, heated by the hot rock, and then brought back to the surface as steam to generate electricity. One test facility near Los Alamos, New Mexico, consists of wells drilled to a depth of 4 kilometers.

Figure 19.23
About 50 percent of the electricity generated in Iceland is from geothermal sources. This is the Blue Lagoon, a warm pool created from the effluent of the hydrothermal power plant visible in the background.

Geopressured brine, yet another form of geothermal energy, is extracted warm subterranean brine water that is high in dissolved methane. Deposits of this brine are found primarily along the coast of the Gulf of Mexico. The methane can be separated on site and then used to heat the brine. Both these technologies are still largely developmental.

Geothermal energy is not without disadvantages. Few gaseous pollutants are emitted, but one of those pollutants is hydrogen sulfide, H_2S, which smells of rotten eggs even at low concentrations. Also, water coming from inside the Earth is often several times more saline than ocean water. This salty water is highly corrosive, and its disposal is a problem. Furthermore, withdrawing water from geologically unstable regions may cause land to subside and possibly trigger earthquakes.

The Energy of Ocean Tides Can Be Harnessed

The moon's gravitational pull on our planet is uneven. The side of the Earth closest to the moon experiences the greatest pull, and the side farthest away experiences the weakest pull. The result of this uneven pull is a subtle, planet-wide elongation of our oceans. As the Earth spins underneath this elongation, Earth-bound observers notice a perpetual rise and fall of sea level. These are our *ocean tides*, and they can be harnessed for energy.

Tidal power is normally obtained from the filling and emptying of a bay or an estuary, which may be closed in by a dam. When tidal waters flow through the dam (in either direction), they cause a paddle wheel or turbine to rotate, and this motion generates electricity.

The prospects for using tidal power on a large scale are not good. In order to be effective, the tides must be relatively great. This severely limits the number of potential sites worldwide. Furthermore, many of these potential sites are praised for their natural beauty. In such cases, public opposition to building a power plant would be strong. Nonetheless, tidal power is an option some communities may wish to consider. One successful site, located in Brittany, a region of France, produces about 240 megawatts of electricity.

Concept Check ✔

Of the four forms of water power discussed in this section, which are solar in origin? Which is lunar? Which is a result of the Earth's hot interior?

Was this your answer? The energy obtained from dams and from ocean thermal energy conversion is solar. Tidal energy is lunar, and geothermal energy is the result of the Earth's hot interior.

19.6 Biomass Is Chemical Energy

Plants use photosynthesis to convert radiant solar energy to chemical energy. This chemical energy comes in the form of the plant material itself—**biomass**. We can use the energy of biomass in two ways: process the

Does biomass sound reminiscent of fossil fuels? It should, because biomass is simply a fossil fuel without the fossil—dead plant material that has yet to turn into coal, petroleum, or natural gas. Just about anything you can do with fossil fuels, you can do with biomass except, of course, deplete it and create as much pollution.

biomass to produce transportable fuels, or burn the biomass at a properly equipped power plant to produce electricity. Biomass can be grown on demand in regions where water supplies are abundant. If biomass is produced at a sustainable rate, the carbon dioxide released when the biomass is burned balances the carbon dioxide consumed during photosynthesis. Biomass combustion products generally contain about one-third the ash produced by coal and about 30 times less sulfur.

Fuels Can Be Obtained from Biomass

The U.S. transportation industry is 97 percent dependent on petroleum and consumes 63 percent of all petroleum stockpiles. Fuels from biomass, such as the alcohols ethanol and methanol, are natural alternatives to petroleum-based fuels. In fact, both these alcohols have a higher octane rating than gasoline, which is why they are the preferred fuels of race-car drivers. They were also the preferred fuels of automotive pioneers Henry Ford and Joseph Diesel, who originally intended biofuels for their automobiles. Ethanol from fermented grains was about to be used by the automobile industry just prior to the passage of the 19th U.S. Constitutional amendment prohibiting its use. Prohibition allowed the petroleum industry to take over.

Today in the United States, government programs requiring the addition of 10 percent alcohol to gasoline are becoming widespread. You can recognize a service station selling *gasohol* by signs such as the one shown in Figure 19.24. Ethanol, also known as *grain alcohol*, can be prepared from the fermentation of food biomass—any grain will do, but simple carbohydrates such as sugars work best. Methanol, also known as *wood alcohol*, can be obtained by heating woody biomass in the absence of atmospheric oxygen. The methanol evaporates and is collected by distillation. The remaining undistilled plant matter produces a charcoal that is the equivalent of sulfur-free coal. Furthermore, some important industrial chemicals, such as acetic acid, acetone, and hydrocarbon-based oils, are also produced.

Ethanol is currently the biofuel used most widely in the United States, accounting for nearly 1 percent of the nation's transportation fuel needs. Production is based in the Midwest, where about 400 million bushels of corn is fermented into more than 4 billion liters of ethanol annually. The country that has the largest program is Brazil, where ethanol is produced from the fermentation of sugar cane. The Brazilian program was an innovative response to the oil crisis of the 1970s. By 1989, some 12 billion liters of ethanol replaced almost 200,000 barrels of gasoline a day in approximately 5 million Brazilian automobiles. The program created 700,000 jobs and significantly improved air quality in polluted metropolises, such as São Paulo and Rio de Janeiro.

Ethanol from fermentation is relatively expensive because of the great financial and environmental costs of growing food biomass, a process

Figure 19.24
Gasohol is gasoline containing an alcohol additive. The alcohol provides an octane boost, allowing an engine to run more efficiently with less pollution. If the alcohol is produced from biomass grown within a nation, there is the added benefit of a reduced dependence on foreign oil.

that requires vast amounts of water and fertilizer. Ethanol derived from petroleum is actually less expensive. (But only because crude oil prices are kept artificially low. If taxpayer subsidies, exemptions from paying for the environmental damage from mining and drilling, and the cost of military protection are factored in, the price of crude oil jumps from $20 a barrel to $89 a barrel. Assuming this latter figure, biomass-derived fuels are extremely cost-competitive.)

Methanol from biomass can also be costly because only relatively small amounts of it are produced through distillation. For this reason, most methanol today is produced from natural gas. The coal industry is even gearing up to produce methanol from its vast coal reserves. However, creating methanol from coal produces more pollution than is saved by burning the cleaner methanol.

On the bright side, technology is being developed for making fermentable sugars out of low-cost woody feed stocks. Ethanol produced from such inexpensive sources is currently cost-competitive with gasoline. If vast supplies of woody plants become available, the wholesale cost of ethanol could drop to as low as 36 cents per gallon.

Biomass Can Be Burned to Generate Electricity

Converting biomass to a transportable fuel is an extra step that decreases energy efficiency. Higher efficiencies are obtained by burning the bio-mass directly. In the United States, electricity produced from biomass has grown from 200 megawatts in the early 1980s to more than 8000 megawatts in the 1990s—a 4000 percent increase. Most biomass power is generated by paper companies and forest products companies, using wood and wood wastes as fuel. Municipalities are experimenting with solid-waste incinerators that provide electricity and waste disposal simultaneously. (On average, about 80 percent of the dry weight of municipal solid waste is combustible organic material.)

The traditional approach to generating electricity from biomass is to burn the biomass in a boiler where water is converted to steam used to drive a steam turbine. Efficiencies can be more than doubled by first converting the biomass to a gaseous fuel, which may be done by applying air and steam at high pressure. Alternatively, gaseous fuels are also produced by mixing the biomass with very hot sand as is done at the Vermont facility shown in Figure 19.25. The gaseous fuel is burned, and the hot combustion products are directed to a gas turbine that generates electricity. In addition, the exhaust gases from the turbine can be used to produce steam for industrial applications or for additional power generation. Whereas steam turbines have shown no improvements in efficiency since the late 1950s, gas turbines have improved continuously. Estimates indicate that a gas turbine running on gas from biomass would be cost-competitive with conventional coal, nuclear, and hydroelectric power both in industrialized countries and in developing countries.

Figure 19.25
The Vermont Biomass Gasification Project supplies more than 50 MW of electricity to the Burlington, Vermont, area. The electricity is generated by gas turbines powered by the combustion of a gaseous fuel mixture created as wood chips are mixed with very hot sand (1000°C).

Concept Check ✓

Concept Check ✓

What do biomass and fossil fuels have in common?

Was this your answer?   They both originate from solar energy.

19.7   Energy Can Be Harnessed from Sunlight

Sunlight can be used directly to heat our homes. Mirrors and lenses can concentrate sunlight onto water to create steam for generating electricity. Sunlight causes winds, which can drive electricity-producing wind turbines. With photovoltaic cells, the energy of sunlight can be converted to an electric current. All these methods of harnessing energy from direct sunlight are advancing rapidly. In the near future, they will be cost-competitive with methods using fossil and nuclear fuels.

Solar Heat Is Easily Collected

Whether you live in a hot or cold climate, you can conserve energy by taking advantage of the sun's rays. Heating water for bathing and washing dishes and clothes consumes a considerable amount of energy. A typical household uses about 15 percent of its total energy consumption to heat water. Energy to heat water could just as well come from the sun as from the local electric power plant or natural-gas supplier.

A solar energy collector for heating water is not much more than a black metal box covered with a glass plate, as Figure 19.26 illustrates.

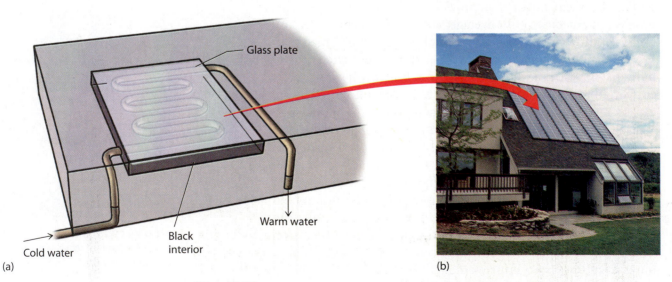

(a)   (b)

Figure 19.26
(a) A solar energy collector is covered with glass to provide a greenhouse effect: sunlight passing into the box is converted to infrared radiation, which cannot escape. (b) Most solar energy collectors are located on rooftops. The collectors are painted black to maximize the absorption of solar heat. The rooftop collector shown here is used for warming an outdoor swimming pool.

Sunlight passes through the glass and into the box, where it is absorbed by the black metal, which becomes hot and so emits infrared radiation. Because glass is opaque to infrared, these rays stay in the box and cause significant warming. Water passing through pipes in the box is heated by this trapped warmth.

Water that has passed through a series of collectors can be scalding hot. This water can be stored in well-insulated containers and used for all washing purposes. Air passing over coils of solar-heated water becomes warm and may be used to heat buildings. Even in cold northern climates, more than 50 percent of all home heating could be obtained from solar collectors. Although installing them can be expensive, over time they more than pay for themselves with energy savings.

Solar Thermal Electric Generation Produces Electricity

A variety of technologies collectively known as *solar thermal electric generation* can be used to produce electricity from sunlight. One technology involves pumping a synthetic oil through a pipe positioned near a mirror-coated trough, as shown in Figure 19.27. The extremely hot oil is then used to generate steam in a steam turbine being used to generate electricity. This design is currently being used in the desert regions of southern California, where more than 360 megawatts of electricity is produced at an operating cost of about 10 cents per kilowatt-hour. A natural-gas burner provides supplemental heat during periods of high demand or cloudy weather.

A second technology involves a large array of sun-tracking mirrors focused so that the sun shines on the top of a central tower, where the temperature soars to about 2200°C. Much of the solar heat is carried away by a molten salt, primarily sodium nitrate, $NaNO_3$. The hot molten salt is piped to an insulated tank where the solar thermal energy can be stored for up to one week. When electricity is needed, the salt is pumped to a conventional steam-generating system for the production of electricity. The molten salt recirculates to the central tower for reheating. An important advantage of this approach is that the salt retains heat for a long time,

Figure 19.27
A solar thermal electric generation unit. A pipe containing synthetic oil is positioned along a mirror-coated trough. Sunlight hitting the mirror is reflected onto the pipe and heats the oil to 370°C. The hot oil is then pumped out and used to convert water to steam in a turbine in an electric power plant.

Hands-On Chemistry: Solar Pool Cover

Which type of cover best keeps the water in a swimming pool warm? Companies claim their "solar pool covers" increase the water temperature by as much as 10 C° above the average outdoor temperature. What is the best material from which to make a solar pool cover? Perform this simple activity to find out.

What You Need

Six identical soup bowls, aluminum foil, colorless plastic food wrap, colorless plastic bubble wrap, black plastic garbage bag, liquid detergent, kitchen thermometer

Procedure

① Cut one piece of aluminum foil, food wrap, bubble wrap, and garbage bag into a circle that completely cover the top of a bowl, with 1 centimeter or so of overhang.

② Set all six bowls outside in the sun (preferably before noon), and fill each with tap water to the brim. Measure and record the temperature in each bowl.

③ Cover one bowl with the aluminum foil (shiny side up), one with the food wrap, one with the bubble wrap (bubble side down), and one with the black garbage bag. Add several drops of liquid detergent to the fifth bowl, and don't do anything to the sixth bowl, which is your control.

④ The temperature of the water will rise as the bowls sit in the sun, but because of the different covers, the temperature increase will vary from bowl to bowl. Guess which water will be warmest at the end of the experiment, which will be second warmest, and so on.

⑤ Allow the bowls to sit in the sun for at least 4 hours. (Good results can also be obtained with skies that are lightly overcast.) Take temperature readings every half hour. Stir the water with the thermometer before taking a reading, and always rinse the thermometer after dipping it into the detergent-containing water. Record your data.

⑥ Plot your data on a graph showing temperature on the vertical axis and time on the horizontal axis. Alternatively, develop a bar graph showing the rise in temperature of the water in each bowl over time.

which allows for the production of electricity even after the sun has set or during periods of inclement weather. Figure 19.28 shows the first commercial central-tower solar thermal energy plant installed in the United States. Located in Barstow, California, it started operating in 1996 and generates about 100 megawatts of electricity.

Figure 19.28
The Solar II power plant near Barstow, California. The central tower is more than 70 meters tall, and the sun-tracking mirrors cover a land area of about eight acres.

Wind Power Is Cheap

Wind power is currently the cheapest form of direct solar energy. This is due, in part, to its simplicity—wind blows on a wind turbine, and the rotation of the turbine blades generates electricity. Most of the early development of wind power took place in California during the 1980s as a result of favorable tax incentives. Many of these early wind machines were not rigorously tested prior to construction because they were merely tax writeoffs. This led to many failures and a marring of the industry's reputation. Costs were also higher than expected because there was little standardizing of procedures and a lack of experience in mass production. Today, most of these problems have been resolved, and the reliability, performance, and cost of wind power have all been improved, as the graph of Figure 19.29 shows. In regions where wind speed averages 13 miles per hour, for example, the operating cost of wind-generated electricity now stands at about 5 cents per kilowatt-hour, which is equal to the operating cost of electricity from conventional sources such as coal.

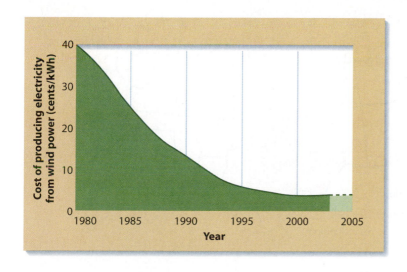

Figure 19.29
Wind is free, but building and maintaining the wind turbines that convert wind energy to electrical energy is not. Over the past 20 years, the reliability and efficiency of wind turbines have improved, resulting in dramatically lower costs for electricity produced from wind power.

Figure 19.30
About 4000 wind turbines at the Tehachapi site in southern California supply 626 megawatts of electricity.

In 1999, the worldwide installed capacity for wind energy totaled about 13,400 megawatts, up dramatically from the worldwide 1990 total of only 2000 megawatts. In the United States, the 1999 installed capacity was about 2400 megawatts, of which 1600 megawatts originated in California. A California installation is shown in Figure 19.30. The production in other states is expected to rise quickly, however, because many new wind projects are currently under development. As is shown in Figure 19.31, most of the wind resources in the United States are concentrated in the northern Great Plains, where more than 10 percent of the nation's electricity needs could be met using existing wind technology. Most of the wind-rich states of the Great Plains could potentially exceed their demand for electricity and either export the excess or use it to create hydrogen fuel through the electrolysis of water.

Improved technology that optimizes the capture of wind energy can bring the cost of wind-generated electricity down to about 3.5 cents per kilowatt-hour. New technologies include improved turbine aerodynamics, a microprocessor to control the turbine, and the introduction of variable-speed turbines that allow optimal blade speeds under a variety of wind conditions. Extended turbine life-spans of 20 to 30 years will also help with maintenance costs.

The major drawback to wind energy is aesthetic: wind turbines are noisy, and many people do not want to see them on the landscape. These objections need to be weighed against the benefit of reduced dependence on fossil fuels and the idea that if people are endeared to the classic wind-

Figure 19.31
Most of the U.S. wind potential is in the Great Plains region.

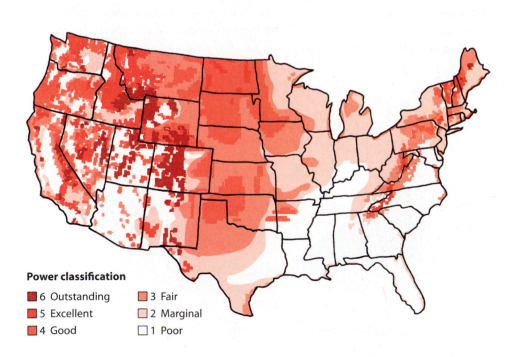

Power classification

- 6 Outstanding
- 5 Excellent
- 4 Good
- 3 Fair
- 2 Marginal
- 1 Poor

mills of Holland, then perhaps these new-age windmills might also one day be looked upon as an attractive sign of prosperity. Fortunately, much of the wind-rich land in the United States is farmland, and wind power and agricultural activities are compatible. At the Altamont wind farms in California, ranchers lost only about 5 percent of their grazing area. Their land prices, however, quadrupled because of royalties they received from wind power companies using their land.

For people who purchase small wind turbines for their home, such as the one pictured in Figure 19.32, unused electricity causes their electric meter to run backwards! As this happens, the wind-generated electricity becomes available for other people connected to the same grid. By law, utility companies are required to either purchase this electricity from the turbine owner or provide credit on his or her electric bill. In this way, the turbine owner is able to "bank" the energy collected on windy days for use on days when the wind speed is too low.

Concept Check ✔

What do fossil fuels have in common with the various forms of direct solar energy just discussed?

Was this your answer? They all originate with the sun.

Photovoltaics Convert Sunlight Directly to Electricity

Photovoltaic cells are the most direct way of converting sunlight to electrical energy. Since their invention in the 1950s, photovoltaics have made remarkable strides. The first major application was in the 1960s with the U.S. space program—space satellites carried photovoltaic cells to power radios and other small electronic devices. Photovoltaics gained further momentum during the energy crisis of the mid-1970s. The cost of photovoltaic electricity dropped from $60 per kilowatt-hour in 1970 to $1 per kilowatt-hour in 1980 to about 20 cents per kilowatt-hour in 2000. Worldwide, sales have grown from less than $2 million in 1975 to more than $1 billion in 2000. More than a billion handheld calculators, several million watches, several million portable lights and battery chargers, and thousands of remote communications facilities are all powered by photovoltaic cells. As Figure 19.33 on page 648 illustrates, the range of their usefulness is huge.

Photovoltaics offer many advantages. Because they need minimal maintenance and no water, they are well suited to remote or arid regions. They can operate on any scale, from multiwatt portable electronic gear to multimegawatt power plants covering millions of square meters. Photovoltaic technology is progressing rapidly and is on the verge of being cost-competitive with fossil and nuclear fuels.

Conventional photovoltaic cells are made from thin slabs of ultrapure silicon. The four valence electrons in a silicon atom allow for four single bonds to four adjacent silicon atoms, as shown in Figure 19.34a. This

Figure 19.32
Children play near a 250-kilowatt wind turbine that provides electricity for elementary schools in Spirit Lake, Iowa. The turbine produces more power than the schools use, and the school district sells the excess to the local utility.

Figure 19.33
Photovoltaic cells come in many sizes, from those in a handheld calculator to those in a rooftop unit that provides electricity to a house.

configuration can be changed by incorporating trace amounts of other elements that have either more or fewer than four valence electrons. Arsenic atoms, for example, have five valence electrons each, also shown in Figure 19.34a. Within the silicon lattice, four of these arsenic electrons participate in bonding with four silicon atoms, but the fifth electron remains free. This is *n-type silicon*, so called because of the presence of free *negative* charges (electrons) brought in by the arsenic. Boron atoms, on the other hand, have only three valence electrons, represented in Figure 19.34b. Incorporating boron into a silicon crystal lattice creates "electron holes," which are bonding sites where electrons should be but are not. This is *p-type silicon*, so called because any passing electron is attracted to this hole as though the hole were a *positive* charge.

What happens when a slice of *n*-type silicon is pressed against a slice of *p*-type silicon? Remember that the *n*-type contains free electrons, and the *p*-type contains electron holes just waiting to attract any available electrons. Sure enough, electrons migrate across the junction from the *n*-type to the

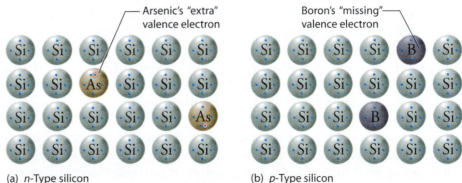

(a) *n*-Type silicon   (b) *p*-Type silicon

Figure 19.34
(a) The four valence electrons in a silicon atom can form four bonds. The fifth valence electron of arsenic is unable to participate in bonding in the silicon lattice and so remains free. Silicon that contains trace amounts of arsenic (or any other element whose atoms have five valence electrons) is called *n*-type silicon. (b) Boron has only three valence electrons for bonding with four silicon atoms. One boron–silicon pair therefore lacks an electron for covalent bonding. Silicon containing trace amounts of boron (or any other element whose atoms have three valence electrons) is called *p*-type silicon.

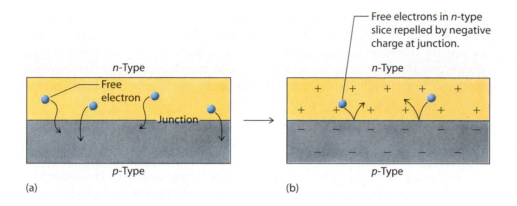

Figure 19.35
(a) Initially, free electrons migrate from the n-type silicon to the p-type silicon. (b) After a while, charge buildup at the junction prevents the continued flow of electrons.

p-type, as shown in Figure 19.35a. This occurs only up to a certain point, however, because losing or gaining electrons upsets the electron-to-proton balance. As the *n*-type silicon loses electrons, it develops a positive charge on its side of the junction. As the *p*-type silicon gains electrons, it develops a negative charge. This charge buildup at the junction serves as a barrier to continued electron migration, as Figure 19.35b illustrates.

Photovoltaic cells rely on the **photoelectric effect**, which is the ability of light to knock electrons away from the atoms in an object. In most materials, the electrons either are completely ejected from the object or simply fall back to their original positions. In a select number of materials, however, including silicon, dislodged electrons roam about randomly in and among neighboring atoms without being locked into any one place, as represented in Figure 19.36. However, *random* electron motion is not an electric current because for every one electron moving to the left, there is another canceling that motion by moving to the right. Greater random motion simply means higher temperatures. This is why solar energy shining onto a piece of silicon produces nothing more than heat.

Once charge builds up, a junction barrier between a slice of *n*-type silicon and a slice of *p*-type, such as the one drawn in Figure 19.35b, is unidirectional. This means that electrons knocked loose in the *n*-type slice are inhibited (by the negative charges in the *p*-type slice) from migrating across the junction to the *p*-type slice, but electrons knocked loose in the *p*-type slice (by solar energy) are drawn into the *n*-type slice by their attraction for the positive charges in the *n*-type slice. As Figure 19.37 shows, a complete

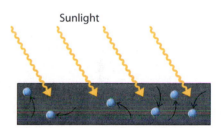

Figure 19.36
The photoelectric effect in silicon. Light rays knock out bonding electrons, which are free to travel about the crystal lattice.

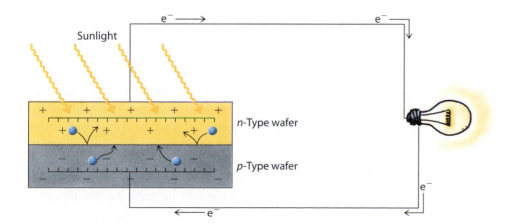

Figure 19.37
Sunlight is converted to electrical energy as it pushes electrons through the external wire from the *n*-type silicon to the *p*-type silicon.

electric circuit can be made by connecting a wire to the outside faces of the two slices of silicon. When light hits either slice, dislodged electrons are forced by the junction barrier to move in one direction: from the *n*-type slice through the wire to the *p*-type slice and back again across the junction to the *n*-type slice. This way, rather than being transformed to heat, energy from sunlight is transformed to electricity.

Concept Check ✓

One wafer of *n*-type silicon and one wafer of *p*-type are pressed together, with no wire connecting the outside faces. An initial flow of electrons from *n*-type to *p*-type results in a charge buildup at the junction. What happens to this charge buildup as light shines on the *n*-type wafer: does it increase, decrease, or stay the same?

Was this your answer? The number of free electrons in the *n*-type silicon is greatly increased because the light knocks electrons out of their atoms. An electric "pressure" thus pushes more electrons through the barrier to the electron holes in the *p*-type wafer. This results in an even greater charge buildup at the junction, which makes the migration of additional electrons even more difficult. When the light is turned off, the charge buildup drops back to its normal level.

The goal of the photovoltaic industry is to create a cell that is highly efficient, inexpensive to manufacture, and easily mass-produced. Traditional photovoltaic cells manufactured from ultrapure crystalline silicon yield efficiencies as high as 15 percent, which is good. However, the great cost of producing ultrapure crystalline silicon prevents these cells from being cost-competitive. Although great strides have been made in the past 20 years, electricity from these cells still remains three to four times more expensive than electricity from conventional sources.

Perhaps the most promising area of photovoltaic research involves thin-film photovoltaic cells. These cells are not produced from expensive crystalline silicon. Rather, they are formed as vaporized silicon or some other photovoltaic material is deposited on a glass or metal substrate. The resulting films are about 400 times thinner than traditional silicon wafers, which saves in material costs. Furthermore, the films are easy to mass-produce.

19.8   Our Future Economy Should Be Based on Hydrogen

One of the most ideal fuels is hydrogen, H_2. By weight, hydrogen packs more energy than any other fuel, which is why it is used to launch spacecraft. It is readily produced from the electrolysis of water, and the electrolysis equipment can be driven by solar-derived electricity. In this way, hydrogen provides a convenient means of storing solar energy.

Hydrogen burns very cleanly, producing almost nothing but water vapor. The vapor can be used to drive a gas turbine for the production of electricity. The waste heat can be used either for industrial purposes or to warm buildings. The water that condenses from the turbine can be used for agriculture or human consumption, or it can be recycled for the production of more hydrogen.

Figure 19.38
A hydrogen-powered test vehicle developed by Mercedes-Benz. The exhaust consists primarily of water vapor.

Because it is a gas, hydrogen is easily transported through pipelines. In fact, pumping hydrogen through pipelines is more energy-efficient than transmitting electricity through power lines. Thus, hydrogen facilities could be located in regions where production is cheapest—desert regions for solar thermal and photovoltaics, windy regions for wind turbines, and moist regions for biomass—and then transported to meet the energy needs of distant communities.

Hydrogen can even be used as a fuel for automobiles, as shown in Figure 19.38. Porous metal alloys have been developed that absorb large quantities of hydrogen gas. Pressing the gas pedal of a car powered by hydrogen fuel sends a warming electric current to the alloy, which is housed in the gas tank. As the alloy heats up, it releases hydrogen, which powers either an internal combustion engine or an electricity-producing fuel cell. Gas stations of the future may truly be *gas* stations!

Concept Check ✔

Why does the combustion of hydrogen produce no carbon dioxide?

Was this your answer? There is no carbon with which to form carbon dioxide! Hydrogen is an ideal fuel that when burned produces almost nothing but water vapor—no carbon dioxide, no carbon monoxide, and no particulate matter.

Fuel Cells Produce Electricity from Fuel

As Section 11.3 discussed, the most efficient way to produce electricity from hydrogen, or any other fuel, is with a *fuel cell*. Utilities could produce megawatts of electricity by stacking fuel cells. Present-day fuel cells have an efficiency of about 60 percent, which is well above the 34 percent efficiency of a typical coal-fired power plant. Interestingly enough, one potential source of hydrogen is coal, which when treated with high-pressure steam and oxygen produces hydrogen gas and methane gas. Burning these gases in a gas turbine results in a coal-to-electricity efficiency as high as 42 percent. If the gases are first run through electricity-producing fuel cells, the efficiency is even higher. Meanwhile, because the coal is not burned directly, fewer pollutants are released into the environment.

Photovoltaic Cells Can Be Used to Produce Hydrogen from Water

The cleanest and most abundant source of hydrogen on our planet is water. Creating hydrogen from water is an energy-intensive process, however, best done using electrolysis, a technique discussed in Section 11.3. Ideally, the electrical energy required to create hydrogen from water could be provided by a solar-driven photovoltaic cell. The electricity generated by the cell is then directed to a second cell, where water is electrolytically cleaved to produce hydrogen, as shown in Figure 19.39.

Over the past several years, there has been great progress toward the realization of a commercially feasible solar-energy-to-hydrogen system. For such a system, the efficiency at which the solar energy is converted to hydrogen needs to be at least 20 percent. Initial systems had efficiencies of no more than 6 percent. One of the main hurdles was developing a photovoltaic system optimized for 1.23 volts, which is the voltage needed for the hydrolysis of water. Such a system was developed by 1998 and provided an efficiency of 12.4 percent. By 2000, researchers were reporting an efficiency of 18.3 percent. With newer technologies now under development, these researchers are predicting that efficiencies on the order of 30 percent may soon be realized. When this level of efficiency is attained, our dependence on fossil fuels may be dramatically reduced.

Figure 19.39
An electric current from a photovoltaic cell is passed through a pair of submerged electrodes, only one of which is shown here. While hydrogen forms at one electrode (shown), oxygen forms at the other (not shown).

In Perspective

The U.S. Department of Energy projects that by 2010, our country will need 200,000 megawatts of new electricity-generating capacity. Furthermore, nearly a quarter of all U.S. facilities that generate electric power will need major renovations, overhauls, or replacement by that time. This represents a grand opportunity for the development of sustainable energy sources. Even greater opportunities exist in developing nations, where there are more than 2 billion people without electricity and many more to come with rapid population growth. An estimated $1 trillion will be spent in the next few years to bring electric power to these people. Immediate choices made about where that energy comes from and how it is produced will affect the environment for a long, long time to come.

All regions of the world will benefit from energy conservation, which is the natural partner of sustainable energy. Conservation measures can be quite effective, as was demonstrated after the world oil embargo imposed by the Organization of Petroleum Exporting Countries (OPEC) in 1973. From the start of this embargo until 1986, energy consumption in the

Table 19.1

The World's Increasing Energy Needs

Year	World Population (billions)	×	Energy per Person (MW)	=	Total Power Consumed (MW)
2000	6.2	×	0.003	=	19 million
2050	10.0	×	0.003	=	30 million
	10.0	×	0.0075	=	75 million
2100	15.0	×	0.003	=	45 million
	15.0	×	0.0075	=	112 million

United States remained constant as a result of increased efficiencies, while the economy grew by 30 percent. Some people argue that we are close to our limit in terms of conservation measures. Others argue that with new technologies and materials we have only scratched the surface. Super-insulated buildings now require only one-tenth as much heating in winter and cooling in summer. Hybrid combustion–electric vehicles can now drive more than 40 kilometers on 1 liter of gasoline and 25 kilowatts of battery power—this corresponds to about 100 miles per gallon! Replace a single 75-watt incandescent bulb with an 18-watt fluorescent one that lasts 10,000 hours, and over that 10,000 hours—about 1 year—you'd save the electricity that a typical U.S. power plant would make from 350 kilograms of coal or 250 liters of oil. As a result, about 730 kilograms of carbon dioxide would not be released into the atmosphere.

Our energy needs are increasing primarily because of our increasing population, as shown in Table 19.1, which assumes an acceptable standard of living can be had at 0.003 megawatt per person. On average, however, the 1.2 billion individuals of developed nations consume about 0.0075 megawatt per person. If conservation measures are not set in place and if the far greater number of individuals in developing nations are brought up to 0.0075 megawatt per person, then the world will use 75 million megawatts in 2050 and 112 million megawatts in 2100, six times the amount consumed in 2000.

As hard as controlling population growth may be, it is likely to be easier than providing increasing numbers of people with energy, food, water, and much else.

Thomas Jefferson described revolution as an extraordinary event necessary to enable all ordinary events to continue. If a decent standard of living for all human beings is to be held as ordinary, what we need is nothing short of an energy revolution. The days of our using nonrenewable fuels are numbered. The sooner we make the necessary transitions, the better.

Key Terms and Matching Definitions

_____ biomass
_____ coal
_____ kilowatt-hour
_____ natural gas
_____ petroleum
_____ photoelectric effect
_____ power
_____ watt

1. The rate at which energy is expended.
2. A unit for measuring power, equal to 1 joule of energy expended per second.
3. The amount of energy consumed in 1 hour at a rate of 1 kilowatt.
4. A solid consisting of a tightly bound network of hydrocarbon chains and rings.
5. A liquid mixture of loosely held hydrocarbon molecules containing not more than 30 carbon atoms each.
6. A mixture of methane plus small amounts of ethane and propane.
7. A general term for plant material.
8. The ability of light to knock electrons out of atoms.

Review Questions

Electricity Is a Convenient Form of Energy

1. What does an electric current generate?
2. What does an electrical wire moving through a magnetic field generate?
3. Is electricity a source of energy?
4. What's a watt?

Fossil Fuels Are a Widely Used but Limited Energy Source

5. Why are fossil fuels such a popular energy resource?
6. Why is coal considered the filthiest fossil fuel?
7. How does a scrubber remove noxious gaseous effluents created in the combustion of coal?
8. Can coal be converted to cleaner-burning fuels?
9. Why is petroleum so convenient to use?
10. What is the prime component of natural gas?

There Are Two Forms of Nuclear Energy

11. Why has the number of operating nuclear fission plants in the United States decreased over the past two decades?
12. For about how long must a fission plant be operating before the initial financial investments are paid off?
13. What are the advantages of having smaller fission reactors that produce on the order of 155 to 600 megawatts of power?
14. Why was most of the radiation from the Chernobyl meltdown released into the environment?
15. What is the source of fuel for nuclear fusion?
16. What are the two methods being explored for extracting energy from nuclear fusion?

What Are Sustainable Energy Sources?

17. What defines an energy source as sustainable?

Water Can Be Used to Generate Electricity

18. What does the acronym OTEC stand for?
19. Why is Hawaii particularly well suited for generating electricity using OTEC technology?
20. Why are hydrothermal vents often quite smelly?
21. How are tides used to generate electricity?

Biomass Is Chemical Energy

22. In what sense is biomass a form of solar energy?
23. Which has a higher octane rating, ethyl alcohol or gasoline?
24. Why is methanol also called wood alcohol?
25. What country named in this chapter currently generates most of its fuel from biomass?
26. What is the most efficient way to convert biomass to electricity?

Energy Can Be Harnessed from Sunlight

27. Why are solar water heaters painted black?
28. How can solar heat be stored for later use?

29. Describe two technologies used to convert solar heat to electricity.

30. What is a major drawback of wind power?

31. How does blending arsenic atoms into a sample of silicon increase the silicon's ability to conduct electricity?

32. How is *p*-type silicon made?

33. What happens when a slice of *n*-type silicon is joined to a slice of *p*-type silicon?

34. Light shining on a slab of pure silicon dislodges electrons from the silicon atoms. What happens to these electrons?

35. When you have *p*-type silicon and *n*-type silicon joined together, why don't electrons dislodged by light in the *n*-type cross the junction barrier and enter the *p*-type?

Our Future Economy Should Be Based on Hydrogen

36. Why is hydrogen such an ideal fuel?

37. How might cars one day store hydrogen fuel?

38. How are fuel cells different from batteries?

In Perspective

39. What is the natural partner of sustainable energy?

40. In what type of nations will there be the greatest increases in energy use in the next few decades?

Hands-On Chemistry Insights

Solar Pool Cover

Evaporation accounts for a significant amount of heat loss from a swimming pool. Therefore just about any covering that inhibits evaporation will help keep a pool warm. This explains why the water covered by aluminum foil gets warmer despite the fact that the aluminum reflects incoming solar radiation. As discussed in Chapter 8, detergents form a thin layer on the surface of water. This layer inhibits evaporation enough to allow the water in the bowl treated with detergent to become slightly warmer than the uncovered control bowl. Some commercial "liquid solar blankets" are nothing more

than nonfoaming detergents that when added to the swimming pool inhibit evaporation and thus cut down on heat loss.

You know that black vinyl car seats get hotter in the sun than do white vinyl seats. Did you then assume that the black covering would result in the warmest water? You may have poked your finger into the water samples to sense the temperature differences. You may have noticed that the water under the black plastic was markedly warmer at the top of the bowl than at the bottom. (This is true of the water in the other bowls, but the effect is not as pronounced.) Although it is true that the black plastic gets the warmest of all the coverings, remember that it's the *water* you want to get warm, not the plastic. The temperature gradient results because only the water directly below the black plastic gets warmed—that is, it's the plastic that heats the water and not the sun.

Transparent covers tend to work best because they both inhibit evaporation and allow the sunlight to heat the water directly. Thus your highest temperatures should have been for the food wrap and the bubble wrap. The food wrap allows the most light to pass through, which explains why the water beneath it tends to reach the highest temperature. After the sun goes down, however, the bubble wrap serves as a better insulator, maintaining a higher temperature over a longer period.

Exercises

1. Why are gas turbines more efficient at generating electricity than steam turbines?

2. Why are airplanes flown by the power of electricity rather than the power of gasoline fuel so impractical?

3. How can fossil fuels be considered a form of solar energy?

4. Why is it impractical to burn any fuel in such a way that no nitrogen oxides are formed as combustion products?

5. Why does burning natural gas produce less carbon dioxide than burning either petroleum or coal? Hint: Consider the chemical formulas of the compounds making up these materials.

6. What are some advantages of deriving energy from nuclear fission?

7. Why would a fission control substance that loses neutron-absorbing power as its temperature increases be dangerous?

8. Why are countries teaming up to develop nuclear fusion energy?

9. In 1973 the Organization of Petroleum Exporting Countries stopped exporting their oil for a time. What effect do you suppose this embargo had on the nuclear power industry?

10. Is it possible for a nuclear fission power plant to blow up like an atomic bomb?

11. What are some advantages of deriving energy from nuclear fusion?

12. Why is it in a country's best national security interests to invest in sustainable energy sources?

13. Why do many environmental groups oppose building more hydroelectric dams when these dams produce virtually no chemical pollutants as they create electric power?

14. At atmospheric pressure, water has a boiling point too high to allow it to be used as the turbine-pushing fluid in an OTEC electric generator. How might the generator be modified to overcome this problem? How might such a modification allow for the production of fresh water from seawater?

15. How is it that here on the Earth we can derive energy from the moon, which is 400,000 kilometers away?

16. What are some disadvantages of producing ethanol by fermenting food biomass?

17. How does a *nonsolar* pool cover increase the efficiency of a pool heater?

18. What is the utility of south-facing windows in northern climates?

19. Suggest how solar energy might help destroy pathogens in drinking water.

20. How might solar heat be used to generate colder temperatures?

21. In the simplified electric generator shown here, a power source is needed at either point A or point B. Where could each of the following power sources be used, at A or at B:

natural gas _____
wind _____
nuclear fusion _____
hydroelectric _____
coal _____

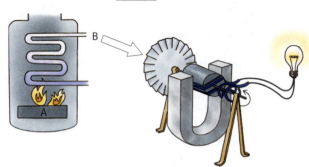

22. How is the junction barrier between *n*-type and *p*-type silicon much like a one-way valve?

23. In an operating photovoltaic cell, electrons move through the external circuit to the negatively charged *p*-type silicon wafer. How can the electrons move to the negatively charged silicon when they themselves are negatively charged? Don't like signs repel each other?

24. What are some major advantages of using seawater to produce hydrogen from electrolysis?

25. A disadvantage of tidal power is that electricity is generated only when tides are coming in or going out. How might a tidal electric power plant be outfitted so as to be able to provide a continuous supply of electricity?

26. What if Mars were inhabitable by humans and we had developed the technology to transport and settle half the world's population there? At present birth rates, how long would it be before we were faced with not one but two world populations of 6 billion? (Hint: See Table 19.1.)

Problems

1. How much money would you save per hour by replacing a 100-watt incandescent light bulb with an equally bright 20-watt fluorescent bulb? Assume the cost for electricity to be 15 cents per kilowatt-hour.

Answers to Calculation Corner

Kilowatt-Hours

Step 1 Total energy consumed in kilowatt-hours:

(10 bulbs)(0.100 kW/bulb)(10 h) = 10 kWh

Step 2 Cost of consuming this much energy:

(10 kWh)(15¢/kWh) = 150¢ = $1.50

Discussion Topics

1. Why might some people consider it a blessing in disguise that fossil fuels are such a limited resource? Centuries from now, what attitudes on the combustion of fossil fuels are our descendants likely to have?

2. The 1986 accident at Chernobyl, in which dozens of people died and thousands more were exposed to radiation that might lead to cancer in the future, caused fear and outrage worldwide and led some people to call for the closing of all nuclear plants. Yet many people choose to smoke cigarettes in spite of the fact that 2 million people die every year from smoking-related diseases. The risks posed by nuclear power plants are involuntary, risks we must all share like it or not, whereas the risks associated with smoking are voluntary because a person chooses to smoke. Why are we so unaccepting of involuntary risk but accepting of voluntary risk?

3. Lithuania is more dependent on nuclear power than any other country in the world. Its two 1500-megawatt reactors produce more than 80 percent of the country's electricity. The reactors, however, are the same design as the unit that caused the Chernobyl disaster. With both reactors working, Lithuania can produce almost twice as much energy as its domestic demand, allowing the country to sell the excess to other nations. If one unit were shut down, however, the country would have to import electricity at a cost of $4 billion a year. A 1998 study determined that Lithuania's older reactor can no longer operate safely. Without foreign aid to assist with the cost of importing electricity once this obsolete plant is shut down, Lithuania intends to operate both facilities for at least 15 more years. What should be done and at whose expense?

Exploring Further

http://www.iea.org

The International Energy Agency was set up in 1974 to focus on issues of energy security, especially oil security, but is today more concerned with how energy production and use can be reconciled with the preservation of our natural environment. An excellent site for obtaining up-to-date statistics on worldwide energy consumption.

http://wwwofe.er.doe.gov

Home page for the U.S. Fusion Energy Sciences Program, whose mission is to acquire the knowledge base needed for an economically and environmentally attractive fusion energy source. Many useful links to worldwide fusion projects are included.

http://www.crest.org/index.html

Solstice, the Center for Renewable Energy and Sustainable Technologies, is an on-line resource for sustainable-energy information.

http://www.rw.doe.gov/homejava/homejava.htm

The Office of Civilian Radioactive Waste Management was established in 1982 to develop and manage a federal system for disposing of spent nuclear fuel from defense activities. It's at this Web site that you'll find the official position of the U.S. government regarding Yucca Mountain as a potential nuclear waste repository.

http://www.iaea.org

The International Atomic Energy Agency is the world's central intergovernmental forum for scientific and technical cooperation in the nuclear field and the international inspectorate for nuclear safeguards.

http://www.awea.org

Home page for the American Wind Energy Association. Since 1974, this organization has advocated the development of wind energy as a reliable, environmentally superior energy alternative.

http://www.nrel.gov

Home page of the National Renewable Energy Laboratory, run by the U.S. Department of Energy.

Appendix A

Scientific Notation Is Used to Express Large and Small Numbers

In science, we often encounter very large and very small numbers. Written in standard decimal notation, these numbers can be cumbersome. There are, for example, about 33,460,000,000,000,000,000,000,000 water molecules in a thimbleful of water, each having a mass of about 0.00000000000000000000002991 gram. To represent such numbers, scientists often use a mathematical shorthand called **scientific notation**. Written in this notation, the number of molecules in a thimbleful of water is 3.346×10^{22}, and the mass of a single molecule is 2.991×10^{-23} gram.

To understand how this shorthand notation works, consider the large number 50,000,000. Mathematically this number is equal to 5 multiplied by $10 \times 10 \times 10 \times 10 \times 10 \times 10 \times 10$ (check this out on your calculator). We can abbreviate this chain of numbers by writing all the 10s in exponential form, which gives us the scientific notation 5×10^7. (Note that 10^7 is the same as $10 \times 10 \times 10 \times 10 \times 10 \times 10 \times 10$. Table A.1 shows the exponential form of some other large and small numbers.) Likewise, the small number 0.0005 is mathematically equal to 5 divided $10 \times 10 \times 10 \times 10$, which is $5/10^4$. Because dividing by a number is exactly equivalent to multiplying by the reciprocal of that number, $5/10^4$ can be written in the form 5×10^{-4}, and so in scientific notation 0.0005 becomes 5×10^{-4} (note the negative exponent).

All scientific notation is written in the general form

$$C \times 10^n$$

where C, called the *coefficient*, is a number between 1 and 9.999 . . . and n is the *exponent*. A positive exponent indicates a number greater than 1, and a negative exponent indicates a number between 0 and 1 (*not* a number less than 0).

Table A.1

Decimal and Exponential Notations

$$
\begin{aligned}
1{,}000{,}000 &= 10 \times 10 \times 10 \times 10 \times 10 \times 10 = 10^6 \\
100{,}000 &= 10 \times 10 \times 10 \times 10 \times 10 = 10^5 \\
10{,}000 &= 10 \times 10 \times 10 \times 10 = 10^4 \\
1000 &= 10 \times 10 \times 10 = 10^3 \\
100 &= 10 \times 10 = 10^2 \\
10 &= 10 = 10^1 \\
1 &= 1 = 10^0 \\
0.1 &= 1/10 = 10^{-1} \\
0.01 &= 1/(10 \times 10) = 10^{-2} \\
0.001 &= 1/(10 \times 10 \times 10) = 10^{-3} \\
0.0001 &= 1/(10 \times 10 \times 10 \times 10) = 10^{-4} \\
0.00001 &= 1/(10 \times 10 \times 10 \times 10 \times 10) = 10^{-5} \\
0.000001 &= 1/(10 \times 10 \times 10 \times 10 \times 10 \times 10) = 10^{-6}
\end{aligned}
$$

Numbers less than 0 are indicated by putting a negative sign *before the coefficient* (not in the exponent):

	Decimal notation	Scientific notation
Large positive number (greater than 1)	6,000,000,000	6×10^9
Small positive number (between 0 and 1)	0.0006	6×10^{-4}
Large negative number (less than -1)	$-6{,}000{,}000{,}000$	-6×10^9
Small negative number (between -1 and 0)	-0.0006	-6×10^{-4}

Table A.2 shows scientific notation used to express some of the physical data often used in science.

Table A.2

Number of molecules in thimbleful of water	3.346×10^{22}
Mass of water molecule	2.991×10^{-23} gram
Average radius of hydrogen atom	5×10^{-11} meter
Proton mass	1.6726×10^{-27} kilogram
Neutron mass	1.6749×10^{-27} kilogram
Electron mass	9.1094×10^{-31} kilogram
Electron charge	1.602×10^{-19} coulomb
Avogadro's number	6.022×10^{23} particles
Atomic mass unit	1.661×10^{-24} gram

To change a decimal number that is greater than $+1$ or less than -1 to scientific notation, you shift the decimal point to the left until you arrive at a number between 1 and 9.999 . . . . For decimal numbers between $+1$ and -1, you move the decimal point to the right until you arrive at a number between 1 and 9.999 . . . . This number is the coefficient part of the notation. The exponent part is simply the number of places you moved the decimal point. For example, to convert the decimal number 45,000 to scientific notation, move the decimal point four places to the left to get a number between 1 and 9.999 . . . :

$$45,000 = 4.5 \times 10^4$$

The number 0.00045, on the other hand, is converted to scientific notation by moving the decimal point four places to the right to arrive at a number between 1 and 9.999 . . . :

$$0.00045 = 4.5 \times 10^{-4}$$

Note that because you moved the decimal point to the right in this case, you must put a minus sign on the exponent.

Your Turn

Express the following exponentials as decimal numbers:

a. 1×10^{-7} b. 1×10^8 c. 8.8×10^5

Express the following decimal numbers in scientific notation:

d. 740,000   e. -0.00354 f. 15

Were these your answers? a. 0.0000001   b. 100,000,000   c. 880,000   d. 7.4×10^5
e. -3.54×10^{-3} f. 1.5×10^1

Appendix B

Significant Figures Are Used to Show Which Digits Have Experimental Meaning

Two kinds of numbers are used in science—those that are *counted or defined* and those that are *measured.* There is a great difference between a counted or defined number and a measured number. The exact value of a counted or defined number can be stated, but the exact value of a measured number can never be stated.

You can count the number of chairs in your classroom, the number of fingers on your hand, or the number of quarters in your pocket with absolute certainty. Thus counted numbers are not subject to error (unless, of course, you count wrong).

Defined numbers are about exact relationships and are defined as being true. The defined number of centimeters in a meter, the defined number of seconds in an hour, and the defined number of sides on a square are examples. Thus defined numbers also are not subject to error (unless you forget a definition).

Every measured number, however, no matter how carefully measured, has some degree of *uncertainty.* This uncertainty (or *margin of error*) in a measurement can be illustrated by the two metersticks shown in Figure B.1. Both sticks are being used to measure the length of a table. Assuming that the zero end of each meterstick has been carefully and accurately positioned at the left end of the table, how long is the table?

The upper meterstick has a scale marked off in centimeter intervals. Using this scale, you can say with certainty that the length is between 51 and 52 centimeters. You can say further that it is closer to 51 centimeters than to 52 centimeters; you can even estimate it to be 51.2 centimeters.

The scale on the lower meterstick has more subdivisions—and therefore greater precision—because it is marked off in millimeters. With this scale, you can say that the length is definitely between 51.2 and 51.3 centimeters, and you can estimate it to be 51.25 centimeters.

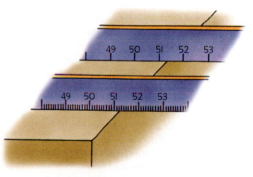

Figure B.1

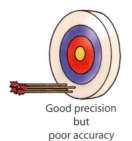

Good precision
but
poor accuracy

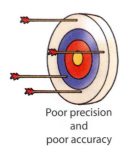

Poor precision
and
poor accuracy

Good precision
and
good accuracy

Figure B.2
Archery as a model for understanding the difference between precision and accuracy. *Precision* means close agreement in a group of measured numbers; *accuracy* means a measured value that is very close to the true value of what is being measured. If you measure the same thing several times and get numbers that are close to one another but are far from the true value (perhaps because your measuring device is not working properly), your measurements are *precise* but not *accurate*.

Note how both readings contain some digits that are *certain*, and one digit (the last one) that is *estimated*. Note also that the uncertainty in the reading from the lower meterstick is less than the uncertainty in the reading from the upper meterstick. The lower meterstick can give a reading to the hundredths place, but the upper one can give a reading only to the tenths place. The lower one is more *precise* than the top one. So, in any measured number, the digits tell us the *magnitude* of the measurement and the location of the decimal point tells us the *precision* of the measurement. (Figure B.2 illustrates the distinction between *precision* and *accuracy*.)

Significant figures are the digits in any measured value that are known with certainty plus one final digit that is estimated and hence uncertain. These are the digits that reflect the precision of the instrument used to generate the number. They are the digits that have experimental meaning. The measurement 51.2 centimeters made with the upper meterstick in Figure B.1, for example, has three significant figures, and the measurement 51.25 centimeters made with the lower meterstick has four significant figures. The rightmost digit is always an estimated digit, and only one estimated digit is ever recorded for a measurement. It would be incorrect to report 51.253 centimeters as the length measured with the lower meterstick. This five-significant-figure value has two estimated digits (the final 5 and 3) and is incorrect because it indicates a *precision* greater than the meterstick can obtain.

Here are some standard rules for writing and using significant figures.

Rule 1   In numbers that do not contain zeros, all the digits are significant:

4.1327	five significant figures
5.14	three significant figures
369	three significant figures

Rule 2   All zeros between significant digits are significant:

8.052	four significant figures
7059	four significant figures
306	three significant figures

Rule 3   Zeros to the left of the first nonzero digit serve only to fix the position of the decimal point and are not significant:

0.0068	two significant figures
0.0427	three significant figures
0.0003506	four significant figures

Rule 4   In a number that contains digits to the right of the decimal point, zeros to the right of the last nonzero digit are significant:

53.0	three significant figures
53.00	four significant figures
0.00200	three significant figures
0.70050	five significant figures

Rule 5 In a number that has no decimal point and that ends in one or more zeros, the zeros that end the number are not significant:

3600	two significant figures
290	two significant figures
5,000,000	one significant figure
10	one significant figure
6050	three significant figures

Rule 6 When a number is expressed in scientific notation, all digits in the coefficient are taken to be significant:

4.6×10^{-5}	two significant figures
4.60×10^{-5}	three significant figures
4.600×10^{-5}	four significant figures
2×10^{-5}	one significant figure
3.0×10^{-5}	two significant figures
4.00×10^{-5}	three significant figures

Your Turn

How many significant figures in

a. 43,384
b. 43,084
c. 0.004308
d. 43,084.0
e. 43,000
f. 4.30×10^4

Were these your answers? a. 5   b. 5   c. 4   d. 6   e. 2   f. 3

In addition to the rules cited above, there is another full set of rules to be followed for significant figures when two or more measured numbers are subtracted, added, divided, or multiplied. These rules are summarized in the appendix of the *Conceptual Chemistry Laboratory Manual.*

Appendix C

Solutions to Odd-Numbered Exercises and Problems

Chapter 1

1. The Exercises are designed to allow you to work with the concepts you learned by reading the chapter and answering the Review Questions. If you have not already learned the concepts, you will have a most difficult time with the Exercises, which are the types of questions often found on exams. To benefit most from the Exercises, either write out your answers or, better yet, explain your answers to a friend.

3. Biology is based on the principles of chemistry applied to living organisms, and chemistry is based on the principles of physics applied to atoms and molecules. Physics is the study of the fundamental rules of nature, which more often than not are simple in design and readily described by mathematical formulas. Thus biology is based on both chemistry *and* physics and so can be considered the most complex science of the three.

5. A good scientist must change his or her mind whenever experimental results disprove a previously held belief. Holding on to hypotheses and theories that are either not testable or have been shown to be wrong is contrary to the spirit of science.

7. A hypothesis is a testable assumption often used to explain an observed phenomenon. A theory is a single comprehensive idea that can be used to explain a broad range of phenomena.

9. This is not unusual at all. As discussed in the answer to Exercise 3, there is much overlap among the sciences. Whereas Baker is interested in how the chemical produced by the sea butterfly may be used for some human purpose, McClintock is interested in how the sea butterfly uses this chemical for self-defense. Here we see two different approaches to the same phenomenon. Studying the system together allows these researchers both to pool their research resources and to learn from each other.

11. If all the material came from the soil, you would expect a large hole to develop in the dirt around the plant. Also, if the plant were grown in a pot, the mass of the soil in the pot should be greater when the plant is young than when the plant is older.

13. The volume of the car and its average density change.

15. Yes, an object can have mass without having weight. This can occur deep in space, where a floating object (which has mass) is "weightless." In order to have weight, however, the object must have mass. So an object cannot have weight without having mass.

17. The mass of a 6-kilogram object is 6 kilograms on the moon, on the Earth, and anyplace else! Mass is always independent of gravity.

19. Yes, a 2-kilogram iron brick has twice the mass of a 1-kilogram block of wood. Volume, however, is a different story. Because the density of iron is much greater than the density of wood, the 1-kilogram wood block occupies more volume than the 2-kilogram iron brick.

21. Yes, your body will possess energy after you die, in the form of chemical potential energy. If you are cremated, however, the amount of this chemical potential energy is reduced substantially.

23. Glass contracts when cooled and expands when warmed. Therefore you should fill the inner glass with cold water while running hot water over the outer glass.

25. The swimming pool has much more total energy. To understand this difference in the amounts of energy in the two bodies of water, consider what your electric utility bill would be after heating them to their respective temperatures.

27. At 25°C, there is a certain amount of energy possessed by each submicroscopic particle of a material. If the attractions between particles are not very strong, the energy each particle has may allow the particles to separate from one another and exist in the gaseous phase. If the attractions are strong, the energy may not be enough to overcome these attractions and keep the particles far away from one another, and so they may be held together in the solid phase. You can assume, therefore, that the attractions among the submicroscopic particles of a material that is a solid at 25°C are stronger than the particle-to-particle attractions in a material that is a gas at this temperature.

29. The different spacings between particles in the left box tell you two phases are present. The very regular stacking of particles on the left indicates a solid, and so this is probably a solid/liquid combination. Taking heat away causes the liquid to freeze so that only solid is present, and that is represented in the middle box—note the regular stacking of particles. Adding heat causes the solid to melt so that only liquid is present, represented in the right box. If these particles represent water molecules, the box on the left represents ice melting, which occurs at 0°C.

31. Density is the ratio of a material's mass to its volume. Because the mass stays the same but the volume decreases, the density of the gas increases.

33. Box a represents the highest density because it has the greatest number of particles packed in the given volume. Because the particles in this box are packed close together but randomly oriented, this box represents a liquid. Box c represents the gaseous phase, which occurs at higher temperatures. This box therefore represents the highest temperature. For most materials, the solid phase is denser than the liquid phase. For the material represented here, however, the liquid phase (box a) is denser than the solid phase (box b). As we explore further in Chapter 8, this is exactly the case for water—the solid phase (ice) is less dense than the liquid phase (liquid water).

Problems

1. Multiply by the conversion factor to arrive at the answer:

$$130 \text{ lb} \times \frac{1 \text{ kg}}{2.205 \text{ lb}} = 59 \text{ kg}$$

3. Multiply by the conversion factor to arrive at the answer:

$$230{,}000 \text{ cal} \times \frac{1 \text{ J}}{0.239 \text{ cal}} = 960{,}000 \text{ J}$$

5. Divide the mass by the volume it displaces to arrive at the density:

$$\text{density} = \frac{\text{mass}}{\text{volume}} = \frac{52.3 \text{ g}}{4.16 \text{ mL}} = 12.6 \text{ g/mL}$$

From Table 1.3, you see that this value is substantially less than the density of pure gold, which is 19.3 g/mL. This evidence indicates that the piece is far from pure.

Chapter 2

1. That this process is reversible suggests a physical change. As you sleep in a reclined position, pressure is taken off the discs in your spinal column, and the reduced pressure allows the discs to expand so that you are significantly taller in the morning. Astronauts returning from extended space visits may be up to 2 inches taller than they were at launch time.

3. In the left box, each particle is made up of one atom of one kind and one atom of another kind, but in the right box, each particle is made up of two identical atoms. This difference tells

you that the particles split apart during the transformation and came together in different combinations, which means a chemical change.

5. The change from A to B represents a physical change because no new types of molecules are formed. The collection of blue/red molecules on the bottom of B represents these molecules in either the liquid or the solid phase after having been in the gaseous phase in A. This phase change means there must have been a decrease in temperature. At this lower temperature, the red/red molecules are still in the gaseous phase, which means that they have a lower boiling point than the blue/red molecules.

7. The oldest known elements are the ones that have atomic symbols that don't match their modern atomic names. Examples include iron, Fe; gold, Au; and copper, Cu.

9. A percentage is converted to a fraction by dividing by 100. To find 50 percent of something, for example, you multiply that something by 50/100 = 0.50. Thus 0.0001 percent is equal to the fraction 0.000001, which when multiplied by 1×10^{24} equals 1×10^{18}. This is certainly a lot of impurity molecules in your glass of water. The number of water molecules, however, far exceeds this number (see Exercise 10), and so these impurity molecules are not a problem. As an analogy, consider that there were about 12 billion pennies minted in 1990. This is certainly a lot of pennies, but they are nonetheless relatively rare in general circulation because the total number of pennies in circulation is far greater—on the order of 300 billion.

11. Study the answer to Exercise 9 to make a strong case that this sample of water is ultra-ultrapure!

13. Table salt (sodium chloride), compound because made of two elements bonded together. Stainless steel, mixture because made of elements (iron and carbon) not bonded together. Tap water, mixture of dihydrogen oxide plus impurities. Table sugar, compound made of one substance— sucrose. Vanilla extract, mixture of water, alcohol, and flavor molecules. Butter, mixture of fat, water, and milk solids. Maple syrup, mixture of water and maple sugar. Aluminum, element. Ice, compound (H_2O) in pure form, mixture when made from impure tap water. Milk, mixture of water and milk solids. Cherry-flavored cough drops, mixture of sugar and flavoring.

15. Element, C; compound, B; mixture of an ele-

ment and compound, A. There are three types of molecules shown: one made of two large atoms, one made of two small atoms, and one made of one large atom and one small atom.

17. The atoms in a compound are chemically bonded together and do not come apart during a physical change. The components of a mixture, however, can be separated from one another by physical means.

19. Taking advantage of differences in chemical properties means you have to make the components of the mixture undergo chemical change. During chemical change, at least one component changes its identity. Thus, you may have separated the component, but now it is something else, which means you need to make it undergo a second chemical change to convert it back to what it was. This can be an energy-intensive and time-intensive process that is much less efficient than a separation based on differences in physical properties.

21. Based on its location in the periodic table, you know that gallium, Ga, is more metallic than germanium, Ge. This means that gallium should be a better conductor of electricity. Gallium chips should therefore operate faster than germanium chips. (Gallium has a low melting point of 30°C, which makes its use in the manufacture of computer chips impractical. Mixtures of gallium and arsenic, however, have found great use in the manufacture of ultrafast, though relatively expensive, computer chips.)

23. Helium is placed at the far right of the periodic table, in group 18, because it has physical and chemical properties similar to those of the other elements of group 18.

25. Calcium is readily absorbed by the body for the building of bones. Because calcium and strontium are in the same group of the periodic table, they have similar physical and chemical properties. The body therefore has a hard time distinguishing between the two, and strontium is absorbed just as though it were calcium.

27. The iron can be separated based on a difference in physical properties. The iron particles are attracted to a magnet, but the pieces of cereal are not. Try this with your next box of iron-fortified cereal.

Chapter 3

1. The cat leaves a trail of molecules across the yard. These molecules leave the ground and mix with the air, where they enter the dog's nose, activating the sense of smell.

3. The atoms in the baby are just as old as those in the elderly person—all appreciably older than the solar system.

5. You are a part of every person around you in the sense that you are composed of atoms that at one time were part of not only every person around you but also every person who ever lived on the Earth!

7. Lavoisier *observed* that tin gained mass as it decomposed into a gray powder and *asked* how that could happen. He *hypothesized* that the tin gained mass because it absorbed something from the air. He then *predicted* that if he could keep track of air volume as this reaction took place, he would find a change in air volume. He *tested* his prediction by heating a piece of tin floating on a wood block enclosed by a glass container. After observing that the volume of air surrounding the tin decreased by 20 percent, Lavoisier *theorized* that 20 percent of the air is made of a gaseous compound that reacts with tin.

9. Oxygen and hydrogen react in an 8:1 ratio by mass to form water. Thus, 1 gram of hydrogen always reacts with only 8 grams of oxygen (no more and no less). If the amount of oxygen available were 10 grams, only 8 grams of this 10 grams would react with 1 gram of hydrogen—2 grams of oxygen would be left over. When they react, the 1 gram of hydrogen and 8 grams of oxygen form a combined mass of 9 grams of water.

11. Both were right. Proust saw oxygen and hydrogen react in an 8:1 ratio by *mass*, and Gay-Lussac saw them react in a 1:2 ratio by *volume*.

13. Iron metal consists only of iron atoms, Fe. Rust is a compound of iron and oxygen (iron oxide, Fe_2O_3). So a sample of iron weighs more after it rusts because it has gained the weight of oxygen atoms. Be sure to recognize that the parts of the sample that have rusted are no longer iron. Rather, those parts are now iron oxide.

15. Avogadro correctly assumed that equal volumes of oxygen gas and water vapor contain equal numbers of particles. To account for the fact that the volume of oxygen has more mass than the volume of water vapor, Avogadro hypothesized that each particle of oxygen gas consists of two oxygen atoms bound together into a single unit that we know today as an oxygen molecule, O_2. With two oxygen atoms per particle, a volume of oxygen gas is more massive than an equal volume of water vapor, which has only one oxygen atom per particle (along with two lightweight hydrogen atoms).

17. For the first reaction, the box on the far right should show the same thing as the box adjacent to it—ten HCl molecules. Each box contains 36.5 grams of HCl. For the second reaction, there is insufficient hydrogen to react with all the chlorine. Thus only 36.5 grams of hydrogen chloride forms. The box on the far right should show the five unreacted Cl_2 molecules. Their combined mass is 35.5 grams.

19. Throughout the development of modern chemistry, there were numerous instances where investigators clung tightly to the ideas of the past. One example presented in this chapter was the reluctance of Dalton and other early chemists to accept Avogadro's hypothesis of the diatomic nature of oxygen and hydrogen. There are many other examples that the text didn't have the time to go into. Boyle, for example, held on to Aristotle's notion that there was only one form of matter and that this "one form" took on different "qualities" to make up the materials around us. Priestley refused to believe Lavoisier's explanation that his new gas was a new element absorbed by materials as they burned. Instead, Priestley used the ideas of his predecessors in stating that the "new air" he had generated was a form of air that lacked an "essence" that materials *release* as they burn. Fire burned brighter in his air because it rapidly absorbed this essence as the essence was released by the burning material.

21. The youngest was Gay-Lussac, who in 1808 at the age of 30 showed that gases react in definite volume ratios. More significantly, you should recognize that most of these investigators were in their 30s at the time of their noted scientific contribution (average age about 37). Thus it is in science that a person's greatest contributions typically are made when she or he is young and not yet fully established in a field. Fresh ideas come from fresh minds.

23. If the particles had a greater charge, they would be bent more because the deflecting force is directly proportional to the charge. (If the particles were more massive, they would be bent less by the magnetic force—obeying the law of inertia.)

25. The neon sign is a fancy cathode ray tube, which means that the ray of light is the result of the flow of electrons through the tube. A magnet held up to these flowing electrons alters their path, which shows up as a distortion of the light ray.

27. The one on the far right, where the nucleus is not visible.

29. The resulting nucleus would be that of arsenic, which is poisonous!

31. The iron atom is electrically neutral when it has 26 electrons to balance its 26 protons.

33. Carbon-13 atoms have more mass than carbon-12 atoms. Because of this, a given mass of carbon-13 contains fewer atoms than the same mass of carbon-12. Look at it this way—golf balls have more mass than Ping-Pong balls. So, which contains more balls: 1 kilogram of golf balls or 1 kilogram of Ping-Pong balls? Because Ping-Pong balls are so much lighter, you need many more of them to get 1 kilogram.

Problems

1. Eight grams of oxygen will react with only 1 gram of hydrogen to produce $8 + 1 = 9$ grams of water. Because only 1 gram of the hydrogen reacts, there is $8 - 1 = 7$ grams of hydrogen left over.

3.

	Mass (amu)		Fraction of Abundance		
Mass of ^{6}Li	6.0151	×	0.0742	=	0.446
Mass of ^{7}Li	7.0160	×	0.9258	=	6.495
					6.941 amu

Chapter 4

1. Any radioactive sample is always a little warmer than its surroundings because the radiating alpha or beta particles impart heat energy to the atoms. (Interestingly enough, the heat energy of the Earth originates from the radioactive decay of the Earth's core and surrounding material.)

3. Alpha and beta particles are deflected in opposite directions in a magnetic field because they are oppositely charged—alpha positive and beta negative. Gamma rays have no electric charge and are therefore undeflected.

5. Alpha radiation decreases the atomic number of the emitting element by 2 and the mass number by 4. Beta radiation increases the atomic number by 1 and does not affect the mass number. Gamma radiation does not affect either atomic number or mass number. So alpha radiation results in the greatest change both in atomic number and in mass number.

7. Only gamma rays are able to penetrate the metal hull of an airplane. Flight crews are required to limit their flying time so as to minimize the potential harm caused by the significant amounts of radiation present at high altitudes. Recall from the text that merely two round-trip flights from New York to California expose each human in the plane to about as much radiation as a chest X ray.

9. Alpha particles are much bigger than beta particles, which makes alphas less able to pass through the "pores" of materials. The smaller beta particles are therefore more effective in penetrating materials.

11. To decay to one-sixteenth the original amount will take four half-lives, which in this case equals 120 years.

13. When radium (atomic number 88) emits an alpha particle, its atomic number decreases by 2, and so the resulting nucleus is radon (atomic number 86). The alpha decay causes the mass number of the radium to decrease by 4. Because the radium here is the isotope radium-226, the radon isotope must have mass number 222, which means its atomic mass is 222 atomic mass units.

15. An element can decay to one that has a higher atomic number by emitting an electron (beta particle). When this happens, a neutron becomes a proton and the atomic number increases by 1.

17. The Earth's natural energy that heats the water in the hot spring is the energy of radioactive decay, which keeps the Earth's interior molten. Radioactivity heats the water but doesn't make the water radioactive. The warmth of hot springs is one of the nicer effects of radioactive decay. You and your friend will most likely encounter more radioactivity from the granite outcroppings of the foothills than you would

encounter near a nuclear power plant. Furthermore, at high altitude you'll both be exposed to increased cosmic radiation. However, these radiations are not appreciably different from the radiation you would encounter in the "safest" of situations. The probability of dying from something or other is 100 percent, and so in the meantime you and your friend should enjoy life!

19. Although there is significantly more radioactivity in a nuclear power plant than in a coal-fired plant, the absence of shielding in the coal plant results in more radioactivity in the environment around the coal plant. All nuclear plants are shielded; coal plants are not.

21. Radioactive decay rates are statistical averages of large numbers of decaying atoms. Because of the relatively short half-life of carbon-14, only trace amounts would be left after 50,000 years—too little to be statistically meaningful.

23. At 3:00 P.M., there will be 1/8 gram left. At 6:00 P.M., 1/64 gram left. At 10:00 P.M., 1/1024 gram left.

25. A neutron makes a better "bullet" for penetrating atomic nuclei because it has no electric charge and is therefore not deflected from its path by electrical interactions, nor is it electrically repelled by the nuclei.

27. Plutonium has a short half-life (24,360 years) relative to the age of the Earth (about 4 billion years), and so any plutonium initially in the Earth's crust has long since decayed. The same is true for any heavier elements that have even shorter half-lives and might have been the elements that decayed to plutonium. Trace amounts of plutonium can occur naturally in uranium-238 deposits, however, as a result of neutron capture. The neutron capture changes uranium-238 to uranium-239, which then beta decays to neptunium-239. The neptunium-239 then beta decays to plutonium-239. (There are elements in the Earth's crust with half-lives shorter than plutonium's, but these are the products of uranium decay and fall between uranium and lead in the periodic table.)

29. Because fissionable plutonium-239 is formed as the uranium-238 absorbs neutrons from the fissioning uranium-235.

31. To predict the energy release of a nuclear reaction, the physicist needs to know the difference between the average mass per nucleon for the beginning nucleus and the average mass per nucleon for the products formed in the reaction (either fission or fusion). This mass difference (called the *mass defect*) can be found from the curve of Figure 4.31 or from a table of nuclear masses. The physicist then multiplies this mass difference by the speed of light squared to determine the energy released—$E = mc^2$.

33. If a uranium nucleus split into three equal-size fragments, each would be smaller than if the nucleus split into only two fragments. The smaller fragments would have lower atomic numbers, more toward iron on the graph of Figure 4.29. The resulting mass per nucleon would be less, which means more mass was converted to energy and so more energy was released.

35. Such speculation could fill volumes. The energy and material abundance that are the expected outcome of a fusion age will likely prompt several fundamental changes. Obvious changes would occur in the fields of economics and commerce, which would be geared to relative abundance rather than scarcity. Our present price system, which is geared to and in many ways dependent on scarcity, often malfunctions in an environment of abundance. Hence we see instances where scarcity is created to keep the economic system functioning. Change at the international level will likely include worldwide economic reform, and change at the personal level might mean a re-evaluation of the idea that scarcity ought to be the basis of value. A fusion age will likely see changes that will touch every facet of our way of life.

Chapter 5

1. Atoms are smaller than the wavelengths of visible light and hence not *visible* in the true sense of the word. We can, however, measure the surface topography of a collection of atoms by scanning an electric current back and forth across the surface. A computer can then assemble the data from such scanning into an image that reveals how the atoms are arranged on the surface. It would be more appropriate to say that with the scanning tunneling microscope we *feel* atoms rather than *see* them.

3. Many objects or systems can be described just as well by a physical model as by a conceptual model. In general, a physical model is used to replicate an object on some convenient scale. A conceptual model is used to represent abstract ideas or to demonstrate how a system behaves. Of the examples given in the exercise, the following are best described by a physical model: brain, solar system, stranger, gold coin, car engine, and virus. The following are best described by a conceptual model: mind, birth of universe, best friend (whose complex behavior you have some understanding of), dollar bill (which represents wealth but is really only a piece of paper), spread of a sexually transmitted disease.

5. The one electron can be boosted to many energy levels and can therefore make many combinations of transitions to lower levels. Each transition is of a specific energy and accompanied by the emission of a photon of a specific frequency. Thus the variety of spectral lines.

7. The sum of the energies of the frequencies emitted during the two-step transition must equal the energy of the frequency emitted during the one-step transition.

9. Blue light has a higher frequency than red light. Therefore the blue light comes from the greater energy transition.

11. An electron not restricted to particular energy levels would release light continuously as it spiraled closer in toward the nucleus. A broad spectrum of colors would be observed rather than distinct lines.

13. It takes *no time at all* for the transition to occur; it is instantaneous. The electron is *never* found between two orbitals.

15. Because of the electron's wave nature, it would be better to say that the electron exists in both lobes at the same time.

17. Electrons ordinarily fill lower orbitals before entering higher ones. The electron configuration for carbon shown on the right, $1s^2 2s^1 2p^3$, shows that one of the $2s$ electrons has been boosted to a (higher-energy) $2p$ orbital. This configuration therefore represents the greater amount of energy.

19. Lowest energy $1s^2 2s^2 2p^5$, highest energy $1s^0 2s^2 2p^5 3s^2$.

21.

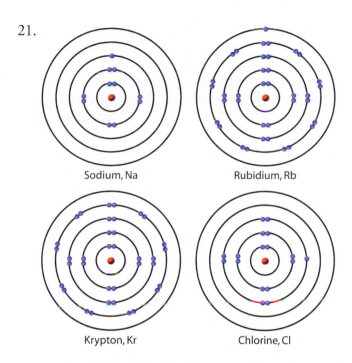

Sodium, Na   Rubidium, Rb

Krypton, Kr   Chlorine, Cl

23. An orbital is just a region of space in which an electron of a given energy resides. This region of space exists with or without the electron. The same logic applies to a shell, which is just a collection of orbitals of similar energy levels. The space defined by the shell exists whether or not an electron is to be found there.

25. Neon's outermost occupied shell is filled to capacity with electrons. Any additional electrons would have to occupy the next shell out, which has an effective nuclear charge of zero.

27. An electron in the outermost occupied shell of sodium experiences the greatest effective nuclear charge. The strength of the electric force diminishes with increasing distance, and the outermost sodium electron is closer to the nucleus than are the outermost electrons of the other atoms discussed in the Exercise.

29. Pb < Sn < As < P.

31. Potassium has one electron in its outermost occupied shell, which is the fourth shell. The effective nuclear charge felt by the electron in this shell is relatively weak (+1), and so this electron is readily lost. If the atom were to lose a second electron, that electron would need to come from the next shell inward (the third shell), where the effective nuclear charge is much stronger (+9). Thus, it is very difficult to pull a second electron away from the potassium atom

because this electron is being held so tightly by this much greater effective nuclear charge.

33. Note that each shell has been divided into a series of finer shells known as *subshells*. Each subshell contains only one type of orbital. In the seventh shell, for instance, working in order of increasing energy levels, the first (innermost) subshell, indicated by the digit 2 in the right-most stack of numbers in the drawing, is for the 2 electrons of the $7s$ orbital. The second subshell of the seventh shell is for the 14 electrons of the seven $5f$ orbitals, the third subshell is for the 10 electrons of the five $6d$ orbitals, and the fourth subshell is for the 6 electrons of the three $7p$ orbitals. Gallium is larger than zinc because gallium has an electron in each of the three subshells of the fourth shell, whereas zinc has an electron only in each of the first two inner subshells of the fourth shell. What you see here is a refinement of the model presented in Section 5.7. Don't worry about fully understanding this refinement, however. Rather, it is more important that you understand that all conceptual models are subject to refinement. We choose whatever level of refinement best suits our needs in a given situation.

Chapter 6

1. This is an example of a chemical change involving the formation of ions, which are different from the neutral atoms from which the ions are made.
3. The nuclear charge experienced by an electron in sodium's third shell is not strong enough to hold this many electrons. As was discussed in Chapter 5, this is because there are 10 inner-shell electrons shielding any third-shell electron from the +11 nucleus. The effective nuclear charge in the third shell is therefore about +1, which means that the shell is able to hold only one electron.
5. Ba_3N_2.
7. There is no more room available in its outermost occupied shell.
9. The hydrogen atom has only one electron to share.
11. There is a gradual change. You can determine which type of bond is likely by noting relative positions in the periodic table. If the elements forming a bond are close together in the peri-

odic table and toward the upper right of the table, the bond is more covalent. If the elements are on opposite sides of the periodic table, the bond is more ionic. For elements between these two extremes, the bonding tends to be a blend, referred to as a *polar covalent* bond.

13. When bonded to an atom that has low electronegativity, such as any group 1 element, the nonmetallic atom pulls the bonding electrons so close to itself that an ion is formed.
15. The chemical formula for phosphine is PH_3, which is similar to the formula for ammonia, NH_3. Note how phosphorus is directly below nitrogen in the periodic table.
17. $:\ddot{Cl}:^{1-}$ Ca^{2+} $:\ddot{Cl}:^{1-}$

19. $\overset{\displaystyle H}{\underset{\displaystyle H}{:\ddot{O}:\ddot{O}:}}$

21. There are four substituents and no nonbonding pairs around the central atom. Therefore the shape of this molecule is the same as the geometry: tetrahedral.
23. The source of an atom's electronegativity is the positive charge of the nucleus. More specifically, it is the effective nuclear charge experienced in the shell the bonding electrons are occupying.
25. The least symmetrical molecule: O=C=S.
27. N–N < N–O < N–F < H–F.
29. Water is a polar molecule because the dipoles in the H_2O molecule do not cancel. Polar molecules tend to stick to one another, which gives rise to relatively high boiling points. Methane is a nonpolar molecule because of its symmetry, which results in no net dipole in the molecule and a relatively low boiling point. The boiling points of water and methane are influenced only a little by molecular mass but a great deal by the degree of intermolecular attraction in each liquid.
31. The greater the polarity of the molecules of a substance, the higher the boiling point of the substance. (a) The compound with the two chlorines on the same side of the molecule is more polar and thus has the higher boiling point. (b) The SCO molecule, with a carbon atom flanked by a sulfur atom and an oxygen atom, is less symmetrical, which means that the dipoles in it do not cancel as well as those in the symmetrical carbon dioxide molecule, CO_2, do. The SCO therefore has the higher boiling point. (c) The chlorine

atoms have a relatively strong electronegativity that pulls electrons away from the carbon. In the $COCl_2$ molecule, this tug by the chlorine atoms is counteracted by a relatively strong tug by the oxygen atom, which tends to decrease the polarity of this molecule. The hydrogens of the $C_2H_2Cl_2$ molecule have an electronegativity that is less than that of carbon, and so they assist the chlorine atoms in allowing electrons to be yanked toward the Cl side of the molecule, which means this molecule is more polar than $COCl_2$. The $C_2H_2Cl_2$ therefore has the higher boiling point.

Chapter 7

1. Ion–dipole attractions are stronger because the magnitude of the electric charge associated with an ion is much greater than the magnitude of the charge associated with a dipole.

3. It sticks to itself by way of induced dipole–induced dipole attractions between molecules.

5. The boiling points go up because of an increase in the number of molecule-to-molecule interactions. The boiling point of a substance refers to a pure sample of that substance. The boiling point of 1-pentanol (the third molecule in the list) is relatively high because 1-pentanol molecules are very strongly attracted to one another (by induced dipole–induced dipole, dipole–dipole, and dipole–induced dipole attractions). Solubility refers to how well a substance interacts with a second substance—in this case, water. Water is much less attracted to 1-pentanol molecules because most of a 1-pentanol molecule is non-polar (the only polar portion is the –OH group). For this reason, 1-pentanol is not very soluble in water. Put yourself in the position of a water molecule and ask yourself how strongly you are attracted to a methanol molecule (the first molecule in the list) compared to how strongly you are attracted to a pentanol molecule.

7. Nitrogen atoms are bigger than helium atoms, and so nitrogen molecules should be more soluble in water because of more possibilities for dipole–induced dipole attractions.

9. The helium is less soluble than nitrogen in body fluids, and so less of it dissolves at a given pressure. Upon decompression, there is less helium to "bubble out" and cause potential harm.

11. One way is to add more sugar and see if it dissolves. If it does, the solution was not saturated.

Alternatively, cool the solution a couple of degrees and see if any sugar precipitates out. If any does, the solution was saturated. Because sugar forms supersaturated solutions so easily, however, neither of these methods is foolproof.

13. The sodium nitrate solution is more concentrated (85 grams/milliliter versus only about 38 grams/milliliter for the sodium chloride solution).

15. Assuming concentration is given as mass (or moles) of solute in a given volume of solution, the concentration decreases with increasing temperature because you have the same mass of solute as before but dissolved in a larger volume of solvent.

17. For a given substance, molecule polarity has a far greater influence on boiling point than does molecule mass. Recall that boiling involves separating the molecules of a liquid from one another. (The more polar the molecules, the more "sticky" they are and hence the higher the boiling point.) The extra neutron in deuterium does not affect the chemical bonding in the D_2O molecule, which means that D_2O has the same chemical structure as H_2O. With the same chemical structure, these two molecules have about equal polarity, making their boiling points very close: 100°C for H_2O and 101°C for D_2O. The extra mass of the deuterium has only a small affect on boiling point.

19. Although oxygen gas, O_2, has poor solubility in water, many examples have good solubility in water. From the concepts discussed in Chapter 6, you should be able to deduce that hydrogen chloride is a somewhat polar molecule. This gaseous material therefore has good solubility in water by virtue of the dipole–dipole attractions between HCl and H_2O molecules.

21. Motor oil is much thicker (more viscous) than gasoline. This difference suggests that molecules of motor oil are more strongly attracted to one another than are molecules of gasoline. A material consisting of molecules of structure A would have greater induced dipole–induced dipole interactions than a material consisting of molecules of structure B. Motor oil molecules therefore are best represented by structure A, and gasoline molecules are best represented by structure B.

23. The arrangement of atoms in a molecule makes all the difference as to physical and chemical

properties. Ethanol contains the −OH group, which is polar. This polarity is what allows the ethanol to dissolve in water. The oxygen of dimethyl ether is bonded to two carbon atoms. The difference in electronegativity between oxygen and carbon is not as great as the difference in electronegativity between oxygen and hydrogen. The polarity of the C−O bond is therefore less than that of the O−H bond. As a consequence, dimethyl ether is significantly less polar than ethanol and not readily soluble in water.

25. In order for you to smell something, the molecules of that something must evaporate and reach your nose. If the new perfume doesn't evaporate, it will not have an odor.

27. These bubbles are gases that were dissolved in the cold water and are now coming out of solution. Because the solubility of gases in water decreases with increasing temperature, a pot of warm water has more bubbles than a pot of cold water.

29. Most home water softeners work by replacing the calcium and magnesium ions of the tap water with sodium ions. The softened water therefore contains increased levels of sodium ions.

Problems

1. Multiply solution concentration by amount of solution to get amount of solute:
(0.5 g/L)(5 L) = 2.5 g.

3. $\dfrac{1 \text{ mole}}{1 \text{ L}} = 1 \text{ M}$

$\dfrac{2 \text{ moles}}{0.5 \text{ L}} = 4 \text{ moles} / \text{L} = 4 \text{ M}$

Chapter 8

1. As discussed in Chapter 6, both hydrogen fluoride, HF, and ammonia, NH_3, are polar molecules. Thus the molecule-to-molecule attractions in either liquid are relatively strong, which translates to relatively high boiling points.

3. As the ice melts, the water level does not change. For insight as to why this is so, think about the answer to Exercise 4.

5. It is important to keep the pipes from freezing because the water in them expands more than the pipe material does, and the expanded volume can fracture the pipes.

7. Any portion of the added heat that goes to the ice increases the vibrations in the H_2O molecules making up the ice. The increased vibrations cause more hydrogen bonds to be broken. Because hydrogen bonds are largely responsible for holding ice crystals together, the increased vibrations mean less ice can form.

9. Graph d.

11. Just as the presence of a solute affects the rate at which ice crystals form, so does it affect the rate at which ice microcyrstals form. As you should recall, it is the formation of microcrystals in fresh water that results in fresh water's expansion as it cools to below 4°C. Without the formation of microcrystals, ocean water continues to contract all the way to its freezing temperature. Ocean water is therefore most dense *just before* it freezes, which makes its freezing behavior most different from that of fresh water. To see for yourself, place a cupful of salt water next to a cupful of fresh water in your kitchen freezer.

13. As it cools to 4°C, the oxygen-rich surface water sinks to the bottom of the lake. This is of benefit to the aquatic organisms living at the bottom. Concurrently, the nutrient-rich deeper water is pushed to the surface. This is of benefit to the aquatic organisms living near the surface.

15. Nutrients are concentrated mostly at lower depths in any body of water. Because tropical surface water never sinks, there is no circulation in the water, with the result that nutrients stay on or near the ocean floor and the water is clear to a considerable depth.

17. The mercury–mercury cohesive forces must be stronger than the mercury–glass adhesive forces.

19. At colder temperatures, water molecules are moving more slowly, which makes it relatively easy for them to cohere to one another, increasing the surface tension. A tablespoon of vegetable oil poured into a pot of cold water typically stays together as a single blob (because the water molecules adjacent to the blob are so attracted to one another). At higher temperatures, water molecules are moving faster, which makes it more difficult for them to cohere to one another, decreasing the surface tension. A tablespoon of vegetable oil poured into a pot of hot water is more inclined to disperse throughout the water, which is of great benefit when someone is cooking spaghetti.

21. Water is not strongly attracted to the wax surface, which is nonpolar. To minimize surface area, the water on the wax surface tends to form a sphere. Sitting on a solid surface, however, the sphere is squashed down into a bead by the force of gravity.

23. When a wet finger is held to the wind, evaporation is greatest on the windy side, which feels cool. The cool side of your finger is windward.

25. A bottle wrapped in a wet cloth cools as the liquid evaporates from the cloth. As evaporation progresses, the average temperature of the water left behind in the cloth can easily drop to below the temperature of the water originally used to wet the cloth. So to cool a bottle of beer, soda, or whatever at a picnic, wet a piece of cloth in a bucket of cool water and wrap the wet cloth around the bottle. As evaporation progresses, the temperature of the water in the cloth drops and cools the bottle to a temperature below that of the water in the bucket.

27. A high temperature and the resulting heat energy given to the food are responsible for cooking. If the water boils at a low temperature (presumably under reduced pressure), there is not enough heat energy to cook the food.

29. The instructor removed most of the air from the flask (which is only partially filled with water). Removing the air caused the air pressure above the water surface to be extremely low. At such low pressure, the small amount of heat that travels from your hand to the water is enough to cause the water to boil. Remember, the boiling point of any liquid decreases as atmospheric pressure decreases.

31. The lid traps heat inside the pot, which shortens the amount of time it takes to bring all the water to its boiling point. The lid also increases the pressure exerted on the water, which raises the boiling point. The hotter water cooks food in a shorter time.

33. You want your radiator liquid to absorb heat from the engine so that the engine doesn't melt. A liquid that has a high specific heat capacity will be more effective at absorbing this heat. Engine efficiency increases with increasing temperature, however, and so keeping the engine too cool is not desirable. Commerically sold radiator liquids are formulated to have a specific heat capacity that helps your engine run at an optimal temperature.

35. The climate of Bermuda, like that of all other islands, is moderated by the high specific heat capacity of water. The climate is moderated by the large amounts of energy given off and absorbed by water for small changes in temperature. When the air is cooler than the water, the water warms the air; when the air is warmer than the water, the water cools the air.

37. Much of the heat from the oven is consumed in changing the phase of the water from liquid to gas. As long as any of the water remains in the liquid phase, the temperature of the oven will not rise much higher than the boiling point of the water—100°C.

39. This heat is radiated outward to the environment. In order to melt the ice, the heat would have to be reflected back into the ice.

Problems

1. heat into water = specific heat capacity × mass × temperature change

$$= (4.184 \text{ J/g} \cdot °C)(100 \text{ g})(+7 \, °C)$$
$$= 2928.8 \text{ J}$$

which to the proper number of significant figures is 3000 J.

3. temperature change

$$= \frac{\text{heat added}}{\text{specific heat capacity} \times \text{mass}}$$
$$= \frac{230 \text{ J}}{(4.184 \text{ J/g} \cdot °C)(5.0 \text{ g})}$$
$$= 11 \, °C$$

Chapter 9

1. a. 4, 3, 2 b. 3, 1, 2 (Remember that, by convention, a coefficient of 1s is not shown in the balanced equation.)

3. H_2O, 18 atomic mass units; C_3H_6, 42 atomic mass units; C_3H_8O, 60 atomic mass units.

5. They have the same number of atoms. The mass of NH_3 represents 1 mole of NH_3, which means 6.02×10^{23} NH_3 molecules. Because there are four atoms per molecule, there are $4(6.02 \times 10^{23}) = 24.08 \times 10^{23}$ atoms in the 17.031 grams. The mass of HCl represents

2 moles of HCl, which means $2(6.02 \times 10^{23})$ molecules. Because there are only two atoms per molecule in this case, you have $2(2)(6.02 \times 10^{23}) = 24.08 \times 10^{23}$ atoms in the 72.922 grams.

7. There is 1 mole of N_2 in 28 grams of N_2, 1 mole of O_2 in 32 grams of O_2; 2 moles of CH_4 in 32 grams of CH_4, and 1 mole of F_2 in 38 grams of F_2. Therefore the answer is c.

9. They assumed incorrectly that one hydrogen atom bonds to one oxygen atom to form water, which would mean that the chemical formula was HO. We know today, however, that two hydrogen molecules (not atoms) react with one oxygen molecule to form water. By a count of molecules—which translates to a count of atoms—the hydrogen and oxygen react in a 2:1 ratio and the formula for water is H_2O. By mass, hydrogen and oxygen still always react in a 1:8 ratio. Because two hydrogen atoms are needed for every one oxygen atom, however, this ratio is better expressed as 0.5 gram of hydrogen to 4 grams of oxygen. Comparing one hydrogen atom to one oxygen atom thus shows us that oxygen is actually 16 times more massive than hydrogen.

11. A single oxygen atom has the very small mass of 16 atomic mass units.

13. Because 1 atomic mass unit equals 1.661×10^{-24} gram, 16 atomic mass units must equal $(16)(1.661 \times 10^{-24}$ gram$) = 26.576 \times 10^{-24}$ gram $= 2.6576 \times 10^{-23}$ gram.

15. No, because this mass is less than that of a single oxygen atom.

17. The water molecules have the greater mass, just as a bunch of golf balls have more mass than the same number of Ping-Pong balls. The water molecules have about nine times more mass because each H_2O molecule ($16 + 1 + 1 = 18$ amu) is about nine times more massive than each H_2 molecule ($1 + 1 = 2$ amu): 18 amu/2 amu $= 9$. The big numbers don't change anything: 1.204×10^{24} molecules of water have a greater mass than 1.204×10^{24} molecules of molecular hydrogen.

19. Chemical reactions, including the biological ones responsible for the spoilage of food, slow down with decreasing temperature. The refrigerator therefore merely delays spoilage.

21. In pure oxygen, there is a greater concentration of one of the reactants (oxygen) for the chemical reaction (combustion). The greater the concentration of reactants, the higher the rate of the reaction.

23. The bubbling is the result of a reaction between the Alka-Seltzer tablet and the water. In the alcohol–water mix, there is a lower proportion of water molecules, which leads to a lower rate of reaction. In terms of molecular collisions, with fewer water molecules around, the probability of collisions between Alka-Seltzer and water molecules is less in the alcohol–water mix.

25. The final result of this reaction is the transformation of three oxygen molecules, O_2, to two ozone molecules, O_3. Although there is no net consumption or production of nitrogen monoxide, NO, nitrogen dioxide, NO_2, or atomic oxygen, O, only the nitrogen monoxide appears to be required for this reaction to begin. Therefore the nitrogen monoxide is best described as the catalyst.

27. Putting more ozone into the atmosphere to replace what's been destroyed is like throwing more fish into a pool of sharks to replace those the sharks have eaten. The solution to the ozone problem is to remove the CFCs that destroy the ozone. Unfortunately, CFCs degrade only slowly, and the ones up there now will remain there for many years to come. Our best bet is to stop producing CFCs and hope we haven't already caused too much damage.

29. $N_2H_4 \longrightarrow 2\,H_2 + N_2$

Energy absorbed as bonds break	Energy released as bonds form
N–N = 159 kJ	H–H = 436 kJ
N–H = 389 kJ	H–H = 436 kJ
N–H = 389 kJ	
N–H = 389 kJ	
N–H = 389 kJ	N≡N = 946 kJ
Total = +1715 kJ absorbed	Total = −1818 kJ released

$$\text{net energy of reaction} = +1715 \text{ kJ} + (-1818 \text{ kJ})$$
$$= -103 \text{ kJ (negative net energy means exothermic reaction)}$$

$$2\ H_2O_2 \longrightarrow O_2 + 2\ H_2O$$

Energy absorbed as bonds break	**Energy released as bonds form**
O–O = 138 kJ	
H–O = 464 kJ	O=O = 498 kJ
H–O = 464 kJ	H–O = 464 kJ
O–O = 138 kJ	H–O = 464 kJ
H–O = 464 kJ	H–O = 464 kJ
H–O = 464 kJ	H–O = 464 kJ
Total = +2132 kJ absorbed	Total = −2354 kJ released

$$\text{net energy of reaction} = +2132\ \text{kJ} + (-2354\ \text{kJ})$$
$$= -222\ \text{kJ (negative net energy means exothermic reaction)}$$

31. The reactions in a disposable battery are exothermic, evidenced by the fact that the battery is producing energy used to operate some device. The reactions taking place as a rechargeable battery is recharged are endothermic, evidenced by the fact that you need to connect the recharger to an external supply of electricity. (Both the disposable battery and the recharging one get warm because of the heat generated as electricity passes through them.)

33. Both are right, provided they recognize that the energy *contained* in the brick is dispersed throughout the brick.

35. Look at Table 9.2 and you'll see some examples of combustible fuels, including hydrogen, H_2, methane, CH_4, and methanol, CH_3OH. You can think of these fuels as sources of relatively concentrated energy, which means their entropies are relatively low. As they burn in oxygen, this concentrated energy is dispersed.

37. Recall from the Hands-On Chemistry of Section 9.5 that this is what happens when a salt, such as sodium chloride, is dissolved in water. As the ions separate from one another, some of the vibrational energy they have when together is transformed to the potential energy of their being apart. The effect is that the ions are no longer vibrating so fast, which we detect as a drop in temperature. Energy from the higher-temperature surroundings then rushes in to contribute to the process. This occurs, however, only because of the large entropy increases that occur as the ions disperse into solution.

Problems

1. $$(0.250\ \text{g})\left(\frac{1\ \text{mole}}{180\ \text{g}}\right)\left(\frac{6.02 \times 10^{23}\ \text{molecules}}{1\ \text{mole}}\right)$$
$$= 8.36 \times 10^{20}\ \text{molecules}$$

3. The chemical equation tells you that 1 mole of 2-propanol yields 1 mole of propene and 1 mole of water. From the chemical formula C_3H_8O, you know that the molar mass of 2-propanol is 60 grams/mole, and so 6.0 grams is 0.10 mole of 2-propanol. Therefore this mass of 2-propanol yields 0.10 mole of propene and 0.10 mole of water. The masses of the two products are therefore

C_3H_6 (42 g/mole)(0.10 mole) = 4.2 g
H_2O (18 g/mole)(0.10 mole) = 1.8 g

5. Absorbed:

1 mole N≡N × 946 kJ/mole = 946 kJ
3 moles H–H × 436 kJ/mole = 1308 kJ
2254 kJ

Released (2 moles NH_3 contain 6 moles N–H bonds):

6 moles N–H × (−436 kJ/mole) = −2334 kJ

net energy of reaction = energy absorbed + energy released
$$= 2254\ \text{kJ} + (-2334\ \text{kJ})$$
$$= -80\ \text{kJ}$$

The negative value means this energy is released.

7. Change in S from products − reactants
= entropy of 2 moles NH_3 − entropies of 1 mole N_2 and 3 moles H_2
= 2(192.5 J/K) − [191.6 J/K + 3(130.7 J/K)]
= 385.0 J/K − 583.7 J/K
= −198.7 J/K

The result in Problem 5, −80 kJ, tells you 80,000 joules of energy is released during this reaction. Because you are now looking at the energy gained by the surroundings, change the sign to positive: +80,000 joules gained by surroundings.

Change in S from energy released during reaction
= +80,000 J/723K
= +110 J/K

Net change in S of universe
= −198.7 J/K + 110 J/K
= −89 J/K

Ammonia, NH_3, is a very important chemical, ranking among the top five chemicals produced worldwide. Its primary use is as a fertilizer, and so its production is directly linked with our food supply. Your calculations should show that the formation of ammonia becomes less favored with higher temperature (negative net entropy change). Industry overcomes this problem by using catalysts and by performing the reaction under high pressure.

Chapter 10

1. Potassium carbonate in the ashes acts as a base and reacts with skin oils to produce a slippery solution of soap.

3. The base accepted one or more hydrogen ions, H^+, and thus gained positive charge. In order to accept H^+ ions, the base had to give up some other positive ions. The base thus formed the positive ions of the salt. The acid donated one or more hydrogen ions and thus lost positive charge. Once the acid lost all its H^+ ions, what remained was only negative ions. The acid thus formed the negative ions of the salt.

5. In the first reaction, the H_3O^+ transforms to a water molecule. This means the H_3O^+ loses a hydrogen ion, which is donated to the Cl^-. The H_3O^+ therefore is behaving as an acid, and the Cl^- is behaving as a base. In the reverse direction, the H_2O gains a hydrogen ion (behaving as a base) to become H_3O^+. It gets this hydrogen ion from the HCl, which in donating is behaving as an acid. You should be able to make similar arguments to arrive at the following answers:
   a. acid, base, base, acid.
   b. acid, base, acid, base.
   c. acid, base, acid, base.

7. When the sodium hydroxide, NaOH, accepts a hydrogen ion from water, the products formed are water and sodium hydroxide! In solution this sodium hydroxide is dissolved as individual sodium ions and hydroxide ions.

9. That the value of K_w is so small tells you that the extent to which water ionizes is also quite small.

11. a. As more hydronium and hydroxide ions form as the temperature increases, the concentration of these ions increases. This means that the product of their concentrations, which is K_w, also increases. Thus, K_w is constant only so long as the temperature is constant. Interestingly, K_w is 1.0×10^{-14} only at 24°C. At 40°C, K_w is 2.92×10^{-14}.

    b. pH is a measure of hydronium ion concentration. The greater the hydronium ion concentration, the lower the pH. According to the information given in this Exercise, the hydronium ion concentration increases (albeit only slightly) as water warms up. Thus, pure water that is hot has a slightly lower pH than pure water that is cold.

    c. As water warms up, the hydronium ion concentration increases, but so does the hydroxide ion concentration—and by the same amount. Thus, the pH decreases and yet the solution remains neutral because the hydronium ion and hydroxide ion concentrations are still equal. At 40°C, for example, the hydronium ion and hydroxide ion concentrations in pure water are both 1.71×10^{-7} mole per liter (the square root of K_w). The pH of this solution is the negative of the logarithm of this number, which is 6.77. This is why most pH meters need to be adjusted for the temperature of the solution being measured. Except for here in this Exercise, which probes your powers of analytical thinking, this textbook ignores the slight role that temperature plays in pH. Unless noted otherwise, please continue to assume that K_w always has the value 1.0×10^{-14}—in other words, assume that the solution being measured is at 24°C.

13. pH = $-\log [H_3O^+]$ = $-\log 1$ = -0 = 0

    This an acidic solution. Yes, pH can be equal to zero!

15. Because pH is defined as the *negative* of the logarithm, you have

$$pH = -\log [H_3O^+] = -3$$
$$\log [H_3O^+] = 3$$
$$[H_3O^+] = 10^3 \ M = 1000 \ M$$

The solution would be impossible to prepare because only so much acid can dissolve in water before the solution is saturated. The highest concentration possible for hydrochloric acid, for example, is 12 *M*. Beyond this concentration, any HCl, which is a gas, added to the water simply bubbles back out into the atmosphere.

17. The concentrated solution of hydrochloric acid becomes more dilute in hydronium ions as the weak acid is added. The solution therefore becomes less acidic.

19. If the chalk is made of calcium carbonate, $CaCO_3$, it is made of the active ingredient in many antacids. Calcium carbonate is a base that neutralizes an acid. Be careful, though, never to take too much calcium carbonate because the stomach is designed to always be somewhat acidic.

21. Limestone (also known as calcium carbonate) is a base. Thus any slight amounts of it dissolved in lake water neutralize some of the acid rain that enters the lake. Granite is not basic, and so there is no such neutralizing possible for acid rain falling into granite-lined lakes.

23. The warmer the ocean, the lower the solubility of any dissolved gases, such as carbon dioxide, CO_2. Thus less CO_2 would be absorbed in a warmer ocean, meaning that more of this gas would remain in the atmosphere to perpetuate global warming.

25. The hydrogen chloride, which behaves as an acid, reacts with the ammonia, which behaves as a base, to form ammonium chloride. The concentration of ammonium chloride increases, and the concentration of ammonia decreases.

27. When the active buffering components are all neutralized.

Problems

1. The concentration of hydroxide ions must be 1×10^{-4} mole per liter because $[H_3O^+][OH^-] = 1 \times 10^{-14}$ mole per liter.

3. The pH of this solution is 4, and the solution is acidic.

5. Doubling the volume of solution means that the hydronium ion concentration is cut in half. The hydronium ion concentration after dilution is therefore 0.05 *M*. The pH is

$$pH = -\log [H_3O^+]$$
$$= -\log 0.05$$
$$= -(-1.3) = 1.3$$

Chapter 11

1. The atom that loses an electron and thus becomes positively charged (the red one) is the one that is oxidized.

3. An oxidizing agent causes other materials to lose electrons. It does so by its tendency to gain electrons. Atoms with high electronegativity tend to have a strong attraction for electrons and therefore behave as strong oxidizing agents. Conversely, a reducing agent causes other materials to gain electrons. It does so by its tendency to lose electrons. Atoms with high electronegativity therefore have little tendency to behave as reducing agents.

5. Fluorine should be the stronger oxidizing agent because it has the greater effective nuclear charge.

7. The electrons flow from the nail to the copper ions.

9. The anode is where oxidation takes place and where free-roaming electrons are generated. The negative sign at a battery anode indicates that this electrode is the source of negatively charged electrons. They run from the anode, through an external circuit, to the cathode, which carries a positive charge to which the electrons are attracted. (When a battery is being recharged, energy is used to force electrons in the opposite direction. In other words, during recharging, electrons move from the positive electrode to the negative electrode— a place where they would not ever go without the input of energy. Electrons are thus gained at the negative electrode, which is now classified as the cathode because the cathode is where reduction—the gain of electrons— occurs. Look carefully at Figure 11.10b to see how this is so.)

11. According to the chemical formula for iron hydroxide, there are two hydroxide groups for

every one iron ion. Each hydroxide group has a single negative charge. This means that the iron ion must carry a double positive charge, which is no different from the free Fe^{2+} ion from which the $Fe(OH)_2$ is formed. This reaction is merely the coming together of oppositely charged ions.

13. The Cu^{2+} ion is reduced as it gains electrons to form copper metal, Cu. The magnesium metal, Mg, is oxidized as it loses electrons to form the Mg^{2+} ion.

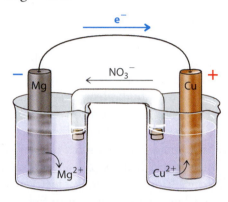

15. How much power a battery can deliver is a function of the number of ions in contact with the electrodes—the more ions, the greater the power. If the lead electrodes are completely submerged both before and after the water is added, diluting the ionic solution in the battery decreases the number of ions in contact with the electrodes and thus decreases the power of the battery. This effect is only temporary, however, because more ions are soon generated as the battery is recharged by the car's generator. If the water level inside the battery is so low that the lead electrodes are not completely submerged, adding water increases the amount of electrode surface area in contact with the solution. This counterbalances the weakening effect of diluting the ionic solution.

17. Aluminum oxide is insoluble in water and thus forms a protective coating that prevents continued oxidation of the aluminum.

19. Combustion reactions are generally exothermic because they involve the transfer of electrons to oxygen, which of all atoms in the periodic table has one of the greatest tendencies for gaining electrons.

21. In water, the oxygen is chemically bound to hydrogen atoms. That chemically bound −O−

is completely different from molecular oxygen, O_2, which is what is required for combustion. Another way to phrase your answer would be to say that the oxygen in water is already "reduced" in the sense that it has gained electrons from the hydrogen atoms to which it is attached. Being already reduced, this oxygen atom no longer has a great attraction for additional electrons.

23. This is very bad news. The iron atoms will lose electrons to the copper atoms, which, being very poor reducing agents, will pass those electrons on to oxygen molecules in the air in contact with the pipe surface, much as is indicated in Figure 11.17.

25. This water is in the gaseous phase and merely floats away from the fire.

Chapter 12

1. A 5-carbon saturated hydrocarbon contains 12 hydrogen atoms. A 5-carbon unsaturated hydrocarbon contains 10 or fewer hydrogen atoms.

3.

5. There are only two structural isomers drawn. The one in the middle and the one on the right are two conformations of the same isomer.

7. The pressure is greatest at the bottom because the temperature is highest here, and the high temperature means the greatest number of vaporized molecules present to cause the high pressure.

9. If the bulk of the alcohol molecule is a nonpolar hydrocarbon chain, the alcohol may be insoluble in water.

11. Ingesting methanol is indirectly harmful to eyes because in the body it is metabolized to formaldehyde, a chemical that damages eyes and can cause blindness. In addition, methanol, just like ethanol, has inherent toxicity and is thus also directly harmful.

13.

Caffeine
(free base)

13. H_3C ... $+ H_2O + Na^+H_2PO_4^-$

15. The HCl would react with the free base to form the hydrochloric acid salt of caffeine, which is soluble in water but insoluble in diethyl ether. With no water available to dissolve this material, it would precipitate out of the diethyl ether as a solid that could be collected by filtration.

17. The caprylic acid reacts with the sodium hydroxide to form a water-soluble salt, which dissolves in the water. The aldehyde is not acidic and so does not form a water-soluble salt.

19. No! This label indicates that the decongestant contains the hydrogen chloride salt of phenylephrine, *not* acidic hydrogen chloride. This organic salt is as different from hydrogen chloride as sodium chloride (table salt) is. (Sodium chloride could go by the name "hydrogen chloride salt of sodium.") Think of it this way: assume you have a cousin named George. You may be George's cousin, but in no way are you George. In a similar fashion, the hydrogen chloride salt of phenylephrine is made using hydrogen chloride, but it is in no way hydrogen chloride. A chemical substance is different from the elements or compounds from which it is made.

21. Structure b. At acidic pH values, the nitrogen atom accepts hydrogen ions from solution and so carries a positive charge as shown in the amine part of structure b. The carboxylic acid, however, remains unchanged and so appears as it does in the carboxylic acid part of structure a. At alkaline pH values, the carboxylic acid part of the molecule loses its hydrogen ion as it reacts with hydroxide ions in solution. As a result, this part of the molecule carries a negative charge as shown in structure b. Because the solution is basic, the nitrogen atom cannot pick up any hydrogen ions and so remains as shown

in the amine part of structure a. At neutral pH, both the concentration of hydronium ions and the concentration of hydroxide ions are quite low. This situation allows the acidic carboxylic acid part of the molecule and the basic amine part of the molecule to react *with each other* to form structure b.

23. 1. ether   2. amide   3. ester   4. amide
    5. alcohol   6. aldehyde   7. amine   8. ether
    9. ketone

25. The combustion of polyacrylonitrile produces hydrogen cyanide. The combustion of polyvinyl chloride and polyvinylidene chloride produces hydrogen chloride gas.

27. The polymer made of long strands is likely to be more viscous because of the tendency of longer strands to get tangled.

29. A fluorine-containing polymer, such as Teflon.

Chapter 13

1. A carbohydrate is made from water and carbon dioxide, but in no way does it "contain" these two materials. Recall from Chapter 2 that any chemical compound produced in a reaction is different from the reactants used to make it.

3. Both cellulose and starch are polymers of glucose. They differ in how the glucose units are linked together. In cellulose, the linking results in linear polymers that align with one another, giving rise to a tough material useful for structural purposes in plants. In starch, the linking permits the formation of alpha helices, with periodic branching along the polymer chain.

5. Lipids are nonpolar molecules containing no polar functional groups. They are made primarily of nonpolar hydrocarbons and therefore do not dissolve in water because they cannot compete with the strong attraction water molecules have for one another.

7. A fat molecule is a triglyceride. So if a product contains no triglycerides, perhaps there is some legitimacy to saying it is "free of fat molecules." When fat (triglyceride) molecules consumed in food are digested, however, they are broken down to fatty acids and glycerol molecules. This "fat-free" product therefore would still offer your body the same type of molecules that fat-containing food does, which means the same number of calories.

9. Your customer's hair is fine because each strand is thin. Because each strand is thin, there is less mass per strand. Less mass per strand means each strand is made of fewer amino acids, which means fewer cysteine amino acids and disulfide cross-linkings. Adding concentrated or even regular-strength reducing agent might cause his hair to fall apart completely. Yikes! You should dilute your reducing agent prior to application. You might even skip the reducing step altogether and move right to the oxidizing step in which disulfide bonds are formed.

11. Ser–Leu–Ser–Leu–Cys

13. At a: dipole–dipole attractions (hydrogen bonding). At b: disulfide bonds. At c: induced dipole–induced dipole attractions (hydrophobic attractions). The primary structure is the sequence of amino acids. The secondary structure is the three alpha-helix regions (the coils) and the one pleated-sheet region (the zig-zags). The tertiary structure is the overall shape of the polypeptide chain. No quaternary structure is observed because this polypeptide is not associated with an adjacent polypeptide.

15. Cytosine without its amine group is uracil. If uracil were present as a normal component of DNA, this "authentic" uracil would not be distinguishable from any uracil generated by cytosine deamination. The methyl group on thymine is apparently the "label" that tells repair enzymes to leave it alone, thereby preventing loss of genetic information.

17. The sequence ATG in DNA gives rise to the codon UAC on mRNA, which codes for the amino acid tyrosine.

19. Asp–Pro–Ala

21. The hydroxyl group on the ribose sugar provides a conformation for the mRNA polymer that doesn't bind well with DNA.

23. Water readily passes through the body. Excess quantities of the water-soluble vitamins are therefore readily excreted. Water-insoluble vitamins, by contrast, tend to build up in nonpolar fatty tissue, where they can remain for extended periods of time.

25. You should tell her that her body needs and will absorb only a certain amount of vitamin C each day. If she loads up with an excessive amount only once a week, her body has no way of storing any excess. Instead, the excess is excreted, leaving her potentially low on this vitamin for days at a time.

27. Sucrose is a disaccharide and must be broken down to its two saccharide units—only one of which is glucose—before it can be used by the body.

29. These polyunsaturated fatty acids came from the vegetarian diet eaten by the cows from which the beef fat came.

31. Every muscle in your body is a storage site of amino acids for yourself (as well as for any organisms that might end up eating you). In times of starvation, your body will access this "stockpile" of amino acids, resulting in a decrease in your muscle mass.

Chapter 14

1. It is the vast diversity of organic chemicals that permits the manufacture of the many types of medicines needed to match the many types of illnesses humans are susceptible to.

3. Whether a drug is isolated from nature or synthesized in the laboratory makes no difference as to "how good it may be for you." There are a multitude of natural products that are harmful, just as there are many synthetic drugs that are harmful. The effectiveness of a drug depends on its chemical structure, not on its source.

5. Typically, cancer chemotherapeutics kill not *all* cancerous cells but rather only a high percentage of them. The remaining cancerous cells are finished off by the immune system. The earlier the cancer is caught, the fewer the number of cancerous cells present in the body, which makes it easier for the chemotherapy and the immune

system to work together. With later-stage cancer, the number of cancerous cells left over after chemotherapy might be too much for the immune system to handle. Also, most cancer chemotherapy agents used today work by killing cancerous cells that are dividing. Young tumors have a greater percentage of dividing cells than do older tumors. Thus, young tumors are quicker to succumb to chemotherapy.

7. They work well together because of the synergistic effect. An invading virus is slowed down by the protease inhibitors and also by the antiviral nucleosides. Combining these agents creates an effect that is greater than the effect of either agent given independently.

9. Of the 6 billion humans on the planet, 3 billion are woman. The percentage of women using birth control pills is therefore (60 million/3 billion) $\times$ 100 = 2 percent. Birth control pills therefore have probably had only a minor impact on the worldwide growth of the human population.

11. Many stimulant drugs work by blocking the re-uptake of excitatory neurotransmitters. Stuck in the synaptic cleft, these neurotransmitters are decomposed by enzymes. By the time re-uptake is no longer blocked by a stimulant, there are few neurotransmitters left to be reabsorbed by the presynaptic neuron, which by this point is also deprived of neurotransmitters. With a lack of neurotransmitters, there is little communication between adjacent neurons, and this has the effect of making the drug user depressed.

13. Gardeners must exercise extreme care when using solutions of nicotine because nicotine is extremely poisonous to humans.

15. The effective dose of MDA is many times greater than the effective dose of LSD. In other words, MDA is taken by the milligram but LSD is taken by the microgram. Because so much MDA is taken, negative side effects are more likely.

The mechanisms of drug withdrawal are quite complex, however. Consider the following. A coffee drinker gets a headache when he or she stops drinking coffee. The headache is a result of a side effect of caffeine. This stimulant component of coffee causes blood vessels to dilate. Over time, the coffee drinker's body becomes accustomed to this dilation and counteracts it by applying a force that causes the dilated blood vessels to constrict. When coffee is no longer

consumed, the body doesn't know to stop counteracting the dilation. Blood vessels therefore become overly constricted, and that gives rise to a headache, which prompts the coffee drinker to drink more coffee.

Likewise, with repeated use of MDA, the body develops a tolerance to its side effects. When a person stops using MDA, the body is still working hard to tolerate the drug. Thus, it is the body's own counteraction to the drug that drives the MDA user to continue using it—a vicious cycle driven by some very real chemistry.

17. The advantage is that the person doesn't run the risk of being sent to jail for smoking marijuana. The disadvantage is that if the person is already nauseated, she or he may not be able to hold the orally taken Marinol down long enough for the antinausea effect to occur.

19. Replace the $-CH_3$ group with a group having a $-C-C-N-$ sequence. With this sequence, there is a nitrogen atom two carbons away from the single-bonded oxygen of the ester (see Figure 14.39). This change gives either procaine or a compound very similar to procaine, and the anesthetic properties of these compounds are greater than those of benzocaine. Of course, once the structure is modified, it is no longer benzocaine but some other chemical.

21. Endorphins likely came first, and then humans discovered how opium contains compounds that mimic the effect of endorphins. Thus, "opioids have endorphin activity" is the more appropriate statement. Why this is so is something you should be sure to discuss with your instructor.

23. Blood clots may form at the plaque site. The clots can then break away and travel through the bloodstream until they reach and clog a blood vessel in the heart, causing a heart attack.

Chapter 15

1. This oxygen is not gaseous O_2 but rather oxygen atoms bound to the cellulose structure of the plants.

3. *Tyrannosaurus rex* was at the top of its food chain (a quaternary consumer), which means the size of its population was limited because of the small food supply available at this level.

5. As it decomposes, the humus releases carbon dioxide, which reacts with moisture in the soil to form carbonic acid.

7. If the open spaces are too large, the soil will not retain water very well. Instead, the water will leach out, taking many plant nutrients with it.

9. Clay is a very compact material containing few open spaces. Because water cannot penetrate clay very well, nutrients are not washed away.

11. In alkaline soil, the ammonium ion, NH_4^+, behaves as a weak acid—it loses a hydrogen ion and thereby becomes ammonia, NH_3. Although ammonia is soluble in water, it is a gas at ambient temperatures and is thus readily lost to the atmosphere.

13. The organic bulk of compost, even though it is chemically inert, provides many open spaces to the soil, and these open spaces are where water and atmospheric oxygen become available to plant roots.

15. Synthetic fertilizers provide no organic bulk to the soil. Without an adequate supply of organic bulk, soil becomes chalky and loses its capacity to hold water, with the result that leaching becomes more significant. The synthetic fertilizers are quickly leached away, which means that greater quantities of them must be applied to compensate for this leaching effect.

17. Sawdust is mostly ground-up cellulose, which, as you learned in Chapter 13, is made of only carbon, oxygen, and hydrogen. The sawdust should therefore have the lowest rating of 0.2–0–0.2. Fish meal is a good source of protein, and proteins are high in nitrogen. The most likely rating for the fish meal therefore is 5–3–3. As discussed in Section 10.1, wood ashes are a source of the alkaline material potassium carbonate, K_2CO_3, and so the most likely rating for the ashes is 0–1.5–8.

19. The plant cannot distinguish between the two ions. An ammonium ion is an ammonium ion no matter where it is from! The main difference for the plant would be that the compost would provide the added benefit of organic bulk.

21. Here, it is all just a matter of semantics. The word *organic* in the phrase *organic farming* refers to an environmentally friendly method of farming. When a pesticide is classified as "organic," the likely meaning is that the pesticide is made of organic molecules, which, as discussed in Chapter 12, are molecules built by the linking together of carbon atoms.

23. One of the most precious resources of a nation is its topsoil because from this topsoil comes the food to feed the nation. One important reason the U.S. government is paying farmers not to farm is to conserve topsoil. At present, food supplies are more than adequate, and so there is an argument to be made that these conservation efforts are prudent. For the millions of starving people in the world, the problem is not an inadequate food supply but rather an inadequate distribution infrastructure for the food.

25. The advances in genetic engineering allow agricultural chemists to modify plants to improve their nutritional qualities and make them more resistant to pest infiltration.

27. Some plants are fairly tolerant of salty soil. The gene or genes that allow for this greater tolerance might be inserted into other plants that normally have a low tolerance for salty soil.

Chapter 16

1. People need fresh water to survive and prosper. Thus it's only natural that communities developed around areas where fresh water was easy to obtain, which means alongside rivers, lakes, and streams. Only with newer technologies has it become feasible for humans to settle in places where aboveground fresh water is not so abundant. In these regions—such as where Denver, Colorado, is now located—the majority of drinking water is obtained from deep wells that tap into groundwater.

3. The ultimate sink for nearly all rainfall on the Earth is the oceans.

5. Most groundwater exists in aquifers, which are underground regions through which water flows relatively slowly. If groundwater removal is stopped, the flow of water through the aquifer will slowly recharge the aquifer. If subsidence has occurred, the underground soil is much more compact than it was originally, which means it is not likely to hold as much water as it once did. If water removal is stopped, therefore, the land may rise some because of the recharging effect, but probably not back to its original level.

7. Groundwater moves only slowly, which means that any pollutants added to it are not flushed away for quite some time. Furthermore, there is no way to clean polluted groundwater while it is in the ground. Our only choice would be to purify the water after we extract it from the ground.

9. A big advantage to using chlorine is that it provides protection from pathogens for several days after treatment. A drawback is that the residual chlorine can adversely affect the taste of the water. Ozone, O_3, is very effective at killing pathogens. However, soon after it is bubbled through the water, it decomposes to oxygen, O_2. Thus, ozone does not remain in the water for as long as chlorine does to provide long-lasting protection. If the ozonated water is consumed fairly soon, however, this may not be a problem. Futhermore, because the ozone decomposes, ozonated water tends to taste better than water treated with chlorine.

11. Reverse osmosis can be used to get fresh water from any solution. The only prerequisite is that the solute particles be larger than the water molecules. When this is the case, only water molecules pass through the semipermeable membrane from the solution side to the fresh-water side.

13. Our mouths are pretty good at discerning the taste of residual components in drinking water—so much so that many of us are willing to pay the 1000 percent markup that water bottlers charge for their products, which are only a fraction of a percentage purer than municipal water. Because the purity of municipal water and the purity of bottled water are comparable, flushing a toilet with municipal water is about as wasteful as flushing it with bottled water. Your sketch should look something like this:

15. When a red blood cell is placed in fresh water, the concentration of solute particles inside the cell is higher than their concentration outside the cell. As a result, water migrates into the cell (moving from a region of low solute concentration to a region of high solute concentration). When enough water has collected in the cell, the cell bursts.

17. As the water freezes, dissolved salts are excluded from the ice crystals that form. Seawater can therefore be desalinated by cooling it to form crystals, which can then be melted to produce fresh water. Before the seawater is completely frozen, however, the mixture of crystals and liquid water needs to be "rinsed" with fresh water to remove the salts. Otherwise, the salts collect in tiny pockets in the ice. Unfortunately, the amount of fresh water required to rinse the freezing seawater is comparable to the amount of fresh water obtained by this process, making the process relatively inefficient.

19. Phosphates are a nutrient for many plants and microorganisms. In the past, the phosphates from used laundry detergents often made their way into rivers, lakes, and ponds. The phosphate-rich water supported such rampant growth of plants and microorganisms that the natural supply of dissolved oxygen was choked off as eutrophication took place.

21. The decomposition is primarily anaerobic because very little oxygen makes it from our mouths to our intestines, where food decomposition takes place. As a consequence, gases that come out the other end are frequently of the odoriferous sort.

23. At the wastewater treatment facility, human waste is extracted from the water and typically ends up in a solid-waste disposal site. So why not use a composting toilet, which allows you to stop wasting water, and send wastes directly to farmlands?

25. To date, advanced integrated pond systems are best suited to small communities that have access to large areas of undeveloped land and lots of sunshine. Even if conditions are right for a particular community, however, still to be overcome is the social inertia almost always associated with doing something different.

Chapter 17

1. The molecules of the gases that make up the atmosphere are held down by the force of gravity.

3. The air pressure exerted on the outside of your eardrums is decreasing more quickly than is the air pressure exerted on the inside of your eardrums. FYI, commercial airplanes maintain a cabin pressure equal to the pressure one experiences at about 2400 meters (8000 feet) up a mountain.

5. Oxygen molecules are more massive and hence heavier than nitrogen molecules. An oxygen molecule therefore requires more kinetic energy to travel to the same altitude as a nitrogen molecule. This is one of the reasons the ratio of nitrogen molecules to oxygen molecules increases (slightly) with increasing altitude.

7. Coal is a fossil fuel, which means it formed from decomposed organic matter. The sulfur in organic matter is primarily in the form of the amino acids cysteine and methionine. Plants obtain the sulfur to make these amino acids from atmospheric sulfur oxides that originated from volcano eruptions and the burning of fossil fuels. Thus, there is no original source of sulfur on the Earth. Rather, like all other elements, the sulfur that is already here recycles through various pathways. Sulfur does have an ultimate origin, however—the nuclear fusion occurring in our sun and all other stars.

9. The airborne sulfur dioxide reacts with oxygen and water to form sulfuric acid, which is carried to the ground by rainwater.

11. Atmospheric nitrogen reacts with atmospheric oxygen under conditions of extreme heat, such as occurs in an automobile engine or in the air surrounding a lightning bolt. The balanced equation for this reaction is $N_2 + O_2 \longrightarrow 2\,NO$.

13. Photosynthesis produces oxygen, O_2, which migrates from the Earth's surface to high up in the stratosphere, where it is converted by the energy of ultraviolet light to ozone, O_3. Plants and all other organisms living on the planet's surface benefit from this stratospheric ozone because of its ability to shade the planet's surface from ultraviolet light coming from the sun.

15. No. Close to the Earth's surface, ozone decomposes as it reacts with various materials, such as plants and airborne hydrocarbons. Thus, the ozone from automobiles doesn't last long enough to make it to distant locations, such as the stratospheric skies over the South Pole.

17. The coating of whitewash keeps visible light out of the greenhouse so that the temperature inside the greenhouse does not get too high.

19. The greenhouse effect works because of the atmosphere's ability to trap terrestrial radiation. If the atmosphere were less able to trap terrestrial radiation, the greenhouse effect would be weaker and global temperatures would drop.

Problems

1. You would need to know the total number of molecules in 1 liter of air. You would divide the number of CFC molecules by that total number and then multiply the quotient by 100 to get the percentage of CFC molecules.

3. (1.25 g/L)(1 mole/28 g) = 0.045 mole

Chapter 18

1. Both paper and cooked spaghetti consist of many overlapping and intertwined strands.

3. In a given length of time, industrial hemp produces much more cellulose fiber per acre than do trees. Hemp plants also contain a much lower proportion of lignins, which means that the cellulose fibers can be extracted from the hemp without the use of harsh chemicals. A disadvantage to using hemp for making paper is that the paper industry is well established at using trees. The short-term costs of outfitting new machinery and other infrastructure would therefore be great. Also, there would be political inertia to overcome because the hemp plant is a close relative of the marijuana plant. Industrial hemp, however, contains insignificant amounts of THC, the active ingredient in marijuana. Furthermore, any cross-breeding between industrial hemp and marijuana would result in a plant that contains smaller amounts of THC and hence would not be so desirable to the illicit drug community.

5. As with any other scientific endeavor, chance played an important role in the development of polymers. In all cases, however, there was a scientist with an open and innovative mind ready to recognize and take advantage of a chance observation. Charles Goodyear accidently tipped sulfur into heated natural rubber to invent vulcanized rubber. Christian Schobein inadvertantly wiped up a nitric acid spill with a cotton rag to create nitrocellulose. Jacques Brandenberger thought of cellophane as he observed stains on tablecloths. Also, as was discussed in Chapter 1, Teflon was first observed when Roy

Plunkett sawed open a gas-storage tank looking for contents that had "disappeared."

7. Celluloid and cellophane are both derived from cellulose. The difference between the two molecules is that in celluloid every hydroxyl group of celluose has been replaced by a nitrate group. Cellophane has the same chemical composition as cellulose but has been transformed to a film by chemical and mechanical processing.

9. They are made of nitrocellulose, which is the highly combustible material used to make guncotton.

11. Collodion, Parkesine, celluloid, viscose, cellophane, PVC, nylon, Teflon.

13. Definitely not! Metal halides are by no means restricted to group 1 metals. In fact, most metals are able to form halides. Iron chloride, $FeCl_3$, and copper chloride, $CuCl_2$, are examples. Figure 18.19 shows only the most common forms of metal compounds. In nature, iron is most commonly found as an oxide, and copper is most commonly found as a sulfide.

15. One process that exploits differences in physical properties takes place in the blast furnace when, because of its greater density, molten iron sinks beneath slag. Another is flotation, a technique that takes advantage of the fact that metal sulfides are relatively nonpolar and therefore attracted to oil. One process that exploits differences in chemical properties is the electrolysis of a metal, such as impure copper. Another is the reduction of a purified metal ore, such as iron ore, using carbon as the reducing agent and the high temperatures found in a blast furnace.

17. The problem is not whether or not we have the metal atoms on this planet—we do! The problem is in the expense of collecting those metal atoms. This expense would be too great if the metal atoms were evenly distributed throughout the planet. We are fortunate, therefore, that there are geologic formations where metal ores have been concentrated by natural processes. Bear in mind that it is only the metal atoms that we can extract from the ground that we are able to recycle. If we don't recycle these metal atoms, then down the road we'll have substantial shortages of new metal ores from which to feed our ever-growing appetite for metal-based consumer goods and building materials.

19. Glass is not fragile when it is molten.

21. Ceramics are made by heating clay to high temperatures. Clay is made up of microcapsules of aluminum oxides and silicon oxides surrounded by water. The heating melts the silicon oxides, and the molten material then surrounds all remaining microcapsules. When the clay cools, the silicon oxides solidify, holding the microcapsules together, much like a hardened glue.

23. Any composite is weakest along any direction in which fibers do not run. Plywood is made by gluing together thin layers of wood such that the grain in one layer runs perpendicular to the grain in the layers above and below. This arrangement provides great strength in two directions. Lay a sheet of plywood flat on the ground, however, and there is no grain that runs in the vertical direction—in other words, in the direction of the sheet's thickness. Plywood is therefore relatively weak in this direction, which is why the first step when an old sheet of plywood deteriorates is the "unstacking" of the once-bound thin layers.

Chapter 19

1. One of the reasons a gas turbine is more efficient at generating electricity is that it provides a more direct method of getting the turbine to rotate. In a steam turbine, fuel is burned to heat water to steam, and it is the steam that cases the turbine to rotate. While the water is being heated, some energy is invariably lost as heat that escapes to the surroundings. In a gas turbine, the hot products of combustion are used to turn the turbine.

3. Fossil fuels are made from plants that lived millions of years ago. These plants got their energy from the sun through photosynthesis. Burning fossil fuels is therefore an indirect way of using solar energy.

5. In order to produce carbon dioxide, there must be carbon. The lower the percentage of carbon in a fuel, the less carbon dioxide in the exhaust when that fuel is burned. Coal is almost nothing but carbon, and so the percentage of carbon is extremely high, close to 100 percent. This means burning coal produces much carbon dioxide. The carbon percentage is also high in petroleum-based fuels. Octane, to take just one example, has eight carbons for every 18 hydrogens, meaning the percentage of carbon by mass is about 84 percent. In the molecules that make up natural gas, on the other hand, there is a lower percentage of carbon. Methane, for example, has

one carbon for every four hydrogens, meaning the mass percentage of carbon is a relatively low 75 percent.

7. If the control substance loses neutron-absorbing power with increasing temperature, then as the reactor gets warmer, there is a proportionate increase in the number of neutrons available to initiate further fission reactions. These are the neutrons that the control substance does not absorb because of the higher temperature. This condition can lead to a runaway reaction. Then once the reactor hits a critical temperature, fission reactions escalate to the point of meltdown. Today's reactors have control substances that become *better* neutron absorbers with increasing temperature, a design that gives the units an effective passive safety mechanism.

9. The 1973 OPEC oil embargo was a boon to the nuclear power industry. The embargo was instrumental in alerting people to the foolishness of depending on one source of energy and prompted the building of many nuclear power plants. Interestingly, there have been no new nuclear facilities constructed in the United States since that time.

11. The fuel supply for nuclear fusion—hydrogen isotopes from the ocean—is orders of magnitude greater than the supply of all other sources of energy combined. A nuclear fusion reactor will produce no air pollutants and fewer radioactive products than present-day fission reactors.

13. Although dams produce virtually no chemical pollutants, they do radically alter ecosystems, to the detriment of many species, including the humans displaced from their homes by the rising waters that come behind a dam.

15. The moon energy available to us here on the Earth is the energy of the tides, which result from the gravitational pull between the Earth and the moon. Interestingly, the energy from the Earth–moon system that goes into creating the tides results in a slowing down of the Earth's rotation. Back during the time of the dinosaurs, a day was only about 19 hours long. In the far, far future, a day will be on the order of 46 hours long. At that time, the Earth's spin rate will exactly match the rate at which the moon orbits the Earth. The result will be that, to future Earthlings, the moon will always appear in the same location in the sky.

17. A nonsolar pool cover helps keep the pool warm by reducing the amount of evaporation that takes place.

19. Pass the drinking water through the hot zone of a solar collector. If contained within pressurized pipes, the water would become superheated, and pathogens would be quickly destroyed. A less expensive use of solar energy for producing drinking water would be to build a solar distillation apparatus like the one shown in Figure 16.11.

21. Natural gas, A or B; wind, B; nuclear fusion, A; hydroelectric, B; coal, A.

23. The electrons do not travel to the negatively charged silicon on their own. They are forced to move in this direction by an external energy source. For a photovoltaic cell, this energy source is sunlight, which literally knocks electrons in the proper direction.

25. The electricity generated during the inflow or outflow of the tides could be used to produce hydrogen from the electrolysis of water. This hydrogen could be stored and used to generate electricity through a fuel cell during times when the tide is reversing its direction.

Problems

1. First calculate the cost of running the 100-watt bulb for 1 hour, understanding that 100 watts is the same as 0.1 kilowatt:

$$0.1 \text{ kW} \times 1 \text{ h} = 0.1 \text{ kWh}$$

$$0.1 \text{ kWh} \times \frac{15 \text{ cents}}{1 \text{ kWh}} = 1.5 \text{ cents}$$

Then calculcate the cost of running the 20-watt bulb for 1 hour, understanding that 20 watts is the same as 0.02 kilowatt:

$$0.02 \text{ kW} \times 1 \text{ h} = 0.02 \text{ kWh}$$

$$0.02 \text{ kWh} \times \frac{15 \text{ cents}}{1 \text{ kWh}} = 0.3 \text{ cents}$$

The savings each hour is therefore 1.5 cents − 0.3 cents = 1.2 cents. This may not seem like much, but if 50 million households changed just one bulb from a 100-watt incandescent to a 20-watt fluorescent, the total annual savings on electric bills would be on the order of 1.2 billion dollars!

Glossary

absolute zero The lowest possible temperature any substance can have; the temperature at which the atoms of a substance have no kinetic energy: $0 \text{ K} = -273.15°C = -459.7°F$.

acid A substance that donates hydrogen ions.

acidic solution A solution in which the hydronium ion concentration is higher than the hydroxide ion concentration.

actinide Any seventh-period inner transition metal.

activation energy The minimum energy required in order for a chemical reaction to proceed.

addition polymer A polymer formed by the joining together of monomer units with no atoms lost as the polymer forms.

adhesive force An attractive force between molecules of two different substances.

aerobic bacteria Bacteria able to decompose organic matter only in the presence of oxygen.

aerosol A moisture-coated microscopic airborne particle up to 0.01 millimeter in diameter that is a site for many atmospheric chemical reactions.

agriculture The organized use of resources for the production of food.

alchemy A medieval endeavor concerned with turning other metals to gold.

alcohol An organic molecule that contains a hydroxyl group bonded to a saturated carbon.

aldehyde An organic molecule containing a carbonyl group the carbon of which is bonded either to one carbon atom and one hydrogen atom or to two hydrogen atoms.

alkali metal Any group 1 element.

alkaline-earth metal Any group 2 element.

alloy A mixture of two or more metallic elements.

alpha particle A helium atom nucleus, which consists of two neutrons and two protons and is ejected by certain radioactive elements.

amide An organic molecule containing a carbonyl group the carbon of which is bonded to a nitrogen atom.

amine An organic molecule containing a nitrogen atom bonded to one or more saturated carbon atoms.

amino acid The monomer of polypeptides, consisting of an amine group and a carboxylic acid group bonded to the same carbon atom.

amphoteric A description of a substance that can behave as either an acid or a base.

anabolism Chemical reactions that synthesize biomolecules in the body.

anaerobic bacteria Bacteria able to decompose organic matter in the absence of oxygen.

analgesic A drug that enhances the ability to tolerate pain without abolishing nerve sensations.

anesthetic A drug that prevents neurons from transmitting sensations to the central nervous system.

anode The electrode where oxidation occurs.

applied research Research dedicated to the development of useful products and processes.

aquifer A soil layer in which groundwater may flow.

aromatic compound Any organic molecule containing a benzene ring.

atmospheric pressure The pressure exerted on any object immersed in the atmosphere.

atomic mass The mass of an element's atoms, listed in the periodic table as an average value based on the relative abundance of the element's isotopes.

atomic nucleus The dense, positively charged center of every atom.

atomic number A count of the number of protons in the atomic nucleus.

atomic orbital A region of space in which an electron in an atom has a 90 percent chance of being located.

atomic spectrum The pattern of frequencies of electromagnetic radiation emitted by the atoms of an element, considered to be an element's "fingerprint."

atomic symbol An abbreviation for an element or atom.

Avogadro's number The number of particles—6.02×10^{23}—contained in 1 mole of anything.

base A substance that accepts hydrogen ions.

basic research Research dedicated to the discovery of the fundamental workings of nature.

basic solution A solution in which the hydroxide ion concentration is higher than the hydronium ion concentration.

beta particle An electron ejected from an atomic nucleus during the radioactive decay of certain nuclei.

bioaccumulation The process whereby a toxic chemical that enters a food chain at a low trophic level becomes more concentrated in organisms higher up the chain.

biochemical oxygen demand A measure of the amount of oxygen consumed by aerobic bacteria in water.

biomass A general term for plant material.

boiling Evaporation in which bubbles form beneath the liquid surface.

bond energy The amount of energy either absorbed as a chemical bond breaks or released as a bond forms.

buffer solution A solution that resists large changes in pH, made from either a weak acid and one of its salts or a weak base and one of its salts.

capillary action The rising of liquid into a small vertical space due to the interplay of cohesive and adhesive forces.

carbohydrate A biomolecule that contains only carbon, hydrogen, and oxygen atoms and is produced by plants through photosynthesis.

carbon-14 dating The process of estimating the age of once-living material by measuring the amount of a radioactive isotope of carbon present in the material.

carbonyl group A carbon atom double-bonded to an oxygen atom, found in ketones, aldehydes, amides, carboxylic acids, and esters.

carboxylic acid An organic molecule containing a carbonyl group the carbon of which is bonded to a hydroxyl group.

catabolism Chemical reactions that break down biomolecules in the body.

catalyst Any substance that increases the rate of a chemical reaction without itself being consumed by the reaction.

cathode ray tube A device that emits a beam of electrons.

cathode The electrode where reduction occurs.

chain reaction A self-sustaining reaction in which the products of one fission event stimulate further events.

chemical change During this kind of change, atoms in a substance are rearranged to give a new substance having a new chemical identity.

chemical equation A representation of a chemical reaction.

chemical formula A notation used to indicate the composition of a compound, consisting of the atomic symbols for the different elements of the compound and numerical subscripts indicating the ratio in which the atoms combine.

chemical property A type of property that characterizes the ability of a substance to change its chemical identity.

chemical reaction Synonymous with chemical change.

chemistry The study of matter and the transformations it can undergo.

chemotherapy The use of drugs to destroy pathogens without destroying the animal host.

chromosome An elongated bundle of DNA and protein that appears in a cell's nucleus just prior to cell division.

coal A solid consisting of a tightly bound three-dimensional network of hydrocarbon chains and rings.

coefficient A number used in a chemical equation to indicate either the number of atoms/molecules or the number of moles of a reactant or product.

cohesive force An attractive force between two identical molecules.

combinatorial chemistry The production of a large number of compounds in order to increase the chances of finding a new drug having medicinal value.

combustion An exothermic oxidation–reduction reaction between a nonmetallic material and molecular oxygen.

composite Any thermoset medium strengthened by the incorporation of fibers.

compost Fertilizer formed by the decay of organic matter.

compound A material in which atoms of different elements are bonded to one another.

concentration A quantitative measure of the amount of solute in a solution.

conceptual model A representation of a system that helps us predict how the system behaves.

condensation A transformation from a gas to a liquid.

condensation polymer A polymer formed by the joining together of monomer units accompanied by the loss of a small molecule, such as water.

conformation One of the possible spatial orientations of a molecule.

consumer An organism that takes in the matter and energy of other organisms.

control test A test performed by scientists to increase the conclusiveness of an experimental test.

corrosion The deterioration of a metal, typically caused by atmospheric oxygen.

covalent bond A chemical bond in which atoms are held together by their mutual attraction for two or more electrons they share.

covalent compound An element or chemical compound in which atoms are held together by covalent bonds.

critical mass The minimum mass of fissionable material needed in a reactor or nuclear bomb that will sustain a chain reaction.

decomposer An organism in the soil that transforms once-living matter to nutrients.

density The ratio of an object's mass to its volume.

deoxyribonucleic acid A nucleic acid containing a deoxy-genated ribose sugar, having a double helical structure, and carrying the genetic code in the nucleotide sequence.

dipole A separation of charge that occurs in a chemical bond because of differences in the electronegativities of the bonded atoms.

dissolving The process of mixing a solute in a solvent.

effective nuclear charge The nuclear charge experienced by outer-shell electrons, diminished by the shielding effect of inner-shell electrons.

electrochemistry The branch of chemistry concerned with the relationship between electrical energy and chemical change.

electrode Any material that conducts electrons into or out of a medium in which electrochemical reactions are occurring.

electrolysis The use of electrical energy to produce chemical change.

electromagnetic spectrum The complete range of electro-magnetic waves, from radio waves to gamma rays.

electron An extremely small, negatively charged subatomic particle found outside the atomic nucleus.

electron configuration The arrangement of electrons in the orbitals of an atom.

electron-dot structure A shorthand notation of the shell model of the atom in which valence electrons are shown around an atomic symbol.

electronegativity The ability of an atom to attract a bonding pair of electrons to itself when bonded to another atom.

element A fundamental material consisting of only one type of atom.

elemental formula A notation that uses the atomic symbol and (sometimes) a numerical subscript to denote how atoms are bonded in an element.

endothermic A term that describes a chemical reaction in which there is a net absorption of energy.

energy The capacity to do work.

energy-level diagram A drawing used to arrange atomic orbitals in order of energy levels.

entropy The total amount of energy contained in a substance at a particular temperature divided by that temperature.

enzyme A protein that catalyzes biochemical reactions.

ester An organic molecule containing a carbonyl group the carbon of which is bonded to one carbon atom and to an oxygen atom bonded to a carbon atom.

ether An organic molecule containing an oxygen atom bonded to two carbons atoms.

eutrophication The process whereby inorganic wastes in water fertilize algae and plants growing in the water and the resulting overgrowth reduces the dissolved oxygen concentration of the water.

evaporation A transformation from a liquid to a gas.

exothermic A term that describes a chemical reaction in which there is a net release of energy.

fat A biomolecule that packs a lot of energy per gram and consists of a glycerol unit attached to three fatty acid molecules.

formula mass The sum of the atomic masses of the atoms in a chemical compound or element.

freezing A transformation from a liquid to a solid.

functional group A specific combination of atoms that behave as a unit in an organic molecule.

gamma ray High-energy radiation emitted by the nuclei of radioactive atoms.

gas Matter that has neither a definite volume nor a definite shape, always filling any space available to it.

gene cloning The technique of incorporating a gene from one organism into the DNA of another organism.

gene A nucleotide sequence in the DNA strand in a chromo-some that leads a cell to manufacture a particular polypeptide.

glycogen A glucose polymer stored in animal tissue and also known as animal starch.

greenhouse effect The process by which visible light from the sun is absorbed by the Earth, which then emits infrared energy that cannot escape and so warms the atmosphere.

group A vertical column in the periodic table, also known as a family of elements.

half reaction One portion of an oxidation–reduction reac-tion, represented by an equation showing electrons as either reactants or products.

half-life The time required for half the atoms in a sample of a radioactive isotope to decay.

halogen Any group 17 element.

heat The energy that flows from one object to another because of a temperature difference between the two.

heat of condensation The energy released by a substance as it transforms from gas to liquid.

heat of freezing The heat energy released by a substance as it transforms from liquid to solid.

heat of melting The heat energy absorbed by a substance as it transforms from solid to liquid.

heat of vaporization The heat energy absorbed by a substance as it transforms from liquid to gas.

heteroatom Any atom other than carbon or hydrogen in an organic molecule.

heterogeneous mixture A mixture in which the various components can be seen as individual substances.

homogeneous mixture A mixture in which the components are so finely mixed that the composition is the same throughout.

humus The organic matter of topsoil.

hydrocarbon A chemical compound containing only carbon and hydrogen atoms.

hydrogen bond A strong dipole–dipole attraction between a slightly positive hydrogen atom on one molecule and a pair of nonbonding electrons on another molecule.

hydrologic cycle The natural circulation of water throughout our planet.

hydronium ion A water molecule after accepting a hydrogen ion.

hydroxide ion A water molecule after losing a hydrogen ion.

impure The state of a material that is a mixture of more than one element or compound.

induced dipole A dipole temporarily created in an otherwise nonpolar molecule, induced by a neighboring charge.

industrial smog Visible airborne pollution, containing large amounts of particulates and sulfur dioxide and produced largely from the combustion of coal and oil.

inner transition metal Any element in the two subgroups of the transition metals.

inner-shell shielding The effect whereby inner-shell electrons diminish the nuclear charge experienced by outer-shell electrons.

insoluble Not capable of dissolving to any appreciable extent in a given solvent.

integrated crop management A whole-farm strategy that involves managing crops in ways that suit local soil, climatic, and economic conditions.

integrated pest management A pest-control strategy that emphasizes prevention, planning, and the use of a variety of pest-control resources.

ion An electrically charged particle created when an atom either loses or gains one or more electrons.

ionic bond A chemical bond in which an attractive electric force holds ions of opposite charge together.

ionic compound Any chemical compound containing ions.

ionization energy The amount of energy required to remove an electron from an atom.

isotope Atoms of the same element whose nuclei contain the same number of protons but different numbers of neutrons.

ketone An organic molecule containing a carbonyl group the carbon of which is bonded to two carbon atoms.

kilowatt-hour The amount of energy consumed in 1 hour at a rate of 1 kilowatt.

kinetic energy Energy due to motion.

lanthanide Any sixth-period inner transition metal.

law of definite proportions A law stating that elements combine in definite mass ratios to form compounds.

law of mass conservation A law stating that matter is neither created nor destroyed in a chemical reaction.

leachate A solution formed by water that has percolated through a solid-waste disposal site and picked up water-soluble substances.

lipid A broad class of biomolecules that are not soluble in water.

liquid Matter that has a definite volume but no definite shape, assuming the shape of its container.

lock-and-key model A model that explains how drugs interact with receptor sites.

mass number The number of nucleons (protons and neutrons) in the atomic nucleus. Used primarily to identify isotopes.

mass The quantitative measure of how much matter an object contains.

matter Anything that occupies space.

melting A transformation from a solid to a liquid.

meniscus The curving of the surface of a liquid at the interface between the liquid surface and its container.

metabolism The general term describing all chemical reactions in the body.

metal An element that is shiny, opaque, and able to conduct electricity and heat.

metallic bond A chemical bond in which the metal ions in a piece of solid metal are held together by their attraction to a "fluid" of electrons in the metal.

metalloid An element that exhibits some properties of metals and some properties of nonmetals.

microirrigation A method of delivering water directly to plant roots.

mineral Inorganic chemicals that play a wide variety of roles in the body.

mixed fertilizer A fertilizer that contains a mixture of the plant nutrients nitrogen, phosphorus, and potassium.

mixture A combination of two or more substances in which each substance retains its properties.

molar mass The mass of 1 mole of a substance.

molarity A unit of concentration equal to the number of moles of a solute per liter of solution.

mole 6.02×10^{23} of anything.

molecule A group of atoms held tightly together by covalent bonds.

monomer The small molecular unit from which a polymer is formed.

natural gas A mixture of methane plus small amounts of ethane and propane.

neuron A specialized cell capable of receiving and sending electrical impulses.

neurotransmitter re-uptake A mechanism whereby a presynaptic neuron absorbs neurotransmitters from the synaptic cleft for reuse.

neurotransmitter An organic compound capable of activating receptor sites on proteins embedded in the membrane of a neuron.

neutral solution A solution in which the hydronium ion concentration is equal to the hydroxide ion concentration.

neutralization A reaction in which an acid and base combine to form a salt.

neutron An electrically neutral subatomic particle of the atomic nucleus.

nitrogen fixation A chemical reaction that converts atmospheric nitrogen to some form of nitrogen usable by plants.

noble gas Any group 18 element.

nonbonding pair Two paired valence electrons that don't participate in a chemical bond and yet influence the shape of the molecule.

nonmetal An element that is located toward the upper right of the periodic table and is neither a metal nor a metalloid.

nonpoint source A pollution source in which the pollutants originate at different locations.

nonpolar bond A chemical bond having no dipole.

nuclear fission The splitting of a heavy nucleus into two lighter nuclei, accompanied by the release of much energy.

nuclear fusion The joining together of light nuclei to form a heavier nucleus, accompanied by the release of much energy.

nucleic acid A long polymeric chain of nucleotide monomers.

nucleon Any subatomic particle found in the atomic nucleus. Another name for either proton or neutron.

nucleotide A nucleic acid monomer consisting of three parts: a nitrogenous base, a ribose sugar, and an ionic phosphate group.

ore A geologic deposit containing relatively high concentrations of one or more metal-containing compounds.

organic chemistry The study of carbon-containing compounds.

organic farming Farming without the use of pesticides or synthetic fertilizers.

osmosis The net flow of water across a semipermeable membrane from a region where the solute concentration is low to a region where the solute concentration is higher.

oxidation–reduction reaction A reaction involving the transfer of electrons from one reactant to another.

oxidation The process whereby a reactant loses one or more electrons.

particulate An airborne particle having a diameter greater than 0.01 millimeter.

period A horizontal row in the periodic table.

periodic table A chart in which all known elements are organized by physical and chemical properties.

periodic trend The gradual change of any property in the elements across a period.

petroleum A liquid mixture of loosely held hydrocarbon molecules containing not more than 30 carbon atoms each.

pH A measure of the acidity of a solution, equal to the negative of the base-10 logarithm of the hydronium ion concentration.

phenol An organic molecule in which a hydroxyl group is bonded to a benzene ring.

pheromone An organic molecule secreted by insects to communicate with one another.

photochemical smog Airborne pollution consisting of pollutants that participate in chemical reactions induced by sunlight.

photoelectric effect The ability of light to knock electrons out of atoms.

photon Another term for a single quantum of light, a name chosen to emphasize the particulate nature of light.

physical change A change in which a substance changes its physical properties without changing its chemical identity.

physical dependence A dependence characterized by the need to continue taking a drug to avoid withdrawal symptoms.

physical model A representation of an object on a different scale.

physical property Any physical attribute of a substance, such as color, density, or hardness.

point source A specific, well-defined location where pollutants enter a body of water.

polar bond A chemical bond having a dipole.

polyatomic ion A molecule that carries a net electric charge.

polymer A long organic molecule made of many repeating units.

potential energy Stored energy.

power The rate at which energy is expended.

precipitate A solute that has come out of solution.

principal quantum number *n* An integer that specifies the quantized energy level of an atomic orbital.

probability cloud The pattern of electron positions plotted over time to show the likelihood of an electron's being at a given position at a given time.

producer An organism at the bottom of a trophic structure.

product A new material formed in a chemical reaction, appearing after the arrow in a chemical equation.

protein A polymer of amino acids, also known as a polypeptide.

proton A positively charged subatomic particle of the atomic nucleus.

psychoactive Said of a drug that affects the mind or behavior.

psychological dependence A deep-rooted craving for a drug.

pure The state of a material that consists of a single element or compound.

quantum hypothesis The idea that light energy is contained in discrete packets called quanta.

quantum A small, discrete packet of light energy.

radioactivity The tendency of some elements, such as uranium, to emit radiation as a result of changes in the atomic nucleus.

reactant A starting material in a chemical reaction, appearing before the arrow in a chemical equation.

reaction rate A measure of how quickly the concentration of products in a chemical reaction increases or the concentration of reactants decreases.

recombinant DNA A hybrid DNA composed of DNA strands from different organisms.

reduction The process whereby a reactant gains one or more electrons.

rem A unit for measuring radiation dosage, obtained by multiplying the number of rads by a factor that allows for the different health effects of different types of radiation.

replication The process by which DNA strands are duplicated.

reverse osmosis A technique for purifying water by forcing it through a semipermeable membrane.

ribonucleic acid A nucleic acid containing a fully oxygenated ribose sugar.

saccharide Another term for carbohydrate. The prefixes *mono-*, *di-*, and *poly-* are used before this term to indicate the length of the carbohydrate.

salinization The process whereby irrigated land becomes saltier.

salt An ionic compound formed from the reaction between an acid and a base.

saturated hydrocarbon A hydrocarbon containing no multiple covalent bonds, with each carbon atom bonded to four other atoms.

saturated solution A solution containing the maximum amount of solute that will dissolve.

science An organized body of knowledge about nature.

scientific hypothesis A testable assumption often used to explain an observed phenomenon.

scientific notation A system for expressing numbers as some value between 1.00 . . . and 9.99 . . . multiplied by an appropriate power of 10.

scientific law Any scientific hypothesis that has been tested over and over again and has not been contradicted. Also known as a scientific principle.

semipermeable membrane A membrane that allows water molecules to pass through its submicroscopic pores but not solute molecules.

significant figure Any of the digits in a measured value that are known with certainty plus one digit that is estimated from the scale shown on the measuring device.

soil horizon A layer of soil.

solid Matter that has a definite volume and a definite shape.

solubility The ability of a solute to dissolve in a given solvent.

soluble Capable of dissolving to an appreciable extent in a given solvent.

solute Any component in a solution that is not the solvent.

solution A homogeneous mixture in which all components are in the same phase.

solvent The component in a solution present in the largest amount.

specific heat capacity The quantity of heat required to change the temperature of 1 gram of a substance by 1 Celsius degree.

spectroscope A device that uses a prism or diffraction grating to separate light into its color components.

steel Iron strengthened by small percentages of carbon.

straight fertilizer A fertilizer that contains only one nutrient.

stratosphere The atmospheric layer that lies immediately above the troposphere and contains the ozone layer.

strong nuclear force The force of interaction between all nucleons, effective only at very, very, very close distances.

structural isomers Molecules that have the same molecular formula but different chemical structures.

sublimation The process of a material transforming from a solid to a gas without passing through the liquid phase.

submicroscopic The realm of atoms and molecules, where objects are smaller than can be detected by optical microscopes.

substituent An atom or nonbonding pair of electrons surrounding a central atom.

superconductor Any material having zero electrical resistance.

surface tension The elastic tendency found at the surface of a liquid.

suspension A homogeneous mixture in which the various components are in different phases.

synaptic cleft A narrow gap across which neurotransmitters pass either from one neuron to the next or from a neuron to a muscle or gland.

synergistic effect One drug enhancing the effect of another.

temperature How warm or cold an object is relative to some standard. Also, a measure of the average kinetic energy per molecule of a substance, measured in degrees Celsius, degrees Fahrenheit, or kelvins.

theory A comprehensive idea that can be used to explain a broad range of phenomena.

thermodynamics The area of science concerned with the role of energy in chemical reactions.

thermometer An instrument used to measure temperature.

thermonuclear fusion Nuclear fusion produced by high temperature.

transcription The process whereby the genetic information of DNA is used to specify the nucleotide sequence of a complementary single strand of messenger RNA.

transgenic organism An organism that contains one or more genes from another species.

transition metal Any element of groups 3 through 12.

translation The process of bringing amino acids together according to the codon sequence on mRNA.

transmutation The conversion of an atomic nucleus of one element to an atomic nucleus of another element through a loss or gain of protons.

trophic structure The pattern of feeding relationships in a community of organisms.

troposphere The atmospheric layer closest to the Earth's surface, containing 90 percent of the atmosphere's mass and essentially all water vapor and clouds.

unsaturated hydrocarbon A hydrocarbon containing at least one multiple covalent bond.

unsaturated solution A solution that will dissolve additional solute if it is added.

valence electron An electron that is located in the outermost occupied shell in an atom and can participate in chemical bonding.

valence shell The outermost occupied shell of an atom.

valence-shell electron-pair repulsion A model that explains molecular geometries in terms of electron pairs striving to be as far apart from one another as possible.

vitamin Organic chemicals that assist in various biochemical reactions in the body and can be obtained only from food.

volume The quantity of space an object occupies.

water table The upper boundary of a soil's zone of saturation, which is the area where every space between soil particles is filled with water.

watt A unit for measuring power, equal to 1 joule of energy expended per second.

wave frequency A measure of how rapidly a wave oscillates. The higher this value, the greater the amount of energy in the wave.

wavelength The distance between two crests of a wave.

weight The gravitational force of attraction between two bodies (where one body is usually the Earth).

Credits

Cover: Photographer, Tim McKenna (tim-mckenna.com);
Surfer: Malik Joyeux

Frontmatter:
Title page: photographer, Tim McKenna (tim-mckenna.com) and
surfer: Malik Joyeux; p. xvii John Suchocki; p. xxx John Suchocki

Chapter 1
Chapter Opener: John Beatty/Getty Images; 1.1(a) Leonard
Lessin/Peter Arnold, Inc., (b) Ray Nelson/Phototake, (c) David
Parker/Photo Researchers,Inc., (d) Ray Nelson/Phototake, (e) Malin
Space Science Systems; 1.2 PhotoDisc, Inc.; 1.4 Todd Gipstein/
CORBIS; 1.6 (right) University of Alabama, (left) Jim McClintock;
1.7 (a, b) Jim Mastro; 1.11 "Re-enactment of the Invention of
DuPont Teflon® flouropolymer in April, 1938." Roy Plunkett is
shown at right. Photo courtesy of DuPont; 1.12 Lisa Jeffers-Fabro;
1.13 John Suchocki; 1.14 National Bureau of Standards/SKA
Photofiles; 1.16 (left, middle, right) Rachel Epstein/SKA; 1.18 John
Suchocki; 1.19 Eric Schrempp/Photo Researchers, Inc.; 1.20 Rachel
Epstein/SKA; 1.26 (left, middle, right) Sharon Hopwood; 1.30 Jim
Goodwin/Photo Researchers, Inc.; p. 29 Tracy Suchocki. **Concept
Check:** p. 15 NASA. **Hands-On Chemistry:** pp. 14, 17 (top,
bottom) Rachel Epstein/SKA; p. 25 Tom Pantages.

Chapter 2
Chapter Opener: (main photo) Astrid & Hanns-Friedler
Michler/Photo Researchers,Inc., (margin photos) top: Charles D.
Winters/Photo Researchers, Inc., second from top: Phillip
Hayson/Photo Researchers, Inc., (third and fourth from top) Tom
Pantages; 2.1 (left) Fundamental Photographs, (center) Definitive
Stock, (right) PhotoDisc, Inc.; 2.2(a) PhotoDisc, Inc., (b) Tom
Pantages; 2.3 (left, middle, right) Tom Pantages; 2.5 John Suchocki;
2.6 Stephen R. Swinburne/Stock, Boston; 2.7 Sharon Hopwood;
2.8 Sharon Hopwood; 2.11 (left, middle) Rachel Epstein/SKA;
2.11 (right) Tony Freeman/PhotoEdit; 2.13 (all) Tom Pantages;
2.14 Getty Images; 2.15 John Suchocki; 2.16 Runk, Scoenberger/
Grant Heilman, Inc.; 2.18 (a) Tom Pantages, (b) Dave Bartruff/Stock,
Boston; 2.19 Georg Gerster/Photo Researchers, Inc.; 2.21 (a) (left)
Kevin Adams/Nature Photography, (middle) Getty Images, (right)
PhotoDisc, Inc., (b) (left) Science Source/Photo Researchers, Inc.,
(middle, right) PhotoDisc, Inc.; 2.22 Brian Yarvin/Photo Researchers,
Inc.; 2.24 (top: left, second from left) PhotoDisc, Inc., (top: third
from left) Peter Arnold, Inc., (top: right) PhotoDisc, Inc., (bottom:
left) Fundamental Photographs, (bottom: second from left) Photo
Researchers, Inc., (bottom: third from left) PhotoDisc, Inc.,
(bottom: right) Fundamental Photographs; 2.25 Mark Martin/Photo
Researchers, Inc.; 2.29 Rachel Epstein/SKA. **Concept Check:** p. 45
John Suchocki. **Hands-On Chemistry:** pp. 42, 49, 54 Tom Pantages.

Chapter 3
Chapter Opener: (background) FPG, (overlay) John Suchocki;
p. 72 The Granger Collection; p. 74 Science Photo Library/Photo
Researchers, Inc.; p. 78 (top) Culver Pictures, (bottom) Science
Photo Library/Photo Researchers, Inc.; pp. 79, 80, 81 Science Photo
Library/Photo Researchers, Inc.; 3.11 The Granger Collection;
3.12 (a) Rachel Epstein/SKA, (b) Richard Megna/Fundamental
Photographs; p. 85 Culver Pictures; p. 86 Courtesy of the Archives,
California Institute of Technology; p. 88 Science Photo Library/
Photo Researchers, Inc.; 3.19 John Suchocki. **Hands-On Chemistry:**
pp. 74, 75, 86 Tom Pantages.

Chapter 4
Chapter Opener: Comstock; 4.1 Eric Schrempp/Photo Researchers,
Inc.; 4.3 (a, b) Science Photo Library/Photo Researchers, Inc.;
4.6 International Atomic Energy Agency; 4.8 Larry Mulvehill/Photo
Researchers, Inc.; 4.9 Richard Megna/Fundamental Photographs;
4.11 Chris Priest/Photo Researchers, Inc.; 4.18 Rich Frishman/Getty
Images; 4.20 CORBIS; 4.27 Comstock; 4.32 (b) Lawrence Livermore
National Laboratory. **Hands-On Chemistry:** p. 107 Rachel Epstein/
SKA.

Chapter 5
Chapter Opener: John Suchocki; 5.3 (a) Volker Steger/Peter Arnold,
Inc., (b) IBM/Peter Arnold, Inc., (c) Alamaden Research Center,
IBM Research Division; 5.4 (a) Tom Pantages, (b) Rachel Epstein/
SKA; 5.6 (a) Phototake, Inc., (b) John Suchocki; 5.7 (a, b) John
Suchocki; 5.8 (flame tests) Tom Pantages, (spectra) Alan J. Jircitano;
fig 5.9 Science Photo Library/Photo Researchers, Inc.; 5.15 (a) John
Suchocki, (b) David Scharf/Peter Arnold, Inc.; 5.16 (a, b, c) John
Suchocki; 5.20 John Suchocki; p. 159 CORBIS; 5.27 (a, b) Material
used by permission of DuPont Central Research and Development;
p. 168 (left) John Suchocki, (right column, top and bottom) Ed
Elliot, (right column, middle) CORBIS. **Hands-On Chemistry:**
pp. 140, 148 Tom Pantages; p. 152 Rachel Epstein/SKA.

Chapter 6
Chapter Opener: Charles M. Falco/Photo Researchers, Inc.;
6.3 Science Photo Library/Photo Researchers, Inc.; 6.8 (a) Rachel
Epstein/SKA, (b) F. Hache/Photo Researchers, Inc.; 6.9 John
Suchocki; 6.10 (top) M. Claye/Photo Researchers, Inc., (bottom)
E. R. Degginger/Photo Researchers, Inc.; 6.11 Dee Breger/Photo
Researchers, Inc.; 6.16 (a, b) Rachel Epstein/SKA; 6.17 Vaughan
Fleming/Photo Researchers, Inc.; 6.25 David Taylor/Photo
Researchers, Inc.; 6.29 Natalie Fobes/Getty Images; p. 202 (molecular
models) Tom Pantages. **Hands-On Chemistry:** pp. 182, 190 Tom
Pantages.

Chapter 7

Chapter Opener: Dr. & T. S. Schrichte/Photo Resource Hawaii; 7.5 Rachel Epstein/SKA; 7.8 (a, b) Tom Pantages; 7.9 Rachel Epstein/SKA; 7.13 (a) Fred Ward/Black Star, (b) Rachel Epstein/SKA, (c) Image Works; 7.18 Leonard Lessin/Peter Arnold, Inc; 7.19 John Suchocki; 7.23 Gordon Baer; 7.25 Sheila Terry/Photo Researchers, Inc. **Hands-On Chemistry:** pp. 212, 218, 223 (margin: middle, bottom) Tom Pantages; p. 223 (margin: top) Rachel Epstein/SKA.

Chapter 8

Chapter Opener: Kirk Weddle; 8.1 John Suchocki; 8.3 (a and b) Tom Pantages; 8.4 Nuridsany et Perennou/Photo Researchers, Inc.; 8.5 (and inset) Wm. R. Sallaz/Duomo; 8.13 (and inset) Rachel Epstein/SKA; 8.15 (a) NASA, (b) Diane Hirsch/Fundamental Photographs; 8.17 (a, b) Tom Pantages; 8.18 Sinclair Stammers/Photo Researchers, Inc.; 8.23 Leonard Lessin/Peter Arnold, Inc.; 8.24 (a) John Suchocki, (b) Pat Crowe/Animals, Animals; 8.26 Charles Cook; 8.27 NASA/Photo Researchers, Inc.; 8.29 Charles D. Winters/Photo Researchers, Inc.; 8.31 Galen Rowell/Mountain Light; 8.33 Richard Megna/Fundamental Photographs; 8.39 Okonewski/The Image Works; 8.40 John Suchocki; 8.41 Michael Dick/Animals, Animals. **Hands-On Chemistry:** pp. 240, 261 Tom Pantages.

Chapter 9

Chapter Opener: Digitalvision; 9.4 (left, middle, right) Tom Pantages; 9.13 CORBIS; 9.15 Rachel Epstein/SKA; 9.16 Photo Researchers, Inc.; 9.17 (top) E. R. Degginger/Photo Researchers, Inc., (bottom) Jon Lemker/Earth Scenes; 9.19 NASA/SKA Photofiles; 9.21 John Suchocki; 9.22 John Suchocki; 9.23 John Suchocki; 9.24 Rachel Epstein/SKA. **Hands-On Chemistry:** p. 293 Rachel Epstein/SKA.

Chapter 10

Chapter Opener: David Harris; 10.1 (a) M. P. Gadomski/Photo Researchers, Inc., (b, c, d) Rachel Epstein/SKA; 10.2 (a) S. Grant/PhotoEdit, (b, c, d) Rachel Epstein/SKA; 10.5 Rachel Epstein/SKA; 10.9 (a, b, c) Tom Pantages; 10.13 (a, b) Tom Pantages; 10.14 (a) (left, right) M. Bleier/Peter Arnold, Inc, (b) Will McIntyre/Photo Researchers, Inc.; 10.16 Tom Pantages; 10.23 Rachel Epstein/SKA. **Hands-On Chemistry:** p. 323 Tom Pantages.

Chapter 11

Chapter Opener: John-Peter Lahall/Photo Researchers, Inc.; 11.1 Tom Pantages; 11.3 John Suchocki; 11.4 (center) John Suchocki; 11.9 Leonard Lesson/Peter Arnold, Inc; 11.12 Xcellis Fuel Cell Engines, Inc.; 11.13 C. Liu, et al., "Single-Walled Carbon Nanotubes at Room Temperature" *Science*, Nov. 5, 1999: 1127–1129; 11.14 John Suchocki; 11.16 Frank Siteman/Stock, Boston; 11.18 Rachel Epstein/SKA; 11.19 Chevron Shipping Co., LLC. **Hands-On Chemistry:** pp. 341, 353 Tom Pantages.

Chapter 12

Chapter Opener: John Suchocki; 12.5 Rachel Epstein/SKA; 12.10 Rachel Epstein/SKA; 12.24 (a) Bob Gibbons/Photo Researchers, Inc., (b) Peter Arnold, Inc.; 12.30 John Suchocki; 12.31 Rachel Epstein/SKA; 12.32 Rachel Epstein/SKA; 12.36 Cambridge Display Technologies. **Hands-On Chemistry:** p. 369 Tom Pantages; p. 392 Rachel Epstein/SKA.

Chapter 13

Chapter Opener: John Suchocki; 13.1 (a) John Suchocki, (b) J. C. Munoz/Peter Arnold, Inc.; 13.2 Mark Chappel/Animals, Animals; 13.3 Rachel Epstein/SKA; 13.4 Leonard Lesson/Peter Arnold, Inc.; 13.5 (a) Rachel Epstein/SKA, (b) Fred Whitehead/Animals, Animals; 13.6 (left) Wally Eberhart/Visuals Unlimited, (right) Manfred Kage/Peter Arnold, Inc.; 13.7 (a) David Leah/Allsport, (b) Cabisco/Visuals Unlimited; 13.9 John Suchocki; 13.10 (left) Scott Camazine/Photo Researchers, Inc., (right) M. Abbey/Visuals Unlimited; 13.12 Johnny Johnson/Animals, Animals; 13.13 (a, b)Rachel Epstein/SKA; 13.18 (clockwise from left) Blair Seitz/Photo Researchers, Inc., David Scharf/Peter Arnold, Inc., Gray Mortimore/Allsport, Science Photo Library/Photo Researchers, Inc., Rachel Epstein/SKA, Science Photo Library/Photo Researchers, Inc., John Suchocki; 13.19 Stan Fleger/Visuals Unlimited; 13.28 (left) Cabisco/Visuals Unlimited, (right) Science Photo Library/Photo Researchers; 13.29 A. Lesk/Photo Researchers, Inc.; 13.32 (a) A. Barrington Brown/Photo Researchers; (b) CORBIS; (c) Science Photo Library/Photo Researchers, Inc.; 13.37 Simon Frazer/Photo Researchers, Inc.; 13.44 Leonard Lesson/Peter Arnold, Inc; 13.46 (a, inset, b) John Suchocki. **Hands-On Chemistry:** pp. 406, 443 Tom Pantages.

Chapter 14

Chapter Opener: Dr. J. Burgess/Photo Researchers, Inc.; 14.3 Kurt Hostettmann; 14.4 (left) David Nunuk/Photo Researchers, Inc., (right) Tom & Pat Leeson/Photo Researchers, Inc.; 14.5 (b) Upjohn; 14.8 Sim/Visuals Unlimited; 14.12 (a) NIBSC/Photo Researchers, Inc.; 14.14 (left, middle, right) Michael Abbey/Photo Researchers, Inc.; 14.16 Scott Camazine/Photo Researchers, Inc.; 14.25 Photo Researchers, Inc.; 14.27 Alan D. Carey/Photo Researchers, Inc.; 14.29 (clockwise from left) Inga Spence/Visuals Unlimited, Adam Jones/Photo Researchers, Inc., Tim Hazael/Getty Images, Matt Meadows/Peter Arnold, Inc., Cristina Pedrazzini/Photo Researchers, Inc.; 14.32 E. R. Degginger/Earth Scenes; 14.33 Science Photo LIbrary/Photo Researchers, Inc.; 14.34 Labat/Jerrican/Photo Researchers, Inc. **Hands-On Chemistry:** p. 469 Tom Pantages.

Chapter 15

Chapter Opener: Monsanto Company; 15.2 Will Trayer/Visuals Unlimited; 15.3 Norm Thomas/Photo Researchers, Inc.; 15.5 (a, b) Nigel Cattlin/Photo Researchers, Inc., (c) SKA Photofiles; 15.7 K. W. Fink/Photo Researchers, Inc.; 15.12 Rachel Epstein/SKA; 15.14 (top: left) G. S. Grant/Photo Researchers, Inc., (right) Norm Thomas/Photo Researchers, Inc., (bottom: left) G. Ochocki/Photo Researchers, Inc., (middle) Nigel Cattlin/Photo Researchers, Inc., (right) Robert Calentine/Visuals Unlimited; 15.16 CORBIS; 15.19 Nigel Cattlin/Photo Researchers, Inc.; 15.24 T. McCabe/Visuals Unlimited; 15.25 CORBIS; 15.26 United States Geological Survey; 15.27 University of New Mexico; 15.28 (left) Grant Heilman, (right) Cavagnaro/Visuals Unlimited; 15.29 Arthur Hill/Visuals Unlimited; 15.30 (left, right) Rachel Epstein/SKA; 15.31 Joe Munroe/Photo Researchers, Inc.; 15.32 A. C. Smith III/Grant Heilman, Inc.; 15.33 Marjorie A. Hoy; 15.34 (left) T. H. Martin/Photo Researchers, Inc.; (right) W. J. Weber/Visuals Unlimited; 15.36 C. S. Prakashi; p. 525 CORBIS. **Hands-On Chemistry:** p. 506 Rachel Epstein/SKA; p. 513 Tom Pantages.

Index